Wildland Fire in

Fire and Nonnative Invasive Plants

Editors

Kristin Zouhar, Ecologist/Technical Information Specialist, Fire Modeling Institute; Fire, Fuel, and Smoke Science Program, Rocky Mountain Research Station, U.S. Department of Agriculture, Forest Service, Missoula, MT.

Jane Kapler Smith, Ecologist, Fire Modeling Institute; Fire, Fuel, and Smoke Science Program, Rocky Mountain Research Station, U.S. Department of Agriculture, Forest Service, Missoula, MT.

Steve Sutherland, Research Ecologist, Forests and Woodlands Ecosystems Science Program, Fire Sciences Laboratory, Rocky Mountain Research Station, U.S. Department of Agriculture, Forest Service, Missoula, MT.

Matthew L. Brooks, Research Botanist, Las Vegas Field Station, Western Ecological Research Center, Biological Resources Division, U.S. Geological Survey, Henderson, NV.

Authors

Alison Ainsworth, NARS Specialist, Hawaii Division of Forestry and Wildlife, Natural Area Reserve System, Hilo, HI 96720

Dawn Anzinger, Research Assistant, Department of Forest Science, Oregon State University, Corvallis, OR 97331

Matthew L. Brooks, Research Botanist, Las Vegas Field Station, Western Ecological Research Center, Biological Resources Division, U.S. Geological Survey, Henderson, NV 89074

Alison C. Dibble, Cooperating Research Ecologist, Northern Research Station, U.S. Department of Agriculture, Forest Service, Bradley, ME 04411

Jonathan P. Freeman, Graduate Research Assistant, Natural Resource Ecology Laboratory, Colorado State University, Fort Collins, CO 80523

James B. Grace, Research Ecologist, National Wetlands Research Center, U.S. Geological Survey, Lafayette, LA 70506

R. Flint Hughes, Ecosystem Ecologist, Institute of Pacific Islands Forestry, Pacific Southwest Research Station, U.S. Department of Agriculture, Forest Service, Hilo, HI 96720

Karen V. S. Hupp, Research Assistant, Department of Agronomy, University of Florida, Gainesville, FL 32611

Molly E. Hunter, Research Associate, Department of Forest, Rangeland, and Watershed Stewardship, Colorado State University, Fort Collins, CO 80523

J. Boone Kauffman, Ecologist and Director, Institute of Pacific Islands Forestry, Pacific Southwest Research Station, U.S. Department of Agriculture, Forest Service, Hilo, HI 96720

Rob Klinger, USGS-BRD, Western Ecological Research Center, Yosemite Field Station Bishop, CA 93515

Anne Marie LaRosa, Forest Health Management Coordinator, Institute of Pacific Islands Forestry, Pacific Southwest Research Station, U.S. Department of Agriculture, Forest Service, Hilo, HI 96720

Erik J. Martinson, Research Associate, Department of Forest, Rangeland, and Watershed Stewardship, Colorado State University, Fort Collins, CO 80523

Guy R. McPherson, Professor, School of Natural Resources and Department of Ecology and Evolutionary Biology, University of Arizona, Tucson, AZ 85721

Gregory T. Munger, Biological Technician, U.S. Department of Agriculture, Forest Service, Fire Sciences Laboratory, Missoula, MT 59808

Philip N. Omi, Professor Emeritus, Department of Forest, Rangeland, and Watershed Stewardship, Colorado State University, Fort Collins, CO 80523

Steven R. Radosevich, Professor, Department of Forest Science, Oregon State University, Corvallis, OR 97331

Lisa J. Rew, Assistant Professor, Land Resources and Environmental Sciences Department, Montana State University, Bozeman, MT 59717

Peter M. Rice, Research Associate, Division of Biological Sciences, The University of Montana, Missoula, MT 59812

Jane Kapler Smith, Ecologist, Fire Modeling Institute; Fire, Fuels, and Smoke Science Program, Rocky Mountain Research Station, U.S. Department of Agriculture, Forest Service, Missoula, MT 59808

Randall K. Stocker, Director, Center for Aquatic and Invasive Plants, Department of Agronomy, University of Florida, Gainesville, FL 32611

Steve Sutherland, Research Ecologist, Forests and Woodlands Ecosystems Science Program, Fire Sciences Laboratory, Rocky Mountain Research Station, U.S. Department of Agriculture, Forest Service, Missoula, MT 59808

J. Timothy Tunison (retired), Chief of Resource Management, Hawai`i Volcanoes National Park, U.S. Department of the Interior, National Park Service, Hawai`i National Park, HI 96718

Robin Wills, Region Fire Ecologist, Pacific West Region, U.S. Department of the Interior, National Park Service, Oakland, CA 94607

Kristin Zouhar, Ecologist/Technical Information Specialist, Fire Modeling Institute; Fire, Fuels, and Smoke Science Program, Rocky Mountain Research Station, U.S. Department of Agriculture, Forest Service, Missoula, MT 59808

Preface

In 1978, a national workshop on fire effects in Denver, Colorado, provided the impetus for the "Effects of Wildland Fire on Ecosystems" series. Recognizing that knowledge of fire was needed for land management planning, state-of-the-knowledge reviews were produced that became known as the "Rainbow Series." The series consisted of six publications, each with a different colored cover (hence the informal title "Rainbow Series"), describing the effects of fire on soil (Wells and others 1979), water (Tiedemann and others 1979), air (Sandberg and others 1979), flora (Lotan and others 1981), fauna (Lyon and others 1978), and fuels (Martin and others 1979).

The Rainbow Series proved popular in providing fire effects information for professionals, students, and others. Printed supplies eventually ran out, but knowledge of fire effects continued to grow. To meet the continuing demand for summary and synthesis of fire effects knowledge, the interagency National Wildfire Coordinating Group asked Forest Service research leaders to update and revise the series. To fulfill this request, a meeting for organizing the revision was held January 1993 in Scottsdale, Arizona. A new, five-volume series was planned, officially named the Rainbow Series, to cover fauna, flora, cultural resources, soil and water, and air. Support for developing the new Series was provided by the Joint Fire Science Program.

Volume 2 of the Rainbow Series, "Effects of fire on flora," was published in December 2002. This volume synthesized information on the relationship between native plant communities and fire, but it provided little coverage of fire's influence on invasions by nonnative plants or nonnative plants' influence on fire regimes. To answer managers' requests for a comprehensive treatment of the relationship between fire and nonnative plants, in 2005 the Joint Fire Science Program provided funding for addition of this volume to the Rainbow Series.

The Rainbow Series emphasizes principles and processes. While it provides many details, examples, and citations, it does not intend to summarize all that is known. The six volumes, taken together, provide a wealth of information and examples to advance understanding of basic concepts regarding fire effects in the United States and Canada. As conceptual background, they provide technical support to fire and resource managers for carrying out interdisciplinary planning, which is essential for managing wildlands in an ecosystem context. Planners and managers will find the series helpful in many aspects of ecosystem-based management, but they also have the responsibility to seek and synthesize the detailed information needed to resolve specific management questions.

— The Editors
July 2007

Acknowledgments

The Rainbow Series was completed under the sponsorship of the Joint Fire Science Program, a cooperative effort of the U.S. Department of Agriculture, Forest Service, and the U.S. Department of the Interior, Bureau of Indian Affairs, Bureau of Land Management, Fish and Wildlife Service, and National Park Service. We thank these sponsors.

The authors and editors would also like to thank the following people for technical reviews, editorial suggestions, sharing of data and insights, and technical assistance:
James Åkerson, Michael Batcher, Ann Camp, Tony Caprio, Nancy Chadwick, Geneva W. Chong, Loa Collins, Steve Cross, Lesley A. DeFalco, Julie S. Denslow, Joseph M. DiTomaso, Lane Eskew, Amy Ferriter, Norm Forder, Neil Frakes, Kelly Ann Gorman, Jim Grace, Mick Harrington, Jeff Heys, John Hom, Todd Hutchinson, Cynthia D. Huebner, Jon E. Keeley, Charles Keller, Rudy King, Rob Klinger, Diane L. Larson, Steven O. Link, Mary Martin, Nanka McMurray, Henry McNab, Guy R. McPherson, Kyle Merriam, Melanie Miller, Scott A. Mincemoyer, David Moore, David J. Moorhead, Ronald L. Myers, Cara R. Nelson, Paul Nelson, Kate O'Brien, Glenn Palmgren, Mike Pellant, Karen J. Phillips, Thomas Poole, Steven R. Radosevich, John M. Randall, Thomas Rawinski, Paul Reeberg, Barry Rice, Julie A. Richburg, Thomas C. Roberts, David Scamardella, Thomas Schuler, Heather Schussman, Dennis Simmerman, David H. Smith, Helen Y. Smith, Marie-Louise Smith, Suzy Stevens, Thomas J. Stohlgren, Neil G. Sugihara, Elaine Kennedy Sutherland, Robert D. Sutter, Mandy Tu, Eric Ulaszek, Peter M. Vitousek, Dale Wade, Andrew Whitman, Troy A. Wirth, Autumn M. Yanzick, Daniel Yaussy.

Contents

Page

Chapter 1: Fire and Nonnative Invasive Plants—Introduction **1**
by Jane Kapler Smith
Kristin Zouhar
Steve Sutherland
Matthew L. Brooks
Scope of This Volume 2
Fire Behavior and Fire Regimes 3
Organization and Use of This Volume 5

Chapter 2: Effects of Fire on Nonnative Invasive Plants and Invasibility of Wildland Ecosystems **7**
by Kristin Zouhar
Jane Kapler Smith
Steve Sutherland
Invasion Ecology 7
Ecosystem Properties and Resource Availability 8
Properties of Native and Nonnative Plants 8
Nonnative Propagule Pressure 9
Influence of Fire on Invasions 9
Influence of Fire on Resource Availability and Interactions Between Plant Species 10
Influence of Fire Severity on Postfire Invasions 12
Influence of Fire Frequency on Postfire Invasions 17
Influence of Spatial Extent and Uniformity of Fire on Postfire Invasions 18
Influence of Fire Season and Plant Phenology on Postfire Invasions 18
Influence of Weather Patterns on Postfire Invasions 19
Generalizations About Fire Effects on Nonnative Invasives 19
Question 1. Does Fire Generally Favor Nonnatives Over Natives? 20
Question 2. Do Invasions Increase With Increasing Fire Severity? 22
Question 3. Does Additional Disturbance Favor Invasions? 22
Question 4. Do Invasions Become Less Severe With Increasing Time After Fire? 25
Question 5. Do Invasions Increase With Disruption of the Presettlement Fire Regime? 27
Question 6. Are Postfire Invasions Less Common in High Elevation Ecosystems? 28
Conclusions 28
Variation 29
Changing Atmosphere and Climate 29
Management in a Changing World 31

Chapter 3: Plant Invasions and Fire Regimes **33**
by Matthew L. Brooks
Fire Behavior and Fire Regimes 33
Biological and Physical Factors that Affect Fire Regimes 34
Effects of Plant Invasions on Fuels 35
The Invasive Plant Fire Regime Cycle 37
Predicting the Effects of Plant Invasions on Fire Regimes 38
Herbaceous Plant Invasions 39
Woody Plant Invasions 42
Preventing or Mitigating Altered Fire Regimes 44
Summary Recommendations 45

Chapter 4: Use of Fire to Manage Populations of Nonnative Invasive Plants **47**
by Peter M. Rice
Jane Kapler Smith
Introduction 47
Use of Fire Alone to Control Nonnative Invasive Plants 48
Prevention of Reproduction by Seed 48
Induced Mortality and Prevention or Delay of Resprouting 51
Burning to Favor Native Species 52
Fire Combined with Other Treatments 53
Treatments that Increase Effectiveness of Prescribed Fire 53
Prescribed Fire to Enhance Efficacy of Other Treatments 54
Altering Fire Severity, Uniformity, Extent, and Frequency to Control Nonnative Invasive plants 56
Prescribed Fire Severity and Uniformity in Relation to Season 56
Prescribed Fire Severity and Uniformity in Relation to Fuel Beds 57
Extent and Uniformity of Burns 58
Fire Frequency 58
Management of a Human Process: Constraints on Use of Prescribed Fire 59
Conclusions 60

Chapter 5: Fire and Nonnative Invasive Plants in the Northeast Bioregion **61**
by Alison C. Dibble
Kristin Zouhar
Jane Kapler Smith
Introduction 61
Fire History in the Northeast Bioregion 61
Nonnative Plants in the Northeast Bioregion 62
Interactions of Fire and Invasive Plants in the Northeast Bioregion 63
Forests 66
Deciduous and Mixed Forests 66
Background 66
Role of Fire and Fire Exclusion in Promoting Nonnative Plant Invasions in Deciduous and Mixed Forests 68
Effects of Nonnative Plant Invasions on Fuel Characteristics and Fire Regimes in Deciduous and Mixed Forests 74
Use of Fire for Controlling Nonnative Invasives in Deciduous and Mixed Forests 75
Coniferous Forests 76
Background 76
Role of Fire and Fire Exclusion in Promoting Nonnative Plant Invasions in Coniferous Forests 77
Effects of Nonnative Plant Invasions on Fuel Characteristics and Fire Regimes in Coniferous Forests 78

Page

Use of Fire to Control Nonnative Invasive Plants in Coniferous Forests ... 78
Grasslands and Early-Successional Old Fields ... 78
Background ... 78
Role of Fire and Fire Exclusion in Promoting Nonnative Invasives in Grasslands and Old Fields ... 79
Effects of Nonnative Plant Invasions on Fuel Characteristics and Fire Regimes in Grasslands and Old Fields ... 80
Use of Fire to Control Nonnative Invasive Plants in Grasslands and Old Fields ... 81
Riparian and Wetland Communities ... 82
Background ... 82
Role of Fire and Fire Exclusion in Promoting Nonnative Invasives in Riparian and Wetland Communities ... 84
Effects of Nonnative Plant Invasions on Fuel Characteristics and Fire Regimes in Riparian and Wetland Communities ... 85
Use of Fire to Control Nonnative Invasive Plants in Riparian and Wetland Communities ... 86
Emerging Issues in the Northeast Bioregion ... 87
Fuel Properties of Invaded Northeastern Plant Communities and Influences on Fire Regimes ... 87
Vulnerability of Forest Gaps to Invasion ... 87
Global Climate Change ... 88
Interactions Between Nonnative Invasive Plants, Fire, and Animals ... 88
Conclusions ... 88
Resources Useful to Managers in the Northeast Bioregion ... 89

Chapter 6: Fire and Nonnative Invasive Plants in the Southeast Bioregion ... 91
by Randall Stocker
Karen V. S. Hupp
Introduction ... 91
Fire in the Southeast Bioregion ... 92
Fire and Invasive Plants in the Southeast Bioregion ... 93
Wet Grassland Habitat ... 94
Background ... 94
Role of Fire and Fire Exclusion in Promoting Nonnative Plant Invasions in Wet Grasslands ... 96
Effects of Plant Invasions on Fuel and Fire Regime Characteristics in Wet Grasslands ... 97
Use of Fire to Manage Invasive Plants in Wet Grasslands ... 98
Pine and Pine Savanna Habitat ... 100
Background ... 100
Role of Fire and Fire Exclusion in Promoting Nonnative Plant Invasions in Pine and Pine Savanna ... 101
Effects of Plant Invasions on Fuel and Fire Regime Characteristics in Pine and Pine Savanna ... 102
Use of Fire to Manage Invasive Plants in Pine and Pine Savanna ... 104
Oak-Hickory Woodland Habitat ... 105
Background ... 105

Page

Role of Fire and Fire Exclusion in Promoting Nonnative Plant Invasions in Woodlands ... 105
Effects of Plant Invasions on Fuel and Fire Regime Characteristics in Woodlands ... 106
Use of Fire to Manage Invasive Plants in Oak-Hickory Woodland ... 106
Tropical Hardwood Forest Habitat ... 107
Background ... 107
Role of Fire and Fire Exclusion in Promoting Nonnative Plant Invasions in Tropical Hardwood Forest ... 107
Cypress Swamp Habitat ... 108
Conclusions and Summary ... 109
Role of Fire and Fire Exclusion in Promoting Nonnative Plant Invasions ... 109
Effects of Plant Invasions on Fuel and Fire Regime Characteristics ... 110
Use of Fire to Manage Invasive Plants ... 110
Additional Research Needs ... 110
Emerging Issues ... 111

Chapter 7: Fire and Nonnative Invasive Plants in the Central Bioregion ... 113
by James B. Grace
Kristin Zouhar
Introduction ... 113
Geographic Context and Chapter Organization ... 114
Fire Regimes ... 115
The Conservation Context ... 115
Overview of Nonnative Plants that Impact or Threaten the Central Bioregion ... 115
Plant-Fire Interactions ... 117
The Mesic Tallgrass Prairie Subregion ... 117
Northern and Central Tallgrass Prairie Formation ... 119
Southern Tallgrass Prairie Formation ... 125
The Great Plains Subregion ... 130
Northern Mixedgrass Prairie Formation ... 131
Southern Mixedgrass Prairie Formation ... 136
Shortgrass Steppe Formation ... 136
Riparian Formation ... 137
Conclusions ... 140

Chapter 8: Fire and Nonnative Invasive Plants in the Interior West Bioregion ... 141
by Peter M. Rice
Guy R. McPherson
Lisa J. Rew
Introduction ... 141
Grasslands ... 145
Desert Grasslands ... 145
Mountain Grasslands ... 145
Effects of Fire and Fire Exclusion on Nonnative Plant Invasions in Interior West Grasslands ... 146
Effects of Nonnative Plant Invasions on Fuels and Fire Regimes in Interior West Grasslands ... 148
Use of Fire to Manage Invasive Plants in Interior West Grasslands ... 152
Shrublands ... 154
Sagebrush Shrublands ... 154
Desert Shrublands and Shrubsteppe ... 155
Chaparral-Mountain Shrub ... 156

Page

Effects of Fire on Nonnative Plant Invasions in Interior West Shrublands....156
Effects of Nonnative Plant Invasions on Fuels and Fire Regimes in Interior West Shrublands....158
Use of Fire to Manage Invasive Plants in Interior West Shrublands....160
Piñon-Juniper Woodlands....161
Effects of Fire on Nonnative Plant Invasions in Interior West Woodlands....162
Effects of Nonnative Plant Invasions on Fuels and Fire Regimes in Interior West Woodlands....163
Use of Fire to Manage Invasive Plants in Interior West Woodlands....163
Open-Canopy Forest....164
Effects of Fire on Nonnative Plant Invasions in Interior West Open-Canopy Forest....164
Effects of Nonnative Plant Invasions on Fuels and Fire Regimes in Interior West Open-Canopy Forest....166
Use of Fire to Manage Invasive Plants in Interior West Open-Canopy Forests....166
Closed-Canopy Forests....166
Effects of Fire on Nonnative Plant Invasions in Interior West Closed-Canopy Forests....167
Effects of Nonnative Plant Invasions on Fuels and Fire Regimes in Interior West Closed-Canopy Forests....168
Use of Fire to Manage Invasive Plants in Interior West Closed-Canopy Forests....168
Riparian Communities....168
Effects of Fire on Nonnative Plant Invasions in Interior West Riparian Communities....169
Effects of Nonnative Plant Invasions on Fuels and Fire Regimes in Interior West Riparian Communities....170
Use of Fire to Manage Invasive Plants in Interior West Riparian Communities....171
Conclusions....171

Chapter 9: Fire and Nonnative Invasive Plants in the Southwest Coastal Bioregion....175
by Rob Klinger
Robin Wills
Matthew L. Brooks
Introduction....175
Grasslands....177
Fire and Nonnative Invasives in Grasslands....180
Chaparral and Coastal Scrub....182
Chaparral....182
Coastal Scrub....183
Fire and Nonnative Invasives in Chaparral and Coastal Scrub....184
Mixed Evergreen Forests....187
Fire and Nonnative Invasives in Mixed Evergreen Forests....188
Conifer Forests....188
Coastal Conifer Forests....189
Klamath and Cascade Range Conifer Forests....189
Sierra Nevada Conifer Forests....189
Fire and Nonnative Invasives in Conifer Forests....190
Wetland and Riparian Communities....191
Fire and Nonnative Invasives in Wetland and Riparian Communities....192
Conclusions....192

Chapter 10: Fire and Nonnative Invasive Plants in the Northwest Coastal Bioregion....197
by Dawn Anzinger
Steven R. Radosevich
Introduction....197
Coastal Douglas-fir Forests....198
Role of Fire in Promoting Nonnative Plant Invasions in Coastal Douglas-fir Forests....199
Effects of Nonnative Plant Invasions on Fuels and Fire Regimes in Coastal Douglas-fir Forests....203
Use of Fire to Manage Invasive Plants in Coastal Douglas-fir Forests....204
Upper Montane Conifer Forests and Meadows....204
Role of Fire in Promoting Invasions of Nonnative Plant Species in Upper Montane Communities....205
Effects of Nonnative Plant Invasions on Fuels and Fire Regimes in Upper Montane Communities....205
Use of Fire to Manage Invasive Plants in Upper Montane Communities....205
Riparian Forests....206
Role of Fire in Promoting Nonnative Plant Invasions in Riparian Forests....206
Effects of Nonnative Plant Invasions on Fuels and Fire Regimes in Riparian Forests....207
Use of Fire to Manage Invasive Plants in Riparian Forests....208
Oregon Oak Woodlands and Prairies....208
Role of Fire in Promoting Nonnative Plant Invasions in Woodlands and Prairies....209
Effects of Nonnative Plant Invasions on Fuels and Fire Regimes in Woodlands and Prairies....212
Use of Fire to Manage Invasive Plants in Woodlands and Prairies....212
Alaska....215
Coastal Hemlock-Spruce Forests....215
Role of Fire in Promoting Nonnative Plant Invasions in Coastal Hemlock-Spruce Forests....215
Effects of Nonnative Plant Invasions on Fuels and Fire Regimes in Coastal Hemlock-Spruce Forests....217
Use of Fire to Manage Invasive Plants in Coastal Hemlock-Spruce Forests....218
Boreal Forests....218
Role of Fire in Promoting Nonnative Plant Invasions in Boreal Forests....218
Effects of Nonnative Plant Invasions on Fuels and Fire Regimes in Boreal Forests....220
Use of Fire to Manage Invasive Plants in Boreal Forests....220
Tundra....220
Role of Fire in Promoting Nonnative Plant Invasions in Tundra....220
Effects of Nonnative Plant Invasions on Fuels and Fire Regimes in Tundra....221
Use of Fire to Manage Invasive Plants in Tundra....221

Page

Summary of Fire-Invasive Plant Relationships in Alaska 221
Conclusions 222

Chapter 11: Fire and Nonnative Invasive Plants in the Hawaiian Islands Bioregion 225
by Anne Marie LaRosa
J. Timothy Tunison
Alison Ainsworth
J. Boone Kauffman
R. Flint Hughes
Introduction 225
Description of Dry and Mesic Grassland, Shrubland, Woodland, and Forest 226
Lowlands 226
Montane and Subalpine 229
Fire History and the Role of Nonnative Invasives on Fire Regimes in Dry and Mesic Ecosystems 230
Fire and Its Effects on Nonnative Invasives in Dry and Mesic Ecosystems 233
Lowland and Montane Wet Forests 237
Fire History and the Role of Nonnative Invasives in Wet Forest Fire Regimes 237
Fire and Its Effects on Nonnative Invasives in Wet Forests 238
Use of Fire to Control Invasive Species and Restore Native Ecosystems 239
Conclusions 240

Chapter 12: Gaps in Scientific Knowledge About Fire and Nonnative Invasive Plants 243
by Kristin Zouhar
Gregory T. Munger
Jane Kapler Smith
Introduction 244
Methods 244
Results 247
Discussion 254
Basic Biology 255
Impacts, Invasiveness, and Invasibility 255
Distribution and Site Information 256
Ecological Information 256
Responses to Fire, Heat, and Postfire Conditions 257
Field Experiments Addressing Fire Effects 257
Representing Information Quality in Literature Reviews: Potential for Illusions of Knowledge 258
Monitoring, Data Sharing, and Adaptive Management 259
Conclusions 259

Chapter 13: Effects of Fuel and Vegetation Management Activities on Nonnative Invasive Plants 261
by Erik J. Martinson
Molly E. Hunter
Jonathan P. Freeman
Philip N. Omi
Introduction 261
High-Frequency, Low-Severity Historic Fire Regime 262

Page

Mixed-Severity Historic Fire Regime 263
Low-Frequency, High-Severity Historic Fire Regime 264
Altered Fire Regime 265
Unaltered Fire Regime 266
Summary 266

Chapter 14: Effects of Fire Suppression and Postfire Management Activities on Plant Invasions 269
by Matthew L. Brooks
Challenges of Identifying Postfire Plant Invasions 269
The Invasion Process 270
Effects of Fire Suppression Activities on Plant Invasions 271
Resource Availability 271
Propagule Pressure 273
Effects of Postfire Management Activities on Plant Invasions 274
Resource Availability 276
Summary 279

Chapter 15: Monitoring the Effects of Fire on Nonnative Invasive Plant Species 281
by Steve Sutherland
Vegetation Monitoring 282
Objectives 282
Stratification 283
Controls 284
Random Sampling 284
Sample Size 285
Statistical Analysis 285
Field Techniques 285
Other Elements of an Effective Monitoring Program 288
Fire Monitoring 290
Fire Behavior 290
Fuel Loading 290
Fire Severity 291
Success and Failure in Monitoring Programs 292

Chapter 16: Fire and Nonnative Plants— Summary and Conclusions 293
by Jane Kapler Smith
Kristin Zouhar
Steve Sutherland
Matthew L. Brooks
Nonnative Invasive Species and Wildland Fire 293
Management Implications 294
Use of Fire to Control Nonnative Invasive Species 295
Management Implications 295
Questions 296

References 297

Appendix A: Glossary 345

Summary

Wildland fire is a process integral to the functioning of most wildland ecosystems of the United States. Where nonnative plant species have invaded wildlands or have potential to invade, fire may influence their abundance and the effects of the nonnative species on native plant communities. This volume synthesizes scientific information regarding wildland fire and nonnative invasive plant species, identifies the nonnative invasive species currently of greatest concern in major bioregions of the United States, and describes emerging fire-invasive issues in each bioregion and throughout the nation. This report can assist fire managers and those concerned with prevention, detection, and eradication or control of nonnative invasive plants. It can help increase understanding of plant invasions and fire and can be used in planning fire management and ecosystem-based land management activities.

The first part of this volume summarizes fundamental concepts regarding relationships between fire and nonnative plant invasions. The introduction sets up a conceptual framework for discussing these relationships, focusing especially on the nature of plant invasions and fire regimes. Chapter 2 summarizes ecological and botanical principles that apply to fires' influences on plant invasions; it also analyzes ways in which the condition of the native plant community affects nonnative species' responses to fire. With this theoretical background, the chapter then examines the applicability of several common generalizations regarding fire and plant invasions. Chapter 3 describes how plant invasions can alter the quantity, spatial distribution, and seasonal availability of fuels and then, in some cases, fire regimes. The invasive plant/fire cycle is examined, in which the invasive species increases or decreases flammability in an ecosystem, then increases or decreases fire frequency, and then increases in abundance, continuing the cycle. Chapter 4 summarizes information on use of fire to control plant invasions.

Part II (chapters 5 through 11) synthesizes information on three topics (effect of fire on nonnative plant invasions, influence of plant invasions on fuels and fire regimes, and use of fire to control nonnative invasives) for seven bioregions of the United States: Northeast, Southeast, Central, Interior West, Southwest Coastal, Northwest Coastal (including Alaska), and the Hawaiian Islands.

The third part of this volume addresses management and research issues of national concern. Chapter 12 describes knowledge gaps regarding fire and nonnative invasive plants, focusing on the urgent need for more information on heat tolerance, postfire establishment, effects of varying fire regimes (severities, seasons, and intervals between burns), and long-term effects of fire. Chapter 13 describes the response of nonnative invasive plants to nonfire fuel treatments. Fuel treatments have reduced wildfire severity in some ecosystems with historically frequent, low-severity fire, and wildfire may pose a greater threat from nonnative invasive species than fuel treatments. However, evidence for this generalization comes mostly from ponderosa pine (*Pinus ponderosa*) forests, so it should be applied cautiously to other vegetation types, especially those with different fire regimes. Chapter 14 analyzes the influence of postfire rehabilitation on invasions. Several procedures can be integrated into postfire rehabilitation and land management plans to minimize new invasions and reduce or avoid increases in existing invasions. Chapter 15 describes the importance of postfire monitoring for invasives and ways to obtain nonnative species information in postfire monitoring. Long-term monitoring may be difficult to plan, implement, analyze, interpret, and integrate into the adaptive management process, but it often provides the best way, and sometimes the only way, to make defensible decisions regarding management of fire and invasives. The final chapter summarizes major findings in this volume and suggests important questions for future research. To manage fire and nonnative invasive plants, managers must integrate many kinds of knowledge while remaining aware of their applications and limitations. Management actions should be implemented with caution, monitored, and adapted as new knowledge develops.

For sale by the Superintendent of Documents, U.S. Government Printing Office
Internet: bookstore.gpo.gov Phone: toll free (866) 512-1800; DC area (202) 512-1800
Fax: (202) 512-2104 Mail: Stop IDCC, Washington, DC 20402-0001

ISBN 978-0-16-081465-5

Jane Kapler Smith
Kristin Zouhar
Steve Sutherland
Matthew L. Brooks

Chapter 1:

Fire and Nonnative Invasive Plants—Introduction

Fire is a process integral to the functioning of most temperate wildland ecosystems. Lightning-caused and anthropogenic fires have influenced the vegetation of North America profoundly for millennia (Brown and Smith 2000; Pyne 1982b). In some cases, fire has been used to manipulate the species composition and structure of ecosystems to meet management objectives, including control of nonnative invasive plant species (DiTomaso and others 2006a; Grace and others 2001; Keeley 2001; Myers and others 2001; Pyke and others, in review). However, fire can also threaten human life, property, and natural and cultural resources. Under some conditions, fire can increase abundance of nonnative invasive plants (Goodwin and others 2002), which may subsequently alter fire behavior and fire regimes, sometimes creating new, self-sustaining, invasive plant/fire cycles (Brooks and others 2004; D'Antonio and Vitousek 1992). These altered fire regimes can reduce native species diversity, alter ecosystem functions, and increase the threat of fire to human communities and wildland ecosystems.

Wildland managers must decide when, where, and for what specific reasons they should use fire to meet management objectives. To develop effective plans and make well-informed decisions, managers need to understand the scientific principles that drive the relationships between fire and nonnative invasive plants. They also need to understand how fire-invasives issues affect management in their geographic regions. Managers have indicated that better interpretation of science, including peer-reviewed synthesis, is essential for "bridging the worlds of fire managers and researchers" (White 2004). Several publications summarize regional and topical aspects of integrated fire and invasive plant management (Brooks and others 2004; Brooks and Esque 2002; Brooks and Pyke 2001; D'Antonio and Vitousek 1992; D'Antonio 2000; DiTomaso and others 2006a). However, a published synthesis of major fire-invasive plant issues on a national scale is lacking. To address this need, this volume reviews the scientific literature regarding relationships between fire and nonnative invasive plants in the United States and presents information useful for improving fire and nonnative species management in wildland ecosystems.

This volume complements Volume 2 of the Rainbow Series, *Wildland Fire and Ecosystems—Effects of Fire on Flora* (Brown and Smith 2000). Readers are referred to that volume for information on autecological relationships between plants and fire, past fire regimes, and successional patterns for forest and grassland ecosystems in the United States. However, *Effects of Fire on Flora* provides only a cursory treatment of nonnative species. In contrast, this volume focuses on nonnative invasive plants and fire. It is intended as a review of knowledge for wildland managers who are interested in using prescribed fire to reduce nonnative invasive plants, and those who are concerned that fire and fire management activities may increase abundance of nonnative species to the detriment of native ecosystems. This volume can be used to inform management plans and actions, although it is not comprehensive or detailed enough to provide prescriptions for management on particular sites. Objectives of this volume are

1. To synthesize scientific information regarding relationships between wildland fire and nonnative invasive plant species; and
2. To identify the nonnative invasive species currently of greatest concern and the wildland communities where fire-invasive issues are of greatest concern in major bioregions of the United States, synthesize information unique to those areas, and describe emerging fire-invasive issues in each bioregion.

Scope of This Volume

Other volumes in the Rainbow Series are framed only in terms of fire effects, but interactions between fire and nonnative invasive plants are more complicated. Nonnative invasive plant species may establish or increase in response to fire, but fire exclusion may also provide opportunities for invaders to establish in some plant communities (chapter 2). Fire can also be used to control invading plant species in some plant communities (chapter 4). Once invading species establish and begin to dominate a site, they may change many properties of fuel beds, which in turn may affect the fire regime (chapter 3). To capture the complex interrelationships between fire and nonnative invasive plant species, we follow three main themes throughout this volume: fire effects on nonnative plant invasions (including effects of fire exclusion policy and fire suppression tactics), changes in fuel characteristics and fire regimes caused by nonnative plant invasions, and the intentional use of fire to control nonnative invasives.

In this volume, "nonnative" refers to a species that has evolved outside the United States and has intentionally or unintentionally been transported and disseminated by humans into and within the United States (adapted from Li 1995). We include a few exceptions to this definition, however. For instance, in chapter 11 we address some species that are native to the mainland United States but not to the Hawaiian Islands; and in chapter 5 we include reed canarygrass (*Phalaris arundinacea*) and common reed (*Phragmites australis*), which have origins in both the United States and Eurasia, and have ecotypes or strains that are invasive in some situations. The patterns and processes described in this volume may also apply to species native to the United States that are spreading outside their historical ranges (for example, black locust (*Robinia pseudoacacia*), mesquite (*Prosopis* spp.), and juniper (*Juniperus* spp.)), but these species are not addressed in detail in this volume.

Only a small subset of nonnative plants is considered "invasive" (Rejmánek and others 2005; Williamson 1996). In this volume, the term "invasive" refers to a species that can establish, persist, and spread in a new area (*sensu* Burke and Grime 1996; Mack and others 2000; Sakai and others 2001) and also cause—or have potential to cause—negative impacts or harm to native ecosystems, habitats, or species. The decision of whether ecological changes or other impacts constitute net "harm" is a function of human values and can be ambiguous (see, for example, reviews by Lodge and others 2006; Mooney 2005). In this volume, harm in natural areas occurs when a species is so abundant that it causes significant changes in ecosystem composition, structure, or function, which are often viewed as harmful (Westbrooks 1998). Randall (1997) states this idea pragmatically: A plant species is considered invasive in natural areas when its occurrence interferes with management goals such as maintenance of native biotic diversity, protection of habitat for rare species, or restoration of ecological processes.

The basic biology of a nonnative species provides some insights regarding its potential invasiveness, so autecology is a useful starting point for understanding the potential impacts of disturbances, including fire, on nonnative species. However, genetic and clinal variation can lead to variable responses (Rejmánek and Richardson 1996; Wade 1997). Therefore, the nonnative species of greatest concern vary from region to region. In this volume, the nonnatives of greatest concern are listed at the beginning of each bioregional chapter, then discussed within the chapter in terms of both autecology and site-specific responses. Throughout this volume, the first reference to a species in each chapter gives both the common name most often used in the literature and the scientific name (from the U.S. Department of Agriculture's Integrated Taxonomic Information System (ITIS Database 2004)); subsequent references in the chapter use only the common name. Readers seeking information on a particular species should refer to the index.

Understanding the interactions among invasive species, native plant communities, and fire requires more than an understanding of autecology. As noted by Bazzaz (1986), "The colonizer and the colonized are partners in the process." Invasibility—the susceptibility of a plant community to invasion (*sensu* Davis and others 2000; Howard and others 2004; Lonsdale 1999; Smith and others 2004; Williamson 1996)—varies among plant communities. Additionally, the responses of plant communities to fire depend on a host of factors, including the frequency and severity of fire, season and spatial extent of burns, preburn vegetation occurrence (including nonnatives) and phenology, site conditions (particularly moisture, available nutrients, light, and disturbance history), and postfire conditions, including weather and availability of seed from invasive plants (Klinger and others 2006a; Pyke and others, in review; Stohlgren and others 2005). Such comprehensive knowledge is rarely available in the scientific literature in relation to invasive species (chapter 12), so managers must integrate incomplete knowledge, gathered at a variety of locations using different methods, for application to invasives in specific plant communities. McPherson (2001) comments that this "creative application of existing knowledge" is "as important, and as difficult, as the development of new knowledge." To assist managers in this formidable task, this volume draws mainly on peer-reviewed reports of primary research, that is, articles published in scientific journals and reports in reviewed scientific publication series, such as USDA Forest Service research papers and general technical reports. We also include information from case studies, informal reports, unpublished research, and personal communications. While this kind of information is probably accurate, its scope of inference is often difficult to determine. Therefore, the reader should use caution in applying that information, especially in locations or under conditions that differ from those described.

Secondary sources, including literature reviews, receive less emphasis in this volume than primary research because (1) reviews do not consistently report the scope of a study or its limits of inference, and (2) reviews occasionally misquote an original source. However, this volume does rely on *Fire Effects on Flora* (Brown and Smith 2000) for background information on fire regimes in the bioregional chapters in Part II. Additionally, reviews from two Internet sources, the Fire Effects Information System (FEIS, at www.fs.fed.us/database/feis) and The Nature Conservancy's (TNC) Element Stewardship Abstracts (ESAs, provided by the Invasive Species Initiative at http://tncweeds.ucdavis.edu/index.html), receive frequent use in this volume. Both sources include extensive information and are written specifically for managers. FEIS species reviews include biological and ecological information as well as information relating specifically to fire. ESAs focus especially on management and control techniques for invasive species. Literature reviews are identified as such the first time they are used in each chapter, so the reader will be aware that these are not primary sources. In regard to FEIS and TNC reviews, this note can also alert the reader that the Internet site may provide further information on fire, impacts, and control methods.

This volume is a survey as well as a synthesis, so it does not include every citation on every nonnative invasive plant species in the United States. Even unlimited resources would make a comprehensive treatment of fire and invasive species infeasible since new nonnative plants continue to invade the United States (Westbrooks 1998) and changing climatic conditions may alter competitive relationships between native and nonnative species (D'Antonio 2000). However, since the patterns described here are based on biological and ecological concepts, this volume may help managers evaluate the potential invasiveness of nonnative species about which little is known, the potential vulnerability of a particular ecosystem to establishment of new invasives, and how fire management activities may affect an ecosystem's susceptibility to invasion.

A literature review can provide perspective on a problem and suggest possible approaches for addressing it, but reviews and research from other areas cannot substitute for careful, long-term monitoring of what is taking place on the particular landscape of concern to a manager. Clear objectives, literature-based knowledge, and field-based knowledge are all essential for effective management. McPherson (2001) describes the complex challenges inherent in managing nonnative invasive species. He urges policy makers to resist applying oversimplified answers to complex issues such as those associated with biological invasions, and encourages managers to "synthesize disparate information for practical use and rely on all relevant knowledge at their disposal." As a synthesis of current knowledge on fire and nonnative invasives, this volume provides a tool that policy makers, managers, and members of the public can use in meeting management challenges.

Fire Behavior and Fire Regimes

We discuss here the key concepts of fire behavior and ecology that are used in this volume. A glossary of technical terms is in the appendix.

Fire behavior—The pattern of fire spread and heat release in an individual fire is referred to as fire behavior. It is affected by the fuel, weather, and topographic conditions at the time of burning. Fires are often described on the basis of the vertical stratum of fuels in which fire spread occurs. *Ground fires* spread in duff or peat (partly decayed organic matter in contact with the mineral soil surface). *Surface fires* spread in litter, woody material on or near the soil surface, herbs, shrubs, and small trees. *Crown fires* spread in the crowns of trees or tall shrubs (National Wildfire Coordinating Group 1996).

Individual fires are described quantitatively by *rate of spread*, *residence time*, *flame length*, and *flame depth* in the flaming front. They can also be described in terms of energy release. *Fire intensity* is the amount of heat released per unit time; this rate is described quantitatively by *fireline intensity*, the rate of heat release in the flaming front, regardless of its depth, and *reaction intensity,* the rate of heat release per unit area of the flaming front. *Total heat release* includes the heat produced in the flaming front and also behind the flaming front, by glowing and smoldering combustion (McPherson and others 1990).

Fire severity—The degree to which a site has been altered by fire is referred to as fire severity (National Wildfire Coordinating Group 1996). Descriptions of fire severity depend on the specific fire effects under study and the measurement methods used. Ecological studies typically examine relationships between fire severity and effects on individuals, populations, communities, and ecosystems. Management documents such as Burned Area Emergency Response plans relate fire severity to the effects of fire on soil stability. If fire severity is measured using aerial or satellite measurements, changes in the vegetation canopy (consumed, scorched, or intact) are the main descriptors. If fire severity is measured in sampling plots on the ground, descriptions of effects on individual organisms, litter and duff cover, and physical change in the soil can be used.

In this volume, we use "low severity" to refer to fires that cause little alteration to the soil and little mortality to underground plant parts or seed banks. "High severity" refers to fires that alter soil properties and/or kill substantial amounts of underground plant tissue. Ideally, primary research and postfire monitoring programs could describe fire severity either in terms of vegetation change, with detail as to the vegetation type and stratum discussed, or to changes in soil properties (DeBano and others 1998) (table 1-1). Relationships between fire severity and plant survival and persistence are discussed more thoroughly in chapter 2 (see "Influence of Fire Severity on Postfire Invasions" on page 12).

Fire regimes—The cumulative effects of fires over time have profound influence on ecosystem components, structure, and processes. The characteristic pattern of repeated burning over large expanses of space and long periods of time is referred to as the fire regime (Lyon and others 2000a; Sugihara and others 2006a). Fire regimes are described for a specific geographic area or vegetation type by the characteristic fire type (ground, surface, or crown fire), frequency, intensity, severity, size, spatial complexity, and seasonality. Fire frequency is described in this volume by the fire-return interval, the average time before fire reburns a given area.

The following categories are used to describe fire regimes in this volume, following Brown (2000):

- *Stand-replacement fire regime* refers to a pattern in which fire kills or top-kills the aboveground parts of the dominant vegetation. Using this definition, forests that routinely experience crown fire or severe surface fire have a stand-replacement fire regime; grasslands and many shrublands also have stand-replacement fire regimes because fire usually kills or top-kills the dominant vegetation layer.
- *Understory fire regime* understory fire regime" applies only to forests, woodlands, and shrublands. In a plant community with this kind of fire regime, most fires do not kill or top-kill the overstory vegetation and thus do not substantially change the plant community structure.
- *Mixed-severity fire regime* also applies only to forests, woodlands, and shrublands. In plant communities with this kind of fire regime, most fires either cause selective mortality of the overstory vegetation, depending on different species' susceptibility to fire, or sequential fires vary in severity.

In this volume, we use the term "presettlement" to describe fire regimes before extensive settlement by European Americans, extensive conversion of wildlands for agriculture, and effective fire exclusion. This term is common in the literature on fire regimes (for example, Bonnicksen and Stone 1982; Brown 2000; Brown and others 1994; Feeney and others 1998; Frost 1998) because settlement by Europeans is generally considered the time at which ecological communities began to deviate dramatically from conditions under

Table 1-1—Examples of fire severity descriptors.

What is being described?	Fire severity	Description
Forest, woodland, shrub canopy	high	Overstory foliage consumed by fire
	moderate	Overstory foliage killed by fire but not consumed
	low	Overstory foliage not altered by fire
Ground and soil surface[a]	high	Consumes or chars organic material below soil surface
	moderate	Consumes all organic material on soil surface
	low	Leaves soil covered with partially charred organic material

[a] Adapted from DeBano and others (1998).

which they evolved or at least persisted for centuries or millennia. As a set of conditions theoretically based in evolutionary time, presettlement conditions are considered more ecologically robust than current conditions and therefore may be used to determine desired conditions for wildlands. While the term "presettlement" is convenient for discussion, it remains imprecise and value-laden: (1) it is difficult to decide just when the "presettlement" era ends for any given location; (2) in many ecosystems, it is difficult to describe the presettlement fire regime precisely; and (3) presettlement conditions may not provide appropriate or achievable management goals. This becomes especially clear when one considers the impossibility of recreating the Native American influences from past eras and eliminating European American influences, including the introduction of thousands of nonnative species. For on-the-ground management of specific plant and animal communities, it may be more useful to refer to the desired fire regime as a "reference" or "baseline" for management, without insisting that past conditions be reconstituted on the landscape (chapter 3).

Organization and Use of This Volume

This volume synthesizes current scientific understanding of relationships between fire and nonnative invasive plant species at a conceptual level in part I and at a bioregional level in part II. In part III this information is synthesized further and related to specific management issues.

Chapters in part I describe botanical and ecological principles that govern relationships between fire and nonnative invasive plants. Chapter 2 describes the role of fire in promoting plant invasions, including traits that enable invasives to establish, persist, and spread after fire, and also the role of changes to plant communities brought about by fire (for example, altered nutrient availability and exposed mineral soil). It also discusses ways in which different plant communities may be more or less susceptible to establishment and spread of nonnative invasives in a postfire environment. Chapter 3 discusses the effects of nonnative invasive plants on fire regimes and includes information on fuel properties of plant communities influenced by invasive species. Chapter 4 addresses the use of prescribed fire to manage invasive plant species and interactions of fire with other control methods.

We have limited ability to predict the species likely to have negative impacts in a particular community (Woods 1997), so it is important for managers to have information about specific invasives and the relative invasibility of particular plant communities. Part II synthesizes this information at a bioregional level. We defined the bioregions according to broad plant community formations (fig. 1-1), based on Bailey's (1995) classification. Each chapter in part II lists the nonnative species of greatest

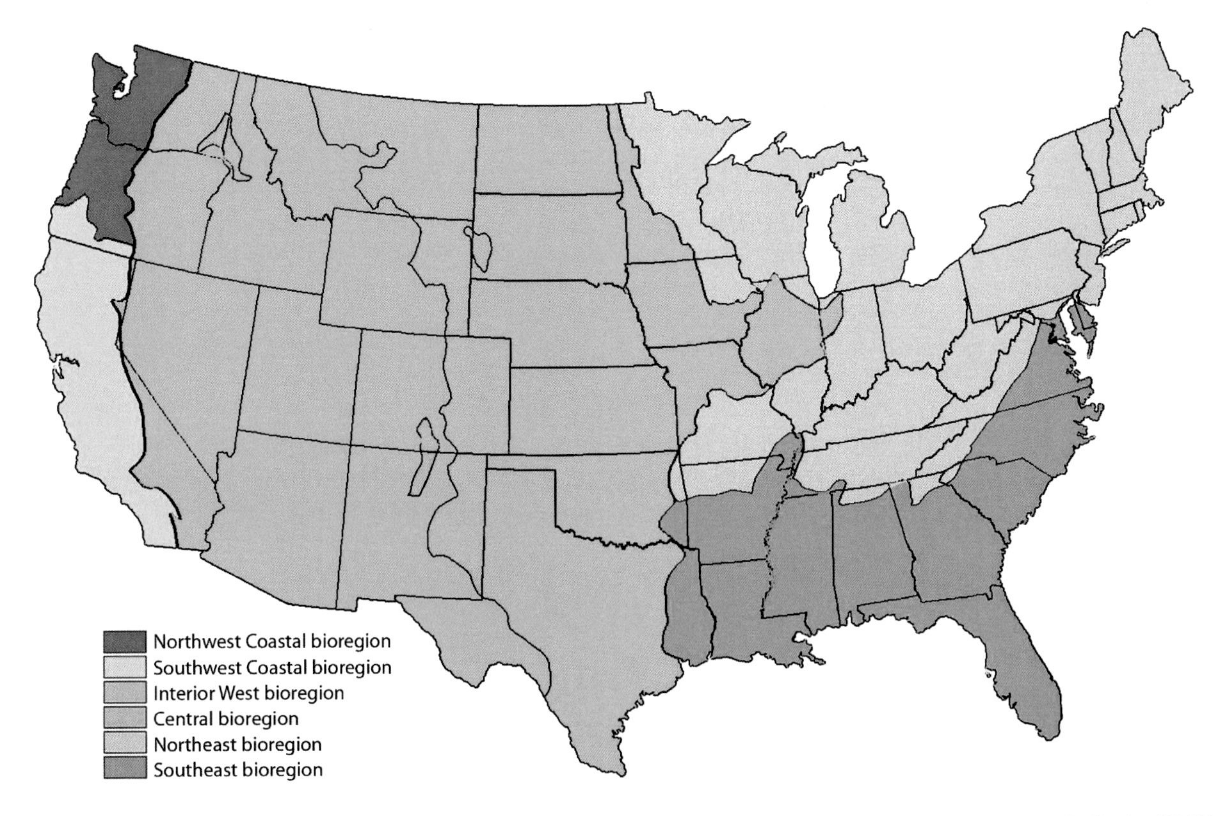

Figure 1-1—Approximate boundaries of bioregions defined for this volume. Bioregions are based on Bailey's (1995) classification. Alaska (included in chapter 10, Northwest Coastal bioregion) and Hawai`i (chapter 11) are not pictured.

concern in that bioregion and the plant communities considered most vulnerable to invasives. Thus, not all plant communities that occur in a bioregion are included in this volume.

The bioregional boundaries shown in figure 1-1 are, in reality, ecotones that may be hundreds of miles wide, and many invasive species occur in several bioregions. Most species are discussed in depth in only one bioregional chapter; in other chapters, discussion of the species focuses only on variations specific to the given bioregion. Each bioregional chapter synthesizes research and management information concerning fire effects on nonnative invasions, invasives' effects on fire regimes, and use of fire to control invasives. Finally, each chapter in part II identifies emerging issues in the bioregion regarding fire and invasives.

In addition to summarizing fire-invasive information for each bioregion, part II of this volume can be used as a reference on specific invasives and plant communities. To assess whether a particular community is likely to be susceptible to invasion by a particular nonnative species, first find the bioregional chapter that discusses that plant community. Look up the plant community in the table at the beginning of the bioregional chapter. Examine the list of invasive species of "high concern." Is the species you are seeking included? Does it have the same life form and regeneration strategies as species that are of high concern? Is it listed as a "potentially" invasive species for that community? The answers to these questions provide some indication of the potential threat.

Part III evaluates the knowledge available to managers on fire and nonnative invasive plants (chapter 12), summarizes effects of nonfire fuel management on nonnative species (chapter 13), describes the effects of fire suppression and postfire emergency stabilization, restoration, and rehabilitation treatments on nonnative invasive plants (chapter 14), and suggests monitoring strategies to evaluate the effects of fire management actions on nonnatives (chapter 15). The final chapter summarizes major issues reviewed in this volume, describes barriers to effective management and possible ways to address these, and lists current burning questions in relation to fire and nonnative invasive species.

This volume is intended as a review useful for managers, policy makers, and the general public. It provides a place to start when designing management plans that involve both fire and invasive plants. Final plans will typically require additional, more detailed, local information than is provided in this volume.

Kristin Zouhar
Jane Kapler Smith
Steve Sutherland

Chapter 2: Effects of Fire on Nonnative Invasive Plants and Invasibility of Wildland Ecosystems

Considerable experimental and theoretical work has been done on general concepts regarding nonnative species and disturbance, but experimental research on the effects of fire on nonnative invasive species is sparse. We begin this chapter by connecting fundamental concepts from the literature of invasion ecology to fire. Then we examine fire behavior characteristics, immediate fire effects, and fire regime attributes in relation to invasion potential. These concepts form the basis for examining the literature that supports or refutes several common generalizations regarding fire effects on nonnative invasives. We conclude with a summary of management implications regarding fire effects on nonnative invasive plants.

Invasion Ecology

Invasion ecology is influenced by interactions of ecosystem properties, properties of native and nonnative plant species, and nonnative propagule pressure (Lonsdale 1999) (fig. 2-1). Ecosystem properties include disturbance regimes and fluctuations in resource availability. In the context of invasion, this is the availability of resources needed by a nonnative species to establish, persist, and spread. Morphological properties, phenological properties, and competitive ability of native species influence resistance to invasion, while the same properties of nonnative species influence potential to invade. Native and nonnative plant responses to fire, such as damage or stimulation from heat and increases or decreases in postfire years, are particularly important for our discussion. Propagule pressure is the availability, abundance, and mobility of propagules in and around a plant community.

In this chapter, we examine several generalizations that have been suggested about wildland invasion by nonnative species after fire. We treat these generalizations as questions that can be examined in light of current research.

- Question 1. Does fire generally favor nonnative over native species?
- Question 2. Do invasions increase with increasing fire severity?

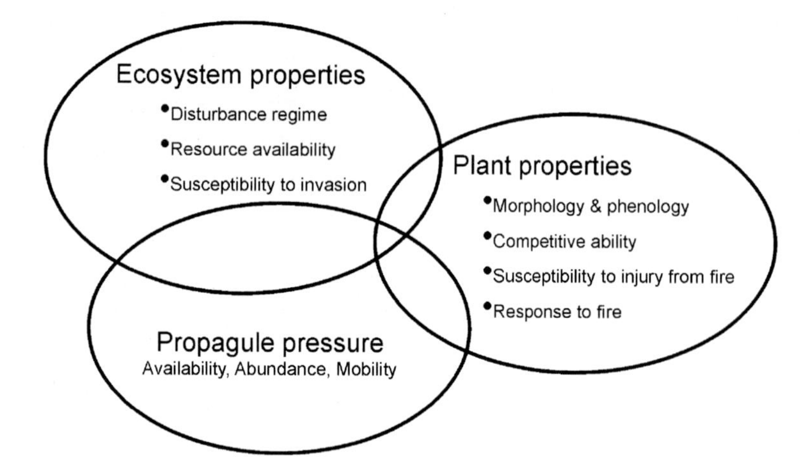

Figure 2-1—Susceptibility of a plant community to invasion by nonnative species after fire depends on properties of the ecosystem itself, properties of plant populations (both native and nonnative) and availability of nonnative plant propagules (following Lonsdale 1999).

- Question 3. Does additional disturbance (before, during, or after fire) favor invasions?
- Question 4. Do invasions become less severe with increasing time after fire?
- Question 5. Do invasions increase with disruption of the presettlement fire regime?
- Question 6. Are postfire invasions less common in high elevation ecosystems?

We will return to these questions after reviewing the connections between invasion concepts and fire.

Ecosystem Properties and Resource Availability

Invading species must have access to resources, including light, nutrients, and water, so community susceptibility to invasion can be explained to some extent by changes in resource availability. A species will "enjoy greater success in invading a community if it does not encounter intense competition for these resources from resident species" (Davis and others 2000). Therefore, a plant community becomes more susceptible to invasion when the amount of unused resources increases. Fire can increase resource availability by reducing resource use by resident vegetation (through mortality or injury) or by altering the form and availability of nutrients. Reports of postfire increases in nonnative species, often attributed to increased light or other resources, are available in the literature (for example, D'Antonio 2000, review; Hunter and others 2006; Keeley and others 2003).

Disturbed areas are often considered vulnerable to invasion (Sakai and others 2001), and burned areas are no exception. However, some plant communities that have evolved with recurring fire, such as California chaparral, are not considered highly invasible under their historic fire regime (Keeley 2001; Keeley and others 2003). Furthermore, divergence from the historic fire regime in the "opposite" direction—with reduced fire frequency or severity—may also increase invasibility. For invasion to actually occur, community susceptibility and resources adequate for the spread of nonnative species must coincide with availability of propagules of nonnative species that can successfully compete with native vegetation for those resources (Davis and others 2000).

Properties of Native and Nonnative Plants

Several plant characteristics influence their susceptibility to fire injury, ability to recover and compete for resources following fire, and changes in cover and dominance over time after fire. Fire has less potential to kill individuals and impact populations if the species' meristem tissues, buds, and seeds are protected from heat-caused damage. Avoidance of heat damage can be based on structural features, location of meristematic tissues, or phenology (see "Influence of Fire Season and Plant Phenology on Postfire Invasions" page 18). Elevated buds, thick bark, and underground vegetative structures can provide protection.

Fire survivors and species that form persistent seed banks, native or nonnative, have early access to resources on a burned site (see "Influence of Fire on Resource Availability and Interactions Between Plant Species" page 10). They can spread by regenerating from surviving structures or establishing from the seed bank, then producing abundant seeds that establish on the exposed mineral soil seedbed (see "Downward heat pulse effects on plant survival" page 14). Many nonnative invasives are annuals or biennials with short generations, ability to self pollinate, and low shade tolerance (Sutherland 2004). A review by

Barrett (2000) highlights the relationship between disturbance and opportunistic or invasive species with short life cycles, well-developed dispersal powers, and high reproductive output. These typically ephemeral species can establish on burned sites only if abundant propagules are available from the soil bank or nearby unburned areas. Unless they alter ecosystem processes to perpetuate early-seral conditions, ephemerals are often replaced by perennials within a few years after fire (see "Question 4" page 25).

Nonnative Propagule Pressure

The spatial distribution of nonnative source populations and their mode of propagule dispersal influence their establishment and spread in new areas (Amor and Stevens 1975; Giessow and Zedler 1996; Wiser and others 1998), including burns (Keeley and others 2003). Some seeds are heat tolerant and therefore may survive a fire onsite (Volland and Dell 1981). Postfire establishment and spread of nonnative species depends, in part, on propagule pressure (*sensu* Colautti and others 2006; Drake and Lodge 2006; Lockwood and others 2005)— the abundance of nonnative propagules occurring onsite and within dispersal distance of the burned area. D'Antonio and others (2001b) contend that variation in propagule supply interacts with the "ecological resistance" of an ecosystem (*sensu* Elton 1958) such that when resistance is low, few propagules are needed for successful invasion, and as resistance increases it takes proportionately more propagules for invaders to establish. In a meta-analysis designed to examine characteristics of invasiveness and invasibility, Colautti and others (2006) found that while propagule pressure was rarely considered in studies of biological invasions, it was a significant predictor of invasibility. More disturbance and higher resource availability are also significant predictors of invasibility (Colautti and others 2006), though field studies rarely isolate these factors and measure their influence quantitatively.

The scientific literature provides numerous examples of a positive relationship between anthropogenic disturbance and nonnative invasive species richness and abundance. Where burns are associated with anthropogenic disturbance, they are likely to be subject to greater propagule pressure and may therefore be more susceptible to postfire invasion than burns in less disturbed areas (see "Question 3" page 22). The generally positive relationship between nonnative invasive species and anthropogenic disturbance (for example, Dark 2004; Johnson and others 2006; McKinney 2002; Moffat and others 2004) may have implications for plant communities and bioregions that currently show relatively little effect of fire on nonnative plant invasions. Fire appears only weakly related to spread of nonnative species in the Northeastern bioregion (chapter 5), but the plethora of nonnative invasive species present in this region (Mehrhoff and others 2003) and the prevalence of anthropogenic disturbance suggest that, if burning increases, impacts from nonnative species may increase as well. Similarly, while fire-caused increases in nonnative invaders are currently uncommon in Alaska (chapter 10), expanding human influences on wildlands coupled with climate change may increase problems with nonnative plants in that state. Similar concerns have been voiced regarding invasive species in Colorado shortgrass steppe (Kotanen and others 1998) and may apply in many areas of the United States.

Influence of Fire on Invasions

The responses of plants to fire depend on both fire attributes and plant attributes relating to survival and establishment (Pyke and others, in review). Nonnative plants that survive on site, establish from the seed bank, or disperse seed into burns soon after fire have early access to resources that are more plentiful or more available after fire. Fire behavior characteristics, immediate fire effects, and fire regime attributes (table 2-1) influence persistence of on-site populations and postfire establishment from on- and off-site sources. While fire behavior characteristics are often measured and recorded on wildfires, they are not as clearly related to invasiveness and invasibility

Table 2-1—Fire attributes that can influence invasion by nonnative plant species. Concepts listed here are defined in greater detail in chapter 1 and the glossary.

Fire behavior attributes	Immediate fire effects	Fire regime attributes
Fire type (ground, surface, and/or crown fire)	Fuel consumption	Fire type (ground, surface, and/or crown fire)
Fireline intensity	Soil heating pattern	Intensity
Rate of spread	Total heat release	Frequency
Residence time	Burn pattern	Severity
Flame length	Crown scorch	Size
Flame depth	Crown	Spatial complexity
Reaction intensity	Consumption	Seasonality
	Smoke production	

as immediate fire effects (most of which relate to fire severity) and fire regime attributes. Fire type, severity, and frequency affect the persistence of invasive populations and their potential for spread within burned areas. Spatial characteristics (fire size, the distribution of burned and unburned patches, and the spatial pattern of fire severity) influence the potential for establishment from unburned areas. Burn season also influences nonnative plant response, especially as it interacts with plant phenology and vulnerability to heat damage.

Influence of Fire on Resource Availability and Interactions Between Plant Species

Superior competitive ability is often used to explain postfire invasions by nonnative species. Explanations for spread of invasives based on competition theory include the natural enemies hypothesis (Elton 1958; Mack and others 2000), the evolution of increased competitive ability hypothesis (Blossey and Notzold 1995), and the novel weapons hypothesis (Callaway and Ridenour 2004). These hypotheses are supported by examples of particular species, but their relevance to fire has not been demonstrated. We focus here on theoretical concepts leading to the expectation that fire will alter the competitive balance between native and nonnative species, and empirical evidence of such impacts.

A review of nonfire competition experiments from world literature suggests that at least some nonnatives are better competitors than native species, with the caveat that "Invaded communities are not random assemblages, and researchers tend to study the most competitive alien plants" (Vilà and Weiner 2004). Reduction of nonnatives can lead to increases in native species, as demonstrated in the Mojave Desert (Brooks 2000). These results are consistent with the hypothesis that nonnatives can outcompete natives for limited resources in early successional environments (MacDougall and Turkington 2004).

If fire increases resource availability on an invaded site, the relative competitive abilities of the species present should theoretically determine which will benefit most from the increased resources. However, competitive interactions between native and nonnative species are poorly understood and difficult to measure in the field. Studies conclusively demonstrating postfire competition between native and nonnative species for a specific resource are lacking. This is not surprising when one considers the scale and methodological differences between fire research and competition research. Most fire research addresses plant communities with high spatial variability and variation in fire severity, many native and sometimes many nonnative species, and several resources altered by fire. In contrast, most competition studies are comparisons of paired species under carefully controlled conditions (Vilà and Weiner 2004).

Availability of Specific Resources—Many disturbances increase the availability of resources for plant growth and thus have the potential to increase a community's susceptibility to invasion (Davis and others 2000). Fire can increase light availability by reducing cover. It can increase water availability by killing vegetation and thus the demand for moisture. It can increase nutrient availability by killing vegetation and also by converting nutrients from storage in biomass to forms that can be absorbed by plants. By consuming surface organic layers, fire increases exposed mineral soil; while not a resource itself, mineral soil exposure affects postfire germination and establishment (reviewed in Miller 2000).

Postfire increases in the availability of specific resources are likely to be interrelated and subtle. For example, by reducing the quantity and vigor of existing vegetation, a canopy fire may increase not only light levels but also moisture availability; however, increased exposure to light and wind and decreased albedo may dry the surface layer of the soil. A surface fire that causes lethal crown scorch may initially increase moisture availability and mineral soil exposure but have little effect on light availability; then, as foliage is cast, light levels will increase and exposed mineral soil will decrease. Most surface and ground fires increase mineral soil exposure, but they may increase or decrease nutrient availability depending on fire severity. Correlations between levels of different resources make it difficult, if not impossible, to isolate plant response to fire-caused changes in a particular resource.

The light available to understory species after fire in forests, woodlands, and some shrublands increases as canopy cover and woody basal area decrease (for example, Keyser and others 2006). Increased light in the understory is generally associated with increased cover and biomass of understory species and sometimes with increased species richness (for example, Battles and others 2001; Messier and others 1998; Son and others 2004). This pattern has been well documented following fire (Miller 2000, review). Nonnative invasive species that are shade-intolerant—the majority (Sutherland 2004)—are likely to benefit from increased light if they survive or establish in forests, woodlands, or shrublands after fire. If the canopy closes with time after fire, decreased light levels may then reduce the abundance of shade-intolerant nonnative species (see "Question 4" page 25). However, nonnative species that persist at low abundance or maintain a viable soil seed bank when the canopy closes may increase rapidly when fire or another disturbance opens the canopy and again increases available light.

Growing-season fires reduce aboveground vegetation, so they are likely to reduce moisture uptake by plants, at least temporarily (Knoepp and others 2005; Neary and Ffolliot 2005). These changes can increase the moisture available to sprouting plants and seedlings, although the increase may be offset by runoff and evaporation from exposed mineral soil (DeBano and others 1998). In shrub-steppe ecosystems of the Great Basin, soil moisture patterns on burned sites differ both spatially and temporally between burned and unburned sites; these differences may affect the success of nonnative species relative to native species (Prater and others 2006). However, research to date has not isolated soil moisture as the cause of postfire spread of nonnative species. The effect can be inferred in a grassland study in which the effect of late spring prescribed fire on Kentucky bluegrass (*Poa pratensis*) was related to postfire moisture. Burning reduced this nonnative grass significantly on sites that experienced subsequent dry growing conditions but not on sites that had abundant postfire moisture (Blankespoor and Bich 1991). Conversely, smooth brome (*Bromus inermis*) decreases when postfire moisture availability is high and increases when available moisture is low. The authors suggest that when soil moisture is high, native warm-season grasses are able to outcompete fire-injured smooth brome for water; and when less soil moisture is available, native grasses are less competitive (Blankespoor and Larson 1994).

Fire mineralizes several plant nutrients, including nitrogen, potassium, and phosphorus, releasing them from complex molecules in tissues and either volatilizing them or depositing them in forms that are more available for plant uptake (Anderson and others 2004; Bauhus and others 1993; Keeley and others 2003; White and Zak 2004). We focus here on nitrogen, since fire research on this plant nutrient is somewhat more complete than on others. Nitrogen often limits plant growth because it is used in many organic molecules essential for life, including proteins and DNA. When plants and litter are burned, some of the nitrogen from organic compounds is volatilized, and the rest remains on site as ammonium and nitrate—small ions that plants can readily absorb with soil water (Knoepp and others 2005). Subsequent changes in soil biota also affect availability of these ions to plants (Blank and others 1996). A meta-analysis of the effect of fire on nitrogen in forests, shrublands, and grasslands (Wan and others 2001) found no significant effect on total nitrogen but a significant short-term increase in available soil nitrogen (ammonium and nitrate). Ammonium usually peaked immediately after fire, while nitrate peaked 7 to 12 months after fire. Fire-caused increases in available nitrogen were transitory. In the 22 studies analyzed, ammonium and nitrate returned to prefire levels within 3 to 5 years after fire.

Increases in available nitrogen generally favor nonnative annual species over native perennials (McClendon and Redente 1992). Cheatgrass (*Bromus tectorum*), for example, effectively uses both patches and early pulses of nitrogen, which may contribute to its successful competition with perennials for available nitrogen (Duke and Caldwell 2001). No studies elaborate on nitrogen's influence on annual-perennial relationships in the postfire environment, though a similar relationship might be assumed with the postfire flush of available nitrogen. Several reviews link postfire increases in nitrogen to increased nonnative plant biomass (for example, Brooks 1998; 2002; Floyd and others 2006; Hobbs and others 1988; Huenneke and others 1990). However, none has demonstrated a link between increased nitrogen and increased nonnative abundance at the expense of native species. Research in Hawai`i demonstrated that nonnative grasses, which convert native Hawaiian woodlands to fire-maintained grasslands, alter the seasonal pattern of nitrogen availability to plants (Mack and D'Antonio 2003); however, this change was described as a likely result of the invasion rather than the cause of it.

Fires consume litter and organic layers, exposing mineral soil, a condition that may favor nonnative invasive species. A meta-analysis of the impact of litter on understory vegetation indicated a generally inhibiting effect on germination, establishment, and productivity (Xiong and Nilsson 1999), though the analysis did not differentiate between nonnative and native plants. Postfire research in northern Arizona ponderosa pine (*Pinus ponderosa* var. *scopulorum*) forests suggested that sites with bare mineral soil and little litter favored nonnative plants, whereas native herbs were more tolerant of litter cover (Crawford and others 2001). In contrast, abundance of nonnative grasses Japanese brome (*Bromus japonicus*), soft chess (*Bromus hordeaceus*), and Italian ryegrass (*Lolium multiflorum*) may be reduced by fire when litter is removed because they rely on the moisture retained in the litter layers for germination and establishment (see "Influence of Weather Patterns on Postfire Invasions" page 19).

A small body of research focuses on establishment of nonnative species on burned versus unburned soil or effects of ash on establishment. Research in piñon-juniper woodlands suggests that some nonnative species have an affinity for burned microsites within larger harvested units (for example, prickly lettuce (*Lactuca serriola*), Japanese brome, and London rocket (*Sisymbrium irio*)). Other species, Dalmatian toadflax (*Linaria dalmatica*), white sweetclover (*Melilotus album*), and red brome (*Bromus rubens*), showed no preference for burned soil (Haskins and Gehring 2004). Scotch broom (*Cytisus scoparius*) establishment may even be reduced by exposure to ash (Regan 2001).

Several nonnative invasives occur on harvested forest sites following broadcast burning and appear to prefer burned microsites (chapter 10). Maret and Wilson (2000) found that several nonnative species in western Oregon prairies showed similar emergence but better survival on burned than unburned plots.

Influence of Fire Severity on Postfire Invasions

Fire severity is a measure of a fire's effects on an ecosystem. Specifically, it is the degree to which a site has been altered by fire (National Wildfire Coordinating Group 1996; chapter 1). Fire severity is complex, difficult to measure and predict, and not directly linked to the difficulty of controlling fire, so it is not monitored or reported as regularly as fire behavior descriptors. Nevertheless, an understanding of fire severity is crucial to understanding differential effects of fire on different plant species.

Fire severity is often described as the result of both an upward heat pulse and a downward heat pulse, which are not necessarily correlated (Neary and others 2005b; Ryan and Noste 1985). The upward heat pulse is formed by the flaming front and described by rate of spread, fireline intensity, and flame length. The downward heat pulse is influenced to some extent by the flaming front, especially flaming zone depth, residence time, and reaction intensity. It is influenced more strongly by total heat production and duration, including smoldering and glowing combustion. Upward and downward heat pulses depend to some extent on fire type. Ground fires may heat the soil substantially without producing a strong upward heat pulse. In contrast, surface and crown fires can produce long flames and strong upward heating but may move too fast to ignite ground fuels or heat the soil appreciably.

Upward Heat Pulse—The upward heat pulse from a fire largely determines survival of aboveground plant tissues. Just as for native plants (Miller 2000), nonnatives with aboveground parts that most often survive surface or crown fire are trees with a high canopy, protected buds, and/or thick bark. For example, a single fire results in little mortality of mature, nonnative melaleuca (*Melaleuca quinquenervia*) trees despite a high occurrence of torching and crown fire. Additionally, melaleuca has an aerial seed bank consisting of canopy-stored seed that survives even severe fire. It is one of the first species to germinate after fire in many habitats in southern Florida and can subsequently establish large seedling populations (Munger 2005b, FEIS review). Herbaceous species that retain mature seeds in inflorescences such as nonnative annuals in the Mojave Desert (Brooks 2002), medusahead (*Taeniatherum caput-medusae*) (Pyke 1994), and diffuse knapweed (*Centaurea diffusa*) (Watson and Renney 1974) may be more susceptible to seed mortality from fire than species with soil-stored seed. Of course, this depends on timing of fire relative to eventual seed dispersal.

Seedlings and saplings of woody plants are more susceptible to mortality from fire than larger individuals because their buds are closer to the ground, and their bark—which can protect the cambium from heat damage—is generally thinner (Morgan and Neuenschwander 1988). For example, while mature common buckthorn (*Rhamnus cathartica*) (Boudreau and Willson 1992), Chinese tallow (*Triadica sebifera*) (Grace and others 2005), and Brazilian pepper (*Schinus terebinthifolius*) (Meyer 2005a, FEIS review) often survive and/or sprout from underground parts after fire, their seedlings and saplings are typically killed by fire.

Downward Heat Pulse—The downward heat pulse from fire influences survival of belowground plant tissue, survival and potential heat scarification of buried seed, consumption of soil organic matter, changes in soil texture and water-holding capacity, and changes in soil nutrient availability—all of which influence the potential for nonnative species to establish, persist, and spread on burned sites, possibly at the expense of native species. An understanding of this aspect of fire severity is crucial for understanding mortality and survival of plants and seeds that are present on a site when it burns. Research investigating the relationship between the downward heat pulse and abundance of nonnative invasive species is discussed below (see "Question 2" page 22). Fundamental aspects of the downward heat pulse are summarized here; more detailed discussion is provided by DeBano and Neary (2005) and Knoepp and others (2005).

The peak temperature reached in soils during fire usually declines rapidly with depth (for example, Beadle 1940; DeBano and others 1979; Neal and others 1965; Ryan and Frandsen 1991). Sites with dry soils that are without heavy fuel loads may burn with no change in temperature 1 to 2 inches (2 to 5 cm) beneath the soil surface (DeBano and others 2005; Whelan 1995). Surface and crown fires may heat the soil relatively little if it is insulated by thick surface organic horizons ("duff") that do not burn (DeBano and others 2005; Hartford and Frandsen 1992). Duff in forests of western Montana and northern Idaho was unlikely to burn if its moisture content exceeded 60 percent, though it burned even without continued heat from surface fire if its moisture content was less than 30 percent (Hartford and Frandsen 1992).

Duff moisture changes in response to long-term weather patterns (Alexander 1982). When duff becomes dry enough and is subjected to sufficient heat to ignite, seeds and plant parts within it are consumed, its insulating value decreases, and it begins to contribute to the fire's downward heat pulse. Because of its high bulk density, duff burns slowly, usually with smoldering rather than flaming combustion, and produces an ash layer that can provide a new form of insulation—preventing heat from dispersing upward (DeBano and others 2005).

The heat produced by fire interacts with soil moisture and soil physical properties in complex ways to influence soil heating. Following are some of the principles governing heat transfer into the soil:

1. More heat input is needed to increase the temperature of moist soils than dry soils (DeBano and others 1998; 2005).
2. The thermal conductivity of moist soils may increase with increasing temperature (Campbell and others 1994).
3. When soil water is heated to vaporization, it absorbs substantial heat from its surroundings. When the resulting steam moves to an area of low temperature, it condenses and releases substantial heat to its surroundings (DeBano and others 1998; 2005).

If managers want to minimize soil heating from prescribed burns, the planning process should address both fuel load and the desired soil moisture range. Frandsen and Ryan (1986) found that increasing soil moisture reduced the maximum temperature and duration of soil heating beneath burning fuel piles. Busse and others (2006) measured temperature regimes in soils under approximately 60 tons/acre of masticated fuels and found that soil moisture greater than 20 percent (by volume) kept mineral soil temperatures below 140 °F (60 °C) at depths greater than 2 inches (5 cm); lower soil moisture allowed for greater soil heating. For a review of soil heating models and heat transfer in soil, see Albini and others (1996).

The effects of soil heating on survival of underground plant parts depend on both temperature and duration of heating (Hare 1961). Lethal temperatures for plant tissues generally range from about 104 to 158 °F (40 to 70 °C); some seeds can survive exposure to much higher temperatures (Hungerford and others 1991; Levitt 1980; Volland and Dell 1981). Fire effects studies sometimes use maximum temperature to describe fire severity, but elevated temperatures lasting only seconds are much less likely to kill or damage living tissue than the same temperatures sustained for minutes or hours. Species differ in their susceptibility to heat, and the time needed to kill plants of a given species decreases exponentially as temperature increases (fig. 2-2). Therefore, a time-temperature profile of the soil is a better indicator of fire effects on underground plant parts than a maximum temperature profile. Time-temperature profiles generally show the surface layers reaching higher temperatures than deeper layers, which is consistent with maximum temperature profiles; in addition, they often show shorter duration of elevated temperatures at the soil surface than in deeper layers (for example, Hartford and Frandsen 1992; Ryan and Frandsen 1991). Figure 2-3 shows time-temperature profiles measured during an August prescribed fire under two mature ponderosa pines in northwestern Montana. Maximum temperatures in litter and duff were higher than in deeper layers, but the duration of elevated temperatures increased with depth.

Fire effects on plant tissues also vary with the moisture content and metabolic state of the tissues, a topic addressed in more detail under "Influence of Fire Season and Plant Phenology on Postfire Invasions" page 18.

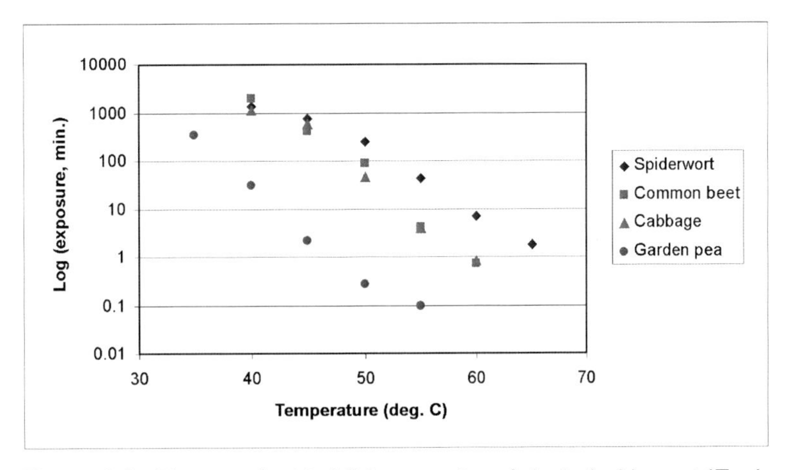

Figure 2-2—Time required to kill four species of plants (spiderwort (*Tradescantia* sp.), common beet (*Beta vulgaris*), cabbage (*Brassica oleracea*), and garden pea (*Pisum sativum*)) at a range of temperatures, adapted from Levitt (1980), from laboratory research conducted in Germany. Note that the y axis is a logarithmic scale. Highest time values for spiderwort, beet, and pea are approximate or the midpoint of a range.

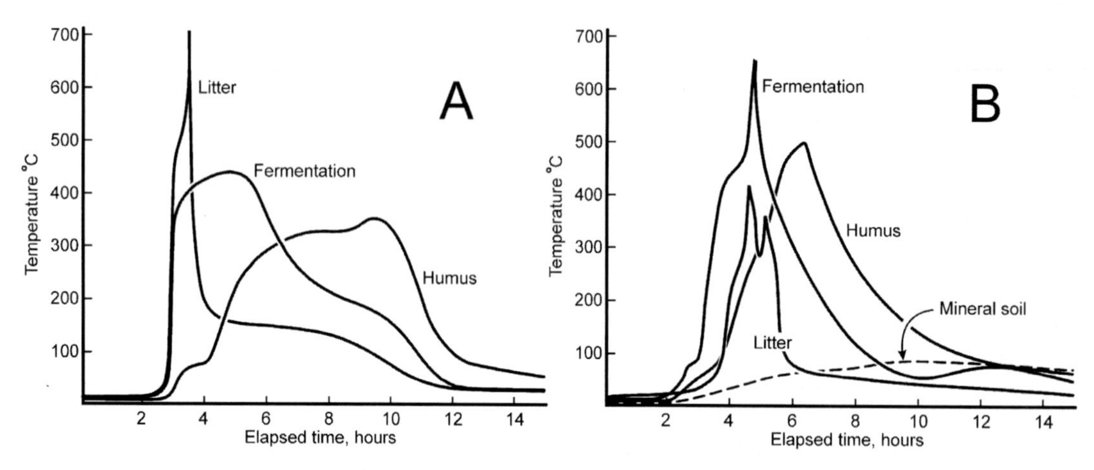

Figure 2-3—Time-temperature profiles under mature ponderosa pine canopy in northwestern Montana (Ryan and Frandsen 1991). (A) Downward-spreading ground fire in litter and fermentation layer (duff) 7 cm deep. (B) Laterally spreading ground fire in litter/duff 17 cm deep. (Adapted with permission from *International Journal of Wildland Fire,* CSIRO Publishing, Melbourne Australia.)

Downward heat pulse effects on plant survival—The ability of individual plants to survive fire depends on the temperature regime at the location of their perennating tissues (Miller 2000) and is thus related to fire severity. Raunkiaer (1934) classified plants according to their means of surviving freezing temperatures, by noting the vertical position of their perennating tissues, or buds, above and below the soil surface (table 2-2). This classification can be adapted to explain plant response to fire by considering not only dormant buds that survive fire but also adventitious buds that sprout after the plant is top-killed. Many plants survive lethal heating of aboveground tissues because their underground parts are capable of producing new stems, roots, and leaves (Smith 2000).

Table 2-2—Effects of fire on Raunkiaer (1934) plant life forms.

Raunkiaer life form	Example	Perennating tissue	Potential injury from fire
Therophytes	Annuals	Seeds that reside on or under the soil surface, or on senesced plants	Depends on where seeds are located during fire
Chamaephytes	Some shrubs & herbs	Perennial tissue and/or adventitious buds just above the soil surface	Often killed by fire due to position within the flaming zone
Hemicryptophytes	Rhizomatous plants, root sprouters	Perennial tissue and/or adventitious buds just above or below the soil surface	Depends on their location in organic or mineral soil; combustion of litter and duff; amount of soil heating from smoldering combustion; can be as well protected as bulbs or corms
Cryptophytes	Plants with bulbs or corms	Perennial tissue and/or adventitious buds well below the soil surface	Protected from all but severe fires due to insulation from soil
Phanerophytes	Trees & tall shrubs	Perennial tissue and/or epicormic buds well above the soil surface	Can be killed by crown fires, which consume the canopy, or by surface fires if severe enough to kill the cambium or perennating buds

Plants with buds located in the combustible organic layers of soil can survive if the organic matter does not burn. Plants with buds in mineral soil have greater potential to survive; the deeper the perennating tissue, the more likely their survival. Dormant and adventitious buds can occur on stolons, root crowns, rhizomes, roots, caudices, bulbs, and corms (fig. 2-4). Stoloniferous plants have stems or branches that grow on the surface and can sprout from buds along their length. Because of their position, stolons are likely to be damaged by fire. In contrast, buds in the root crown, the transition area between stem and root, are somewhat better protected from fire because of their position at or beneath the surface, possible insulation from bark, and thermal mass. Rhizomes (usually horizontal) and caudices (vertical) are plant stems growing within the organic or mineral soil, and roots also grow in these layers (fig. 2-5). Duff and mineral soil may insulate their buds from heat damage, especially when they are located in mineral soil, well below the surface. Bulbs and corms—underground plant storage organs bearing roots on their lower surfaces—usually grow below the organic layer in mineral soil and are well protected from all but severe ground fires.

Figure 2-5—Sulfur cinquefoil sprouting, probably from surviving caudex, within a month after September wildfire in a western Montana mountain grassland. (Photo by Peter Rice.)

Several species of nonnative invasive trees sprout after fire. Examples include melaleuca (Munger 2005b), Chinese tallow (Meyer 2005b, FEIS review), and tamarisk (*Tamarix* spp.) (Zouhar 2003c, FEIS review). However, severe fire can kill both Chinese tallow (Grace and others 2005) and tamarisk (Ellis 2001). Fire can top-kill most nonnative invasive shrubs, but many persist via underground tissues, with survival depending on fire severity. For example, gorse (*Ulex europaeus*) occurs in heathlands in its native range, where it responds to low-severity surface fire by sprouting from the basal stem region. Under these circumstances, postfire vegetative regeneration of gorse can be prolific and rapid. However, severe ground fire, which consumes most or all of a deep organic surface horizon, typically kills gorse (Zouhar 2005d, FEIS review). Mortality of Scotch broom in Australia (Downey 2000) and Washington (Tveten and Fonda 1999), and French broom (*Genista monspessulana*) in California (Boyd 1995; 1998) also appears to be related to fire severity. Gorse and brooms also establish prolifically from on-site seed in the postfire environment, with abundance dependent on fire severity (see "Downward heat pulse effects on seed" page 16).

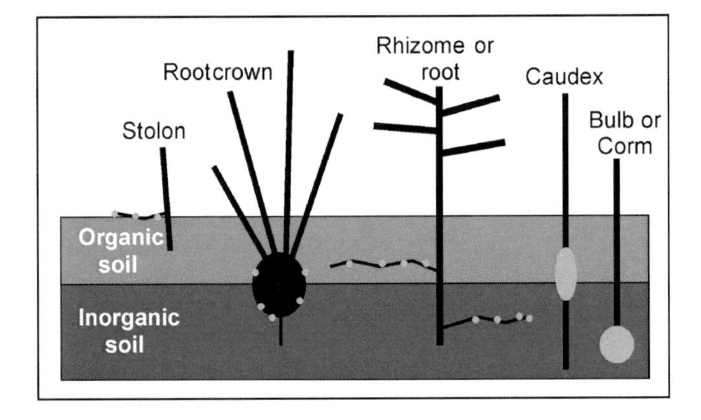

Figure 2-4—Organs that may enable perennial plant species to survive fire. Green circles indicate meristem tissues. Native species may have any of these organs; none of the nonnative species discussed in this volume have bulbs or corms. Organic soil horizons are not always distinct, as suggested by this diagram; they often intergrade with top layer of mineral soil. For more discussion of plant organs in relation to fire, see Miller (2000).

Most grasses are top-killed by fire. However, perennial grasses sprout seasonally, so removal of aboveground biomass in itself is not a factor affecting postfire survival. Rhizomatous grasses have an extensive underground network of rhizomes that are likely to survive and sprout after fire. There are many examples of invasive perennial grasses that sprout after fire, especially in the Central (chapter 7), Interior West (chapter 8), and Hawaiian Islands (chapter 11) bioregions. These species typically respond with rapid sprouting and high fecundity in the postfire environment (see "Question 1" page 20). However, some nonnative perennial grasses may be killed by high-severity fire. Redtop (*Agrostis gigantea*), for example, a rhizomatous grass introduced from Europe, generally survives fire (Carey 1995, FEIS review) but may be killed by ground fires in peat (Frolik 1941). Crested wheatgrass (*Agropyron cristatum*) and desert wheatgrass (*Agropyron desertorum*) are nonnative bunchgrasses that generally burn quickly, transferring little heat into the soil. However, fires that smolder in the dense clusters of stems in these bunchgrasses, burning for extended periods after the fire front has passed, are likely to kill them (Skinner and Wakimoto 1989).

Fires may kill seedlings and top-kill adult nonnative perennial forbs, but adult plants typically survive and sprout from perennial underground parts after fire. This is probably because even the most severe fires rarely damage plant tissues below 2 inches (10 cm) in the soil, while perennial rhizomes and roots with dormant or adventitious buds on some plants can penetrate the soil to a depth of several feet. Information on differential effects of fire severity was found for several invasive perennial forbs, but results from these studies are inconclusive due to incomplete information. A study from Australia suggests that St. Johnswort (*Hypericum perforatum*) mortality increases with increased fire severity (Briese 1996); and two studies from central Illinois suggest that garlic mustard (*Alliaria petiolata*) may be sensitive to severe fire (Nuzzo 1996; Nuzzo and others 1996).

Where information on response to fire is lacking for most nonnative species, cautious inferences can be made based on plant morphological traits. Species with subterranean dormant and adventitious buds are likely to survive and sprout following fire (Goodwin and others 2002). Perennial woody and herbaceous species known to sprout following mechanical damage or top-kill by means other than fire may be capable of similar responses to fire if their perennating tissues are protected from the downward heat pulse. However, fires and mechanical disturbances alter a site in different ways, so biological responses cannot be assumed to be equivalent.

Downward heat pulse effects on seed—Like plants that survive a fire onsite, residual colonizers (species that leave viable seed onsite even if mature plants are killed by fire) have early access to resources in the postfire environment. Seed survival depends on seed location relative to the occurrence of lethal temperature regimes. Because grasses produce fine fuels with high surface-to-volume ratios, fuel consumption in grassland communities is often rapid, residence times are short, and lethal temperatures may not occur at the soil surface (Daubenmire 1968a). Invasive grasses and grassland invaders with seed that frequently survives fire include medusahead (Blank and others 1996), cheatgrass (Evans and Young 1987), yellow starthistle (*Centaurea solstitialis*) (Hastings and DiTomaso 1996), and filaree (*Erodium* spp.) (chapter 9). Seeds in soil organic layers may be killed or consumed by fire if the organic material burns, but seed at the mineral soil surface may survive even where litter is burned (for example, Japanese brome (Whisenant 1985)). It should be noted that grassland fires do not always produce mild temperature regimes. For example, in prescribed fires on the Texas plains, the maximum temperature recorded at the soil surface was 1260 °F (682 °C), and temperatures exceeded 150 °F (66 °C) for as long as 8.5 minutes (Stinson and Wright 1969).

Buried seeds are most likely to survive fire. For example, tumble mustard (*Sisymbrium altissimum*) has tiny seeds that fall into fire-safe microsites such as soil crevices (Howard 2003b, FEIS review), and cutleaf filaree (*Erodium cicutarium*) seed is driven into the soil by the styles (Felger 1990)—traits that may protect these seeds from fire. However, seeds buried too deep in the soil may fail to establish if they require light for germination (for example, bull thistle (*Cirsium vulgare*) and Lehmann lovegrass (*Eragrostis lehmanniana*)) or if endosperm resources are depleted before the seedlings emerge from the soil. The effect of seed burial depth on germination is demonstrated by several nonnative invasives. Rattail sixweeks grass (*Vulpia myuros*), a nonnative annual grass, germinated more successfully from 0.5-inch (1-cm) depth than from 2-inch (5-cm) depth in a greenhouse study; seedlings emerging from 5 cm weighed significantly less than seedlings from 1 cm (Dillon and Forcella 1984). Optimum germination of spotted knapweed (*Centaurea biebersteinii*) seed occurs with the seeds at the soil surface and decreases with depth, with little germination below 2 inches (5 cm) (Spears and others 1980; Watson and Renney 1974). Scotch broom germination rates are highest in the top inch (2 cm) of soil, and seedlings do not emerge from below 3 inches (8 cm) (Bossard 1993). St. Johnswort seed germination is limited in the dark, and seedlings emerging from seed buried as little as 0.5 inch (1 cm) rarely survive (Zouhar 2004, FEIS review).

Seed bank formation is complex and depends on many factors, including (1) seed rain, dormancy, predation, longevity, and size; (2) soil texture, structure, deposition, and compaction; and (3) movement of seeds by wind, earthworms, insects, and animals (Baskin and Baskin 2001). Because of this complexity, seed longevity under field conditions is rarely known accurately and is often estimated from field observations and laboratory studies. Several nonnative forbs have been reported to regenerate from a soil seed bank, including bull thistle (Doucet and Cavers 1996), St. Johnswort (Zouhar 2004), tansy ragwort (*Senecio jacobaea*) (chapter 10), and common groundsel (*S. vulgaris*) (Zammit and Zedler 1994). Establishment and spread of these species is triggered by disturbances that remove existing vegetation. Gorse, Scotch broom, and French broom all form seed banks (Zouhar 2005a,c,d). The ability of these shrubs to establish large numbers of seedlings after fire is related to prolific seed production, longevity of viable seed, and a scarification requirement for germination.

Seeds of several nonnative species are stimulated to germinate by exposure to heat or fire. Brooms and gorse seed germination is stimulated by heat scarification (Zouhar 2005a,c,d). Mimosa (*Albizia julibrissin*) seeds exposed to flame in the laboratory had higher germination rates than unheated seeds (Gogue and Emino 1979). Lehmann lovegrass seeds are dormant at maturity, but seed on the soil surface can be scarified either by fire or by high summertime seedbed temperatures (Sumrall and others 1991). Field observations (Briese 1996; Sampson and Parker 1930; Walker 2000) and laboratory tests (Sampson and Parker 1930) suggest that fire stimulates germination of St. Johnswort seed. Most yellow sweetclover (*Melilotus officinalis*) and white sweetclover seeds can remain viable in the seed bank for 20 to 40 years (Smith and Gorz 1965; Smoliak and others 1981; Turkington and others 1978) and have hard seed coats that require scarification for germination (Smith and Gorz 1965). Fire aids establishment of sweetclover in grasslands, probably because it scarifies seed and simultaneously creates openings in which sweetclover can establish (Heitlinger 1975). Soil heating by fire may promote kudzu (*Pueraria montana* var. *lobata*) germination by scarifying the seedcoat, allowing water to penetrate (Munger 2002b, FEIS review).

While heat stimulates seed of some nonnative invasives to germinate, it inhibits others. Examples from laboratory tests include spotted knapweed (Abella and MacDonald 2000), bull thistle, woodland groundsel (*Senecio sylvaticus*) (Clark and Wilson 1994), common velvetgrass (*Holcus lanatus*) (Rivas and others 2006), and Johnson grass (*Sorghum halepense*) (Mitchell and Dabbert 2000). Other species show reduced establishment following fire in the field. Menvielle and Scopel (1999, abstract) report that the surface seed bank of chinaberry (*Melia azedarach*) is completely killed by fire, although there was "some" emergence from buried seed. Brooks (2002) found that nonnative annuals (red brome, Mediterranean grass (*Schismus* spp.), and cutleaf filaree) in the Mojave Desert responded to different temperature regimes in different microsites. The highest temperatures occurred under the canopies of creosotebush (*Larrea tridentata*) shrubs (where the most fuel was consumed), and these microsites had reduced biomass of the nonnative annuals for 4 years after fire. At the canopy dripline, where temperatures were lower, annual plant biomass was reduced for 1 year, while negligible postfire changes occurred in interspace microsites, where fire produced little soil heating (Brooks 2002).

Influence of Fire Frequency on Postfire Invasions

The relationship between nonnative species and fire frequency has received little attention outside the context of control efforts (chapter 4). A plant's response to fire frequency should theoretically be related to its life history, morphology, and maturity. Many annuals can persist under a regime of frequent, even annual, burning if their seeds are protected from heat and subsequent growing conditions are favorable (see table 2-2). Examples among nonnative species that persist under a regime of frequent fire include many annual grasses and forbs in the Great Basin and California (chapters 8 and 9). Exceptions include medusahead and ripgut brome (*Bromus diandrus*), which showed a significant decrease in abundance after two consecutive burns (DiTomaso and others 2006b), probably because their seeds are not protected from heat. Another exception is prickly lettuce in Central bioregion tallgrass prairie (Towne and Kemp 2003). In fact, native prairie species tend to be adapted to frequent fire and can often resist invasion by nonnatives under a regime of frequent fire (chapter 7). This is especially evident in large, intact ecosystems with low propagule pressure, as compared to fragmented landscapes with large pools of nonnatives present (Smith and Knapp 2001).

The ability of perennial species to persist through repeated fires depends on protection of their meristem, buds, and seed from heat and their ability to replenish energy stores and buds after fire (Whelan 1995). Most perennial herbs are vulnerable to fire as seedlings, so repeated fires at short intervals are likely to reduce establishment. Ability to withstand fire is likely to increase with maturity if underground structures expand (Gill 1995). Unfortunately, literature describing responses of nonnative plants to differing fire intervals is rarely available (chapter 12). Results of studies in different locations can be compared, but it is difficult to ascertain whether differing results are caused by different fire frequencies or by other variables, such as community properties and fire severity and seasonality. For example, spotted knapweed abundance tends to increase after single fires in ponderosa pine communities in the Interior West bioregion (chapter 8), while in Michigan, annual spring prescribed burning under severe conditions (when humidity and dead fine fuel moisture are as low as possible) reduces spotted knapweed populations and increases the competitiveness of the native prairie vegetation (J. McGowan-Stinski, personal communication 2001).

Woody species seedlings also tend to be susceptible to fire, though fire resistance for many increases with age as bark thickens, underground structures expand, and bud-bearing stems become taller. Even though mature melaleuca trees are very resistant to damage from repeated fires (Geary and Woodall 1990), most seedlings (up to 12 inches (30 cm) tall) are killed by fire (Timmer and Teague 1991). Similarly, Grace and others (2005) describe prescribed fires that killed all Chinese tallow less than 4 inches (10 cm) tall and 40 percent of those 4 inches to 3 feet (10 cm to 1 m) tall (Grace and others 2005). It is not surprising that research is lacking on the effects of varying fire frequencies on invasive trees, since this information can only be obtained from long-term studies. Considering the potential for interactions among carbohydrate reserve patterns, fluctuating resources in the ecosystem due to fire, heat damage to plants and secondary damage from insects and pathogens, and competitive interactions among species, it is difficult to accurately predict the effects of varying fire frequencies on long-lived woody species without field research, and even then results are likely to be specific to the plant community studied. Long-term research is needed on how varying fire intervals and their interactions with fire severity and seasonality affect nonnative plants.

Influence of Spatial Extent and Uniformity of Fire on Postfire Invasions

The availability of propagules within a burn and from nearby unburned sites depends on fire size, patchiness, and uniformity of fire severity. Giessow and Zedler (1996) found that rates of establishment of nonnative species declined with distance from source populations. If a burned area is large, species establishing from off-site are likely to be represented by long-distance seed dispersers. Several nonnative species with small, wind-dispersed seed are reported in early postfire communities (see "Question 1" page 20). Animal dispersal of invasive plant seeds after fire has not been documented in the literature, but this mode of establishment seems likely for many nonnative invasives, such as Brazilian pepper in the Southeast (Ewel and others 1982) and numerous shrubs and vines in the Northeast (chapter 5).

When burned areas occur in patchy vegetation or a highly fragmented landscape, rates of postfire establishment of nonnative species can be high (Allen 1998; Minnich and Dezzani 1998). After comparing the establishment of nonnative invasive species from small and large species pools in Kansas tallgrass prairie, Smith and Knapp (2001) suggest that increasing fragmentation of ecosystems will increase invasibility. Cole (1991) notes that sweetclover may persist despite repeated burns to control it if the fires are patchy, leaving some of the seed bank intact and enabling second-year shoots to survive. Keeley and others (2003) found that nonnatives were uncommon in unburned chaparral but persistent in adjacent blue oak (*Quercus douglasii*) savannas. Because these two communities occurred in a mosaic, nonnatives rapidly established in patches of burned chaparral from the savanna. Nonnatives in chaparral constituted 8 percent of the plant species present 1 year after fire, 23 percent the second year, and 32 percent the third year.

Variation in fire severity (which may result from patchy vegetation, variation in fuel structure and moisture, or other factors) may also increase the susceptibility of a site to spread of invasives (see "Question 2" page 22).

Influence of Fire Season and Plant Phenology on Postfire Invasions

Fire effects on plant tissues vary with the moisture content and metabolic state of the tissues themselves (Hare 1961; Volland and Dell 1981). More heat is required to raise the temperature of large, thick tissues than fine ones (Hungerford and others 1991; Levitt 1980; Whelan 1995), so lignotubers and thick rhizomes are generally less susceptible to fire damage than root hairs and mycorrhizae at the same depth in the soil. In addition, actively growing plants generally suffer damage at lower temperatures than seeds or dormant plants of the same species (Volland and Dell 1981). Kentucky bluegrass, for example, flowers early and is dormant by mid-summer. The species is not usually damaged by late-summer fire unless it occurs during drought (Uchytil 1993, FEIS review). This variation in fire response may be related to the higher water content of growing than dormant plants (Zwolinski 1990) or the lack of stored carbohydrates available for regrowth if plants are burned during the growing season (Whelan 1995). Phenological patterns may interact with soil moisture patterns to influence a species' susceptibility to heat damage, since plant and soil moisture may vary together through the seasons.

Influence of fire season on invasions by nonnative species is not often described in the scientific literature (chapter 12). Since temperate herbs die back to the ground at the end of the growing season, dormant season fires usually have little impact on their survival. Growing season fires are more likely to cause direct mortality, damage actively growing tissues, deplete resources, and increase postfire recovery time (Miller 2000) in herbaceous plants and woody species as well. A review by Richburg and others (2001) suggests that prescribed burns conducted in the Northeast bioregion during the dormant season ultimately increase the density of invasive woody species. Similarly, dormant season and growing season burns do not differ in immediate damage to Chinese tallow, but growing season fires result in weaker recovery and greater long-term impacts to this species (Grace and others 2005). Additionally, season of burning may indirectly affect postfire response of a particular species due to its relationship to fire severity. For example, higher severity of fall fires versus spring fires may account for the significantly higher mortality and lower basal sprouting of Scotch broom following fall burning (Tveten and Fonda 1999). Managers can take advantage of differences in phenology between nonnative invasives and desired native species in planning burns to increase dominance of desired species (chapter 4).

Influence of Weather Patterns on Postfire Invasions

Weather patterns, especially timing and amount of precipitation, may be decisive in determining the ability of nonnative invasive species to establish, persist, and spread. This may be particularly evident in arid and semiarid communities. Abundance of nonnative annuals in desert shrublands, for example, is strongly affected by precipitation patterns (chapter 8). Increased fuel loads and continuity in years with above-average precipitation can increase the probability that an area supporting nonnative annual grasses will burn in the following dry season (Knapp 1995, 1998). When these annual grasses persist and spread after fire, creating conditions favorable for more fire, a grass/fire cycle may result (chapter 3).

Postfire weather conditions affect the ability of nonnative invasive species to persist and spread after fire (D'Antonio 2000). Melaleuca seedling establishment, for example, is affected by timing and amount of precipitation relative to burning (Munger 2005b; chapter 6). In a central Utah sagebrush (*Artemisia tridentata* ssp. *wyomingensis*) community, postfire abundance of cheatgrass over a 20-year period seems closely tied to precipitation patterns—declining during drought and increasing during wet periods (Hosten and West 1994; West and Hassan 1985; West and Yorks 2002).

Species exhibiting a reduced abundance after fire coincident with lower than average postfire precipitation include yellow starthistle, sulfur cinquefoil (*Potentilla recta*), and Japanese brome. A single burn typically increases germination and density of yellow starthistle, but fire eliminated yellow starthistle on a site that experienced drought after fire (DiTomaso and others 2006b). For 2 years following an August wildfire in grasslands dominated by bluebunch wheatgrass (*Pseudoroegneria spicata*) and Sandberg bluegrass (*Poa secunda*) in Idaho, yellow starthistle canopy cover increased significantly—probably aided by substantial precipitation the month after the fire (Gucker 2004). In a northwestern Montana rough fescue (*Festuca altaica*) grassland, small prescribed burn treatments were followed by increased density of small sulfur cinquefoil plants, but the population then decreased under the drought conditions that prevailed during the 5-year study (Lesica and Martin 2003). The current view of fire effects on Japanese brome is based on its requirement for sufficient moisture to establish and the role of plant litter in retaining soil moisture (Whisenant 1989). Fire kills the majority of Japanese brome plants and much of the seed retained by the plant and also removes the litter layer, so populations of Japanese brome are often substantially reduced following fire (for example, Ewing and others 2005; Whisenant and Uresk 1990). When fall precipitation is plentiful, however, litter is not required for successful establishment and populations can rebound immediately (Whisenant 1990b). Moisture availability may influence Japanese brome population dynamics more than fire (chapter 7).

In contrast to the above examples, burned perennial Lehmann lovegrass may increase under postfire drought conditions. This species exhibited no reduction in biomass production during an experimental drought (Fernandez and Reynolds 2000) and had greater reproductive output on burned versus unburned plots during 2 years of lower than average precipitation after fire in the High Plains of Texas (McFarland and Mitchell 2000).

Generalizations About Fire Effects on Nonnative Invasives

In the previous sections, we applied concepts of invasion ecology, fire behavior, fire regimes, and competition to the potential effects of fire on nonnative invasive species. In this section, we use that conceptual basis to examine several generalizations about postfire invasion that are often suggested. We treat these generalizations as questions and explore their applicability and scope using examples from the scientific literature. While several of the generalizations are supported by examples from the literature, each one also has exceptions. The take-home message of this analysis is that, while generalizations are useful for *describing and explaining* fire's relationship to nonnative invasive species, they have limited usefulness for *predicting* what will happen on a given site after a given burn. Generalizations can alert the manager to what might happen after fire; but local knowledge of plant communities, the status of nonnative species, and the burn itself (especially severity and uniformity) are essential for managers to select and prioritize management actions that will minimize ecosystem impacts from nonnative species after fire and to avoid management actions that are unnecessary and could themselves cause environmental damage.

One problem with most scientific literature on nonnative species is that the species considered have been selected for study because they are problematic. They are usually among the ~1 percent of nonnative species that become invasive (causing ecological or economic harm) or are otherwise considered pest species (Williamson 1993; Williamson and Brown 1986). Because researchers tend to study the most invasive nonnative species (for example, see Vilà and Weiner 2004), it is worthwhile to keep in mind that (1) not all nonnative species are invasive, (2) no invasive species causes harm in every native plant community in which it occurs, and (3) some nonnatives currently considered innocuous may eventually cause ecological damage to a native community. As mentioned frequently in this volume, local knowledge is as important as an understanding of general concepts relating to nonnative species and fire.

Question 1. Does Fire Generally Favor Nonnatives Over Natives?

Generally speaking, if a fire occurs in a plant community where nonnative propagules are abundant and/or the native species are stressed, then nonnative species are likely to establish and/or spread in the postfire environment. To what degree they will dominate, and for how long, is less clear. Chapter 12 points out the lack of long-term studies on nonnative species after fire.

The interaction between fire and nonnative species is complex and research results are limited and variable. A review of recent research on fire and nonnative species (D'Antonio 2000) supports the contention that accidental and natural fires often result in increases in some nonnative species. However, the scope of the review is limited with regard to North American plant communities: It includes studies from five habitats in California, two in the Great Basin Desert, three in the Sonoran Desert, one in Canada, and one in Hawai`i (D'Antonio 2000). The bioregional discussions in this volume provide a more comprehensive review of nonnative-fire interactions in North American plant communities. Community-level information supporting postfire increases in nonnatives is available for California grasslands and shrublands (chapter 9), desert shrublands (chapter 8), wet grasslands invaded by melaleuca (chapter 6), and closed-canopy forests (chapter 10). In addition, a growing body of literature describes postfire increases of nonnatives in other communities, including forests dominated by ponderosa pine (for example, Cooper and Jean 2001; Phillips and Crisp 2001; Sackett and Haase 1998; Sieg and others 2003;) piñon-juniper (*Pinus* spp. – *Juniperus* spp.) woodlands (chapters 8 and 9), and Hawaiian shrublands and grasslands (chapter 11). These bioregional discussions also present exceptions. Fires in grasslands and prairies, which have evolved with frequent fire, often favor native species over nonnatives (chapters 6, 7, 8, 10). In other plant communities (for example, Oregon white oak (*Quercus garryana*) woodlands) and other bioregions (for example, the Northeast and Alaska), information on interactions between fire and invasives does not follow a consistent pattern. Because research is limited and results are variable, the generalization that fire favors nonnatives over natives cannot be applied to all nonnative species or all ecosystems. A breakdown based on postfire regeneration strategies of the nonnative species may be more helpful.

Survivors—Most nonnative perennial species studied have the ability to sprout from root crowns, roots, or rhizomes following top-kill or damage. For several of these species, the literature reports postfire sprouting. Some reports also note that these species spread in the postfire environment, although information regarding their effects on native communities, especially over the long term, tends to be sparse.

Nonnative invasive woody species are most common in the Northeast, Southeast, and Northwest Coastal bioregions, and in riparian communities in the Interior West and Central bioregions. Melaleuca (Munger 2005b; chapter 6) and tamarisk (Busch 1995; Ellis 2001) are known to sprout after fire with greater vigor than associated native species and tend to dominate postfire communities. The woody vine kudzu, known to dominate plant communities in the Southeast bioregion to the detriment of native species, sprouts from the root crown after fire and may return to previous levels of dominance by the second postfire growing season (Munger 2002b). For other woody species, however, such as tree-of-heaven (Gibson and others 2003), Russian-olive (USDA Forest Service 2004), and autumn-olive (chapter 5), reports of postfire sprouting tend to be anecdotal, and postfire consequences to ecosystems are not described.

Nonnative shrubs such as chinaberry, bush honeysuckles (*Lonicera* spp.), and glossy buckthorn (*Frangula alnus*) are known to sprout after fire, but information on postfire response in invaded communities is limited and sometimes conflicting. For example, chinaberry exhibited vigorous crown and root sprouting from adventitious buds after fire in Argentina (Menvielle and Scopel 1999, abstract). It is speculated that postfire sprouting of this type can lead to spread of chinaberry (Tourn and others 1999), but to date, no fire research on this species has been published from North America. Several studies indicate limited mortality and basal sprouting in bush honeysuckles after fires in spring, summer, and fall (for example, Barnes 1972; Kline and McClintock 1994; Mitchell and Malecki 2003), but none provide information beyond the first postfire year. Glossy buckthorn was reported to increase after fire in a calcareous fen in Michigan (chapter 5) and in an alvar woodland in Ontario (Catling and others 2001), but this species was also strongly associated with unburned alvar woodland (Catling and others 2002), so the specific effects of fire are unclear. In the Northwest Coastal bioregion, several nonnative woody species including blackberries (*Rubus* spp.), Scotch broom, sweetbriar rose (*Rosa eglanteria*), and common pear (*Pyrus communis*) sprout from underground parts after fire, often with increased stem density; however, effects of these species on the native community after fire are not well described. Additionally, these species also spread with fire exclusion in some communities (chapter 10).

Nonnative perennial herbs such as Canada thistle (*Cirsium arvense*) (Zouhar 2001d, FEIS review), spotted knapweed (MacDonald and others 2001; Zouhar 2001c, FEIS review), Dalmatian toadflax (*Linaria dalmatica*) (Jacobs and Sheley 2003a), St. Johnswort (Zouhar 2004), and sulfur cinquefoil (Lesica and Martin 2003) tend to survive fire and may spread in postfire communities (see "Downward heat pulse effects on plant survival" page 14). But postfire dominance is likely to vary with plant community, fire frequency, and fire severity. For example, spotted knapweed and Canada thistle may increase in abundance in ponderosa pine and closed-canopy forests after fire, while in native prairies, where the dominant native species are well adapted to frequent fire, their abundance may be reduced by fire (see "Influence of Fire Frequency on Postfire Invasions" page 17, and "Question 5" page 27).

In Hawai`i, nonnative perennial grasses and nonnative Asian sword fern (*Nephrolepis multiflora*) survive fire and can respond with increased cover at the expense of native species (chapter 11). For example, fountain grass (*Pennisetum setaceum*) can sprout rapidly following top-kill and set seed within a few weeks (Goergen and Daehler 2001). In another study, total nonnative grass cover was about 30 percent higher and total native species cover lower in burned than unburned transects 2 to 5 years after fire (D'Antonio and others 2000; Tunison and others 1995). Asian sword fern is observed to sprout shortly after fire and quickly dominate the understory in mesic `ōhi`a forest (Ainsworth and others 2005; Tunison and others 1995).

Seed Bankers—Residual colonizers with surviving viable seed in the soil after fire have early access to resources and may dominate the postfire environment, at least in the short term. Several examples are presented above (see "Downward heat pulse effects on seed" page 16).

Flushes of seedlings from heat-scarified seed in the soil seed bank can be dramatic, so these species tend to dominate immediately after fire. Examples include brooms (Zouhar 2005a,c), St. Johnswort (Sampson and Parker 1930; Walker 2000), and lovegrasses (*Eragrostis* spp.) (Ruyle and others1988; Sumrall and others 1991). Dense populations of these species can persist in some communities. For example, Scotch and French broom form dense thickets in California grasslands. Flushes of broom seedlings after fire (for example, see Haubensak and others 2004) are likely to maintain populations of these species indefinitely (for example, Boyd 1995, 1998).

Other species that establish from the soil seed bank include annual grasses and forbs, though they may not dominate until the second or third postfire season, and may or may not persist. Density and timing of postfire dominance by these species may depend on precipitation (see "Influence of Weather Patterns on Postfire Invasions" page19). Once established, populations can persist for many years. For example, in a Wyoming big sagebrush shrub-steppe community on the Snake River Plain south of Boise, Idaho, cover of nonnative annual grasses was sparse in control plots, which were dominated by predominantly native species, while nonnative annuals dominated burned plots 10 years after fire (Hilty and others 2004).

Seed Dispersers—Dramatic postfire increases in nonnative species with wind-dispersed seed are commonly described in the literature, although the seed source is rarely indicated so establishment may be from the soil seed bank in some cases. Nonnative species with small, wind-dispersed seed often occur and sometimes dominate burned forest sites in the early postfire environment in the Interior West and Northwest Coastal bioregions. Examples include Canada thistle (Floyd and others 2006; MacDougall 2005; Turner and others 1997; Zouhar 2001d), bull thistle (MacDougall 2005; Zouhar 2002b, FEIS review), musk thistle (*Carduus nutans*) (Floyd and others 2006), wild lettuces (*Lactuca* spp. and *Mycelis* spp.) (Agee and Huff 1980; Sutherland, unpublished data 2008; Turner and others 1997), tansy ragwort (Agee and Huff 1980), hairy catsear (*Hypochaeris radicata*) (Agee and Huff 1980), common velvetgrass (Agee 1996a,b), and dandelion (*Taraxacum officinale*) (Wein and others 1992). These species tend to be absent from adjacent undisturbed forest. Their abundance usually peaks 2 to 4 years after fire, after which their numbers decline (see "Question 4" page 25). However, there are exceptions to this pattern. For example, piñon-juniper communities have supported populations of Canada and musk thistle for over 13 postfire years (Floyd and others 2006), and many species can survive on site through viable seed in the soil seed bank from which seedlings can establish after another disturbance (Clark and Wilson 1994; Doucet and Cavers 1996). Other invasives that establish after fire via long-distance seed dispersal include princesstree (*Paulownia tomentosa*) in the Northeast bioregion (Reilly and others 2006), cogongrass (*Imperata cylindrica*) (Mishra and Ramakrishnan 1983) in the Southeast bioregion, and fountain grass in Hawai`i (Nonner 2006).

Species With Increased Fecundity After Fire—Several species produce unusually large seed crops in the postfire environment. For instance, the August following a stand-replacing fire at Lees Ferry, Arizona, 69 percent of burned tamarisk plants were blooming heavily, while on adjacent unburned sites 11 percent of tamarisk plants were blooming (Stevens 1989). Other perennials showing an increase in flowering and seed production after fire include Dalmatian toadflax (Jacobs and Sheley 2003a) and St. Johnswort (Briese 1996). Annual species often produce more seed in burned than unburned sites, allowing the annuals to spread rapidly during the time when resource availability may be high (Brooks and Pyke 2001). Examples include cheatgrass in diverse habitats (Mojave desert, sagebrush (*Artemisia* spp.) grasslands, and dry ponderosa pine and grassland in Idaho) (Zouhar 2003a, FEIS review), yellow starthistle in California grasslands (Hastings and DiTomaso 1996), and annual vernal grass (*Anthoxanthum aristatum*) in Oregon white oak woodlands (Clark and Wilson 2001).

Exceptions—While many studies support the generalization that nonnatives increase after fire, the above discussion illustrates substantial variation. Additionally, impacts from postfire invasions are not well documented, especially over the long term.

Some species can be reduced by fire (chapter 4), and some research demonstrates that fire exclusion contributes to invasion of native plant communities that have evolved with frequent fire. For example, native prairies are invaded by woody species and cool-season grasses in the Central bioregion (chapter 7). Oregon white oak woodlands and Idaho fescue (*Festuca idahoensis*) prairies are invaded by nonnatives Scotch broom and Himalayan blackberry (*Rubus discolor*) in the Northwest Coastal bioregion (chapter 10). Wet grasslands and pine habitats are invaded by nonnative woody species in the Southeast (chapter 6), and oak forests and savannas are invaded by nonnative shrubs and vines in the Northeast (chapter 5).

Question 2. Do Invasions Increase With Increasing Fire Severity?

Several researchers report greater abundance of nonnative species following high-severity fire compared with unburned or low-severity burned sites. Definitions of fire severity vary in these accounts, with some relating severity to canopy removal and others relating it to litter or fuel consumption and/or ground char. On conifer sites in California, abundance of nonnative species was low in virtually all burned sites, but was greatest in areas with high-severity fire (Keeley and others 2003). Similarly, nonnative species cover in ponderosa pine forests of Colorado, New Mexico, and Arizona was positively correlated with fire severity and reduction of tree cover (Crawford and others 2001; Hunter and others 2006). High-severity burn patches were associated with establishment of nonnative invasive species such as tansy ragwort and common velvetgrass in closed-canopy forests in the Northwest Coastal bioregion (Agee 1996a,b). Establishment of prickly lettuce was greatest in high-severity burn patches in forests of Yellowstone National Park (Turner and others 1997) and ponderosa pine forests in Idaho (Armour and others 1984). Much of the literature on burning of slash piles, which produces high-severity patches in an otherwise unburned site, indicates that ruderal species (native and nonnative) establish readily in burned patches, but persistence is variable (see Question 3 below).

Although several studies support the generalization that severe fire leads to increased establishment and spread of nonnative species, fire obviously has the potential to consume all living tissue if it is severe enough, and high fire severity has also been associated with decreases in nonnative species abundance. For example, seed banking species may show lower establishment in microsites that experience high temperatures for long durations (for example, see Brooks 2002) (also see "Downward heat pulse effects on seed" page 16). Similarly, sprouting species may have greater mortality after high-severity fires (see "Downward heat pulse effects on plant survival" page 14). Where burning is severe enough to kill both sprouters and seed bankers, postfire invasion depends on propagule pressure from outside the burned area.

Question 3. Does Additional Disturbance Favor Invasions?

Postfire establishment of nonnative species may be exacerbated by other types of disturbance. This is related to the observations that postfire species composition is strongly related to prefire composition, and disturbance tends to increase nonnative abundance in communities that are already severely invaded (Harrison and others 2003). This section first demonstrates that nonnative species are often associated with nonfire disturbances, then examines the evidence that fire exacerbates establishment and spread of nonnatives—whether nonfire disturbances occur before, during, or after fire. Postfire establishment of nonnative species may also be enhanced in areas subjected to postfire rehabilitation activities (see chapter 14).

The scientific literature is rich with examples of relationships between site disturbance and nonnative species richness and abundance. At regional or landscape scales, richness and abundance of nonnative invasive plants tend to be lower in protected or undeveloped areas than in human-dominated landscapes or landscapes fragmented by human use (Barton and others 2004; Ervin and others 2006; Forcella and Harvey 1983; Huenneke 1997; McKinney 2002; Pauchard and Alaback 2004), although exceptions to this pattern are noted in some locations (for example, Fornwalt and others 2003; also see chapter 13). High nonnative species abundance and richness often occur in areas of high road density (for example, see Dark 2004), large human populations, a history of human occupation, and agricultural use of surrounding areas (Johnson and others 2006; McKinney 2002; Moffat and others 2004). Regional variation in the number of nonnative plant species is positively correlated with human population density ($R^2 = 0.58$, $P = 0.01$) (fig. 2-6). An analysis of nonnative species richness using broad geographic regions (the Geographic Area Command Centers for fire management) shows a 10-fold difference in the number of nonnative species between the South (1,981) and Alaska (193), with areas of highest human population density (California, the South, and the East) having the most nonnative species (data from Kartesz and Meacham 1999).

At local scales, nonnative invasive species richness and abundance are generally highest in and around disturbed patches, corridors, and edges such as small animal disturbances (for example, Larson 2003), riparian corridors (for example, DeFerrari and Naiman 1994), and transportation corridors (roadsides, old road beds, and/or trails) (Benninger-Truax and others 1992; Flory and Clay 2006; Frenkel 1970; Gelbard and Belnap 2003; Gelbard and Harrison 2003; Harrison and others 2002; Larson 2003; Parendes and Jones 2000; Parker and others 1993; Reed and others 1996; Tyser and Worley 1992; Watkins and others 2003; Weaver and others 1990). Forest edges typically have higher nonnative plant abundance than forest interiors (Ambrose and Bratton 1990; Brothers and Spingarn 1992; Fraver 1994; Hunter and Mattice 2002; Ranney and others 1981; Robertson and others 1994; Saunders and others 1991; Williams 1993). Features common in logged areas such as skid trails are also likely to support populations and propagules of nonnative plants (Buckley and others 2003; Lundgren and others 2004; Marsh and others 2005; Parendes and Jones 2000). Similarly, areas with fuel treatments, including forest thinning (chapter 13) (fig. 2-7), fuel breaks (Giessow and Zedler 1996; Keeley 2006b; Merriam and others 2006), and firelines (for example, Benson and Kurth 1995; Sutherland, unpublished data 2008), often support higher abundance of nonnatives than nearby untreated areas.

While there is concern regarding the effects of livestock grazing on changes in community composition including effects on the abundance of nonnative plants, relatively few quantitative studies are available on this topic. In a review of the literature on disturbance and biological invasions, D'Antonio and others (1999) found that a majority of the available studies suggest a correlation between livestock grazing and nonnative species abundance. A small number of case studies from

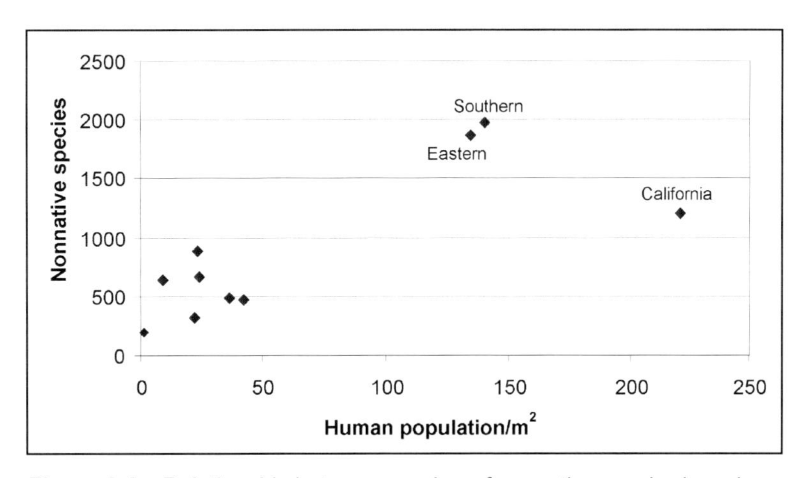

Figure 2-6—Relationship between number of nonnative species in various regions of the United States and human population density. The three most populous areas are labeled. The remaining areas are the Northern and Central Rocky Mountains/plains, East and West Great Basin, Northwest, Southwest, and Alaska (which has the lowest population density). Hawai`i is not included. Regions used here are based on the national Geographic Area Coordination Centers for managing wildland fire and other incidents (information available at http://www.nifc.gov/fireinfo/geomap.html).

Figure 2-7—Effects of fuel reduction treatment on a closed-canopy ponderosa pine-Douglas-fir forest, western Montana. (A) Before treatment, understory is comprised of sparse clumps of native grasses and limited spotted knapweed. (B) Same photo point 3 years after thinning to reduce canopy fuels, followed by prescribed fire. Spotted knapweed (forb with gray-green foliage) and flannel mullein (forb with tall brown inflorescence) dominate understory. (Photo by Mick Harrington.)

western North America suggests that grazing plays an important role in the decrease of native perennial grasses and an increase in dominance by nonnative annual species; however, invasion has been found to occur with and without grazing in some areas. While it is difficult to discern the relative importance of grazing, climate, and fire on nonnative plant abundance (D'Antonio and others 1999), areas with a history of livestock grazing often support a variety of nonnative species, especially in areas where nonnatives have been introduced to increase forage value of rangelands or pastures (see, for example, chapters 7, 8).

When fire occurs in an area with a large number of nonnative plants in and around the burned area, one might expect establishment and spread of nonnatives within that burned area because (1) resources become more available after fire, and (2) nonnative propagules are available to establish and spread on that site. Research demonstrating this pattern is available from several areas in the central and western United States. Nonnative species abundance often increases, sometimes dramatically, in postfire plant communities in southwestern ponderosa pine forests with a history of anthropogenic disturbance (for example, see Crawford and others 2001; Griffis and others 2001). This contrasts with postfire dominance by native species and occurrence of very few nonnative species in relatively undisturbed mixed conifer and ponderosa pine communities at Grand Canyon National Park, Arizona, and Bandelier National Monument, New Mexico (Foxx 1996; Huisinga and others 2005; Laughlin and others 2004). This is likely due to lower nonnative propagule pressure in less disturbed landscapes. A seed bank study conducted in northern Arizona ponderosa pine communities representing "a historical land use disturbance gradient" found that the soil seed bank on sites with high and intermediate disturbance had many nonnatives, while sites with low levels of disturbance had only two nonnative species in the seed bank: cheatgrass and annual canarygrass (*Phalaris canariensis*) (Korb and others 2005). In high elevation Rocky Mountain lodgepole pine (*Pinus contorta* var. *latifolia*) forests in West Yellowstone, establishment and spread of nonnatives in forests was significantly enhanced along roadsides (where nonnative species richness was highest) but not along the edges of burns or clearcuts (Pauchard and Alaback 2006). Christensen and Muller (1975) also noted that nonnative plants were most common after fire in heavily disturbed parts of their California chaparral study area, such as along roadsides. In tallgrass prairie, postfire increases in nonnatives were greater in areas where the landscape is fragmented and nonnative propagule pressure is higher than in less fragmented areas with fewer nonnatives (Smith and Knapp 2001).

Fuel reduction efforts using fire may enhance the invasibility of treated forests, although long-term studies are needed to determine if established nonnatives will persist (Keeley 2006b, review) (also see chapter 13). Several studies have found that a combination of thinning and burning resulted in greater abundance of nonnatives than either thinning or burning alone in ponderosa pine forests of Arizona (Fulé and others 2005; Moore and others 2006; Wienk and others 2004) and western Montana (Dodson and Fiedler 2006; Metlen and Fiedler 2006). Similarly, cheatgrass, Japanese brome, North Africa grass (*Ventenata dubia*), and prickly lettuce were more strongly associated with plots that were thinned and burned than plots that were only burned or thinned in low-elevation forests of northeastern Oregon (Youngblood and others 2006).

Fire suppression activities (including construction of fire lines, temporary roads, fire camps, and helicopter pads) may increase nonnative species in the postfire environment by disturbing soil, dispersing propagules (Backer and others 2004, review), and altering plant nutrient availability (chapter 14). For example, following wildfire in a mixed conifer forest in Glacier National Park, nonnative species were more diverse in bulldozed (23 species) than burned (5 species) or undisturbed plots (3 species) (Benson and Kurth 1995). One year after wildfire in dense ponderosa pine/Douglas-fir forest in western Montana/northern Idaho, nonnative species richness was 7 on bulldozed plots and 1.7 on adjacent burned and unburned plots (Sutherland, unpublished data 2008). In an eastern Ontario alvar woodland dominated by northern white-cedar (*Thuja occidentalis*), many nonnative species were associated exclusively with bulldozed tracks and did not occur on sites that were undisturbed or burned within the previous year (Catling and others 2002). Nonnatives in areas disturbed during fire suppression may provide propagules for spread into adjacent native communities (see "Nonnative Propagule Pressure" page 9). Although no studies are available, there is concern that because fire retardant supplies nitrogen and phosphorus to the soil, the establishment and spread of invasive species may increase in the nutrient-rich environment where it is applied.

Livestock grazing before or after fire is another disturbance that can influence nonnative species establishment, persistence, and spread. Interactions of grazing with invasive species and fire, however, are complex (Collins and others 1995, 1998; Fuhlendorf and Engle 2004; Stohlgren and others 1999b), and studies that incorporate all three topics are rare. Plant communities that are in poor condition due to prolonged or excessive grazing may be more susceptible to nonnative plant invasions (chapters 7, 8, 9). Similarly, when livestock grazing occurs soon after a fire, the potential for animals to disperse nonnative propagules while possibly stressing desirable species must be considered. On the other hand, grazing has occasionally been used in conjunction with prescribed fire to reduce invasive species (see "Treatments That Increase Effectiveness of Prescribed Fire," chapter 4).

Question 4. Do Invasions Become Less Severe With Increasing Time After Fire?

Traits that allow nonnative species to exploit disturbed sites may also make them dependent on disturbance in some plant communities. As vegetation recovers after fire, canopy cover increases and sunlight reaching the soil surface decreases. Nutrients and soil moisture are taken up by the dominant vegetation. Nonnative species that are not adapted to these new conditions are likely to decline. This pattern is demonstrated to some extent in plant communities where fires are infrequent and postfire communities succeed to forest (for example, closed-canopy forests in the Northwest, Southwest Coastal, and Interior West bioregion) or shrubland (for example, chaparral in the Southwest Coastal bioregion). However, it is not consistent among all studies reviewed, and the duration of most studies on postfire succession is too short to demonstrate or refute this generalization. This generalization is usually examined using chronosequence studies, which assume that conditions on a site are consistent through time. This assumption is unlikely to hold true in regard to nonnative species. The nonnative portion of a plant community is unlikely to be constant in species or abundance over many decades. For example, in 1959, there were fewer than 800 nonnative species in California (Munz and Keck 1959), but by 1999 there were 1,200 nonnative species (Kartesz 1999). If most invasives establish soon after disturbance, a plant community burned in 1959 in California would not have been exposed to the same suite of species or the same degree of nonnative propagule pressure as a plant community burned in 1999.

In coniferous forests of the Northwest Coastal bioregion, information on postfire persistence of nonnative species comes primarily from studies on the effects of timber harvest and associated slash burning. Ruderal herbs, mostly native but including some nonnatives, are the dominant vegetation during the first few years after slash burning. Nonnative species in this group include woodland groundsel, tansy ragwort, bull thistle, Canada thistle, St. Johnswort, and wild lettuces. Non-ruderal native species typically regain dominance within about 4 to 5 years after slash burning (chapter 10). A wildfire chronosequence from this bioregion supports this pattern for woodland groundsel and wall-lettuce (*Mycelis muralis*). These two species dominated the herb layer 3 years after fire and are not mentioned in any other postfire year by this study, which covered stands 1 to 515 years after fire (Agee and Huff 1987). Conversely, St. Johnswort was present in mature (80 to 95 years old) and old growth (200 to 730 years old) stands in Oregon and California (Ruggiero and others 1991), indicating that it can establish and persist in closed-canopy forests. Woody nonnatives such as Scotch broom and Himalayan blackberry typically invade disturbed forests and sometimes form dense thickets. While these species are not shade-tolerant, and therefore may not persist after canopy closure, they may prevent or delay reforestation (chapter 10).

In closed-canopy forests of the Interior West bioregion that have burned, postfire invasion of nonnative species is not well studied or well documented, although two studies provide some support for this generalization, and a third demonstrates this pattern in ponderosa pine forest. Doyle and others (1998) observed an initial increase in Canada thistle abundance followed by a steady decline after fire in a mixed conifer forest in Grand Teton National Park. Turner and others (1997) document prickly lettuce densities of around 100 stems/ha 3 years after fire in Yellowstone National Park, followed by a 50 percent decrease in density by the fifth postfire year; however, Canada thistle density increased from 2 to 5 years after fire. Similarly, in ponderosa pine forest in western Montana, prickly lettuce reached nearly 4 percent average cover the second year after stand-replacing fire (fig. 2-8) but declined substantially in the next 2 years to near preburn levels (Sutherland, unpublished data 2008).

Stands of chaparral and coastal scrub with intact canopies are relatively resistant to invasion by nonnative plants, and postfire succession by resprouting dominants follows a relatively predictable but highly dynamic pattern in these communities when fire-return intervals occur within the range of 20 to 50 years (chapter 9; Keeley and Keeley 1981; Keeley and others 2005). Herbaceous species, including some nonnatives, dominate in the first few years after fire, then gradually diminish as succession proceeds, shrub cover increases, and the canopy closes (for example, Guo 2001; Horton and Kraebel 1955; Keeley and others 1981, 2005; Klinger and others 2006a). However, when fire intervals decline to 15 years or less, shrub dominance declines, and nonnative annual grasses and forbs are more likely to dominate and initiate a grass/fire cycle in which it is extremely difficult for woody and herbaceous native species to establish and regenerate (chapter 9).

Resprouting dominants in mountain shrub communities of Mesa Verde National Park reduce invasibility. Dominants in these communities include Gambel oak (*Quercus gambelii*) and Utah serviceberry (*Amelanchier utahensis*), which sprout rapidly after fire, apparently utilizing available resources so efficiently that nonnative species have limited opportunity to become established. Dominants in adjacent piñon-juniper communities do not resprout; consequently, these communities recover their prefire structure slowly, which provides open conditions favoring nonnative species after fire. Eight and 13 years after fire in Mesa Verde National Park, mountain shrub communities were less invaded than adjacent piñon-juniper communities, based on density and species richness measures (Floyd and others 2006); comparisons to unburned sites were not provided.

Research on nonnative species in piñon-juniper woodlands is not clear in regard to this generalization. A study of six fires over 15 years indicates that musk thistle, Canada thistle, and cheatgrass have persisted for at least 13 years after wildfire in piñon-juniper communities in Mesa Verde National Park. Conversely, prickly lettuce and prickly Russian-thistle (*Salsola tragus*) were common 3 years after fire but were not recorded 8 and 13 years after fire (Floyd and others 2006). Chronosequence studies from piñon-juniper woodlands in Mesa Verde, Colorado (Erdman 1970), Nevada and California (Koniak 1985),

Figure 2-8—Dense prickly lettuce establishment the second year after stand-replacing fire in ponderosa pine forest in western Montana. Red-stemmed plants are native fireweed (*Chamerion angustifolium*). (Photo by Steve Sutherland.)

and west-central Utah (Barney and Frischknecht 1974) suggest that nonnative annuals are most abundant in early postfire years and decline in later successional stages.

Even if long-term research eventually demonstrates that nonnative invasive species decline during succession as native species increase and a closed canopy develops, one cannot assume that the invasives have disappeared from the site. Seeds of many nonnative invasives can remain viable in the soil seed bank for many years or decades, and nonnative perennials may persist in suppressed, nonflowering form at low densities under closed canopies. Another fire is likely to again produce conditions favoring their development and dominance, but the long-term successional outcome may be different. Many factors, such as reduced abundance and vigor of native species, different postfire precipitation patterns, or presence of additional nonnative species, could alter successional trends and make it more difficult for native species to regain dominance. Unfortunately, research on the influence of multiple burns is lacking for most nonnative invasive species (chapter 12).

Question 5. Do Invasions Increase With Disruption of the Presettlement Fire Regime?

If ecological processes that have shaped a plant community are altered, the vigor and abundance of native plants may decline, theoretically making the community more invasible. Application of this concept to fire regimes leads to the generalization that disruption of a plant community's fire regime increases its invasibility (Huenneke 1997). This generalization may apply to changes in any aspect of the fire regime, but the primary aspects treated in the literature to date are fire severity and fire frequency. Examples include ecosystems where fire exclusion or, conversely, increased fire frequency have stressed native species adapted to fire regimes of different frequencies and severities. Fire exclusion from grasslands, for example, may stress native species adapted to frequent fire and favor nonnative species that are intolerant of frequent fire. Exclusion of fire from open-canopy forests, on the other hand, has led to increased surface and ladder fuels and subsequent increases in fire severity in some areas, when the forests eventually burn. Native plant communities are likely to be adversely impacted by fire under these fuel conditions, so nonnative species may be favored in the postfire environment. Ecosystems where fire frequency has increased, either due to increases in anthropogenic ignitions or changes in fuel structure brought about by invasive species themselves, also support this generalization (chapter 3).

Fire exclusion from grasslands and savannas adapted to frequent fires may favor nonnative invasive grasses, forbs, or woody species. Tallgrass prairie ecosystems, for example, tend to support more nonnative grasses and forbs under a regime of infrequent fire than with frequent burning (chapter 7). Many ecosystems are invaded by woody plants when fire is excluded: honeysuckles, buckthorns (*Rhamnus cathartica* and *Frangula alnus*) and barberries (*Berberis* spp.) occur in oak savannas of the Northeast and Central bioregions (chapters 5 and 7); melaleuca, Chinese tallow, Brazilian pepper and chinaberry invade wet grasslands of the Southeast bioregion (chapter 6); Chinese tallow increases in southern tallgrass prairie (chapter 7); and brooms and gorse may spread in oak savannas and grasslands in the Northwest and Southwest Coastal bioregions (chapters 9 and 10). Most of these woody invasives are fire-tolerant and continue to reproduce and thrive even after fire is reintroduced. In some cases, they shade herbaceous species, reducing the cover and continuity of fine fuels such that they are difficult to burn. Chinese tallow, Brazilian peppertree, and common buckthorn are examples of invasive species for which this pattern has been suggested.

In open-canopy forests, such as ponderosa pine forests in the Interior West bioregion, fire exclusion has led to changes in structure, species composition, and fuel accumulation such that, when wildfire occurs, it may be more severe than was common in presettlement times. Several nonnative forbs and grasses increase after fire in these successionally altered plant communities. Canada thistle, bull thistle, and knapweeds are the most frequently recorded nonnative forbs during the early postfire years (Cooper and Jean 2001; Crawford and others 2001; Griffis and others 2001; Phillips and Crisp 2001; Sackett and Haase 1998; Sieg and others 2003). This is in contrast to conifer forests where fire intervals and fire severity have not increased substantially (Foxx 1996; Huisinga and others 2005; Laughlin and others 2004). For example, few nonnative species were present at any site (burned or unburned) after a low-severity fire in remote ponderosa pine forests on the North Rim of Grand Canyon National Park, Arizona, where fire regimes have not been disrupted, grazing has been minimal, and logging has not occurred (Laughlin and others 2004).

Interactions between fire exclusion and grazing have influenced invasion of piñon-juniper woodlands and sagebrush grasslands by nonnative species. At many contemporary piñon-juniper sites, perennial grass cover has declined and tree cover has increased following decades of livestock grazing and fire exclusion (for example, Laycock 1991; Ott and others 2001). In sagebrush grasslands, livestock grazing has reduced native grasses while fire exclusion has allowed trees, especially juniper, to spread (M. Miller, personal communication 2007). As piñon-juniper stands increase in density and approach crown closure, native herbaceous cover (Tausch and West 1995), seed production, and seed bank density decline (Everett and Sharrow 1983; Koniak and Everett 1982). Nonnative species, especially cheatgrass, are typically present in and around these sites and are likely to establish and dominate early successional stages after fire under these conditions. Dominance of cheatgrass, in turn, may lead to increases in fire size and frequency, thus initiating an annual grass/fire cycle (chapter 3). Successional trajectories in piñon-juniper stands are further complicated by recent widespread tree mortality caused by extended, severe drought interacting with insects, root fungi, and piñon dwarf mistletoe (*Arceuthobium divericatum*) (Breshears and others 2005; Shaw and others 2005) (see chapter 8 for more information).

In some ecosystems, fire frequency has increased and favors nonnative species. These increases may be due to increases in anthropogenic ignitions or changes in fuel structure brought about by the invasive species themselves. The latter case is best exemplified by invasions of nonnative grasses in Hawai`i and in southwestern and Great Basin desert shrublands and the resulting grass/fire cycle (chapter 3). An example of invasive species' response to increased fire frequency due to anthropogenic ignitions is found in Fort Lewis, Washington, on a 2,500 to 3,000 acre (1,000 to 1,200 ha) area called Artillery Prairie. Here broadcast burns ignited by artillery fire have occurred nearly annually for about 50 years, resulting in a plant community dominated by nonnative forbs and annual grasses. The natural fire cycle is less frequent, and a prescribed fire regime of burning every 3 to 5 years maintains native prairies and oak woodlands (Tveten and Fonda 1999).

Question 6. Are Postfire Invasions Less Common in High Elevation Ecosystems?

Several studies indicate a negative correlation between elevation and nonnative species richness or abundance; this pattern has been observed in California (Dark 2004; Frenkel 1970; Randall and others 1998; Keeley and others 2003), the northern Rocky Mountains (Forcella and Harvey 1983; Sutherland, unpublished data 2008; Weaver and others 1990), and the Southwest (Bashkin and others 2003; Fisher and Fulé 2004). Only a few of these studies relate to fire, and no research has illuminated the reasons for these correlations. Here we discuss possible explanations for and management implications of this generalization.

Invasive species richness may decline with increasing elevation because fewer species (native as well as nonnative) thrive in the shorter growing seasons, cooler temperatures, and generally more stressful environment of subalpine and alpine ecosystems than at lower elevations. Fire would further limit the number of invasives to species that can survive a burn or disperse into burned sites.

Nonnative species that can persist at high elevations may show relatively low abundance because, like native species, they grow and spread more slowly in severe conditions. This factor suggests that, while high-elevation ecosystems may currently be less invaded than lower-elevation sites, they have no intrinsic immunity to invasion and could be impacted as severely as any other community type in time. Insofar as fire increases resource availability and mineral soil exposure and reduces native species dominance and vigor, it could accelerate invasions; however, the ruderal species most favored in recent burns are unlikely to persist in high-elevation environments, which favor slow-growing, perennial species with persistent underground structures.

Another explanation for lower invasion levels at high elevations is that human-caused disturbance is generally less and propagules are less likely to be introduced in large numbers in high-elevation ecosystems (Klinger and others 2006b). This is supported by observations that, with increased disturbance such as roads and clearcuts, nonnative species occurrence extended to higher elevations (Forcella and Harvey 1983; Weaver and others 1990). Thus increases in accessibility, use, and mechanical disturbance of high-elevation plant communities—including activities related to fire management or fire suppression—have potential to increase propagule pressure from nonnative invasive species and invasibility of these sites.

Climate change, expressed at high elevations by longer growing seasons and milder temperature regimes, is likely to simultaneously increase stress on native plants and favor more nonnative invasive species. Fire frequency may increase at high (as at low) elevations, occurring at intervals shorter than the regeneration time for some native plants and creating more disturbed sites for establishment of nonnatives. See "Changing Atmosphere and Climate" page 29.

Conclusions

Generalizations that explain patterns across a wide range of systems are elusive in invasion ecology in general (for example, see review by Rejmánek and others 2005), a principle that certainly applies to fire. Nonnative invasive species show some patterns in their responses to wildland fire. The generalization that fire favors nonnatives over natives is supported by the literature for some nonnative species in some plant communities under some conditions. Postfire invasions can be intense and lead to severe impacts on native communities, so vigilance is warranted. However, invasions also vary with numerous site and climatic factors, depend on the nonnative propagules within and near the burn, and can be short-lived. Information about fire effects on specific plant communities with specific invasive species is the best knowledge base for making management decisions. Second best is knowledge of nonnative species in similar environments. The more conditions in the area of concern diverge from conditions in published research or other known areas, the less reliable predictions will be.

Examination of the literature provides insights regarding other common assumptions about fire and nonnative species:

- Nonnative species establishment increases with increasing fire severity. This pattern depends on fire resistance of onsite species, propagule pressure, and the uniformity and size of high-severity burn patches.
- Additional disturbance favors invasions in most circumstances, though the influence of grazing-fire interactions on nonnative species are complex and may not follow this pattern consistently.
- Invasions become less severe with increasing time since fire in some plant communities, particularly where ruderal species invade closed-canopy forests and chaparral after fire. However, there are few long-term studies investigating this pattern and there are many exceptions. Without local, long-term knowledge, this generalization may not be reliable as a predictive tool.
- Invasions increase in some plant communities with disrupted fire regimes, whether the disruption relates to fire regime characteristics that increase or decrease relative to baseline conditions. Communities that have developed an invasive grass/fire cycle support this generalization (chapter 3), as do native grasslands from which fire has been excluded (chapter 6, 7, 8, 10). However, a "disrupted fire regime" may be too complex an ecological property to use as a predictive tool. Specific stresses on the native plant community arising from disrupted fire regimes may be more helpful.
- Postfire invasion is currently less likely to occur and persist in high-elevation than low-elevation ecosystems, but elevation *per se* does not provide the only explanation for this pattern. Differences in human-caused disturbance, resource availability, and propagule pressure should also be considered; where these influences are increasing, high-elevation ecosystems are likely to become more vulnerable to postfire invasions.

It is important to keep in mind that exceptions to these patterns are common. Our knowledge about fire and nonnative plants is not extensive enough in space and time for use in widely applicable predictive tools. Site-specific knowledge is essential for management to successfully meet objectives.

Variation

There are many reasons for variation in nonnative species' responses to fire. The invasive potential of nonnative species varies throughout their range (Klinger and others 2006a). For the majority of nonnative plants in the United States, the distribution of the species is much larger than the states that have declared it a noxious species (Kartesz 1999). Fire itself also varies tremendously in severity, size, spatial complexity, frequency, and seasonality. Finally, invasibility of a community is also influenced by site history, the condition of native plant populations, and postfire precipitation patterns, so postfire spread of nonnatives may be inconsistent even within a plant community.

Site-specific knowledge about fire effects on nonnative plants depends to some extent on the monitoring techniques used and the length of time monitored (chapter 15). Data collected only within the first 2 or 3 postfire years usually cannot be used to project long-term patterns; this is particularly true of forests and woodlands where tree regeneration occurs over many decades, changing stand structure and the availability of resources. Research has not been particularly helpful with this problem. Despite the need for long-term studies, 70 percent of recent literature reviews covering fire effects on nonnative invasive species in the Fire Effects Information System contain no postfire response information beyond the first 2 postfire years (chapter 12).

Changing Atmosphere and Climate

Our understanding of the responses of nonnative species to fire is based on the premise that the community of native species occurring on a site has been shaped by natural selection to be well suited to local site conditions and climate. Confidence in the robustness of the native community is part of the rationale for using "natural" conditions as a baseline for management. Past conditions are useful as a reference for desired future conditions, not because of a hope to return to the past, but because past conditions capture a range of variability that was sustained over long periods (Klinger and others 2006a). This rationale may not apply in a world where the climate is changing substantially from historic patterns, as is currently occurring. Earth's air and oceans are warming to levels not seen in the past 1,300 years, possibly in the past 10,000 years. According to the International Panel on Climate Change, this warming is now unequivocal and could continue for centuries. Global climate change is expressed in many areas as earlier snowmelt, longer warm seasons, and changes in extreme weather; however, these effects vary geographically, and some areas are likely to be colder or more moist than in past centuries (Alley and others 2007). In the Western United States, climate change appears to be contributing to longer fire seasons and more frequent, more extensive fires (McKenzie and others 2004; Westerling and others 2006), which may increase the vulnerability of many ecosystems to invasion and spread of nonnative plants.

An example of potential impact of climate change is the die-off of piñon over 4,600 square miles (12,000 km^2) in the southwestern United States. On an intensively studied site in northern New Mexico, Breshears and others (2005) found that greater than 90 percent of the dominant overstory tree, Colorado piñon (*Pinus edulis*), and greater than 50 percent of the dominant understory herb, blue grama (*Bouteloua gracilis*), died after a 4-year drought with unusually high temperatures. These results were supported by four regional studies in Arizona, Colorado, New Mexico, and Utah, reporting piñon die-off ranging from 40 to 80 percent. Not only is the die-off a major ecosystem disturbance in its own right, but also the increase in dead fuel loading increases wildfire hazard in this area. Although the study does not address invasives, the disturbance resulting from die-off and the possibility of subsequent high-severity wildfires may increase this area's susceptibility to invasion and spread of nonnative species.

Increased atmospheric carbon dioxide alters not only climate but also plant properties and the balance between species (Huenneke 1997). Laboratory research on cheatgrass biology demonstrates changes in plant properties. Cheatgrass grown at carbon dioxide levels representative of current conditions matures more quickly, produces more seed and greater biomass, and produces significantly more heat per unit biomass when burned (associated with reduced mineral and lignin concentrations) than cheatgrass grown at "pre-industrial" carbon dioxide levels (fig. 2-9). These responses to increasing carbon dioxide may have increased flammability in cheatgrass communities during the past century (Blank and others 2006; Ziska and others 2005). Research on cheatgrass has not addressed the possibility that native species biomass has increased along with that of cheatgrass in response to increasing carbon dioxide, but a study in the Mojave Desert has addressed that possibility in regard to red brome, another nonnative annual grass. In an environment with elevated carbon dioxide, red brome density increased while density of four native annuals decreased. Red brome showed greater increases in biomass and seed production in response to elevated carbon dioxide than did the four native species (Smith and others 2000). At the community level, nonnative species that respond favorably to increased carbon dioxide may thrive at the expense of native species with lower nutrient requirements; this change in the balance between species favors many fast-growing plants, including nonnative invasives (Huenneke 1997).

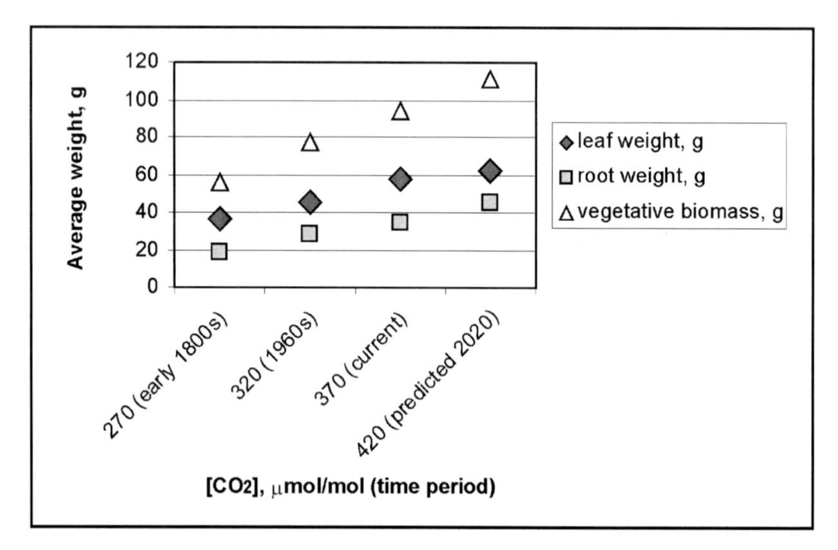

Figure 2-9—Average weight of cheatgrass seedlings grown for 87 days under varying concentrations of carbon dioxide. Carbon dioxide level was a significant predictor of leaf weight, root weight, and total vegetative biomass. In addition, increasing CO_2 reduced the time from germination to flowering. (Graphed from data in Ziska and others 2005, table 1.)

Increased carbon dioxide is not the only atmospheric change affecting wildland ecosystems and, potentially, fire regimes. Nitrogen deposited from air pollution can increase available soil nitrogen (for example, see Baez and others 2007; Padgett and others 1999), altering the relative abundance of native and nonnative species. In the Mojave Desert, artificial nitrogen addition significantly increased the density and biomass of nonnative annual herbs and, in 1 of 2 years, significantly reduced the density and biomass of native species. Increased biomass is likely to increase fire frequency on these desert sites (Brooks 2003). In more mesic areas, greater depletion of surface soil moisture by invading C_3 grasses may favor deep-rooted shrubs, and the increase in nitrogen fixation that occurs with the increased metabolic activity triggered by increased carbon dioxide could favor leguminous shrubs (Dukes 2000).

Not all nonnative species will benefit more than native species from atmospheric changes (Dukes 2000). However, the dispersal capability of many nonnatives, plus their rapid growth in disturbed areas, makes them well suited to the conditions accompanying climate change. Climate change will be expressed as local changes in growing conditions and fire regimes, which may interact synergistically to increase invasions (Barrett 2000). Local knowledge regarding nonnative invasive species, monitoring, and adaptive use of the knowledge gained will be increasingly important for successful management.

Management in a Changing World

The relationships between nonnative invasive species and fire described here are based on information from the past century or two, and they hold true only to the extent that conditions that shaped this relationship continue. Management of wildlands must be based on current conditions and likely future conditions (Klinger and others 2006a). How can managers prepare and respond? A decision framework such as that presented by Pyke and others (in review) can be a useful tool. To use a decision framework or model effectively, knowledge about local patterns of invasion, problematic nonnative species, and highly invasible sites is critical. This information should be readily available to fire managers and postfire rehabilitation specialists. Communication between local fire managers and local botanists is important. Careful monitoring of burned areas is crucial, and the monitoring must extend over decades rather than years. Knowledge gained from monitoring prescribed burns may be helpful for projecting effects of wildfires, although wildfires are likely to be larger, more severe, and occur in a different season than prescribed fires.

Prevention and early eradication are daunting tasks. The number of native species within a particular plant community on a specific site is finite, whereas the number of species from around the world that could potentially grow on the site may be greater by orders of magnitude (Randall and others 1998; Williamson 1993). Prevention of invasion is always the best strategy, since control and eradication are costly and may never be complete (Klinger and others 2006a). Even if a nonnative species is eradicated, any invasion leaves a legacy of subtle alterations in the site and gene pool of the remaining species.

Control of invasives already present on a burn, combined with early detection and treatment of new invasives and regular monitoring after treatment will be essential for preventing dominance of postfire habitats by nonnative invasive species. And continued research is needed, especially long-term studies addressing effects of various fire severities, frequencies, intervals, and seasons.

Notes

Matthew L. Brooks

Chapter 3:
Plant Invasions and Fire Regimes

The alteration of fire regimes is one of the most significant ways that plant invasions can affect ecosystems (Brooks and others 2004; D'Antonio 2000; D'Antonio and Vitousek 1992; Vitousek 1990). The suites of changes that can accompany an invasion include both direct effects of invaders on native plants through competitive interference, and indirect effects on all taxa through changes in habitat characteristics, biogeochemical cycles, and disturbance regimes. Effects can be far-reaching as they cascade up to higher trophic levels within an ecosystem (Brooks and others 2004; Mack and D'Antonio 1998).

Direct interference of invaders with native plants can be mitigated by removing or controlling the invading species. In contrast, when invaders cause changes in fundamental ecosystem processes, such as disturbance regimes, the effects can persist long after the invading species are removed. Restoration of native plant communities and their associated disturbance regimes may be necessary to restore pre-invasion landscape conditions. In this chapter, I describe ways in which invasions by nonnative plant species can change fuel conditions and fire regimes, and discuss what can be done to prevent or mitigate these effects.

Fire Behavior and Fire Regimes

Fire behavior is described by the rate of spread, residence time, flame length, and flame depth of an individual fire (chapter 1). This behavior is affected by fuel, weather, and topographic conditions at the time of burning. Individual fires can have significant short-term effects, such as stand-level mortality of vegetation. However, it is the cumulative effects of multiple fires over time that largely influence ecosystem structure and processes. The characteristic pattern of repeated burning over large areas and long periods of time is referred to as the fire regime.

A fire regime is specifically defined by a characteristic type (ground, surface, or crown fire), frequency (for example, return interval), intensity (heat release), severity (effects on soils and/or vegetation), size, spatial complexity, and seasonality of fire within a given geographic area or vegetation type (Sugihara and others 2006b; chapter 1). Fire regimes can be described quantitatively by a range of values that are typically called the "natural range of variation" or "historical range of variation." The term "natural" requires a value judgment that can be interpreted in many different ways, and the term "historical" requires a temporal context that is often poorly understood, as does the term "presettlement" (chapter 1). Any of these concepts may be inappropriate as a reference set for management decisions if conditions have changed so much that the landscape cannot support a fire regime that is either natural or historical. The term "reference fire regime" is used in this chapter to describe a range of fire regime conditions under which native species have probably evolved, or which are likely to achieve specific management objectives such as maximizing native species diversity and sustainability. Current patterns of fire regime characteristics can be objectively compared to reference conditions, and shifts outside of the reference range of variation can be used to identify "altered" fire regimes.

Biological and Physical Factors that Affect Fire Regimes

Fire regimes are influenced by topographic patterns, climatic conditions, ignition sources, and fuels (fig. 3-1). Topographic factors are extremely stable at the scale of thousands to millions of years. Climate factors are relatively stable at the scale of hundreds to thousand of years, although rapid changes have occurred during the 1900s due to anthropogenic influences on the atmosphere (Houghton and others 1990). Ignition rates from natural sources such as lightning are related to climate and have a similar degree of stability. However, ignition rates from anthropogenic sources are related to human population levels, which can increase dramatically at the scale of tens to hundreds of years. In contrast, fuels are created by vegetation, which can change over a period of days to weeks due to land cover conversion to agriculture, some other human use, or other disturbance. Vegetation can also be changed over a period of years to decades due to plant invasions that can displace native vegetation and change fuel properties. Thus, alterations in anthropogenic ignitions and vegetation characteristics are the two primary ways in which fire regimes can change rapidly. In this chapter, I focus on fire regime changes due to vegetation changes caused by plant invasions.

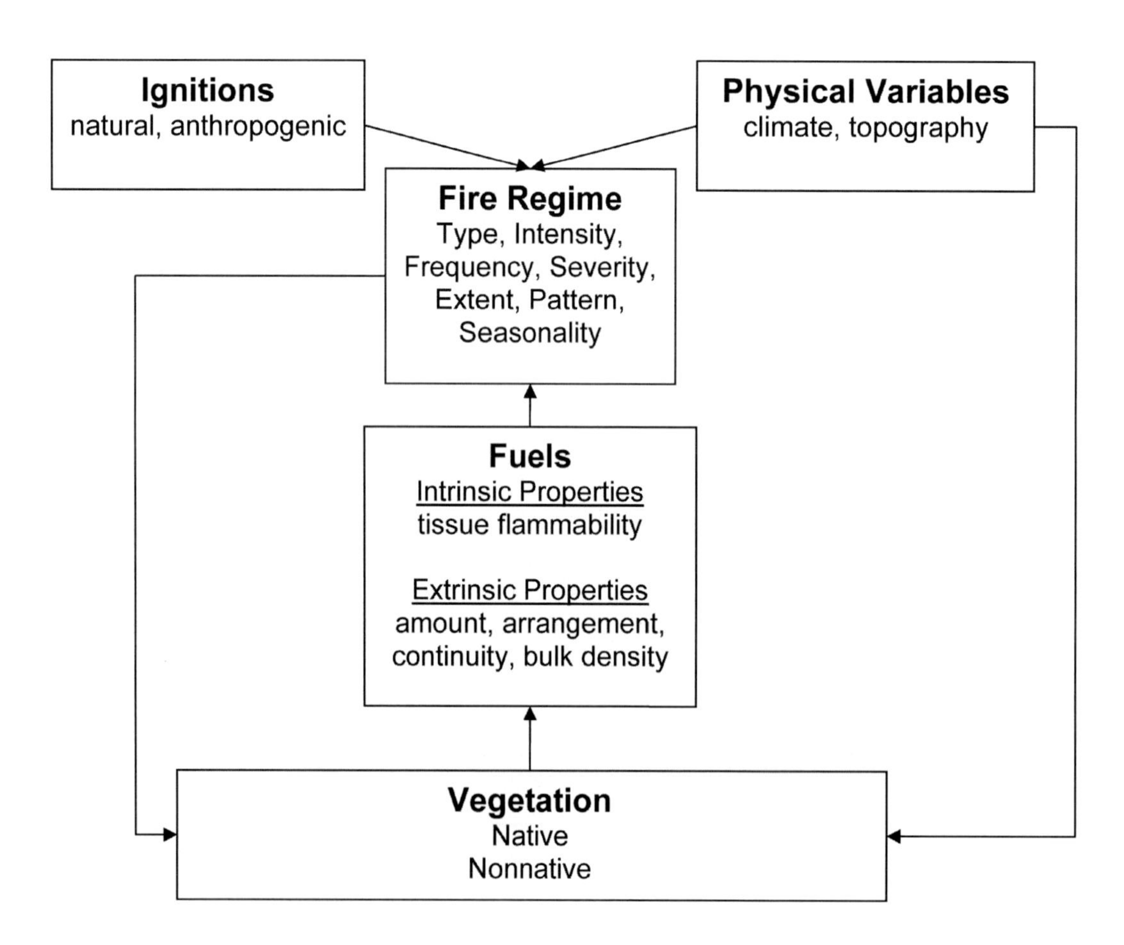

Figure 3-1—Biological and physical factors influencing fire regimes.

Effects of Plant Invasions on Fuels

Fuels can be classified into layers based on their vertical arrangement on the landscape. The fuel layers typically used to describe fire spread are ground, surface, and canopy fuels (chapter 1). Since plant invasions are less likely to alter ground fuels (organic duff and peat layers) in most ecosystems, surface and canopy fuels are emphasized in this chapter.

Surface fuels are typically dominated by litter plus herbaceous plants and shrubs, comprised mostly of fine (1-hour timelag: <0.25 inch (<0.6cm)) and medium-size (10-hour timelag: 0.25 to 1 inch (0.6 to 2.5cm)) fuels (Deeming and others 1977). In forested ecosystems, coarse (100-hour timelag: 1 to 3 inches (2.5 to 7.5 cm)) and larger diameter surface fuels may also be plentiful. Horizontal continuity of surface fuels is generally high in productive ecosystems such as riparian zones but may be very low in low-productivity ecosystems such as desert uplands (fig. 3-2).

Canopy fuels include fine to very coarse fuels, have low to high horizontal continuity, and have relatively low ignitability. The vertical spread of fire from surface to canopy fuels often requires ladder fuels that provide vertical continuity between strata, thereby allowing fire to carry from surface fuels into the crowns of trees or shrubs (fig. 3-3). Ladder fuels help initiate crown fires and contribute to their spread (National Wildfire Coordinating Group 1996). They may consist of vines or be comprised of the tallest surface fuels and the shortest canopy fuels at a given site.

In addition to fuel particle size and the horizontal and vertical distribution of fuel layers, other extrinsic fuel properties related to the way fuels are arranged on the landscape can influence fire behavior (fig. 3-1; table 3-1). The amount of fuel, or fuel load, primarily affects the intensity of fires. The fuel bed bulk density, or amount of fuel per unit volume of space, affects the rate of combustion, which influences the frequency, intensity, and seasonal burning window of fires. These properties also affect residence time and the

Figure 3-2—(A) High horizontal fuel continuity created by the nonnative grasses ripgut brome (*Bromus diandrus*) and cheatgrass (*Bromus rubens*) in a southwestern riparian ecosystem. (B) Low horizontal fuel continuity in a native Mojave Desert shrubland. (Photos by Matt Brooks, USGS, Western Ecological Research Center.)

Figure 3-3—High vertical continuity (ladder fuels) created by the nonnative giant reed (*Arundo donax*) in a southern California riparian woodland. (Photo by Tom Dudley, UC Santa Barbara.)

Table 3-1—Primary effects of fuelbed changes on potential fire regimes.[a]

Fuelbed change	Fire regime change
Increased amount (load)	Increased fire intensity and seasonal burn window; increased likelihood of crown fire; increased fire severity
Decreased amount (load)	Decreased fire intensity and seasonal burn window; decreased likelihood of crown fire; decreased fire severity
Increased horizontal continuity	Increased fire frequency and size, increased spatial homogeneity
Decreased horizontal continuity	Decreased fire frequency and size, decreased spatial homogeneity
Increased vertical continuity	Increased fire intensity and likelihood of crown fire, which could increase size and spatial homogeneity
Decreased vertical continuity	Decreased fire intensity and likelihood of crown fire, which could reduce size and homogeneity
Change in bulk density	Change in fire frequency, intensity, and seasonality; change in fire severity
Increased plant tissue flammability	Increased fire frequency, intensity, and seasonal burn window; possible increase in fire frequency or severity
Decreased plant tissue flammability	Decreased fire frequency, intensity, and seasonal burn-window; possible decrease in fire frequency or severity

[a] Modified from Brooks and others (2004) table 1.

duration of smoldering combustion, which influence fire severity. Intrinsic fuel properties such as plant tissue flammability, influenced by moisture content or chemical composition (for example, the presence of salts or volatile oils), also affect fire behavior.

Nonnative plants can alter fuelbeds directly, based on their own extrinsic and intrinsic properties as fuels, or indirectly by altering the abundance and arrangement of native plant fuels. If fuelbed characteristics are changed to the extent that fire type, frequency, intensity, severity, size, spatial complexity, or seasonality is altered, then the invasive plant has altered the fire regime. Plant invasions that alter fire regimes typically do so by altering more than one fuel or fire regime property (Brooks and others 2004). For example, grass invasions into shrublands increase horizontal fuel continuity and create a fuel bed bulk density more conducive to the ignition and spread of fire, thereby increasing fire frequency, size, and spatial homogeneity (in other words, completeness of burning). At the same time, the replacement of shrubs with grasses generally decreases the total fuel load, resulting in less heat release and decreased fire residence time and possibly reduced fire severity on the new, grass-dominated fuelbed. Reduced severity allows nonnative annual grasses to recover quickly following fire and establish fuel and fire regime conditions that could persist indefinitely.

The Invasive Plant Fire Regime Cycle

A conceptual model was recently developed describing the four general phases that lead to a shift in fuelbed composition from native to nonnative species, culminating in a self-sustaining invasive plant/fire regime cycle in which nonnatives dominate (Brooks and others 2004) (fig. 3-4). This process begins with *Phase 1*, before nonnative species disperse into and establish in the region of interest. Potential invader propagules are located in adjacent geographic areas or associated with potential vectors of spread (for example, contaminants of hay mulch or seeding mixes) and are poised to disperse into the region. The significance of this phase is that management can be focused solely on the prevention of propagule dispersal and the early detection and eradication of individuals or groups of individuals before populations can become established. *Phase 2* is characterized by the (1) establishment, (2) persistence, and (3) spread of a nonnative species (chapter 1). This requires that the species overcome local environmental and reproductive barriers, for example, the absence of an appropriate pollinator. Initial populations typically establish and persist in areas with substantial anthropogenic disturbances, such as urban, agricultural, or roadside sites. *Phase 3* is marked by the emergence of substantial negative ecosystem effects, or "ecological harm" as described in chapter 1; examples include reduction of native species abundance and diversity and deterioration of habitat for native animal species (Mack and D'Antonio 1998). In *Phase 4*, the negative effects of the nonnatives on native plants, and the presence of the nonnatives themselves, combine to alter fuel properties sufficiently to shift at least one characteristic of the subsequent fire regime outside of the reference range of variation. If the new fire regime favors the dominance of the invasive species causing the new fuel conditions and negatively affects the native species, an invasive plant/fire regime cycle becomes established (fig. 3-5).

The establishment of an invasive plant/fire regime cycle is of concern to land managers for two primary reasons. First, it may alter fuels and fire regimes in ways that impact human health, safety, or economic well-being. For example, invasions may increase "hazardous fuels," increasing the potential for fire-caused damage to human life, property, or economic commodities, particularly in wildland/urban interface areas. The second major concern for land managers, and the one most pertinent to management of

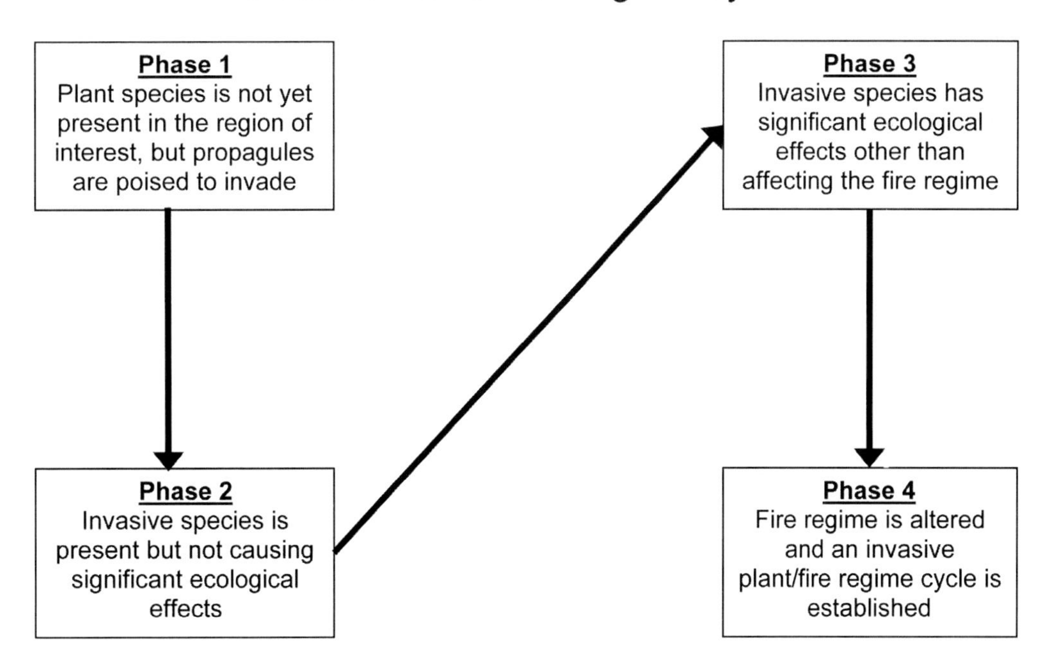

Figure 3-4—Phases leading to the establishment of an invasive plant/fire regime cycle,modified from Brooks and others (2004).

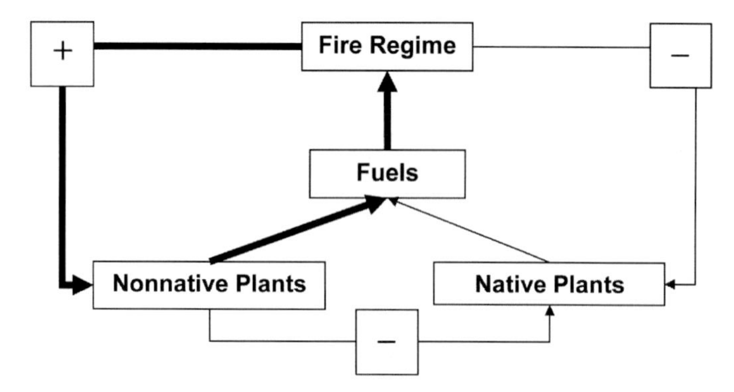

Figure 3-5—The invasive plant/fire regime cycle. Modified from Brooks and others (2004).

wildlands, is that altered fire regimes may have substantial negative effects on native plant and animal populations and ecosystems. The invasive plant/fire regime cycle may increase fuels or increase the rate of fuel replenishment after fire, leading to increased fire intensity or frequency; or it may reduce fuels in ways that suppress the spread of fire in ecosystems where fire is desirable. Disturbance regimes are strong forces driving the evolution of species, and the shifting of fire regime variables outside the range of variation to which they are adapted can tip the balance in plant communities toward new suites of dominant species. The resulting changes in vegetation may affect higher trophic levels. For example, cheatgrass (*Bromus tectorum*) and medusahead (*Taeniatherum caput-medusae*) now dominate 25 to 50 percent of sagebrush steppe in the Great Basin (West 2000), reducing habitat for the greater sage-grouse (*Centrocercus urophasianus*), a species experiencing declining population levels (Sands and others 2000). This change in vegetation has also reduced black-tailed jackrabbit (*Lepus californicus*) populations, which in turn has decreased the carrying capacity for golden eagles (*Aquila chrysaetos*) at the Snake River Birds of Prey National Conservation Area in Idaho. In the same area, Piute ground squirrel (*Spermophilus mollis idahoensis*) populations fluctuate more dramatically when sagebrush cover is lost, which may adversely affect prairie falcon (*Falco mexicanus*) populations (Sands and others 2000; Sullivan 2005). In the Mojave Desert, the replacement of native shrublands with nonnative annual grasslands dominated by red brome (*Bromus rubens*) reduces forage quality and habitat structure for the federally threatened desert tortoise (*Gopherus agassizii*) (Brooks and Esque 2002). When habitats are altered to the degree that existing wildlife populations cannot survive, new populations may invade or increase in dominance, and ecosystem properties may be further altered.

Predicting the Effects of Plant Invasions on Fire Regimes

Plant invasions that only affect the quantitative magnitude of pre-existing fuel properties (continuous-trait invaders, Chapin and others 1996) are likely to have less impact than those that alter the more fundamental qualitative properties of the existing fuelbed (discrete-trait invaders, Chapin and others 1996). For example, a nonnative grass that invades an existing grassland may quantitatively change the fuel load and/or the continuity of fuels somewhat, but it does not represent a qualitatively new fuel type. In contrast, a woody plant invading the same grassland represents a qualitatively different fuel type and is likely to have a much greater impact on fuel conditions and the reference fire regime. For example, in prairies in the Willamette Valley of Oregon and the Puget Lowlands of Washington the invasion of woody plants has shifted the fire regime from a frequent, low-severity regime fueled by grasses and herbaceous vegetation, to a mixed-severity regime with longer fire-return intervals and accumulations of woody fuels (chapter 10).

Although many publications and models relate fuel characteristics to fire behavior, relatively little has been documented relating plant invasions, fuelbeds, and fire behavior to fire regimes. To determine that an invasive plant/fire regime cycle has been established one must (1) document that a plant invasion or set of invasions has altered fuelbed characteristics, (2) demonstrate that these fuelbed changes alter the spatial and/or temporal distribution of fire on the landscape, and (3) show that the new regime promotes the dominance of the fuels that drive it (fig. 3-6). This requires evidence that spans multiple fire-return intervals. Evidence can accumulate rapidly if the change results in shortened return intervals. However, if the change results in longer fire-return intervals, many decades to centuries may need to elapse before there is enough evidence to show that an invasive plant/fire regime cycle has in fact established. Although it may be very difficult to obtain the information necessary to definitively document that an invasive plant/fire regime cycle has established, reasonable inferences can be made based on a comparison of fuel and fire behavior characteristics of the invading species (fig. 3-6, steps 1 and 2) and estimated reference conditions in the habitats they are invading.

An excellent example of how indirect inference can be used to evaluate the potential for an invasive plant/fire regime cycle is provided by Rossiter and others (2003). These authors tested two assumptions of D'Antonio and Vitousek's (1992) grass/fire model:

1. Nonnative grass invasions alter fuel loads and fuelbed flammability.

Steps of Inference

1) Document that a plant invasion or set of invasions has altered fuelbed characteristics.

2) Demonstrate that these fuelbed changes alter the spatial and/or temporal distribution of fire on the landscape.

3) Show that the new fire regime promotes the dominance of the fuels that drive it.

Relative Difficulty to Accomplish

Figure 3-6—Steps required to determine the establishment of an invasive plant/fire regime cycle.

2. These changes increase fire frequency and/or intensity compared to uninvaded vegetation.

Their results indicate that a perennial grass invader from Africa, Gamba grass (*Andropogon gayanus*), can create fuelbeds with seven times more biomass than those created by native Australian savanna species. This higher fuel load led to a fire that was eight times more intense than other fires recorded in native fuelbeds during the same time of year, and produced the highest temperatures of any early dry season fire ever recorded in Northern Territory, Australia. Although this study did not demonstrate that the invading species preferentially benefited from the fire behavior it created, numerous examples from other ecosystems suggest that African grasses typically benefit from frequent, moderate to high intensity fires (Brooks and others 2004; D'Antonio and Vitousek 1992).

Another good example where indirect inference was used as evidence for an invasive plant/fire regime cycle comes from the Mojave Desert of western North America. Fires were historically infrequent and relatively small in this ecosystem that is dominated by sparse native desert vegetation (Brooks and Minnich 2006). Invasion by nonnative annual grasses from Eurasia has increased the amount, continuity, and persistence of fine fuels (Brooks 1999a; Brooks and Minnich 2006). The nonnative species that have caused these fuelbed changes include cheatgrass at higher elevations, red brome at middle elevations, and Mediterranean grass (*Schismus arabicus*, *S. barbatus*) at lower elevations (Brooks and Berry 2006). Fire often spreads exclusively in these fine nonnative fuels (Brooks 1999), allowing fires to occur under conditions where they otherwise would not. In addition, fires are larger following years of high rainfall that stimulate the growth of these nonnative annual grasses (Brooks and Matchett 2006), and these nonnnative species often dominate postfire landscapes within a few years after burning, setting the stage for recurrent fire (Brooks and Matchett 2003; Brooks and Minnich 2006). Some areas in the northeastern part of the Mojave Desert have recently experienced fires in excess of 100,000 ha (250,000 acres) (Brooks and Matchett 2006). Historical photographic and other anecdotal evidence suggest that since red brome invaded the region around 1900, fire-return intervals may have become as short as 20 years in some places, compared to historical estimates of greater than 100 years (Brooks and Minnich 2006). U.S. Bureau of Land Management photographs from the region clearly show continuous fuelbeds of red brome within a few years following fires that occurred during the 1940s. Thus, these invaders have altered fuelbeds, influenced fire behavior, recovered quickly following fire, and may have led to fire-return intervals on the order of 20 years in some places, all results that support the assumptions of both D'Antonio and Vitousek's (1992) grass/fire model and the invasive plant/fire regime cycle presented here. A similar example can be cited for cheatgrass in the Great Basin Desert of western North America (see discussion on page 40 and in chapter 8).

Herbaceous Plant Invasions

Herbaceous plants produce mostly fine fuels (1-hour timelag), which typically contain a high proportion of dead tissue late in the fire season. This is because the aboveground parts of herbaceous plants characteristically die back each year, either completely (for example, annuals, biennials, geophytes) or partially (for example, some perennial grasses, herbaceous shrubs).

Fine fuelbeds dominated by dead fuels respond rapidly to atmospheric conditions, especially with increased or decreased moisture content following changes in ambient relative humidity. Some herbaceous species create fuelbeds that do not readily carry fire because they have fuel moisture contents that are too high, horizontal continuities that are insufficient, or fuel bed bulk densities that are less than ideal for combustion. However, other species produce fuelbeds that can readily carry fire because they have low fuel moisture, high horizontal continuity, and fuel bed bulk densities that are conducive to combustion. Based on these parameters, forbs are generally the least flammable herbaceous invaders, and grasses (especially annual species) the most flammable.

Where highly flammable nonnative herbaceous plants have increased horizontal continuity of fine fuels, fires may be larger and more uniform than in uninvaded sites. Easy ignition and high fuel continuity, coupled with rapid recovery of nonnative herbs following fire, result in lengthy annual fire seasons and short fire-return intervals. Successive years of high rainfall and accumulation of dead herbaceous fuels can be associated with increased area burned (for example, Rogers and Vint 1987). The herbaceous fuels at the start of a given fire season consist mainly of litter from the past year's growth; later in the fire season, they are mostly comprised of the current year's growth (Jim Grace, personal communication 2006; M. Brooks personal observation, Mojave Desert uplands, spring and summer 1993). During drought periods, when herbaceous production is low, fuel load and continuity, and resulting probability of fire, tend to be low as well.

Perhaps the best known example of an invasive plant/fire regime cycle caused by an herbaceous plant is the invasion of the nonnative annual cheatgrass into sagebrush-steppe regions of the Intermountain West of North America (Brooks and Pyke 2001). Cheatgrass creates a type of fuel bed that was not previously present in this region (fig. 3-7). The altered fuel bed facilitates the ignition and spread of fire between adjacent shrubs (fig. 3-8). After fires occur, cheatgrass recovers rapidly, producing a fuel bed that can carry fire after as few as 5 years (Whisenant 1990a). Many native plants in this community, especially the subspecies of big sagebrush, are adapted to a longer fire-return interval and cannot persist where cheatgrass invasion substantially reduces the time between fires. Thus, cheatgrass both promotes frequent fire and recovers soon following fire, creating an invasive plant/fire regime cycle that has converted vast landscapes of native sagebrush-steppe to nonnative annual grasslands (chapter 8; Menakis and others 2003). These vegetation changes can cascade to higher trophic levels, affecting wildlife prey and predator species as well (reviewed by Brooks and others 2004).

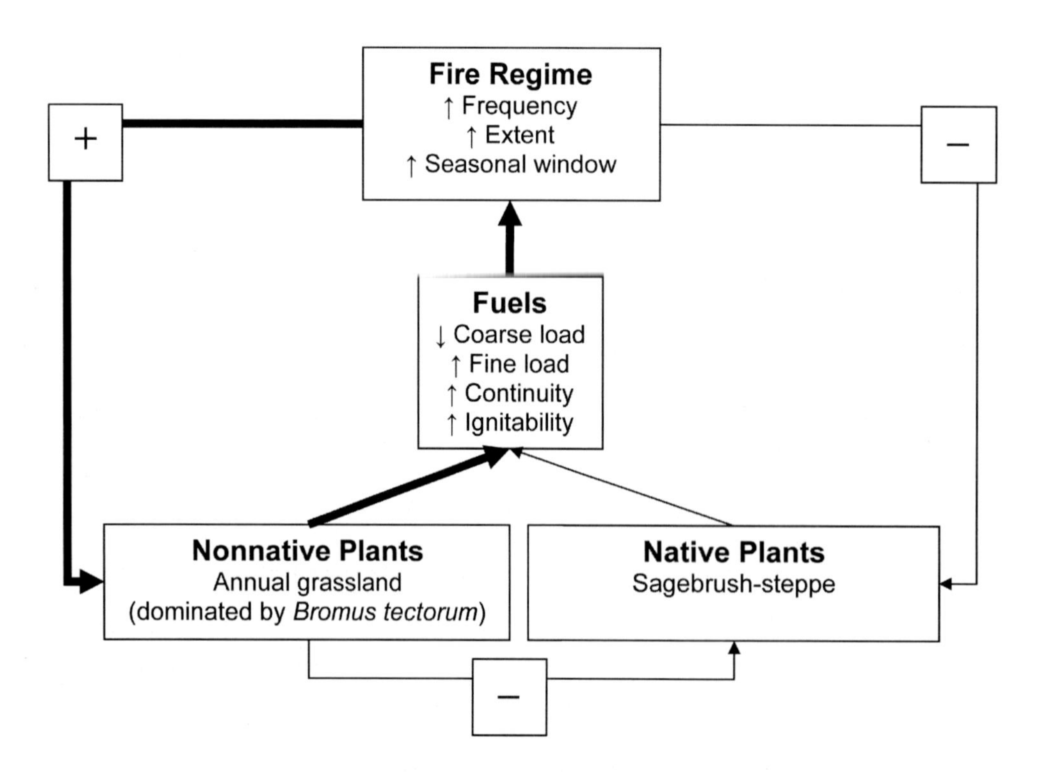

Figure 3-7—Changes in fuelbed and fire regime properties caused by the invasion of nonnative annual grasses into native sagebrush-steppe in the Intermountain West of North America.

Figure 3-8—Cheatgrass leading to an invasive plant/fire regime cycle in the Intermountain West of North America. (A) Initial invasion filling interspaces between shrubs creating fuelbeds of fine and woody fuels; (B) initial fires; (C) subsequent fuelbed dominated by fine fuels with few woody fuels; (D) recurrent fire that perpetuates fine fuelbeds. (Photos A, C, and D by Mike Pellant, BLM; photo B by J. R. Matchett, USGS Western Ecological Research Center.)

The fuel changes caused by cheatgrass invasion include increased fine fuel loads, horizontal continuity, and ignitability of fuels (figs. 3-7 and 3-8). These fuelbed characteristics lead to increased fire frequency, size, and seasonal window of burning. As shrublands have been replaced by grasslands, coarse woody fuels have been replaced by fine grasses. This may lead to decreased fire residence times and possibly reduced fire intensities, although these changes have not been specifically documented.

Other examples of herbaceous plant invaders that may increase fire frequency include other annual grasses (for example, red brome) and perennial grasses (for example, buffelgrass (*Pennisetum ciliare*), fountain grass (*Pennisetum setaceum*), Johnson grass (*Sorghum halepense*)) in desert shrublands (chapter 8) and the Pacific Islands (chapter 11), Japanese stiltgrass (*Microstegium vimineum*) in northeastern forests (chapter 5), climbing ferns (*Lygodium* spp.) in the Southeast bioregion (chapter 6), and Japanese knotweed (*Polygonum cuspidatum*) in the Northwest Coastal bioregion (chapter 10). While empirical evidence for fire regime change is lacking in many of these examples, differences in fuel characteristics between invaded and uninvaded areas imply that there is potential for such changes.

There are cases when herbaceous perennial plants may decrease fire frequency and intensity, especially when the invaders have high live fuel moisture. The invasion of iceplant species (for example, *Carpobrotus* spp., *Mesembryanthemum* spp.) into coastal sage-scrub in California appears to have reduced the likelihood of fire in some areas (M. Brooks, personal observation, coastal southern California, Ventura County, summer 2000). Invasion of cactus species (for example, *Opuntia* spp.) into fire-adapted matorral shrubland in Spain may be having a similar effect, as fires have been noted to be less frequent and intense where cacti have invaded (Vilà and others 2005). This could have deleterious consequences for "fire followers," species that depend on the fire regeneration niche to complete their life cycles.

Herbaceous species that have allelopathic effects on other plants, such as diffuse knapweed (*Centaurea diffusa*) (Callaway and Aschehoug 2000; Callaway and Ridenour 2004), may reduce fine fuel loads and thereby reduce the frequency and size of fires. In some cases, plant invasions such as these may suppress extant nonnatives that previously altered the fire regime (for example, *Bromus* spp.), and dynamics between the two nonnatives, native plants, and the fire regime may settle into a new cycle.

Woody Plant Invasions

Woody shrubs and trees produce coarse fuels (10-hour timelag and larger) in addition to fine fuels from twigs and foliage. As woody plant canopies approach closure, herbaceous surface fuels may be suppressed. The coarser structure of woody fuelbeds compared to herbaceous fuelbeds can narrow seasonal burning windows and lengthen fire-return intervals, especially if this change is accompanied by reduction in fine surface fuels. Horizontal continuity of fuels is highly variable, leading to highly variable size and occurrence of unburned patches. Fuel loads can be relatively high and contain a high proportion of dead tissue, especially in old stands with deep accumulations of duff and litter. These conditions can lead to intense, stand-replacing fires and significant soil heating (thus increased fire severity). Examples include mesic shrublands and evergreen and deciduous forests.

The replacement of continuous, fine, grassland fuels by more patchy, coarse shrubland and woodland fuels (fig. 3-9) has reduced the frequency of fire in a number of places worldwide (Bruce and others 1997; Drewa and others 2001; Gordon 1998; Grace and others 2001; Miller and Tausch 2001; van Wilgen and Richardson 1985). Fire exclusion often facilitates the early stages of these invasions by extinguishing fires before they spread and/or reducing fine fuels through intensive livestock grazing. However, at some point the altered fuelbeds reduce the chance of fire due to their inherent characteristics. The longer fire is excluded and woody species persist with their patchy distribution, the greater the chance that topsoil and the underlying distribution of soil nutrients will shift from homogeneous patterns characteristic of grasslands to patchy patterns

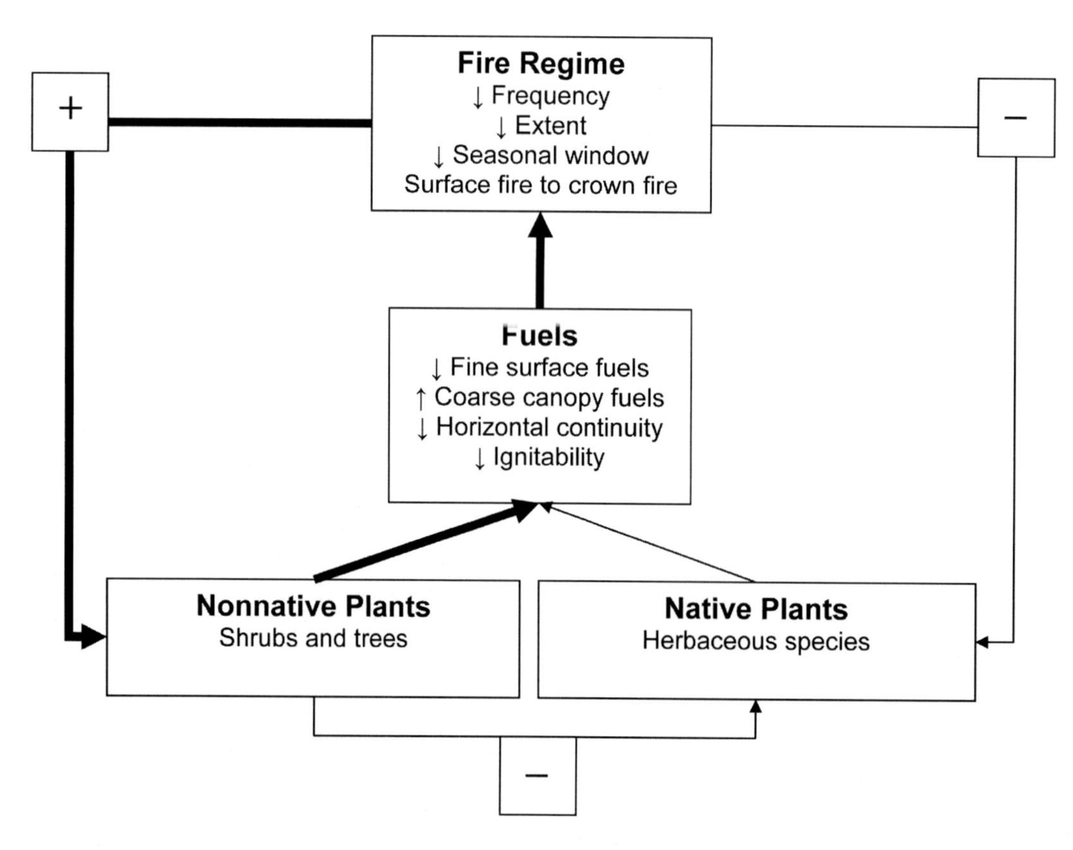

Figure 3-9—Changes in fuelbed and fire regime properties caused by the invasion of nonnative woody shrubs or trees into native herbaceous plant assemblages.

characteristic of shrub- and woodlands (Schlesinger and others 1990). Although positive feedback cycles between invading woody fuels and native grassland fuels have not been established over multiple fire-return intervals, it seems likely that these changes can be characterized as an invasive plant/fire regime cycle. This may be especially true if soil conditions, such as the spatial distribution of soil nutrients, change to the point where their patchiness does not support the continuous vegetation cover (Schlesinger and others 1996) needed to re-establish a reference regime of frequent, low-intensity fire.

Invasion of Chinese tallow (*Triadica sebifera*) into coastal prairie is a good example of how a tree invasion can alter fuels and fire regime properties (Bruce and others 1997; Grace and others 2001; chapter 7). As tallow trees invade, they overtop and shade out surface vegetation, resulting in reduced fine grass and forb surface fuels and increased coarse woody fuels (fig. 3-10). The fuelbed that results from Chinese tallow invasion is comprised mostly of canopy fuels, with few surface or ladder fuels. As a result, fires are difficult to start. If they do start, the high live fuel moisture content of Chinese tallow usually precludes spread into their crowns (Jim Grace, personal communication 2006). Although there are currently no data documenting a positive feedback loop between long fire-return intervals and dominance by Chinese tallow, the species' capacity for vigorous resprouting and rapid growth, and relatively low postfire mortality rates of mature trees, suggest that it can recover rapidly after fire and continue promoting a long fire-return interval. Theoretically, grasslands invaded by woody species that have reduced fine fuels and lengthened

Figure 3-10—Chinese tallow creating a closed-canopy forest that drives out native surface fuels and may result in an invasive plant/fire regime cycle in the Gulf States of North America. (A) Uninvaded prairie of fine fuels. (B) Early invasion adding woody fuels. (C) Late invasion creating a fire resistant stand of closed-canopy woody fuels and few fine surface fuels. (D) To allow native prairie to recover, fire may be needed to control Chinese tallow at early stages of its invasion before closed-canopy stands become established. (Photos by Larry Allain, USGS, National Wetlands Research Center.)

fire-return intervals could support crown fire under very hazardous fire weather conditions. However, thus far no studies have described this phenomenon.

Saltcedar (*Tamarix ramosissima, T. chinensis*) provides an example of how a woody plant invader can alter fuel characteristics of native riparian assemblages comprised largely of woody species (chapter 8). Saltcedar produces a nearly continuous litter layer that is highly flammable (M. Brooks, personal observation, lower Colorado River, San Bernardino County, California, summer 2000). Fires that start in these surface fuels can easily carry through mature saltcedar trees and up into the canopies of native riparian trees. This may result in a frequent, high intensity, crown fire regime where an infrequent, low to moderate intensity, surface fire regime previously existed (Brooks and Minnich 2006). Saltcedar can resprout readily after burning and benefit from nutrients released by fire, whereas native woody riparian plants often do not resprout as vigorously (M. Brooks, personal observation, Virgin River, Clark County, Nevada, summer 2005; Ellis 2001). Although there are no quantitative descriptions relating saltcedar invasions to fire regime changes, numerous anecdotal observations and accounts suggest that this species can establish an invasive plant/fire regime cycle in riparian ecosystems in western North America (Dudley and Brooks 2006).

Preventing or Mitigating Altered Fire Regimes

Exclusion of potentially threatening species before they invade and early detection and rapid response to eradicate populations at the very early stages of invasion are the most cost-effective and successful approaches to preventing the establishment of an invasive plant/fire regime cycle. These approaches focus on *Phase 1* (dispersal) and *Phase 2* (establishment, persistence, spread) of the cycle (fig. 3-11). The cost of control is lowest and probability of successful management is highest during *Phase 1*. There may be economic costs associated with exclusion of plant species that are used in ornamental horticulture or as livestock forage, but these short-term costs would be eclipsed by the long-term costs of inaction if the species moves into phases 2 to 4. After a species has established multiple local populations during *Phase 2*, management costs begin to rise and the probability of successful prevention or mitigation of negative effects begins to decline, but management can still be focused entirely on the invasive plant species. In contrast, once *Phase 3* begins, management must focus on controlling the invader, revegetating native plant communities, and possibly also restoring ecosystem processes (other than fire regime) that have allowed invasion or have

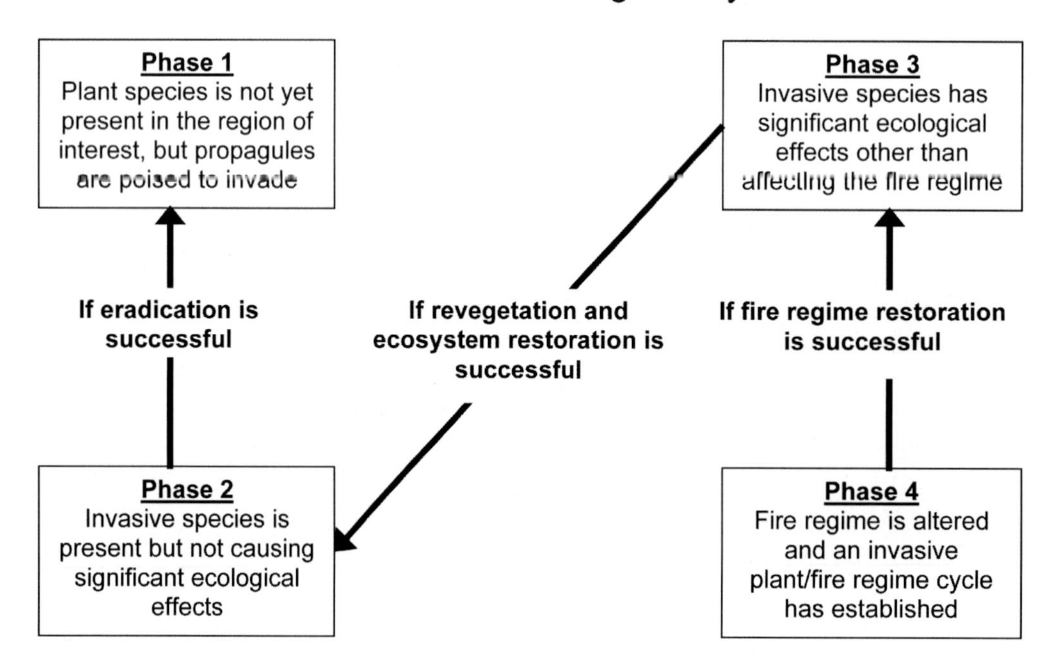

Figure 3-11—Steps toward breaking the invasive plant/fire regime cycle and reversing the effects of plant invasions on native plant communities and ecosystem properties, modified from Brooks and others (2004).

been altered by the invader. When *Phase 4* is reached and an invasive plant/fire regime cycle has been established, the invader needs to be controlled, native vegetation needs to be re-established, and various ecosystem processes, including the fire regime, must be restored. Thus, as each successive phase of the cycle is reached, additional management considerations are added, costs increase, and the probability of successful management decreases (Brooks and others 2004).

Summary Recommendations

Recommendations for management can be related to the steps of invasion and fire regime change shown in figure 3-11. Regarding *Phase 1,* plant species that have not yet invaded a region need to be evaluated for their potential to establish, become invasive, cause ecological impact, alter fuels, and eventually alter fire regimes. Species with high potential to alter fire regimes should be prioritized for exclusion from a region. Regarding *Phase 2, s*pecies that have persisted or begun to expand need to be evaluated for their potential to cause significant ecological impact and alter fuels. Species with a high potential to cause negative impact or alter fuels need to be prioritized for control. Regarding *Phase 3,* species that have already caused significant ecological impact need to be evaluated for their potential to alter fuels and fire regimes under any of the environmental conditions that occur in the region. Species with high potential to alter fire regimes should be prioritized for control, and restoration of pre-invasion plant community and ecosystem properties may be necessary. Species that introduce qualitatively novel fuel characteristics should in general be considered greater threats than those that only change fuel conditions quantitatively. Regarding *Phase 4,* when a species has already changed one or more fire regime characteristics, the altered regime needs to be evaluated for its potential to have negative effects on public safety, property, local economies, natural resources, and wildland ecosystems. Ecosystem effects may include reduced biodiversity of native species, loss of wildlife habitat, promotion of subsequent invasions by other nonnative species, altered watershed functioning, loss of tourist appeal, and increased fire-associated hazards.

In some cases, it may not be possible to restore communities to their pre-invasion state. For example, fire-enhancing tropical grasses from Central America and Africa have invaded seasonally dry habitats in the Hawaiian Islands, increased fire frequency, and reduced the amount of native forest. Restoration of pre-invasion native vegetation was found to be difficult, and managers have instead created "replacement communities" of native grassland species that are more fire tolerant than forest species and can coexist with the nonnative grasses (Tunison and others 2001; chapter 11).

Nonnative species that are not invasive may also be used in postfire revegetation to compete with invaders and recreate pre-invasion fuel characteristics to help restore altered fire regimes. For example, the nonnative bunchgrass crested wheatgrass (*Agropyron desertorum*) has been seeded into postfire landscapes in the Great Basin desert of North America to suppress growth of the nonnative annual cheatgrass and reduce fuel continuity and flammability (Hull and Stewart 1948). The idea is that nonnative bunchgrasses can help restore a pre-invasion fire regime of infrequent, low intensity fire, which will allow native plants to re-establish more easily. This process has been referred to as "assisted succession" (Cox and Anderson 2004). Proponents anticipated that, if the desired fire regime can be maintained, over time the original native plant species may be gradually reintroduced. It is difficult to say how successful this management approach has been because effectiveness of past postfire emergency stabilization, rehabilitation, and restoration in the United States has not been monitored (GAO 2003). However, current studies are in progress to provide some of this information.

In this chapter I have presented a number of examples of how plant invasions can alter fire regimes. Although the ecological implications of these changes can be significant, one must remember that few plant invasions will result in fire regimes shifted beyond their reference conditions. Even so, the potential effects of invaders on fire regimes must be considered along with other potential effects when prioritizing plant invaders for management (for example, Warner and others 2003). To help in this task, I have described key elements linking fuel conditions with fire regimes that can help in screening plant invaders for their potential effects. This task would further benefit from additional examples of cases where invasions by nonnative plants have altered fire regimes. There are very likely many examples that await discovery, especially in contexts where these relationships may not be expected or otherwise cause people to take notice.

Notes

Peter M. Rice
Jane Kapler Smith

Chapter 4:

Use of Fire to Manage Populations of Nonnative Invasive Plants

Introduction

It may be impossible to overstate the complexity of relationships among wildland ecosystems, fires, and nonnative invasives. Strategies for managing these relationships are similarly complex; they require information on local plant phenology, ability to produce various levels of fire severity within burns, willingness to combine fire with other management techniques, and systematic monitoring to improve effectiveness. Oversimplification and short-sightedness in planning can lead to unintended degradation of the ecosystem; lack of monitoring may leave such consequences unnoticed and unaddressed. An inventory of the knowledge needed for planning an effective burning program could begin with the topics listed in table 4-1; managers need to understand the regeneration strategies and phenology of both target and desired species and their respective sensitivity to fire regime characteristics. Extensive information like this is currently available for only a few invasive species. If the information is not

Table 4-1—Inventory of species-specific knowledge needed to assess potential for using prescribed fire to control nonnative invasive plants. This information is needed not only for the invasive species but also for desired native species.

Topic	
Postfire regeneration from seed (production, dispersal, mobility, use of seed bank)	
Postfire vegetative regeneration strategies & location of perennating tissues	
Season	Most vulnerable to fire Least vulnerable to fire
Fire interval	Most favorable to regeneration Least favorable to regeneration
Probable fire effect	Low-severity High-severity

available to managers, they must monitor treatment sites carefully and learn from experience.

To assess the potential for managing nonnative invasive species with prescribed fire, managers must integrate knowledge about individual species. This includes understanding the condition of the plant community to be treated and altering conditions on the site that favor nonnatives (Brooks 2006; Keeley 2006b). If disturbance favors the invasive more than desired natives, fire alone is probably inappropriate (Keeley 2006b). If desired species are unable to establish dominance soon after a treatment, the target species or another undesired species is likely to take over (Goodwin and others 2002). Finally, ecological considerations must be integrated with practical aspects of fire management:

- What fire season(s), severities, and intervals seem most desirable for meeting treatment objectives?
- What fuels, weather conditions, and firing techniques are needed to produce the needed fire behavior?
- During what seasons, and how long after previous fire, are these conditions present?
- What other treatments might enhance the benefits of fire? Can fire be used to enhance the benefits or reduce the negative impacts of other treatments?

Treatments that prove successful in one place may not succeed in another (McPherson 2001). Garlic mustard (*Alliaria petiolata*), a nonnative biennial herb, provides an example of a species with fire responses that vary with fire regime characteristics and with the plant community being treated. In an Illinois oak (*Quercus* spp.) forest, spring fires with flame lengths up to 6 inches (15 cm) and fairly uniform burning reduced the density of both seedlings and mature garlic mustard plants (Nuzzo 1991); lower-intensity, less uniform fires had no appreciable effect. In another study of an oak forest in Illinois, a series of three annual dormant-season fires (flame heights up to 4 feet (1.2 m)) maintained garlic mustard at low percent cover, whereas it increased substantially without fire. Increased native species cover and richness accompanied decreases in garlic mustard (Nuzzo and others 1996). Conversely, in hardwood forests of Kentucky dominated by sugar maple (*Acer saccharum*) and white ash (*Fraxinus americana*), repeated prescribed fires had no appreciable effect on garlic mustard abundance or richness of native species (Luken and Shea 2000). Perhaps the only generalization that can be applied to management of invasive species with fire is that results of any treatment should be monitored and evaluated so management programs can be improved with time and experience (chapter 15).

Scientific literature on control of nonnative invasives with prescribed fire is limited. A recent comprehensive literature search and case history review found only 235 references on this topic (Rice 2005, review). Many of these were proceedings, abstracts, or managers' reports without supporting data. Relatively few publications report studies with replicated treatments and controls that meet the standards of peer reviewed journals. Details on fuel loads and fire behavior are generally lacking. But the biggest challenge to applying research on burning to control invasives is the variability of plant invasions themselves—the apparently limitless potential interactions of target invasives, desirable competitors, fuel properties, fire behavior, climate, and other ecosystem properties.

Despite the limitations of the knowledge available, a survey of studies conducted in North America provides insights about the use of fire for controlling nonnative invasive species. There are many ways to examine this subject. In the first section below, we discuss the effects of fire alone for managing invasives. Second, we look at fire combined with other management tools. In the third section, we examine the potential for manipulating three aspects of the fire regime—fire severity, uniformity, and frequency—to control undesired species and move the ecosystem toward a desired condition. Finally, in the last section we take a brief look at political and logistical aspects of managing nonnative species with prescribed fire. While this chapter looks at several facets of use of prescribed fire for controlling nonnative invasive species, it does not attempt to describe management of individual species in depth; for that information, see discussions of individual species in the bioregional chapters (chapters 5 through 11) and other sources, especially the Fire Effects Information System (FEIS, www.fs.fed.us/database/feis) and The Nature Conservancy's Element Stewardship Abstracts (tncweeds.ucdavis.edu/esadocs.html).

Use of Fire Alone to Control Nonnative Invasive Plants

To achieve long-term control of a nonnative invasive population with fire, managers must consider the species' regeneration strategies (by seed and vegetative means), phenology, and site requirements (reviews by Brooks 2006; DiTomaso 2006a; Rice 2005). To favor native or other desired species, the same considerations apply (table 4-1).

Prevention of Reproduction by Seed

Preventing Flowering or Seed Set—Prescribed fire can be used to prevent seed production in nonnative invasive species by killing aboveground tissues prior to flowering or seed maturation. The fire must be severe enough to damage target plants, so success may depend on quantity and quality of fuel on the site (see "Prescribed Fire Severity and Uniformity in Relation to Fuel Beds" page 57). Two examples highlight the importance of evaluating effectiveness over the long term and the difficulty of using fire to both reduce the target species and enhance native species.

Multiyear burning in California grasslands provided only temporary control of yellow starthistle (*Centaurea solstitialis*), a nonnative invasive annual forb. Yellow starthistle was associated with a variety of nonnative annual grasses and also native species, including purple needlegrass (*Nassella pulchra*), blue wildrye (*Elymus glaucus*), and beardless wildrye (*Leymus triticoides*). Conducting prescribed burns during late floral bud stage or early flowering stage of yellow starthistle prevented seed production while allowing for seed dispersal by associated vegetation (Hastings and DiTomaso 1996). Reduction in yellow starthistle vegetative cover following 3 consecutive years of burning corresponded to a 99 percent depletion of yellow starthistle seeds in the soil seed bank. During the same period, species richness and native forb cover increased on burned sites compared to unburned controls (Hastings and DiTomaso 1996; Kyser and DiTomaso 2002). Four years after cessation of annual burning on these sites, however, yellow starthistle cover and seed bank density had increased to near pretreatment levels, and native forbs, total plant cover, and diversity had declined (Kyser and DiTomaso 2002).

Use of prescribed fire to control biennial species is complex, but some success was achieved at a 45-year-old restored tallgrass prairie site in Minnesota. Prescribed burning of second-year biennial sweetclover (*Melilotus* spp.) prevented seed formation and reduced sweetclover frequency. However, the optimal burning schedule for reducing sweetclover (a sequence of early spring burning one year, followed by a May burn the next year) reduced native forb frequencies. Dormant season burns were least successful at controlling sweetclover but increased native forb frequencies (Kline 1983).

Destroying Seeds in Inflorescences—Burning the seeds of nonnative invasive annual grasses before they disperse is a goal of many restoration programs (for example see Allen 1995; Kan and Pollack 2000; Menke 1992). Important considerations for success include burning when seeds are most vulnerable to heat and before they are dispersed to the soil surface, and producing fires severe enough to kill the seed.

Seeds of many species are most vulnerable to heat damage before they are fully cured (DiTomaso and others 2001; Furbush 1953; McKell and others 1962). Backfiring has been recommended (McKell and others 1962; Murphy and Turner 1959) to maximize fuel consumption and thus heat produced, and to increase the duration of heating. McKell and others (1962) burned California grasslands dominated by medusahead (*Taeniatherum caput-medusae*), a nonnative annual grass, at different times as the seed developed and dispersed. Stands burned with a slow-moving backfire while medusahead seed was in the soft dough stage (highest moisture content) had the lowest density of mature medusahead the next growing season. Similarly, early June burning of medusahead infestations in northern California rangelands when desirable annual grasses had cured but medusahead seed was still in the milk to early dough stage "effectively removed" medusahead and increased desirable forage for at least 3 years (Furbush 1953). DiTomaso and others (2001) completely burned a barbed goatgrass (*Aegilops triuncialis*) infestation in California in 2 consecutive years when goatgrass seed was still in the soft dough stage. The barbed goatgrass seedling density in this pasture was reduced to 16 percent of the control in the year after the first burn and to zero the year after a second burn. Density reduction was less in a second pasture, which did not have enough fuel for a complete burn in the second year.

In fires carried by fine grasses and forbs, fire temperatures may be higher in the fine fuel canopy than at the soil surface (Brooks 2002; DiTomaso and others 1999). Under these conditions, seeds retained in inflorescences are likely to be more susceptible to fire than seeds on or in the soil. Brooks (2002) reports this phenomenon for red brome (*Bromus rubens*), a nonnative annual grass invading many sites in the Mojave Desert; Kan and Pollak (2000) report the same phenomenon for medusahead. However, a grassland fire may consume most of the standing dead biomass and still not produce enough heat to kill seeds in intact seedheads. For example, Sharp and others (1957) measured 87 percent germination of medusahead seeds collected after a prescribed fire that burned the culms nearly to the head but did not scorch the heads. Kan and Pollak (2000) comment that August burns (after seed dispersal) may actually increase medusahead abundance.

Laboratory studies confirm that extending the magnitude and duration of heating can greatly increase seed mortality for some nonnative species—for example, jointed goatgrass (*Aegilops cylindrica*) (Willis and others 1988; Young and others 1990) and spotted knapweed (*Centaurea biebersteinii*) (Abella and MacDonald 2000). Duration of heating in prescribed burns can be manipulated by planning the burn for a time when fuels are abundant, scheduling for the time of day with the highest temperatures, and manipulating ignition patterns. Deferring grazing increases fine fuels and can thus increase heat release (George

1992). Deferring grazing also avoids livestock-caused seed dispersal, which deposits seed on the ground and may make it less vulnerable to fire (Kan and Pollak 2000; Major and others 1960).

Destroying Seeds in the Litter and Soil—Grassland and surface fires may kill seed in the litter layer (Daubenmire 1968a; DeBano and others 2005, review), but it is often difficult to produce high enough fire severity at the soil surface to cause mortality. Species that release seed rapidly after maturation are especially difficult to eradicate with fire. For example, cheatgrass (*Bromus tectorum*) seeds begin to disperse shortly after culms cure enough to carry a fire. Consequently, for most of the year, almost all cheatgrass seeds are in the litter or on the soil. Fire may consume most of this seed, but some is likely to survive and establish highly fecund plants the following year. Thus fire is unlikely to cause long-term reduction of cheatgrass (Zouhar 2003a, FEIS review).

Mortality of invasive grass seed may be higher under the canopy of burned shrubs, where woody fuels increase heat release, than in open areas. Fires often destroy cheatgrass seed located directly under the sagebrush (*Artemisia* spp.) canopy (Young and Evans 1978). These areas must be planted with desirable species the year of the fire, however, or they will be reinvaded by seed from cheatgrass growing in the shrub interspaces, where sagebrush and woody fuels are lacking and hence burning is less severe (Evans and Young 1987).

Seeds in the soil are unlikely to be damaged by grassland fires (Daubenmire 1968a,b; Vogl 1974), and fires in shrublands and woodlands do not generally produce enough heat to kill seed buried deeper than about 2 inches (5 cm), since soil temperatures at this depth may not change at all during fire (Whelan 1995). However, the soil may experience temperatures lethal to seeds when heavy fuels, such as large woody fuels or deep duff, burn for long periods (chapter 2). Peak temperatures from spring and summer prescribed fires in the Mojave Desert varied with aboveground fuels and vertical location (fig. 4-1). Temperatures in areas with sparse grass/forb cover and along the edge of native creosote bush (*Larrea tridentata*) plants were not lethal to seeds, but temperatures under the shrubs, as deep as <1 inch (2 cm) below the soil surface, were lethal to red brome and native annual seeds. The spatial variability in fire severity led to complex fire effects on the plant community. While burning reduced red brome and native annuals under the burned shrub canopy for the 4 years of the study, two nonnative perennials—Mediterranean grass (*Schismus* spp.) and cutleaf filaree (*Erodium cicutarium*), which occur more commonly near the dripline than under the shrub canopy—recovered to preburn levels by the second postfire year. Also under the drip line, native annuals increased during the first 3 postfire years (Brooks 2002).

Ground fires in habitats with heavy litter and duff could produce lethal temperature regimes below the soil surface, but there are no reported cases where this type of burn has been employed to control invasive plants. Ground fires severe enough to destroy buried seed would probably kill perennating tissues of desired native plants, a negative consequence likely to outweigh the benefits of reducing the nonnative seed bank.

Depleting the Seed Bank by Fire-stimulated Germination—Synchronous germination of significant portions of the seed bank can increase target populations' vulnerability to follow-up treatments, including repeat burning. French broom (*Genista monspessulana*) and Scotch broom (*Cytisus scoparius*) are nonnative woody plants that invade grasslands, woodlands, and open forests; these species are currently most problematic in Washington, Oregon, and

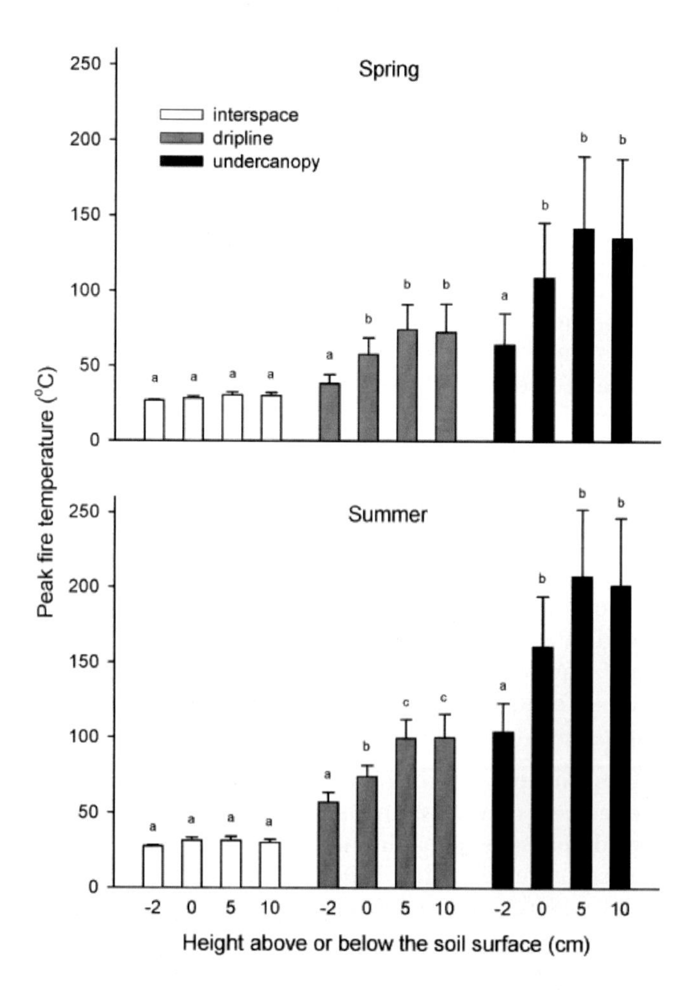

Figure 4-1—Peak fire temperature for three microhabitats and four heights from the soil surface in two seasons. Significant differences are indicated by different lower case letters. (Adapted from Brooks 2002, with permission from the Ecological Society of America.)

California (Zouhar 2005a,c, FEIS reviews). Standing, herbicide-treated, or cut broom stands in California are burned to kill aboveground tissues and to kill seed or encourage germination from the persistent seed bank (Bossard 2000a,b; Boyd 1995). Follow-up treatments with prescribed fire, propane torch, hand pulling, brush cutter, or herbicide within 2 to 3 years can kill sprouts and seedlings before new seeds are produced. With appropriate timing, seedlings may also die during seasonal drought periods following germination. Repeated prescribed burning is most effective if grasses are present to carry the fire (see "Prescribed Fire Severity and Uniformity in Relation to Fuel Beds" page 57). Removal sites should be monitored annually for 5 to 10 years to locate and kill new seedlings (Bossard 2000a,b).

Fire has also been used to deplete the seed bank of white and yellow sweetclover (*Melilotus alba, M. officinalis*) in Minnesota prairies (Cole 1991; Kline 1983,1986), but the benefit of seed reduction may be offset by reduction in native species cover (see "Preventing Flowering or Seed Set" page 48).

Induced Mortality and Prevention or Delay of Resprouting

Control of invasive biennial and perennial species requires either direct mortality of perennating tissues or depletion of carbohydrate reserves in these tissues (Whelan 1995).

Direct Mortality—It is generally not feasible to kill belowground perennating tissues with prescribed fire. Daubenmire (1968a) and Whelan (1995) reviewed studies of surface fires and found that soil temperatures below 1 inch (2.5 cm) are unlikely to reach 212 °F (100 °C) during a fire, even with high surface fuel loads in shrublands and forests. Temperatures decline rapidly with small increases in soil depth, reaching temperatures no higher than 120 °F (50 °C) 2 inches (5 cm) below the surface (Whelan 1995). Plant cell death begins when temperatures reach 120 to 130 °F (50 to 55 °C) (Hare 1961).

Direct mortality from fire has been achieved for some woody species by use of cutting or herbicides to increase fuel loads (see "Treatments that Increase Effectiveness of Prescribed Fire" page 53 and "Prescribed Fire Severity and Uniformity in Relation to Fuel Beds" page 57). In addition, prescribed fire may be effective for controlling species with shallow perennating buds. Steuter (1988) burned a mixedgrass prairie in South Dakota to suppress absinth wormwood (*Artemisia absinthium*), a nonnative invasive subshrub that has perennating buds at or near the soil surface. A series of four early May fires within a 5-year period reduced density of absinth wormwood by 96 percent.

Prescribed Burning to Deplete Carbohydrate Reserves—Repeated fires have been used to reduce postfire sprouting in some woody species, probably by reducing the plants' carbohydrate reserves. Managers of fire-adapted midwestern and eastern plant communities often suggest annual or biennial burning to control sprouting of nonnative shrubs if the burn treatment can be repeated for periods as long as 5 or 6 years (Heidorn 1991, review). In Alabama, annual burning during very dry periods eliminated European and Chinese privet (*Ligustrum vulgare* and *L. sinense*) (Batcher 2000a, TNC review), whereas a single burn treatment in northwestern Georgia (Faulkner and others 1989) caused no significant change in Chinese privet.

Use of repeated fire has produced equivocal results for several woody species. Glossy buckthorn (*Frangula alnus*) and common buckthorn (*Rhamnus cathartica*) are nonnative shrubs that form dense thickets in native grasslands. They have been reported both to decrease (Grese 1992, review) and increase (Post and others 1990) after repeated prescribed fire. In wet prairies of the Willamette Valley, Oregon, neither a single fall burn nor two consecutive fall burns significantly altered the density of nonnative invasive shrubs (sweetbriar rose (*Rosa eglanteria*), Himalayan blackberry (*Rubus discolor*), and cutleaf blackberry (*R. laciniatus*)) or trees (oneseed hawthorn (*Crataegus monogyna*) and cultivated pear (*Pyrus communis*)) (Pendergrass and others 1998). Nevertheless, the authors comment that repeated burning may gradually reduce the density and retard the expansion of woody species.

Prescribed burns are often conducted during the dormant season to protect vulnerable wildlife species (for example, Mitchell and Malecki 2003; Schramm 1978), but some researchers suggest that growing season burns would offer better control of woody nonnatives. Dormant season burns in the Northeast are followed by profuse sprouting from the roots and rhizomes of many nonnative woody species. Total nonstructural carbohydrate reserves are lowest during the growing season, so burning late in the growing season may deplete root carbohydrates of nonnatives more effectively than dormant-season burns (Richburg and others 2001; Richburg and Patterson 2003b). The same principle may apply in coastal prairies and forests in the southern United States being invaded by Chinese tallow (*Triadica sebifera*), a nonnative tree. The species is difficult to control with fire, but prescribed fires have reduced sprouting of small trees and prevented tallow from gaining dominance (Grace 1998; Grace and others 2001). Growing-season burns were more effective than dormant-season burns. Intense fires can damage large tallow trees, but stands of Chinese tallow suppress herbaceous species needed to carry fire, so frequent burning is not usually feasible (Grace and others 2001, 2005).

Burning to Favor Native Species

Taking Advantage of Varying Plant Phenology—Burning while an invasive species is actively growing and desired native species are dormant can reduce the invasive and simultaneously enhance the productivity of native species. This technique has been studied mainly in grassland ecosystems. In an Iowa prairie remnant burned 1 to 3 times in 3 years, native vegetation began growing earlier, matured earlier, and produced more flower stalks on burned than unburned sites. Repeated fires reduced the density of the nonnative invasive Kentucky bluegrass (*Poa pratensis*), which was beginning growth at the time of burning while native grasses were still dormant (Ehrenreich 1959). Willson (1992) found that burning in mid-May reduced smooth brome (*Bromus inermis*) and increased big bluestem (*Andropogon gerardii*), the desired native dominant, in a Nebraska tallgrass prairie. Smooth brome tillers were elongating at the time of burning. Earlier burning, when tillers were emerging, did not reduce smooth brome, as demonstrated by this study and also research by Anderson (1994).

Since plant communities are usually complex mixtures of species, it is not surprising that burn treatments that reduce nonnative invasive species often have mixed effects on the native plant community. It may be very difficult to use fire to reduce a nonnative species if a desired species has similar phenology. Two Wisconsin studies on prairie restoration and maintenance demonstrate this point. In one study native tallgrass prairie species were planted in abandoned fields dominated by nonnatives Kentucky bluegrass and Canada bluegrass (*Poa compressa*), and the sites were burned the following spring and 1 year later. Burns occurred while nonnative bluegrasses and natives Canada wildrye (*Elymus canadensis*) and Virginia wildrye (*Elymus virginicus*) were actively growing, but before other natives (indiangrass (*Sorghastrum nutans*), switchgrass (*Panicum virgatum*), big bluestem, and little bluestem (*Schizachyrium scoparium*)) commenced growth. One year after the second burn, the bluegrasses and wildryes had declined; most other native grasses had increased (Robocker and Miller 1955). A second study examined the effects of 8 burns in 10 years on a prairie remnant (Henderson 1992). Repeated late-spring fires reduced Kentucky bluegrass but also reduced native sedges (*Carex* spp.), bunchgrasses, and some forbs. Native porcupinegrass (*Hesperostipa spartea*) increased. In the same area, late fall and early spring burns reduced Kentucky bluegrass and had less effect on native species.

Where a site is infested with multiple nonnative invasive species, differences in their phenologies limit the benefits of burning. The manager may need to determine which species is most detrimental to the ecosystem and focus resources on controlling it (Randall 1996). Reports of fire effects in tallgrass prairie demonstrate the complexity of scheduling prescribed burns to maximize benefits. Late April burns on an Illinois prairie eliminated Kentucky bluegrass, which began growth in early April and reached peak production in mid-May. But the burns did not eliminate smooth brome, which began growth in mid-April and reached peak production about 3 months after the burns, although its productivity was reduced (Old 1969). Becker's (1989) research on repeat spring burning of prairie in southwestern Minnesota also had complex results. Five consecutive spring burns reduced cover of Kentucky bluegrass, smooth brome, and Canada thistle (*Cirsium arvense*), and native prairie species were favored in locations near prairie remnants. However, large patches of quackgrass (*Elymus repens*) persisted and expanded slightly into areas where invading woody species were killed, probably because quackgrass was not actively growing at the time of the fires.

Use of fire in California grasslands demonstrates the need for flexibility in scheduling burns to take advantage of susceptible phenological stages. Prescribed fire reduced nonnative invasive grasses (red brome, mouse barley (*Hordeum murinum*), slender oat (*Avena barbata*), and wild oat (*Avena fatua*)) and increased native plant cover in California grasslands if it was applied just before the invasive grasses set seed. However, the time of seed set in these grasslands can vary by as much as 2 months from year to year, depending on precipitation (Meyer and Schiffman 1999). More detail on this experiment is provided in "Burning Litter to Manipulate Species Composition" below. Thus monitoring of grass phenology and flexibility in management are both needed to use fire effectively. Flexibility and detailed scheduling may also be needed for use of fire in treatment areas that are large or cover complex terrain, where plant phenology and burning conditions may vary across the area.

If native species are sparse or low in vigor, burning will probably not shift dominance from nonnative invasives to native species, as demonstrated by pasture restoration efforts in Iowa (Rosburg and Glenn-Lewin 1992) and studies of "abused" rangeland in southern Nebraska (Schacht and Stubbendieck 1985).

Burning Litter to Manipulate Species Composition—Burning may stimulate fire-adapted native species by removing dead stems and litter, and may reduce nonnatives that grow well in litter. However, this technique is ineffective where litter removal leaves many surviving invasives or favors establishment of new invasives.

Season-dependent success with burning to remove litter is illustrated by research in the California grasslands described above (Meyer and Schiffman

1999). Initially, cover by nonnative annual grasses was greater than 97 percent. Litter was removed by burning and clipping/raking. Late spring burns and fall burns significantly increased cover and diversity of native vegetation and reduced cover and seed viability of nonnative grasses. Neither winter burns, which were less severe, nor partial mulch removal enhanced native cover or reduced nonnative cover significantly. Litter removal by fire can reduce other nonnative species, including soft chess (*Bromus hordeaceus*) (Heady 1956), cheatgrass (Evans and Young 1970, 1972; Young and others 1972a), and medusahead (Evans and Young 1970).

The usefulness of litter removal to reduce nonnative annual grass abundance is compromised if individual nonnatives respond with increased vigor and fecundity in the postfire environment. Examples from Sierra Nevada foothills and Midwestern prairie sites illustrate this point. At a Sierra Nevada foothills ponderosa pine site infested with nonnative annual grasses, a fall wildfire consumed the 1- to 2-inch (3- to 5-cm) litter layer. Cheatgrass and soft chess were reduced but not eliminated on burned plots; seedling density was 16 percent of that on adjacent unburned plots in the growing season after the fire. The reduction in annual grass density was not significant in the second postburn growing season, however, and by the third year the burned and unburned plots had near equal abundance of nonnative annual grasses. Effects on native species were not reported (Smith 1970). Whisenant and Uresk (1990) burned plots in Badlands National Park, South Dakota, that were dominated by Japanese brome (*Bromus japonicus*) and western wheatgrass (*Agropyron smithii*). Burning reduced Japanese brome density and standing crop in the first postfire growing season and favored native grasses, but Japanese brome density returned to preburn levels by the second growing season as litter began to accumulate (Whisenant 1990b; Whisenant and Uresk 1990).

While litter removal may reduce some nonnative annual grasses, nonnative forbs with regeneration facilitated by seed-to-soil contact often increase after litter removal. Cutleaf filaree seeds have a twisted, awn-like structure that forces the seed deep into the soil as it wets and dries, thus favoring establishment of this nonnative forb on bare soil (Bentley and Fenner 1958); germination is inhibited by a litter layer (Howard 1992a, FEIS review). Pickford (1932) noted a high abundance of cutleaf filaree in the Great Salt Lake Valley, Utah, in sagebrush and cheatgrass areas subject to frequent burning. Meyer and Schiffman (1999) measured a tenfold increase in cutleaf filaree on late spring burn plots in contrast to unburned control plots in a California grassland. In creosote bush and blackbrush (*Coleogyne ramosissima*) communities, cutleaf filaree cover was greater on burned than control plots 2 to 14 years after burns (Brooks 2002; Brooks and Matchett 2003).

Fire Combined with Other Treatments

If invasive species are generally promoted by fire, it does not make sense to attempt to use fire alone to reduce them (Keeley 2006b); however, fire is sometimes effective when used in combination with other treatments. Use of fire to deplete the seed bank, when mature plants will be controlled by other means, is discussed above (see "Depleting the Seed Bank by Fire-stimulated Germination" page 50), as is use of fire to suppress target species by litter removal. Mechanical and herbicide treatments can be useful to prepare for prescribed burning, especially on sites with sparse fuels. In addition, fire can be used to increase herbicide efficacy, prepare for other disturbance treatments, prepare a site for introduction of desired native species, and promote expansion of biocontrol organisms.

Treatments that Increase Effectiveness of Prescribed Fire

Mechanical, cultural, and chemical treatments can be used to increase the effectiveness of prescribed fire. Melaleuca (*Melaleuca quinquenervia*) is a nonnative tree that rapidly establishes, spreads, and eventually dominates southeastern wetland coastal communities and is well adapted to fire (Molnar and others 1991). A control program in southern Florida integrates cutting mature melaleuca, treating stumps with herbicide, and prescribed burning 6 to 12 months later to kill seedlings (Myers and others 2001). Complex programs of cutting, herbicide treatment, prescribed fire, and hand pulling have been combined to reduce French broom and Scotch broom in California (see "Depleting the Seed Bank by Fire-stimulated Germination" page 50). Research in the Northeastern states showed that cutting in early summer followed by burning in late summer prevented full recovery of nonstructural carbohydrates for at least 2 years in common buckthorn and another nonnative woody species, Japanese barberry (*Berberis thunbergii*). Growing season treatments were more effective than dormant season treatments (Richburg and Patterson 2003a).

Fire has been combined with grazing to restore tallgrass prairie in Oklahoma (Fuhlendorf and Engle 2004), based on the assumption that the native plant community evolved under a grazing-fire regime. This study compared plant community composition on unburned sites with sites managed in a patchy burn pattern, where one-third of the area was burned each year. Cattle were "moderately stocked" in both treatments.

In patch-treated areas, livestock devoted 75 percent of their grazing time to the most recently burned area. Treatment had little effect on native tallgrasses during the 4 years of the study. Abundance, diversity, and structural complexity of native forbs increased in patch-burned areas but did not change in untreated areas. Cover of sericea lespedeza (*Lespedeza cuneata*), a nonnative invasive forb, showed no net change on patch-burned areas but increased steadily on unburned areas.

Prescribed Fire to Enhance Efficacy of Other Treatments

Burning to Increase Herbicide Efficacy—Fire can be used to prepare a site for herbicide treatment, and combining herbicides with prescribed fire may reduce the amount of herbicide needed or the number of applications required. Herbicides may be more effective after fire in part because postfire herbaceous growth tends to be more succulent and have a less-developed cuticle than unburned herbage, resulting in more efficient absorption of herbicide (DiTomaso and others 2006a). It is important to note, however, that some herbicides cannot be applied immediately after burning, lest charcoal bind the active ingredient and make it unavailable for plant uptake (DiTomaso, personal communication 2004). Burning cheatgrass stands before emergence in preparation for applying herbicide may increase efficacy and reduce the herbicide required for control (Vollmer and Vollmer, personal communication 2005). DiTomaso's (2006b) review reports that yellow starthistle control usually requires 3 years of prescribed burning or clopyralid treatment when either method is used alone, but a similar level of control can be accomplished in only 2 years when a prescribed burn is conducted in the summer of the first year and clopyralid is applied the following winter or early spring. Fire has been used prior to herbicide application to enhance control of many other invasive species, including:

- Grasses (Lehmann lovegrass (*Eragrostis lehmanniana*), ripgut brome (*Bromus diandrus*), medusahead, and tall fescue (*Lolium arundinaceum*)) (Rice 2005; Washburn and others 2002)
- Forbs (fennel (*Foeniculum vulgare*), Sahara mustard (*Brassica tournefortii*), and perennial pepperweed (*Lepidium latifolium*)) (Rice 2005)
- Shrubs and trees (Macartney rose (*Rosa bracteata*), French broom, Scotch broom, gorse (*Ulex europaeus*), and tamarisk) (Gordon and Scifres 1977; Gordon and others 1982; Rice 2005)
- Vines (Japanese honeysuckle (*Lonicera japonica*) and kudzu (*Pueraria montana* var. *lobata*)) (Rice 2005)

Some success with this approach has also been reported for controlling medusahead (Carpinelli 2005) and squarrose knapweed (*Centaurea triumfettii*) (Dewey and others 2000). However, fire did not enhance herbicide effectiveness for controlling spotted knapweed (Carpenter 1986; Rice and Harrington 2005a), St. Johnswort (*Hypericum perforatum*), leafy spurge (*Euphorbia esula*), or Dalmatian toadflax (*Linaria dalmatica*) in western Montana (Rice and Harrington 2005a) (fig. 4-2).

Figure 4-2—Mountain grassland at National Bison Range, Montana, 3 months after April burning to assess fire effects on Dalmatian toadflax. (A) Left side: Burn-only treatment produced no changes in Dalmatian toadflax cover or cover of native grasses relative to control plots. (A) Right side: Spray-only treatments reduced Dalmatian toadflax and enhanced native grass cover. (B) Spray-burn combination reduced Dalmatian toadflax, but native grass cover did not increase until the second growing season after burning. (Photos by Mick Harrington.)

Burning has been used in the southern states to prepare sites dominated by kudzu for efficient herbicide application, and fire is also used after herbicides to promote germination of native plants from the seed bank and encourage kudzu seed germination; seedlings can then be eliminated with herbicides (Munger 2002b, FEIS review).

The National Park Service has been using prescribed burning for over a decade to prepare tamarisk-invaded sites in the Lake Mead area for herbicide treatment (Curt Deuser, personal communication 2004). Prescribed crown fires are used to consume as much aboveground tamarisk biomass as possible; extreme fire weather conditions are usually necessary to initiate these fires, and yet they reduce tamarisk by 10 percent or less. Within 6 to 12 months of the burn, tamarisk sprouts are treated with low volume basal spray, which increases mortality to over 95 percent.

Even if combinations of fire or herbicide treatments control the target species, they may not enhance the native plant community. Research on Santa Cruz Island, California, found that although fire increased the effectiveness of herbicide for reducing fennel, the native plant community did not recover. The most substantial change that followed herbicide treatment, with and without fire, was an increase in other nonnative forbs and nonnative annual grasses (R. Klinger, personal communication 2006; Ogden and Rejmánek 2005).

Burning Before Flood Treatment—In wetland management, top-killing nonnative invasives with fire before flooding may allow water to cover sprouts, which may in turn reduce regrowth. Bahia grass (*Paspalum notatum*) was top-killed with fire in Florida wetlands, then flooded; percent cover declined from 25 percent before treatment to 11 percent after flooding. The time elapsed between treatments and observations was not reported, so success of the program is difficult to gauge (Van Horn and others 1995).

Fire is used with flooding to restore native woody species and animal habitat in the Bosque del Apache National Wildlife Refuge, New Mexico. Friederici (1995) reports that late summer burning followed by flooding reduced tamarisk. Taylor and McDaniel (1998a) describe combinations of herbicide treatment, mechanical removal, and burning that killed or top-killed tamarisk and disposed of residual biomass. Planting of native tree and shrub species on treated sites met with limited success, but natural recruitment of natives was very successful in areas flood-irrigated after tamarisk removal. These changes in habitat composition and structure were accompanied by increases in animal diversity: During the 5 years following treatment, the number of bird, small mammal, and reptile/amphibian species increased in the restored area.

Burning Before Seeding or Planting—Burning may be used to prepare a site dominated by nonnative invasive species for planting desired species. This approach has met with success in grasslands, especially those with deep litter. In the central Great Plains, herbicides were applied in the fall to a mixedgrass prairie infested with leafy spurge, Kentucky bluegrass, and smooth brome. Residual litter was burned the following spring, and native tallgrasses (big bluestem, indiangrass, and switchgrass) were then drill seeded. Nonnative grasses declined and native tallgrass production increased following these treatments. Where native tallgrass productivity was high, leafy spurge productivity was reduced. Litter removal by burning was considered an important part of the treatment, although results were not compared to an unburned control (Masters and Nissen 1998; Masters and others 1996).

Burning cheatgrass in sagebrush steppe has proven useful in preparation for seeding of desired grass species (Rasmussen 1994). Seeded perennial grasses established successfully after summer burning of cheatgrass-infested rangeland in the Palouse of eastern Washington. Burning reduced the cheatgrass seed crop and facilitated soil contact by planted native seeds. Cheatgrass seedlings emerged (90 stems/m^2) along with the desired perennial grasses, but cheatgrass density was less than on untreated sites or sites treated with herbicide or disking (all more than 170 stems/m^2). Fall herbicide application on burned sites reduced cheatgrass seedling density to less than 40 stems/m^2 (Haferkamp and others 1987).

The success of burning/seeding programs is limited if seed from target species is abundant adjacent to treated areas (Maret and Wilson 2000).

Burning to Enhance Biocontrol Efficacy—A recurring theme in this volume is the importance of managing **for** desired conditions or species as well as managing **against** invasives that have negative impacts on the ecosystem. Desired species may include introduced biological control agents already present on a site.

Prescribed fire may seem incompatible with use of biocontrol agents, especially insects; however, integration of knowledge about the invasive species, the biocontrol agent, and the fire regime may lead to a successful management program (Briese 1996) (fig. 4-3). Many factors listed in figure 4-3 have already been considered in this chapter, but some merit specific discussion here: The scale and uniformity of burns are important because they influence the availability of refugia for biocontrol agents and the ability of the biocontrol agent to recolonize the burned area. Fire season and frequency may need adjustment to accommodate the life cycle and reproductive capacity of the control agent. If the biocontrol agent passes some

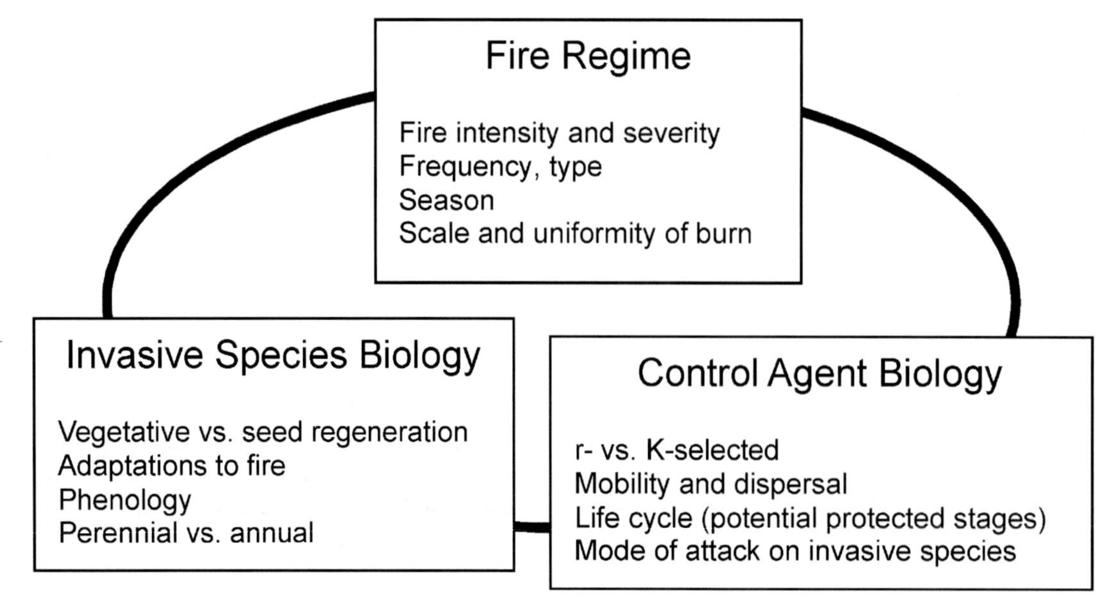

Figure 4-3—Interactions that need to be considered for management combining fire and biocontrol. (Adapted from Briese 1996.)

of its life cycle in a protected location (for example, root-boring larvae), the protected phase may provide a good season for burning. High-intensity wildfires in western Montana did not eliminate populations of *Agapeta zoegana,* a biocontrol agent that feeds on roots of spotted knapweed and has been introduced in many locations in western Montana (Sturdevant and Dewey 2002). Briese (1996) suggests that '*r*-selected' biocontrol agents could be preferable to '*K*-selected' agents if burning is planned, since the former could rapidly establish and increase after fire and would be more likely to have an impact on target plant density before another burn.

Research on grasslands in North Dakota demonstrates successful use of prescribed fire during seasons when a biocontrol agent is below ground. Areas invaded by leafy spurge were burned in mid-October and mid-May before the introduction of leafy spurge flea beetles (*Aphthona nigriscutis*). The beetles established successfully on 83 percent of burned plots, more than twice the establishment rate on unburned plots—possibly because litter removal favored establishment. Plots where flea beetle colonies established were then burned again in mid-October or mid-May. The adults were not active and juveniles were below ground at the time of both burns, and beetle populations were not affected. No reduction in leafy spurge density was attributed to the flea beetles in this short-term, small-plot study, but a large release of the beetles in a different area led to reduction of leafy spurge. The authors caution that spring burns of established colonies must be timed to allow leafy spurge regrowth before adult beetles emerge and need a food source (Fellows and Newton 1999).

Altering Fire Severity, Uniformity, Extent, and Frequency to Control Nonnative Invasive Plants

Thus far, this chapter has mentioned several aspects of the fire regime, but only seasonality of burning is discussed in detail (see "Prescribed Burning to Deplete Carbohydrate Reserves" page 51 and "Taking Advantage of Varying Plant Phenology" page 52). This section addresses several other aspects of fire regimes in relation to use of fire to control populations of nonnative invasive plants.

Prescribed Fire Severity and Uniformity in Relation to Season

Severity of prescribed fire varies with fuel moisture, which varies with season. For example, spring fires were too patchy to reduce density or cover of Scotch broom thickets in western Washington because grass cover was sparse and fuel moisture was high, whereas fall fires burned more continuously, produced higher maximum temperatures, and reduced Scotch broom significantly (Tveten and Fonda 1999). As mentioned in "Depleting the Seed Bank by Fire-stimulated Germination" (pg. 50), summer may be the optimal time to burn Scotch broom because seedlings germinating after fire will be exposed to harsh, dry conditions, increasing mortality.

The severity of fire prescribed for tamarisk control may also vary with season. One year after July burns on the Ouray National Wildlife Refuge, northeastern

Utah, 64 percent of tamarisk plants failed to sprout. Significantly fewer tamarisk plants were killed by September and October treatments, ranging from 4 to 9 percent. Fuels were drier and wind speeds were lower during the July burns than during fall burns (Howard and others 1983), possibly contributing to greater fire residence time and thus greater severity in July burns.

Prescribed Fire Severity and Uniformity in Relation to Fuel Beds

Some invasive forbs and trees reduce the amount and continuity of fine surface fuels and thus may reduce the ability of fires to spread, limit the time available for prescribed burning, or reduce fire severity. In a management review, Glass (1991) comments that two nonnative invasive forbs, cutleaf and common teasel (*Dipsacus laciniatus* and *D. sylvestris*), can be controlled in sparse, open grasslands by late spring burns. After teasel cover becomes dense, however, fire does not carry well so other treatments are needed—though they can perhaps be combined with fire. In western Montana, a discontinuous, nonuniform fuel bed forms as spotted knapweed density increases and displaces fine grasses. The coarse knapweed stems do not carry fire well under mild weather conditions, so the range of conditions that will produce effective but safe burns is narrow in knapweed-infested sites (Xanthopoulos 1986) (fig. 4-4). Some invasive woody species also reduce fine fuels. For example, beneath the canopy of Brazilian pepper (*Schinus terebinthifolius*), a nonnative invasive shrub-tree in Florida, grasses—and hence fine fuels—decrease as the plants increase in size. Doren and Whiteaker (1990) found that small Brazilian pepper plants with heavy grass fuel accumulations could be killed or severely retarded in growth by repeated biennial spring burning, but larger plants on sites with less grassy fuel either recovered rapidly from less severe burning or did not burn at all.

Numerous techniques have been used for increasing fine fuels to make prescribed burning feasible or increase its effectiveness. These include adding dead fine fuels to a site, cutting or mowing, planting noninvasive annual grasses, deferring grazing, and using herbicides to increase dead fuels. In Illinois prairies being converted from dominance by invasive cool-season grasses and forbs to native prairie species, Schramm (1978) recommends adding dry straw to facilitate spread of spring fires the first year after seeding. A common practice for reducing Scotch broom and French broom in the West is to cut the mature stems and let them cure prior to burning, thus increasing fire severity to discourage postfire sprouting and encourage germination from the soil seed bank (see

Figure 4-4—Fall prescribed burn in a mountain grassland in western Montana following herbicide treatment to reduce spotted knapweed. Due to low wind speeds, fire did not spread readily, as indicated by short flame lengths and patchy burn pattern. (Photo by Mick Harrington.)

"Depleting the Seed Bank by Fire-stimulated Germination" page 50). Boyd (1995) found that cut fuels in a site dominated by French broom in California were sufficient for one burn, but fine fuels were insufficient to fuel a second burn severe enough to kill broom resprouts and seedlings. Introduction of two annual grasses (soft chess and rattail sixweeks grass (*Vulpia myuros*)) increased fuels and effectiveness of fire. A previous seeding of grain barley (*Hordeum vulgare*) to increase fine fuel had failed to establish.

Litter and standing dead biomass must be present in a burned area before it can be reburned successfully. In rangelands, deferral of grazing can increase the fuel load and thus the heat produced by burning (George 1992; Rice 2005). In ecosystems where productivity is low, either due to site conditions or fluctuating weather patterns, several years' growth may be needed for accumulation of enough litter and dead fuel to carry an effective fire. Young and others (1972b) initially burned medusahead stands near Alturas, California, using a backing fire. Less litter and increased poverty weed (*Iva axillaris*), a native subshrub with succulent leaves, prevented second- and third-year fires from carrying with backfires and necessitated use of head fires. The three annual burns did not reduce medusahead.

Extent and Uniformity of Burns

Treatments to reduce nonnative invasive plants generally must cover a large enough area to prevent immediate re-establishment of the invasive. Many invasive species annually produce copious amounts of easily-dispersed seed (Bryson and Carter 2004). Such species cannot be eradicated from a small area if a propagule source is nearby. Thus fire size, uniform severity, and the condition of adjacent areas are important considerations. Research in sagebrush grasslands (Young and Evans 1978) and creosote bush scrub communities (Brooks 2002) demonstrates how spatial variation in fuels and fire severity can limit the effectiveness of fire for controlling nonnative invasive annual grasses (see details under "Destroying Seeds in the Litter and Soil" page 50). In Sierra Nevada forests, management strategies to reduce postfire invasion by nonnative species into fire-created gaps include elimination of nonnative seed sources from roadsides and other disturbed areas adjacent to burn sites and increasing the size of prescribed burns to increase the distance from seed sources (Keeley 2001).

Another issue related to the size and uniformity of burns is the regenerative capacity of desired native species: Can they establish in a treated area rapidly enough to attain dominance before the target invasive species re-establishes or is replaced by other invasives? Ogden and Rejmánek (2005) compared small- and large-scale treatments to restore grasslands on Santa Cruz Island, California. Fire-herbicide treatments at both scales reduced fennel but also led to substantial increases in nonnative invasive grasses (oats (*Avena* spp.), soft chess, Italian ryegrass (*Lolium multiflorum*), mouse barley, and ripgut brome). In the small-scale treatments, native species increased in cover and diversity, but this effect did not occur in the large-scale study site—probably because the small-scale plots were embedded in a more diverse plant community and, because of a higher ratio of edge to treated area, natives could spread readily into the treated area.

Fire Frequency

Where invasive species are susceptible to fire, a single burn usually only provides short-term control followed by recovery of the invasive in subsequent growing seasons or invasion by other undesired species (Rice 2005). Repeated burning is usually needed to sustain dominance of native species and suppression of invasives—a pattern that is not surprising in native communities that evolved with frequent fire. For example, one-time burning provided only short-term control of nonnative cool-season grasses in mixedgrass prairie in South Dakota (Whisenant and Bulsiewicz 1986), and of spotted knapweed in prairie remnants in Michigan (Emery and Gross 2005). In contrast, frequent burning of tallgrass prairie reduced abundance of nonnative cool-season grasses while stimulating native warm-season grasses (Smith and Knapp 1999, 2001; Svedarsky and others 1986). Parsons and Stohlgren (1989) found that repeated spring and fall fires reduced the diversity and dominance of nonnative invasive grasses in Sierra Nevada foothills grasslands, but this effect lasted less than 2 years. "Prescribed Burning to Deplete Carbohydrate Reserves" (pg. 51) presents some examples of the use of repeated burns to control nonnative invasive woody species.

Even where a plant community seems adapted to frequent fire, fuels may not accumulate rapidly enough to support fires frequent enough, severe enough, or uniform enough to accomplish management objectives. An example comes from California grasslands dominated by nonnative annual grasses, where DiTomaso and others (2001) conducted late spring burns for two consecutive years to reduce barbed goatgrass. The first burn was complete on all study sites, but fuels were too sparse to support a complete second burn on two of the three sites. Control of barbed goatgrass after the second burn was proportional to the completeness of the burns; a burn that covered about half of one study site did not reduce barbed goatgrass cover at all (fig. 4-5).

High-frequency burning, even in ecosystems with short presettlement fire-return intervals, may increase the likelihood of new invasions, cause unwanted erosion, or reduce desired native species. DiTomaso's

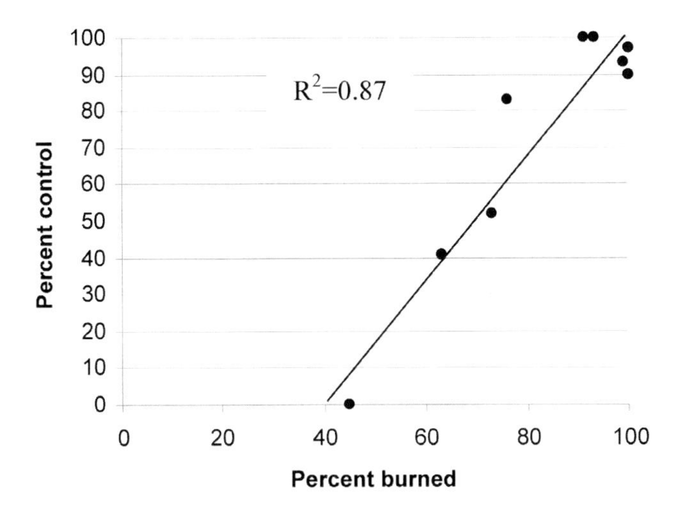

Figure 4-5—Relationship between completeness of burn and control (percent reduction in cover) of barbed goatgrass in a California grassland. (Adapted with permission from DiTomaso and others 2001, California Agriculture 55(6):47-53; copyright UC Regents 2001.)

(2006b) review notes that repeated burning may accelerate establishment and spread of invasive species not targeted by the original treatment, especially producers of abundant windblown seed. Repeated burning also exposes the soil to repeated heating and postfire raindrop impact, increasing the risk of erosion (Brooks and others 2004). In western Washington, 50 years of annual broadcast burning converted a community dominated by native perennial bunchgrasses, especially Idaho fescue (*Festuca idahoensis*), to one dominated by nonnative invasive grasses and forbs. On the other hand, fire exclusion allowed woody species, including the nonnative Scotch broom, to establish and persist (Tveten and Fonda 1999). A northern prairie grassland in the aspen parkland of east-central Alberta was burned each spring for at least 24 years. Annual burning significantly reduced smooth brome cover, but native rough fescue (*Festuca altaica*) cover was also 50 percent lower, and several other native cool-season grasses declined under this regime (Anderson and Bailey 1980). These results suggest that burning with variable frequency should be considered for controlling nonnatives and promoting native species. Variable fire-return intervals no doubt characterized many historic fire regimes and may be important for maintaining desired plant community composition and structure (Wills 2000).

In ecosystems that have not evolved with frequent fire, fire-return intervals short enough to suppress one nonnative invasive species may favor another or may cause other negative impacts. For example, multiple burn treatments are likely to select against native animals that have young at the time of burning (DiTomaso 1997). Keeley (2001) observes that frequent understory burns could reduce nonnative bull thistle (*Cirsium vulgare*) in ponderosa pine forests of the Sierra Nevada, but they would also severely reduce ponderosa pine seedlings. For prescribed fire to control Scotch broom effectively in Oregon white oak (*Quercus garryana*) woodlands, it must be applied frequently enough to prevent fuel buildup but not so frequently that nonnative herbaceous species are favored over native species (Zouhar 2005a).

Management of a Human Process: Constraints on Use of Prescribed Fire

Prescribed fire can be used to control some invasions by nonnative plant species, especially when integrated with other control methods in a long-term program. Fire is often seen as a means of treating a large area in a cost effective manner (Minnich 2006). However, use of fire is accompanied by concerns about safety and effects on other resources. The political and logistical obstacles to use of fire are not necessarily related to the objective. Responsibilities for safety and protection of property apply to use of fire for any purpose, and these challenges have been discussed by many authors. Minnich (2006) presents a thorough discussion of issues in regard to using prescribed fire for invasive species control, summarized here.

Any group or agency using prescribed fire is responsible for safety and protection of property. Operational challenges include staffing with a qualified program coordinator and fire manager, completing agreements with partners, and obtaining necessary training and equipment. Other obstacles include

- Restrictions on allowable burn area or season due to smoke impacts
- Lack of a suitable time window for completing the burn
- Opposition from neighbors and the community
- Unwillingness of employees to assume additional work or responsibility
- Lack of commitment at higher levels of an organization
- Lack of support from regulatory agencies

Use of prescribed fire may also conflict with other management needs or resource objectives, another issue not unique to use of fire to control invasive species. While the optimum time for a prescribed burn may be summer or fall, resources may be unavailable during these seasons due to wildfire activity, competing projects, or limited funds (Minnich 2006).

A common ecosystem-related problem is that timing of burns interferes with wildlife needs. DiTomaso (1997) notes that fire's potential impact on small animals and insects may be the most overlooked risk of burning. Spring burning is prohibited on many wildlife refuges because of impacts on nesting birds (Rice 2005). Illinois grasslands, for example, can be burned from mid-March to mid-April, but after that, burning may disrupt nesting birds and cause mortality to reptiles (Schramm 1978). The author's description of the difficulty of accomplishing a successful burn within a limited time is apt. One must be "poised and ready to burn at the proper moment" since usually there is only one chance for a "good" burn. In some areas, such as bush honeysuckle and buckthorn stands, nonnative species have formed dense monospecific communities that native songbirds now depend upon. In such cases, the nonnative species may need to be removed incrementally, in coordination with restoration of native shrubs, to provide continuous nesting habitat (Whelan and Dilger 1992). These few examples demonstrate the importance of developing clear objectives regarding nonnative invasive species and integrating all management programs to meet management goals.

Conclusions

To determine if fire can be used to reduce invasions by nonnative species, precise knowledge of invasive plant morphology, phenology, and life history must be combined with knowledge of the invaded site, its community composition, condition, and fire regime. Nonnative species that survive and/or reproduce successfully in burned areas are not likely to be suppressed by fire alone unless some aspect of the fire regime (usually season, frequency, or severity) can be manipulated to stress the nonnative without stressing the native species. This kind of treatment is most likely to succeed in ecosystems where the native plant community responds well to fire. Burning has been used with some success in grasslands and to prepare a site dominated by nonnative invasive species for planting of desired species.

It is possible to combine fire with other treatments to reduce plant invasions. In wetlands of Florida and riparian areas of the Southwest, fire has been used to top-kill nonnative species before flooding, a treatment combination that may reduce bahia grass in Florida and tamarisk in New Mexico. Fire has been used to prepare invaded sites for herbicide treatment, and herbicides have been used before fire to increase dead fuels, thus increasing fire intensity and severity. While success has been reported with these techniques in a variety of plant communities, there have also been failures, and long-term studies are few. The order, number, and timing of treatments influence success, so monitoring, follow-up over the long term, and an adaptive approach are essential components of a treatment program.

Treatments to reduce invasions by nonnative plants must cover a large enough area to prevent immediate re-establishment of the invasive. Even then, success will likely be limited if seed from target species is abundant adjacent to treated areas or if other conditions (soil disturbance or climate change, for instance) prevent desired native species from increasing and dominating on treated sites.

Alison C. Dibble
Kristin Zouhar
Jane Kapler Smith

Chapter 5:
Fire and Nonnative Invasive Plants in the Northeast Bioregion

Introduction

The Northeast bioregion extends from Maine to Maryland and northern Virginia, south along the northwest slope of the Appalachians to Tennessee, and west to the ecotone between prairie and woodland from Minnesota to northeastern Oklahoma. It is composed of a wide variety of landforms and vegetation types. Elevation ranges from sea level along the Atlantic coast, to 243 to 600 feet (74 to 183 m) at the Great Lakes, to over 6,000 feet (1,800 m) in the northern Appalachian Mountains.

Much of the native vegetation in the Northeast bioregion was historically closed-canopy coniferous and deciduous forest. Coniferous forests are characterized by spruce (*Picea* spp.) and balsam fir (*Abies balsamea*) in the north and by eastern white, red, pitch, and jack pines (*Pinus strobus*, *P. resinosa*, *P. rigida*, *P. banksiana*) in the northeastern coastal and Great Lakes areas. Deciduous forests in the northern part of the bioregion include those dominated by maple (*Acer* spp.), American beech (*Fagus grandifolia*), and birch (*Betula* spp.), others dominated by aspen (*Populus* spp.) and birch, others dominated by oak (*Quercus* spp.), and lowland and riparian forests dominated by elm (*Ulmus* spp.), ash (*Fraxinus* spp.), and cottonwood (*Populus* spp.). In the central and southern portions of the bioregion, oak and hickory (*Carya* spp.) are codominant species. Scattered stands of mixed oak and pine become more common toward the transition zone between oak-hickory and southern pine forests. Oak savannas, barrens, and tallgrass prairie remnants occur in the transitional area between eastern deciduous forests and central prairie and in other isolated locations. Early successional grasslands and woodlands occur in scattered areas and on abandoned farm land (old fields) (Garrison and others 1977; Smith and others 2001). Stands of eastern white pine often occupy former agricultural fields.

Fire History in the Northeast Bioregion

Information on fire history and fire regimes in the Northeast bioregion is given here and within sections on general plant communities to provide a context in which to discuss relationships between fire and nonnative invasive plant species. This information is derived largely from literature reviews, such as those in Brown and Smith (2000), except where otherwise indicated.

In northeastern plant communities, natural stand-replacing disturbances are more often caused by hurricanes, catastrophic wind events (Dey 2002a), and ice storms (for example, see Fahey and Reiners 1981) than by fire. However, fire has played a role in shaping the structure and composition of the vegetation in many areas. Fire has been a recurring disturbance in parts of the Northeast bioregion both before and after European settlement (Cronon 1983; Patterson and Sassaman 1988). At the landscape scale, there was substantial heterogeneity in fire regimes, and the relative evolutionary importance of fire varies among plant communities (Wade and others 2000). While estimates of presettlement fire regimes are difficult to confirm (Clark and Royall 1996), recent attempts have been made for vegetation types in the Northeast bioregion as part of the nationwide LANDFIRE Rapid Assessment (2005c). Evidence indicates that some northeastern communities burned more regularly than others (Parshall and Foster 2002). In general, fire regimes varied from almost no fires in beech-maple forests; to infrequent, high-severity, stand-replacement fires in northern coniferous forests dominated by spruce and fir; to frequent surface fires in oak-hickory forests, savannas, barrens, and prairie remnants (Wade and others 2000).

Lightning-caused fires are rare in the Northeast bioregion (Ruffner and Abrams 1998). People have been and continue to be the primary source of ignition (Leete 1938; Wade and others 2000). Before European contact, Native Americans used fire to manage landscapes for hunting, gathering, agriculture, and travel (Cronon 1983; Day 1953; Delcourt and Delcourt 1997; Dey 2002a; Pyne 1982a), and fire frequency was correlated with Native American occupancy. Fires continued to occur after European settlement and were ignited purposely and accidentally by both settlers and natives. The frequency and extent varied spatially and temporally based on factors such as topography, fuel loads, population levels, land use and fragmentation, and cultural values (for example, see Dey and Guyette 2000; Guyette and others 2002, 2003).

Fire exclusion efforts in the 20th century reduced fire frequency and extent in many fire-adapted plant communities (Shumway and others 2001; Sutherland 1997). On the New England sandplains, for example, several large fires occurred in the early 1900s but relatively few fires have occurred since, and these have been of smaller extent, due in part to suppression activities (Motzkin and others 1996). As forest succession and fire exclusion have proceeded, early-successional habitats have been reduced, and oak-dominated forests are gradually being replaced by forest dominated by a mix of maples and beech (review by Artman and others 2005). In an old-growth forest in western Maryland, for example, the overstory is currently dominated by oaks, but the recruitment layer has shifted from oaks to maple and birch; this shift corresponds with a lack of major fires since 1930 (Shumway and others 2001).

Since the 1980s, the Northeast bioregion has seen increasing use of fire as a management tool. In the last 10 years, prescribed fire, alone or in combination with silvicultural treatments, has been advocated to restore presettlement fire regimes or reference conditions in the Northeast bioregion, particularly in savannas and oak-dominated forests (Brose and others 2001; Healy and McShea 2002; Lorimer 1993; Van Lear and Watt 1993). Today prescribed burns are used routinely on public lands and lands managed by The Nature Conservancy for hazard fuel reduction, maintenance of fire-adapted ecosystems, promotion of oak regeneration, restoration of savannas, retention of early successional vegetation for breeding birds (reviews by, Artman and others 2005; Mitchell and Malecki 2003; Vickery and others 2005), and protection of rare plants (for example, see Arabas 2000; Patterson and others 2005; Trammell and others 2004). Burning may be a useful tool to aid in American chestnut (*Castanea dentata*) recovery in eastern oak forests (McCament and McCarthy 2005).

The actual use of prescribed fire, however, has been limited, and the spatial extent of burning has been relatively small. Nearly 70 percent of forest land in the region is owned by non-industrial private landowners (Smith and others 2001) who seldom use prescribed fire (Artman and others 2005). However, several states (for example, Ohio, Virginia, and North Carolina) have initiated programs to certify public land managers and private citizens in the use of prescribed burns. Thus prescribed burning may be used more frequently on private lands in the future (Artman and others 2005).

Nonnative Plants in the Northeast Bioregion

Current and presettlement vegetation types may have little in common in the Northeast bioregion because most forests were harvested or cleared for agriculture by the early 20th century. European settlers vastly increased the amount of open grassland and introduced many species of nonnative grasses, forbs, and shrubs to the bioregion. Many nonnative plants were introduced as contaminants in crop seed or other imported products, while others were introduced intentionally for agricultural and horticultural purposes. Most plant community types in this bioregion are invaded by nonnative plants in some areas, and the spread of these species is an increasing problem today (Mehrhoff and others 2003; Richburg and others 2001).

Most large infestations of nonnative species occur in or near settled areas, agricultural lands, roads and trails, or on public lands where they were deliberately introduced (for example, see Barton and others 2004; Ebinger and McClain 1996). "Conservation plantings" previously advocated by federal agencies (Knopf and others 1988) included invasive species such as Japanese barberry (*Berberis thunbergii*), multiflora rose (*Rosa multiflora*), autumn-olive (*Elaeagnus umbellata*), bush honeysuckles (*Lonicera* spp.), and buckthorns (glossy, *Frangula alnus*, and common, *Rhamnus cathartica*). Relatively fewer infestations occur in remote, upland natural areas; however, spread of several nonnative species into more remote areas is facilitated by ongoing development, propagule dispersal along roads, rivers, and other corridors (Barton and others 2004; Buckley and others 2003; Lundgren and others 2004), and especially seed dispersal by birds (for example, see White and Stiles 1992). Many invasive shrubs and vines in the Northeast have bird-dispersed seed (Mack 1996) (fig. 5-1). Nonnative plant species recorded from traps, feces, feeding observations or stomach contents of birds in a study in New Jersey include Japanese barberry, multiflora rose, Oriental bittersweet (*Celastrus orbiculatus*), winged euonymus (*Euonymus alatus*), common buckthorn, European privet (*Ligustrum vulgare*), Japanese honeysuckle (*Lonicera japonica*), Amur honeysuckle (*L. maackii*), and Tatarian honeysuckle (*L. tatarica*) (White and Stiles 1992). Additionally, efforts to assist native wild turkey recovery in the Northeast bioregion include planting nonnative honeysuckles (*Lonicera* spp.) and Oriental bittersweet, seeds of which are subsequently dispersed by wild turkeys (Poole, personal communication 2005).

Figure 5-1—Glossy buckthorn fruits are bird-dispersed and sometimes taken deep into the shady forest (Bradley, Maine). (Photo by Alison C. Dibble.)

Interactions of Fire and Invasive Plants in the Northeast Bioregion

Managers in the Northeast share with other bioregions a concern about the interactions of fire and invasive species. With increasing use of prescribed fire for a variety of management objectives, managers need information on the effects of fire on nonnative plants present in areas to be burned, and on the potential establishment and spread of those plants in the postfire environment.

Of particular concern in the northeast are the effects of nonnative plants on fuel characteristics. Changes in fire regimes due to the presence of nonnative invasive plants in the Northeast bioregion were discussed by Richburg and others (2001), but this topic has otherwise received little attention in the scientific literature, and data are insufficient for making generalizations. Observations and data from other bioregions indicate that changes in fuel characteristics brought about by nonnative species invasions can lead to changes in fire behavior and alter fire regime characteristics such as frequency, intensity, extent, type, and seasonality of fire, and thus impact native plant and animal communities (chapter 3). Invaded forest communities in the Northeast studied by Dibble and others (2003) and Dibble and Rees (2005) often had substantially higher cover of shrubs than uninvaded communities, resulting in increased height and density of surface fuels and suggesting an increased potential for fire to carry into the tree canopy. Additionally, the authors found higher percent cover of nonnative grasses on several invaded forest sites. If nonnative grasses differ in fuel loading, spatial distribution, phenology, or other characteristics from desired native understory species, they may affect fire frequency and seasonal burning window (Dibble and others 2003; Dibble and Rees 2005). Heat content, measured in the cone calorimeter for 42 plant species, differed between some native and nonnative invasive plants found in the Northeast bioregion, with no trend exclusive to one or the other group. For example, plants of fire-adapted ecosystems including black huckleberry (*Gaylussacia baccata*), pitch pine, bear oak (*Quercus ilicifolia*), barberry (*Berberis* spp.), and reindeer lichen (*Cladonia* spp.) had especially high heat content while nonnatives black locust (*Robinia pseudoacacia*), Norway maple (*Acer platanoides*), Japanese stiltgrass (*Microstegium vimineum*), sheep sorrel (*Rumex acetosella*), and glossy buckthorn (*Frangula alnus*) had low heat content (Dibble and others 2007).

There is growing interest in use of prescribed burning to control nonnative invasive plants (Bennett and others 2003); however, little is known about the effects of fire on nonnative invasives in this bioregion. Additionally, the use of prescribed fire in the Northeast is constrained by a highly reticulated wildland urban interface (WUI) in which the human population is high, and habitat fragmentation and new development are proceeding at a rapid pace. Prescribed burning in the Northeast is also difficult because it can be too moist and cool in most years for prescribed fires to carry or be effective. Additionally, policy restricts use of prescribed fire to particular seasons in many areas. For example, on Nantucket Island, Massachusetts, prescribed fires can only be conducted during the dormant season between October and April (personal communication cited in Vickery and others 2005), despite the fact that growing-season burns are probably more effective for controlling shrubs (Richburg 2005; Rudnicky and others 1997). Use of fire for controlling invasive plants is most effective when combined with other control methods (Bennett and others 2003).

In this chapter, we review the available literature on the interactions between fire and nonnative invasive plants in seven broad vegetation types in the Northeast bioregion: deciduous forest, coniferous forest, mixed forest, grasslands and early successional old fields, fresh wetland, tidal wetland, and riparian zone (fig. 5-2). Established vegetation classifications were not used because the limited data and literature available on fire and invasives in the Northeast bioregion makes using more specific classifications unrealistic. A brief description of the vegetation, presettlement fire regimes, and management issues is presented for each vegetation type. Consult Wade and others (2000) and Duchesne and Hawkes (2000) for greater detail about presettlement fire regimes and fire management considerations in the absence of nonnative invasive plants. The role of fire and/or fire exclusion in promoting nonnative plant invasions, fire regime changes brought about by nonnative plant invasions, and use of fire to control nonnative invasive species is discussed for each vegetation type. The focus is on nonnative species of concern for which some information is available regarding their relationship to fire or their response to other disturbances (table 5-1). This is only a subset of problematic nonnative species in the Northeast bioregion; many nonnatives of concern were excluded from this discussion due to lack of information. Interactions of fire and invasive plants can vary by species, vegetation type, and location, so the information presented in this chapter must be adapted for site-specific applications.

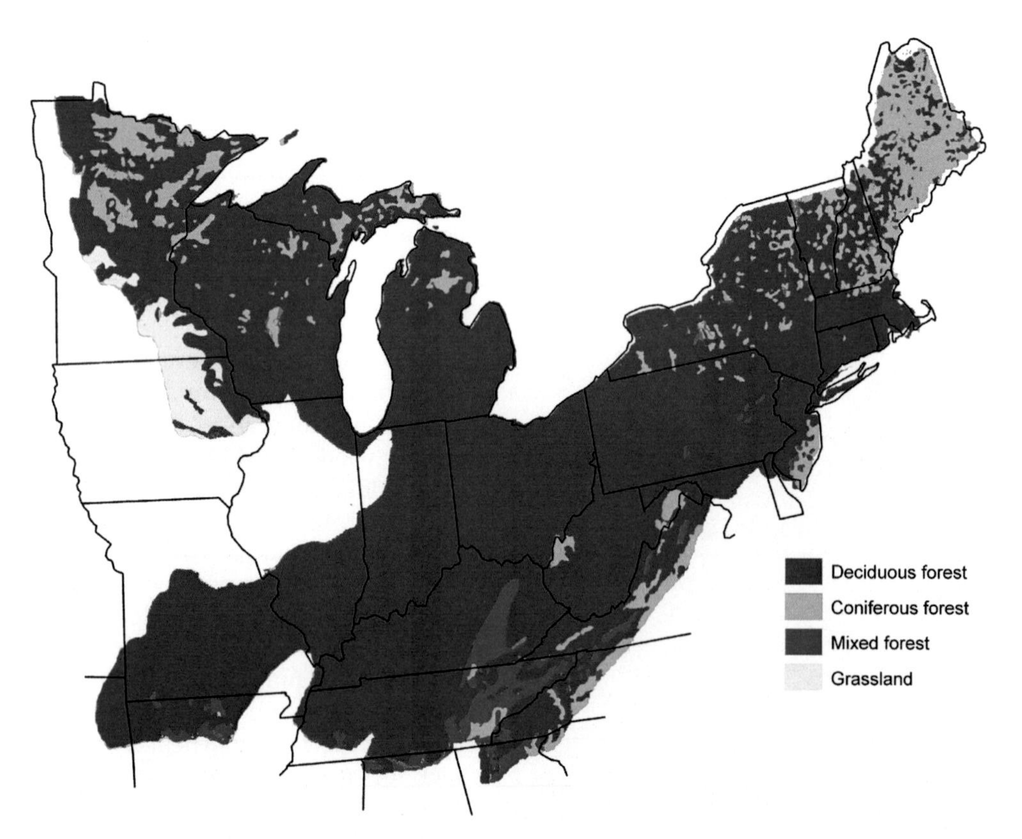

Figure 5-2—Approximate distribution of broad vegetation types in the Northeast bioregion. Riparian areas, wetlands, small grassland patches, and old fields are not shown. (Adapted from Garrison and others 1977.)

Table 5-1—General vegetation types in the Northeast bioregion and perceived threat potential of several nonnative plants in each type (L= low threat, H = high threat, P = potentially high threat, N= not invasive, U = unknown). Designations are **approximations** based on information from state and regional invasive species lists and an informal survey of 17 land managers and researchers in the bioregion.

Species		Plant communities						
Scientific name	Common name	Deciduous forest	Coniferous forest	Mixed forest	Grassland/ old field	Riparian	Fresh wetlands	Tidal wetlands
Acer platanoides	Norway maple	H	P	P	L	H	P	U
Ailanthus altissima	Tree-of-heaven	H	P	P	P	H	P	P
Alliaria petiolata	Garlic mustard	H	H	H	L	H	U	U
Ampelopsis brevipedunculata	Porcelainberry	P	P	P	H	H	P	N
Berberis thunbergii	Japanese barberry	H	H	H	H	H	H	U
Celastrus orbiculatus	Oriental bittersweet	H	H	H	H	H	P	P
Cynanchum louiseae	Black swallow-wort	P	P	P	H	H	P	U
Cynanchum rossicum	Pale swallow-wort	P	P	P	H	H	P	U
Cytisus scoparius	Scotch broom	L	L	L	H	U	U	N
Elaeagnus umbellata	Autumn-olive	P	P	P	H	P	L	N
Euonymus alatus	Winged euonymus	H	H	H	H	H	U	U
Frangula alnus	Glossy buckthorn	H	P	H	H	H	H	H
Glechoma hederacea	Ground-ivy	P	P	P	U	H	P	N
Ligustrum spp.	Privet	H	H	H	P	H	H	U
Lonicera japonica	Japanese honeysuckle	H	P	H	P	H	P	U
Lonicera spp.	Bush honeysuckles	H	H	H	P	H	H	N
Lythrum salicaria	Purple loosestrife	N	N	N	P	H	H	L
Microstegium vimineum	Japanese stiltgrass	H	H	H	H	H	H	U
Paulownia tomentosa	Princesstree	H	H	H	P	P	P	U
Phalaris arundinacea[a]	Reed canarygrass	P	U	U	P	H	H	U
Phragmites australis[a]	Common reed	N	N	N	N	H	H	H
Polygonum cuspidatum	Japanese knotweed	P	P	P	P	H	H	P
Polygonum perfoliatum	Mile-a-minute	P	P	P	H	H	H	U
Pueraria montana var. lobata	Kudzu	P	P	P	H	P	P	P
Rhamnus cathartica	Common buckthorn	H	H	H	H	H	P	N
Robinia pseudoacacia[a]	Black locust	P	H	P	H	P	L	U
Rosa multiflora	Multiflora rose	P	P	P	H	L	P	N

[a] Indicates species that are native in North America, but considered nonnative in all or part of the Northeast bioregion.

Forests

Nonnative invasive plants, insects, and pathogens pose a significant threat to forest integrity in eastern North America, especially in conjunction with forest fragmentation and climate change (Luken 2003; Vitousek and others 1997). In pre-colonial times, inland vegetation in the Northeast bioregion was probably dominated by forests with closed canopies and an accumulation of organic matter on the forest floor. These forests had relatively small areas of edge habitat resulting from disturbances such as the death of a large tree, severe storm damage, or fire (Marks 1974, 1983). These openings succeeded rapidly to thickets, young woodland and, in less than a century, forest (Marks 1974). While relatively stable, largely undisturbed forests in the Northeast can be susceptible to establishment and persistence of some shade-tolerant nonnative species such as Japanese stiltgrass, Japanese barberry, privets (*Ligustrum* spp.), bush honeysuckles, Japanese honeysuckle, and Norway maple (Brothers and Spingarn 1992; Ehrenfeld 1997; Fraver 1994; McCarthy 1997; Webb and others 2000). These forests tend to resist invasion by other nonnative plants (Auclair and Cottam 1971; Barton and others 2004; Fraver 1994).

Many nonnatives are restricted to edges and disturbed patches within forests, such as travel corridors including firelanes (Patterson and others 2005), recreation areas (Pyle 1995), sites associated with timber harvest (Buckley and others 2003; Lundgren and others 2004), and areas impacted by severe storm damage (for example, see Taverna and others 2005). These edges and patches typically have a higher abundance of nonnative plant species than forest interiors (Ambrose and Bratton 1990; Brothers and Spingarn 1992; Fraver 1994; Hunter and Mattice 2002; Ranney and others 1981; Robertson and others 1994). Nonnative species that thrive in and can quickly dominate forest edge habitat include Oriental bittersweet, porcelainberry (*Ampelopsis brevipedunculata*), Japanese honeysuckle, kudzu (*Pueraria montana* var. *lobata*), tree-of-heaven (*Ailanthus altissima*), and princesstree (Brothers and Spingarn 1992; McDonald and Urban 2006; Ranney and others 1981; Robertson and others 1994; Saunders and others 1991; Williams 1993). Edges may function as "safe sites" for nonnative invasives, where they can establish, reproduce, and disperse to additional locations including the forest interior (Fraver 1994).

Invasibility of a particular forest site is strongly influenced by its disturbance history, fragmentation of the surrounding landscape, and spatial relationship to propagule sources. Widespread forest clearing for agricultural land use and subsequent abandonment have resulted in secondary forest sites that contain many nonnative plants and propagules (for example, Ashton and others 2005; Bellemare and others 2002; Fike and Niering 1999; Vankat and Snyder 1991). Contemporary forested natural areas and preserves may be an assemblage of forest remnants, abandoned agricultural fields in later stages of succession, woodlots, and streamside corridors embedded in an agricultural and suburban matrix, producing substantial edge habitat (review by Robertson and others 1994). Gaps and edge habitat in northeastern forests are also created by extensive die-off of important canopy trees such as happened with chestnut blight (*Cryphonectria parasitica*) (Myers and others 2004) and Dutch elm disease (*Ophiostoma ulmi*). In canopy gaps resulting from high mortality of eastern hemlock (*Tsuga canadensis*) from the nonnative hemlock wooly adelgid (*Adelges tsugae*), several nonnative species had high cover; these included Japanese stiltgrass, Oriental bittersweet, Japanese barberry, and tree-of-heaven (Orwig and Foster 1998). Similarly, canopy gaps created by wild or prescribed fire could provide seed beds and edge habitat for nonnative invasive populations, although little research is available on this topic. Propagule pressure from existing nonnative invasive populations, coupled with establishment opportunities provided by ongoing disturbances and forest fragmentation, will likely lead to continued spread of these species.

Deciduous and Mixed Forests

Background

Two major deciduous forest types in the Northeast bioregion include the maple-beech-birch and oak-hickory ecosystems described by Garrison and others (1977). These types are treated separately here based on differences in fire ecology. The mixed forest type described below is typically dominated by oaks and pines, especially northern red oak and eastern white pine. Maples, birches, beech, and hemlock are common associates.

Maple-beech-birch Forests—Vegetation in this ecosystem includes northern hardwood forests; southward it transitions into mixed mesophytic hardwoods, as discussed by Wade and others (2000). The northern hardwoods occur on mesic and fire protected sites in the Lake States and farther east. The dominant hardwood species include sugar maple, yellow birch (*Betula alleghaniensis*), American beech (*Fagus grandifolia*), and basswood (*Tilia americana*) in the Midwest. Northern hardwoods mix with boreal spruce (including *Picea glauca*, *P. mariana*, and *P. rubens*) and balsam fir to the northeast, and with eastern hemlock, eastern white pine and oaks to the north, south, and west. Mixed mesophytic hardwoods occupy the transition zone between northern hardwood forest and oak-hickory forest, and contain a large diversity of canopy tree species. This type transitions into sugar maple-beech-birch forest in northern West Virginia, southwestern Pennsylvania, and southern Ohio in the north, and into the oak-hickory-pine type in northern Alabama in the south (Wade and others 2000). Dominant species include sugar maple, beech, basswood, white oak (*Quercus alba*), northern red oak (*Q. rubra*), buckeye (*Aesculus octandra*), and tulip tree (*Liriodendron tulipifera*) (Küchler 1964).

Presettlement fire regimes in eastern deciduous forests varied among forest types. Charcoal evidence suggests that fires were more common in mixed mesophytic forests than in northern hardwood forests (Wade and others 2000). Northern hardwood forests are not very flammable and if fires penetrate the forest, they tend to burn as patchy, creeping surface fires. Crown fires are unusual in eastern deciduous forests (Lorimer 1977; Turner and Romme 1994). Fire return intervals are estimated to exceed 1,000 years throughout the northern hardwoods type and are estimated at 35 to over 200 years in the mixed mesophytic type. Both types are characterized by a mixed-severity fire regime (Wade and others 2000); that is, if fire does occur in these forests it would cause selective mortality in dominant vegetation, depending on the susceptibility of different tree species to fire (Brown 2000). Fires may have occurred more frequently in areas that were burned by Native Americans and where conifers occur as substantial components of the hardwood forests (Wade and others 2000). Although northern hardwood species are generally thought to have little resistance to fire, maple and birch sprout vigorously from the stump, and beech suckers from the root system. See reviews on individual species in the Fire Effects Information System (FEIS) for more information on fire ecology of dominant species in these ecosystems. Also see Wade and others (2000) and Parshall and Foster (2002) for more information on fire regimes in northern hardwood and mixed mesophytic forest types.

Oak-hickory Forests and Oak Savanna—The oak-hickory ecosystem is extensive in the Northeast bioregion, reaching from southern Maine, southwest along the Appalachian Highlands to the northern part of Georgia and Alabama, and westward to the oak savannas and central grasslands. The oak-hickory ecosystem varies from open to closed woods with a sparse to dense understory of shrubs, vines, and herbaceous plants. Associated species vary with latitude and location (Garrison and others 1977). Over three dozen species of oak and almost two dozen species of hickory are possible in the overstory of this ecosystem. It includes oak-hickory and Appalachian oak ecosystems as described by Küchler (1964) and becomes an oak-hickory-pine type in the Mid-Atlantic States, including stands that can be classified as mixed mesophytic forest (Wade and others 2000). At its western extent, this forest type grades into open oak woodlands and oak savannas. Oak savannas are associated with prairies and are generally dominated by prairie grasses and forbs, with widely spaced groves or individual trees (review in LANDFIRE Rapid Assessment 2005a).

According to a review by Wade and others (2000), the presettlement fire regime in the oak-hickory ecosystem was characterized by high frequency, understory fires often ignited by Native Americans. Presettlement fire frequencies are not known but are estimated between 3 and 35 years. Subsequent settlement by Euro-Americans, who used fire for many of the same reasons as Native Americans, increased the frequency and extent of burning in oak-hickory forests. Fire intervals decreased to less than 10 years, and many sites burned annually. Frequent fire maintained open oak-hickory woodlands with large, old, fire-resistant trees and a groundcover of grasses and forbs. Shrubs, understory trees, and woody debris were rare in oak-hickory forests and savannas. Where present, ericaceous shrubs such as mountain laurel (*Kalmia latifolia*) and rhododendron (*Rhododendron* spp.) could burn with extreme fire behavior, resulting in mixed-severity or stand-replacing fires (Wade and others 2000). Fire regimes in oak savanna are characterized primarily as frequent surface fires occurring at about 4-year intervals (LANDFIRE Rapid Assessment 2005a).

Exclusion of fire has profoundly changed oak-hickory forests and oak savannas and, in many areas, has led to dominance by mixed mesophytic and northern hardwood species and allowed the mid-story canopy to close and shade out herbaceous plants (LANDFIRE Rapid Assessment 2005a; Wade and others 2000). Managers in Virginia note replacement of some oak-hickory stands by maple and beech where fire has been excluded (Gorman, personal communication 2005).

Surface fires enhance regeneration of oak and hickory, and there has been much recent research on use of prescribed fire to promote establishment of oak (Boerner and others 2000a, b; Dey 2002b; Kuddes-Fischer and Arthur 2002; Lorimer 1993; Wade and others 2000). In upland oak-hickory forests at Quantico Marine Base and Fort Pickett Military Reservation, frequent fires associated with training activities enhance the oak-hickory community, including an endangered shrub, Michaux's sumac (*Rhus michauxii*) (Virginia Department of Conservation and Recreation 2005). In areas where fire has been excluded and fuel loads are high, reintroduction of fire might need to be phased in with a series of fuel reduction treatments.

Mixed Forests—Presettlement fire regimes in mixed forests were characterized by a range of fire frequencies and fire severities. Estimates given by Wade and others (2000) for mixed forest types range from understory fires with return intervals of less than 10 years in shortleaf pine-oak communities, to mixed-severity fires with return intervals between

10 and 35 years in Virginia pine-oak communities, to stand-replacement fires with return intervals greater than 200 years in northern hardwoods types with components of spruce or fir. Fire regimes for white pine-red oak-red maple communities are thought to consist of stand-replacement fires with return intervals of less than 35 to 200 years (Duchesne and Hawkes 2000).

Nonnative invasive plants that threaten forests in the Northeast bioregion are similar among deciduous and mixed forests (table 5-1), so they are discussed together here. Distinctions are made among forest types where possible.

Role of Fire and Fire Exclusion in Promoting Nonnative Plant Invasions in Deciduous and Mixed Forests

Managers are concerned about the potential for establishment and spread of nonnative plants after fire. Evidence is sparse regarding postfire response of nonnatives for northeastern deciduous and mixed forests, although inferences may be possible based on life history and reproductive traits (table 5-2). Species such as Japanese honeysuckle and Oriental bittersweet can persist in low numbers in the understory and spread following canopy or soil disturbance, while others such as tree-of-heaven may establish in open areas via long-distance seed dispersal. Several invasive species (for example, tree-of-heaven, autumn-olive, Japanese barberry, privet, honeysuckles, and kudzu) are able to reproduce vegetatively and sprout following top-kill. These species could spread and possibly dominate postfire communities. While there is little hard evidence of seed banking for invasive plants, observations suggest that some species (for example, princesstree, buckthorn, multiflora rose, kudzu, ground-ivy, and Japanese stiltgrass) may establish from the soil seed bank after fire (table 5-2). Some invasive species (for example, common buckthorn, bush honeysuckles) have established and spread in areas such as oak-hickory forests and oak savannas, where fire has been excluded from plant communities adapted to a regime of frequent surface fires.

Many woody invasives in the Northeast bioregion have some traits in common, including the ability to sprout following top-kill (table 5-2). Managers at Virginia-area national parks note that autumn-olive sprouts following aboveground damage, and sprouts are especially vigorous following dormant season burning in oak woodlands (Gorman, personal communication 2005; Virginia Department of Conservation and Recreation 2002a). Similarly, at Fort Devens, Massachusetts, autumn-olive established from both root sprouts and seedlings following a single fire (Poole, personal communication 2005). Observations indicate that a related nonnative species, Russian-olive (*Elaeagnus angustifolia*), sprouts after fire in the Central and Interior West bioregions (chapters 7 and 8). Russian-olive occurs in northeastern deciduous forests but is considered less invasive than autumn-olive in this bioregion (Mehrhoff and others 2003). While available literature does not describe postfire response of Norway maple, Simpfendorfer (1989) lists it among species that regenerate by coppicing following fire. It is also likely that, if Norway maple saplings and seedlings survive fire, they would respond favorably to gap formation (Munger 2003a, FEIS review). Tree-of-heaven produces abundant root sprouts after complete top-kill from fire (Howard 2004a, FEIS review). Japanese barberry sprouted after cutting and/or burning treatments in a deciduous forest site in western Massacusetts, although total cover of this species was reduced 2 years after burning (Richburg 2005). Observations in a mixed forest in northwestern Georgia indicate that Chinese privet responds to aboveground damage from fire by vigorously sprouting from the root crown (Faulkner and others 1989), and an anecdotal account suggests that Japanese privet (*Ligustrum japonicum*) can "resprout following fire" (Louisiana State University 2001, review). Other privet species are likely to sprout from roots and/or root crowns following fire; however, documentation is lacking (Munger 2003c, FEIS review). Glossy buckthorn sprouted from roots or root crowns after wildfire in a mixed alvar woodland near Ottawa. Sprouts were 3 to 5 feet (1 to 1.5 m) tall after 100 days, but no prefire data were available for comparison. Two nonnative grasses, Canada bluegrass and redtop (*Agrostis gigantea*), occurred in burned areas in these studies at 22 and 18 percent frequency, respectively (Catling and others 2001). While only anecdotal evidence is available suggesting postfire sprouting in multiflora rose (Virginia Department of Conservation and Recreation 2002d), in a deciduous forest in southeastern Ohio germination and recruitment of multiflora rose was higher on open-canopy plots and on plots treated with high-severity prescribed fire than in control plots (Glasgow and Matlack 2007).

Some invasive vines can occur in the forest understory in small numbers and spread via vegetative regeneration or recruitment from the soil seed bank following disturbance. Oriental bittersweet and Japanese honeysuckle often occur under closed canopies, and when disturbance creates canopy gaps, these vines can grow and spread rapidly (Howard 2005a; Munger 2002a, FEIS reviews). Several sources indicate that Japanese honeysuckle sprouts after damage from fire, and postfire sprouting can lead to rapid recovery of preexisting populations. Scattered subpopulations of Japanese honeysuckle can also persist with frequent fire, possibly within fire refugia or via continued recruitment from bird-dispersed seed (Munger 2002a). Kudzu stems and foliage are likely to resist fire damage during

Table 5-2—Biological and ecological characteristics of nonnative species selected from table 5-1 that occur in forested habitats in the Northeast bioregion. Information is derived from a comprehensive review of available literature for each species. Superscripted letters following plant names refer to citations for biological and ecological information in footnotes. Superscripts in right-hand column indicate information type: * indicates anecdotal information; ** indicates information from experiments; *** indicates a literature review in which additional fire information is available.

Species	Seed dispersal/ establishment	Seed banking	Vegetative reproduction	Shade tolerance	Available fire information
Trees					
Norway maple[a]	Wind-dispersed	No information	Anecdotal evidence of regeneration by coppicing following fire; cut trees can sprout from stumps	Yes	Simpfendorfer 1989*
tree-of-heaven[b]	Abundant, wind-dispersed seeds	Transient	Abundant, rapidly growing sprouts from roots, root crown, and/or bole following complete top-kill or bole damage	Low	Gorman, personal communication 2005*; Hoshovsky 1988*
princesstree[c]	Abundant, wind-dispersed seed (up to 4 km); establishes best on exposed mineral soil with high light levels	Viable seeds found in forest soil seed banks	Sprouts from roots and/or stumps after pulling or cutting, and may also do so after fire	No	Reilly and others 2006*; Langdon and Johnson 1994*
Shrubs					
Japanese barberry[d]	Gravity-, bird- and possibly animal-dispersed seed; high rates of seedling recruitment	No information	Sprouted after cutting and burning in deciduous forest and forested swamp sites; cover substantially reduced two years after cutting and/or burning treatments	Yes	Richburg 2005**; D'Appollonio 2006**
autumn-olive[e]	Abundant bird-dispersed seed	No information	Anecdotal evidence suggests that it can sprout following damage from fire	Low to moderate	Gorman, personal communication 2005*; Virginia Department of Conservation and Recreation 2002a*; Poole, personal communication 2005*

Table 5-2—(Continued)

winged euonymus[f]	Large quantities of gravity- and bird-dispersed seed	No information	Anecdotal evidence suggests that it spreads by root suckers and can sprout following damage to aboveground parts	Very	none
buckthorns[g]	Bird- and animal-dispersed seed; recruitment best on sites with ample light and exposed soil	Seed requires scarification for germination. Research on seed banking is not available	Glossy buckthorn sprouts from roots or root crowns after fire; common buckthorn sprouted following prescribed fire in a grassland habitat; in an oak savanna, spring prescribed fire killed common buckthorn seedlings while mature buckthorn were top-killed and sprouted after fire	Low to moderate	Apfelbaum and Haney 1990*; Converse 1984a, TNC review*; Heidorn 1991*; Larson and Stearns 1990*; McGowan-Stinski 2006*; Boudreau and Willson 1992**; Catling and others 2001**; Neumann and Dickmann 2001**; Post and others 1990**; Richburg 2005**
privets[h]	Bird- and animal-dispersed seed	Transient	Anecdotal evidence suggests that privets can sprout following damage (including fire) to aboveground parts; and observations indicate that Chinese privet responds to fire by vigorously sprouting from the root crown	Moderate	Gorman, personal communication 2005*; Louisiana State University 2001* (Japanese privet); Faulkner and others 1989** (Chinese privet)
bush honeysuckles[i]	Abundant, wind-dispersed seed	Not likely to form a persistent seed bank, but information is lacking	Postfire sprouting of bush honeysuckles has been observed; spring prescribed burning may kill bush honeysuckle seedlings and top-kill larger plants, although results have been mixed	Yes	Apfelbaum and Haney 1990*; Kline and McClintock 1994*; Mitchell and Malecki 2003**; Maine Department of Conservation, Natural Areas Program 2004*; Solecki 1997*
multiflora rose[j]	Gravity and bird-dispersed seed	Persistent seed bank	Reproduces by root suckering and layering; sprouting is reported following top-kill, including fire; germination and seedling growth promoted in experimentally burned microsites	Low to moderate	Hruska and Ebinger 1995**; Virginia Department of Conservation and Recreation 2002d*; Glasgow and Matlack 2007**
Woody vines					
porcelainberry[k]	No info	No information	Spreads by "extensive underground growth"; able to reproduce vegetatively from stem or root segments	Low	None

Table 5-2—(Continued)

Oriental bittersweet[l]	Bird-dispersed seed	Conflicting info; may have small, persistent seed bank	Deeply buried perennating buds; capable of sprouting from the root crown, roots, root fragments, and runners; damage to branches, root crowns, runners, or roots encourages sprouting	Yes	None
Japanese honeysuckle[m]	Bird-dispersed seed	Possible, but information is lacking	Sprouts from root crowns after damage from fire. Repeated burning can have negative impacts	Yes	Munger 2002a***; Faulkner and others 1989**; Schwegman and Anderson 1986**
kudzu[n]	No information	Mature seeds are dormant and require scarification for germination so seeds are likely to persist, although information on longevity is lacking	Sprouts from the root crown after fire; seed dormancy may be broken by high temperatures	No	Munger 2002b***; Rader 2000**; Miller 1988*
Herbaceous vines					
swallow-worts[o]	Abundant gravity- and wind-dispersed seed	No information	Mature plants of both species sprout readily from perennating buds on the root crown occurring below the soil surface and likely to be protected from fire; fire thought to be ineffective for controlling these species	Populations on shaded sites may persist for years and expand rapidly if a gap is created	Lawlor 2002, TNC review*; Lawlor 2000*; DiTommaso and others 2005*; Sheeley 1992*; Richburg 2005**
mile-a-minute[p]	Bird-, animal-, and water-dispersed seed	Seeds can remain dormant in the soil seed bank for at least three years	No info; annual species	Tolerates light shade	None
Herbs					
garlic mustard[q]	Primarily short-distance, gravity and ballistic dispersal; (long-distance dispersal by animals is possible)	Mostly transient, though some seeds may persist for several years	Top-killed when exposed to fire, but can survive by sprouting from the root crown	Moderate	Hintz 1996**; Luken and Shea 2000**; Munger 2001***; Nuzzo and others 1996**

Table 5-2—(Continued)

ground-ivy[r]	No information	Seeds likely to persist in soil seed bank	Spreads by stolons with roots at each node; displayed rapid vegetative growth after a spring prescribed fire in a mixed forest	Somewhat	Chapman and Crow 1981*
Japanese stiltgrass[s]	Water-, wind- and animal-dispersed seed	Soil-stored seed may remain viable for 3 to 5 years	An annual plant that can grow back from tillers and stolons following top-kill from early-season fire; can establish from the soil seed bank after fire; germination and seedling growth promoted in experimentally burned microsites	Very; but shows significant increase in biomass following canopy removal	Barden 1987*; Gorman, personal communication 2005*; Tu 2000, TNC review*; Glasgow and Matlack 2007**
reed canarygrass[t]	No information	Extensive germination from the soil seed bank after fire and herbicide application	Vigorous, rapidly spreading rhizomes	No	Howe 1994b**; Apfelbaum and Sams 1987*; Hutchison 1992b*; Preuninger and Umbanhowar 1994*

[a] Munger 2003a; Webb and others 2001; Martin and Marks 2006.
[b] Howard 2004a, FEIS review; Hutchinson and others 2004; Marsh and others 2005.
[c] Williams 1993; Dobberpuhl 1980; Hyatt and Casper 2000; Johnson, K. 1996, review.
[d] Mehrhoff and others 2003; Rhoads and Block 2002, TNC review; Ehrenfeld 1997; Ehrenfeld 1999; Richburg and others 2001, review; Johnson, E. 1996; Silander and Klepeis 1999; D'Appollonio 2006.
[e] Munger 2003b, FEIS review.
[f] Mehrhoff and others 2003, review; Miller 2003, review; Ebinger 1983; Martin 2000, TNC weed alert; Ma and Moore 2004.
[g] Converse 1984a, TNC review; Chapter 7, this volume.
[h] Mehrhoff and others 2003; Munger 2003c, FEIS review; Miller 2003, review.
[i] Munger 2005a, FEIS review; Bartuszevige and Gorchov 2006.
[j] Munger 2002c, FEIS review; Szafoni 1991; Richburg 2005.
[k] Antenen 1996, review; Antenen and others 1989; Yost and others 1991; Robertson and Antenen 1990.
[l] Howard 2005a, FEIS review.
[m] Munger 2002a, FEIS review.
[n] Munger 2002b, FEIS review; Susko and others 2001.
[o] Lawlor 2002, TNC review; Cappuccino and others 2002; Lumer and Yost 1995; Smith and others 2006; Sheeley and Raynal 1996.
[p] Kumar and DiTommaso 2005.
[q] Munger 2001, FEIS review.
[r] Hutchings and Price 1999; Mitich 1994.
[s] Cole and Weltzin 2004; Cole and Weltzin 2005; Flory and others 2007; Howard 2005c, FEIS review; Oswalt and others 2007; Winter and others 1982 .
[t] Hoffman and Kearns 2004; Leck and Leck 2005; Snyder 1992b, FEIS review.

the growing season because they typically maintain high water content and are relatively unflammable except after frost-kill in autumn (Wade, personal communication 2005). When large kudzu plants do burn, they can sprout from the root crown after top-kill and reestablish soon after dormant-season fire, returning to prefire abundance by the second postfire growing season. Additionally, soil heating by fire may promote kudzu seed germination by scarifying the seedcoat. However, dormant season fire can kill root crowns of small, newly established kudzu plants (Munger 2002b, FEIS review).

Fire information is available for only a few species of the nonnative herbs occurring in northeastern forests. Researchers report that garlic mustard can establish and persist after fire in northeastern deciduous forests. Establishment of garlic mustard from the soil seed bank may be facilitated by postfire conditions (Munger 2001, FEIS review). Repeated fall burning (2 to 3 annual burns) did not reduce abundance or relative importance of garlic mustard in an eastern mesophytic forest understory in Kentucky (Luken and Shea 2000). In a white pine-mixed deciduous forest in New Hampshire, ground-ivy displayed rapid vegetative growth after a spring prescribed fire but did not occur on fall-burned plots (Chapman and Crow 1981). Addtionally, some evidence suggests that ground-ivy seeds might survive in the soil seed bank for a number of years (Hutchings and Price 1999) and therefore may be capable of recruitment after fire. Japanese stiltgrass can establish in forest understories (for example, see Dibble and Rees 2005; Ehrenfeld 2003), in gaps created by overstory mortality (Orwig and Foster 1998), and in disturbed areas created by hurricanes (Taverna and others 2005). In oak-hickory woodland sites, Japanese stiltgrass established after mechanical thinning and prescribed fire in southern Illinois (Anderson and others 2000) and established from the soil seed bank after an accidental fire on a North Carolina floodplain (Barden 1987). In deciduous forest in southeastern Ohio, germination and growth of Japanese stiltgrass was higher in open-canopy plots and in plots treated with high-severity prescribed fire than in untreated controls (Glasgow and Matlack 2007). Observations indicate that it also grows back from tillers and stolons following top-kill from early-season fire (Tu 2000, TNC review).

Fire exclusion affects forest types differently depending on the extent to which they are dominated by fire-adapted species or have gap formation. Many nonnative invasive species in the Northeast bioregion are shade-tolerant (table 5-2) and thus may invade forests and savannas where fire has been excluded. Hobbs (1988) suggested that common buckthorn, known for its shade tolerance, may have spread in northeastern deciduous forests in part because of fire exclusion, as well as in gaps that occurred with the demise of American elm. Common buckthorn and bush honeysuckles have infested oak savanna remnants where fire has been excluded. In a study of 24 oak savanna remnants in northern Illinois, Indiana, and southern Wisconsin, Apfelbaum and Haney (1990) suggested that, without fires about every 35 to 100 years, native and nonnative woody species, including common buckthorn and bush honeysuckles, establish in oak understories and interfere with oak regeneration. In 24 oak savannas that varied in soil type and management history, periodic fires afforded some control of mesic shrub infestations and promoted oak regeneration (Apfelbaum and Haney 1990). Glossy buckthorn is typically associated with unburned woodland in mixed forests (Catling and Brownell 1998; Catling and others 2002). A relatively shade-tolerant herbaceous species, garlic mustard, occurred in areas of low ambient light where reduced fire frequency resulted in increased tree canopy cover in a northern Illinois oak savanna remnant (Bowles and McBride 1998). Invasive populations of Japanese honeysuckle apparently do not occur in communities with frequent, low-severity fires (Munger 2002a). More data are needed to uncover whether this pattern can be related to fire exclusion.

Fire exclusion from fire-seral communities such as oak-hickory forests or oak savannas diminishes opportunities for maintenance of the dominant species. Instead, establishment and growth of shade-tolerant species are enhanced, and fire-seral species can be replaced. Once shade-tolerant nonnative species establish in closed-canopy forests, they may persist, spread, and possibly dominate the understory. Evidence suggests that Norway maple establishes, persists, and grows in forest understories in the absence of fire or other stand-level disturbances. In a New Jersey piedmont forest, for example, Norway maple, American beech, and sugar maple are gradually replacing white oak, northern red oak, and black oak (*Q. velutina*), which were formerly dominant (Webb and Kaunzinger 1993; Webb and others 2000). Nonnative grasses (sweet vernal grass (*Anthoxanthum odoratum*), fineleaf sheep fescue (*Festuca filiformis*), Japanese stiltgrass (Dibble and Rees 2005), and Canada bluegrass (*Poa compressa*) (Swan 1970)) often occur in the understory of deciduous forests and seem to spread in the absence of fire. Of these, Japanese stiltgrass is the most studied and is one of the most invasive grasses in forests and riparian areas, especially along trails and roadsides but also in undisturbed, shaded sites (Cole and Weltzin 2004). It can invade woodlands with incomplete canopy closure (Winter and others 1982) and persist after the canopy closes completely (Howard 2005c). If fire is reintroduced to these communities, restoration of native plant communities is not certain; shade-tolerant nonnatives that can sprout after top-kill or establish from the seed bank may dominate the postfire community.

Effects of Nonnative Plant Invasions on Fuel Characteristics and Fire Regimes in Deciduous and Mixed Forests

Nonnative plant invasions may change fuel properties, fire behavior, and possibly fire regimes in several ways (Richburg and others 2001). Individual species could affect fire behavior due to differences from native species in heat content (Dibble and others 2007), moisture content, volatility, fuel packing, and phenology. Nonnative grasses have the potential to increase biomass and continuity of fine fuels on invaded sites (Dibble and Rees 2005). Invasive shrubs and vines may affect biomass and flammability of the shrub and herb layers or act as ladder fuels. While increased fuel loads due to nonnative invasive plants might not be a concern to fire managers in wet years, hazard fuels must be considered if drought occurs. However, properties of individual species might be less important than fuel moisture, topography, and wind velocity (Ducey 2003) during a wildfire.

There is concern that encroachment by nonnative grasses, vines, and shrubs could increase flammability and fuel continuity in deciduous forests. Fuels in five invaded mid-successional deciduous forest stands dominated by oak-hickory (Maryland), poplar (Maine), oak-bigtooth aspen (Maine), oak-yellow poplar (New Jersey), and mixed hardwoods (Vermont) were studied by Dibble and others (2003) and by Dibble and Rees (2005). Invaded stands were compared to nearby uninvaded stands. Under invaded conditions, graminoid and shrub cover were greater because of the frequency and height of the nonnative plants (including fineleaf sheep fescue, sweet vernal grass, Japanese stiltgrass, bush honeysuckles (fig. 5-3), Japanese barberry, and others). If fire occurs in invaded stands, patches of fine fuels represented by nonnative grasses and shrubs could increase fire intensity (Dibble and Rees 2005). For example, Japanese stiltgrass forms large ($\geq$0.2 ha), dense patches with hundreds to thousands of stems per square foot (Dibble, unpublished data 2005; Dibble and Rees 2005). This species produces large amounts of litter and fine fuels, and stems lie down soon after they die in autumn, creating a continuous fuelbed of matted straw (Barden 1987) that may constitute an increase in biomass and continuity of fine fuels compared to uninvaded sites. More information on fuel properties of several native and nonnative grasses is available (Dibble and others 2007).

Invasive vines such as Oriental bittersweet (fig. 5-4), Japanese honeysuckle, kudzu, Chinese wisteria (*Wisteria sinensis*), porcelainberry, and English ivy (*Hedera helix*) have potential to alter fuel characteristics of invaded communities. They could increase fuel loading and continuity by growing up and over supporting vines, shrubs, and trees, and by killing the vegetation beneath them. Invasive vines could increase the likelihood of crown fire, especially under drought conditions, by acting as ladder fuels. Such changes have not been quantified. In the southern Appalachians, Oriental bittersweet contributes substantial vine biomass (Greenberg and others 2001). It can also support later-successional vines and lianas (Fike and Niering 1999), possibly enabling other species to become ladder fuels

Figure 5-3—Bush honeysuckle in the understory of a mixed forest that developed on an abandoned agricultural field (Bradley, Maine). (Photo by Alison C. Dibble.)

Figure 5-4—Oriental bittersweet has completely overtaken this eastern white pine (Rockland, Maine). (Photo by Alison C. Dibble.)

(Howard 2005a). In a deciduous forest in New York, several gaps were occupied by porcelainberry growing over Amur honeysuckle. In some quadrats, the cover of these two species combined was well over 100 percent, and few tree seedlings and herbs grew beneath the tangled canopy (Yost and others 1991).

In deciduous forests where invasive plants are prevalent, the morphology and stand structure of invasive Japanese barberry may alter fuel characteristics (Dibble and Rees 2005). Individual plants consist of multiple stems originating from the root collar and varying in length and morphology. Stems die after a few years, as new stems sprout from the base (Ehrenfeld 1999; Silander and Klepeis 1999). Japanese barberry populations can become dense, nearly impenetrable thickets within 15 years of initial establishment (Ehrenfeld 1999), even under closed canopies. Populations may become so thick that they shade out understory species (Johnson, E. 1996, review).

In some cases, nonnative vegetation might decrease the potential for ignition and spread of fire, although there are no studies documenting this in northeastern deciduous forests. Mile-a-minute (*Polygonum perfoliatum*), Japanese barberry, privet, kudzu, Japanese honeysuckle, and Oriental bittersweet are thought to reduce flammability on some oak-hickory sites in Virginia-area national parks. For example, mile-a-minute vine produces a dense mass of succulent, almost nonflammable vegetation. Where mile-a-minute dominates, managers are concerned that the use of prescribed fire to promote regeneration of desirable native species may not be possible. Japanese barberry and privet displace native, flammable ericaceous species including mountain laurel and blueberry (*Vaccinium* spp.), and there is concern that dominance by nonnative species may reduce flammability of the invaded community (Gorman, personal communication 2005). In dense thickets of Chinese privet in northwestern Georgia, prescribed fire was spotty and erratic. Lack of fire spread in privet infestations might be explained by moist and compacted privet litter or by the affinity of Chinese privet for moist, low-lying soils (Faulkner and others 1989). Because kudzu stems and foliage maintain high water content, flammability of invaded sites may be reduced even during drought, when desired native plants become susceptible to fire due to desiccation (Munger 2002b). Similarly, it has been suggested that dense stands of garlic mustard may inhibit the ability of a forest understory to carry surface fire (Nuzzo 1991).

Differences in phenology between native and nonnative species could theoretically affect fire seasonality, rendering a community more or less flammable during particular seasons. For example, buckthorns (Converse 1984a, TNC review) and bush honeysuckles (Batcher and Stiles 2000) leaf out earlier than native vegetation and retain their leaves later into autumn. This topic deserves further study.

Use of Fire for Controlling Nonnative Invasives in Deciduous and Mixed Forests

Fire alone is probably not sufficient to control most invasive species in deciduous forests of the Northeast bioregion because high fuel moisture and insufficient fuel accumulation limit both fire severity and the frequency with which burning can be conducted. In general, a long-term commitment and some combination of control treatments will likely be more effective for controlling invasive species than any single approach (Bennett and others 2003). Additionally, the usefulness and effectiveness of prescribed fire differ among forest types and depend to some extent on the fire types and frequency to which native plant communities are adapted. In this sense, prescribed fire is more likely to be an effective tool for controlling invasive species and promoting native vegetation in oak-hickory forests and oak savannas than in other forest types.

In maple-beech forests, a lack of dry surface fuels and/or a brief weather window for burning make the use of prescribed fire difficult. Additionally, if managers seek to maintain an overstory of fire-intolerant species such as maple and beech, burning under conditions where fires are severe enough to kill nonnative species will likely kill desired species as well. Using prescribed fire to control Norway maple, for example, would probably be detrimental to sugar maple and American beech.

Repeated prescribed burning may be more effective in oak-hickory forests and oak savannas than in maple-beech forests because dominant native species in these plant communities are adapted to relatively frequent fires. In this case, fire may be appropriate where management goals include controlling nonnative species or reducing fuels, accompanied by maintenance of native seral species.

Where conditions are appropriate for carrying a surface fire, nonnative invasive trees such as Norway maple and tree-of-heaven may be top-killed by fire, but both species can sprout following top-kill (Howard 2004a; Webb and others 2001). Observations indicate that seedlings of tree-of-heaven are killed by fire, but larger individuals tend to survive and sprout after fire (Gorman, personal communication 2005), even following heat-girdling (Hoshovsky 1988, TNC review). No experimental information is available on the effects of fire on Norway maple; however, cutting Norway maple resulted in sprouting from both seedlings and larger trees the following summer (Webb and others 2001). Additionally, removal of Norway maple from the canopy of a mixed maple forest in New Jersey resulted in a dramatic floristic and structural change in some areas, with establishment of both native and nonnative plant species not previously seen in the forest. Among the new arrivals were tree-of-heaven, Japanese barberry, winged euonymus, Japanese honeysuckle, wineberry (*Rubus phoenicolasius*), black locust, and garlic mustard. It is unclear whether these species established from the seed bank or from off-site sources. Removal of Norway maple seedlings also resulted in a large pulse of Norway maple recruitment (Webb and others 2001).

Sprouting is reported in many invasive woody species following top-kill (table 5-2). Prescribed fire during the dormant season is generally ineffective for controlling invasive shrubs in the Northeast bioregion. These fires reduce shrub cover temporarily and may kill seedlings and smaller plants, but populations are not controlled as shrubs resprout (Richburg 2005; Richburg and others 2001). However, on a study site in a mature deciduous forest in western Massacusetts, cover of Japanese barberry was significantly reduced 2 years after both cutting and burning (conducted in April and November), with the greatest reductions in areas where treatments were combined (Richburg 2005). Results of a study on the use of fire to control Japanese honeysuckle, Chinese privet, and native poison ivy in an oak-hickory-pine forest in northwestern Georgia are relevant to forests in the Northeast bioregion. Both fall and winter burns significantly ($P < 0.05$) reduced Japanese honeysuckle biomass. However, sprouting from buds protected by unburned litter was evident as early as 1 month following fire. Chinese privet showed no significant response to fire or season of burning, and many plants sprouted from root crowns. The response of privet to fire was unclear because fire did not spread well in privet thickets (Faulkner and others 1989).

It has been suggested that repeated prescribed fire may be effective for controlling species such as bush honeysuckles (Munger 2005a, FEIS review; Nyboer 1992), privets (Batcher 2000a, TNC review), and multiflora rose (Virginia Department of Conservation and Recreation 2002d). However, little empirical evidence is available to support these suggestions for northeastern forests. Additionally, repeated burns may be limited by insufficient fuel accumulation to carry fires that are scheduled close together (for example, see Richburg 2005). Repeated prescribed fire has been used with some success for controlling nonnative shrubs such as common buckthorn in oak savannas (chapter 7).

Japanese honeysuckle was reduced by repeated prescribed burning in North Carolina shortleaf pine forest and in an Illinois barren remnant (Munger 2002a). However, cessation of prescribed fire treatments, even after multiple consecutive or near-consecutive years of burning, can lead to reinvasion (Schwegman and Anderson 1986). Anecdotal evidence suggests that kudzu may also be controlled by repeated prescribed fire under certain conditions. Managers in Virginia observed that 3 to 4 years of prescribed fire late in the growing season can eliminate kudzu in the treated area (Gorman, personal communication 2005).

Little information is available on the use of fire to control invasive grasses in deciduous forests. It has been suggested that Japanese stiltgrass is not controlled by spring burning or mowing in oak-hickory forests because seeds germinate from the soil seed bank after treatment, and plants may grow rapidly enough to set seed that same year (Virginia Department of Conservation and Recreation 2002b). More effective control of Japanese stiltgrass might be achieved by timing prescribed fire before seeds ripen but late enough in the season to prevent a second flush of seed production (Gorman, personal communication 2005). In an oak savanna in Wisconsin, early April burning was not effective for controlling reed canarygrass (*Phalaris arundinacea*); fire appeared to enhance its spread. Burning in mid to late May weakened reed canarygrass and prevented seed production, though it did not eliminate the infestation and was detrimental to desired native herbs such as shooting star (*Dodecatheon media*) (Henderson 1990). For more information on control of reed canarygrass using prescribed fire, see the "Riparian and Wetland Communities" section, page 82.

Prescribed fire can be used to temporarily control garlic mustard under some conditions. However, garlic mustard has a moderately persistent seed bank and rapid population growth, and some individuals are likely to survive understory and mixed-severity fires in deciduous and mixed forests due to the patchiness of these fires (Munger 2001). Three consecutive years of prescribed burning in a central Illinois black oak forest failed to eradicate garlic mustard. One reason was that individuals survived in protected, unburned microsites such as the lee of a downed log or a patch of damp litter, and these survivors were successful in producing seed (Nuzzo and others 1996). Additionally, removal of garlic mustard may lead to proliferation of other undesirable species, so caution is warranted to avoid interventions that may be detrimental to the native community (McCarthy 1997).

Coniferous Forests

Background

Coniferous forests in the Northeast bioregion include white-red-jack pine ecosystems in the Great Lakes area; pitch pine communities in parts of the New England coast, the New Jersey Pine Barrens, and upstate New York; and spruce-fir (*Picea-Abies*) ecosystems in the Lake and New England States and at high elevations in the Appalachian Mountains. Other coniferous forest types include Virginia pine (*Pinus virginiana*), shortleaf pine (*Pinus echinata*), and Table Mountain pine

(*Pinus pungens*) within the loblolly-shortleaf pine and oak-pine ecosystems described by Garrison and others (1977). Some of these coniferous forest types also occur in the Southeast bioregion. More information is available on interactions between fire and invasive species in pine forests and savannas in chapter 6.

The white-red-jack pine ecosystem occurs on plains and tablelands of the northern Lake States and parts of New York and New England. In the Lake States these forests are used principally for timber and recreation, while large urban areas fragment this ecosystem in the northeast (Garrison and others 1977). Prior to Euro-American settlement, eastern white pine and red pine associations were generally fire-maintained seral types and existed occasionally as self-perpetuating climax under mixed fire regimes in the Great Lakes area. Fire exclusion can alter plant community structure and composition in these forest types, with shade-tolerant species becoming widespread. These stands may respond well to prescribed burning; however, understory invasion by shade-tolerant species could make burning difficult by developing a layer of less flammable surface material (Duchesne and Hawkes 2000). No information was found regarding nonnative species invasions in these ecosystems.

Pitch pine is well adapted to frequent fire, with presettlement fire regimes characterized by surface fires at intervals less than 10 years where burning by Native Americans was common and mixed-severity fires at intervals of about 10 to 35 years (Wade and others 2000). In the absence of disturbance, pitch pine is replaced by various hardwoods, especially oak and hickory, or by eastern white pine if present. Fire exclusion has also led to conversion of pitch pine forests to black locust-dominated stands (Dooley 2003). Black locust is an early-successional tree that colonizes old fields and burned areas in its native range (Converse 1984b, TNC review), from Pennsylvania southward. It is considered nonnative to the north and east (Fernald 1950). Today prescribed fire is used to reduce fuel loads and maintain or restore fire-adapted vegetation in some pitch pine communities (Patterson and Crary 2004).

Virginia pine and shortleaf pine types are estimated to have relatively frequent presettlement fire-return intervals (~2 to 35 years). Table Mountain pine fire regimes are characterized by stand-replacement fires at intervals of <35 to 200 years (Wade and others 2000). Little information is available regarding invasive species in these forest types. At Manassas National Battlefield Park in northeastern Virginia, some old agricultural fields have succeeded to Virginia pine and support spreading populations of Japanese stiltgrass, Japanese honeysuckle, privet, and winged euonymus (Dibble and Rees 2005). Princesstree occurs in Table Mountain pine-pitch pine forests in the southern Appalachian Mountains (Williams 1998). With the exception of princesstree, there is no information in the literature regarding invasive species and fire in these forest types.

Northeastern spruce-fir forests are characterized by a presettlement fire regime of stand-replacement fires with long return intervals (35 to over 200 years) (Duchesne and Hawkes 2000; Wade and others 2000). Spruce-fir stands are presumed to be less vulnerable than some other vegetation types to encroachment by nonnative invasive plants, though exceptions can occur where seed sources are available. In Maine, spruce-fir stands can have persistent, spreading populations of bush honeysuckles, Norway maple, Japanese barberry, and/or winged euonymus (Dibble and Rees 2005). Loss of dominant trees to nonnative insect pests such as balsam wooly adelgid and hemlock wooly adelgid (Dale and others 1991, 2001) provides openings for establishment of Japanese barberry, Oriental bittersweet, tree-of-heaven, and Japanese stiltgrass (Orwig and Foster 1998).

Role of Fire and Fire Exclusion in Promoting Nonnative Plant Invasions in Coniferous Forests

Few studies in northeastern coniferous forests discuss the establishment or spread of nonnative invasive species following fire or a period of fire exclusion.

Princesstree is widely planted in the eastern United States as an ornamental and a source of high-value export lumber. It is an early successional species that produces large numbers of wind-dispersed seed. It is not shade-tolerant, and seed germination and seedling establishment are restricted to disturbed areas such as exposed mineral soil, where light levels are high and leaf litter is absent (Williams 1993). Princesstree established after wildfire in forests dominated by Table Mountain pine, pitch pine, and Virginia pine in the southern Appalachians (Reilly and others 2006). Managers report postfire establishment of princesstree after "several wildfires" in Great Smoky Mountains National Park (Langdon and Johnson 1994). Invasion of native forests by princesstree is facilitated by other large-scale disturbances such as timber harvest, construction, gypsy moth defoliation, hurricanes, and floods (Johnson, K. 1996, review; Miller 2003; Williams 1993). In debris avalanches following Hurricane Camille in Virginia, princesstree established at densities ranging from 75 to 310 stems/ha on 3 of 4 study sites. Other species that occupied these sites included Japanese honeysuckle and tree-of-heaven, but these species were rarely found in the canopy of a mature forest (Hull and Scott 1982). Viable princesstree seeds have been found in the soil seed bank of some forest communities (Dobberpuhl 1980; Hyatt and Casper 2000). This includes the Pine Barrens of southern New Jersey (Matlack and Good 1990), where prescribed fire is sometimes used (Patterson and Crary 2004). Managers should be aware of the possibility of princesstree establishment from the seed bank. Princesstree sprouts from roots and/or stumps after pulling or cutting (Johnson, K. 1996) and may also do so after fire.

Sheep sorrel might be promoted by fire in conifer stands that have developed in agricultural openings or in openings over bedrock. In a stand of white pine on an old field in southeastern Nova Scotia that was clearcut and burned in June, sheep sorrel had a high stem density relative to other vegetation 1 year after fire (Martin 1956). In a 4-year study of a spruce-fir stand in southwestern New Brunswick, Canada, which was clearcut and burned twice, Hall (1955) found that sheep sorrel established immediately after the first burn and persisted after the second, though at a low stem count.

Responses of nonnative buckthorn species to fire may vary depending on frequency of burning. Four years after a low-severity spring burn in white and red pine plantations of Michigan, common and glossy buckthorns less than 0.8 inch (2 cm) DBH were present on plots burned only once but absent from plots burned three times in 5 years. Larger buckthorns (0.8 to 2.3 inches (2.0 to 5.9 cm) DBH) occurred on unburned plots but not on any burned plots, suggesting that the larger size class had been eliminated by fire (Neumann and Dickmann 2001). It is not clear to what extent the two nonnative buckthorn species differed in their response to repeated burning in the pine plantations.

Effects of Nonnative Plant Invasions on Fuel Characteristics and Fire Regimes in Coniferous Forests

There is no documentation that fire regimes have been changed by nonnative invasive species in northeastern coniferous forests, although studies suggest that fuel characteristics may be altered on invaded sites.

A study by Dibble and Rees (2005) suggests that nonnative species have altered fuel characteristics in coniferous forests in southern Maine. Invaded stands support a shrub layer dominated by nonnative honeysuckle species, Japanese barberry, Oriental bittersweet, common buckthorn, and/or glossy buckthorn, and have significantly greater shrub cover and frequency than nearby, relatively uninvaded stands. Fineleaf sheep fescue and wood bluegrass (*Poa nemoralis*, native to Eurasia and recently added to Maine's list of nonnative invasive plants) are abundant in the herb layer and may increase fine fuel loads and continuity on invaded conifer sites (Dibble and Rees 2005).

Black locust has been observed to reduce potential fire spread in pitch pine stands on sandy outwash plains, especially on old farm fields. Black locust litter on the forest floor tends to lie flat and stay relatively damp due to closed-canopy conditions created by black locust clones. The higher live-to-dead fuel ratios and higher fuel moistures effectively slow surface fires compared to uninvaded pitch pine stands (Dooley 2003). Native plants and animals in these fire-dependent plant communities can be adversely impacted by black locust dominance. When black locust encroached in dunes of Indiana, decline in native plant diversity was often accompanied by an increase in nonnative cheatgrass (Peloquin and Hiebert 1999). Larvae of the federally endangered Karner blue butterfly (*Lycaeides melissa sameulis*) of northeastern and upper midwestern North America feed solely on blue lupine (*Lupinus perennis*) in fire-adapted pitch pine woodlands and oak savannas (King 2003; Kleintjes and others 2003), which may be degraded by black locust invasion. Note that the host plant is not bigleaf lupine (*Lupinus polyphyllus*) of the Pacific Northwest, though that species dominates some roadsides and openings in Maine and is listed as a nonnative invasive species in that state.

Use of Fire to Control Nonnative Invasive Plants in Coniferous Forests

Fire is not typically used to control nonnative invasive plants in coniferous forests in the Northeast bioregion where fire suppression is usually the fire management priority. Prescribed fire is sometimes used to maintain fire-adapted ecosystems such as pitch pine, where black locust may be controlled by frequent, severe burning in late spring (Dooley 2003). However, it is difficult to obtain fires of sufficient severity to kill black locust, which typically responds to burning, cutting, and girdling by resprouting and suckering (Converse 1984b).

Grasslands and Early-Successional Old Fields

Background

True grasslands in the Northeast bioregion are sparse and discontinuous compared to their counterparts to the west and south. This section includes barrens and sandplains (as described by Curtis 1959 and Dunwiddie and others 1996) and early-successional old fields—areas that were initially cleared for agriculture and are currently maintained in early successional stages (Richburg and others 2004). While material covered here also pertains to disjunct populations of prairie communities extending east as far as Pennsylvania, Kentucky, and

Tennessee (Transeau 1935), discussion of prairie ecosystems is presented more fully in chapter 7.

The grasses and forbs that comprise northeastern grasslands include unique plant assemblages and numerous rare plants and animals (for example, see Dunwiddie 1998; Mitchell and Malecki 2003) and thus are of particular concern to managers. The ecology and much of the fire history of northeastern grasslands are summarized in Vickery and Dunwiddie (1997). Fire, primarily anthropogenic in origin, has been identified as one factor contributing to the origin and persistence of these plant communities (for example, Niering and Dreyer 1987; Parshall and Foster 2002; Patterson and Sassaman 1988; Transeau 1935; Winne 1997). Prescribed fire is currently used in some areas to maintain early successional species. For example, in a sandy outwash plain in southern Maine where native grassland has persisted for more than 900 years (Winne 1997), fire is used to maintain habitat for grasshopper sparrow and a large population of a rare herb, northern blazing star (*Liatris scariosa* var. *novae-angliae*) (Vickery 2002). Xeric blueberry barrens in southeastern Maine have been an open grassland-pine/shrub type for at least 1,700 years; many of these areas are now maintained by burning in alternate years (Winne 1997).

The area of old fields in the Northeast bioregion is extensive. From the early days of the colonial period, forests were converted to pasture and cropland; by the mid-19th century, less than 40 percent of Vermont, Massachusetts, Rhode Island, and Connecticut was forested. Conversion of forest to agriculture resulted in intentional and accidental introduction of many nonnative plants. When agricultural fields were abandoned, seeds of nonnative plants were no doubt present, and other species may have been introduced with "conservation plantings" (Knopf and others 1988). Thus propagules of nonnative invasive plants are likely to be more abundant in old fields than in grasslands of other origins. As agriculture declined in the Northeast during the late 19th and 20th centuries, forests reclaimed much of the landscape (review by Vickery and Dunwiddie 1997), while some areas are maintained in early succession for bird habitat (for example, see Vickery and others 2005) or other conservation purposes.

Grasslands and old fields in the Northeast bioregion are early successional communities capable of supporting woody vegetation. The most problematic invasives are woody species, both native and nonnative, which alter the structure as well as the species composition of these habitats. Techniques for controlling woody invasives in grasslands typically include cutting or mowing, herbicides (Barnes 2004), and/or fire (for example, Dunwiddie 1998). Grassland burns are commonly conducted in fall or spring, though control of woody species may be more effective if burns occur during the growing season (Mitchell and Malecki 2003, Richburg 2005). In the wildland urban interface, early successional vegetation is more commonly maintained by mowing than by burning. Because these habitats require disturbance to remain in an early successional stage, they may be especially vulnerable to establishment and spread of nonnative plants (for example, see Johnson and others 2006).

Nonnative species that seem especially problematic in northeastern grasslands include Scotch broom (*Cytisus scoparius*), multiflora rose, porcelainberry, and swallow-wort (*Cynanchum louiseae, C. rossicum*). Numerous other nonnative species are invasive or potentially invasive in northeastern grasslands (table 5-1), but research on their relationship to fire in grasslands or old fields is lacking. Mehrhoff and others (2003) state that most nonnative species of concern in this bioregion are common in old fields, but many are also problematic in forest or riparian communities, and some of these are discussed in other sections of this chapter.

Role of Fire and Fire Exclusion in Promoting Nonnative Invasives in Grasslands and Old Fields

Nonnative herbs may increase after fire in northeastern grasslands and old fields. Swan (1970) quantified vegetation response to wildfires of the early 1960s in goldenrod-dominated fields in south-central New York. Two years after fire, three nonnative species showed higher relative frequency on burned than unburned areas: Canada bluegrass (81 percent vs. 56 percent), redtop (39 percent vs. 25 percent), and sheep sorrel (44 vs. 33 percent). Burning appears to enhance germination of sheep sorrel, possibly by removing the litter layer (Kitajima and Tilman 1996). However, Dunwiddie (1998) reports no effect of fire on sheep sorrel, and Niering and Dreyer (1989) report equivocal results: In Connecticut old fields dominated by little bluestem, relative frequency of sheep sorrel decreased after 17 years of annual burning but also decreased on unburned plots. Fire alone increased stem density of spotted knapweed for 3 years after spring burning in old fields in Saratoga National Historical Park, Saratoga Springs, New York (Gorman, personal communication 2005).

Nonnative shrubs are likely to survive all but severe, growing-season fires, though information specific to northeastern grasslands is limited. Most research on Scotch broom comes from the Pacific Northwest and is summarized briefly here; this species is covered more thoroughly in chapter 10. Scotch broom spreads from abundant seeds and can sprout from stumps or root crowns following damage to aboveground parts. Scotch broom seeds can survive in the soil for at least 5 years and possibly as long as 30 years. Laboratory studies and postfire field observations indicate that heat scarification induces germination (Zouhar 2005a, FEIS review). Prescribed fires in a northeastern grassland (on one site, August cut, then burn; on another, April burn) reduced Scotch broom cover significantly, but numerous seedlings established (Richburg 2005).

Fire can top-kill bush honeysuckle plants and is likely to kill seedlings and stressed plants. However, perennating tissues on roots and root crowns are often protected by soil from fire damage, so there is potential for postfire sprouting. Sprouting of Bell's honeysuckle (*Lonicera* × *bella*) was observed after spring and late-summer fires at the University of Wisconsin Arboretum (Munger 2005a). After spring burning in an oak forest, sprouts of Bell's honeysuckle were described as "not very vigorous" (Kline and McClintock 1994).

Multiflora rose is a shrub that can reproduce by root suckering and layering, and forms dense thickets that displace herbaceous plants, especially in early successional habitats such as old fields. Both multiflora rose and common buckthorn produced sprouts following stem removal by cutting, with or without burning, in the dormant and growing season (Richburg 2005). Native rose species are typically top-killed by fire; with increasing fire severity, they may be subject to root crown and rhizome damage (Munger 2002c, FEIS review). Thus it seems likely that multiflora rose will survive fire and, because seeds remain viable in the soil for 10 to 20 years (Szafoni 1991), possibly regenerate from seed after fire.

Several nonnative vines are invasive in northeastern grasslands; however, information on responses of nonnative vines to fire in these communities is lacking. We could find no peer-reviewed accounts of invasion or spread of porcelainberry or Oriental bittersweet after fire. However, the ability of these vines to regenerate vegetatively, produce abundant seed (table 5-2), and establish in openings suggests that fire may favor their spread. Black swallow-wort cover increased after dormant season cutting and burning of invasive shrubs in a New York grassland (Richburg 2005), although significance of differences from untreated sites was not reported. Lawlor (2002, TNC review) reports that swallow-wort recovered and reproduced the season following prescribed fires in New York and Wisconsin (see "Use of Fire to Control Nonnative Invasive Plants in Grasslands and Old Fields," page 81).

In the absence of fire or other disturbance (for example, mowing), woody species generally increase in northeastern grasslands. Two studies illustrate how the presence and abundance of nonnative species in old fields may change over time without disturbance. In New Jersey, 40 years of vegetation data from old fields were used to evaluate changes in nonnative species abundance and diversity over time. Invasions were initially severe, with nonnative species comprising over 50 percent of the cover and species in each field. After 20 or more years of abandonment, the abundance and richness of nonnative species had declined significantly without management intervention. As woody cover increased, many nonnative herbaceous species that had dominated earlier in succession, particularly annuals and biennials, became much less abundant. Some shade-tolerant invasive species (garlic mustard, bush honeysuckles, Norway maple, Japanese stiltgrass, and Japanese barberry) are currently increasing on these sites and may present the next invasion challenge to the managers of the grassland community (Meiners and others 2002). A site in southeastern Connecticut that was abandoned and burned 40 years earlier became partially dominated by Oriental bittersweet, which increased in cover along with Japanese honeysuckle, Morrow's honeysuckle, and multiflora rose during the last decades of the study (Fike and Niering 1999).

Effects of Nonnative Plant Invasions on Fuel Characteristics and Fire Regimes in Grasslands and Old Fields

Presettlement and even postsettlement fire regimes for northeastern grasslands are not well described in the literature (but see Vickery and Dunwiddie 1997), so departure of current patterns from past fire regimes is difficult to determine, and the influence of nonnative species on fire regime changes is difficult to estimate. Old fields have no reference fire regime because they are a recent anthropogenic vegetation type. It is more fruitful, in this section, to discuss how nonnative species may alter fuels in northeastern grasslands and thereby influence the fire regimes desired for maintenance of these plant communities or for protecting property in the wildland urban interface. Scotch broom is the only nonnative species for which information on fuel characterisitcs and fire behavior in northeastern grasslands has been published (Richburg 2005).

Scotch broom establishes in old fields and grasslands, where it can eventually replace native plants with a dense, monospecific stand. As Scotch broom stands age, the ratio of woody to green material increases and dead wood accumulates (Waloff and Richards 1977). During experimental fires intended to control this species in old fields on Naushon Island, Massachusetts, where the effects of cutting and burning were studied, Scotch broom was observed to be highly flammable, even when green (Richburg 2005; Richburg and others 2004). Cutting reduced non-woody fine fuels and increased 1-hour and 10-hour woody fuels. Fuel bed depth did not change. Subsequent burns showed flame lengths of approximately 20 feet (6 m) on uncut plots burned with a headfire in April, and 3 feet (1 m) on cut plots burned in August with a backfire under "very dry conditions."

Due to their growth form and habits, nonnative vines can affect fuel load and distribution, as described above in the section on deciduous and mixed forests. Porcelainberry vines, for example, can cover the ground in sunny openings, such as old fields, and grow up into trees and shrubs at the forest edge. These vines can eventually kill the supporting vegetation (Yost and others 1991), and festooned trees are also susceptible to wind damage, further increasing mortality of supporting species. Similarly, pale swallow-wort and black swallow-wort, which are viny and twining herbs (Gleason and Cronquist 1991), can form large, monospecific stands in open areas and can over-top and smother shrubs (Lawlor 2002).

Use of Fire to Control Nonnative Invasive Plants in Grasslands and Old Fields

The nonnative shrubs and vines that are most problematic in northeastern grasslands are all able to sprout after fire, at least to some extent, and several persist in the soil seed bank (table 5-2). While fire may be a desirable tool for promoting desired grasses and forbs in grasslands, its effectiveness may be confounded by the ability of nonnative species to survive and thrive after fire. Research in grasslands in south-central New York and Naushon Island, Massachusetts, indicated that a single dormant-season burn is unlikely to reduce nonnative woody species. Combining growing-season prescribed fire with other treatments may improve control. A treatment that reduced common buckthorn and Scotch broom in grasslands consisted of a late spring mowing, allowing cut fuels to cure, and then a late summer burn. Growth rate of common buckthorn sprouts was slower on August-burned plots than on unburned plots or spring-burned plots. Effects may be short-lived, however. Nonstructural carbohydrates in common buckthorn and multiflora rose declined after cutting, mowing, or burning treatments but recovered within 1 year (Richburg 2005).

Scotch broom may be susceptible to heat damage from fire, but regeneration from the seed bank complicates the use of fire to control this species. Several researchers provide evidence that Scotch broom seed germination is stimulated by fire (Zouhar 2005a), although results vary among locations (for example, see Parker 2001). Chapters 4 and 10 cover this species in some detail, but one study is relevant here. In old fields being maintained as grasslands on Naushon Island, Massachusetts, Richburg (2005) found that prescribed fires, whether in the dormant season or growing season, killed Scotch broom but led to copious recruitment of Scotch broom germinants from the soil seed bank and/or from nearby untreated plants. Cover of native graminoids and herbs was low within Scotch broom patches and decreased within a year after burning. When prescribed fire is used to stimulate Scotch broom germination from the seed bank, follow-up treatments such as subsequent controlled burns, spot burning, revegetation with fast growing native species, herbicide treatments, grazing, and hand-pulling can be used to kill seedlings and thus reduce the seed bank (Zouhar 2005a).

Multiflora rose seems able to survive fire but does not usually increase immediately after burning. Thus repeated fires may be useful in controlling this species. In a savanna restoration project on an old agricultural field in Illinois, Hruska and Ebinger (1995) significantly reduced stem density of multiflora rose and autumn-olive following March fires in 2 successive years. They were concerned that desired native oak seedlings were adversely impacted. In plant communities comprised of fire-adapted grasses and forbs, periodic prescribed burns will likely retard multiflora rose invasion and establishment (Munger 2002c). The Virginia Department of Conservation and Recreation (2002d) recommends spring prescribed fire to reduce cover of multiflora rose, with follow-up burns in subsequent years for severe infestations.

Bush honeysuckle species (Morrow's honeysuckle, Bell's honeysuckle, and others) may be controlled with prescribed fire in fire-adapted grassland or old-field communities. According to several sources, spring prescribed burning may kill bush honeysuckle seedlings and top-kill larger plants, although results have been mixed (Munger 2005a). Morrow's honeysuckle was not reduced by dormant-season prescribed fire in old fields of western New York, but growing-season fires preceded by growing-season mowing reduced this species (Mitchell and Malecki 2003). The Maine Department of Conservation, Natural Areas Program (2004) recommends burning during the growing season. Regardless of season, a single prescribed fire is usually not sufficient to eradicate bush honeysuckles. Annual or biennial burns may be needed for several years (Munger 2005a). Solecki (1997, review) recommends annual or biennial spring burning for 5 or more years to control bush honeysuckles in prairie ecosystems.

Fire may be useful for controlling Japanese honeysuckle in grasslands, but only with repeated use and long-term commitment to monitoring and follow-up treatments. Cessation of prescribed fire treatments, even after multiple consecutive or near-consecutive years of burning, often leads to reinvasion. Following spring burns in 4 out of 5 years, fire was excluded from a southern Illinois barren. Japanese honeysuckle frequency decreased following the fires. However, shade increased during fire exclusion years, and 11 years after the last fire, frequency of Japanese honeysuckle was nearly four times preburn levels (Schwegman and Anderson 1986).

Numerous sources agree that fire is not effective for reducing swallow-wort populations (DiTommaso and others 2005; Lawlor 2000, 2002; Sheeley 1992). Perennating buds on the root crowns generally occur a centimeter or more below the soil surface and are thus likely to be protected from fire (Sheeley 1992; Lawlor 2000, 2002). At Montezuma National Wildlife Refuge, western New York, a large swallow-wort infested area was burned in late spring to reduce woody debris in grasslands (Lawlor 2002). The swallow-worts recovered and reproduced as usual the following season. Similar results were observed after prescribed burning in Wisconsin. Lawlor (2002) suggests that burning or flaming could be used to control seedlings after mature growth has been killed with herbicides; swallow-wort seedlings lack the well-developed root crown of more mature plants.

Riparian and Wetland Communities

Background

Riparian and wetland communities in the Northeast bioregion vary in native plant community composition, site characteristics, and fire regime; however, several species of nonnative invasive plants are common among these community types, so they are discussed together here.

Riparian plant communities in the Northeast bioregion may be dominated by hardwoods, conifers, or mixed stands, and a dense layer of shrubs and vines can occur beneath the tree canopy. Many native plants grow almost exclusively in riparian areas and may be adapted to intense disturbance from seasonal flooding and scour by water and ice. Disturbance by fire is unusual in Northeastern riparian communities, so riparian plants may not be fire-adapted; however, adaptations that allow these plants to recover after flooding and scour could aid in their recovery after fire (chapter 2).

The bottomland hardwood vegetation type described by Wade and others (2000) includes the elm-ash-cottonwood ecosystem (*sensu* Garrison and others 1977) that occurs in riparian areas along major streams or scattered swamp areas throughout the eastern United States and includes several forest cover types. The historical role of fire in these ecosystems is unclear, although many of the dominant riparian species are sensitive to fire and especially intolerant of repeated burning. Presettlement fire regimes were thought to be of mixed-severity or stand-replacement types, with intervals of about 35 to 200 years. Fuel loads were generally low due to rapid decomposition, so large, severe fires probably occurred only during extended drought or in heavy fuels caused by damaging wind storms (Wade and others 2000). Conditions in spring and fall are often too wet for prescribed burning.

Freshwater wetlands in the Northeast bioregion include forested wetlands such as red maple swamps, silver maple (*Acer saccharinum*) floodplain forests, alder thickets, conifer bogs, Atlantic white cedar (*Chamaecyparis thyoides*), black gum (*Nyssa sylvatica*), and bay forests; and also fens and marshes dominated by sedges and grasses (Garrison and others 1977). Forested wetlands such as conifer bogs are probably only susceptible to fire in severe drought years due to their typically humid environment. Ground fires are possible with severe drought or drainage; with strong winds, conifer bogs can sustain crown fires. Presettlement, stand-replacement fire intervals are estimated between 35 and 200 years (Duchesne and Hawkes 2000). Prescribed fire is probably not appropriate in forested wetlands, and fire is typically excluded from these communities.

Wet grasslands in the Northeast bioregion include freshwater and salt or brackish tidal wetlands along the Atlantic coast, as well as freshwater inland marshes. Frost (1995) provides information on dominant vegetation along gradients of salinity and fire frequency. Consistent differences in species composition and fire behavior occur between saltwater and freshwater wetlands. Freshwater wetlands support a high diversity of species and a variety of plant associations. These are typically dominated by herbaceous species but may also support woody associations, although woody plant development is impeded by factors including ice scour, wave action, and periodic fires. Saltwater wetlands of the Northeast include the northern cordgrass prairie described by Küchler (1964), which is dominated by cordgrasses (*Spartina* spp.), saltgrass (*Distichlis* spp.), and rushes (*Juncus* spp.) (Wade and others 2000). Sedges (especially *Scirpus,* but also *Schoenoplectus* and *Bolboschoenus*) are also common. Forbs may be present where fresh water mixes into the system. Woody plants are typically intolerant of the salinity and the twice-daily inundation that characterize tidal wetlands, but they may occupy hummocks or outcrops. Since presettlement times, the assumption is that woody plants have extended into the marsh, vegetation is taller, and native plants have been displaced by tall, dense stands of common reed (*Phragmites australis*) (LANDFIRE Rapid Assessment 2005b).

Most information on fire regime characteristics in herbaceous wetlands comes from the southeastern United States (for example, Frost 1995), with relatively little information on wetlands in the Northeast bioregion. Fires are common in southeastern wetlands, which support large quantities of flammable, herbaceous vegetation that is well-adapted to frequent fires. Occurrence of woody plants can alter fire behavior, and groundwater levels influence both fire behavior and fire effects on soils and vegetation. In freshwater marshes, flammability varies due to the large diversity of plant communities, but species such as sawgrass (*Cladium*), cattail (*Typha*), common reed, maidencane (*Amphicarpum purshii, Panicum hemitomon*), and switchgrass (*Panicum virgatum*) provide flammable fuels that can support continuous, intense fires. Cordgrass species that dominate tidal marshes are also quite flammable. Presettlement fire frequency for northern cordgrass prairie communities is estimated at 1- to 3-year intervals (Wade and others 2000).

Prescribed fire is used more extensively in salt marshes than in freshwater marshes. Conditions in spring and fall are often too wet for prescribed burning in freshwater wetlands, although fire is sometimes used to reduce fuel loads, control invasive plants, and promote native species. Prescribed fire is frequently used in saltwater grasslands to enhance productivity and to reduce plant cover, fuel loadings, and woody species (Wade and others 2000).

Few nonnative invasive plants pose a high threat potential in tidal wetlands (table 5-1). Along margins and in areas where tidal influence or salinities have been altered by land use and development, woody species such as tree-of-heaven (Kiviat 2004) and Oriental bittersweet (Bean and McClellan 1997, review) may be invasive. Common reed, a large, perennial, rhizomatous grass with nearly worldwide distribution, is the invasive species of most concern in tidal wetlands (for example, see Leck and Leck 2005; Niering 1992; Weis and Weis 2003) and also invades fresh wetlands and riparian areas in the Northeast bioregion (table 5-1). Literature reviews (for example, D'Antonio and Meyerson 2002; Marks and others 1993, TNC review) suggest that, although common reed is native to North America, invasive strains may have been introduced from other parts of the world; and while there is evidence that common reed is native in the Northeast bioregion, many marshes are occupied by a European genotype (Saltonstall 2003). Common reed is regarded as aggressive and undesirable in parts of the eastern United States, but it may also be a stable component of a wetland community that poses little or no threat in areas where the habitat is undisturbed. Examples of areas with stable, native populations of common reed include sea-level fens in Delaware and Virginia and along Mattagodus Stream in Maine. In areas where common reed is invasive, large monospecific stands may negatively impact native plant diversity and create a fire hazard (D'Antonio and Meyerson 2002; Marks and others 1993).

Riparian areas often support more invasive species than upland habitats (for example, see Barton and others 2004; Brown and Peet 2003). This is attributed to high levels of propagule pressure (that is, abundance of seeds or vegetative fragments), a high-frequency disturbance regime, and water dispersal of propagules (Barton and others 2004; Robertson and others 1994). Several nonnative invasive plant species occur in fresh wetlands and/or riparian areas in the Northeast bioregion (table 5-1), including widespread species such as Japanese stiltgrass, garlic mustard, tree-of-heaven, Norway maple, Japanese barberry, bush honeysuckles, privets, multiflora rose, common buckthorn, Oriental bittersweet, and ground-ivy, which are covered in more detail in other sections of this chapter. Species that may be common in old fields and other areas of anthropogenic disturbance, such as porcelainberry and swallow-worts (covered in the "Grasslands and Early-Successional Old Fields" section, page 78), can be invasive along rivers and streams where scouring spring floods occur. Swallow-worts, for example, occur in areas subject to hydrologic extremes such as alvar communities of the eastern Lake Ontario region or New England coastal areas (Lawlor 2002). Species that seem to have a particular affinity for wetland and riparian communities include common reed, reed canarygrass, purple loosestrife (fig. 5-5), glossy buckthorn, Japanese knotweed (*Polygonum cuspidatum*), and mile-a-minute.

Figure 5-5—Reed canarygrass and purple loosestrife grow in dense patches on the typically rocky shore of the Penobscot River (Eddington, Maine). (Photo by Alison C. Dibble.)

Role of Fire and Fire Exclusion in Promoting Nonnative Invasives in Riparian and Wetland Communities

There is little published literature on the role of fire or fire exclusion in promoting plant invasions in riparian communities or wetlands in the Northeast bioregion. However, managers should be alert to the possibility of invasion by nonnative species after wild or prescribed fires, and the possibility that wetland areas adapted to frequent fires could be invaded in the absence of fire by nonnative woody species such as glossy and common buckthorn (Moran 1981).

Studies in the north-central United States and adjacent Manitoba, Canada, indicate that common reed is not typically damaged by fire because it has deeply buried rhizomes that are often under water, and the heat from most fires does not penetrate deeply enough into the soil to injure them. When fire consumes the aboveground foliage of common reed, new top growth is initiated from the surviving rhizomes. Rhizomes may be damaged by severe fire when the soil is dry and humidity low (Uchytil 1992b, FEIS review). Fires of this severity are likely to occur only under conditions of artificial drainage and/or severe drought.

Reed canarygrass is a cool-season, rhizomatous grass that can form dense, monotypic stands in marshes, wet prairies, wet meadows, fens, stream banks, and swales (Hutchison 1992b) (fig. 5-6). It is native to North America and also to temperate regions of Europe and Asia (Rosburg 2001; Solecki 1997). In the United States, cultivars of the Eurasian ecotype have been developed for increased vigor and thus may be more invasive than native ecotypes (Wisconsin Department of Natural Resources 2004). Reed canarygrass is considered a threat to native wetlands because of its rapid early growth, cold hardiness, and ability to exclude desired native plants (Hutchison 1992b; Lyons 1998, TNC review; Wisconsin Department of Natural Resources 2004). Anthropogenic disturbance and alteration of water levels encourage its spread (Wisconsin Department of Natural Resources 2004). Reed canarygrass seems well adapted to survive and reproduce after fire, but its response to wildfire has not been described in the literature. Burning of a Minnesota wetland followed by repeated herbicide application led to extensive germination of reed canarygrass from the seed bank, probably because of increased light at the soil surface (Preuninger and Umbanhowar 1994).

Figure 5-6—Reed canarygrass quickly filled this low conifer forest when the hydrology changed and the overstory died. The forest had previously contained only a sparse understory layer. (Photo by Alison C. Dibble.)

Glossy buckthorn is similar to common buckthorn (see table 5-2) in its reproductive biology and sometimes invades similar woodland habitats, but it more commonly invades moist to wet sites that are not fully flooded (Andreas and Knoop 1992; Frappier and others 2003; Taft and Solecki 1990). Reviews indicate that it grows best in drier parts of wetlands, in wetlands where some drainage has occurred, and possibly where fires have been excluded (Converse 1984a; Larson and Stearns 1990). Glossy buckthorn recruitment is most successful with ample light and exposed mineral soil. Burning to maintain vigor of the native plant community may prevent glossy buckthorn seedling establishment; however, if seed sources occur near burned areas, seedlings can establish readily on exposed soils. Glossy buckthorn also sprouts from roots or the root crown after fire (Catling and others 2001; Post and others 1990). In a calcareous fen in Michigan burned in the fall, glossy buckthorn stem density was twice as great the summer after burning as the summer before burning, and stems were one-third the height of preburn stems (unpublished report cited in review by Converse 1984a). On a prairie site in northwest Indiana, prescribed fire in October resulted in complete top-kill of glossy buckthorn, yet 1 year after fire there was a 48 percent increase in total stems of glossy buckthorn. The site was burned again the following April and sampled the following September with similar results. Overall stem numbers increased 59 percent. The authors suggest that prescribed burning may be used to prevent seed set but that plants will resprout (Post and others 1990).

Purple loosestrife is one of the most invasive species of freshwater wetlands and riparian areas in North America. It is an herbaceous perennial forb with buds that overwinter on the root crown about 0.8 inches (2 cm) below the soil surface (DiTomaso and Healy 2003). Surface fires are unlikely to provide enough heat or burn long enough to cause substantial damage to roots or the root crown of purple loosestrife (Munger

2002d, FEIS review; Thompson and others 1987). Information describing interactions between purple loosestrife and fire are lacking, although it is likely that purple loosestrife can survive fire by sprouting from buds located below the soil surface. Fire may also lead to recruitment of purple loosestrife seedlings due to exposure of bare substrate containing a substantial seed bank (Munger 2002d).

Two members of the buckwheat family (Polygonaceae), Japanese knotweed and mile-a-minute, are especially invasive in riparian and freshwater wetland communities. Japanese knotweed is an herbaceous perennial that is widely distributed in much of the eastern United States, where it spreads primarily along river banks (fig. 5-7) but also occurs in wetlands, along roadways, and in other disturbed areas (Seiger 1991, TNC review). Japanese knotweed reproduces from seed (Bram and McNair 2004; Forman and Kesseli 2003) and perennial rhizomes that can extend 18 inches (46 cm) below ground, are 50 to 65 feet (15 to 20 m) long, and can survive repeated control attempts. It can also establish from rhizome and stem fragments. Once established, Japanese knotweed spreads via rhizomes to form virtual monocultures (Child and Wade 2000) that are extremely persistent and difficult to control (Seiger 1991). Given its extensive root system and its response to repeated cutting, it seems likely to survive even frequent, severe fire, though no peer-reviewed reports are available on this topic.

Figure 5-7—Japanese knotweed spread from a nearby house site to occupy at least 50 m along both sides of this stream in Blue Hill Falls, Maine. (Photo by Alison C. Dibble.)

Mile-a-minute is a prickly, annual, scrambling vine that is especially prevalent along roadsides, ditches, stream banks, wet meadows, and recently harvested forest sites (Virginia Department of Conservation and Recreation 2002c). A review by Kumar and DiTommaso (2005) indicates that mile-a-minute grows best in sunny locations on damp soil but can also tolerate light shade. The prickly stem and leaves allow it to climb over neighboring vegetation and to form dense, tangled mats that cover small trees and shrubs to a height of about 26 ft (8 m) along forest edges. Mile-a-minute reproduces by seed that is dispersed by birds and mammals, as well as by water transport. Seeds can remain dormant in the soil seed bank for at least 3 years (Kumar and DiTommaso 2005). Since mile-a-minute thrives in gaps and disturbed areas and its seed is widely dispersed by birds (Okay 2005, review), fire could contribute to its increase; however, this has not been documented.

Effects of Nonnative Plant Invasions on Fuel Characteristics and Fire Regimes in Riparian and Wetland Communities

We found no studies that specifically address changes in fuel characteristics and fire behavior in riparian or wetland communities in the Northeast bioregion. Existing reports do not indicate that nonnative plant invasions have altered the fire regimes in these communities. The discussion of fuel properties here is based on morphology and phenology of nonnative invasives.

Common reed is perceived as a fire hazard where it occurs in dense stands in wetlands. It produces substantial amounts of aboveground biomass each year, and dead canes remain standing for 3 to 4 years (Thompson and Shay 1985). It has been suggested that common reed colonies increase the potential for marsh fires during the winter when aboveground portions of the plant die and dry out (Reimer 1973). Thompson and Shay (1989) observed that, even when common reed stands are green, the typically abundant litter allows fires to burn. Additionally, head fires in common reed stands may provide firebrands that ignite spot fires more than 100 feet (30 m) away (Beall 1984, as cited by Marks and others 1993).

Glossy buckthorn branches profusely from the base, with dead stems often found among smaller, live stems (Taft and Solecki 1990). Herbaceous fuels are usually sparse beneath large glossy buckthorn shrubs or in dense thickets (Packard 1988). Where glossy buckthorn invasion has reduced fine herbaceous fuels and increased dead woody fuels, fire behavior may be altered.

Purple loosestrife is difficult to burn, based on reports from managers who attempted to use prescribed fire to control it (Munger 2002d). Such attempts are commonly described as being confounded by moist soil conditions and patchy fuel distribution. A persistent stand of purple loosestrife could alter fuel conditions and fire behavior if it displaces native vegetation that is more flammable, and could thus further alter plant community composition. There is, however, currently no empirical evidence of such effects from purple loosestrife invasion.

Use of Fire to Control Nonnative Invasive Plants in Riparian and Wetland Communities

Prescribed fire is not likely to be a useful control measure for invasive species in plant communities where fires are typically rare and native species are not fire-adapted. Many forested wetlands, for example, are typically too wet to burn except during drought. Conversely, herbaceous wetlands commonly support native species that are adapted to frequent fire (Frost 1995; Wade and others 2000), and prescribed fire may be useful for controlling nonnative invasives in these communities.

Prescribed fire and herbicides are often used, alone and in combination, to manage common reed in wetlands in the Northeast bioregion and adjacent Canada. For example, fire has been used to reduce common reed in marshes at Prime Hook National Wildlife Refuge near residential areas of Delaware (Vickers 2003). However, little published quantitative information is available regarding the efficacy of fire. Prescribed burning alone removes accumulated litter and results in a temporary decrease in aboveground biomass of common reed, but fire does not kill plants unless rhizomes are burned and killed. This seldom occurs because the rhizomes are usually covered by a layer of soil, mud, and/or water (Marks and others 1993).

Season of burning may influence postfire response of common reed. Researchers in Europe found that burning common reed in winter caused little damage, while burning during the emergence period killed the majority of common reed shoots (Toorn and Mook 1982). Spring burning at the Delta Marsh in Manitoba removes litter and promotes a dense stand of even-aged canes, whereas summer burning results in stunted shoots and may control vegetative spread (Thompson and Shay 1985; Ward 1968). Both spring and fall burning of common reed resulted in greater shoot biomass, and summer burns resulted in lower shoot biomass in comparison with controls, while total shoot density on all burned plots was higher on controls. Similarly, belowground production in common reed was higher by mid-September on spring and fall burns than on controls, but was not higher on summer burns. Summer burns resulted in increased species diversity, richness, and evenness, while these community characteristics were not altered by spring and fall burns (Thompson and Shay 1985).

Burning is sometimes used in conjunction with herbicide treatments and manipulation of water levels to control common reed. Clark (1998) found that herbicide applied late in the growing season, followed by dormant season prescribed fire and a second herbicide application the following growing season, was more effective than spraying alone. A significant decrease in density and frequency of common reed was recorded in spray-burn treatments compared to pretreatment measures, untreated controls, and spray only treatments (Clark 1998). At Wertheim National Wildlife Refuge in New York, common reed was eliminated from a freshwater impoundment that was drained in the fall, burned the following winter, and then reflooded. Common reed remained absent for at least 3 years following treatment (Parris, personal communication cited by Marks and others 1993). The same TNC review presents several additional case studies documenting attempts to control common reed using prescribed fire.

Reed canarygrass is difficult to control because it has vigorous, rapidly spreading rhizomes and forms a large seed bank (Hoffman and Kearns 2004, Leck and Leck 2005); in addition, control efforts could reduce native ecotypes of this species or harm other native species (Lavergne and Molofsky 2006, review; Lyons 1998). Effects of prescribed fire on reed canarygrass vary. Moist bottomlands in Wisconsin undergoing restoration from agriculture to tallgrass prairie were burned on a 3-year rotation—one group of plots in late March and another group in mid-July. Neither frequency nor cover of reed canarygrass changed significantly in any of the treatments (no burn, spring burn, summer burn) (Howe 1994b). A review by Apfelbaum and Sams (1987) included an account of burning of wet prairie in Illinois every 2 to 3 years. This treatment appeared to restrict reed canarygrass to disturbed sites and prevent spread into undisturbed wetland. The effects of burning reed canarygrass at different seasons have not been studied for wetlands. Hutchison's (1992b) management guidelines suggest that late spring or late autumn burning for 5 to 6 consecutive years may produce "good control" of reed canarygrass in wetlands, but that treatment will be ineffective unless desired species are present or seeded in. Prescribed fires may be difficult to conduct in stands dominated by reed canarygrass because of high water levels and vegetation greenness. Management guidelines from the Wisconsin Department of Natural Resources (2004) suggest that treatment with glyphosate could make fall burning more feasible. Because reed canarygrass alters water circulation, increases sedimentation, and may increase the uniformity of wetland microtopography (Zedler and Kercher 2004), restoration of invaded ecosystems is likely to require restoration of physical structure of the habitat and seeding.

The literature on glossy buckthorn, largely anecdotal, suggests that prescribed fire may be used to control this species, especially in communities adapted to frequent fire (for example, Heidorn 1991; Larson and Stearns 1990). According to a management guideline by Heidorn (1991), regular prescribed fire (annual or biennial burns for 5 or 6 years or more) can control both glossy and common buckthorns in communities adapted to frequent fire such as fens, sedge meadows, and marshes. A review by Converse (1984a, TNC review) suggests glossy buckthorn can be reduced by cutting in the spring at leaf expansion and again in the fall, followed by spring burning the next 2 years. Postfire sprouts of glossy buckthorn may be more susceptibile to herbicides (Converse 1984a) or other control measures. McGowan-Stinski (2006, review) indicates that the season after mature buckthorn shrubs have been removed from an area, large numbers of seedlings are likely to germinate; in addition, untreated saplings and/or resprouts are likely to occur. He suggests controlling seedlings, saplings and sprouts by burning them with a propane torch in the first growing season after removal of adults. It is most efficient to torch seedlings and saplings at the stem base until wilting occurs. Repeat treatment could be needed. Seedlings are usually not capable of resprouting if torched before August (McGowan-Stinski 2006).

In a forested swamp dominated by white ash and red maple in the Berkshire Hills of western Massachusetts, Morrow's honeysuckle dominated a dense shrub understory, and Japanese barberry was a common associate. Richburg (2005) compared treatments to reduce the ability of these species to store root carbohydrates. A growing season cut followed by fall burning and a cut the next year had the greatest effect on reducing nonstructural carbohydrates for Morrow's honeysuckle. For Japanese barberry, this treatment and the dormant-season cut led to the lowest root carbohydrate levels, which was interpreted as a decrease in plant vigor. Plots cut in the dormant season had taller sprouts and greater growth rates by late summer 2003 than plots treated during the growing season.

For three additional species, use of fire as a control method is ineffective or not well-known. The use of fire as a control measure for purple loosestrife has been largely dismissed as ineffective. Attempts to burn residual biomass following cutting or herbicide treatments may merely result in recruitment of purple loosestrife seedlings where burning exposes soil containing a substantial seed bank (Munger 2002d). Burning Japanese knotweed when it is actively growing is not recommended as an effective control method according to a control manual published in England (Child and Wade 2000). There is no information in the literature regarding the use of fire to control mile-a-minute. Control efforts should focus on eliminating or reducing seed output, especially near waterways, and avoiding disturbance and the creation of gaps in existing vegetation (Okay 2005).

Emerging Issues in the Northeast Bioregion

As nonnative invasive plants continue to spread into previously uninvaded areas and managers gain experience with their control, questions and concerns about the relationship of invasive species to fire will also change. Some of the following matters are under active discussion in the region:

Fuel Properties of Invaded Northeastern Plant Communities and Influences on Fire Regimes

Ducey (2003) pointed out the inadequacy of fuels information specific to the Northeast, especially regarding heat content of dead fuels. Currently fuel models must be extrapolated from models developed in western vegetation types. The Photo Series for the Northeast (http://depts.washington.edu/nwfire/dps/) will be an important resource but will not focus on invaded fuel beds. Dibble and others (2007) assessed the relative flammability of native versus nonnative fuels for 42 species, but more research is needed.

Fuel accumulations that may exceed reference conditions in forested areas have resulted from fire exclusion, extensive mortality of dominant tree species, severe weather events such as the region-wide ice storm in January 1998, and encroachment by nonnative invasive plants. In some locations in the Northeast, nonnative invasive grasses form a more continuous fine surface fuel layer than occurred in nearby uninvaded conditions (Dibble and Rees 2005). At other locations, invasive vines have become common and can act as ladder fuels. However, it is not known if these changes in fuel bed characteristics will result in an increase in fire size, frequency, and/or severity.

Vulnerability of Forest Gaps to Invasion

Just as millions of American chestnut trees succumbed in the 1900s to chestnut blight (*Cryphonectria parasitica*), so we are likely to see continued tree mortality in the northeastern bioregion associated with insects and diseases. Agents of tree mortality include:

- Hemlock wooly adelgid (*Adelges tsugae*)
- Gypsy moth (*Lymantria dispar*)
- Sudden oak death (*Phytophthora ramorum*)
- White pine blister rust (*Cronartium ribicola*)
- Balsam wooly adelgid (*Adelges piceae*)
- Dutch elm disease (*Ophiostoma ulmi*)
- Spruce budworm (*Choristoneura fumiferana*)
- Asian long-horned beetle (*Anoplophora glabripennis*)
- European wood wasp (*Sirex noctilio*)
- Emerald ash borer (*Agrilus planipennis*)

Of these, only spruce budworm is native to North America. The impact of high tree mortality probably exceeds the impact of fire in promoting invasive plants in this bioregion. Additionally, the effects of salvage operations following insect kill, and timber harvest in general, may introduce and promote invasive species. In salvage operations, log yards and skid trails are often the sites and corridors for new infestations of nonnative invasive plants. Tree-of-heaven has invaded harvested stands in Virginia (Call and Nilsen 2003, 2005) and West Virginia (Marsh and others 2005), and Japanese stiltgrass established after timber harvest in eastern Tennessee (Cole and Weltzin 2004). Openings in infested stands might be invaded by nonnative honeysuckles, Oriental bittersweet, Japanese barberry, tree-of-heaven, invasive grasses, or other nonnative plants. The presence of these species could alter fuelbed structure and possibly biomass and seasonal drying patterns.

Global Climate Change

Population expansions by nonnative plants in the Northeast are likely to be facilitated by a warming climate, which is expected to continue to increase the frequency and intensity of disturbances and thus opportunities for invasion. Recent models (Adger and others 2007; Intergovernmental Panel on Climate Change 2001) indicate that climate change will reduce snowfall and alter streamflow in eastern forests. These effects would be accompanied by greater uncertainty in weather. Ice storms, hurricanes, and episodes of drought are expected to increase in frequency, intensity, or duration. These events may result in more frequent wildfire, accompanied by increases in nonnative invasives favored by fire.

Interactions Between Nonnative Invasive Plants, Fire, and Animals

Information is needed on changes in wildlife nutrition that come about when nonnative plants are burned in the Northeast bioregion. Lyon and others (2000b) reviewed changes in nutritional content of wildlife foods when vegetation is burned, but they focused on native plant communities.

Invertebrate species may affect the relationships between fire and plant communities. In New Jersey hardwoods (oaks, yellow poplar, maple), areas invaded by Japanese barberry and Japanese stiltgrass differed from uninvaded areas not only in plant composition and structure but also in forest floor properties. Invaded areas had higher pH, thinner litter and organic layers (Kourtev and others 1998), and higher nitrate concentrations accompanied by greater nonnative earthworm density (Kourtev and others 1999). Earthworms are an important wildlife food (for example, for American robin, and woodcock) and their abundance could lead to altered behavior and habitat use. Because they consume the litter layer, they may influence the potential for surface fires.

Conclusions

The highly fragmented landscape, proximity of the wildland urban interface, and large number of nonnative species that occur in the Northeast bioregion complicate land management decisions, including fire and fuel management. A relative lack of peer-reviewed literature on the relationships between fire and invasive plants for this bioregion further challenges the manager to make informed decisions. Managers must consider the possibility of nonnative species establishing or spreading after wild or prescribed fire. Ideally, monitoring for invasive species and far-sighted mitigation will be included in their fire management plans. Available information suggests that some nonnative invasive plants have potential to alter fuel characteristics and that these differ from reference conditions (for example, Dibble and Rees 2005).

When planning prescribed fire with the objective of controlling invasives, fire impacts on all species must be considered and efforts made to prescribe a fire or fire regime that will favor native vegetation over invasive plants. Use of fire to control invasives in a plant community where the fire is outside reference conditions could produce undesired effects on the native community. Additional considerations for the use of prescribed fire for controlling invasive plants in the Northeast bioregion include:

1. It is important to prioritize safety and compliance with air quality and other regulations within the wildland-urban interface.
2. Cooperation among adjoining landowners is key.
3. Multiple control methods and repeated treatments are likely to be needed to reduce most invasive plant populations.

4. Implemention of high-quality, long-term monitoring, archiving of data, and information sharing are essential components of a successful control project.

Resources Useful to Managers in the Northeast Bioregion

Because peer-reviewed literature is limited regarding the relationship between fire and nonnative invasive plants in the Northeast bioregion, information sharing by managers can be especially effective:

- A listserve maintained by the Mid-Altantic Exotic Pest Plant Council enables managers to relate their successes and failures using control treatments, including prescribed burning (www.ma-eppc.org).
- Spread of nonnative invasive plants in six New England states is tracked by county in an online atlas (Invasive Plant Atlas of New England, http://invasives.uconn.edu/ipane/), based on herbarium specimens. Information on weed control is also included (Mehrhoff and others 2003).
- The Virginia Native Plant Society offers fact sheets about nonnative invasive plants at http://www.dcr.state.va.us/dnh/invlist.htm. These fact sheets cover use of prescribed fire as a management tool, though fire effects are rarely noted and few references are given.
- Rapid assessment reference condition models are available for several "potential natural vegetation groups" in the Northeast bioregion through the LANDFIRE website (http://www.landfire.gov). Model descriptions can be downloaded and compared to existing conditions. This can aid in estimating fuel loads and fire regime characteristics that are desirable in restoration projects and hazard fuels management.

Notes

Randall Stocker
Karen V. S. Hupp

Chapter 6:

Fire and Nonnative Invasive Plants in the Southeast Bioregion

Introduction

This chapter identifies major concerns about fire and nonnative invasive plants in the Southeast bioregion. The geographic area covered by this chapter includes the entire States of Louisiana, Mississippi, and Florida; all except the northernmost portions of Delaware and Maryland; the foothill and coastal ecosystems of Virginia, North Carolina, South Carolina, Georgia, and Alabama; and the lower elevation plant communities of Arkansas, southeastern Missouri, southeastern Oklahoma, southwestern Tennessee, and eastern Texas. This area coincides with common designations of the Atlantic Coastal Plain and the Piedmont (the plateau region between the Atlantic and Gulf of Mexico Coastal Plain and the Appalachian Mountains). Soils are generally moist year-round, with permanent ponds, lakes, rivers, streams, bogs, and other wetlands. Elevations vary from 2,407 feet (734 m) on Cheaha Mountain, Alabama, to –8 feet (–2.4 m) in New Orleans, Louisiana (USGS 2001). Westerly winds bring winter precipitation to the bioregion, and tropical air from the Gulf of Mexico, Atlantic Ocean, and Caribbean Sea brings summer moisture. Southward through this region the contribution of winter rainfall decreases, as does the frequency of freezing temperatures. Tropical conditions occur at the southern tip of Florida. The percentage of evergreen species and palms (*Serenoa* spp., *Sabal* spp.) increases along this climate gradient (Daubenmire 1978).

Plant communities within this portion of the temperate mesophytic forest are complex and subject to a long history of natural and anthropogenic disturbance. Various methods have been used to estimate the dominant presettlement forest types. Plummer (1975) reported that pine (*Pinus* spp.) and post oak (*Quercus stellata*) were the dominant trees on historical survey corner tree lists in the Georgia Piedmont, and Nelson (1957) used soil type to estimate that 40 percent of the Piedmont was dominated by hardwood species, 45 percent was in mixed hardwood and pine stands, and 15 percent was predominantly pine. On the southeastern Coastal Plain, pine savannas may have covered between two-thirds and three-fourths of the area (Platt 1999).

Currently, forests include a mosaic of mostly deciduous angiosperms that form a dense canopy of tall trees with a "diffuse" layer of shorter, shade-tolerant trees, interspersed with disturbance- (mostly fire-) derived pine stands (Daubenmire 1978). Vines are common and frequently include native grape (*Vitis* spp.) species. Large streams often have extensive floodplains and oxbows, and areas where the water table occurs at or near the surface year-round usually support stands of bald cypress (*Taxodium distichum*). Coastal dunes are often populated by American beachgrass (*Ammophila breviligulata*) from North Carolina northward and by sea-oats (*Uniola paniculata*) throughout the region. Salt water-influenced wetlands occur landward of coastal dunes and are dominated by a variety of species including inland saltgrass (*Distichlis spicata*), needlegrass rush (*Juncus roemerianus*), smooth cordgrass (*Spartina alterniflora*), and saltmeadow cordgrass (*Spartina patens*) (Daubenmire 1978).

While nonnative plants can be found throughout this region, the highest proportion of nonnative plants is found in southern Florida (Ewel 1986; Long 1974). Prior to this century's increase in transport and trade, the southern Florida peninsula had geographical and geological barriers to plant species introductions from the north, and surrounding waters provided barriers to tropical species introductions. More recently, southern Florida has become especially vulnerable to nonnative plant invasions because of a large number of temperate and tropical species introductions for horticulture (Gordon and Thomas 1997), the proximity of the introduction pathways to potentially invasible habitats, and the relatively depauperate native flora (Schmitz and others 1997). Additional human-caused changes in hydrology, fire regime, and salinity have combined to increase the vulnerability of the vast low elevation freshwater wetlands south of Lake Okeechobee (Hofstetter 1991; Myers 1983).

Discussion of fire and nonnative plant interactions is complicated by the limited number of experimental field studies, the lack of a complete understanding of presettlement fire regimes, and the unknown effects of increasing atmospheric carbon dioxide and nitrogen and other aspects of climate change (Archer and others 2001). Complications notwithstanding, a better comprehension of the factors and forces at work is critical to developing fire and other habitat management practices that provide a more effective means of achieving ecological and societal objectives (D'Antonio 2000).

Fire in the Southeast Bioregion

Naturally occurring fires are, and were, common in this region (for example, Chapman 1932; Harper 1927; Komarek 1964; Platt 1999; Stanturf and others 2002). The Southeast includes locations with some of the highest lightning incidence levels on earth. Six of the eight highest lightning-strike rates in the United States are found in the southeastern region (Tampa-Orlando, Florida; Texarkana, Arkansas; Palestine, Texas; Mobile, Alabama; Northern Gulf of Mexico; and Gulf Stream-East Carolinas).

Little is known about "natural" or prehistoric fire regimes in the Southeast. During the interval between the retreat of the ice 18,000 years ago and the initial influence of Native Americans beginning around 14,000 years ago, variations in soil moisture, lightning strikes, fuel accumulation, and disturbance history likely resulted in a wide range of fire-return intervals, from as short as one year to as long as centuries. Similarly, fire severities probably ranged from minor fires in the understory to stand-replacement events (Stanturf and others 2002).

Fire frequency and severity were important factors in the evolution of southeastern plant communities (Komarek 1964, 1974; Platt 1999; Pyne 1982a; Pyne and others 1984; Snyder 1991; Van Lear and Harlow 2001; Williams 1989), and fire contributes to the high diversity of communities such as pine and shrub (pineland) communities of south Florida (Snyder 1991) and pine savannas (Platt 1999). Estimates of presettlement fire regime characteristics are summarized in reviews by Wade and others (2000) and Myers (2000) for major vegetation types in the Southeast bioregion.

Substantial evidence from many disciplines supports the contention that fire was widespread prior to European arrival (Stanturf and others 2002). Fires induced by native peoples created and maintained open woodlands, savannas, and prairies (McCleery 1993; Williams 1989), and kept forests in early successional plant communities. Native peoples often burned up to twice a year and extended the fire season beyond summer lightning-induced fires (Van Lear and Harlow 2001).

After adopting the practices and utilizing the clearings made by native people, European settlers influenced fire patterns and plant communities by expanding areas of agricultural clearing and repeated burning (Brender and Merrick 1950; Stoddard 1962; Williams 1992), maintaining permanent fields (Stanturf and others 2002), introducing large herds of hogs and cattle (McWhiney 1988; Stanturf and others 2002; Williams 1992), and heavily logging coastal pine forests, bald cypress, and bottomland hardwood stands (Stanturf and others 2002; Williams 1989). Frequent anthropogenic burning, in combination with grazing cattle and feral pigs, eliminated regeneration of pine and other woody species in large areas (Brender and Merrick 1950; Frost 1993).

Subsequent land and fire management practices and policies oscillated between periods of controlled burning and fire exclusion, and varied from place to place (Brueckheimer 1979; Johnson and Hale 2000; Paisley 1968; Stoddard 1931). This range of fire practices was not the result of carefully planned and organized management strategies but instead was a reaction to political and social influences at a variety of geographical scales, local to regional. Little scientific information was available, especially in the early years, to inform the ongoing debate over fire exclusion and controlled burning (Frost 1993). Intentional burning practices rarely attempted to mimic presettlement fire conditions (Doren and others 1993; Drewa and others 2002; Platt 1999; Platt and Peet 1998; Slocum and others 2003) but were conducted mainly for agriculture and land clearing.

Contemporary objectives of controlled burns in the Southeast bioregion include hazard fuels reduction, wildlife habitat improvement, and range management (Wade and others 2000). Increasing numbers of acres are being burned for ecosystem restoration and maintenance (Stanturf and others 2002) and to sustain populations of rare and endangered plants (Hessl and Spackman 1995, review; Kaye and others 2001; Lesica 1996). Contemporary fire management practices often strive to recreate presettlement fire regimes, assuming that this will promote maximum diversity (Good 1981; Roberts and Gilliam 1995). Because presettlement fire regimes are not always well understood, however, it is difficult to design a fire management program to meet this objective (Slocum and others 2003).

The negative consequences of past fire management practices have been interpreted as an "ecological disaster" (Brenner and Wade 2003). Exclusion of fire from southeastern pine savannas, for instance, has been blamed for loss of fire-adapted, species-rich herbaceous ground cover and subsequent increase in less fire-tolerant native and nonnative woody species (DeCoster and others 1999; Heyward 1939; Platt 1999; Slocum and others 2003; Streng and others 1993; Walker and Peet 1983).

Fire and Invasive Plants in the Southeast Bioregion

Fire can contribute to the establishment and spread of nonnative invasive plants under some circumstances (Mack and D'Antonio 1998). Melaleuca (*Melaleuca quinquenervia*), for instance, invades fire-cleared mineral soils in south Florida (Myers 1975).

Fire exclusion has also been blamed for reducing native species in favor of nonnatives in fire-adapted communities. For example, Chinese tallow (*Triadica sebifera*) invades fresh marshes (Grace 1999), and Brazilian pepper (*Schinus terebinthifolius*) invades subtropical pine habitats (Myers 2000) in the absence of fire. Exclusion of fire from longleaf pine (*Pinus palustris*) communities generally results in woody species overtopping herbs, thicker duff layers, and changes in nutrient availability that "all favor extrinsic species at the expense of endemic residents" (Wade and others 2000, page 66).

Nonnative plant invasions can affect fuel and fire characteristics in invaded communities (Brooks and others 2004; Chapter 3) and may subsequently reduce native plant density and diversity. Altered fuel characteristics associated with some invasive species may result in fires that kill native plants but not fire-resistant invasive species (Drake 1990; Pimm 1984). For example, cogongrass (*Imperata cylindrica*) invasions in Florida sandhills increase biomass, horizontal continuity, and vertical distribution of fine fuels, compared to uninvaded pine savanna. Fires in stands invaded by cogongrass have higher maximum temperatures than fire in uninvaded stands (Lippincott 2000) and may therefore cause greater mortality in native species than fires fueled by native species. Melaleuca invasion can alter the vertical distribution of fuels such that communities that typically experienced low-severity surface fires have a greater incidence of crown fire in invaded communities (Myers 2000). Conversely, Brazilian pepper and Chinese tallow develop dense stands that suppress native understory grasses, resulting in lower fine fuel loads than the fire-maintained plant communities being replaced (Doren and others 1991; Grace and others 2001). Lower fuel loads may lead to reduced fire frequency and lower fire severity, which may favor the fire sensitive seedling stages of the invasives (Mack and D'Antonio 1998).

Controlled burning is sometimes used in an effort to manage invasive plants in the Southeast. However, D'Antonio's review (2000) suggests that fire-versus-invasives results are highly variable and depend on fire intensity, time of burning (Hastings and DiTomaso 1996; Parsons and Stohlgren 1989; Willson and Stubbendieck 1997), weather, and the status of the remaining seed bank (Lunt 1990; Parsons and Stohlgren 1989). It is also important to note that, while dormant season fires have been recommended to control invasive shrubs in grasslands, they may result in increases in nonnatives (Richburg and others 2001, review). It has been recommended that, if fire is used to reduce populations of nonnative invasive plants, burning should be timed to reduce flowers and/or seed production, or at the young seedling/sapling stage (chapter 4). Spot-burning very small populations of invasive plants has also been recommended as "cheaper and easier than implementing a prescribed fire" (Tu and others 2001).

The limited number of replicated, long-term, field experiments on fire and invasive plants reflects the very difficult nature of conducting the needed studies. Even where detailed measurements of the effects of fire on native and nonnative plant species have been collected, the studies are typically short-term and do not necessarily reflect longer-term changes (Freckleton 2004; Freckleton and Watkinson 2001). More research is needed over longer periods of time to better understand the relationships between fire and invasive species in the Southeast bioregion.

The remainder of this chapter presents information on the known relationships of fire and invasive plant species for five major plant habitats: wet grassland, pine and pine savanna, oak-hickory (*Quercus – Carya*) woodland, tropical hardwood forest, and a brief treatment of cypress (*Taxodium distichum*) swamp (fig. 6-1). For each habitat except cypress swamp, a summary is provided of the role of fire and fire exclusion in promoting invasions by nonnative plant species, fire regimes changed by plant invasions, and use of fire to manage invasive plants, with an emphasis on those species included in table 6-1. The final section presents general conclusions and emerging issues relating to fire and invasive species management in the Southeast bioregion. All parts of the Southeast have been and continue to be affected by management practices including fire exclusion, controlled burning, or both (Brenner and Wade 2003; Freckleton 2004). Therefore, we have made no attempt to make a distinction between "more managed" (for example, pine plantations) and "less managed" (for example, conservation areas) ecosystems in this section.

Wet Grassland Habitat

Background

The term "wet grasslands" is used here to include the Everglades region of southern Florida, grassland savannas with cabbage palm (*Sabal palmetto*) and cypress in Florida (Küchler's (1964) palmetto prairie and cypress savanna, respectively), and the coastal grass-dominated wetlands from Virginia to Texas (Küchler's (1964) northern and southern cordgrass (*Spartina* spp.) prairie). Native species in these wetlands include smooth cordgrass dominating tidally flushed saltmarshes; smooth and gulf cordgrass (*Spartina spartinae*), needlegrass rush, pickleweed (*Salicornia spp.*), inland saltgrass, saltmeadow cordgrass, and saltmeadow rush (*Juncus gerardii*) in less frequently flooded more inland marshes. Aquatic species in fresh marshes include pond-lily (*Nuphar* spp.), waterlily (*Nymphaea* spp.), wild rice (*Zizania aquatica*), cutgrass (*Zizaniopsis miliacea*), pickerelweed (*Pontedaria cordata*), arrowhead (*Sagittaria* spp.), cattail (*Typha* spp.), maidencane (*Panicum hemitomon*), spikerush (*Eleocharis* spp.), and sedges (*Carex* spp.) (Wade and others 2000).

While wet grassland communities border many different habitats, some of the smallest non-graminoid dominated types include 1-to-several acre (0.5-to-several hectare) hardwood forest sites found within marsh, prairie, or savanna in southern Florida. Locally termed "tree islands," fire in these locations is usually driven by processes in the adjacent plant community, and special features related to the spread of fire into the tree islands from adjacent wet grasslands are discussed in the "Tropical Hardwood forest" section of this chapter.

Wet grassland plant communities tend to be flammable and are adapted to an environment of frequent wet season (summer) fires (Leenhouts 1982; Schmalzer and others 1991; Wade 1988; Wade and others 1980). The following information on fire regimes in these plant communities comes from Wade and others (2000) and Myers (2000) (see these reviews for more detail). Presettlement fire regimes in wet grasslands in much of the coastal region in the Southeast are classified as stand-replacement types with 1- to 10-year return intervals (Myers 2000; Wade and others 2000).

Fire behavior differs among wet grassland types due to differences in flammability of dominant species, which vary with groundwater levels and salinity. Cordgrass communities in coastal salt marshes tend to be quite flammable, with green tissues of saltmeadow cordgrass and gulf cordgrass capable of burning several times during a growing season. Fire in these communities will carry over standing water. Flammability in fresh and brackish marshes is more variable due to considerable plant diversity. Grass dominated stands generally experience more intense and continuous fires than forb and sedge dominated stands with important exceptions, including cattail and sawgrass stands (*Cladium jamaicense*) (Myers 2000; Wade and others 2000).

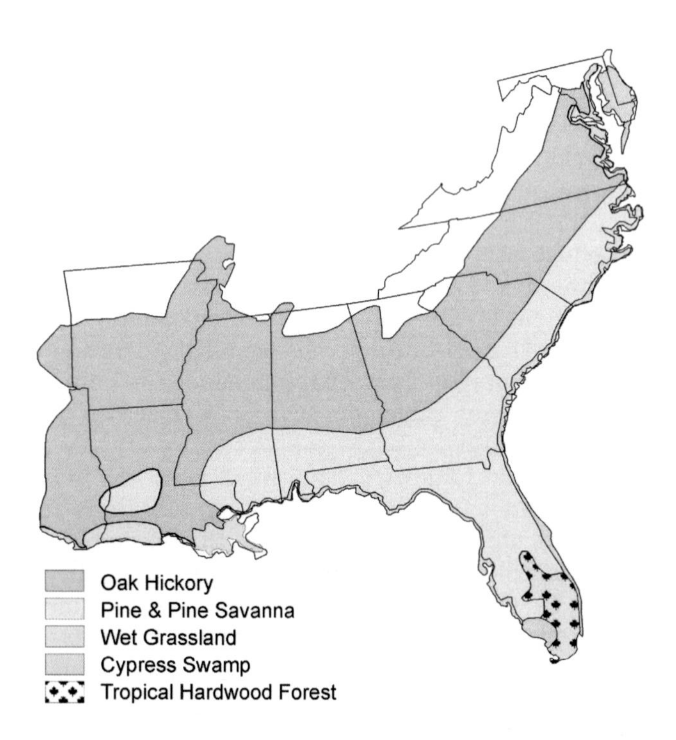

Figure 6-1—Approximate distribution of major plant habitats in the Southeast bioregion. Upland grasslands and palmetto prairie are not identified here.

Table 6-1—Major plant habitats in the Southeast bioregion and perceived threat potential of several important nonnative plants in each habitat (L= low threat, H = high threat, P = potentially high threat, N= not invasive, U = unknown). Designations are approximations based on available literature.

		Grassland			Forest and woodland			
Scientific name	Common name	Wet grassland	Upland grassland	Palmetto prairie	Pine and pine savanna	Oak-hickory woodland	Tropical hardwood forest	Cypress swamp
Ailanthus altissima	Tree-of-heaven	N	N	N	L	H	N	N
Albizia julibrissin	Mimosa	N	N	N	U	P	N	N
Alliaria petiolata	Garlic mustard	N	N	N	P	P	N	N
Dioscorea alata	Water yam	N	N	N	L	P	P	L
Dioscorea bulbifera	Air potato	N	N	N	H	H	H	H
Dioscorea oppositifolia	Chinese yam	U	U	N	U	P	U	U
Elaeagnus pungens	Thorny-olive	N	N	N	U	P	N	N
Elaeagnus umbellata	Autumn-olive	U	H	N	U	U	N	N
Euonymus alatus	Winged euonymus	U	U	N	U	P	N	N
Euonymus fortunei	Winter creeper	U	U	N	U	P	N	N
Hedera helix	English ivy	N	N	N	U	P	U	U
Imperata cylindrica	Cogongrass	N	L	U	H	L	N	N
Lespedeza bicolor	Lespedeza	N	H	N	L	L	N	N
Lespedeza cuneata	Sericea lespedeza	N	L	L	L	L	L	N
Ligustrum spp.	Privet species	U	N	U	L	H	N	N
Lolium arundinaceum	Tall fescue	U	H	N	U	N	N	N
Lonicera japonica	Japanese honeysuckle	N	H	N	H	H	N	N
Lonicera spp.	Bush honeysuckles	N	H	N	U	L	U	N
Lygodium japonicum	Japanese climbing fern	U	U	U	H	P	U	U
Lygodium microphyllum	Old World climbing fern	H	N	H	H	N	H	H
Melaleuca quinquenervia	Melaleuca	H	N	H	H	N	H	H
Melia azedarach	Chinaberry	H	H	H	U	P	U	N
Microstegium vimineum	Japanese stiltgrass	P	H	P	U	H	U	U
Paspalum notatum	Bahia grass	U	L	U	L	L	N	N
Pueraria montana var. lobata	kudzu	U	U	U	H	H	U	U
Schinus terebinthifolius	Brazilian pepper	H	P	H	H	N	P	P
Solanum viarum	tropical soda apple	U	H	U	P	U	N	N
Triadica sebifera	Chinese tallow	H	H	H	P	P	U	U

Postfire succession patterns in wet grasslands are influenced by season of burning and the interplay of hydroperiod and fuels, which together determine whether fires are lethal or nonlethal to the dominant species. Hydroperiod factors that influence the effect of fire on belowground plant parts and substrate include proximity to the water table, tidal conditions, and drought cycles. In most cases, the aboveground vegetation is consumed by fire, and the dominant species that make up the fuel sprout from underground buds, tubers, or rhizomes after fire. Peat fires and postfire flooding are two disturbance events that can kill both above- and belowground organs of existing vegetation. Severe peat fires can occur in organic substrates when severe drought coincides with low water table levels. Vegetation can also be killed by water overtopping recovering vegetation after a fire. Successional species would be expected to be primarily those represented in the seed bank (Myers 2000; Wade and others 2000).

In some areas, native wet grassland communities have been altered by fire exclusion, allowing invasion of woody species. Fire is now being reintroduced in many areas to restore native species compositions (for example, see Leenhouts 1982).

Role of Fire and Fire Exclusion in Promoting Nonnative Plant Invasions in Wet Grasslands

Fire exclusion in wet grasslands during the past century has resulted in less frequent but more severe fires than occurred prior to European settlement. These fires have opened up wet grasslands to invasion by nonnative plant species (Bruce and others 1995).

Melaleuca is well adapted to survive fire and to establish and spread in the postfire environment. There may not be many better fire-adapted tree species in the world than melaleuca. Nicknamed the "Australian fireproof tree" (Meskimen 1962), its native habitats include fire-shaped ecosystems in Australia (Stocker and Mott 1981). It produces serotinous capsules (le Maitre and Midgley 1992) that release as many as 20 million seeds per tree (Woodall 1981) following fire. Complete capsule dehiscence can occur as quickly as a few days following a crown fire (Woodall 1983). Seedlings establish on fire-cleared mineral soil and can survive fire within a few weeks or months after germination (Meskimen 1962; Myers 1975, 1983). Melaleuca sprouts from roots and from epicormic trunk buds to resume growth after fire. Although the outer bark can easily burn (fig. 6-2), the trunk wood is protected from fire damage by spongy inner bark that is saturated with water (Turner and others 1998).

Fire is not necessary for melaleuca establishment (Woodall 1981); in fact, melaleuca spreads readily with changes in hydrology and mechanical damage to habitats (Cost and Carver 1981). Nevertheless, its ability to capitalize on burned areas is remarkable (Hofstetter 1991; Myers 1983). "If melaleuca were managed as a desired species, prescribed fire would be the single most important tool available to the resource manager" (Wade 1981).

Melaleuca seedlings that establish after fire at the height of the dry season may have 5 to 8 months' advantage over native tree species such as south Florida slash pine (*Pinus elliottii* var. *densa*) and bald cypress, which release seed during the wet season (Wade 1981). Other species that may establish and/or spread following fire in wet grasslands include climbing ferns (*Lygodium* spp.) and the shrub chinaberry (*Melia azedarach*). According to a review by Ferriter (2001), Old World climbing fern (*Lygodium microphyllum*) occurs in sawgrass marsh in southern Florida and may spread following fire. No published information was found related to the response of chinaberry to fire in the Southeast; however, it reproduces vegetatively from both stumps and roots following fire in Argentina (Menvielle and Scopel 1999; Tourn and others 1999).

In areas where fire has been excluded from wet grassland communities, native species are often

Figure 6-2—Melaleuca's papery trunk. (Photo by Forest & Kim Starr, United States Geological Survey, Bugwood.org.)

replaced by a dense woody overstory composed of nonnative invasive trees and/or shrubs (for example, melaleuca, Chinese tallow, Brazilian pepper and chinaberry). When this occurs, native species numbers and diversity are dramatically reduced (for example, Bruce and others 1995). It has been suggested that fire exclusion can contribute to the invasion of Chinese tallow into coastal prairies (Bruce and others 1995; chapter 7; D'Antonio 2000). Similarly, reduced fire frequency due to human-induced changes to hydrology is blamed for invasion of Brazilian pepper into sand cordgrass (*Spartina bakeri*) and black rush (*Juncus roemerianus*) dominated salt marshes in Florida's Indian River Lagoon (Schmalzer 1995).

The relationship between fire and fire exclusion and the increase in Brazilian pepper in south Florida wetlands is not well understood. Brazilian pepper is not a fire-adapted species (Smith, C. 1985) and is generally kept out by fire in adjacent pinelands (Loope and Dunevitz 1981). The response to fire in wetlands may not be the same as in pinelands, although the field studies have not been conducted on "typical" wetland plant communities. The limited published research includes assessment of the effect of repeated fire (generally every two years) on experimental plots in highly altered "rock-plowed" limestone substrate. Much of this formerly agricultural area was originally sawgrass marsh, although many woody species invaded the rock-plowed portions after the fields were abandoned. Fire exclusion (control plots in this study) resulted in increased Brazilian pepper stem density, but Brazilian pepper stem density also increased in burned plots (Doren and others 1991). The effects of fire were related to the size of the Brazilian pepper plant. Smaller, apparently younger plants were badly damaged or killed by fire, while larger plants either recovered completely or did not burn. The conclusion of the authors was that Brazilian pepper invasion progressed with or without fire, and that fire is not an appropriate management tool for this unique area (Doren and others 1991). Dry season wildfires are thought to contribute to increasing Brazilian pepper populations in "tree islands," which are typically tree-dominated areas within the wet grasslands of southern Florida (Ferriter 1997, FLEPPC review) (see "Tropical Hardwood Forest Habitat" section page 107).

Chinaberry may also invade wet grasslands where fire has been excluded. This shrub occurs primarily in disturbed areas but is also said to invade relatively undisturbed floodplain hammocks, marshes, and upland woods in Florida (Batcher 2000b, TNC review), although no additional information is available. In Texas, riparian woodlands and upland grasslands have also been extensively invaded by chinaberry (Randall and Rice unpublished, cited in Batcher 2000b).

Tall fescue (*Lolium arundinaceum*) is found on disturbed upland grassland sites or where "the natural fire regime has been suppressed (Eidson 1997)" (as cited in Batcher 2004, TNC review).

Effects of Plant Invasions on Fuel and Fire Regime Characteristics in Wet Grasslands

Several nonnative invasive plants are thought to change fire regimes in wet grasslands in the Southeast bioregion by changing the quantity and/or quality of fuels in invaded communities. Changes in fuels may subsequently reduce or increase fire frequency and severity. Examples of both cases are evident in wet grassland plant communities.

Observational and limited experimental evidence suggests that invasive hardwoods such as Chinese privet (*Ligustrum sinense*), Chinese tallow, and Brazilian pepper shade out and/or replace native plant species in southeastern marshes and prairies such that fine fuel loads and horizontal continuity are reduced (for example, Doren and Whiteaker 1990; Doren and others 1991; Grace 1999; Platt and Stanton 2003). When this occurs, fire frequency and intensity may be reduced and fire patchiness increased. However, there is little experimental evidence to support these conjectures, and more research is needed to better understand the implications of these vegetation changes on fire regimes.

Melaleuca invasion can have variable effects on fuels and fire behavior. Large amounts of litter under melaleuca stands (Gordon 1998) promote intense and severe fires (Flowers 1991; Timmer and Teague 1991) that are difficult to control and have large potential for economic damage, loss of human life and property, and negative ecological consequences (Flowers 1991; Schmitz and Hofstetter 1999, FLEPPC review; Wade 1981). These high intensity fires promote melaleuca establishment and spread, and reduce cover of native species. Severe fire also removes the outer, highly-flammable melaleuca bark layers, thus reducing the probability of damage to mature melaleuca from subsequent fires (Wade 1981). Thus, a positive feedback loop of fire-promoting-melaleuca and melaleuca-promoting-fire is created (Hofstetter 1991; Morton 1962). Conversely, intense fires fueled by melaleuca may reduce the chances of subsequent fires when organic soils are consumed and the elevation of the soil surface lowered (Schmitz and Hofstetter 1999). Small changes in water level can then theoretically reduce the likelihood of fire by flooding these formerly unflooded sites.

Old World climbing fern alters plant community fuel structure in wet grasslands and associated tree islands with extensive, dry-standing frond "skirts" that create ladder fuels that facilitate fire spread into tree canopies (Ferriter 2001, FLEPPC review) (see section on "Tropical Hardwood Forest Habitat" page 107). Roberts (D. 1996) also reports that fire penetrates into wet grasslands from the margins of forested communities where Old World climbing fern has invaded and provides a novel source of additional fuel (see section on "Pine and Pine Savanna Habitat" page 100). Fire spread may also be promoted by pieces of burning fern frond blowing aloft into grasslands from tree islands (Roberts, D. 1996).

While the potential exists for invasive plant species to influence abiotic factors that affect fire behavior, including water table elevation and surface hydrology, this relationship has not yet been shown to be important in wet grasslands in the Southeast bioregion. It is logical to assume, for example, that invasive species that lower the water table through evapotranspiration could reduce soil moisture and thus affect subsequent fire characteristics. It has been suggested that melaleuca increases the amount of water lost to the environment in sites with standing water by adding its evapotranspiration to the water surface evaporation (for example, Gordon 1998; Schmitz and Hofstetter 1999; Versfeld and van Wilgen 1986; Vitousek 1986). In the southwestern United States, transpiration of dense stands of nonnative tamarisk (*Tamarix* spp.) can result in the loss of large quantities of water on sites where the water table is just below the soil surface (Sala and others 1996). In many wet grasslands in the southeastern United States, however, the water table is at or above the soil surface, and evaporation and evapotranspiration are both driven and limited by solar energy. Simply adding another species to the system does not increase the available energy and therefore does not increase the amount of water lost to the atmosphere, although it may increase the available evaporative surface (Allen and others 1997).

Other invasive species-induced changes in hydrology may have some effect on wet grasslands in the Southeast. The thick (over 1 m) rachis mat formed by decades of Old World climbing fern growth may have diverted shallow stream meandering of the Loxahatchee River in east-central Florida by a distance of about 164 feet (50 m) (R. Stocker, personal observation, fall 1997). At this scale only very small portions of the invaded habitat would be affected. Additional study of the relationships among invasive plants and abiotic factors that affect fire regime is warranted.

Use of Fire to Manage Invasive Plants in Wet Grasslands

Controlled burning has been used extensively to manage invasive plants in the Southeast bioregion, with varied results. Only a small portion of the literature describes research on the use of fire to control invasives in wet grasslands.

Many wet grassland sites are on organic soils, and fires occurring when the organic surface soil is dry can consume the peat and affect the type of vegetation that subsequently develops on the site (Ferriter 2001; Myers 2000; Schmitz and Hofstetter 1999). Therefore, it may be possible to prescribe fires that could substantially damage plant roots in wet grasslands during dry periods (Nyman and Chabreck 1995). Documented success using such burns to control invasive nonnative plant species, however, is lacking (Wade and others 2000), and care must be taken to avoid substantial damage to desirable species.

Frequent fires in wet grasslands during historic and prehistoric times are thought to have maintained grasslands with very little woody vegetation (Schmalzer 1995). It follows that prescribed fires with a frequency and seasonality within the reference range of variation experienced in these habitats might favor native wet grassland species over nonnative woody species. Controlled burning following flooding or plant flowering has been suggested as particularly effective in reducing "unwanted woody vegetation" in salt marshes of the St. Johns National Wildlife Refuge (Leenhouts 1982). Fire is not, however, effective for controlling all invasive species in wet grasslands, some of which are well adapted to frequent fires.

Controlled burning has been promoted as a means to reduce woody vegetation in salt marshes (Leenhouts 1982) but has not been effective in controlling melaleuca (Belles and others 1999, FLEPPC review) and has provided mixed results for Chinese tallow (Grace 1999; Grace and others 2001) and chinaberry (Tourn and others 1999).

Controlled burning alone is not effective at controlling melaleuca (Wade 1981) and will not eliminate mature stands (Belles and others 1999). Fires timed to consume seedlings after most germination has occurred have the best potential to control melaleuca (Woodall 1981). Fire can kill melaleuca seedlings less than 6 months old (Belles and others 1999); however, it is difficult to achieve the needed degree of soil surface dryness and fuel load to carry a fire severe enough to prevent postfire sprouting. Melaleuca seedlings less than 1 year old may sprout from root collars after fire damage (Myers 1984). Susceptibility to fire-induced mortality is reduced as seedlings and saplings grow taller, with more than 50 percent of 1.5-foot-(0.5 m) tall saplings surviving in one study (Myers and others 2001). Because melaleuca seeds are able to survive in flooded organic soils for about 1.5 years and in unflooded sandy soils for 2 to 2.3 years (Van and others 2005), postfire establishment from the soil seed bank is also a concern.

Some resource managers maintain that melaleuca control can be achieved with proper timing of prescribed burns (Belles and others 1999; Maffei 1991; Molnar and others 1991; Pernas and Snyder 1999), but the success of this approach depends on postfire rainfall, which often does not follow anticipated patterns. Two seasonal windows of opportunity may exist, depending on rainfall patterns (Belles and others 1999). (1) Burning during the late wet season, when surface soils are likely to be moist but not flooded, would encourage melaleuca seed germination just prior to soil dry-down during the dry season. With average dry-season rainfall, melaleuca seedlings are likely to die before the wet season returns the following May or June. If dry-season rainfall is above average, however, melaleuca recruitment is likely to be high. (2) Burning at the beginning of the wet season also encourages seed germination, and normal rainfall patterns might provide sufficient flooding to kill seedlings. Fluctuating rainfall patterns or less than average quantity during the wet season could result in substantial melaleuca recruitment (Belles and others 1999). Because melaleuca has very small wind- and water-dispersed seeds, reproductive and outlying individuals must be killed if long-term reduction of populations is to be achieved (Woodall 1981).

Repeated fires may have potential for controlling melaleuca; however, fuel loads may be insufficient to carry fire in consecutive years. Some wet grasslands might be capable of providing sufficient fuel for a second fire within 2 or 3 years after the first fire. Nearly all melaleuca seedlings were killed in a second fire 2 years after a wildfire in a wet grassland dominated by muhly grass (*Muhlenbergia capillaris*) (Belles and others 1999).

Recommendations for controlling mature melaleuca stands include using fire only after first killing reproductive individuals with herbicide (Myers and others 2001) (fig. 6-3). Herbicide-treated trees release large quantities of viable seed. Prescribed burning should then be conducted within 2 years (6 to 12 months recommended for Big Cypress Preserve; Myers and others 2001) of the herbicide-induced seed release and subsequent germination (Belles and others 1999). Mature melaleuca stands burned by wildfire should have high priority for management because of the potential for postfire spread following seed release from fire-damaged melaleuca or adjacent stands of unburned melaleuca. Recommendations include additional specifications for herbicide use and careful monitoring for several years after fire (Belles and others 1999).

Repeated burning, especially in combination with other control methods, can effectively control Chinese tallow under some circumstances. Chinese tallow is difficult to manage with fire because fuel loads under tallow infestations are often insufficient to carry fire (Grace 1999). See chapter 7 for more information on the use of fire to control Chinese tallow.

The effect of repeated fire (generally every two years) to control Brazilian pepper has been evaluated in highly altered "rock-plowed" limestone substrate in south Florida. Fine fuel supply was insufficient to carry annual fires, which the authors attributed to the replacement of graminoid species with Brazilian pepper (Doren and others 1991). While density and coverage of Brazilian pepper had increased on both burned and unburned plots at the end of the 6 years of evaluation, increases on burned plots occurred more slowly than on unburned plots (Doren and others 1991). Control of Brazilian pepper using herbicide followed by prescribed burning has also been attempted (fig. 6-4), though results are not reported in the literature.

Figure 6-3—Burning herbicide-killed melaleuca at South Florida Water Management District. Melaleuca stand was treated with herbicide in spring of 1996 and burned in winter 2001. The objective was to consume a large portion of the standing dead biomass with the fire, but only the tops of trees were burned. (Photo by Steve Smith.)

Figure 6-4—Using a helitorch to ignite herbicide-killed Brazilian pepper in spring 2006. Herbicide was applied one year before the fire. The fire was not effective at consuming dead pepper trees due to standing water and low fuel loads. (Photo by Steve Smith.)

Published reports were not found that document attempts to manage chinaberry with fire in the Southeast bioregion. Chinaberry recovered fully from a single autumn fire in South America, reproducing vegetatively from both stumps and roots (Menvielle and Scopel 1999; Tourn and others 1999). A single surface fire killed all seeds in the seed bank, and fruit production was 90 percent less than in unburned control plots. Chinaberry seedling emergence following the fire was 5 to 20 times greater in unburned control plots than in burned plots; however, the seasonal pattern of seedling emergence and survivorship was not affected by fire (Menvielle and Scopel 1999). The South American studies suggest that a single fire is not effective in controlling chinaberry, with populations quickly returning to prefire levels or even expanding (Tourn and others 1999). Additional research is needed to determine if fire in different seasons, multiple-year fires, or a combination of fire and herbicide application are effective for controlling Chinaberry in the Southeast.

The use of fire alone is not likely to cause enough damage to kill Old World climbing fern plants and prevent postfire sprouting and rapid recovery in wet grasslands. This is due to the high moisture content of most wet grassland fuels resulting in low-severity fire (Stocker and others 1997). Spot burning prior to herbicide application can reduce the amount of herbicide needed to control Old World climbing fern by about 50 percent (Stocker and others, In press). See the "Pine and Pine Savanna Habitat" section for more information on the use of fire to control climbing ferns.

Autumn-olive (*Elaeagnus umbellata),* sericea lespedeza (*Lespedeza cuneata*), shrubby lespedeza (*L. bicolor*), and tall fescue commonly occur on disturbed sites near southeastern grasslands. While no specific studies are available that examine the relationship between these species and fire in Southeast grassland habitats, studies conducted in other areas suggest that prescribed fire has a limited potential for controlling these species under certain circumstances and in combination with other control methods. Autumn-olive may respond to fire damage by sprouting, but empirical information on the relationship of this species to fire and fire management is lacking (Munger 2003b, FEIS review). See FEIS reviews by Munger (2004) and Tesky (1992) and TNC reviews by Stevens (2002) and Morisawa (1999a) for more information on the use of fire for management of lespedeza species. Recent introductions of tall fescue can be controlled by spring burning, and combinations of prescribed burns and herbicide applications have "moderate to high potential for restoration" (Batcher 2004).

Several resource management organizations have suggested that fire can be used successfully to reduce south Florida silkreed (*Neyraudia reynaudiana*) populations prior to spraying regrowth with herbicide. They caution, however, that silkreed is a highly combustible fuel source and, because of that, a special burning permit may be required (Rasha 2005, review). Burning without follow-up herbicide or mechanical control is ineffective in controlling silkreed and may enhance its growth and spread (Guala 1990, TNC review).

Other species that can be found in drier or upland portions of the wet grassland habitat include Japanese honeysuckle (*Lonicera japonica*), bush honeysuckles (Amur honeysuckle (*L. maackii*), Morrow's honeysuckle (*L. morrowii*), and tatarian honeysuckle (*L. tatarica*), Japanese stiltgrass (*Microstegium vimineum*) and tropical soda apple (*Solanum viarum*) (table 6-1). Bush honeysuckles are typically top-killed by fire, and fire may kill seeds and seedlings. Adult plants probably survive by postfire sprouting from roots and/or root crowns. Studies conducted in the Southeast bioregion are not available; however, field work in the Northeast bioregion suggests that repeated prescribed fire may be useful in controlling bush honeysuckles (chapter 5; Munger 2005a, FEIS review). Fire research on Japanese stiltgrass in the Southeast bioregion is also needed; however, studies outside of the Southeast suggest that prescribed fire prior to seed set might aid in controlling this species (Howard 2005c, FEIS review). No published information is available on the relationship of tropical soda apple to fire.

Pine and Pine Savanna Habitat ____

Background

Pine and pine savanna habitats covered here include southern mixed forest, oak-hickory-pine forest, and subtropical pine forest associations as described by Küchler (1964). Pine and oak species are the dominant trees, including longleaf pine, shortleaf pine (*Pinus echinata*), loblolly pine (*P. taeda*), slash pine (*P. elliottii*), pond pine (*P. serotina*), southern red oak (*Quercus falcata*), turkey oak (*Q. laevis*), sand-post oak (*Q. margaretta*), bluejack oak (*Q. incana*), blackjack oak (*Q. marilandica*), post oak, and water oak (*Q. nigra*). Pond cypress and palms are the dominant trees in some wetter and more southern sites. Shrub species are common, including runner oaks (*Q. minima* and *Q. pumila*), sumac (*Rhus* spp.), ericaceous shrubs (for example, *Vaccinium*), palms, wax myrtle (*Myrica cerifera*), and hollies (*Ilex* spp.). Understory species include grasses such as wiregrass (*Aristida stricta* and *A. beyrichiana*), little bluestem (*Schizachyrium scoparium*), and numerous forbs. When Europeans first arrived in the Southeast, pine stands, and especially pine savannas, may well have been the dominant vegetation in most of this area, extending from southeastern

Virginia to eastern Texas and from northern Georgia and Alabama to the Florida Keys (Platt 1999).

Presettlement fire regimes are poorly understood, but it is inferred that the high number of lightning strikes resulted in a fire-return interval of less than 13 years in pine forests and savannas. Larger and more intense fires probably occurred in May and June, after the start of the lightning/rain season but before large amounts of rain had fallen. Summer fires were probably more frequent but less intense and smaller in area. Ignitions by Native Americans probably increased fire frequency in many locations, shaping the savannas seen by early explorers (Wade and others 2000). Fire intervals may have been 1 to 4 years (1 to 5 years for subtropical pine forest; Myers 2000) before the arrival of European settlers, and then 1 to 3 years until fire exclusion became the norm in the early 1900s (Wade and others 2000).

Fires in these habitats were historically understory fires. Short return-interval (<10 years), understory fires predominated in most of the southern mixed forest and oak-hickory-pine types (*sensu* Küchler 1964). Slash pine and loblolly pine habitats experienced understory and mixed-severity fire regimes, with presettlement fire-return intervals estimated between 1 and 35 years (Myers 2000; Wade and others 2000).

The once-common southern pine forests were dramatically reduced by invasions of native hardwood species when fire exclusion policies were adopted in the 1920s and 1930s. Also affected were the populations of native plant and wildlife species, nutrient cycling, fuel reduction, and range management objectives that were associated with the historical fire regime. It has been suggested that savanna ecosystems that are not too seriously degraded can be restored if the appropriate fire regime (short return-interval, understory, spring and summer fires) is re-introduced, because the native plant species are adapted to this regime (Wade and others 2000).

Among the most threatening nonnative invasive plant species found in these habitats are cogongrass, Japanese honeysuckle, Brazilian pepper, and melaleuca. These species are fire-tolerant, thus reducing the effectiveness of fire in controlling their establishment and dispersal. Additionally, populations of climbing ferns appear to be increasing rapidly in Florida pine habitats and are spreading in Alabama, Florida, Georgia, and Mississippi pine habitats. Japanese climbing fern (*Lygodium japonicum*) has become particularly troublesome where pine straw is collected for sale as mulch. Spores of this fern have been found in straw bales, and the distribution of mulch bales throughout the Southeast has spread Japanese climbing fern into new areas. There is no specific information on fire and management of this species in this vegetation type. Old World climbing fern and melaleuca are also serious problems in wet grasslands and are discussed in the "Wet Grasslands Habitat" section.

Role of Fire and Fire Exclusion in Promoting Nonnative Plant Invasions in Pine and Pine Savanna

Frequent surface fires typical of presettlement fire regimes in most pine habitats promote some invasive species such as cogongrass. When fire is excluded, pine habitats are especially vulnerable to invasions by nonnative plants such as Brazilian pepper and Japanese honeysuckle.

Cogongrass and closely related Brazilian satintail (*Imperata brasiliensis*) are perennial, rhizomatous grasses that are well adapted to frequent fire. Both are early-seral species in wet-tropical and subtropical regions around the world. Discussion about management of these species is complicated by difficulty in distinguishing between the two species, lack of consensus among taxonomists about whether they actually are separate species, and determination of their native ranges (Howard 2005b, FEIS review). Among the more recent treatments, Wunderlin and Hansen (2003) describe them as distinct species, distinguished by anther number, and suggest that, while cogongrass is not native to the United States, Brazilian satintail is native to Florida.

Cogongrass requires some type of disturbance, such as fire, to maintain its dominance in southeastern pine understory. Frequent fire typically favors cogongrass over native species including big bluestem (*Andropogon gerardii*), Beyrick threeawn (*Aristida beyrichiana*), golden colicroot (*Aletris aurea*), and roundleaf thoroughroot (*Eupatorium rotundifolium*). Cogongrass flowering and seed production may be triggered by burning and other disturbances, although flowering has also been observed in undisturbed populations. In the absence of fire, vegetative growth from rhizomes (fig. 6-5) allows expansion of populations. Rhizomes also sprout easily after burning (Howard 2005b). In Mississippi wet pine savannas, cogongrass seedlings had higher levels of survival (for at least two months) in burned than in unburned study plots (King and Grace 2000).

Only limited information is available on two invasive shrub species. Brazilian pepper invades pine rockland (southern Florida habitat on limestone substrate) where fire has been excluded (Loope and Dunevitz 1981). It is suggested that Japanese honeysuckle is intolerant of frequent, low-severity fire and is typically absent from plant communities with this type of fire regime, such as longleaf pine. Therefore exclusion of fire from these communities may promote its establishment and spread (Munger 2002a, FEIS review).

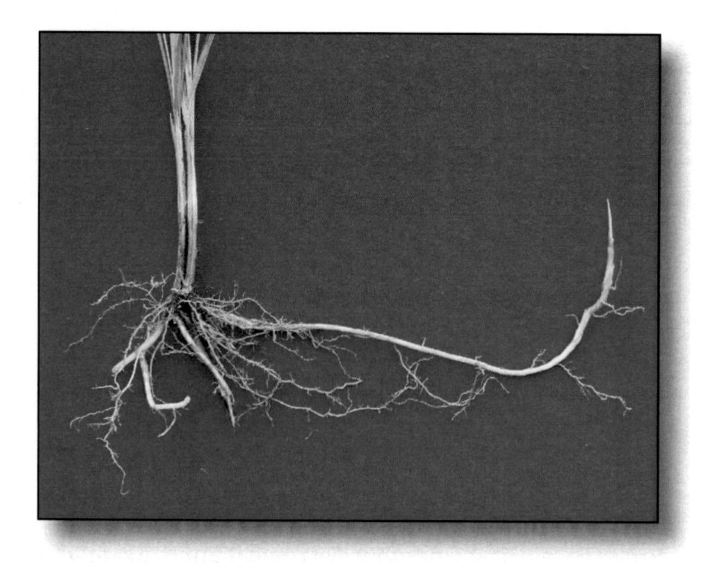

Figure 6-5—Cogongrass rhizomes. (Photo by Chris Evans, River to River CWMA. Bugwood.org.jpg.)

Effects of Plant Invasions on Fuel and Fire Regime Characteristics in Pine and Pine Savanna

Nonnative species life-forms that have invaded southeastern pine habitats and altered fuel and fire regime characteristics include trees (melaleuca), shrubs (Brazilian pepper), grasses (cogongrass), and ferns (climbing ferns) (fig. 6-6). In some situations these species replace native species and fill similar forest strata, but in other cases the invasive plants completely alter the horizontal structure and fuel characteristics of the invaded plant community.

Figure 6-6—Old World climbing fern climbing and overtopping vegetation in a pine habitat at the Jonathan Dickinson State Park in central Florida. (Photo by Mandy Tu, The Nature Conservancy.)

Old World climbing fern invasions provide a novel source of fuel in pine habitats (Roberts, D. 1996) and alter fire behavior by altering plant community fuel structure with extensive dry-standing frond "skirts" that ladder fire into the canopies of trees (Ferriter 2001) (fig. 6-7). Resulting canopy fires often kill trees that are adapted to low-severity surface fires, as well as native bromeliads (for example, wild pine (*Tillandsia fasciculata*)) resident on tree trunks. Fire spread may also be promoted when pieces of burning fern frond are kited into adjacent areas (Roberts, D. 1996). Increased fuel loads, altered fuel structure, and spotting from Old World climbing fern are blamed for tree mortality and escape of prescribed fires in pine stands at Jonathan Dickinson State Park in Florida. The park's fire management plan has been revised to no longer depend on wetland buffers to act as fire breaks if they contain climbing fern (Ferriter 2001).

Figure 6-7—Old World climbing fern on a slash pine, burning during a routine prescribed burn in pine flatwoods at the Reese Groves Property in Jupiter, Florida. Large slash pine and cypress have been killed by fire when climbing fern "ladders" carry fire into the canopy. (Photo by Amy Ferriter, South Florida Water Management District, Bugwood.org.)

Cogongrass invasion changes fuel properties in southeastern pine communities (Howard 2005b) (fig. 6-8). In a study of fuels and fire behavior in invaded and uninvaded pine stands in Florida, Lippincott (2000) found that invasion by cogongrass may lead to changes in fire behavior and fire effects in these communities. Native plant and cogongrass fuels have similar energy content; however, fire behavior is driven by factors other than energy content of fuels. Sites invaded by cogongrass had greater fine-fuel loads, more horizontal continuity, and greater vertical distribution of fuels. The resulting fires were more horizontally continuous and had higher maximum temperatures and greater flame lengths than fires in adjacent plots not invaded by cogongrass, and they resulted in higher subsequent mortality to young longleaf pine (Lippencott 2000). Similarly, Platt and Gottschalk (2001) found fine fuel and litter biomass were higher in cogongrass and nonnative silkreed stands than in adjacent pine stands without these grasses. The authors suggest that increases in fine fuels attributed to cogongrass could increase fire intensity at heights of 3 to 7 feet (1 to 2 m) above the ground (Platt and Gottschalk 2001).

Figure 6-8—(A) Infestation of cogongrass in a slash pine plantation in Charles M. Deaton Preserve, Mississippi. (Photo by John M. Randall, The Nature Conservancy.) (B) Cogongrass among planted pines in Mitchell County, Georgia, forms large accumulation of fine fuels around the base of trees. (Photo by Chris Evans, River to River CWMA, Bugwood.org.)

Bahia grass (*Paspalum notatum*) is not commonly found within existing pine stands but occupies heavily disturbed pine habitat that has been converted to pasture and rangeland, and interferes with efforts to restore pine communities. Bahia grass forms a continuous "sod fuel layer" in the previously patchy pine community, thus increasing fuel continuity (Violi 2000, TNC review). Because bahia grass is important forage for livestock, rangeland managers use controlled burns in winter to stimulate its growth. Winter burns negatively affect some native understory species, including wiregrass, which responds better to late summer and fall burns (Abrahamson 1984).

Melaleuca invasion in pine flatwoods can alter the fire regime from frequent (1- to 5-year return interval), low-severity surface fires to a mixed regime with less frequent (<35 to 200 year return interval) fires and greater incidence of crown fires. Crown fires are typically nonlethal to melaleuca trees but usually result in pine mortality. This combination of high-intensity fire and crown-fire survival is uncommon in North America (Myers 2000).

Low levels of fuel under mature Brazilian pepper (Doren and others 1991) and the difficulty of burning Brazilian pepper wood and leaves due to their high moisture content (Meyer 2005a, FEIS review) probably reduce fire intensity and fire spread in areas of dense infestation. Similarly, invasion by kudzu (*Pueraria montana* var. *lobata*) may reduce flammability of invaded pine habitats during the growing season due to its luxuriant, moist foliage. Conversely, the large amount of fuel biomass contributed by kudzu (fig. 6-9) and by plants killed by its invasion may increase the potential for dormant-season fires by increasing fuel loads, and its vining nature may increase the chance of fire crowning (Munger 2002b, FEIS review). These conjectures have not, however, been tested empirically.

Figure 6-9—Kudzu infestation at Travelers Rest, South Carolina in (A) summer, and (B) winter. Green kudzu foliage in summer may reduce potential for fire spread, while the opposite may be true in winter. (Photos by Randy Cyr, GREENTREE Technologies, Bugwood.org.)

Use of Fire to Manage Invasive Plants in Pine and Pine Savanna

Fire has become an important tool of natural area managers for removal of nonnative invasive species and maintenance of fire-adapted pine communities (Rhoades and others 2002). A study in a pine flatwoods community found that annual winter (non-growing season) burning increased native species richness and provided habitat for rare and listed plant species. Nonnative invasive species were found only in unburned plots. This may, however, be attributed to microsite differences within the treatment areas, as there was greater moisture availability in burned plots than in unburned plots (Beever and Beever 1993).

Old World climbing fern aerial fronds burn easily, and individual fronds can be completely consumed by a fire of sufficient intensity, but there are several reasons why prescribed fire is not expected to provide a major role in management of this species (Ferriter 2001). Pieces of burning climbing fern fronds get caught in fire-induced updrafts, reducing the ability to manage the fire perimeter when they are transported to adjacent areas. Climbing fern spores are very small and probably travel great distances by wind, including fire currents. Old World climbing fern plants in a south Florida slash pine stand were observed to sprout and recover rapidly after low-severity fire was applied using a hand-held propane torch (Stocker and others 1997).

Brazilian pepper is another species that is unlikely to be eliminated from pine stands by fire. Low-severity fire does not kill adult pepper trees, as girdling of the stem results in profuse sprouting from aboveground stems and root crowns (Woodall 1979). Brazilian pepper seeds can be killed by heat (70 °C for 1 hour; Nilsen and Muller 1980), and young seedlings can be killed by fire (Ferriter 1997). However, the intense crown fires necessary to kill adult plants (Doren and others 1991; Smith, C. 1985) do not commonly occur in pine stands with dense Brazilian pepper infestation. Fire does not carry well in mature Brazilian pepper stands, and fire rarely penetrates dense stands (Meyer 2005a). Brazilian pepper litter decomposes rapidly, leaving little litter for fuel, and moisture levels of branches, leaves, and litter are typically high (Doren and others 1991).

Prescribed fire may be more effective for controlling young Brazilian pepper stands. In areas where the water table lies below the soil surface for at least part of the year, grasses should provide sufficient fuels to carry fire of sufficient severity to kill young Brazilian pepper seedlings, and may also kill seeds (Nilsen and Muller 1980). Maintaining fire programs that killed seedlings prior to reaching unspecified "fire-resistant heights" has resulted in pepper-free areas (Ferriter 1997), and it has been noted that fire with a 5-year fire-return interval in Everglades National Park has excluded Brazilian pepper (Loope and Dunevitz 1981). On sites where either higher or lower water tables reduce the development of herbaceous fuels, prescribed fire may not be of sufficient severity to kill young Brazilian pepper plants (Ferriter 1997). In a study in south Florida pinelands, for example, most Brazilian pepper saplings over 3 feet (1 m) tall survived fire by coppicing (Loope and Dunevitz 1981). In any case, Brazilian pepper seed is readily dispersed from nearby stands by animals (Ewel and others 1982). The conclusion of a group of resource managers and scientists is that repeated burning may slow invasions of this species by killing seeds and seedlings, but fire "is not an effective control method for mature Brazilian peppertree stands" (Ferriter 1997).

It has been suggested that Japanese honeysuckle can be controlled by prescribed burning in pine plantations or in fire-dependent natural communities. Prescribed burns in Virginia are recommended to reduce Japanese honeysuckle cover and to "inhibit spread" for 1 to 2 growing seasons (Williams 1994, Virginia Department of Conservation and Recreation review). Two annual fires in a pine-hardwood forest resulted in an 80 percent reduction in Japanese honeysuckle crown volume and a 35 percent reduction of ground coverage. While these treatments do not eliminate Japanese honeysuckle from the site, the authors suggest that they may reduce the amount of herbicide required in an integrated management program (Barden and Matthews 1980).

Fire by itself does not control cogongrass, and in fact frequent fire promotes cogongrass. Fire can, however, improve the success of an integrated management approach using tillage and herbicides. Fire is also important for maintaining native plant diversity in pine habitat, and restoration of native plant species may be a critical factor in longer term control of cogongrass (Howard 2005b).

Controlled burning has not been effective in killing kudzu, but it can be used to remove vines and leaves to permit inspection of root crowns for population and stand monitoring. Fire also promotes seed germination in kudzu, after which seedlings can be effectively controlled with herbicides. Spring burns are recommended to reduce soil erosion by winter rainfall (Moorhead and Johnson 2002, Bugwood Network review). When removed from a portion of its occupied area, kudzu can re-invade from water- and bird-disseminated seed (Brender 1961). Similarly, controlled burning has not been effective in managing bahia grass because it sprouts readily after fire (Violi 2000).

Oak-Hickory Woodland Habitat ____

Background

The distribution of oak-hickory woodlands in the Southeast bioregion has depended on historical fire management practices. Limited to the most mesic and protected sites during periods of shortened fire-return intervals, oak-hickory woodland habitats have increased in area during the fire-exclusion decades of the early 1900s and continue to occupy many parts of the Southeast region today (Daubenmire 1978).

Plant communities in this type are dominated by a variety of oaks and hickories, with a mixture of other tree species, including maple (*Acer* spp.), magnolia (*Magnolia* spp.), sassafras (*Sassafras* spp.), and ericaceous shrubs. Several vines commonly occur, including grape and greenbrier (*Smilax* spp.). Many pine species are found in areas with edaphic and/or fire disturbances (Daubenmire 1978).

Fire regimes in oak-hickory habitats are classified as understory types with return intervals estimated between 2 and 35 years. Presettlement fire regimes are poorly understood, although estimates based on dendrochronology indicate a fire-return interval of 7 to 14 years in the mid-Atlantic region. After European settlement, fire-return intervals were reduced to 2 to 10 years, with some sites burned annually. At the present time, the fire regime of oak-hickory forests is infrequent, low-severity surface fires occurring principally during spring and fall. They are mainly human-caused and only burn small areas (Wade and others 2000).

Oak-hickory woodland habitats in the Southeast are heavily invaded by aggressive nonnative vines, shrubs, and trees including kudzu, Japanese honeysuckle, privet (*Ligustrum* spp.), bush honeysuckles, and tree-of-heaven (*Ailanthus altissima*). Other invasive species are found in oak-hickory woodland habitats, but much less information is available for them. Mimosa (*Albizia julibrissin*) is a small tree found throughout the Southeast bioregion in many types of disturbed areas, including old fields, stream banks, and roadsides (Miller 2003). While it is a common species, little published information describes its relationship with fire. Giant reed (*Arundo donax*) is commonly found in riparian areas in much of the United States and has been reported as invasive in Georgia, Virginia, and Maryland (Swearingen 2005). Thorny-olive (*Elaeagnus pungens*) is found as an ornamental escape in the Southeast (Miller 2003), but there are no published reports on the relationship of this species to fire. Winged euonymus (*Euonymus alatus*) is reported in a variety of east coast habitats, including forests, coastal scrublands and prairies (USFWS 2004, review), but no information is available on the relationship of this species to fire. No information is available concerning a related species, winter creeper (*Euonymus fortunei*). It is found in many states throughout the East (Swearingen and others 2002), but no specific information identifies habitats where it commonly occurs.

Role of Fire and Fire Exclusion in Promoting Nonnative Plant Invasions in Oak-Hickory Woodland

Only limited information is available on the role of fire and fire exclusion as they affect invasive plant species in Southeast oak-hickory woodland habitat. In some cases, fire exclusion seems to promote establishment and spread of nonnatives, while in other cases the canopy gaps created by fire may increase the likelihood of establishment and spread of nonnatives. In more tropical parts of the Southeast, it may be assumed that increased light penetration into the plant community will promote establishment and spread of shade-intolerant nonnative invasive species.

A study of the consequences of hurricane damage in conservation lands of south Florida demonstrates the effects of canopy gaps on nonnative plant invasions. Air potato (*Dioscorea bulbifera*) and other nonnative vine species increased after the tree canopy was damaged by Hurricane Andrew (Maguire 1995). This species may respond in a similar fashion to canopy gaps created by fire. Similarly, fire is one of many types of disturbance that creates canopy gaps that improve the chances for establishment by tree-of-heaven in old-growth woodland. Additionally, tree-of-heaven seed germination is delayed and reduced by leaf litter and may therefore be enhanced by fire when litter is consumed. On the other hand, fire may produce a flush of herbaceous growth that could inhibit tree-of-heaven germination (Howard 2004a, FEIS review). More information is needed on the effects of fire on seed germination in this species.

Soil heating may promote kudzu establishment by scarifying kudzu seedcoats and stimulating germination (Miller 1988). Similarly, mimosa seeds exposed to fire for 1 to 3 seconds had higher germination rates than unheated seeds (Gogue and Emino 1979). It has also been suggested that fire exclusion may promote Japanese honeysuckle spread (Munger 2002a).

Effects of Plant Invasions on Fuel and Fire Regime Characteristics in Oak-Hickory Woodland

The role of invasive plants in altering fire regimes in the Southeast bioregion is complicated by the existing mosaic of fire exclusion and controlled burning. Tree-of-heaven, for instance, is found in many types of woodlands in North America where presettlement fire regimes have been disrupted in many different ways. This makes it difficult to make definitive statements about the potential effect of tree-of-heaven on more natural fire regimes. The large amount of litter produced by tree-of-heaven from large leaves and broken branches, and its tendency to form dense thickets, may contribute to fire spread and crown fires in invaded areas (Howard 2004a).

A FEIS review speculates that the abundant moist foliage of kudzu could inhibit fire, effectively lengthening the time between fires in woodland habitats. On the other hand, the large amount of kudzu biomass may increase the potential for dormant-season fires by increasing fuel loads, and its vining nature may increase the chance of fire crowning. Additionally, increases in standing and surface fuels formed by plants killed following kudzu invasion may increase both fire intensity and frequency. The author points out that studies needed to test these hypothetical statements have not been conducted (Munger 2002b).

Use of Fire to Manage Invasive Plants in Oak-Hickory Woodland

Fire has not been recommended as a sole management tool to control tree-of-heaven because of this species' potential to burn in crown fire, its ability to sprout from the root crown and/or roots following top-kill from fire, and the potential for fire to promote seed germination. Fire has been used to reduce aboveground biomass of tree-of-heaven (Howard 2004a). A flame-thrower or weed burning device has been suggested to kill lower limbs (Hoshovsky 1988, TNC review), but this is not a population reduction measure.

Fire has been used to reduce cover of Japanese honeysuckle but does not kill plants. Japanese honeysuckle sprouts from subterranean buds, roots, and stems, recovering to various levels after fire (Munger 2002a). Japanese honeysuckle remained a site dominant after two consecutive annual fires in a pine-hardwood forest in North Carolina (Barden and Matthews 1980). Experimental plot (abandoned agricultural field) burns 5 years apart near Nacogdoches, Texas, resulted in Japanese honeysuckle plants with fewer and shorter prostrate shoots than in unburned plots 1 year after the last burn, but plants were not killed (Stransky 1984). Because prostrate shoots are an important part of this species' ability to invade native plant communities (Larson 2000), reduction in numbers of these shoots could theoretically slow the invasion process.

Seasonality of burns can affect postfire response of Japanese honeysuckle. Prescribed burns in October in a Tennessee oak-hickory-pine forest with a maple and dogwood (*Cornus* sp.) understory reduced Japanese honeysuckle coverage by 93 percent; burns in January or March reduced Japanese honeysuckle by 59 percent. Vegetation measurements were taken at the end of the growing season (September) about 1.5 years after burning (Faulkner and others 1989). The Nature Conservancy recommends fall, winter, or early spring prescribed burning to control Japanese honeysuckle in northern states, when Japanese honeysuckle maintains some leaves and most native plants are leafless (Nuzzo 1997, TNC review). This improved ability to target a particular species may have some applicability in southeastern habitats, but more often other native species retain leaves through the winter and may therefore be more subject to damage by fire at those times.

The Nature Conservancy also suggests that integrating fire and herbicide treatments to control Japanese honeysuckle may be more effective than either approach alone, with herbicides applied about a month after sprouting occurs following a late fall or winter burn (Nuzzo 1997). Application of herbicide about 1 year after a burn was not effective, possibly because postfire increases in herbaceous vegetation resulted in less herbicide contacting Japanese honeysuckle (Faulkner and others 1989). Fire is also helpful in controlling fire-intolerant Japanese honeysuckle seedlings and young plants. Efforts should be made to avoid soil disturbance as much as possible to reduce subsequent germination of Japanese honeysuckle seeds in the seed bank (Nuzzo 1997).

Prescribed fires have been suggested for controlling bush honeysuckles (Tatarian honeysuckle, Morrow's honeysuckle, Bell's honeysuckle (*Lonicera X bella*), and Amur honeysuckle) in fire-adapted communities (Nyboer 1990, Illinois Nature Preserves Commission review). Spring burns kill bush honeysuckle seedlings and top-kill mature plants; however, plants sprout readily after fire. Effective control may come from annual or biennial fires conducted for 5 years or more (Nyboer 1990).

It has been suggested that Chinese privet is intolerant of fire (Matlack 2002) and can be managed successfully by repeated fire, especially on sites with low stem density and high fine fuel loads (Batcher 2000a, TNC review). A single fire does not result in sufficient kill of mature plants (Faulkner and others 1989) but instead promotes sprouts from root crowns and/or roots (Munger 2003c, FEIS review). Chinese privet burns poorly without additional fuel. However, if sufficient low-moisture fuels are available (Batcher 2000a), annual fires may substantially reduce or kill aboveground portions of Chinese privet, although they will not eliminate it from a site. Three annual prescribed burns did not eradicate Chinese privet from areas where fire had been excluded for more than 45 years (Munger 2003c). Platt and Stanton (2003) suggest that dominance by Chinese privet cannot be reversed, but increases in population size can be prevented with short return interval, lightning-season fires. Japanese privet (*Ligustrum japonicum*) and European privet (*L. vulgare*) also occur in this vegetation type, but no specific information on management and fire for these species in this vegetation type is available.

Prescribed burns have been suggested as part of a strategy to manage kudzu. Information on the limitations of prescribed fire and effects of kudzu removal on native vegetation (presented in the "Pine and Pine Savanna Habitat" section page 100) is relevant to this habitat as well.

Air potato (*Dioscorea bulbifera*) invades woodland habitats throughout Florida (Schmitz and others 1997). While only a limited amount of research has been conducted, prescribed fire may be useful in killing stem growth (Morisawa 1999b, TNC review) and bulbils (Schultz 1993, TNC review) of air potato in woodlands. A related species, Chinese yam (*Dioscorea oppositifolia*), has been reviewed by The Natural Conservancy (Tu 2002b). Chinese yam is found in mesic bottomland forests, along streambanks and drainageways in many states of the Southeast. Only very limited information is available about the use of fire to manage this species. It was noted that reduced amounts of Chinese yam were present the year following a fall wildfire in Great Smoky Mountains National Park (Tu 2002b), but the specific habitat information is not available.

Tropical Hardwood Forest Habitat

Background

Scattered throughout the grassland and savanna plant communities of south Florida are "islands" of tropical hardwood species, often called hammocks, and typically found on somewhat drier sites. Common species include gumbo limbo (*Bursera simaruba*), black ironwood (*Krugiodendron ferreum*), inkwood (*Exothea paniculata*), lancewood (*Ocotea coriacea*), marlberry (*Ardisia escallonoides*), pigeon plum (*Coccoloba diversifolia*), satinleaf (*Chrysophyllum oliviforme*), poisonwood (*Metopium toxiferum*), and white stopper (*Eugenia axillaris*). While limited in extent compared to the other habitats discussed in this chapter, they are important because of the plant diversity they provide within the grassland landscape.

Fire regime in this habitat varies from low-intensity surface fires to crown fires, with an estimated presettlement return interval of 35 to over 200 years. Fire has not been the dominant force in shaping hardwood hammock plant communities because they are usually difficult to burn. Although many of the hardwood species sprout following top-kill from fire, the stands can be destroyed by fire during periods of drought if the organic soil is consumed (Myers 2000).

Many hardwood hammocks are being aggressively invaded by air potato, melaleuca (fig. 6-10), and Old World climbing fern. Water yam (*Dioscorea alata*) has been reported in coastal hammocks in Florida (FLEPPC 1996). This species is related to air potato and may act in a similar manner to air potato in relation to fire, although no specific information is available for either species.

Role of Fire and Fire Exclusion in Promoting Nonnative Plant Invasions in Tropical Hardwood Forest

Limited information is available regarding the role of fire in promoting nonnative plant invasions in tropical hardwood hammocks. Melaleuca is capable

Figure 6-10—Melaleuca saplings (green trees) ringing a Cypress clump at the Loxahatchee National Wildlife Refuge around 1994. The melaleuca eventually overtook the cypress (Ferriter, personal communication 2007). (Photo by Amy Ferriter, South Florida Water Management District, Bugwood.org.)

of spreading quickly into this habitat following fires that remove most of the vegetation and expose bare mineral soil (Bodle and Van 1999).

Old World climbing fern is rapidly invading hardwood hammocks, but it is not known to what extent invasion is dependent on or retarded by fire. Anecdotal evidence suggests that the fern appears to grow especially well in hammocks where native vegetation was either damaged or killed by fire during a drought. Researchers and natural area managers in southern Florida suspect that Old World climbing fern is particularly robust in areas where native tree island vegetation was damaged or killed by a fire during the drought of 1989–1990. It has also been suggested, although not demonstrated, that convection currents generated by burning fern growth that has formed a trellis up into tall trees could increase the dispersal of spores (Ferriter 2001). Old World climbing fern plants sprout and recover rapidly after low-severity fire (Stocker and others 1997).

The influence of melaleuca and Old-World climbing fern on fuels and fire regimes has been discussed in previous parts of this chapter, as has the use of fire for managing these species. Fire by itself is not an effective method to manage melaleuca or Old World climbing fern, although suggestions have been made for ways in which fire could be incorporated into an integrated management approach with other control techniques. Since native species in this habitat probably did not evolve in a regime of frequent fire, increasing fire frequency with prescribed burning might have unintended effects on native plant species (Ferriter 2001).

Cypress Swamp Habitat

Very limited information is available about the relationship between fire and invasive species in cypress swamp habitat. Depressional wetlands in central and south Florida dominated by bald cypress (Küchler's (1964) Southern Floodplain Forest) had a presettlement stand-replacement fire-return interval estimated at 100 to 200 years or greater (Wade and others 2000). This fire regime has been altered by invasions of melaleuca and Old World climbing fern. Both species increase the probability of more frequent stand-replacement fires because of the ease with which they burn and because Old World climbing fern can form a fuel bridge between adjacent, more frequently burned habitats and the much wetter cypress swamp

(Langeland 2006). Information on the role of these species in relation to fire is also presented in the "Wet Grassland Habitat," "Pine and Pine Savanna Habitat" and "Oak-Hickory Woodland Habitat" sections.

Catclaw mimosa (*Mimosa pigra*) occurs on about 1,000 acres (400 ha) in Florida, including cypress swamp habitat. While fire has been used experimentally in Australia to clear mature plants, enhance seed germination for subsequent herbicide application, and kill some seeds (Lonsdale and Miller 1993), no similar studies have been conducted in the United States.

Conclusions and Summary

The importance of fire and fire management in influencing species composition and dynamics of plant communities in the Southeast bioregion is often stated, if not completely understood. Incomplete or contradictory information, for instance, describes the plant communities prior to human influence (Stanturf and others 2002). With its high number of lightning strikes, its many fire-adapted communities, and its historical human dependence on fire-maintained habitats, fire has probably been a more important factor in the Southeast than in any other broadly defined region of the country. Plant communities such as longleaf pine once covered millions of acres when human populations supplemented naturally occurring fires with intentional blazes (Chapman 1932). Even wet grasslands in the Southeast tend to be flammable and are adapted to frequent fires (Leenhouts 1982; Schmalzer and others 1991; Wade 1988; Wade and others 1980).

Human influence on fire regime was accompanied by a large number of intentional introductions of plants for agricultural, horticultural, medicinal, and religious purposes, as well as many accidental imports. The large number of intentional introductions and the escape of these introduced plants led to very high levels of invasive plants in the Southeast bioregion, especially Florida. More than 25,000 species and cultivars have been introduced to Florida (D. Hall, personal communication, cited by Gordon 1998), a state with around 2,523 native species (Ward 1990). While many of these introductions have served their intended purposes, a small proportion (about 10 percent (Gordon 1998)) has caused unintended damage to forest and conservation lands. It is not clear whether the large number of nonnative plant invasions in this bioregion is due to the large number of introductions (in other words, propagule pressure), or whether habitats in the Southeast are more susceptible to invasion. The high number of plant species introductions in the Southeast may be related to the diversity of cultural origins of the human populations and a range of climates from tropical to temperate. This latter factor could also influence the susceptibility of the region by providing a wider range of potentially suitable conditions for establishment and spread.

The causes and effects of invasive plant establishment and spread are related to fire and fire management to varying degrees. The role of fire and fire exclusion in promoting nonnative plant invasions is very different among the five habitats discussed in this chapter. Fire exclusion policy, of course, does not mean the absence of fire. Historical and current efforts to exclude fire have often led to less frequent but more severe fires, which then can lead to substantial changes in native and invasive species populations (Wade and others 2000).

Role of Fire and Fire Exclusion in Promoting Nonnative Plant Invasions

The complex and dramatic relationship between invasive species and both fire and fire exclusion in the Southeast bioregion may be best exemplified by melaleuca invasion in pine, wet grassland, and tropical hardwood habitats in south Florida. Melaleuca has spread into thousands of acres apparently without need of fire, but has spread most dramatically following natural fire, controlled burns, and uncontrolled fire following decades of fire exclusion efforts.

Wet grasslands may be the habitat type most affected by fire and fire exclusion. Expansion of melaleuca following fire has been more thoroughly reported for wet grassland habitats (Ferriter 1999). Fire in wet grasslands is also possibly responsible for increases in Old World climbing fern (Langeland 2006) and chinaberry (Menvielle and Scopel 1999; Tourn and others 1999), although the evidence is principally anecdotal. Successful exclusion of fire has been blamed for Brazilian pepper invasions of salt marshes (Schmalzer 1995) and for Chinese tallow (Bruce and others 1995) and tall fescue invasions (Eidson 1997) in wet grassland habitat.

Fire exclusion has clear ecological impacts in pine habitats, which frequently accumulate additional woody shrub species in the absence of fire and ultimately become hardwood dominated habitats with a minor pine component (DeCoster and others 1999; Heyward 1939; Platt 1999; Slocum and others 2003; Streng and others 1993; Walker and Peet 1983). Less has been studied, however, about these longer-term changes and the interplay of fire and invasives. Brazilian pepper (Loope and Dunevitz 1981) and Japanese honeysuckle (Munger 2002a) are among the woody shrubs that are promoted in pine habitats when fire is excluded. Conversely, cogongrass in pine habitats is promoted by frequent surface fires and, in fact, requires some regular disturbance to maintain its dominance (Howard 2005b).

Effects of Plant Invasions on Fuel and Fire Regime Characteristics

Some of the most obvious visual changes in fire occur when melaleuca invasions fuel crown fires in wet grassland communities. But while these conflagrations are dramatic, the removal of aboveground plant material is not a substantial change to the wet grassland fire regime. It is the secondary effects of these fires, such as consumption of the surface of organic soils followed by flooding, that can lead to major changes in plant communities (Wade and others 2000).

Some evidence suggests that when species such as Chinese privet, Chinese tallow, and Brazilian pepper replace native plant species in wet grasslands, fine fuel loads and horizontal continuity are reduced (for example, Doren and Whiteaker 1990; Doren and others 1991; Grace 1999; Platt and Stanton 2003). Change in fine fuels may lead to reduced fire frequency and intensity, and increased fire patchiness. There is, however, little experimental evidence to support these suggestions.

The most substantial changes in fire regime caused by nonnative plant invasion are probably in pine habitats. Invasions of Old World climbing fern, and possibly its congener Japanese climbing fern, increase incidence of crown fires, carry fire across wetland barriers that would have stopped the fire if they had not contained Old World climbing fern, and possibly "kite" fire to new locations (Langeland 2006). Cogongrass changes fire behavior and effects in pine habitats. Cogongrass invasions lead to increased biomass, horizontal continuity, and vertical distribution of fine fuels when compared with uninvaded pine savanna, and higher maximum fire temperatures have been reported (Lippincott 2000). In this particular example, the detailed studies necessary to show actual replacement of native species have not been conducted. Melaleuca changes the fire regime in pine stands from frequent, low-severity surface fires to a mixed fire regime with less frequent fires and greater incidence of crown fires. These crown fires are often lethal to pines but not to melaleuca (Myers 2000). It is possible that Brazilian pepper invasions have opposite effects and reduce fire intensity and fire spread where it is densely distributed (Doren and Whiteaker 1990; Doren and others 1991).

Use of Fire to Manage Invasive Plants

Among the habitats covered in this section, pine and pine savanna habitats are probably the most conducive to use of fire to manage invasive plants, in large part because of the role of fire in maintaining these pyric communities. Controlled burns in oak-hickory woodland have resulted in reductions of Japanese honeysuckle (Barden and Matthews 1980; Williams 1994) and reduced spread of Chinese privet (Batcher 2000a). Since frequent prescribed fire in oak-hickory woodland will favor pine species at the expense of young hardwoods, it will probably be more difficult to maintain as aggressive a burning practice in oak-hickory woodland than in pine and pine savanna.

In the presence of propagule sources from invasive species such as melaleuca and Old World climbing fern, prescribed fire in wet grasslands must be conducted with extreme care and may result in expansion of these species. The margin between use of fire for successful reduction of melaleuca seedlings following a seed-release event and accidental expansion of melaleuca into fire-cleared seed beds is very small and is affected by weather and water management patterns out of the manager's control (Belles and others 1999).

Additional Research Needs

Among the general information needs related to fire and invasive plants, several specific needs stand out. While we have case studies on short-term effects of various fire related practices for individual species, we don't know which fire management practices in which habitats will provide the most effective means of reducing existing invasive plant populations or preventing future invasions. This is not an easy area of research, in part because we do not know which of the tens of thousands of novel species that could be introduced to southeastern habitats will become management problems, making it nearly impossible to know what practices will provide the most future benefit. There are many nonnative species already present in Southeast bioregion habitats for which no information is available. Included in this category are giant reed, field bindweed (*Convolvulus arvensis*), Chinese silvergrass (*Miscanthus sinensis*), princesstree (*Paulownia tomentosa*), golden bamboo (*Phyllostachys aurea*), multiflora rose (*Rosa multiflora*), sericea lespedeza, five-stamen tamarisk (*Tamarix chinensis*), French tamarisk (*T. gallica*), smallflower tamarisk (*T. parviflora*), saltceder (*T. ramosissima*), bigleaf periwinkle (*Vinca major*), common periwinkle (*V. minor*), Japanese wisteria (*Wisteria floribunda*), and Chinese wisteria (*Wisteria sinensis*).

At a minimum we need longer-term studies that document broad species changes in population size and distribution. For instance, it has been suggested that replacement of wet grassland species by invasive hardwood shrubs (Brazilian pepper) and trees (Chinese tallow) results in reduced fine fuel load and horizontal continuity (Doren and Whiteaker 1990; Doren and others 1991; Grace 1999). These fuel changes could logically lead to changes in fire frequency, severity, and patchiness, but that has yet to be documented. There is also potential for subsequent changes in native plant species coverage and/or diversity—a topic that deserves further study.

Because so much of the Southeast is relatively low-lying, the effect of natural and artificially-manipulated hydrology on the relationship between fire and invasive plants needs to be examined. For instance, we know that artificially lowered surface-water elevations in south Florida wet grasslands lead to increased fires and exposure of mineral soil that facilitate melaleuca invasions, but management practices that could prevent such invasions have not been determined.

Emerging Issues

The Southeast has more fire and more invasive plants than most other parts of the country. To further complicate the situation, the Southeast is also rapidly increasing its human population. The "sunbelt" is the fastest growing part of the United States, with a 21 percent increase in population between 1970 and 1980, and an 18 percent increase between 1980 and 1997 (NPA 1999). It remains to be seen whether prescribed fire practices can be implemented and maintained with more and more urban incursions into forest and natural areas.

Global warming and related increases in carbon dioxide may well affect the relationships among invasive plants, native plant communities, and fire. Lightning frequency is expected to increase with global warming. The U.S. Global Change Research Program's National Assessment Synthesis Team has predicted that the "seasonal severity of fire hazard" will increase about 10 percent for much of the United States, but a 30 percent increase in fire hazard is predicted for the Southeast (NAST 2000).

Major changes in plant communities have occurred in the Southeast because of the interaction of invasive nonnative plants and fire management policy and practice. However, it is not a given that the Southeast would have avoided its serious invasive plant problems if presettlement fire regimes had been maintained, nor that reinstating presettlement fire regimes in the 1800s would have prevented problems during the next two centuries.

Better understanding of the relationship between invasive plants, fire, and fire management is certainly needed if resource managers are to maximize their ability to prevent further nonnative plant invasions and adjust fire policy to both reduce the existing invasive plant populations and achieve other management objectives.

Notes

James B. Grace
Kristin Zouhar

Chapter 7:

Fire and Nonnative Invasive Plants in the Central Bioregion

Introduction

The Central bioregion is a vast area, stretching from Canada to Mexico and from the eastern forests to the Rocky Mountains, dominated by grasslands and shrublands, but inclusive of riparian and other forests. This bioregion has been impacted by many human-induced changes, particularly relating to agricultural practices, over the past 150 years. Also changed are fire regimes, first by native peoples who used fire for a variety of purposes and then by European settlers, who directly and indirectly contributed to a great reduction in the frequency of fire on the landscape. Perhaps of even greater importance has been the introduction of nonnative plant species, which have come to impact every community type to some degree.

Nonnative plants have a wide array of impacts on native ecosystems and populations in the Central bioregion, and these impacts continue to mount and evolve. Many long-time invaders, such as smooth brome (*Bromus inermis*), and leafy spurge (*Euphorbia esula*), have already spread to large areas, and their ranges may still be expanding. Others, such as tamarisk or saltcedar (*Tamarix* spp.) and buffelgrass (*Pennisetum ciliare*), are rapidly spreading at the present time, while still others have likely not yet shown their full potential for expansion. In this volume, as well as in this chapter, our emphasis is on the interaction of nonnatives with fire, how it affects them and how they affect it.

The ecosystems of the Central bioregion have been shaped by fire, including fires associated with natural ignitions and those deliberately set by humans. Both grasslands and shrublands in this bioregion experienced frequent and widespread fires during their evolution (Stewart 2002). Prescribed fire is now widely used to manage some areas for their natural characteristics. Thus, while changed in character, both by conditions that now limit wildfire occurrence and spread and by prescribed burning, the Central bioregion remains one with a high fire frequency (Wade and others 2000).

Fire interactions with nonnative plants can have important impacts. In some cases, fire can be a means of reducing impacts of nonnative species (chapter 4). In other cases, fire may facilitate the establishment and spread of nonnatives (chapter 2). Some nonnative species can radically change the fire regime itself (chapter 3). Because of the widespread use of prescribed fire in this bioregion, it is important to know how nonnative species interact with fire and whether there are means whereby these interactions can be controlled.

Geographic Context and Chapter Organization

Grasslands characterize most of the Central bioregion, much of which has varying amounts of shrubland and woodland components. Two major subdivisions occur within the Central bioregion: the mesic tallgrass prairie subregion in the east and the drier Great Plains subregion in the west. A wide variety of conditions and species exist within each of these two major subregions. To capture some of this variability, these subregions are further divided into formations based on dominant life forms and north-south gradients. Within each of these formations, some nonnatives are conspicuously associated with riparian zones, which are typically forested even if surrounded by grassland. The boundaries of vegetation communities in the Central bioregion are not usually strictly delimited; rather, many types occur scattered throughout the eastern deciduous biome and western and southwestern forests. The central grasslands intergrade from open expanses into arid grasslands and savannas in the west and southwest and into grassland inclusions in the east.

A great range of climate, topography, soil conditions, and historic land use practices occur within the Central bioregion. In the east, most of the pre-Columbian tallgrass prairies have been converted to agricultural fields, some of which have been abandoned and allowed to succeed to various states that include a variety of nonnative and potentially invasive species. In the west, grasslands have also been used for crop production but are used more often for livestock grazing, which means that much larger expanses of grassland exist in at least a quasi-native state in the Great Plains subregion. The embedded wetland and riparian habitats also vary in their condition. In areas where wetlands were small and isolated, there was a tendency for them to be eliminated. In the more extensive coastal wetlands, the communities generally retain more of their native character and are often comparatively pristine.

We would like to be able to predict individual species' interactions with fire within ecoregional boundaries. What is most predictive is information about the biology of individual species. Consistent botanical traits lead to relatively consistent outcomes. Less reliable is our ability to predict how a species will respond in the presence of other species. Competitive ability varies with environment, and complex interactions with the varying native flora and fauna ensure that outcomes will be somewhat conditional. The inadequacy of available information and the complexity of particular situations make it imperative that we keep in mind the possibility that particular situations may deviate from general guidelines.

This chapter presents individual species and their relation to fire within the mesic tallgrass prairie and Great Plains subregions, which match up approximately with the tallgrass prairie in the east (covered by Wade and others 2000) and the plains grasslands in the west (covered by Paysen and others 2000). The treatment is further divided by formations where individual species seem to be most problematic. At this level, we follow the system presented by Risser and others (1981). Using this approach, we recognize two formations in the mesic tallgrass prairie subregion, (1) the northern and central tallgrass prairie and (2) the southern tallgrass prairie. Within the Great Plains subregion, we recognize (1) the northern mixedgrass prairie, (2) the southern mixedgrass prairie, and (3) the shortgrass steppe (fig. 7-1). Finally, while not uniform throughout, we recognize a riparian formation as distinct in character. These classifications recognize the strong role that climate plays in nonnative species distributions and impacts. The reader is reminded that these classifications are approximations and that nonnative invasive species' ranges are not always precisely known and may be expanding. Within-formation variation, which can also be of great importance, will be discussed for each species within this framework when evidence of such variation is available.

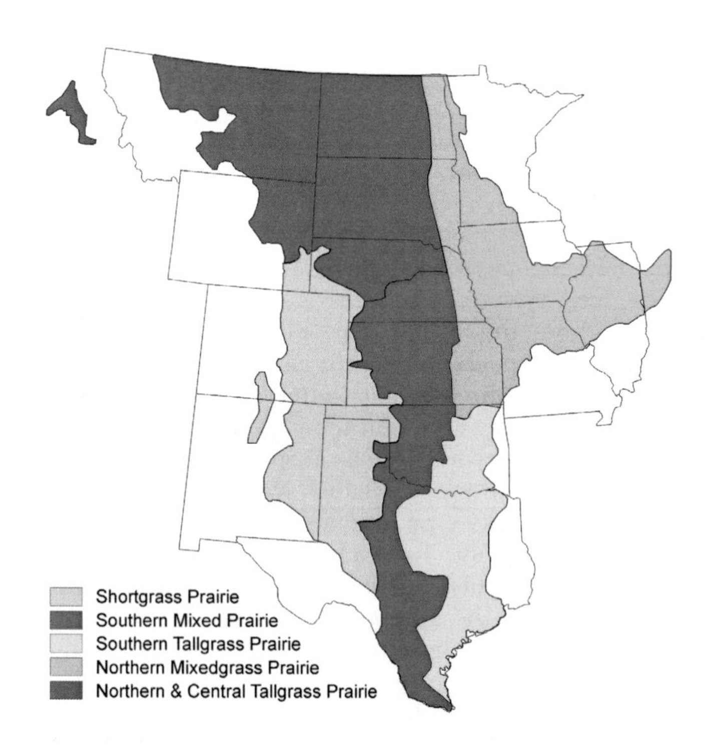

Figure 7-1—Approximate distribution of major vegetation formations in the Central bioregion. Riparian areas are not shown.

Fire Regimes

Grassland and shrubland fire regimes have many similarities. Presettlement fires were frequent and the natural or reference fire regime was dominated by growing-season fires, while the anthropogenic fire regime during the past few millennia seems to be heavily slanted toward dormant season burning (Stewart 2002). Most of these plant communities have now been either cultivated or grazed, and many common features follow from these major influences. To provide useful information, the challenge is to present a picture that is general enough to supply broad guidance to burning programs and decision making. At the same time, we must recognize that there is a great deal of complexity hidden by general descriptions. Our purpose is to focus on some of the major nonnative plants in this bioregion and exemplify the kinds of information known and unknown relating to their interactions with fire. In this vein, in the Effects of Fire on Flora volume (Brown and Smith 2000) figure 1-2 (page 7) provides a useful depiction of fire regime types, including a representation of the average conditions in the Central bioregion.

Wildfires in grasslands are generally warm-season fires that consume most of the aboveground herbaceous growth; that is, they are "stand-replacement" fires as defined by Brown (2000). Presettlement fire regimes in the mesic subregion (northern, central, and southern tallgrass prairies) are characterized as stand-replacement fires with average return intervals of 10 years or less (Wade and others 2000). The Great Plains subregion contains communities with fire regimes that fall into the categories of either stand-replacing or mixed-severity fires (due to the presence of shrubs) and have fire-return intervals of 35 years or less (Paysen and others 2000). Fire severity patterns depend largely on the continuity and abundance of fuels, which in turn is influenced by rainfall and local factors such as grazing. Fire exclusion has a profound impact on the current fire regime. While many managed areas are subject to routine prescribed burning, the majority of the landscape is now managed for fire exclusion. Also of great importance is grazing. Where livestock grazing is intensive, the availability of fuels can be greatly reduced, strongly impacting the likelihood and character of fire.

The Conservation Context

Temperate grasslands, which predominate in the Central bioregion, include some of the most threatened ecosystems in the world (Ricketts and others 1999). In North America, their widespread use for agriculture and livestock grazing, in addition to the effects of urbanization and other human activities, have led to dramatic alterations in their extent and condition. Nonnative plants constitute a major additional threat to conservation, rehabilitation, and restoration of temperate grasslands (Smith and Knapp 1999; Stohlgren and others 1999a; Westbrooks 1998). Studies indicate that the success and consequences of an invasion depend on many factors, including fire. Because of the historic importance of fire in this bioregion, the interactions between invasives and fire is likely to be of critical importance in these systems.

A variety of interactions between invading species and fire is possible. In some cases, fire may act as an environmental filter that eliminates or reduces nonnative invaders. In other cases, fire-adapted invaders will be quite impervious to burning, and fire may facilitate the establishment and spread of certain nonnative plants. Alternatively, fire exclusion may provide a window of opportunity for the establishment of certain nonnatives that may not be easily displaced once they have established. Species that invade and become dominant may drastically change the fire regime and, through that change, have detrimental effects on the native community. Both the natural characteristics of a landscape and anthropogenic modifications can be expected to influence the interactions between invaders and native communities. Also, interactions between fire and invasives can be complicated by additional factors such as grazing and other disturbances (Collins and others 1995, 1998; Stohlgren and others 1999b). More information is needed in order to manage natural and seminatural systems using fire in ways that are of greatest utility to conservation, particularly in the face of nonnative invasions.

Overview of Nonnative Plants that Impact or Threaten the Central Bioregion

Many nonnative plants occur in the Central bioregion (table 7-1). Some species have very broad distributions and have demonstrated invasiveness in many, but not all, ecosystems within their ranges. These species include cheatgrass (*Bromus tectorum*), smooth brome, tamarisk, leafy spurge, spotted knapweed (*Centaurea biebersteinii*), and Canada thistle (*Cirsium arvense*). Other species are of increasing concern, such as Caucasian bluestem (*Bothriochloa bladhii*), itchgrass (*Rottboellia cochinchinensis*), and guineagrass (*Urochloa maxima*). Still others (such as Angleton bluestem (*Dichanthium aristatum*)) are restricted to a limited portion of the range of grassland types; and the status of some nonnatives is not well documented.

The degree of concern associated with these nonnatives varies depending on climate and other factors. Of greater importance in the Northern and Central States are leafy spurge, cheatgrass, smooth brome, spotted knapweed, Canada thistle, musk thistle (*Carduus nutans*), sweetclovers (*Melilotus* spp.), Dalmatian

Table 7-1—Nonnative plants of major concern in the central United States (from Grace and others 2001). Scientific nomenclature is from the ITIS Database (http://www.itis.usda.gov/).

Common name	Scientific name
Grasses	
Angleton bluestem	*Dichanthium aristatum* (Poir.) C.E. Hubbard
Bahia grass	*Paspalum notatum* Fluegge
Bermudagrass	*Cynodon dactylon* (L.) Pers.
Buffelgrass	*Pennisetum ciliare* var. *ciliare* (L.) Link
Caucasian bluestem	*Bothriochloa bladhii* (Retz.) S.T. Blake
Cheatgrass	*Bromus tectorum* L.
Cogongrass	*Imperata cylindrica* (L.) Beauv.
Crested wheatgrass	*Agropyron cristatum* (L.) Gaertn.
Fountain grass	*Pennisetum glaucum* (L.) R. Br.
Giant reed	*Arundo donax* L.
Giant sugarcane plumegrass	*Saccharum giganteum* (Walt.) Pers.
Guineagrass	*Urochloa maxima* (Jacq.) R. Webster
Itchgrass	*Rottboellia cochinchinensis* (Lour.) W.D. Clayton
Japanese brome	*Bromus japonicus* Thunb. ex Murr.
Johnson grass	*Sorghum halepense* (L.) Pers.
Kentucky bluegrass	*Poa pratensis* L.
Kleberg bluestem	*Dichanthium annulatum* (Forsk.) Stapf
Lehmann's lovegrass	*Eragrostis lehmanniana* Nees
Orchard grass	*Dactylis glomerata* L.
Quackgrass	*Elymus repens* (L.) Gould
Ryegrass spp.	*Lolium* spp. L.
Smooth brome	*Bromus inermis* Leyss.
Vaseygrass	*Paspalum urvillei* Steud.
Yellow bluestem	*Bothriochloa ischaemum* (L.) Keng
Forbs	
Brazilian vervain	*Verbena brasiliensis* Vell.
Canada thistle	*Cirsium arvense* (L.) Scop.
Common mullein	*Verbascum thapsus* L.
Crown vetch	*Coronilla varia* L.
Dalmatian toadflax	*Linaria dalmatica* (L.) P. Mill.
Diffuse knapweed	*Centaurea diffusa* Lam.
Garlic mustard	*Alliaria petiolata* (Bieb.) Cavara & Grande
Japanese climbing fern	*Lygodium japonicum* (Thunb. ex Murr.) Sw.
Kochia	*Kochia prostrata* (L.) Schrad.
Leafy spurge	*Euphorbia esula* L.
Missouri bladderpod	*Lesquerella filiformis* Rollins
Musk thistle	*Carduus nutans* L.
Oxeye daisy	*Leucanthemum vulgare* Lam.
Purple loosestrife	*Lythrum salicaria* L.
Red-horned poppy	*Glaucium corniculatum* (L.) J.H. Rudolph
Russian knapweed	*Acroptilon repens* (L.) DC.
Scotch thistle	*Onopordum acanthium* L.
Sericea lespedeza	*Lespedeza cuneata* (Dum.-Cours.) G. Don
Spotted knapweed	*Centaurea biebersteinii* DC.
Squarrose knapweed	*Centaurea triumfettii* All.
White Sweetclover	*Melilotus alba* Medikus
Whitetop	*Cardaria draba* (L.) Desv.
Yellow sweetclover	*Melilotus officinalis* (L.) Lam.
Yellow starthistle	*Centaurea solstitialis* L.
Yellow toadflax	*Linaria vulgaris* P. Mill.
Woody Species	
Chinese privet	*Ligustrum sinense* Lour.
Chinese tallow	*Triadica sebifera* (L.) Small
Common buckthorn	*Rhamnus cathartica* L.
Japanese honeysuckle	*Lonicera japonica* Thunb.
Macartney rose	*Rosa bracteata* J.C. Wendl.
Multiflora rose	*Rosa multiflora* Thunb. ex Murr.
Russian-olive	*Elaeagnus angustifolia* L.
Tamarisk	*Tamarix* spp. L.

toadflax (*Linaria dalmatica*), and Kentucky bluegrass (*Poa pratensis*). In the southern portion of the Central bioregion, buffelgrass, guineagrass, Bermudagrass (*Cynodon dactylon*), sericea lespedeza (*Lespedeza cuneata*), Johnson grass (*Sorghum halepense*), Chinese tallow (*Triadica sebifera*), Macartney rose (*Rosa bracteata*), Caucasian bluestem, and several other escaped pasture grasses are the most often mentioned species. Riparian and wetland habitats in many parts of the Central bioregion have been invaded by tamarisk, Russian-olive (*Elaeagnus angustifolia*), and purple loosestrife (*Lythrum salicaria*). In southern wetlands, nonnative invaders include giant reed (*Arundo donax*).

The species covered in this chapter represent only a fraction of the nonnatives currently present in temperate grasslands. In a review of invasive species in the USGS Central Region (Burkett and others 2000), resource management agencies throughout the Central States were polled to obtain their views on which nonnative invaders (plants and animals) were of greatest concern. Some 69 plant taxa were reported to be of concern, which gives an indication of the magnitude of the problem. A preliminary assessment of the flora of the southern tallgrass prairie formation (Allain and others, unpublished report, 2004) has identified over 150 nonnative species in this type of prairie alone. Also, of the 304 grass species in Texas coastal prairies and marshes, 85 (26 percent) are introduced (Hatch, S. and others 1999). Our current knowledge about the full suite of introduced species in the Central bioregion and how they relate to fire is incomplete. While certain generalizations may apply to nonnative invaders, experience tells us that species-specific and site-specific effects are often highly important and difficult to anticipate without a great deal more information than is currently available.

Plant-Fire Interactions

The conceptual model presented by Grace and others (2001) represents a framework for summarizing and evaluating how an invading nonnative may interact with the native community and the fire regime. In this framework, the major categories of influences are the native community characteristics, the fire regime, the growth conditions for both nonnative and native species, and the influences that disturbances, human impacts, and landscape characteristics have had in the past and will have in the future. According to this model, several major relationships will determine whether an invasive species can successfully establish and persist in a habitat in the absence of fire. These include whether an invading species can survive and spread when fires occur and the degree to which the invader may ultimately alter that system. The process of invasion can be broken into four stages: (1) establishment, typically from seed, though long-distance dispersal of vegetative propagules may occur in some cases; (2) survival and reproduction (that is, persistence); (3) density increase within a site, which includes spread within a site, either through seed or vegetatively, as well as increases in abundance in surrounding areas; and (4) dominance, which implies not only the establishment of substantial abundance but also the suppression of other species or other types of ecological harm.

Several questions about how an invasive species relates to fire need to be addressed in order to predict the species relationship to fire in a particular setting:

1. Does fire enhance establishment by the nonnative?
2. Does fire result in the mortality of the nonnative?
3. Are burned plants able to regrow following fire and, if so, how rapidly do they recover?
4. How important is competition with native species to the response by a nonnative to fire? and
5. What effects does a nonnative species have on the characteristics of the fire regime?

The answers to these questions, though approximate, are used to frame the discussion of each of the invasive species discussed below and can be used to classify them into different functional types as described by Grace and others (2001).

The goal of this chapter is not to produce an exhaustive treatment of all major nonnatives in the Central bioregion, which can be found in Grace and others (2001). Rather, we summarize available information relating to nonnatives of greatest concern with regard to interactions with fire in these ecosystems. Following an introduction to the subregion and formations, three main topics will be discussed for species or groups of species within each formation in that subregion: (1) the role of fire or fire exclusion in promoting nonnative species invasions, (2) the effects of nonnative invasives on fuels and fire regimes, and (3) the use of fire to control nonnative invasives. Table 7-2 lists the species, organized by the formations in which they pose the greatest problem, that will be discussed in detail in this chapter.

The Mesic Tallgrass Prairie Subregion

This subregion is represented by tallgrass prairie ecosystems that extend, though not continuously, from Canada to Mexico (Risser and others 1981). Both characteristic grasses (for example, prairie cordgrass (*Spartina pectinata*) in the north and brownseed paspalum (*Paspalum plicatulum*) in the south) and

Table 7-2—Approximate threat potential of several major nonnative species in the Central bioregion within seven broad vegetation formations. L= low threat, H = high threat, P = potentially high threat, N= not invasive

Species		Formations					
Scientific name	Common name	Northern & Central tallgrass prairie	Southern tallgrass prairie	Northern mixedgrass prairie	Southern mixedgrass prairie	Shortgrass steppe	Riparian
Arundo donax	Giant reed	N	N	N	N	N	H
Bothriochloa ischaemum	Yellow bluestem	N	H	N	H	P	N
Bothriochloa bladhii	Caucasian bluestem	H	H	H	P	P	N
Bromus inermis	Smooth brome	H	N	P	N	N	N
Bromus japonicus	Japanese brome	N	N	H	N	H	N
Bromus tectorum	Cheatgrass	N	N	H	P	H	N
Centaurea spp.	Knapweeds	L	N	H	H	H	N
Cirsium arvense	Canada thistle	H	L	H	L	H	N
Coronilla varia	Crown vetch	H	N	L	N	N	N
Elaeagnus angustifolia	Russian-olive	N	N	H	L	H	H
Euphorbia esula	Leafy spurge	H	N	H	N	N	N
Imperata cylindrica	Cogongrass	N	H	N	N	N	N
Lespedeza cuneata	Sericea lespedeza	H	H	L	N	L	N
Melilotus alba	White sweetclover	H	N	H	N	L	N
Melilotus officinalis	Yellow sweetclover	H	N	H	N	L	N
Poa pratensis	Kentucky bluegrass	H	N	H	N	N	N
Pennisetum ciliare var. *ciliare*	Buffelgrass	N	L	N	H	H	N
Rhamnus cathartica	Common buckthorn	H	N	N	N	N	H
Rosa bracteata	Macartney rose	N	H	N	N	N	N
Sorghum halepense	Johnson grass	N	H	N	N	N	N
Tamarix spp.	Saltcedar, tamarisk	N	H	N	H	H	H
Triadica sebifera	Chinese tallow	N	H	N	N	N	H
Urochloa maxima	Guineagrass	N	N	N	H	N	N

forbs (for example, Canadian milk vetch (*Astragalus canadensis*) and prairie cinquefoil (*Potentialla arguta*) in the north, and Texas coneflower (*Rudbeckia texana*) and tropical puff (*Neptunia pubescens*) in the south) are regionally restricted to the northern versus southern lobes of the tallgrass prairie (Kucera 1992; Smeins and others 1992).

Historically, Native Americans were probably the primary source of ignition for fires in the mesic prairie subregion during normal weather conditions. Evidence regarding fire frequencies from this subregion is most reliably taken from scars embedded in trees. Such data indicate that prior to 1870, when European influences had clearly come to dominate, fires occurred at roughly 5-year return intervals, while for the period after that, return intervals of around 20 years are more characteristic (Wade and others 2000). Because of the strong influence of human-caused fires on the fire regime, predominant fire season has varied depending on the practices of the peoples living in the area more than on the influences of lightning strikes and dry frontal passages. Fuel loads in mesic grassland are characteristically high, in the range of 5 to 10 mt/ha. Such fuels characteristically produce flame lengths of 10 to 13 feet (3 to 4 m) and typically precluded the establishment and persistence of woody plants in areas where moisture conditions were adequate for forest development. The exclusion of fire leads to rapid conversion of many areas of mesic prairie to woodland (Wade and others 2000). Species of oak (*Quercus* spp.) and juniper (*Juniperus* spp.) are among the native woody plant groups whose ranges have expanded in recent times.

Successional responses to fire among prairie natives are influenced by season and frequency of burning. Prescribed burning in tallgrass prairie has been most commonly applied in spring or fall, with late spring or fall fires having a tendency to favor warm season grasses compared to early spring fires. Generally, late spring burns are recommended to foster the density of warm-season (C_4) grasses and prevent the establishment of nonnatives (Willson and Stubbendieck 2000). However, it is possible that while such a strategy may be successful against cool-season nonnatives (for example, Kentucky bluegrass) it may be less successful in cases of warm-season nonnatives (for example, Johnson grass).

Native prairie species respond differentially to fire frequency. Generally, big bluestem (*Andropogon gerardii*) shows no response to time since fire, while little bluestem (*Schizachyrium scoparium*), indiangrass (*Sorghastrum nutans*), and switchgrass (*Panicum virgatum*) all decrease with time since fire. Gibson (1988) found that perennial forbs and cool-season grasses increased with time since fire, while annual forbs and warm-season grasses decreased. Annual or frequent burning tends to reduce herbaceous plant diversity in tallgrass prairie (Wade and others 2000).

The management of wildlife species may also influence burn times in order to avoid direct mortality of individuals, particularly during nesting season. Integration of livestock grazing and fire management is constrained by the needs of the animals and by the effects of grazing on fuels. Results from various studies suggest that for the conservation of native diversity, a variety of burn times should be used (Howe 1995). However, when nonnatives are involved, many such recommendations must be reconsidered to avoid promotion of nonnative dominance.

Northern and Central Tallgrass Prairie Formation

In the northern and central tallgrass prairie, from Canada to Oklahoma, the topography is mostly gently rolling plains. Some areas are nearly flat, while other areas have high rounded hills. Elevation ranges from 300 to 2,000 feet (90 to 600 m). Summers are usually hot and winters cold, with the frost-free season ranging from 120 days in the northern portion to 235 days in the central portion. Average annual precipitation ranges from 20 to 40 inches (510 to 1,020 mm), and falls primarily during the growing season (Bailey 1995).

Most of the area is cultivated, and little of the native vegetation remains. Where it does occur, native vegetation in this formation is predominantly prairie dominated by bluestems (big bluestem, little bluestem, sand bluestem (*Andropogon gerardii* var. *paucipilus*)) and variously codominated by switchgrass, indiangrass, prairie sandreed (*Calamovilfa longifolia*), needle-and-thread grass (*Hesperostipa comata*), and hairy grama (*Bouteloua hirsuta*) (Küchler 1964). Many flowering forbs are also present. Most of these prairie species are classified as warm-season plants. Woody vegetation is uncharacteristic in native prairie, except within the eastern ecotone where the prairie grades into oak-hickory (*Quercus* spp.-*Carya* spp.) forest, and on floodplains and moist hillsides (Garrison and others 1977; Küchler 1964). Additionally, in places where fire is excluded and grazing is limited, deciduous forest is encroaching on the prairies (Bailey 1995). Where conditions are more mesic or otherwise favor tree and shrub growth, savannas (for example, oak savanna) are common. In these savannas, the vegetation between trees comprises species typical of tallgrass prairie (Nuzzo 1986).

The effects of long-term, frequent fires on central tallgrass prairie plant communities and the relationship between nonnative species and fire in these communities was examined in a 15-year study at Konza Prairie in northeastern Kansas (Smith and Knapp 1999). Long-term annual spring burning resulted in 80 percent to 100 percent reductions in number and abundance of nonnative plant species compared with infrequently burned plots. Nonnative species were absent from sites that had been burned 26 of 27 years, and nonnative species richness steadily increased as the number of times a site was burned decreased. The highest nonnative species richness occurred on sites burned fewer than 6 times over the 27-year period. Thus the cumulative effects of fire seem to be important in controlling invasion by nonnative species in tallgrass prairie. This effect may be due more to the increased productivity of dominant native C_4 grasses under a regime of frequent fire rather than to direct negative impacts of fire on nonnative species (Smith and Knapp 1999). Some species such as Caucasian bluestem appear to be less easily controlled by frequent fire (USDA, NRCS 2006).

The effects of frequent fire on plant community composition in tallgrass prairie varies with burn season (Howe 1995; Towne and Kemp 2003), other land management practices (for example, grazing by livestock and bison), climatic variation, topographic position, and the impact of management practices on soil, moisture availability, fire patchiness, and propagule pressure (Coppedge and others 1998; Hartnett and others 1996; Trager and others 2004; Vinton and others 1993). For example, while annual burning appears to reduce invasibility of areas studied at Konza Prairie, there is evidence that nonnative species richness increases 2-fold in annually burned sites that are grazed when compared to similar ungrazed sites (Smith and Knapp 1999). Similarly, the percent of species that were nonnative in bison wallows on Konza Prairie was higher than in surrounding grazed prairie sites whether sites were burned annually or on a 4-year rotation. In both wallow and prairie sites, most nonnative species (smooth brome, common pepperweed (*Lepidium densiflorum*), Kentucky bluegrass, and prostrate knotweed (*Polygonum aviculare*)) had higher cover on annually burned sites than on those burned every 4 years, although differences were not always significant (Trager and others 2004).

When there is a source of nonnative propagules in the area surrounding an annually burned site, the site may be more susceptible to establishment and spread of these species in the postfire environment (Grace and others 2001). Further studies at Konza Prairie by Smith and Knapp (2001) suggest that the size and composition of local species pools surrounding a target community is as important as the effects of fire in determining invasibility at this tallgrass prairie site. So, while fire is an important management tool for promoting dominance of native C_4 grasses and reducing nonnative species invasions, frequent fire may not sufficiently limit nonnative species invasions if the local nonnative species pool is relatively large. This is an important consideration as prairie ecosystems become increasingly fragmented and nonnative species establish in surrounding areas (Smith and Knapp 2001).

Important nonnative species in the tallgrass prairie formation include several introduced perennial pasture grasses such as smooth brome, Kentucky bluegrass and Caucasian bluestem; perennial forbs such as Canada thistle, crown vetch (*Coronilla varia*), sericea lespedeza, biennial sweetclovers, and leafy spurge; and the shrub common buckthorn (*Rhamnus cathartica*). Information on fire relationships for these species follows, although Caucasian bluestem is covered in the section on southern tallgrass prairie and leafy spurge is covered in the section on northern mixedgrass prairie.

Smooth brome—Smooth brome is a widespread nonnative grass that has been widely planted to increase forage or to reduce erosion following wildfires; it readily escapes into other habitats, particularly in the northern portion of its range (Sather 1987a, TNC review). It generally invades after disturbance and is a common invader of prairie habitat throughout the Great Plains (Howard 1996, FEIS review). Smooth brome is a cool-season, perennial grass that begins its growth in early spring and continues growth late into fall. It is a prolific seed producer (figure 7-2), setting seed mid-summer through fall. Studies indicate that it has a high level of drought resistance while not being particularly flood tolerant (Dibbern 1947). Most smooth brome cultivars are rhizomatous, though some northern cultivars have bunchgrass morphology. Because of the magnitude of morphological variation among cultivars, Howard (1996) suggests that postfire recovery may differ among cultivars, although such differences have not been documented in the literature.

Figure 7-2—Smooth brome in flower. This widely distributed species, characteristic of the northern and central tallgrass prairie subregions, is a major invader of disturbed native prairie and restoration sites. (Photograph by Mike Haddock, with permission, (see www.lib.ksu.edu/wildflower).)

Exclusion of fire from tallgrass prairie sites may promote smooth brome invasion, and smooth brome can be adversely affected by burning in some cases (Blankespoor and Larson 1994; Kirsch and Kruse 1973). According to Masters and Vogel (1989), smooth brome is usually found in tallgrass prairie in areas with a history of overgrazing and/or fire exclusion. A comparison of burned and unburned prairie remnants in northwestern Illinois indicates that fire exclusion favors nonnatives such as smooth brome and Kentucky bluegrass over native prairie species such as little bluestem and porcupinegrass (*Hesperostipa spartea*), while periodic spring burning over a 20-year period increased mean native species richness and reduced nonnative species richness (Bowles and others 2003).

Smooth brome invasion may sometimes be enhanced by fire, and it appears to suffer little mortality when burned, responding to fire by sprouting from rhizomes and possibly by tillering. Fire in early spring or fall may promote smooth brome by removing litter from sod-bound plants (Howard 1996). Over a 15-year period, annual spring burning did not reduce this species in tallgrass prairie in northeastern Kansas (Smith and Knapp 1999). Smooth brome's phenological stage at the time of burning may be more important in determining its response to fire than the date of burning. Burns during the spring growth period may reduce smooth brome vigor, and several researchers have reported reductions in smooth brome after late-spring fires (Howard 1996). Grilz and Romo (1994, 1995) found no significant effect of fall or spring burning on smooth brome stem density in tallgrass prairie in Saskatchewan. The authors speculate that a single dormant-season burn in this C_3-dominated fescue prairie is not expected to reduce smooth brome, and may actually increase smooth brome density if native species are suppressed (Grilz and Romo 1994).

Willson and Stubbendieck (2000) have proposed a model for managing smooth brome using prescribed fire in northern tallgrass prairie. One fundamental aspect of their proposed approach relates to the plant's phenology. Tiller emergence in smooth brome occurs twice: once after flowering in early summer and again in fall (Lamp 1952). Fall tillers overwinter and then elongate in early May in the central Great Plains and Midwestern States. Willson (1991) found a 50 percent reduction in the density of tillers following prescribed burning in eastern Nebraska in early May, presumably because of reduced carbohydrate reserves during the time of tiller elongation. Kirsch and Kruse (1973) and Old (1969) also reported that burns in April and May have negative effects on smooth brome. Additional studies by Willson and Stubbendieck (1996, 1997) in Minnesota and Nebraska have examined burning during (1) tiller emergence, (2) tiller elongation, (3) heading (initiation of morphological changes associated with sexual reproduction), and (4) flowering. Burning during any of these periods resulted in reductions in plant growth.

A second aspect of control of smooth brome relates to native competitors. Field observations indicate that burning when smooth brome tillers are elongating in the spring shifts the competitive balance to favor warm-season tallgrass species such as big bluestem. In the absence of native warm-season grasses, the adverse effects of fire on smooth brome are not sufficient to control smooth brome (Willson and Stubbendieck 2000). Fire and competition in combination may inhibit this species, though in the absence of native warm-season competitors, even fire precisely timed to have maximum impact on smooth brome will not be sufficient to lead to sustained population reductions (Willson and Stubbendieck 2000). Other examples of this pattern have been observed involving Kentucky bluegrass, quackgrass, and possibly also crested wheatgrass and Canada thistle.

There is no indication that smooth brome alters the fire regime of systems it invades. In the absence of specific information, we might hazard a guess that this results partly from the fact that the grasslands that smooth brome invades are typically high in herbaceous cover and naturally prone to high fire frequencies. Another factor that might contribute is its perennial nature and the fact that its aboveground tissues do not dry and become highly flammable early in the summer. Since the literature suggests that smooth brome can be burned almost any time of year, it is unclear how important the difference in fuel drying may be.

Kentucky bluegrass—Kentucky bluegrass, like smooth brome, is a perennial, cool season, rhizomatous invader of native grasslands that is widely planted, in this case primarily for lawns, as well as for pastures and erosion control. Kentucky bluegrass is a significant invader in more mesic sites in the upper Great Plains as well as in eastern prairies and it is considered a major problem for tallgrass and mixedgrass prairies (Hensel 1923; Sather 1987b, TNC review; Stohlgren and others 1998). Due to its strongly rhizomatous nature, Kentucky bluegrass is capable of rapid vegetative spread (Etter 1951), and populations are persistent.

Exclusion of fire from tallgrass prairie sites seems to favor nonnative, cool season grasses such as Kentucky bluegrass (Benson 2001; Bowles and others 2003; Smith and Knapp 1999), and many studies have shown that burning has negative effects on Kentucky bluegrass production in tallgrass prairie. Fire impacts on Kentucky bluegrass are most pronounced when burning takes place during tiller elongation in the spring (for example, Archibold and others 2003; Curtis and Partch 1948; Ehrenreich 1959) and when fires are repeated at relatively frequent intervals. Kentucky bluegrass cover tends to be lower on annually burned prairie sites than on sites burned at 4-year or longer intervals (Collins and others 1995; Hartnett and others 1996; Vinton and others 1993). Kentucky bluegrass is not always adversely affected by burning. Kirsch and Kruse (1973) found it to be unaffected by a May burn in mixedgrass prairie in North Dakota.

The effect of fire on Kentucky bluegrass is strongly influenced by available soil moisture, which may either enhance or nullify any detrimental impact of fire (Anderson 1965; Blankespoor and Bich 1991; Zedler and Loucks 1969). Late-spring burning on a tallgrass prairie remnant in South Dakota resulted in a significant ($P < 0.01$) reduction in Kentucky bluegrass biomass on plots with low water content and an insignificant reduction on plots with high water content. Kentucky bluegrass biomass increased on unburned plots regardless of soil moisture conditions (Blankespoor and Bich 1991).

Because Kentucky bluegrass is a cool-season grass that elongates early in the growing season (an attribute that helps to make it a desirable forage species), prescribed burning designed to shift the competitive balance to native grasses is used widely for controlling this species. In addition to the importance of timing and frequency of fire and available moisture, native competitors make a critical difference in the impact of burning on Kentucky bluegrass (Schacht and Stubbendieck 1985). Where Kentucky bluegrass grows with warm-season native grasses (as would be typical in tallgrass prairie), repeated spring burning offers a substantial opportunity for shifting the competitive balance toward native species (Owensby and Smith 1973; Towne and Owensby 1984). Indeed, native warm-season grasses dominated and Kentucky bluegrass occurred less frequently in annually-burned, mesic, tallgrass prairie sites at the Konza Preserve in Kansas, compared to infrequently burned (20-year burn cycle) or unburned sites (Benson 2001; Smith and Knapp 1999). In more arid regions of the western Great Plains, where native cool-season species are more common, only a narrow window of opportunity exists for burning while Kentucky bluegrass is elongating but before native species (for example, western wheatgrass (*Pascopyrum smithii*) and needlegrasses (*Achnatherum* and *Hesperostipa* spp.)) elongate (Sather 1987b). The use of early spring burning becomes somewhat more complex when both smooth brome and Kentucky bluegrass occur together because the optimum time for controlling each of these species with spring burning does not appear to be the same. At present, there are insufficient examples for drawing firm conclusions about whether Kentucky bluegrass can be eliminated from tallgrass prairie with fire.

Canada thistle—Canada thistle is a perennial forb that spreads by seed and by creeping, horizontal roots. It occurs throughout much of Canada and the United States in nearly every upland herbaceous community within its range, particularly prairie communities and riparian habitats. In the Central bioregion Canada thistle threatens northern mixedgrass prairie and shortgrass steppe formations, in addition to northern and central tallgrass prairie. It is an early successional species, establishing and developing best in open, moist, disturbed areas. Canada thistle often grows in large clonal patches, and individual clones may reach 115 feet (35 m) in diameter, interfering with native species (Zouhar 2001d, FEIS review).

Canada thistle is adapted to establish on exposed bare soil on recently burned sites and to survive fire. Several examples in the literature indicate Canada thistle establishment from wind-deposited seed, anywhere from 2 to 9 years after fire, on sites where it was absent from the prefire plant community and adjacent unburned areas (Zouhar 2001d). Additionally, Canada thistle may dominate the soil seed bank where it occurs, as was observed on mixedgrass prairie sites in North Dakota (Travnicek and others 2005). Its extensive root system allows Canada thistle to survive major disturbances and fires of varying severity by producing new shoots from adventitious buds on roots. The response of Canada thistle to fire is variable, however, depending on vegetation and site characteristics, as well as frequency, severity, and season of burning (Zouhar 2001d).

Results from studies on prairie and riparian sites demonstrate Canada thistle's variable response to fire. There were no significant differences ($P < 0.05$) in Canada thistle cover after spring burning in the prairie pothole region of Iowa (Messinger 1974). Prescribed burning in spring either reduced or did not change canopy cover of Canada thistle on bluestem prairie sites in Minnesota. Results varied among sites that differed in plant community composition and in time and frequency of burning (Olson 1975). In a prairie site at Pipestone National Monument, Minnesota, 5 years of annual spring burning in mid- to late April, with fires of low to moderate severity, reduced the frequency of Canada thistle over time until it was absent after the fifth year (Becker 1989). Similarly, observations in tallgrass prairie sites in South Dakota indicate that late spring prescribed burning (when native species are still dormant) on a 4 to 5 year rotation (as per the historic fire regime) encourages the growth of native plants and discourages the growth of Canada, bull, and musk thistles. Livestock use must be carefully timed following burning, since grazing early in the growing season can potentially negate beneficial effects of prescribed fire (Dailey, personal communication, 2001). In a study conducted on a mesic prairie site in Colorado, plots that were burned frequently (5 times over 7 years) had lower density of Canada thistle than did an area that was burned only twice during the same period. Results were inconclusive, however, since the final season of the study saw increased spread of Canada thistle from the surrounding area, probably due to clonal growth from existing plants (Morghan and others 2000). On a common reed marsh in Manitoba, Canada, thistle response to burning varied with burn season. Aboveground biomass, stem density, and seedling density were unchanged on spring burns but increased on both summer and fall burns (Thompson and Shay 1989). A Canada thistle clone in a mid-boreal wetland site in northeastern Alberta was not noticeably changed when burned in the spring with a propane torch to simulate both "light" and "deep" burns (Hogenbirk and Wein 1991). The authors concluded that there exists a moderate to high probability that Canada thistle and other Eurasian xerophytic species will dominate these wet meadows in the short term after fire, and that they will continue to dominate small areas for longer periods (Hogenbirk and Wein 1995).

Canada thistle may change the fuel characteristics of invaded sites with its abundant aboveground biomass. Hogenbirk and Wein (1995) suggest that in boreal wet-meadows, Canada thistle has the potential to increase fire frequency and perhaps severity as a result of its abundant and readily ignited litter. No additional information on the potential for Canada thistle to alter fuel characteristics or fire regimes in the Central bioregion is available in the literature.

Season of burn is an important consideration for determining species composition and cover in the postfire tallgrass prairie community (Howe 1994a, b). According to Hutchison (1992a) prescribed burning is a "preferred treatment" for control of Canada thistle, and late spring burns effectively discourage this species, whereas early spring burns can increase sprouting and reproduction. During the first 3 years of control efforts, he recommends that burns be conducted annually. At a mixedgrass prairie site at Theodore Roosevelt National Park, North Dakota, fall burning appeared to increase Canada thistle stem density because it emerged more rapidly on burned sites than on unburned sites. By the second postfire growing season, however, Canada thistle stem densities were similar on burned and unburned sites. Grass foliar cover was unaffected by burning, while forb cover increased to 4 percent in the burned area compared to a 1 percent increase in the unburned area. The number and variety of species in the soil seed bank were not affected by prescribed burning, and Canada thistle and Kentucky bluegrass accounted for over 80 percent of germinants from the soil seed bank (Travnicek and others 2005).

Crown vetch—Crown vetch is a perennial forb that has been widely planted in the Northern United States, primarily for erosion control but also for pasture, mine reclamation, and ornamental ground cover. It tolerates a broad range of conditions and has been reported to escape from cultivation in every state except Alaska, California, North Dakota, and Louisiana. In the Central bioregion, it occurs primarily in northern and central tallgrass and northern mixedgrass prairies, growing best in areas with more than 18 inches (460 mm) annual precipitation. A review by Tu (2003, TNC review) indicates that crown vetch infestations commonly begin along roads and rights-of-way and spread into adjacent natural areas such as grasslands and dunes in Missouri, Minnesota, Illinois, and Iowa. It sometimes forms monotypic stands and can exclude native plant species by covering and shading them. Crown vetch spreads by abundant seed production, a multi-branched root system, and strong rhizomes. Observations suggest that crown vetch seed can remain viable in the soil for many years, although maximum longevity is not known (Tu 2003).

Little research is available regarding the invasiveness, impacts, longevity, and control of crown vetch in any plant community, particularly in relation to fire. One may speculate that as a rhizomatous, perennial legume it is likely to occur in the initial postfire community by sprouting from rhizomes or establishing from seed in the soil seed bank. Crown vetch withstands some grazing pressure and sprouts when aboveground tissue is removed by pulling, cutting, or mowing. Observations by managers indicate both sprouting and stimulation of germination following burning (Tu 2003).

Some managers use fire as part of their management program for crown vetch. Solecki (1997, review) suggests that burning in late spring controls crown vetch seedlings but that mature plants are only top-killed. A manager in central Minnesota uses a 3-year burning cycle with the objective of exhausting the crown vetch seed bank by stimulating germination. Burning kills crown vetch seedlings and slows spread of mature plants. Where crown vetch dominates, burning alone is not likely to control it. In large infestations with sufficient fuel, fire can reduce crown vetch on the periphery of stands but leave infestation interiors unaffected. Frequent (annual) prescribed fire in late spring may control sparse infestations of crown vetch in ecosystems dominated by species adapted to such fire regimes (Tu 2003).

Crown vetch infestations may alter fuel characteristics on invaded sites, especially where it occurs in single-species stands. Managers report that dense infestations of crown vetch do not carry fire well (Tu 2003). More information is needed on the fire ecology of this species.

Sericea lespedeza—Sericea lespedeza is a warm-season, perennial, erect, multi-stemmed forb that grows 2 to 7 feet (0.5 to 2 m) tall. It spreads by seed and can sprout following damage to aboveground tissues. Sericea lespedeza has been widely planted in the United States and has escaped cultivation in many areas. It occurs from southern New England west to southern Wisconsin, Iowa and Nebraska, and south to Texas and Florida. Within this range it may be invasive on open sites such as prairie, grassland, and pasture, especially near areas where it was previously planted. It grows best in areas with more than 30 inches (760 mm) of annual precipitation, although its deep, well-developed root system allows sericea lespedeza to grow on relatively dry or infertile sites (Munger 2004, FEIS review).

No published information specifically describes adaptations of sericea lespedeza to fire. Perennating buds on sericea lespedeza are located 0.5 to 3 inches (1 to 8 cm) below the soil surface; therefore, individual plants are likely to survive the heat of most surface fires despite damage or destruction of aboveground tissue. Sericea lespedeza is also known to sprout in response to mechanical damage of aboveground tissue. It may develop a seed bank and thereby establish from on-site seed sources after fire. Sericea lespedeza seedlings may be favored in postfire environments (Munger 2004). Conversely, frequent fire may exclude sericea lespedeza from plant communities adapted to such fire regimes. For example, on three prairie remnants in eastern Arkansas, sericea lespedeza was absent from two sites that were annually burned for over 60 years and occurred on one site where fire was excluded for 17 years prior to study (Irving and others 1980). Although inconclusive, these results suggest that frequent fire in prairie communities may exclude sericea lespedeza. Conversely, Griffith (1996) implicates annual burning among reasons for spread of sericea lespedeza based on field observations.

Some studies suggest that fire effects on sericea lespedeza populations depend on a number of factors including fire conditions, climatic conditions before and after fire, plant community composition, and interactions with other disturbances or management practices. In a study designed to evaluate interactive effects of fire and grazing in an area primarily characterized as tallgrass prairie (dominated by little bluestem, big bluestem, and indiangrass) in central Oklahoma (Fuhlendorf and Engle 2004), sericea lespedeza occurred throughout the study area. Treatments included (1) burning of spatially distinct patches within a treatment unit (one sixth of each unit burned each spring (March to April) and one sixth burned each summer (July to October), representing a 3-year return interval for each patch), and free access by moderately stocked cattle; and (2) no burning with free access by moderately stocked cattle. Sericea lespedeza occurred in all treatment units and increased overall during the study. Cover and rate of increase of sericea lespedeza were significantly greater in the unburned units during the 4 years of the study. Sericea lespedeza cover fluctuated at low values in the units with burned patches. These results suggest that the use of prescribed fire in conjunction with grazing in a shifting mosaic pattern may help reduce the importance and spread of sericea lespedeza, while grazing in the absence of burning leads to continued increases in sericea lespedeza cover (Fuhlendorf and Engle 2004).

Few studies document the results of prescribed fire used to control sericea lespedeza populations, although it is suggested that spring fires may promote seed germination (Segelquist 1971) and sprouting (Stevens 2002, TNC review). Under these circumstances, a useful strategy might be to promote spring growth of native species with a prescribed burn and follow up with control measures that focus on eradication of sericea lespedeza sprouts and seedlings. Sericea lespedeza seedlings may be susceptible to mortality from spring burning (Cooperative Quail Study Association 1961), although it is unclear at what age newly established plants can survive fire damage by sprouting. Prescribed fire may be useful for reducing density of sericea lespedeza seed banks. Evidence from laboratory studies on heat tolerance of sericea lespedeza seed suggests that seeds exposed to fire may sustain lethal damage (Munger 2004).

No information is available on the effects of sericea lespedeza invasion on fuel or fire regime characteristics in invaded communities. More research is needed to understand its response to fire in tallgrass prairie.

Biennial sweetclovers—White sweetclover (*Melilotus alba*) and yellow sweetclover (*Melilotus officinalis*) are widely planted annual or, more typically, biennial leguminous forbs that have escaped cultivation and are established in disturbed areas, along roadsides, and in many grasslands and prairies (Turkington and others 1978). Sweetclover sometimes occurs in dense patches that exclude native vegetation (for example, Heitlinger 1975; Randa and Yunger 2001). Sweetclover reproduces by seed, with peak germination in early spring. Seeds are hard upon dispersal and may remain viable in the soil for many years and possibly decades. Biennial sweetclovers produce a vegetative shoot in the first season of growth that dies back in fall. Most root development occurs in late summer, and during this time crown buds form just below or at the soil surface. The taproot and crown buds overwinter; the following spring and early summer, one or more flowering shoots emerge from the buds and rapidly elongate (Turkington and others 1978).

Literature reviews indicate that burning enhances sweetclover establishment in grasslands (for example, Heitlinger 1975; Turkington and others 1978; Uchytil 1992a), and evidence suggests that burning helps to break seed dormancy and enhance postfire germination (Kline 1986). The effect of fire on sweetclover populations depends on timing and frequency. Sweetclover plants are killed when burned during the growing period (Heitlinger 1975), though they survive if burned before shoots begin to elongate (Eckardt 1987, TNC review; Kline 1986). A fire on a tallgrass prairie site in Minnesota in early May resulted in an increased number of first-year sweetclover plants and a decreased number of second-year plants, compared to an unburned area. Fire on the same site in early July reduced the number of both first- and second-year sweetclover plants (Heitlinger 1975). Infrequent burning, especially in combination with grazing (Hulbert 1986, review), actually promotes spread of these species (Kline 1986; Randa and Yunger 2001).

Prescribed fire is often used as a part of management plans for controlling sweetclover. Eckardt (1987) presents information from managers who use prescribed fire as part of their management plan for prairie sites where sweetclover occurs. Based on work in Minnesota, Heitlinger (1975) suggests three possible strategies for using prescribed fire to reduce soil seed reserves and prevent production of additional seed: (1) burning annually in early May, or when second year shoots are clearly visible; (2) burning every other year in early July, before second-year plants ripen seed; and (3) burning annually in early September near the beginning of the "critical growth period" when roots are growing rapidly. Any surviving second-year plants must be hand-pulled or clipped at the base before they produce seed (Heitlinger 1975). In Wisconsin, a combination of an April burn followed the next year by a May burn was more successful in reducing white sweetclover than other burning combinations. Heavily infested prairie stands where this burning combination was conducted twice, separated by 2 years without burning, became almost completely free of white sweetclover (Kline 1986). Problems with this method may arise if the burn is either too early or patchy, leaving viable seeds or undamaged second-year shoots (Cole 1991). These relatively aggressive fire prescriptions designed to control sweetclover will have variable impacts on desirable native species that must be considered before implementing management plans. In arid regions, there may be insufficient fuels for frequent burning as well. Combinations of mowing or herbicides with prescribed burning are also suggested for sweetclover control (Cole 1991; Eckhardt 1987).

No information is available regarding the potential for sweetclover to alter fuel characteristics or fire regimes in tallgrass prairie.

Common buckthorn—Common buckthorn is a dioecious, large shrub or small tree that attains heights up to 20 feet (6 m). Although it has the ability to produce vegetative sprouts, it reproduces primarily by seed. While it can successfully establish in undisturbed sites, there is evidence that excess shading reduces seedling growth (Converse 1984a, TNC review).

Common buckthorn invades oak savanna, pastures, fens, and prairies. Savanna and prairie communities where fire has been excluded may be especially vulnerable to common buckthorn invasion (for example, Apfelbaum and Haney 1990; Packard 1988[1]). Boudreau and Willson (1992) report that where it has invaded oak savanna habitat at Pipestone National Monument in Minnesota, native midstory and understory species have been virtually eliminated. Godwin (1936) reported an example in Europe for a related buckthorn species, glossy buckthorn (*Frangula alnus*), where seedlings established and developed into a continuous shrub thicket in 20 years in a mixed sedge marsh. The most detailed description of stand development in common buckthorn yet presented is by Archibold and others (1997). In this study, initial establishment of a few individuals on a site near Saskatoon, Saskatchewan, was followed by a localized increase in seedling recruitment once the initial trees became reproductive.

Common buckthorn often forms dense, even-aged thickets in both wetlands and woodland understories (for example, Archibold and others 1997; Packard 1988). These stands may reduce or alter understory herbaceous density and composition, thus altering fuel characteristics on invaded sites. This topic deserves further study.

Common buckthorn reproduces from seed. Seeds are dispersed by birds and mammals, and require scarification for germination. Seedling recruitment is most successful where there is ample light and exposed soil (Converse 1984a). These qualities imply that common buckthorn may be adapted to seedling establishment in the postfire environment, although research on this topic is unavailable.

Common buckthorn also sprouts vigorously following removal of or damage to topgrowth (Heidorn 1991), and a limited amount of information suggests that large common buckthorn plants are able to survive fire but that small individuals and seedlings are killed. Boudreau and Willson (1992) report that spring burning in an oak savanna in Minnesota killed common buckthorn seedlings, which were unable to sprout after fire. However, mature trees were only top-killed and survived by sprouting. Converse (1984a) suggests that prescribed fire treatments do not control *Rhamnus* species, in part because species in this genus suppress herbaceous fuel beneath the canopy.

Repeated fires may facilitate development of the herbaceous layer and lead to eventual control of common buckthorn, with the caveat that adequate fuels must exist for control to be obtained through burning. Heidorn (1991) suggests that regular prescribed fire kills buckthorn (*R. cathartica*, *Frangula alnus*, and *R. davurica*) seedlings and larger stems in fire-adapted upland and wetland sites, and that annual or biennial burns may be required for 5 to 6 years or more to reduce the population. Prescribed spring burns conducted annually for 5 years from mid- to late April (when cool-season grasses had initiated growth) at Pipestone National Monument resulted in no appreciable change in common buckthorn cover (Becker 1989). In a dry oak forest dominated by hybrid black oak (*Quercus velutina* x *Q. ellipsoidalis*) and heavily infested with bush honeysuckle (*Lonicera* x *bella*) and common buckthorn, prescribed burns in 2 consecutive years followed by a year with no burn resulted in a decrease in cover of buckthorn in each of the burn years, followed by a slight increase in the no-burn year. Most individuals sprouted from the base, but sprouts were not vigorous (Kline and McClintock 1994).

Southern Tallgrass Prairie Formation

The southern tallgrass or coastal prairie formation occurs in east-central Texas and along the Gulf coastal plain in Texas and Louisiana. It is composed of gently rolling to flat plains, with elevations ranging from sea level to 1,300 feet (400 m). Most of the Coastal Plain streams and rivers are sluggish, with numerous wetland areas along the coast. Winters are warmer and there is more precipitation than in the more temperate northern and central prairies. Average annual precipitation ranges from 35 to 55 inches (890 to 1,410 mm) from west to east along the coast. The length of the frost-free season varies from 300 days to an almost entirely frost-free climate along the coast (Bailey 1995; Garrison and others 1977).

[1] Packard (1988) provides a particularly poignant quote, "An especially sad landscape features forlorn, aristocratic old oaks in an unbroken sea of buckthorn – the understory kept so dark by the dense, alien shrubbery that not one young oak, not one spring trillium, not one native grass can be found"

Vegetation in the southern tallgrass formation consists of savannas, prairies and coastal wetlands. Post oak and blackjack oak dominate the cross timbers habitat that abuts the prairie from northeastern Texas to southeastern Kansas, and oak and hickory are common on the eastern ecotone near the forest region. At higher elevations, little bluestem and indiangrass dominate. Switchgrass and eastern gamagrass (*Tripsacum dactyloides*) become dominant in low areas. In areas where prairie and salt marsh intergrade (salty prairie), gulf cordgrass becomes the dominant graminoid. Big bluestem is present in moist, sandy soils but is less common than in the northern and central tallgrass prairies. Brownseed paspalum is an important and conspicuous part of the vegetation of the upper coastal prairie north of the San Antonio River, while meadow dropseed (*Sporobolus asper* var. *asper*) replaces it in the lower coastal prairie (Smeins and others 1992).

The coastal prairie has been largely converted to agricultural fields and urban areas and is severely infested with invasive species. Only about 1 percent of the original coastal prairie remains in a seminatural condition (USGS 2000). Research in coastal prairie at the Brazoria National Wildlife Refuge, Angleton, Texas, provides some preliminary observations about the relationship between fire and invasive species in this landscape. Factors such as previous soil disturbance and soil salinity influence both variety and abundance of native and nonnative species. For example, nonnative invasive species such as Chinese tallow and Macartney rose are well established in abandoned rice fields and previously overgrazed areas, respectively. Areas with high-salinity soils are generally free of invasive species, with the exception of tamarisk. Several other invasive species occur on roadsides and in disturbed areas of the refuge but have not yet invaded native communities. This suggests that these species may be inhibited by the competitive effects of native dominant species, and that practices that support native species' dominance may discourage further invasions. For example, since fire has been reintroduced to this landscape, native species seed production has increased and recolonization of former agricultural fields by native species has been observed. Further research is needed to evaluate the threat potential of nonnative species and the role that fire plays (Grace and others, unpublished report, 2005).

The following discussion provides information on the biology and fire ecology of some important nonnative invasive species in the southern tallgrass formation. Common nonnative invasive woody species include Chinese tallow, tamarisk, and Macartney rose. Several introduced pasture grasses have become invasive in southern tallgrass ecosystems and mesquite savannas of southern Texas. Invasive grass species in these ecosystems include yellow and Caucasian bluestems, guineagrass, and Johnson grass. Cogongrass (*Imperata cylindrica*) is just beginning to invade areas of southern tallgrass prairie, and itchgrass is also potentially invasive in southern tallgrass prairie ecosystems. In addition to the species covered in this section, species such as Canada thistle, sericea lespedeza, and buffelgrass are also invasive in southern tallgrass prairies, but are covered in other sections of this report.

Chinese tallow—Chinese tallow is an early successional tree (Bruce and others 1995) that can reach heights as great as 60 feet (18 m) and diameters in excess of 3 feet (1 m). The maximum life span is not known, but after a few decades trees typically become senescent. This species has exceptional growth rates and precocious sexual reproduction, producing seed within only a few years of establishment. Chinese tallow is also capable of vigorous sprouting from the roots following top-kill or damage, either at the base of the tree or at distances up to 16 feet (5 m) from the trunk. North American populations of Chinese tallow may be susceptible to herbivory, but the predators that control them in their native and other habitats do not occur in southern tallgrass prairie (Rogers and Siemann 2002, 2003, 2004, 2005; Siemann and Rogers 2003a, b).

Seeds of Chinese tallow are dispersed primarily by birds or via surface water flow. Seed germination typically takes place during fall through spring. Cameron and others (2000) indicate that seeds can survive up to 7 years when stored in refrigerated conditions. Chinese tallow appears to develop at least a short-lived seed bank (Renne and Spira 2001), though it is doubtful the seeds last more than a few years in the field. Preliminary analyses indicate that in the field, tallow seeds germinate within their first year without a light requirement if moisture is available (Billock, personal communication, 2003). There is also evidence that germination is stimulated by fluctuating temperatures (Donahue and others 2004, Nijjer and others 2002), but there is no indication that germination and establishment are particularly promoted by fire; information is lacking on this point.

The distribution of Chinese tallow extends from South Carolina in the east to southern Texas, and it has recently established in areas of southern California. Studies indicate that it has a broad tolerance of soil and moisture conditions (Butterfield and others 2004), although it can be limited by excess salinity (Barrilleaux and Grace 2000). Chinese tallow is likely to be a problem invader only in the southernmost parts of the Central bioregion. Chinese tallow invades a variety of habitats, but the principal grasslands it occupies are coastal marsh and coastal prairie. Here it establishes in abandoned rice fields, riparian zones and fence lines, and then spreads into the prairie (fig. 7-3). It appears

Figure 7-3—Chinese tallow is a fast growing tree that invades both disturbed and undisturbed prairie (as well as other habitats) in the southern tallgrass prairie formation. (Photo by Jim Grace.)

to be able to establish in undisturbed sod at some sites (J. Grace, personal observations, Brazoria National Wildlife Refuge, Texas, 1997-2002), though competition from native vegetation can inhibit persistence (Siemann and Rogers 2003a). It is likely that this species poses the greatest single threat to the conservation and restoration of the southern tallgrass prairie due to its capacity for rapid invasion (Bruce and others 1995) and its dramatic impacts (Grace 1998).

Studies indicate that Chinese tallow is vulnerable to fire when fuels are adequate (Grace 1998). Seedlings, in particular, suffer substantial mortality when burned within their first year. Because of its well-developed capacity for regrowth, it has been hypothesized that Chinese tallow is able to survive burning once it reaches a sufficient size. The critical size would depend in part on available fuel and related fire intensity/severity. Therefore, if burning can be conducted at sufficient frequencies and fuel conditions are sufficient for complete burns to be maintained, fire can be used to keep trees below the critical size and maintain the pyrogenicity of the system. Several experiments (Grace and others, unpublished report, 2005) indicate that the relationship between fire effect and tree size is actually complex and depends on season of burn as well as other factors. Available evidence indicates that burns have long-term detrimental effects on even the largest Chinese tallow trees. Thus, for isolated trees surrounded by a sufficient layer of herbaceous fuels, even the largest individuals can be top-killed by repeated fires (J. Grace, personal observations, Brazoria National Wildlife Refuge, Texas, 1997-2002).

What makes Chinese tallow invasions more problematic than invasions by other woody species is the change in fuel properties of tallow-invaded communities. When trees grow in closed stands, herbaceous fuel biomass and continuity are rapidly reduced, making the system less flammable. Currently, only widespread chemical applications are capable of returning these ecosystems to a condition where grass cover is sufficient to carry fire.

Substantial effort is underway to determine whether a prescribed burning regime can be used to prevent Chinese tallow from invading native prairie (Grace and others, unpublished report, 2005). Many factors influence successful management, including (1) growth conditions for tallow, (2) growth conditions for fuel species, (3) site history, (4) site management, and (5) seed sources and site conditions for tallow recruitment. Results are preliminary and in need of verification. Nonetheless, there are several tentative findings from this work:

1. Fire results in substantial mortality of tallow trees only when they are small (less than 1 foot tall).
2. Top-killing of Chinese tallow can be accomplished for trees of any size as long as there is sufficient fuel around the tree to expose the base of the tree to prolonged lethal temperatures. Tallow cannot be managed successfully using fire alone when fuel conditions are insufficient.
3. Resprouting rates are high (greater than 50 percent) for established tallow trees, and regrowth can be rapid (within weeks following a growing-season fire).
4. Equal rates of mortality or top-kill can be obtained using either growing- or dormant-season burns. However, regrowth rates are greater following dormant-season burns. Thus overall, growing-season burns are more effective in controlling tallow, though dormant-season burns can be of some use.
5. The presence of individual trees does not suppress herbaceous growth sufficiently to render fire ineffective for management. However, on sites previously under cultivation, fuel load and continuity are often poor and individual trees are not likely to be controlled by fire.
6. Dense seed sources can overwhelm even mature prairie. Only small gaps in grass cover are needed for tallow to establish successfully. Thus, unmanaged prairie is highly vulnerable to complete replacement by tallow when seed sources are within bird-dispersal range.
7. It appears that when fuel conditions are sufficient to obtain complete burns, growing-season burns approximately every 3 years can keep tallow from invading and taking over prairie. However, single burns that are not repeated can actually increase the vulnerability of prairie to tallow invasion (Hartley and others 2007).

Chinese tallow has ability to establish and grow rapidly, reducing fine fuels in the understory. Frequent burning and vigilance will be needed to prevent further conversion of great tracts of southern mesic grasslands to tallow forests. Many areas have already succumbed and now require substantial rehabilitation to recover any sort of native character.

Tamarisk—Tamarisk refers to a complex of taxa that are of disputed affinity, of which "saltcedar" typically refers to *Tamarix ramosissima* or *T. chinensis*. In practice and in this report, these entities are generally combined for ecological and management discussions. Tamarisk is a long-lived (50 to 100 or more years), deciduous shrub or tree that attains heights of up to 26 feet (8 m), develops a deep taproot, and can spread by adventitious roots (Zouhar 2003c, FEIS review).

Tamarisk is most widespread and invasive along riparian areas of the Southwestern United States (chapter 8). As a facultative phreatophyte that is tolerant of high salinity (Zouhar 2003c), tamarisk is also well adapted to invade coastal prairie sites with saline soils, and is spreading along the Gulf of Mexico into the southern tallgrass prairie (Grace and others 2001). In the coastal prairie, sites occupied by tamarisk are generally upland and part of the fire-dependent ecosystem (J. Grace, personal observations, San Bernard National Wildlife Refuge, Texas, 1999-2000).

Reproductive strategies of tamarisk (evident from research in other bioregions) may allow it to establish and spread after fire in southern tallgrass and coastal prairie. As of this writing, no information is available in the literature regarding the response of tamarisk to fire in these ecosystems. However, it appears to be able to withstand a great deal of damage without incurring death (J. Grace, personal observation, San Bernard National Wildlife Refuge, Texas, summer 2001). Additionally, research in other ecosystems indicates that tamarisk is highly tolerant of burning and usually sprouts vigorously after fire (see section on "Riparian Formation" for more details page 137).

While changes in fuel and fire regime characteristics are sometimes attributed to tamarisk invasion in southwestern riparian areas (chapter 8), the effects of tamarisk invasion are difficult to disentangle from the many impacts of modern development that have enhanced its spread (Glenn and Nagler 2005; Zouhar 2003c). Fine fuels are much less abundant beneath tamarisk canopies in southern tallgrass prairies than in uninvaded prairie sites (J. Grace, personal observations, San Bernard National Wildlife Refuge, Texas, summer 2001), suggesting a lower likelihood for an invaded community to carry surface fire than an uninvaded grassland. Researchers in other areas note tamarisk plants with many stems and high rates of stem mortality, resulting in dense accumulations of dead, dry branches that catch leaf litter above the ground surface and enhance the crowns' flammability (Busch 1995; Racher and others 2002). Attempts to burn tamarisk in southern tallgrass prairie suggest that it is little affected by prescribed fire because its foliar moisture reduces flammability, thereby reducing fire continuity (J. Grace, personal observations, San Bernard National Wildlife Refuge, Texas, summer 2001).

The use of fire to control tamarisk has been investigated in riparian areas in shortgrass prairie and southern mixedgrass prairie in the Central bioregion (see the "Riparian Formation" section for details page 137). Tamarisk appears to be relatively unaffected by fire or competition. This is not to say that fire has no effect on tamarisk; rather, fire alone does not appear adequate as a control measure.

Macartney rose—This nonnative is an evergreen, perennial, thorny vine or shrub (Vines 1960), introduced from China primarily for erosion control and as a natural fence for pasture. Widely planted in the southern states, it has escaped in various habitats from Texas to Virginia. Macartney rose is a widespread invader in southern Texas and has been estimated to represent a problem on over a quarter million hectares (Scifres and Hamilton 1993). It is unpalatable and increases under grazing, forming dense, impenetrable mounds up to 20 feet (6 m) in height. It is an especially troublesome species in pastures as well as in prairies that have been formerly overgrazed and that are now being rehabilitated.

Existing evidence does not indicate that either fire or fire exclusion promotes the establishment and spread of Macartney rose (Scifres and Hamilton 1993). Rather, it seems that Macartney rose is very well adapted to surviving fire once it is established. Scifres and Hamilton (1993) suggest that Macartney rose may possess volatile oils, and we have observed that it burns readily (Grace and Allain, personal observations, San Bernard National Wildlife Refuge, Texas, 2001-2003). Fires generally top-kill the plants regardless of the season of burn. However, regrowth following winter or summer burns is rapid, and complete recovery may occur as early as the following season (Scifres and Hamilton 1993). Because of low mortality and rapid regrowth following burning, repeated burning is required to prevent Macartney rose from spreading further. We have observed substantial growth of Macartney rose following repeated growing-season burns that were followed by periods of extreme drought.

Macartney rose is not likely to be controlled using fire once it has established. Probably the best approach to controlling this species is through preventive management, such as by avoiding overgrazing and through early detection and eradication. Combined use of burning and herbicide treatments can kill individual plants, though the herbicide treatment is the critical element.

Figure 7-4—(A) Caucasian bluestem (in foreground), with inflorescence highlighted against the sky, invading tallgrass prairie. This species is a recent addition to the list of species of concern to mixedgrass and tallgrass prairie in the central and southern part of the central grasslands. (Photo by Mike Haddock, with permission, (see www.lib.ksu.edu/wildflower).) (B) A near monoculture of yellow bluestem at Barton Creek Preserve, Texas. (Photo by John M. Randall, The Nature Conservancy.)

Perennial grasses—Several nonnative perennial grasses have been introduced to increase forage value of rangelands or pastures in the Central bioregion. Among these, cogongrass, yellow bluestem (*Bothriochloa ischaemum*), Caucasian bluestem (fig. 7-4), guineagrass, and Johnson grass may pose some threat to southern tallgrass prairies. All of these species are at least partially adapted to fire, and most are capable of remaining in late-successional grasslands. Several of the invasive grasses found in the coastal tallgrass prairie are actually a greater threat in other formations, particularly the southern mixedgrass prairie. Here we shall treat those of greatest concern to coastal prairie, cogongrass and Johnson grass. Others will be discussed in the "Southern Mixedgrass Prairie Formation" section.

Information on the relationships of nonnative perennial grasses to fire and fire regimes in southern tallgrass prairie is limited. Cogongrass is the most studied with regard to fire ecology and the impacts of invasion on native fire regimes (Howard 2005b, FEIS review). Cogongrass is invasive in coastal systems from Florida west to Louisiana, with an outlying population in eastern Texas (fig. 7-5). It is not yet known to have invaded coastal prairie, although given its westward spread along the Gulf Coast, it is expected to establish in this ecosystem (Grace and others 2001). In parts of the eastern Gulf of Mexico, cogongrass is noted for

Figure 7-5—Cogongrass in eastern Texas. This grass is invasive along the coast of the Gulf of Mexico from Florida to Louisiana, and there is concern regarding its westward spread that it will become invasive in southern tallgrass prairie. (Photo by Jim Grace.)

the formation of monocultures once established. However, its small seeds and need for outcrossing appear to slow, though not stop, its rate of spread (King and Grace 2000). See Howard (2005b) and chapter 6 for more information on this species.

Johnson grass survives fire by sprouting from rhizomes that occur 8 inches (20 cm) or more below ground. As a seed banking species that establishes in open, disturbed sites, it is likely that Johnson grass is capable of postfire seedling establishment, although this is not documented. In some areas, Johnson grass may produce more persistent dry-matter biomass than associated native species, thereby altering fuel characteristics of invaded sites. Fire may be a useful tool for controlling Johnson grass when used in conjunction with postfire treatments to control rhizome sprouts (Howard 2004b, FEIS review). More information is needed to understand the relationship of Johnson grass to fire and fire regimes in southern tallgrass ecosystems.

The Great Plains Subregion

The Great Plains subregion represents the area of the central states that lies between the mesic tallgrass prairie subregion and the Rocky Mountains. It is generally characterized by mid- and shortgrass species (Coupland 1992).

This subregion coincides with the plains grassland and Texas savanna ecosystems described by Garrison and others (1977). The plains grassland ecosystem occurs on a broad area of high land that slopes gradually eastward and down from an altitude of 5,500 feet (1,700 m) near the eastern foothills of the Rocky Mountains to an altitude of 1,500 feet (460 m) in the Central States, where it merges into the prairie ecosystem. The prairie in the Great Plains is characterized by periodic droughts, with mean annual precipitation ranging from 10 inches (250 mm) in the north to more than 25 inches (630 mm) in the south (Garrison and others 1997). Droughts are less frequent and severe along the eastern edge, where annual precipitation approaches 30 inches (770 mm) from Oklahoma to Nebraska, and about 20 inches (510 mm) in North Dakota. The frost-free season ranges from less than 100 days in the north to more than 200 days in Oklahoma and Texas. Winters are cold and dry, and summers are warm to hot (Bailey 1995; Garrison and others 1977). In the Texas savanna, elevation ranges from sea level to 3,600 feet (1,100 m) on the Edwards Plateau. Precipitation ranges from 20 to 30 inches (510 to 770 mm) annually, and the frost-free season ranges from about 250 to well over 300 days (Garrison and others 1977). Three formations are described for this subregion, based on differences in climate, soils, and vegetation: (1) northern mixedgrass prairie, (2) shortgrass steppe, and (3) southern mixedgrass prairie.

Presettlement fire frequencies in the Great Plains varied, with estimates ranging from 4 to 20 years (Paysen and others 2000). Before Euro-American settlement, fires likely burned over very large areas in this subregion, and dry lightning storms perhaps played a greater role than in the mesic subregion. In areas where fuel might have been strongly limited by precipitation, fire regime characteristics would have varied over time. When cattle grazing became widespread, fuel loads were probably greatly reduced and there was an accompanying reduction in fire frequency. Generally, fuel loads in the Great Plains are substantially lower (about 1 t/ha) than in the mesic subregion. For this reason, continuity of fuels and the presence of flammable shrubs are even more critical to sustained ignition. Fire intensities associated with the herbaceous component in the Great Plains would generally be on the low side, but the presence of flammable shrubs could increase fire severity. More detail about historic changes in fire regime can be found in Paysen and others (2000).

Wide temperature ranges are found in the Northern Great Plains, though a somewhat reliable moisture supply makes the native grasses more hardy. Caution must be exercised to avoid overgrazing, particularly at times of peak evapotranspiration. In the Southern Great Plains, moisture tends to be predominantly available during mid growing season (though this can be variable), and fires during drought lead to conditions deleterious to vegetation recovery. Kucera (1981) offered an interesting theory about the relative differences in successional development following fire in semi-arid versus mesic grasslands, in which the development of thatch plays a central role. Thatch development can occur rapidly in mesic grasslands, leading to reduced production for grazers. In contrast, in semi-arid grasslands thatch is slow to develop and helps conserve moisture in the system. Thus, to maintain productivity, the optimum fire interval would probably be longer in more arid grasslands, particularly in shortgrass steppe, than in mesic grasslands.

The interaction of nonnative species with the fire regime in the Central bioregion is perhaps more dramatic and widespread in the Great Plains than in the mesic tallgrass subregion. Across the central grasslands, vegetation types high in native plant diversity have been heavily invaded by nonnative plant species (Stohlgren and others 1998, 1999a, b). In general, both native and nonnative species richness and cover are positively correlated with soil fertility and soil water holding capacity (Stohlgren and others 1999a). In shortgrass steppe and northern mixedgrass prairie, significantly more nonnative species occur in riparian zones than in adjacent upland sites (Stohlgren and others 1998). While grazing is thought to play a role in the spread of invasive species, a study in temperate and montane grasslands found nonnative species richness and cover were no different in long-term grazed areas than in long-term ungrazed areas (Stohlgren and others 1999b). We might expect that cessation of grazing would increase fuel loads and fire frequency.

Northern Mixedgrass Prairie Formation

The northern mixedgrass prairie occurs from eastern Wyoming and parts of Nebraska north through Montana, South and North Dakota, and into Canada. Vegetation in this formation is dominated by herbaceous species such as western wheatgrass, blue grama, needle-and thread grass, Idaho fescue, rough fescue, green needlegrass, big bluestem and buffalo grass (Küchler 1964). It is characteristic of mixedgrass prairies that they contain elements of both tallgrass and shortgrass formations. In most cases, the major invaders of the shortgrass systems can probably also impact mixedgrass prairies. This is expected because tolerances are asymmetric. Drought tolerant species grow quite well with abundant moisture supplies while mesic species do not tolerate drought in most cases. We can also expect invaders that are very competitive in arid lands to be somewhat less competitive in more mesic areas compared to the highly productive, mesic natives. For these reasons, variations in competitiveness and threat level for invaders will likely vary depending on moisture supply. Latitudinal and elevational influences of a similar nature are expected. It should be noted that in this subregion, many factors can limit the application of prescribed burning (Larson, personal communication, 2006). Spring is often wet, and dominant cool-season grasses green up too quickly to carry fire in many cases. Also, by the time that cool-season grasses cure sufficiently to permit burning, dry conditions bring restrictions on burning. An additional difficulty is that both natives and nonnatives are frequently cool-season grasses, preventing the ability to time burning to selectively impact the nonnatives. Despite these restrictions, fire is a common feature of the landscape, and complex effects are frequently observed, often depending on community type (Larson and others 2001).

Cheatgrass—Cheatgrass occurs throughout the United States and southern Canada, though it is less common in the Southern States. It is highly invasive and occupies large areas of shrubland in the Interior West bioregion (chapter 8). It is also a significant and increasing component of grasslands along the eastern front range of the Rocky Mountains (Carpenter and Murray 1999, TNC review). In the Central bioregion, cheatgrass is problematic in northern mixedgrass prairie and even more troublesome in shortgrass steppe formations.

Cheatgrass is an annual that is able to complete its life cycle in the spring before the dry summer weather begins. The phenology of this species is particularly important to both its desirability as an early season forage grass and to its ability to compete with native species, survive fire, and alter fire regimes. It is able to germinate from fall through spring whenever moisture conditions are suitable, and growth proceeds throughout the winter irrespective of cold temperatures. Rapid early growth allows it to set seed before most other species. Once it begins to senesce, its persistent litter produces a highly flammable fuel (Zouhar 2003a, FEIS review).

Studies of cheatgrass in the Great Basin indicate that it is well adapted to fire and thrives under a regime of relatively frequent fires (3- to 10-year return intervals) in these ecosystems (Zouhar 2003a). Little published information is available regarding the relationship of cheatgrass with fire in the Central bioregion. Cheatgrass is favored by occasional burning at study sites within shortgrass steppe and mixedgrass prairies (Grace and others 2001). However, Smith and Knapp (1999) provide evidence that cheatgrass and other nonnative species are less frequent on annually burned tallgrass prairie sites at Konza Prairie, Kansas than they are on less frequently burned and unburned sites.

Because it dries 4 to 6 weeks earlier than native perennial species and also retains its dead leaf tissues for up to 2 months longer in fall, cheatgrass infestations may increase opportunities for fire. Fire has been shown to reduce densities of cheatgrass the following year; however, the plants that establish generally have especially high seed production (perhaps due to increased soil resources and reduced intraspecific competition). Because of this, cheatgrass recovery after fire is rapid (Young and Evans 1978).

In mesic grasslands, cheatgrass does not appear to be an especially successful competitor against fire-adapted native perennial grasses, and it does not appear to pose as great a threat to native communities. In semiarid and arid grasslands and shrublands, the tolerance of cheatgrass to dry conditions, its ability to fill in the spaces between shrubs and other grasses, and its retention of dead leaf tissue gives it the potential to change fuel characteristics and ultimately alter fire regimes. This property probably does not have as dramatic an effect in the Central bioregion as in more arid regions, where cheatgrass can completely alter fire regimes (chapter 8).

Japanese brome—Japanese brome (*Bromus japonicus*) is a cool-season annual grass that has been widely planted for rangeland improvement and is highly palatable to deer and bison. It is a prolific seed producer, and germination appears to be enhanced by a layer of litter that helps to retain moisture (Howard 1994, FEIS review). Most successful seed germination occurs in fall when sufficient rainfall occurs (Whisenant 1990b). Because of dormancy mechanisms, fall germinants are typically from the previous year's seed crop (Baskin and Baskin 1981).

In the Western United States, Japanese brome occurs in prairie, sagebrush (*Artemisia* spp.) steppe, piñon pine (*Pinus edulis*, *P. monophylla*) woodland, and arid grasslands. It is most common on disturbed sites and grazed areas (Nagel 1995; Smith and Knapp 1999; Stohlgren and others 1999b), but is also reported to invade native communities. It has been found to invade shortgrass prairie in eastern Wyoming (Fisser and others 1989), mixedgrass prairie in southwestern South Dakota (Cincotta and others 1989), tallgrass prairie in central Oklahoma (Ahshapanek 1962) and Kansas (Smith and Knapp 1999), mesquite (*Prosopis* spp.) savanna in Texas (Heitschmidt and others 1988), and riparian zones in south-central Oklahoma (Petranka and Holland 1980). This species is considered a threat to native diversity in rangelands and prairie because of its ability to out-compete many native perennials for water and nutrients. Disturbance appears to enhance Japanese brome populations. For example, mowing in a Kansas tallgrass prairie in August resulted in a major increase in its abundance (Gibson 1989). The absence of disturbance may lead to the disappearance of Japanese brome from a site when woody plants invade (Fitch and Kettle 1983), though it appears to persist even into late successional conditions when woody plants do not dominate (Fitch and Kettle 1983; Hulbert 1955).

The current view of the effects of fire on Japanese brome population dynamics is based on the need for this species to have sufficient moisture for successful establishment from seed and the role of plant litter in retaining soil moisture (Whisenant 1989). Fire not only kills the majority of Japanese brome plants and much of the seed retained by the plant, but also removes the litter layer. Thus, populations of Japanese brome may be substantially reduced following fire (for example, Ewing and others 2005; Whisenant and Uresk 1990). The rate at which they recover depends on precipitation; when there is ample fall precipitation, litter is not required for successful establishment and populations can rebound immediately (Whisenant 1990b). When moisture is less available, a litter layer is typically required before population recovery occurs. An example consistent with this hypothesis comes from Anderson (1965), who observed that annually burned, native prairie in the Flint Hills of Kansas remained free of Japanese brome while grazed prairie that was not burned was invaded by both this species and Kentucky bluegrass. Similarly, after 17 years of periodic burning (about once every 3 years, on average) and resting from grazing at Willa Cather Prairie in Nebraska, relative abundance of Japanese brome and cheatgrass was reduced from about 4.5 percent to about 0.03 percent (Nagel 1995). A similar pattern was observed in tallgrass prairie in Kansas, where Japanese brome occurred at higher frequency on unburned sites (25 percent) than on annually burned sites (14 percent) (Smith and Knapp 1999).

Moisture availability may be more important to Japanese brome population dynamics than the effects of fire. For example, Japanese brome density increased dramatically on both burned (from 12,141 to 194,253 plants per acre 1 year after fire) and control plots (from 6,295 to 482,036 plants per acre during the same period) in the Hill Country of Texas following unusually high winter rainfall (Wimmer and others 2001). In circumstances where Japanese brome can overcome moisture limitation, recovery from fire permits it to develop substantial dominance.

Knapweeds—Several species of knapweeds occur in temperate grassland ecosystems, although they tend to be more problematic in the Western United States than in the Central bioregion. A partial list of knapweeds that are important in semi-arid regions of the Central bioregion includes spotted knapweed, diffuse knapweed (*Centaurea diffusa*), Russian knapweed (*Acroptilon repens*), and yellow starthistle (*C. solstitialis*). Except for spotted knapweed, very little information is available on the ecology and impacts of these species in the Central bioregion.

Spotted knapweed—Spotted knapweed is a short-lived perennial that has spread throughout most of the United States except for the Southeastern States. The highest concentrations are in the Interior West bioregion (chapter 8) and western Canada, though it is also of major importance in the upper Great Plains and North Central States. Spotted knapweed reproduces from seed and forms a persistent seed bank. Seeds are dispersed by gravity and by human activities, and they typically germinate in fall or early spring (Schirman 1981; Spears and others 1980). Established plants are reported to have an average lifespan of 3 to 5 years and a maximum lifespan of at least 9 years (Boggs and Story 1987). As plants age, they may develop short lateral shoots and multiple rosettes (Watson and Renney 1974, review).

Fires create the type of disturbance that may promote establishment and spread of knapweeds by creating areas of bare soil and increasing the amount of sunlight that reaches the ground surface. Spotted knapweed seedlings may emerge from the seed bank or establish on bare ground from off-site seed sources following fire, and mature spotted knapweed can survive on-site and sprout from root crowns following fire (Zouhar 2001c, FEIS review). It is generally presumed that a single, low-severity prescribed fire will not control spotted knapweed, and may lead to increased population density and spread of this species. This is supported by research and observations in the Interior West bioregion (chapter 8; Sutherland, personal communication, 2006; Zouhar 2001c).

Research and observations from the Central bioregion, however, suggest that prescribed fire can be used to control spotted knapweed under some conditions. In a greenhouse study using seeds from a spotted knapweed-infested site in Michigan and temperatures simulating controlled burns, germination of spotted knapweed seeds was significantly ($P < 0.05$) reduced when exposed to 392 °F (200 °C) for 120 seconds or more, and when exposed to 752 °F (400 °C) for 30 seconds or more (Abella and MacDonald 2000). Additionally, results from a greenhouse experiment suggest that burning shortly after seed germination (mid-April to early May) (using fuel loads near the average biomass for temperate grasslands, and intended to simulate both low- and high-severity fires) may reduce spotted knapweed seedling establishment. Burning significantly reduced seedling establishment, regardless of whether the burn occurred prior to seed germination, during the cotyledon stage (1 week), or after seedlings had initiated primary leaves (2 weeks). Timing, rather than fuel load, had a greater effect on seedling establishment, with the greatest reduction occurring when burning occurred 2 weeks after seeding (MacDonald and others 2001). These results suggest that spring burns (timed to kill recently emerged knapweed seedlings) in combination with measures to remove surviving knapweed plants between burns could result in reduced density of spotted knapweed in plant communities dominated by warm-season grasses and forbs that are stimulated by burning during this season (Howe 1995; MacDonald and others 2001).

Emery and Gross (2005) compared the effects of spring, summer, and fall burns at two frequencies (annually from 2000 to 2003 and biennially, 2001 and 2003) on survival, growth, and reproduction of spotted knapweed populations in prairie remnants in southwestern Michigan. Annual summer burning was the only treatment that significantly reduced overall population growth rates. Results further indicated that attempts to control spotted knapweed should focus on combinations of treatments that target both reproductive and non-reproductive adults.

On prairie sites in Michigan with low to moderate spotted knapweed density and sufficient fine fuels to carry a fire, annual spring prescribed burning under severe conditions (when humidity and dead fine fuel moisture are as low as possible) serves to reduce spotted knapweed populations and increase the competitiveness of the native prairie vegetation. On some sites, 3 years of this regimen reduced spotted knapweed to the point where it could be controlled by hand-pulling individuals and increasing the fire-return interval to 3 to 5 years (McGowan-Stinski, personal communication, 2001).

Prescribed burning of spotted knapweed can be difficult, especially if no fine grass fuels are present, because fire does not easily carry through spotted knapweed stems (Xanthopoulos 1988). In dense infestations (>60 rosettes/m^2) in Michigan, broadcast burning is ineffective due to lack of adequate fuel to carry the fire. In this case, spotted knapweed plants can be killed with repeated spot-burning (using a propane torch) of individuals and sprouts 3 to 4 times during the growing season until root reserves are depleted. Seedlings emerge in the time between burning treatments and are killed with subsequent burning, thus depleting the seed bank. This treatment does not seem to harm existing prairie natives; however, newly germinating natives may be at risk, and seeding after the last burn treatment may help them to recover (McGowan-Stinski, personal communication, 2001).

Spotted knapweed infestations can alter fuel characteristics and may reduce flammability in invaded communities (Xanthopoulos 1988), although spotted knapweed fuel loading varies between sites (Xanthopoulos 1986). Existing grass fuel models work poorly for spotted knapweed unless associated grasses exceed 40 to 50 percent cover (Xanthopoulos 1988), so a specific fuel model was developed for this species. In order for a stand of spotted knapweed to carry fire, both modeling and field tests indicate that burns should be conducted in early spring prior to grass and forb growth (because of the high moisture content of grasses and forbs in the spring). Moreover, the sparse foliage and discontinuous nature of spotted knapweed biomass suggest that, under low wind speeds and flat or declining slopes, fires will likely fail to spread (Xanthopoulos 1986).

Diffuse knapweed—Diffuse knapweed functions as a biennial or short-lived, semelparous perennial. It has a well-developed taproot but is not capable of vegetative spread, relying solely on seed for population growth (Zouhar 2001b, FEIS review). Most germination takes place in spring, and seedlings develop into rosettes that persist for 2 to several years before flowering and dying (Watson and Renney 1974). Reproductive success depends on plant size, and the proportion of diffuse knapweed plants that survive to flower and produce seed each year is largely a function of competition for moisture and increases with available growing space (Powell 1990). Diffuse knapweed currently occurs largely in the Western United States and Southwestern Canada. It is not widespread in the Great Plains or tallgrass prairie region, although it does occur in Iowa, Missouri, Minnesota, Indiana, Michigan, Kentucky, and Tennessee. There are no published studies on the relationship of diffuse knapweed to fire; however, its capacity to withstand repeated mowing (Roche and Roche 1999, review; Watson and Renney 1974) and its possession of a taproot suggest that it may be able to tolerate moderate-severity fire.

Russian knapweed—Russian knapweed is a perennial herb that occurs throughout the Western and Central United States, including the Great Plains, but is largely absent from eastern and southeastern areas (Watson 1980, review). This species is used by bighorn sheep and white-tailed deer, although it is not palatable to livestock and causes a neurological disease in horses (Zouhar 2001a, FEIS review). Botanical descriptions indicate that Russian knapweed has an extensive root system, is capable of rapid vegetative spread by root buds, and typically forms dense, often monotypic colonies that may persist more than 75 years (Watson 1980). Russian knapweed is also capable of establishing from seed; however, seed dispersal over any distance larger than the height of the plant requires a dispersal agent.

No information regarding fire adaptations of Russian knapweed is available in the literature, although its deep, extensive root system is likely to survive even severe fire. The tolerance of Russian knapweed seeds to heating is unknown. Russian knapweed probably sprouts from root buds after fire and may establish from on-site seed or from seed brought on site by animals or vehicles (Zouhar 2001a). Researchers burned plots of Russian knapweed in Wyoming after first mowing them to a height of 3 to 5 inches (8 to 12 cm). Observations following these treatments suggest that Russian knapweed plants were injured to a depth of 1 to 2 inches (2.5 to 4 cm) below the soil surface, and that lateral roots at the 3- to 6-inch (7.5 to 15 cm) depth did not appear to be injured. Russian knapweed seedheads were also burned but the seed "appeared to be viable"; however, Russian knapweed seedlings were not observed after burning (Bottoms, personal communication, 2002).

Yellow starthistle—Yellow starthistle, an invasive winter annual, is most problematic and best studied in California (chapter 9) and Interior West (chapter 8) ecosystems. Infestations of yellow starthistle in the eastern two-thirds of the United States are sporadic and localized, where populations apparently fail to establish and persist on a year-to-year basis, possibly because of unfavorable growing conditions (Great Plains Flora Association 1986; Maddox and others 1985). More research is needed to understand the role of yellow starthistle invasion in ecosystems of the Central bioregion. For information on the use of fire to control yellow starthistle and the potential for yellow starthistle to impact fire regimes, see chapter 9, or the FEIS review for this species (Zouhar 2002a).

Leafy spurge—Leafy spurge is a perennial forb that quickly develops an extensive root system reaching depths of 15 to 30 feet (4.5 to 9 m) (Simonin 2000, FEIS review). This species has a prolific capacity for vegetative spread and regrowth, which is facilitated by the production of a basal crown of tissue just beneath the soil surface (Biesboer and Eckardt 1987, TNC review). This basal crown persists for many years, and develops a large number of buds that produce aboveground shoots as well as new roots. In addition, leafy spurge produces rhizomes and root buds that are capable of forming independent plants, which is likely to occur when disturbed (Selleck and others 1962). A notable characteristic of leafy spurge is its capacity to recover from damage by sprouting from rhizomes and root buds as early as 7 to 10 days after germination. Another important characteristic is prolific seed production in explosive capsules that can disperse seed up to 15 feet (4.5 m). Seeds may remain viable in the soil for up to 8 years, though it has also been reported that most seeds germinate or perish within 2 years of production (Selleck and others 1962). Based on this information, it appears that leafy spurge relies on its bud bank as much as its seed bank for regeneration.

The highest concentrations of leafy spurge in the United States are currently in Oregon, Idaho, Montana, Wyoming, Colorado, North Dakota, South Dakota, Nebraska, Minnesota, and Wisconsin. Leafy spurge occupies a broad range of ecological conditions from xeric to subhumid, though its most rapid growth is in more mesic sites. In particular, it invades disturbed areas with exposed mineral soil (Belcher and Wilson 1989; Wilson 1989), but it can also establish and spread in pastures and ungrazed native grasslands (Selleck and others 1962). Forbs and grasses may be completely displaced where leafy spurge invades (Biesboer and Eckardt 1987) and forms dense patches (fig. 7-6). In addition to adverse effects on native diversity and restoration efforts (Belcher and Wilson 1989; Butler and Cogan 2004), it has been found to be toxic to cattle and humans, though goats and sheep are able to consume it without adverse effects (Landgraf and others 1984; Stoneberg 1989). In balance, leafy spurge perhaps poses the greatest problems in mixedgrass prairies, though it is also a problem in semiarid and mesic conditions.

One of the more frequently cited examples of adverse effects of fire on leafy spurge is attributed to Dix (1960). In this study, he compared three pairs of unburned and previously burned sites in North Dakota native grasslands to evaluate the effects of fire on community composition. Leafy spurge was found in only one of the six sites examined, and this site happened to be one of the unburned grasslands. Dix drew no conclu-

Figure 7-6—Meadow infested with leafy spurge in Rist Canyon, west of Fort Collins, Colorado. (Photo by William M. Ciesla, Forest Health Management International, Bugwood.org.)

sions about the effects of fire on leafy spurge from this study, and given the absence of information about its distribution prior to the burns, we feel that no firm conclusions can be drawn from this example.

Other studies have generally found prescribed fire to have limited effects on leafy spurge. Prosser and others (1999) examined the effects of fall burning on leafy spurge in North Dakota and found no effects on stem densities the next year compared to unburned controls. Wolters and others (1994) conducted studies of the effects of spring and fall burns on leafy spurge in mixedgrass prairie at the Little Missouri National Grassland in North Dakota. They found that both spring and fall burns reduced seed germination but increased stem density through vigorous postfire sprouting. Similarly, Masters (1994) found that stem densities increased with late spring burning in mixedgrass prairies in Nebraska. Fellows and Newton (1999) found that burning in late spring or early fall produced immediate reduction of leafy spurge densities, but these effects were not evident the following growing season. In grasslands at the Tewaukon National Wildlife Refuge in North Dakota, leafy spurge canopy cover was not changed 2 months after fire, but canopy cover was greater in burned versus unburned plots 1 year after fire (Olson 1975). These studies suggest that leafy spurge is especially well adapted to sprouting following fire, and that burning alone is ineffective in controlling leafy spurge.

While there is evidence that spurge can be impacted by competition with commercial grass cultivars (Lym and Tober 1997), there is no evidence that fire can be used to tip the competitive balance toward native species to a sufficient degree to negatively impact leafy spurge.

Prescribed burning in conjunction with herbicide treatments may be more effective than either treatment alone for reducing leafy spurge (Masters and others 2001; Nelson and Lym 2003; Winter 1992; Wolters and others 1994). For example, Wolters and others (1994) found that herbicides with or without burning were most effective in reducing leafy spurge stem density, while spring burning, with or without fall herbicide application, was the most effective treatment for reducing leafy spurge seed germination. They concluded that a fall application of picloram plus 2,4-D followed by spring burning would most effectively reduce both germination and stem density (Wolters and others 1994). Prescribed fire was used to enhance chemical control efforts at Bluestem Prairie in Minnesota. According to Winter (1992), prescribed fire effectively removed plant litter in the most heavily infested leafy spurge areas, thereby increasing visibility of leafy spurge plants, especially small shoots, and enabling more chemical to reach leafy spurge foliage and roots (Winter 1992).

Success in the control of leafy spurge will likely involve biocontrol agents (fleabeetle species), though it is not clear at this point the degree of success that can be expected (Larson and Grace 2004). Preliminary evidence suggests that the use of herbicide treatments can interfere with biocontrol success (Larson and others 2007) because the initial decline in spurge density following herbicide treatment can result in a decline in biocontrol agents. Fellows and Newton (1999) examined the effects of prescribed burning on the flea beetle *Aphthona nigriscutis*, a potential control agent, at the Arrowwood National Wildlife Refuge in North Dakota. They found that burning prior to beetle release enhanced establishment of most beetles. The insects did not persist on all plots, but on plots where they did persist, spring and fall burns 2 years later did not affect their population size.

Leafy spurge has a tremendous capacity to sprout following top-kill or damage to aboveground tissues. Even extensive control programs that attempt to eliminate leafy spurge using a combination of mechanical removal, biological control, chemical treatments, and prescribed fire have found that this species must be treated repeatedly for many years to achieve satisfactory reductions.

Southern Mixedgrass Prairie Formation

The southern mixedgrass prairie occurs in eastern Kansas and Oklahoma, south through central Texas and into Mexico. This formation is dominated by little bluestem, side-oats grama, and blue grama. Shrubs become more common in the southernmost part of the formation, in the area described by Garrison and others (1977) as the Texas savanna ecosystem. Here the vegetation is characterized by dense to very open synusia of deciduous and evergreen low trees and shrubs. The grass varies from short to medium tall, and the herbaceous vegetation varies from dense to open (Garrison and others 1977). Common species include little bluestem and seacoast bluestem (Küchler 1964). Mesquite is the most widespread woody plant. Other woody species include *Acacia* species, oaks, juniper, and ceniza along the Rio Grande valley and bluffs (Garrison and others 1977; Küchler 1964). At present, there is a relatively small list of important nonnative invaders in southern mixedgrass prairie (table 7-2); these are primarily introduced grasses and trees such as tamarisk (covered in Riparian section).

Yellow and Caucasian Bluestems—Yellow and Caucasian bluestems, perennial, C_4 bunchgrasses, are among several species of Old World bluestems that have been seeded throughout much of the southern Great Plains for stabilizing marginal cropland or increasing forage production on rangelands (Berg 1993; Marietta and Britton 1989; McCoy and others 1992; Phillips and Coleman 1995). Spring burning and nitrogen fertilization are common management practices to improve productivity of Old World bluestem pastures (for example, Berg 1993; Phillips and Coleman 1995).

Spring burning of uniform stands of yellow bluestem in established pastures on former cropland in northwestern Oklahoma significantly reduced ($P < 0.01$) herbage yields by 6 to 30 percent per year. Soils were calcareous and noncalcareous loams (Berg 1993). Winter burning (February) of artificial grasslands (chaparral sites seeded to Lehmann lovegrass (*Eragrostis lehmanniana*) and yellow bluestem) in central Arizona resulted in a 16 percent increase in bluestem production following fire and rootplowing (Pase 1971). In central Texas, there is some indication that yellow bluestem has a high tolerance of burning and a wide tolerance of soil conditions (Gabbard 2003), though many details are lacking at present.

Caucasian bluestem appears to be well adapted to frequent fire. Research in India indicates that annual and biennial burning in either winter or summer resulted in increased abundance of Caucasian bluestem over a 2-year period (Gupta and Trivedi 2001). According to an NRCS Plant Fact Sheet, however, Caucasian bluestem "does not respond as strongly to fire as do our native species," although it is unclear what plant community or native species are being compared (USDA, NRCS 2006). Furthermore, at Konza Prairie, when compared to native C_4 grass species such as big bluestem, Caucasian bluestem had greater plant biomass, lower pools of soil N, lower rates of decay and carbon cycling, and higher foliar and root tissue C:N ratio in response to spring burning. Additionally, areas dominated by Caucasian bluestem had significantly lower plant species diversity. The authors conclude that the threat of invasion by nonnative C_4 species such as Caucasian bluestem raises a dilemma for using prescribed fire to promote native C_4 grasses in these prairies (Reed and others 2005).

Guineagrass—Guineagrass, an important invader of the western portion of coastal prairie, has short, creeping rhizomes and produces wind-dispersed seed, although seeds are often nonviable. It often forms dense stands in open pastures and disturbed areas and may displace natives on fertile soils. It is drought tolerant but will not survive long periods of desiccation. According to a review by the Invasive Species Specialist Group (2006), guineagrass survives grass fires (quick-moving fires that do not harm the rhizomes), regenerates rapidly from rhizomes, and may dominate the postfire plant community. More information is needed on this species' relationship to fire in the Central bioregion.

Shortgrass Steppe Formation

The shortgrass steppe occurs from eastern Colorado and New Mexico into western Kansas and the panhandles of Oklahoma and Texas (fig. 7-1). This formation is characterized by tablelands and rolling lands, with numerous inclusions of eroded hills, buttes, and valleys, abutting the Rocky Mountains to the west. Presettlement fire regimes were variable in this ecosystem depending on ignition sources and fuel loads, with fire frequencies estimated at 4 to 20 years (Paysen and others 2000). The shortgrass steppe formation is dominated by blue grama (*Bouteloua gracilis*) and buffalo grass (*Buchloe dactyloides*) (Küchler 1964). According to Lauenroth and Milchunas (1992), about 60 percent of the shortgrass steppe remains in native vegetation and is used for grazing, while the remaining 40 percent is largely used for cropland. The shortgrass steppe is still considered to be largely comprised of native species and not yet significantly impacted by nonnative invasive species (Kotanen and others 1998). The list of invaders of major importance is short at present (table 7-2), with primary concern surrounding buffelgrass in the southern range and tamarisk in riparian zones.

Buffelgrass—Buffelgrass is a drought-tolerant, perennial bunchgrass, ranging in height from 4 to 60 inches (10 to 150 cm) at maturity, and reproducing both vegetatively and by wind- and animal-dispersed seed (Tu 2002a, TNC review). It is well established in disturbed and intact desert shrub communities in the Southwestern United States (chapter 8). It has been widely planted and established as a pasture grass in northern Mexico and western and southern Texas. It thrives on sandy soils where annual precipitation ranges between 8 and 47 inches (200 to 1,200 mm) (Hanselka 1988; Ibarra-F. and others 1995), and tends to form dense swards that exclude native vegetation (Tu 2002a).

Buffelgrass is well adapted to survive fire. Cool-season prescribed fires are sometimes used in southern Texas to maintain and rejuvenate buffelgrass pastures by removing litter and controlling woody invaders (Hamilton and Scifres 1982; Mayeux and Hamilton 1983; Scifres and Hamilton 1993). Research in India indicates that annual and biennial burning in either winter or summer increased abundance of buffelgrass over a 2-year period (Gupta and Trivedi 2001). A study in the west-central portion of the South Texas Plains produced similar results, where burning in February resulted in greater buffelgrass production (kg/ha) by mid-spring. However, buffelgrass production was lower on burned than unburned plots as soil water was depleted during summer (Hamilton and Scifres 1982). Prescribed winter (December or February) burning of a buffelgrass pasture in the lower Rio Grande Valley of southern Texas similarly resulted in small, temporary increases in buffelgrass standing crop (kg/ha), while burning combined with broadleaf herbicides to control associated invasive forbs increased buffelgrass standing crop as much as 3-fold (Mayeux and Hamilton 1983). Summer prescribed burning of buffelgrass in Sonora, Mexico, produced a mixed response, depending on plant phenology at the time of burning and possibly related to soil water availability. Burning in summer, after the accumulation of 2 inches (50 mm) of precipitation and between the second leaf stage and early culm elongation in buffelgrass, appeared to stimulate buffelgrass growth for 3 to 4 years. Conversely, burning later in the growing season, at the peak of buffelgrass live biomass production, reduced buffelgrass production by almost 50 percent for 4 years (Martin-Rivera and others 1999).

Where buffelgrass invades desert shrublands in the Southwestern United States and Mexico, it appears to increase fire frequency and spread (chapter 8) by introducing a novel life form (in other words, fuel type) in invaded communities (D'Antonio 2000). No information is available regarding the impacts of buffelgrass on fuel characteristics and fire regimes in invaded communities in shortgrass steppe, but one might speculate that it would have less of an impact on fuel characteristics in a community already dominated by grasses.

Prescribed burning is not likely to be an effective control method for buffelgrass, especially when conducted in winter or other times when soil water availability is high (Tu 2002a), but might be more effective under drier conditions. However, no reports were found in the literature discussing application of prescribed burning in summer as a potential control method for buffelgrass in shortgrass steppe ecosystems.

Riparian Formation

The dominant variable influencing the distribution of species in the central grasslands is the availability of water. For species associated with riparian zones, this constraint is lifted, at least part of the time. Characteristics of riparian zones can vary substantially, as can the degree of constancy of water supply and edaphic influences. Although riparian zones in the Central United States are a small part of the total area, they are ecologically important and distinct from the surrounding communities in their species composition. The vegetation of the riparian zones, for example, is often of special importance as wildlife habitat (Smith 2000). As a group, plants of the riparian zones are distinctly different from plants of upland habitats, so the invaders of riparian areas are treated separately here.

Riparian zones are often dominated by trees, contrasting with the surrounding landscape (Shelford 1963), and range from narrow fringes of vegetation to broad floodplains kilometers wide. Little is known about fire regimes in riparian communities, though it is suggested that they differ from the surrounding communities in fire regimes and adaptations. For some riparian zones, fire frequency is quite low (Wright and Bailey 1982), though in other cases fires can be frequent. Periodic flooding, along with forest canopy development, can reduce herbaceous fuels and make fire less likely to spread into riparian vegetation.

Riparian zones are particularly prone to invasion by nonnative invaders (Kotanen and others 1998), even when phylogenetically independent comparisons are made. Because of their rich alluvial soils, floodplains along large rivers have often historically been converted to cropland and consequently support propagules of many nonnative invasive species. On the Rio Grande Delta, for example, restoration of native woody species is often impeded by a number of invasive grasses including bermudagrass, guineagrass, Johnson grass, buffelgrass, Kleberg bluestem (*Dichanthium annulatum*), Angleton bluestem, and

Natal grass (*Rhynchelytrum repens*). Although research from these areas is unavailable, there is concern that these grasses not only prevent establishment of native trees and shrubs but also increase wildfire risk (for example, Twedt and Best 2004). Many riparian invaders are widespread, capable of transforming ecosystems through their complete domination, and difficult to control. Here we discuss two species in some depth, tamarisk and Russian-olive, and mention a third of lesser importance in the Central bioregion, giant reed.

Tamarisk—In the Central bioregion, tamarisk is most common in riparian areas in southern mixedgrass prairie and shortgrass steppe ecosystems in eastern New Mexico and western Texas, and it is spreading along the coastal plain of the Gulf of Mexico (see "Southern Tallgrass Prairie Formation" section page 125).

Tamarisk establishment and spread appear to be facilitated by fire, although experimental evidence is limited. Tamarisk produces abundant, wind-dispersed seed throughout the growing season that can germinate and establish on bare, saturated soil (Zouhar 2003c). Additionally, burned tamarisk trees that are not killed may show an enhanced flowering response (Stevens 1989). Tamarisk is also capable of vigorous and abundant sprouting from roots and root crown following fire or other disturbance that kills or injures aboveground portions of the plant (Zouhar 2003c). Even when completely top-killed, it may produce sprouts 7 to 10 feet (2 to 3 m) tall in the next year. For example, growth of surviving tamarisk plants after a July wildfire at Lake Meredith National Recreation Area, Texas, exceeded 6 feet (1.8 m) at the end of that growing season (Fox and others 1999).

Postfire response of tamarisk depends on season of burning, fire frequency, fire severity, and postfire plant community composition. Burning during the peak of summer appears to have the strongest adverse effect on tamarisk, presumably due to ensuing water stress (Grace and others 2001). Tamarisk mortality exceeded 60 percent 12 months following a July wildfire at Lake Meredith National Recreation Area, Texas (Fox and others 1999). In cases where annual burning can be achieved for several years, it is suggested that tamarisk can be controlled with fire alone (Duncan 1994). Similarly, control of tamarisk using prescribed burning has been reported by land managers when several years of fire treatments were applied in succession with high fuel loads (Racher and Mitchell 1999).

Several factors seem to influence flammability of tamarisk:

- High water and salt content make tamarisk difficult to burn under some conditions (Busch 1995; Racher and others 2002).
- Flammability of tamarisk increases with the build-up of dead and senescent woody material within the plant, such that old, dense stands of tamarisk can be highly flammable (Hohlt and others 2002; Racher and others 2002).
- Tamarisk in dense stands that have not burned in 25 to 30 years exhibit extreme fire behavior and crowning due to the closed canopy, regardless of time of year. They can have flame lengths exceeding 140 feet (43 m), resulting in consumption of a majority of the woody material.
- Stands reburned within 5 to 6 years show much less intense fire behavior, carrying fire only if there is adequate fine fuel load and continuity. Few trees torch during burning, though some trees are top-killed (Hohlt and others 2002; Racher and others 2001).

Tamarisk stands in the Pecos Valley, New Mexico, can burn "hot," with erratic fire behavior and numerous firebrands transported over 500 feet (150 m) downwind, regardless of whether the stand has burned before (Racher and others 2001, 2002).

Use of fire alone to control tamarisk is generally ineffective. In some areas, prescribed fire can be used to manage tamarisk by eliminating the closed canopy, slowing the rate of invasion, and allowing desirable vegetation to respond, thereby increasing diversity in monotypic tamarisk stands. Burning these communities under controlled conditions can also reduce the potential for costly wildfires that must be suppressed to avoid property loss (Racher and others 2001). Typically control of tamarisk requires some combination of mechanical and chemical control methods, sometimes integrating prescribed fire (Zouhar 2003c). Tamarisk plants that are stressed following defoliation by biocontrol beetles may also be more susceptible to mortality following fire (Brooks personal communication, 2007b). See the tamarisk review in FEIS for more details on the ecology and management of tamarisk and related species.

Much current research on prescribed burning of tamarisk has been conducted in the Pecos Valley, New Mexico, and in the Texas panhandle. Abstracts summarizing the status of this research are available in Texas Tech University's *Research Highlights*. Research includes studies of fire behavior and spotting potential of fires in tamarisk-dominated communities (Racher and others 2001; Racher and others 2002); effects of season of burning on response of tamarisk to prescribed fire (Bryan and others 2001; Hohlt and others 2002; Racher and Mitchell 1999); response of herbaceous species to prescribed fires in tamarisk-dominated communities (Bryan and others 2001); effectiveness of using herbicides in conjunction with prescribed burning on tamarisk control (Racher and Mitchell 1999); and tamarisk response to wildfire followed by mechanical and chemical control efforts (Fox and others 2000; Fox, R. and others 2001).

Russian-olive—Russian-olive is a deciduous shrub or small tree that has been cultivated for a variety of purposes both in its native range and in North America. It is widely distributed from the west coast to the Dakotas, Nebraska, Kansas, Oklahoma, and Texas (Knopf and Olson 1984). Russian-olive readily establishes and spreads on disturbed floodplains, stream banks, and in some situations, marshy areas and wet grasslands (Olson and Knopf 1986). Once established, it appears to have a substantial capacity to withstand periodic drought and out compete native woody species, and is considered the successional climax species in many cases (Howe and Knopf 1991). Russian-olive was introduced to many of the Great Plains States by the early 1900s and it was planted extensively in windbreaks in the 1930s and 1940s in association with government programs. It has persisted and spread in many areas in the Central bioregion, especially in the understory of riparian communities in northern mixedgrass prairies in the western part of the Great Plains (Zouhar 2005b, FEIS review). It is also invasive in riparian areas in much of the Interior West bioregion (chapter 8).

Transplanted Russian-olive seedlings typically produce fruit about 3 to 5 years after transplanting; however, its average reproductive age varies with latitude (Borell 1971). On the Marias and Yellowstone rivers in eastern Montana, the average age of first reproduction for Russian-olive is around 10 years (Lesica and Miles 2001). Seeds are typically dispersed by birds and small mammals, and probably also by water and ice. Longevity of Russian-olive seed under field conditions is unknown, though stored seeds retain viability for about 3 years. Mature Russian-olive seed is typically dormant and may require scarification for germination. Russian-olive may be able to exploit suitable germination conditions over a relatively long time compared to native taxa with which it commonly associates (Shafroth and others 1995). It can establish on disturbed sites in full sun, in shade, or in intact ground cover, though it grows best in full sun. Shade tolerance, reproduction, growth, and recruitment rates may vary with latitude, making Russian-olive less invasive at the upper and lower latitudinal limits of its North American range (Borell 1971; Lesica and Miles 2001; Zouhar 2005b). Russian-olive can produce adventitious roots on buried stems (Brock 1998), and Bovey (1965) suggested that Russian-olive spreads by "underground rootstalks"; however, there is no evidence in the literature that indicates that Russian-olive spreads by asexual reproduction under field conditions, except following injury or top-kill.

No information in the literature specifically addresses Russian-olive regeneration or establishment from seed after fire, although several researchers and managers report that Russian-olive sprouts from the trunk, root crown, and/or roots after top-kill or damage (Zouhar 2005b), and some report sprouting from roots and root crown following fire (Brock 1998; Caplan, personal communication, 2005; Winter, personal communication, 2005). Mixed species stands along the Rio Grande often become monospecific stands of Russian-olive following fire due to vigorous Russian-olive root sprouting (Caplan 2002). Managers should be alert to the possibility of Russian-olive establishment and spread in the postfire environment when it occurs onsite or in the vicinity of burn areas.

The growth habit (in other words, fuel arrangement) of Russian-olive varies among plant communities and site conditions. It may have a single trunk or multiple stems branching at ground level. On many sites, Russian-olive grows in dense thickets with close spacing, sometimes with scattered mature cottonwood in the canopy. Stands may be so dense that other riparian species are excluded entirely. The lower 6.5 feet (2 m) of vegetation may contain a tangle of dense, often dead, branches with little live foliage (Zouhar 2005b). Katz and Shafroth (2003) present data describing density and cover of several established Russian-olive populations in western North America.

Dense growth of Russian-olive may be more fire-prone than the native communities that it invades, although this has not been studied or reported in the literature and deserves further investigation. Additionally, there is little quantitative information on prehistoric frequency, seasonality, severity and spatial extent of fire in North American riparian ecosystems, where Russian-olive is commonly invasive. Many interrelated factors, such as urban and agricultural development, groundwater pumping, dams, and flood suppression, contribute to altered disturbance regimes in these ecosystems.

There is little information available on the use of fire to control Russian-olive. Observational evidence indicates that Russian-olive is top-killed by prescribed fire in tallgrass prairie (Winter, personal communication, 2005) and by wildfire in riparian communities on the Rio Grande River (Caplan, personal communication, 2005). Fire in tallgrass prairie generally does not top-kill trees greater than 2-inch (5 cm) DBH, so Russian-olive is maintained at "brush height" with regular burning in tallgrass prairie sites in Minnesota (Rice and Randall 2004; Winter, personal communication, 2005). According to Deiter (2000, review), stump burning may successfully control sprouting in Russian-olive, but it is time-consuming and therefore likely to be costly. Prescribed fire is probably most effective as a control method for Russian-olive when combined with other control techniques. The species' occurrence in riparian zones may make the use of fire problematic in cases where adverse impacts on fire-intolerant natives may be greater in magnitude. This kind of information is needed to determine the role fire may play in management of Russian-olive.

Giant reed—Giant reed is an escaped ornamental grass that occurs in riparian and wetland areas throughout the Southern United States, from Virginia to California. Giant reed is clump-forming and can grow up to 20 feet (6 m) tall. In deep water, shoots of this species can be semi-floating. Giant reed spreads locally by rhizomes and is reported to flower only once every few years (McWilliams 2004, FEIS review), though in southern Louisiana it appears that some plants flower every year (J. Grace, personal observation, Brazoria National Wildlife Refuge, summer 2002). This species establishes and spreads in riverbanks, marshes, and floodplains and has been observed to replace common reed (*Phragmites communis*) on some sites. The aboveground material of giant reed is quite flammable once leaves have dried. It is capable of surviving fire because of extensive storage organs belowground (McWilliams 2004). At present, there appears to be no information on its effects on fire regimes or succession, nor information about whether fire aids in its establishment.

Conclusions

In general, native species of the central grasslands are adapted to frequent fires. Many native species are promoted by fires in appropriate seasons, and many nonnatives decline. According to Solecki (1997), fire and competition from native prairie plants eliminate most nonnative invaders in prairie ecosystems. However, a fraction of nonnative invaders can tolerate, avoid, or alter the fire regime and pose major threats to central grassland ecosystems. Management of habitats that contain these species or are exposed to them will benefit from considering the interaction of these invaders with fire.

The interaction of fire and nonnatives is an important topic that we are only beginning to address, and additional work is needed. The literature describing how particular nonnatives react to burning and alter fuels and fire regimes is satisfying in its details for only a few species. The few examples where precise applications of burning can be used to favor native species over nonnatives provide encouragement that there is considerable potential for such uses of fire. On the other hand, the degree to which such potentials exist or whether they can be successfully accommodated in burn programs is not known at present.

Several areas are in need of research. One emerging topic of particular interest is how fire impacts biocontrol. Biocontrol agents are being used increasingly as a primary means of controlling nonnative plants, a trend that is likely to continue. How this will relate to the use of fire is largely unknown. Preliminary results for select cases, such as for fleabeetles that feed on leafy spurge, indicate that biocontrol agents can persist in the presence of periodic fire. It seems unlikely that this will universally be the case. Detailed studies will be required for each situation where biocontrol is used in combination with burning.

Peter M. Rice
Guy R. McPherson
Lisa J. Rew

Chapter 8:

Fire and Nonnative Invasive Plants in the Interior West Bioregion

Introduction

The Interior West bioregion is bounded on the east by the eastern slope of the Rocky Mountains from Canada south to Mexico and on the west by the eastern foothills of the Cascade Range in Washington and Oregon and the eastern foothills of the Sierra Nevada in California. The bioregion includes the Chihuahuan, Sonoran, and Mojave hot deserts and the Great Basin cold desert, as well as the forested Rocky Mountains and associated grasslands and meadows in the valleys, foothills, and mid-slopes.

Much of the Interior West consists of wildlands that currently undergo limited active management. Major vegetation types can be summarized along general south to north and/or low to high elevation trends. This bioregion includes the following ecosystems: desert grasslands, desert shrub, southwestern shrubsteppe, chaparral-mountain shrub, piñon-juniper (*Pinus-Juniperus*), sagebrush (*Artemisia* spp.), mountain grasslands, ponderosa pine (*Pinus ponderosa*), Rocky Mountain Douglas-fir (*Pseudotsuga menziesii* var. *glauca*, hereafter Douglas-fir, unless another variety is specified), western white pine (*Pinus monticola*), fir-spruce (*Abies-Picea*), western larch (*Larix occidentalis*), and Rocky Mountain lodgepole pine (*Pinus contorta* var. *latifolia*, hereafter lodgepole pine) (Garrison and others 1977). Reviews by Arno (2000) and Paysen and others (2000) offer detailed descriptions of major vegetation types found in the Interior West and survey knowledge about fire regimes in this region, which vary substantially among vegetation types. Wildfire exclusion, extensive logging, and grazing practices over the past centuries have led to changes in fuel loads and canopy structures such that large deviations from past fire regimes are occurring in many areas (Arno 2000; Bock and Bock 1998; Gruell 1999; Mutch and others 1993).

Major problems with nonnative invasive plants in the Interior West occur in grasslands, shrublands, woodlands, riparian areas, and open-canopy ponderosa pine and Douglas-fir forest. Annual grasses, perennial forbs, and a few biennial forbs are the most invasive life forms in dry ecosystems. Invasion by nonnative woody species has been limited except in riparian areas, where tamarisk (*Tamarix* spp.) and Russian-olive (*Elaeagnus angustifolia*) have altered plant communities substantially. Nonnative plant invasions have apparently been less successful in mesic, high elevation, closed-canopy forests to date. Disturbances that thin or remove the overstory in these dense forests allow invasion by nonnative species such as St. Johnswort (*Hypericum perforatum*) and nonnative hawkweeds (*Hieracium aurantiacum*, *H. caespitosum*), but research to date does not indicate whether these nonnative populations persist as the tree overstory closes.

Although some earlier introductions were observed, the nonnative plant invasions with the most severe impacts in this region did not begin until the later part of the 19th century. By the 1890s much of the regional native vegetation was impacted by agriculture, livestock grazing, and timber harvest (Holechek and others 1998). Intense anthropogenic disruption of native plant communities and soil surface integrity continued in many areas through the Second World War and until more sustainable management practices began to be adopted (Holechek and others 1998; Knapp 1996). Although land use practices have improved, the number of nonnative plants that have established self-maintaining populations has continued to increase (fig. 8-1) (Rice 2003).

Actual documentation of nonnative plant impacts is limited compared to their apparent and presumed effects on native plant communities, ecosystem functions, recreation, and the economics associated with resource use (Duncan and Clark 2005). Some of the more serious consequences attributed to nonnative plant invasions in the Interior West include conversion of native shrub-dominated communities to near monotypic annual grasslands and an increasing trend for mountain bunchgrass communities to be dominated by perennial nonnative forbs.

There is substantial evidence that the widespread and overwhelming abundance of cheatgrass (*Bromus tectorum*), red brome (*Bromus rubens*), and other nonnative annual grasses has resulted in the establishment of annual grass/fire cycle in areas that had relatively low fire frequency prior to invasion (Brooks and others 2004; Whisenant 1990a). Cheatgrass has become the primary forage species in the Great Basin, but its ten-fold annual variability in production and its low palatability when cured make it an unreliable food resource compared to the native perennials it has replaced (Hull and Pechanec 1947). Furthermore, the ecological consequences of cheatgrass invasion in the Great Basin are severe (see "Sagebrush Shrublands" section page 154).

Nonnative forbs that have invaded native bunchgrass communities and open-canopy forests have altered the herbaceous canopy structure and successional pattern in these communities with cascading effects on other trophic levels and possibly ecosystem functions. Spotted knapweed (*Centaurea biebersteinii*) displacement of native herbaceous species in the northern Rockies demonstrates a wide variety of impacts from a single invasive. Knapweed infestation lowers the carrying capacity of elk winter ranges where bunchgrasses

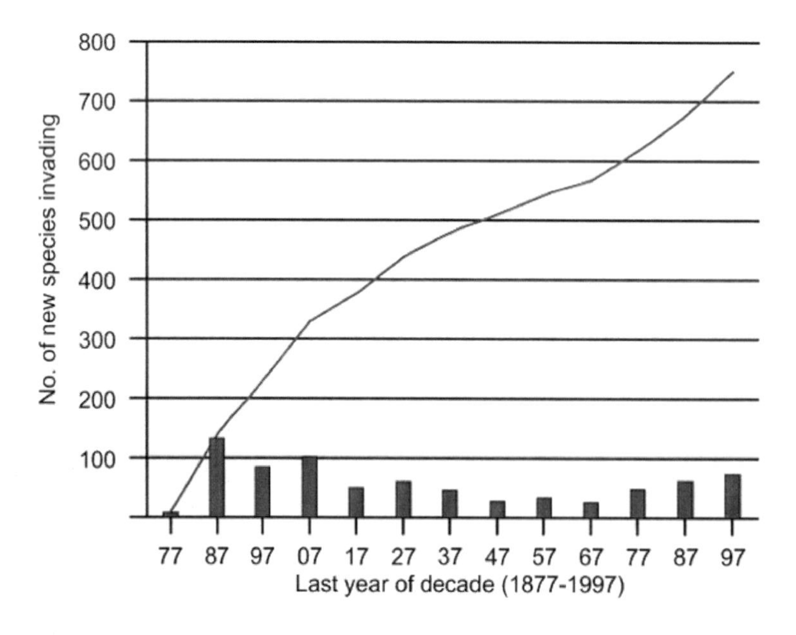

Figure 8-1—Number of nonnative plant species newly established as self-sustaining populations in five northwestern states (WA, OR, ID, MT, WY). The line represents the cumulative number of species over time. (Adapted from Rice 2003.)

could provide late winter and early spring forage (Rice and others 1997). Ortega and others (2004) found that the abundance of deer mice was higher in knapweed dominated bunchgrass habitat than in uninfested bunchgrass communities because introduced biocontrol agents, the knapweed seedhead gall flies (*Urophora affinis* and *U. quadrifasciata*) provided abundant, high quality winter food for the mice. However, summer breeding success of deer mice was lower in knapweed-dominated sites than native plant-dominated sites, perhaps because native seed and insect availability was lower in the knapweed sites (Pearson and others 2000). Songbird nest initiation, nesting site fidelity, and modeled reproductive success were lower in knapweed infested sites. Grasshoppers, an important food source for songbirds, were also less abundant in the knapweed sites (Ortega and others 2006). Knapweed flower stalks provide ideal anchor points for *Dictyna* spider webs so the spiders can double their web size, allowing higher prey capture rates and large increases in the spider population (Pearson, unpublished data, 2006). Carabid beetle species composition and structure were more homogeneous in knapweed sites than in native plant sites, and evenness and beta diversity were significantly lower for carabid functional groups and species assemblages in knapweed sites (Hansen and others 2006). In addition to effects of knapweed invasion on animal populations, Lacey and others (1989) noted that bunchgrass plots dominated by spotted knapweed produce more surface runoff and greater sediment yields than adjacent uninfested plots.

This chapter discusses the vegetation formations in figure 8-2. A brief description is presented of the vegetation within each formation, the historic and current fire regimes, and management issues. This is followed by a summary of the available literature on the interactions between fire and nonnative invasive species in that formation (table 8-1). The fire and nonnative invasive interactions within a formation are addressed in three subsections: (1) the effects of fire and fire exclusion on nonnative plant invasions, (2) changes in fuel properties and fire regimes caused by nonnative plant invasions, and (3) the intentional use of fire to manage nonnative invasives.

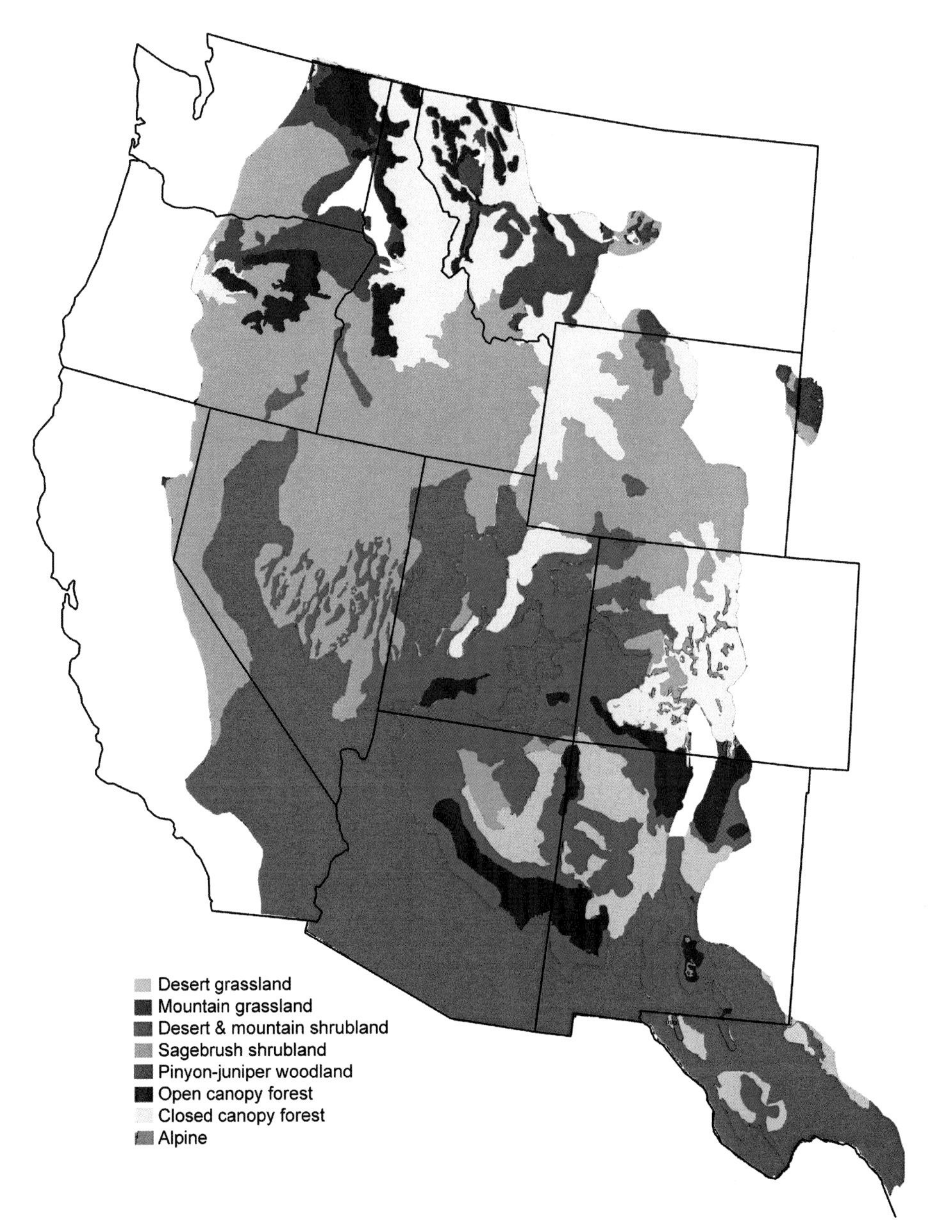

Figure 8-2—Approximate distribution of vegetation formations in the Interior West bioregion. Riparian areas are not shown.

Table 8-1—Vegetation formations in the Interior West and approximate threat imposed by nonnative plants occurring in each formation (L= low threat, H = high threat, P = potentially high threat, U = unknown threat). Designations are approximations based on information from the literature, state and regional invasive species lists, and reports from field scientists and managers in the bioregion.

Species		Formation							
Scientific name	Common name	Desert grassland	Mountain grassland	Desert shrubland	Sagebrush shrubland	Pinyon-juniper woodland	Open-canopy forest	Closed-canopy forest	Riparian
Acroptilon repens	Russian knapweed	L	H	L	H	L	L	L	P
Brassica tournefortii	Sahara mustard	P	L	H	L	L	U	U	U
Bromus japonicus	Japanese brome	L	H	L	H	L	H	L	P
Bromus rubens	red brome	L	L	H	L	L	L	L	P
Bromus tectorum	cheatgrass	L	H	H	H	H	H	L	P
Carduus nutans	musk thistle	L	L	L	L	H	P	P	H
Centaurea biebersteinii	spotted knapweed	L	H	H	H	H	H	L	P
Centaurea diffusa	diffuse knapweed	L	H	H	H	H	H	L	P
Centaurea solstitialis	yellow starthistle	L	H	P	L	L	H	L	H
Chondrilla juncea	rush skeletonweed	L	H	L	H	L	H	L	H
Cirsium arvense	Canada thistle	L	H	L	L	L	L	P	H
Cirsium vulgare	bull thistle	L	L	L	L	L	P	P	P
Convolvulus arvensis	field bindweed	P	L	H	L	L	L	L	U
Elaeagnus angustifolia	Russian-olive	L	N	L	L	L	L	N	H
Erodium cicutarium	cutleaf filaree	L	L	H	L	L	L	L	U
Euphorbia esula	leafy spurge	L	H	L	L	L	H	L	H
Euryops multifidus	sweet resinbush	H	L	H	L	L	U	U	U
Eragrostis curvula	weeping lovegrass	H	L	L	L	L	L	L	L
Eragrostis lehmanniana	Lehmann lovegrass	H	L	L	L	L	L	L	L
Hypericum perforatum	St. Johnswort	L	H	L	L	P	H	P	H
Lactuca serriolata	prickly lettuce	L	L	L	L	L	L	L	P
Linaria dalmatica	Dalmatian toadflax	L	H	L	H	H	H	L	P
Linaria vulgaris	yellow toadflax	L	H	L	U	U	H	L	H
Pennisetum ciliare	buffelgrass	P	L	H	L	L	L	L	L
Pennisetum setaceum	fountain grass	P	L	H	L	L	U	U	U
Poa pratensis	Kentucky bluegrass	L	H	L	L	L	P	L	H
Potentilla recta	sulfur cinquefoil	L	H	H	H	L	H	L	P
Salsola kali	Russian-thistle	H	L	H	H	H	L	L	L
Schismus arabicus and *S. barbatus*	Mediterranean grasses	L	L	H	L	L	L	L	P
Sisymbrium altissimum	tumble mustard	P	H	P	H	H	H	L	U
Sisymbrium irio	London rocket	H	L	H	L	L	U	U	U
Sorghum halepense	Johnson grass	H	L	L	L	L	L	L	U
Taeniatherum caput-medusae	medusahead	L	H	H	H	H	H	L	L
Tamarix spp.	tamarisk	L	L	L	L	L	L	L	H
Tribulus terrestris	puncture vine	H	L	H	L	L	L	L	U

Grasslands

Two major grassland types occur within the Interior West bioregion: desert grasslands in the south and mountain grasslands in the north. Fire regime characteristics of these two types are discussed together by Paysen and others (2000). They are separated in this report due to differences in nonnative species occurrence and impacts.

Wildfires in grasslands are generally warm-season fires that consume most of the aboveground herbaceous growth; that is, they are "stand-replacement" fires as defined by Brown (2000). Native grassland species are generally fire-tolerant; many perennial species sprout from belowground meristematic tissue following fire (Wright and Bailey 1982).

Desert Grasslands

Desert grasslands occur on tablelands from moderate to considerable relief in Arizona, New Mexico, and Utah, and on plains in southwestern Texas. Elevations range from 5,000 to 6,900 feet (1,500 to 2,100 m). Precipitation ranges from 8 to 12 inches (200 to 300 mm), with a frost-free season of about 120 days in Utah to over 200 days in Texas. Grasses dominate the intermediate elevations, and shrubs dominate at higher and lower elevations. In transition zones, shrub cover grades to grasses, from galleta (*Pleuraphis jamesii*) to black grama (*Bouteloua eriopoda*) to blue grama (*B. gracilis*); single-species stands are most common. Tobosa (*Pleuraphis mutica*) dominates in the southern extensions in Texas, and threeawn (*Aristida* spp.) in the northern extensions in Utah (Garrison and others 1977). Desert grassland ecosystems include grama-galleta (*Bouteloua-Pleuraphis*) steppe, grama-tobosa prairie, and galleta-threeawn (*Pleuraphis-Aristida*) shrubsteppe habitats as described by Küchler (1964).

Historical accounts suggest that extensive fires may have occurred frequently, at a frequency of about 5 to 15 years, in most desert grasslands (Humphrey 1958; Humphrey and Mehrhoff 1958; McPherson 1995; Wright and Bailey 1982)—although there is debate about this issue. The time between successive fires at a specific location undoubtedly varied considerably, perhaps ranging from 2 years to 30 years (see, for example, McPherson 1995; Wright and Bailey 1982). Variation in time between fires, probably a result of variation in precipitation, was probably important for maintaining biological diversity, since different species were favored by different fire frequencies and different variation in fire frequency. Fire spread apparently was constrained primarily by fuel continuity and abundance, especially of fine fuels, but this constraint was alleviated when 2 or 3 years of above-average precipitation occurred (McPherson 1995; Wright and Bailey 1982). On the other hand, below-average precipitation, in combination with the stochastic nature of lightning, may have precluded fire from some sites for a few decades.

Most fires occurred in late June or early July when the first summer thunderstorms moved into the region after the extended hot, dry period in May and June. Summer fires were (and probably are) particularly important for sustaining grasses at the expense of woody plants. Most perennial plants, including grasses, are susceptible to mortality from summer fires, but woody plants are especially susceptible to summer fires (Cable 1965, 1973; Glendening and Paulsen 1955; Martin 1983; Pase 1971). It seems likely that seasonality and variation in time between fires were critical components of the fire regime with respect to maintenance of biological diversity, by contributing to both spatial and temporal variability of species composition.

The complex interactions between fire and biological invasions in desert grasslands are confounded by land-use history. Desert grasslands in the southwestern states have been grazed by livestock since the 1500s (Humphrey 1958). In addition, nonnative Lehmann lovegrass (*Eragrostis lehmanniana*) and weeping lovegrass (*E. curvula*) were extensively seeded in the Southwest during the mid-20th century (Bock and Bock 1992b; Cable 1973). Given the poorly documented history of land use on specific sites, it is not surprising that research in these systems often reports varied responses of nonnative plant species to fire (for example, Bock and Bock 1992b; Cable 1965).

Mountain Grasslands

In the Interior West bioregion, mountain grasslands occur in Montana, Idaho, Colorado, and northern New Mexico. They may have occurred historically in Arizona as well. The grasslands occur mainly in open, untimbered areas intermingled with ponderosa pine, Douglas-fir, or lodgepole pine forests at moderate elevations. At higher elevations, grasslands occur on slopes and faces adjacent to subalpine spruce-fir forests and patches of whitebark pine or subalpine fir (*Abies lasiocarpa*). Mountain grasslands also occur on the rich, well-drained soils in valley-like areas and in foothills, tablelands, and lower mountain areas, generally bounded by forest communities. The average frost-free period is 120 days, but it can be fewer than 80 days at high elevations in the northern Rocky Mountains. Precipitation is approximately 13 to 20 inches (500 mm) annually but can exceed 30 inches (760 mm) in the higher mountains. Mountain grasslands are characterized throughout the range by bunchgrasses such as fescue (*Festuca* spp.) and wheatgrass species (for example, bluebunch wheatgrass (*Pseudoroegneria spicata*)) (Garrison and others 1977). The mountain grassland ecosystem includes fescue-oatgrass (*F. idahoensis-Danthonia californica*), fescue-wheatgrass (*F. idahoensis-Pseudoroegneria spicata*), wheatgrass-bluegrass (*P. spicata-Poa secunda*), and foothills prairie (*P. spicata-F. idahoensis, F. altaica-Hesperostipa comata*) habitats as described by Küchler (1964). See individual species reviews in FEIS for information on dominant species' fire ecology.

Historically, encroachment by woody species into mountain grasslands was an ongoing process kept in check by repeated fires. Presettlement fire frequency in mountain grasslands is estimated at less than 10 and up to 35 years (Paysen and others 2000). As Euro-American settlement proceeded, grazing by livestock, cessation of ignitions by Native Americans, and fire exclusion greatly reduced fire frequency in mountain grasslands. As a result, cover of woody species increased substantially on many sites (for example, Bakeman and Nimlos 1985), especially along ecotonal boundaries. Elimination of periodic burning has apparently also reduced diversity of herbaceous species in some areas (Wright and Bailey 1982). Contemporary fires tend to be less frequent in mountain grasslands; they may be less severe where livestock grazing and other activities have reduced fuel loads and continuity (Paysen and others 2000).

Effects of Fire and Fire Exclusion on Nonnative Plant Invasions in Interior West Grasslands

Desert Grasslands—Active fire exclusion in desert grasslands began around 1900 (Paysen and others 2000). There is general agreement that native shrubs have increased in density and invaded new areas in these grasslands over the last century (Bahre and Shelton 1993; Hastings and Turner 1965; Wright and Bailey 1982).

In the late 19th and early 20th centuries, nonnative species were planted in desert grasslands, and in many cases they spread to adjacent areas. The most successful of these were three species of lovegrass from southern Africa—Lehmann lovegrass, weeping lovegrass, and Boer lovegrass (*Eragrostis chloromelas*) (Crider 1945). These grasses were intentionally introduced to stabilize soils throughout the region that had been heavily grazed by livestock, and to provide livestock forage during the fall, winter, and spring when they are green and palatable (Cable 1971; Ruyle and others 1988). As of 2005, the potential distribution of Lehmann lovegrass was modeled to be at least 27,000 square miles (71,000 square km), with much of that area in monotypic stands (Mau-Crimmins and others 2005). Spread of these nonnative grasses is stimulated by fire, as described in the following paragraphs.

Fire promotes lovegrasses in desert grasslands by stimulating seed germination (Ruyle and others 1988; Sumrall and others 1991) (fig. 8-3). Established weeping lovegrass plants are fire tolerant because of their deep root system and densely packed crown that resists burning (Phillips and others 1991). Bock and Bock (1992b) recorded an increase in nonnative lovegrasses after a July 1987 wildfire in Arizona grassland. Lehmann and weeping lovegrass had been seeded at least 40 years previously, and they increased in the first 3 postfire years (1988 through 1990) on plots that had been dominated by native grasses before the wildfire. The virtual absence of lovegrasses from unburned plots precluded statistical comparisons. On plots dominated by nonnative grasses before fire, lovegrass increased after fire although it did not reach prefire levels within 4 years. Forbs initially responded positively to fire, and the nonnative forbs redstar (*Ipomoea coccinea*) and spreading fanpetals (*Sida abutifolia*) initially increased after fire but declined in postfire years 3 and 4.

Johnson grass (*Sorghum halepense*) is a warm-season, perennial, typically rhizomatous grass that was planted as forage in wetlands in the southern portion of the Interior West bioregion. In hot deserts and desert grasslands, it has a scattered, patchy distribution in riparian areas and on relatively moist sites (Anderson and others 1953; McPherson 1995). Once established, Johnson grass is extremely resistant to control efforts, in part because the relatively deep rhizomes survive and sprout following fire (Holm and others 1977).

Figure 8-3—Lehmann lovegrass in desert grassland at Fort Huachuca Military Reservation, Arizona, in June 2001 after a March 2001 fire and ample precipitation. The percent of total biomass on the plot represented by Lehmann lovegrass increased from 53% before the fire to 91% in postfire year 2. (Photo by Erika L. Geiger.)

Several other nonnative species are common invaders of desert grasslands, including sweet resinbush (*Euryops multifidus*), London rocket (*Sisymbrium irio*), puncture vine (*Tribulus terrestris*), and Russian-thistle (*Salsola kali*). Fire effects on these species are not described in the literature, although all establish readily in the high-light environments typically found after fire. According to a FEIS review, Russian-thistle can dominate burned sites 1 to 3 years after fire (Howard 1992b). Sweet resinbush establishes in undisturbed stands throughout desert grasslands (Munda, personal communication, 2001). Sweet resinbush, puncture vine, and Russian-thistle persist for several decades in the absence of disturbance (McPherson, personal observations, July 2003 and 2004).

Many native woody plants associated with desert grasslands are susceptible to fire, especially as seedlings (Bock and Bock 1992b), and relatively few species of woody plants sprout after they are burned or otherwise top-killed as seedlings (Glendening and Paulsen 1955; Wright and Bailey 1982). Fires also kill many seeds of woody plants found on the soil surface (Cox and others 1993). Many of this region's woody plants produce seeds only after they are at least 10 years old (Burns and Honkala 1990a, b; Humphrey 1958), which suggests that relatively frequent fires prevent the establishment of woody plants. In contrast to the decade or more required by woody plants to recover from fire, most grasses experience a period of reduced productivity shorter than 3 years (Wright and Bailey 1982) and generally recover to prefire levels of foliar cover within 3 years (McPherson 1995; Wright and Bailey 1982). These differences in postfire response likely gave grasses a postfire competitive advantage and helped prevent shrub encroachment before Euro-American settlement.

When livestock were introduced widely throughout the region in the late 1800s, the concomitant reduction in fine fuels (especially grasses) reduced the probability of fire spread in desert grasslands, thus contributing to a rapid decline in fire occurrence and increased dominance of woody plants (McPherson 1995, 1997). On many sites, woody plants now interfere with grass production to the extent that fires rarely spread, even under extremely hot, dry, and windy weather conditions. Further, the dominant woody plant on most of the region's former grassland and savannas is mesquite (*Prosopis* spp.). Once this native, long-lived tree or shrub has established, as it has throughout much of the American Southwest, it is remarkably resistant to mortality via fire or other means. Removal of the aboveground portion of the plant, even with recurrent high-intensity fires, rarely induces mortality in mesquite plants that exceed a few years in age (for example, Drewa 2003; Weltzin and others 1998). Personal observations indicate that two high-intensity, early-summer fires within a period of 5 years will cause about 10 percent mortality of established mesquite plants. This appears to represent an approximate upper bound on fire-induced mortality (McPherson 1995). The physiognomic shift from grassland dominated by warm-season (C_4) grasses to shrubland dominated by C_3 woody plants undoubtedly has had consequences for nonnative, invasive species. It seems likely that sites dominated by nitrogen-fixing woody plants such as mesquite will favor species adapted to shady, nitrogen-rich conditions, rather than those adapted to relatively open grasslands that formerly characterized these plant communities. Given the decrease in fire occurrence, spread, and intensity associated with mesquite shrublands relative to grasslands, nonnative species that are poorly adapted to recurrent fires likely will be favored by the physiognomic shift to shrubland. However, these seemingly logical consequences have not been described empirically, and there are no known examples of nonnative species that likely would benefit from altered conditions associated with mesquite dominance.

Mountain Grasslands—Many nonnative species occur in mountain grasslands, and some of these have the potential to establish and/or spread following fire. Documentation in the literature of fire effects on nonnative plant invasions in mountain grasslands is sparse; however, some inferences are suggested based on available experimental evidence as well as life history traits of these species.

Invasive annual grasses such as cheatgrass, Japanese brome (*Bromus japonicus*), and medusahead (*Taeniatherum caput-medusae*) occur in mountain grasslands but are poorly studied in these communities. While invasive grasses tend to increase after fire in plant communities where they dominated before fire, they are not always favored by fire in communities dominated by native species before burning.

Cheatgrass's relationship to fire is well studied in sagebrush ecosystems, and it seems to respond in a similar manner in mountain grasslands. Cheatgrass often increases after fire, but sites dominated by native grasses before fire may be more resistant to cheatgrass spread than sites where cover of native species is sparse. Cheatgrass occurred in all burned plots, but occurred in only 33 percent of unburned bluebunch wheatgrass-Idaho fescue (*Festuca idahoensis*) plots in western Montana 1 year following wildfires. During the next 2 years, cheatgrass frequency significantly increased on burned but not on unburned grassland plots (Sutherland, unpublished data, 2007). Also, Rice and Harrington (2005a) found that single, low-severity, early-spring prescribed burns in western Montana did not cause ecologically significant changes in cheatgrass abundance on bluebunch wheatgrass and fescue-bunchgrass sites infested with broadleaved nonnatives. Cheatgrass abundance did not change following early-spring prescribed burning in combination with herbicide spraying at three study sites that had high frequency of native perennial grasses. However, cheatgrass abundance increased significantly following herbicide spraying alone and in combination with early-spring prescribed burning at a fourth site where native bunchgrasses were lacking. A similar response pattern was observed following combinations of herbicide spraying and low-severity, late-fall prescribed burning in an 11-year restoration study of spotted knapweed-infested rough fescue (*F. altaica*)—Idaho fescue, and Idaho fescue—bluebunch wheatgrass habitat types in western Montana (Rice and Harrington 2003, 2005b).

No research is available on Japanese brome or medusahead in mountain grasslands, although they often occur with cheatgrass in these communities. Chapter 7 provides additional information on the relationship of Japanese brome to fire in prairie and plains grasslands. Medusahead is adapted to survive frequent fires and produces abundant, flammable litter (Archer 2001, FEIS review).

Annual forbs such as tumble mustard (*Sisymbrium altissimum*) and yellow starthistle (*Centaurea solstitialis*) are common in mountain grasslands and tend to establish or increase following fire in these communities. Tumble mustard establishes from soil-stored seed and seed blown or transported in after fire, and is most frequent in early seral stages (Howard 2003b, FEIS review). For example, in an Idaho fescue-prairie Junegrass (*Koeleria macrantha*) community of northeastern Oregon, tumble mustard and native mountain tansymustard (*Descurainia incana*) pioneered on severely burned sites but were absent by the fifth postfire year (Johnson 1998). Tumble mustard cover increased immediately after a June 1977 fire in a rough fescue-Idaho fescue-bluebunch wheatgrass community in western Montana and continued to increase through postfire year 2 (Antos and others 1983; McCune 1978). Most literature on yellow starthistle comes from studies conducted in California, where carefully timed, consecutive annual fires have been used to control this species (Zouhar 2002a, FEIS review). For 2 years following an August 2000 wildfire in canyon grassland communities dominated by bluebunch wheatgrass and Sandberg bluegrass (*Poa secunda*) in Idaho, yellow starthistle canopy coverage increased significantly. Favorable postfire weather conditions may have aided postfire recovery (Gucker 2004).

Invasive biennial forbs such as musk thistle (*Carduus nutans*), bull thistle (*Cirsium vulgare*), and houndstongue (*Cynoglossum officinale*) sometimes occur in mountain grasslands and adjacent communities. The thistles often establish following fire, either from abundant, wind-dispersed seed or from seed in the soil seed bank. No information is available from the literature on any of these species' relationship to fire in mountain grasslands.

Several nonnative perennial forbs occur in mountain grasslands, and limited experimental evidence describes their relationship to fire (table 8-2). The perennial habit and rooting characteristics of plants such as Dalmatian toadflax (*Linaria dalmatica*), yellow toadflax (*L. vulgaris*), spotted knapweed, Russian knapweed (*Acroptilon repens*), diffuse knapweed (*C. diffusa*), hoary cress (*Cardaria* spp.), rush skeletonweed (*Chondrilla juncea*), leafy spurge (*Euphorbia esula*), Canada thistle (*Cirsium arvense*), St. Johnswort, and sulfur cinquefoil (*Potentilla recta*) suggest that they persist following fire by sprouting from rhizomes, root crowns, or adventitious buds in the root system. Several of these species, such as spotted knapweed, are also known to produce a long-lived seed bank from which they may establish after fire.

Fire did not seem to influence abundance of spotted knapweed in bluebunch wheatgrass-Idaho fescue communities in western Montana. There was no difference in spotted knapweed frequency, density, or cover between burned and unburned plots 1 year after wildfires in 2000. Spotted knapweed abundance did not change significantly on either burned or unburned plots by postfire year 3, but had increased on both burned and unburned plots in postfire year 5 (Sutherland, unpublished data, 2008).

Effects of Nonnative Plant Invasions on Fuels and Fire Regimes in Interior West Grasslands

Presettlement fire frequency for western grasslands is not well known, so evaluations of the effects of nonnative plant invasions on fire regimes are primarily based on inferences from changes in fuel characteristics (Brooks and others 2004).

Nonnative annual grass invasions may increase fire frequency and spread in grasslands by forming a more continuous horizontal distribution of fine fuels relative to native bunchgrass community structure, which has little fuel in the interstices. The fire season may also be extended and the probability of ignition increased by the rapid drying of the fine culms of invasive annual grasses. However, this relationship has not been documented in the literature for mountain grasslands.

The herbaceous dicots that have invaded western grasslands are surface fuels, as are the native grasses, but they differ morphologically from the grasses in ways that may influence fire behavior. For example, the coarse stems of spotted knapweed may reduce potential for fire spread. The higher moisture content of leafy spurge could also reduce fire spread, but deep standing litter might facilitate fire spread while grasses are still green (fig. 8-4).

Table 8-2—A summary of available information on fire-related topics relevant to invasive perennial forbs in mountain grasslands. Where individual references are cited, other than those from FEIS, these are the only available literature on that topic for that species.

Species	Postfire vegetative response	Postfire establishment from seed	Heat tolerance of seed
Russian knapweed (*Acroptilon repens*)	Roots undamaged by prescribed fire; postfire sprouting observed (Bottoms, personal communication, 2002)	No information	No information
spotted knapweed (*Centaurea biebersteinii*)	Increase in frequency, cover, or stem density 3 years after wildfire in Montana (Sutherland, unpublished data, 2008). Abundance not changed by early-spring prescribed burning in Montana (Rice and Harrington 2005a). Cover and density reported to increase after a single prescribed fire in Washington (Sheley and others 1998, review; Sheley and Roche 1982, abstract), although no data are given	No information	Laboratory tests showed reduced germination in seeds exposed to 200 °C for 120 seconds, and those exposed to 400 °C for 30 seconds (Abella and MacDonald 2000)
diffuse knapweed (*Centaurea diffusa*)	Cover and density reported to increase after prescribed fire in Washington (Sheley and others 1998; Sheley and Roche 1982), although no data are given	No information	No information
hoary cress (*Cardaria* spp.)	"Severe postfire weed spread" in Wyoming is suggested but not documented (Asher and others 1998, abstract)	No information	An undocumented assertion that heat breaks dormancy in hoary cress seed (Parsons and Cuthbertson 1992, review)
rush skeletonweed (*Chondrilla juncea*)	No information	Produces abundant wind-dispersed seed (Panetta and Dodd 1987, review); "serious rush skeleton infestations" observed 1 year after fire in Idaho (Asher and others 1998)	No information
Canada thistle (*Cirsium arvense*)	Survives fire and sprouts from extensive perennial root system after fire (Romme and others 1995). Most fire research conducted in forest and prairie communities; responses to fire are variable (Zouhar 2001d, FEIS review).	Numerous examples in the literature where Canada thistle established after fire from wind-dispersed seed (Zouhar 2001d, FEIS review)	No information
leafy spurge (*Euphorbia esula*)	Abundance not changed by early spring prescribed burning (Rice and Harrington 2005a). Most fire research conducted in plains grassland and prairies (chapter 7)	No information	No information

(continued)

Table 8-2—(Continued).

Species	Postfire vegetative response	Postfire establishment from seed	Heat tolerance of seed
toadflax species (*Linaria* spp.)	Seed production and biomass increased 1 season after fire; biomass and canopy cover were significantly higher 1 and 2 years after fire in big sagebrush-bluebunch wheatgrass communities. Density did not change significantly during 3 postfire years (Jacobs and Sheley 2003a,b, 2005). Abundance not changed by early spring prescribed burning (Rice and Harrington 2005a)	Several studies from southwestern ponderosa pine communities indicate postfire establishment of Dalmatian toadflax (Zouhar 2003b, FEIS review). No information from mountain grasslands.	No information.
St. Johnswort (*Hypericum perforatum*)	Sprouts from roots and root crown following fire, with variable levels and timing of response (Zouhar 2004, FEIS review). Abundance not changed by early spring prescribed burning (Rice and Harrington 2005a)	Observational (Sampson and Parker 1930; Walker 2000) and experimental (Briese 1996) evidence suggests postfire establishment from seed; however, no experimental evidence is available for mountain grasslands.	Seeds exposed to 212 °F (100 °C) for varying time periods had higher germination rates than untreated seed (Sampson and Parker 1930)
sulfur cinquefoil (*Potentilla recta*)	Large sulfur cinquefoil plants survived and sprouted following fire in a fescue grassland in western Montana (Lesica and Martin 2003)	Postfire seedling establishment recorded in a fescue grassland in western Montana; seedling survival may be limited by moisture stress (Lesica and Martin 2003)	No information

Figure 8-4—Fire behavior on April prescribed burn in a western Montana mountain grassland dominated by leafy spurge. Maximum flame length reached 5 to 6 feet (1.5 to 2 m). Fuel was mostly dead leafy spurge stems and grass litter. (Photo by Mick Harrington.)

Desert Grasslands—Livestock grazing, fire suppression, and fragmentation associated with anthropogenic activities have greatly limited fire spread in desert grasslands since the late 1880s (McPherson and Weltzin 2000). In addition, the establishment and subsequent rise to dominance by native mesquite on many sites limits fire spread because grasses are sparsely distributed in the understory of stands that exceed 30 percent cover of mesquite (Gori and Enquist 2003; McPherson 1997).

Nonnative perennial lovegrasses increase after fire and provide more fine fuel to carry subsequent fires than the native warm-season grasses they replace (Cox and others 1984) (fig. 8-5). Anable and others (1992) speculated that Lehmann lovegrass, which can increase biomass production four-fold compared to native grasses (Cox and others 1990) as well as having increased germination after fire, would increase fire frequency and establish a positive feedback loop in desert grasslands. A stand of Lehmann lovegrass in the northern Chihuahuan Desert, for example, had more litter and more continuously distributed fuel than a stand of native black grama. Prescribed burns in the lovegrass-dominated stand spread more quickly than fire in an adjacent stand of black grama (McGlone and Huenneke 2004). The difference in biomass production between nonnative lovegrasses and native grasses is particularly evident during years of below-average precipitation (Cable 1971; Robinett 1992). For example, in contrast to native grasses, Lehmann lovegrass exhibited no reduction in biomass production during an experimental drought (Fernandez and Reynolds 2000). High C:N ratios in many nonnative grasses, including African lovegrasses (*Eragrostis* spp.), likely reduce decomposition rates relative to native grasses and therefore contribute to persistent litter in stands of lovegrass (Cox 1992). Because fire spread in these systems is primarily constrained by fuel abundance, it seems likely that the increased biomass production and decreased decomposition of nonnative lovegrasses relative to native grasses will favor increased fire occurrence and spread, although this conclusion has not been demonstrated empirically (Geiger and McPherson 2005).

Mountain Grasslands—Limited experimental and observational evidence suggests that perennial forbs that displace native grasses in mountain grasslands, such as spotted knapweed and Dalmatian toadflax, may reduce fire frequency and spread. The coarser stems and higher moisture content of spotted knapweed lead to lower plant tissue flammability and thus might be expected to reduce fire frequency, intensity, and length of fire season in mountain bunchgrass communities (Xanthopoulos 1986, 1988). Moisture content of spotted knapweed during the August dry season in western Montana ranged from 120 percent of dry weight at the beginning of the month to 30 percent at the end of August. Native bunchgrasses during this time are desiccated as a consequence of summer drought-induced dormancy (Xanthopoulos 1986). Dalmatian toadflax is another coarse-stemmed

Figure 8-5—Grassland sites at Fort Huachuca Military Reservation, Arizona. (A). Dense stand of Lehmann lovegrass during hot, dry early summer. (B) Site dominated by native species. Gravelly loam soils support widely spaced perennial grasses and diverse wildflowers (Hartweg's sundrops (*Calylophus hartwegii*) seen here). (Photos by Erika L. Geiger.)

nonnative forb that remains green well into the historic mountain grassland fire season (Rice, personal observation, August 2003, Sawmill Research Natural Area, Bitterroot National Forest, Montana). There is no direct evidence that these nonnative invasive forbs have altered mountain grassland fire regimes.

Use of Fire to Manage Invasive Plants in Interior West Grasslands

A review by Rice (2005) suggests that prescribed burning of interior western grasslands has limited potential for directly controlling most nonnative invasive plant species. Permitted burning is usually restricted to the cool seasons (spring and fall) when moisture at the soil-litter interface is high. High surface fuel moisture retards litter consumption and limits fire severity at the ground surface and below (Rice 2005). Grassland fuel loads are typically too low to produce substantial belowground heating even with total fuel consumption (Bentley and Fenner 1958), let alone when fine fuel moisture is high. Information on fire-caused mortality of seeds and underground plant parts is described in chapters 2 and 4.

Desert Grasslands—Fire is generally ineffective for controlling nonnative herbaceous species in desert grasslands. Nearly all research on this topic has been conducted with spring-ignited prescribed fires. Since historical fires generally occurred in summer, spring prescribed fires could threaten biological diversity of native species. However, personal observations (McPherson, September 2004 and January 2008) and very limited data suggest that summer fires are also ineffective for controlling nonnative species, although fires at the appropriate season and frequency are critical to sustaining native plant community composition and structure.

Prescribed burning of weeping lovegrass stands at high fireline intensities (12,603 kW/m) during the winter did not suppress lovegrass recovery in the following growing season (Roberts and others 1988). Lehmann lovegrass was similarly unaffected by spring prescribed burning in a Chihuahuan Desert grassland site undergoing shrub encroachment, but native grasses were reduced by the fire in the short term (McGlone and Huenneke 2004).

Mountain Grasslands—Prescribed burning has been used in mountain grasslands to produce short term reductions in two nonnative annual grasses, cheatgrass and medusahead (Furbush 1953; Wendtland 1993; White and Currie 1983). Optimal phenological timing is critical for suppressing these species. The goal is to burn the seeds or at least heat them enough to kill the embryos. The treatment must be executed while the seeds remain on the culm. Once the seed disperses to the ground it is below the lethal heating zone of most grassland fires (Daubenmire 1968a; Vogl 1974). Unfortunately, cheatgrass plants that establish after sagebrush fires tend to be more fecund, possibly due to less intra-specific competition and a postfire nutrient flush (Young and Evans 1978), thus resulting in increases in the second or third postfire year (Hassan and West 1986). This may also be the case in mountain grasslands.

Burning of medusahead promoted more desirable grasses in cis-montane California grasslands in the short term (Furbush 1953; Major and others 1960; McKell and others 1962; Murphy and Lusk 1961), but attempts to control medusahead by burning in Idaho grasslands have been unsuccessful in the long term, as the medusahead recovers within a few years after the burn (Sharp and others 1957; Torell and others 1961). Additionally, repeat burning (2 and 3 sequential years) of medusahead-dominated grassland in northeastern California, with only limited remnant perennial grasses, resulted in increased abundance of medusahead (Young and others 1972b).

Prescribed burning may be more successful for controlling nonnative annual grass species when combined with other treatments. Native perennial grasses were successfully established on a degraded Palouse Prairie site in Washington by seeding followed by mechanical imprinting of small soil depressions after cheatgrass density was reduced by prescribed fire. Summer burning, when cheatgrass seed was still in inflorescences, had reduced cheatgrass density to less than 90 seedlings/m^2, and summer burning combined with fall herbicide treatment to kill emerging cheatgrass seedlings had reduced cheatgrass density to less than 40 plants/m^2. Burning exposed mineral soil that facilitated seed-to-soil contact by planted native species (Haferkamp and others 1987). Managers in central Oregon have reported success in suppressing medusahead with a combination of prescribed fire, herbicide treatment, and seeding of desirable species. Burning was done in late spring to early summer, while the seed was still on the culms. The following spring, after remnant medusahead seeds had germinated, the area was treated with glyphosate and then seeded with desirable grasses and shrubs (Miller and others 1999).

There is extensive documentation of the successful use of fire to suppress three nonnative, rhizomatous perennial grasses—Kentucky bluegrass (*Poa pratensis*), Canada bluegrass (*Poa compressa*), and smooth brome (*Bromus inermis*), in tallgrass prairies (Rice 2005). However, attempts to suppress these species in western bunchgrass communities have generally been unsuccessful (Grilz and Romo 1995). The difference in response is likely due to the lack of phenological separation between the nonnative perennial grasses and native bunchgrasses in mountain grasslands. Both natives and nonnatives tend to be actively growing during the cool season. Many native bunchgrasses in these communities are susceptible to fire injury because their reproductive growing points are located at or slightly above the soil surface where they may be killed by smoldering dense fuels in the base of the bunch (Antos and others 1983; Redmann and others 1993; Weddell 2001; Wright and Klemmedson 1965). In a disjunct grassland community, Gartner and others (1986) reported suppression of Kentucky bluegrass during the first growing season after fire in western wheatgrass—needle-and-thread grass (*Agropyron smithii - Hesperostipa comata*) mixedgrass prairie in the Black Hills of western South Dakota. Large amounts of herbaceous fuel were present due to long-term fire exclusion and a moderate (rather than severe) grazing regime. Under dry conditions, the early spring strip-headfire consumed most of the litter (Gartner and others 1986). Longer term results are not available.

Perennial nonnative forbs have a tolerance to grassland fires similar to that of the rhizomatous perennial grasses. The perennating tissues are typically at or below the soil surface and thus minimally influenced by low-severity burns; limited annual fuel production in mountain grasslands constrains the opportunities for burning in sequential years. There are few published papers on the use of prescribed fire to control perennial nonnative forbs in the Interior West. Abundance of spotted knapweed, leafy spurge, Dalmatian toadflax, and St. Johnswort did not change following a single, low-severity prescribed fire in early spring at four fescue-bluebunch wheatgrass sites in western Montana (Rice and Harrington 2005a). In a review on spotted and diffuse knapweeds, Sheley and others (1998) stated, "A single low intensity fire increased the cover and density of both weeds in northern Washington without altering the residual, desirable understory species." A citation is given to an abstract of a control study using various treatments (Sheley and Roche 1982); however, the abstract does not present any data, does not directly discuss response to prescribed burning, and just states, "Treatments which did not include an herbicide generally yielded the greatest amount of weeds and least amount of forage." Small test plots of sulfur cinquefoil in a northwestern Montana rough fescue grassland were burned in fall (October) or spring (April) (Lesica and Martin 2003). The burn treatments did not reduce the abundance of mature sulfur cinquefoil plants. The density of small sulfur cinquefoil plants on burn plots increased the first postfire year, presumably due to seed germination. The initial increase in small plants did not lead to a sustained increase in sulfur cinquefoil under the drought conditions that prevailed during the 5-year study (Lesica and Martin 2003). Overall it seems unlikely that fire alone could be used to effectively control many of the perennial nonnative forbs that have invaded Interior West grasslands.

At the time of this writing (2007), only a few trials have examined the effects of integrating other invasive plant control techniques with prescribed burning in mountain grasslands. Burning was combined with herbicide treatments in the rough fescue grassland study described above. Sulfur cinquefoil was more abundant at the end of the 5-year study in plots that were sprayed and then burned than in plots that were just sprayed (Lesica and Martin 2003). Spotted knapweed-infested Idaho fescue grassland plots in western Montana were subjected to a backing fire in April to remove litter. In May these burn plots were sprayed with picloram, clopyralid, and metsulfuron methyl. Community response was measured at the end of two growing seasons after spraying. Combining burning with spraying did not improve knapweed control over that obtained by the same rate of herbicides without burning (Carpenter 1986). Rice and Harrington (2005a) examined combinations of burn and spray treatments in different sequences on bunchgrass sites invaded by spotted knapweed, leafy spurge, Dalmatian toadflax, and/or St. Johnswort. Three years of posttreatment data indicate that various combination treatments did not increase efficacy or extent of target species control obtained by spraying alone with standard herbicide treatments (Rice and Harringtom 2005a).

Herbicide treatments applied both before and after prescribed fire reduced cover, biomass, and density of Dalmatian toadflax on a big sagebrush-bluebunch wheatgrass site in southwestern Montana. Fire alone increased Dalmatian toadflax biomass, and spring prescribed fire did not enhance control over that obtained by herbicide alone (Jacobs and Sheley 2003a, b, 2005). A factorial experiment using spring burning and two herbicides (chlorsulfuron and picloram), applied the fall before burning and 2 weeks after burning, was conducted by Jacobs and Sheley (2003b). Responses were measured in late summer following the spring treatments. Biomass and cover of Dalmatian toadflax doubled following burning. All herbicide treatments reduced biomass, cover, and density of toadflax by 90 percent, compared to controls. If there was a first-year interaction of fire with the herbicide treatments, it may have been masked by the high suppression obtained by the herbicides alone. The high rates of herbicide used to suppress the toadflax also greatly reduced nontarget forb abundance. The investigators conclude these treatments may have left the sites susceptible to reestablishment of toadflax from the soil seed bank (Jacobs and Sheley 2003b). Additional results reported 3 years after fire-herbicide treatments indicate that chlorsulfuron applied in fall or spring and picloram applied in spring suppressed Dalmatian toadflax for up to 3 years (Jacobs and Sheley 2005).

Shrublands

Shrublands considered in this section include sagebrush, desert shrublands, southwestern shrubsteppe, and chaparral-mountain shrub, as defined in Garrison and others (1977). Given the similarity of hot-desert systems and the paucity of information about them, the following discussion will consider the desert shrublands and southwestern shrubsteppe ecosystems together and refer to them, collectively, as "desert shrublands."

Sagebrush Shrublands

Shrublands in the Great Basin are dominated primarily by big sagebrush (*Artemisia tridentata*), although blackbrush (*Coleogyne ramosissima*) and saltbush (*Atriplex* spp.)—black greasewood (*Sarcobatus vermiculatus*) communities also occur in the Great Basin Desert region. The sagebrush ecosystem occurs primarily on the Columbia Plateau of Washington and Oregon; the central portion of the Great Basin, in Utah, Nevada, and southern Idaho; the Wyoming Basin; and the Colorado Plateau and some of the lower reaches of adjacent mountains. Elevations range between about 600 and 10,000 feet (180 to 3,080 m). The length of the frost-free season ranges from 120 days in most areas to 80 days at some mountain sites. Annual precipitation averages 5 to 12 inches (130 to 300 mm), and up to 20 inches (510 mm) in some locations (Garrison and others 1977). The ecosystem is characterized by shrubs, principally of the genus *Artemisia*, which range from about 1 to 7 feet (0.3 to 2 m) tall. Great basin sagebrush, sagebrush steppe, wheatgrass-needlegrass shrubsteppe, and juniper steppe habitats (as described by Küchler 1964) occur within the sagebrush ecosystem. Dominant species include big sagebrush, bluebunch wheatgrass, western wheatgrass, Sandberg bluegrass (*Poa secunda*), needle-and-thread grass, and western juniper (*Juniperus occidentalis*), with several additional shrub and grass components that vary among habitats (Küchler 1964). Understory herbs tend to occur in discontinuous patches with varying amounts of bare soil (Paysen and others 2000). Native species that occur in sagebrush ecosystems vary in their responses to fire. See FEIS reviews on individual species for details on their respective fire ecology.

Plant communities in sagebrush ecosystems vary in structure and composition and in fire regime characteristics. The exact nature of presettlement fire regimes in these ecosystems is not well understood, and a high degree of variability may have occurred. Presettlement fire regimes in big sagebrush/bunchgrass ecosystems have been characterized by mixed-severity and stand-replacement fires, with estimates of fire-return intervals ranging between 10 and 70 years (Arno and Gruell 1983; Burkhardt and Tisdale 1976; Miller and Rose 1995; Paysen and others 2000; Sapsis 1990; Vincent 1992; Young and Evans 1981). These fires occurred primarily between July and September (Acker 1992; Knapp 1995; Young and Evans 1981), with the middle to end of August being the period of the most extreme fire conditions (Bunting and others 1994).

Fuel loading in sagebrush ecosystems varies depending on species composition, site condition, and precipitation patterns. Some sites support fuels that burn readily in some years, and other sites generally cannot carry fire (Paysen and others 2000). For example, work by Miller and Heyerdahl (2008) indicates a high degree of spatial variability in historical fire regimes in mountain big sagebrush (*Artemisia tridentata* ssp. *vaseyana*) communities, even within a relatively small area (4,000 ha, 10,000 acres). In the arid mountain big sagebrush/western needlegrass association, high-severity fires occurred at intervals greater than 200 years, while the more mesic mountain big sagebrush/Idaho fescue associations experienced low-severity fires at 10- to 20-year intervals.

The sagebrush ecosystem and adjacent shrublands have received considerable attention due to the rapid degradation of these systems since Euro-American settlement. Threats to sagebrush shrublands include grazing, changes in fire regimes, juniper invasion, nonnative species invasion, conversion to agriculture, and recreation (Roberts 1996; Sparks and others 1990; Vail 1994). Overstocked rangelands in the late 1800s led to a depletion of perennial grasses and other palatable forage (Paysen and others 2000). The subsequent introduction and spread of cheatgrass and perennial forage grasses (crested and desert wheatgrass (*Agropyron cristatum, A. desertorum*)), compounded by increases in wild and prescribed fire frequency, led to a reduction of big sagebrush and other native plants over vast areas (Billings 1990, 1994; Menakis and others 2003; Paysen and others 2000; Peters and Bunting 1994). Nonnative plant species now dominate many plant communities in this type (Brooks and Pyke 2001 review; Sparks and others 1990; Vail 1994).

Nonnative plant species invasion has been more intense in some areas of the sagebrush shrubland range than in others. For example, Bunting and others (1987) reported that cheatgrass was more likely to invade big sagebrush communities in basin and Wyoming big sagebrush types than in mountain big sagebrush types. Similarly, lower elevation sagebrush shrublands in the Snake River Plain of Idaho have been more affected by nonnative plant species invasion than higher elevation sites, which have higher precipitation and historically had more fine grasses and more frequent wildfires (Gruell 1985; Peters and Bunting 1994).

Desert Shrublands and Shrubsteppe

Desert shrubland communities occur throughout the Interior West bioregion, scattered within all four major deserts—Mojave, Sonoran, Chihuahuan, and Great Basin (Garrison and others1977; Paysen and others 2000). Precipitation patterns, temperature variables, and vegetation structure vary among the four deserts (Brown 1982; Paysen and others 2000). Annual precipitation varies from year to year, averaging 5 to 12 inches (130 to 300 mm) per year in the dry, low elevation plant communities and increasing slightly with elevation. The frost-free season ranges from 120 days in northern desert shrubland communities to 300 days in the Southwest (Garrison and others 1977). Vegetation composition and structure are highly variable, with a large diversity of native plant species. See reviews in FEIS for information on the fire ecology of dominant species or other species of interest in desert shrublands.

Shrubland vegetation in parts of the Great Basin, Mojave, and Sonoran deserts is characterized by xeric shrubs varying in height from 4 inches (10 cm) to many feet. Common dominant species in these areas include blackbrush, saltbush, black greasewood, creosotebush (*Larrea tridentata*), bursage (*Ambrosia* spp.), mesquite, Joshua tree (*Yucca brevifolia*), paloverde (*Parkinsonia* spp.), and a variety of cacti (Cactaceae) (Küchler 1964). Stands are generally open, with a large amount of bare soil and desert pavement (small, nearly interlocking rocks at the surface). However, some stands may be relatively dense. Understory vegetation is generally sparse; however, during years of above-average rainfall, annuals may be conspicuous for a short time (Garrison and others 1977).

At the southern end of the Rocky Mountains, on plains in the Chihuahuan Desert and along the northern edge of the Sonoran among low mountain ranges in southern Arizona, New Mexico, and Texas, vegetation is characterized by shrubs and short grasses. Yucca, mesquite, creosotebush and tarbush are common shrub dominants. Sideoats grama (*Bouteloua curtipendula*), black grama, threeawns, tobosa, and curlymesquite (*Hilaria belangeri*) are dominant grasses (Garrison and others 1977). Some areas have a larger grass component and others a larger shrub component (Paysen and others 2000). Elevation in these communities ranges from about 1,600 to 7,050 feet (500 to 2,100 m), and mean annual precipitation varies from 10 to 18 inches (250 to 460 mm) (Garrison and others 1977).

Presettlement fire regimes in desert shrub communities are characterized by relatively infrequent, stand-replacement fires with return intervals in the range of 35 years to several centuries (Brooks and Minnich 2006; Dick-Peddie 1993; Drewa and Havstad 2001; Paysen and others 2000). Fire-return intervals were at the lower end of this range in Chihuahuan desert scrub with a substantive grass component (Dick-Peddie 1993; Drewa and Havstad 2001) and in mid-elevation communities of the Mojave desert characterized by Joshua tree woodlands, blackbrush scrub, and upper-elevation creosotebush scrub (Brooks and Minnich 2006). In contrast, much of the Sonoran and Mojave deserts had sparse vegetation interspersed with considerable bare soil; the vegetation rarely burned, except after years of above-average winter precipitation when fine fuels were sufficient to permit fire spread (Brooks and Minnich 2006; Rogers and Vint 1987). These fires undoubtedly generated significant long-term consequences for the dominant long-lived woody and succulent plants, few of which are adapted to survive fire (Brooks and Minnich 2006; Schmid and Rogers 1988). Fire spread is now facilitated in particular by the nonnative invasive red brome in Mojave Desert shrublands (Brooks 1999a), especially where it is most abundant at elevations between 2,600 and 3,300 feet (800 to 1,000 m) (Brooks and Berry 2006). Contemporary fires often occur at the ecotone between desert shrublands and desert grasslands, particularly during years of above-average precipitation (Humphrey 1974; McLaughlin and Bowers 1982; Rogers and Vint 1987) and at higher elevations where more mesic conditions allow for higher fuel loads (Brooks and Minnich 2006). Forest Service fire records in the Arizona Upland Subdivision of the Sonoran Desert showed that fire occurrence increased between 1955 and 1983; the authors speculated that increased fuel from nonnative plant species, improved fire detection and reporting, and increased human-caused ignitions contributed to the reported increase in fire starts (Schmid and Rogers 1988). Similarly, the number of fires between 1980 and 1995 in the Mojave Desert increased significantly with time; the authors concluded that the trend was due to increased human-caused fires and increased fuelbed flammability caused by nonnative annual grasses (Brooks and Esque 2002, review).

Desert shrublands have been manipulated for centuries using fire, along with numerous other tools, to meet management objectives. For example, during the era of Euro-American settlement, settlers introduced nonnative species and deliberately overgrazed desert shrubland and desert grassland to release woody plants from interference from herbaceous growth and to reduce fire frequency (Leopold 1924). Due to gradual reductions in livestock numbers during the 20th century, high fuel loading and contiguous herbaceous fuels are now common on many sites. Nonnative grasses commonly occupy areas between native plants and contribute to increased fine fuel biomass and continuity. Later, during the mid-20th century, fire was one of many techniques used to reduce woody plants in desert shrublands. It has been estimated that 20 percent of all blackbrush shrublands in the northeastern Mojave

Desert were burned during this period to promote perennial grasses and other species that provided better livestock forage than shrubs (Brooks and others 2003). Shrub eradication became a common managerial goal during the 1950s, for which a diverse array of tools was developed, including prescribed fire, chemical, and mechanical control methods (for example, Vallentine 1971).

The use of fire in desert shrublands is controversial because natural fire cycles and patterns are not well known for these ecosystems. Fire may be used to achieve desired objectives in many desert shrubland communities; however, fire also may contribute to the loss of desirable fire-intolerant species that are sometimes replaced by less desirable fire-tolerant species, including invasive annual grasses (Brooks and others 2004; Brooks and Minnich 2006).

Chaparral-Mountain Shrub

Chaparral-mountain shrub ecosystems occur on mountains over 3,000 feet (900 m) in the middle Rocky Mountains, low elevations in the Gila Mountains, low and high tablelands in the southern part of the Great Basin, and central Arizona. Precipitation averages 10 to 28 inches (250 to 710 mm) per year. Vegetation is characterized by dense to open deciduous, semideciduous, and evergreen brush or low trees. Understory vegetation is sparse in stands with dense canopy cover, while more open stands have a highly productive understory (Garrison and others 1977). Oak-juniper woodland and mountain-mahogany-oak scrub habitats, as described by Küchler (1964), occur in the Interior West portions of these ecosystems. Dominant native species include alligator juniper (*Juniperus deppeana*), oneseed juniper (*J. monosperma*), Emory oak (*Quercus emoryi*), Mexican blue oak (*Q. oblongifolia*), Gambel oak (*Q. gambelii*), and curlleaf mountain-mahogany (*Cercocarpus ledifolius*) (Küchler 1964).

Young or otherwise sparse chaparral stands with a grass component may or may not experience stand-replacement fires depending on weather and the amount of heat transferred from the grass component to the shrub overstory. Fully developed chaparral stands can be difficult to ignite unless there is some component of dead material and fuels are continuous. However, given an ignition and some wind, fully developed stands will propagate a moving fire even when virtually no dead material exists in them. These are usually stand-replacement fires. Some species sprout following top-kill, while others produce seed that is stimulated by fire to germinate. While chaparral may be considered a fire climax community, fire frequency and timing can be such that chaparral is overtaken by herbaceous vegetation types, such as annual grasses (Paysen and others 2000). Approximately 30 percent of the chaparral at the ecotone between the northeastern Mojave Desert and the western Colorado Plateau is thought to be dominated by nonnative brome grasses and have significantly shortened fire-return intervals (Brooks and others 2003). However, chaparral in Arizona that has been seeded with Lehmann lovegrass does not exhibit this pattern; burned sites tend to be overgrown by chaparral species in about 7 years (Schussman, personal communication, 2006).

Very little information is available on fire and nonnative invasive species in chaparral-mountain shrub communities. In Mesa Verde National Park, mountain shrub communities are dominated by species that resprout after most fires, including Gambel oak and Utah serviceberry (*Amelanchier utahensis*). Eight years after wildfire, Floyd and others (2006) found that density of nonnative plants in burned sites in mountain shrub communities was approximately 1 percent of the density found in piñon-juniper communities, in which few native plants are able to sprout after fire.

Effects of Fire on Nonnative Plant Invasions in Interior West Shrublands

Most of the information on nonnative plant species' responses to wildfire in western shrubland ecosystems comes from studies in sagebrush vegetation types (for example, Cook and others 1994; Hosten and West 1994; Humphrey 1984; Ratzlaff and Anderson 1995; West and Hassan 1985; West and Yorks 2002; Whisenant 1990a). Studies in desert shrubland ecosystems are predominantly in blackbrush habitats in the transition zone between the Great Basin and Mojave Deserts (Beatley 1966; Brooks and Matchett 2003; Callison and others 1985), creosotebush habitats in the Mojave and Sonoran Deserts (Brooks 2002; Brooks and Esque 2002; Brooks and Minnich 2006; Brown and Minnich 1986; Cave and Pattern 1984; O'Leary and Minnich 1981; Rogers and Steele 1980), and upland Sonoran Desert ecosystems dominated by large succulents (Brooks and others 2004; Tellman 2002).

Sagebrush Shrublands—The most common nonnative plant species in sagebrush shrublands is cheatgrass, which has become dominant in many areas during the last century due to grazing, agriculture, fire (Brooks and Pyke 2001; Peters and Bunting 1994; Pickford 1932; Piemeisel 1951; Vail 1994), and possibly increases in atmospheric carbon dioxide (Ziska and others 2005). Medusahead, another winter germinating annual, is also common (Peters and Bunting 1994).

The tendency for wildfire to promote initial replacement of basin big sagebrush by cheatgrass was recognized as early as 1914 (Kearney and others 1914). It is commonly reported that cheatgrass density decreases the first postfire year and approximately equals preburn density by the second or third postfire year in sagebrush shrublands (Zouhar 2003a, FEIS review). Vast areas of sagebrush shrublands have been converted to cheatgrass in the past century (about 80,000 km^2 in the Great Basin alone) (Menakis and others 2003). Low-elevation sites, which are relatively dry and experience wide variation in soil moisture, appear to be more vulnerable to cheatgrass invasion than higher elevation sites with more stable soil moisture. Cheatgrass plants tend to be larger and more fecund in the postfire environment than on unburned sites, potentially leading to subsequent increases in density with favorable climatic conditions (Zouhar 2003a). A few studies and modeling efforts suggest that cheatgrass may decline in the long term after fire on some sagebrush sites, as the increased fire interval provides more opportunity for perennial species to establish and reproduce (for example, Humphrey 1984; Mata-Gonzalez and others 2007; Whisenant 1990a). Increasing the fire-return interval should enhance the native flora; this could be achieved by reducing the size and frequency of fires and sowing native species in depauperate sites (Whisenant 1990a).

The longer-term abundance of cheatgrass after fire appears to be related to precipitation patterns. At a site in Utah that burned in 1981, short-term increases in cheatgrass occurred in both burned and unburned sites, coinciding with above average precipitation (West and Hassan 1985); however, cheatgrass cover declined to a trace 11 years after fire on all sites, coinciding with drought (Hosten and West 1994). Cheatgrass cover then increased during a wetter period over the following 7 years on burned and unburned sites. Thus, cheatgrass dominated the postfire community for a few years, after which the perennial grasses (primarily bluebunch wheatgrass, Indian ricegrass, and Sandberg bluegrass) recovered and began to dominate, especially in ungrazed areas (West and Yorks 2002). It should be noted that cheatgrass has also been recorded to increase with increased precipitation in the absence of fire: At the Jordan Crater Research Natural Area in southern Oregon, a pristine sagebrush site with no fire or other disturbance, cheatgrass abundance increased from 0 to 10 percent over a 14 year period; the author attributed this increase to abundant precipitation in the final year of the study (Kindschy 1994).

Many other nonnative species have been described as postfire invaders in sagebrush ecosystems, but little information is available on their responses to fire. These species include Kentucky bluegrass, Russian-thistle, tumble mustard, flixweed tansymustard (*Descurainia sophia*), leafy spurge, rush skeletonweed, knapweeds (*Centaurea* spp.) (Sparks and others 1990), jointed goatgrass (*Aegilops cylindrica*), Mediterranean sage (*Salvia aethiopis*), and medusahead (Vail 1994). In one study, Kentucky bluegrass had 6 percent cover on 2 sagebrush sites 18 years after fire but was negligible on older sites (Humphrey 1984). At a sagebrush steppe site in central Utah, Russian-thistle cover was negligible until postfire year 10 but increased in year 11, although there was considerable variability between sites suggesting that factors other than fire could be involved (Hosten and West 1994). Tumble mustard and flixweed tansymustard are most frequent in the first years after fire (Howard 2003a,b, FEIS reviews).

Desert Shrublands—Research in California annual grasslands (for example, Allen-Diaz and others 1999) suggests that annual grasses from the Mediterranean region are well adapted to periodic fires, as are some nonnative perennial grasses and annual forbs. These species recruit readily into bare (including recently burned) sites in desert shrublands as well. Nonnative annual grasses commonly observed in postfire studies in desert shrubland communities of the Mojave and Sonoran deserts include

- Bromes (*Bromus* spp.) (Beatley 1966; Brooks and Esque 2002; Brooks and Matchett 2003; Brooks and Minnich 2006; Brown and Minnich 1986; Callison and others 1985; Cave and Patten 1984; O'Leary and Minnich 1981; Rogers and Steele 1980)
- Mediterranean grasses (*Schismus* spp.) (Brooks and Esque 2002; Brooks and Matchett 2003; Brooks and Minnich 2006; Brown and Minnich 1986; Cave and Patten 1984; O'Leary and Minnich 1981)

Nonnative perennial grasses and annual forbs considered "weedy" and observed to increase immediately after fire include

- Perennial grasses, fountain grass (*Pennisetum setaceum*) and buffelgrass (*P. ciliare*) (Brooks and Esque 2002; Brooks and Minnich 2006)
- Annual forbs, Sahara mustard (*Brassica tournefortii*), London rocket, and cutleaf filaree (*Erodium cicutarium*) (Brooks and Matchett 2003; Brown and Minnich 1986; Callison and others 1985; Cave and Patten 1984; O'Leary and Minnich 1981; Rogers and Steele 1980)
- Annual forb, flixweed tansymustard (Howard 2003a)
- Annual forb, halogeton (*Halogeton glomeratus*) (Pavek 1992, FEIS review)

Responses of nonnative annuals to wildfire in desert shrublands are inconsistent. In creosotebush communities in the western Sonoran Desert region, one study reported that cover of nonnative plant species as a group increased 3 to 5 years after wildfire, but the data were not analyzed statistically (Brown and Minnich 1986). Rogers and Steele (1980) described the annual postfire communities at two different sites in south-central Arizona shrublands as dominated by nonnatives, especially cutleaf filaree at one site and red brome at the other. In a study in southern California desert scrub, however, little difference was observed in cover of nonnative plant species, including cutleaf filaree, red brome, and common Mediterranean grass (*Schismus barbatus*), between unburned sites and sites burned 5 years earlier (O'Leary and Minnich 1981). Cave and Patten (1984) studied communities of paloverde-cactus (mostly *Opuntia* and *Carnegiea gigantea*) in the Upper Sonoran Desert after a wildfire and a prescribed fire, and recorded a significant density increase for common Mediterranean grass and a decrease in annual brome grasses 1 and 2 years after fire. No significant difference was found in density of cutleaf filaree 2 years after fire. The differential response among species is probably due to their individual microhabitat affinities, which associate them with areas of differing fuel loads, fire intensities, and subsequent rates of seed bank mortality. For examples from the Mojave Desert, see Brooks (1999b) and Brooks (2002). Callison and others (1985) observed that nonnative annual grasses, nonnative annual forbs, and seeded crested wheatgrass, a nonnative perennial grass, dominated the understory at all burned and unburned sites in their study in a blackbrush community in southwestern Utah. Brooks and Matchett (2003) recorded significantly higher species richness of nonnatives in burned than unburned desert scrub communities at sites burned 6, 8, and 14 years prior, but the specific nonnative dominating species varied from site to site.

At least two species of nonnative perennial grass, buffelgrass and fountain grass (McPherson, personal observations, July 2002 through present), and a forb, Sahara mustard (Matt Brooks, personal observation, Coachella Valley, California, spring 1998) can increase in cover following fire in desert shrublands, and these species quickly dominate many sites in the Sonoran and Mojave Deserts after fire (Matt Brooks, personal observation, Littlefield Arizona, spring 2001). Documentation of response to fire is lacking for the many other species of nonnative plants in these shrubland systems.

Effects of Nonnative Plant Invasions on Fuels and Fire Regimes in Interior West Shrublands

Sagebrush Shrublands—Annual nonnative grass invasions have greatly shortened fire-free intervals in Great Basin sagebrush communities by increasing biomass and horizontal continuity of fine fuels (Peters and Bunting 1994). By 1930 Pickford (1932) had documented "promiscuous" fire-linked degradation of big sagebrush/bunchgrass steppe in the Great Salt Lake Valley and an increase in cheatgrass on ungrazed, burned sites. By the 1940s it was apparent that cheatgrass invasion in sagebrush/bunchgrass steppe was altering the fire regime (Stewart and Hull 1949).

Cheatgrass invasion promotes more frequent fires by increasing the biomass and horizontal continuity of fine fuels that persist during the summer lightning season and by allowing fire to spread across landscapes where fire was previously restricted to isolated patches (Beatley 1966; Billings 1994; Brooks and Pyke 2001; D'Antonio 2000; Knick and Rotenberry 1997; Stewart and Hull 1949; Whisenant 1990a). As cheatgrass spreads in sagebrush communities, community structure shifts from a complex, shrub-dominated canopy with low fuel loads in the shrub interspaces, to one with continuous fine fuels in the shrub interspaces, thus increasing the probability of ignition and fire spread (Billings 1990; Knapp 1996; Knick 1999; Knick and Rotenberry 1997; Whisenant 1990a).

Cheatgrass is better adapted to recover and thrive in the postfire environment (Hassan and West 1986; Young and Evans 1975; Young and Evans 1978; Young and others 1976) than are most native species in Great Basin sagebrush communities. Sagebrush does not resprout after burning, and many other native perennial plants are top-killed and have slow postfire recovery (Wright and Bailey 1982). There are fewer native species competing for resources following an initial fire, so cover of cheatgrass may increase in the relatively bare soil environment. Cheatgrass density may be reduced from prefire levels in the first growing season after fire, but its fecundity is likely to increase, contributing to increased density in subsequent postfire years (Young and Evans 1978).

With cheatgrass infestation, fire-return intervals are as short as 5 years (Billings 1994; Peters and Bunting 1994) on some Great Basin sagebrush sites, where presettlement fire-return intervals were in the range of 30 to 110 years (Whisenant 1990a; Wright and Bailey 1982). The wildfire season also begins earlier and may extend later into the fall because cheatgrass cures by early July and remains flammable throughout the summer dry season (Young 1991a). The size of fires has also increased with the spread of cheatgrass (Menakis and others 2003; Peters and Bunting 1994). Climatic conditions favoring the growth of nonnative annual grasses provide fuels that contribute to extensive fire spread the subsequent fire season (Brooks and Berry 2006; Knapp 1995). Fires ignited in cheatgrass stands may spread to adjacent sagebrush-bunchgrass steppe and forests (Stewart and Hull 1949).

Medusahead also forms continuous fine fuels and near-monotypic stands such that these infestations are altering fuel structure and continuity in sagebrush steppe in a way similar to cheatgrass (Blank and others 1992; D'Antonio and Vitousek 1992; Knapp 1998;

Peters and Bunting 1994; Torell and others 1961). The high silica content of this nonnative annual grass slows litter decay, leading to increases in fine fuel loads and thus increasing fire hazard (Brooks 1999a; McKell and others 1962; Mutch and Philpot 1970; Swenson and others 1964).

Plant and animal species diversity has declined in western shrublands as invasive grasses have simplified the canopy structure. Many species in sagebrush steppe and bunchgrass habitat are not adapted to frequent burning. These include bluebunch wheatgrass, Idaho fescue, and rough fescue (Antos and others 1983; Wright and Klemmedson 1965), which may be susceptible to fire injury and lack persistent seed banks (Humphrey and Schupp 2001). Unfortunately, the presence of nonnative annual grasses that thrive after fire constrains the options for managers attempting to reestablish presettlement fire regimes in these ecosystems (Brooks and Pyke 2001; Brooks and others 2004).

Russian-thistle may also facilitate fire spread in degraded sagebrush desert and other low elevation communities in the Great Basin (it was a fire hazard on the Great Plains by the late 1800s). The spacing of dried stems allows rapid combustion; and burning, tumbling plants can cause new ignitions beyond the fire front (Young 1991b). The persistent flammable stems may also extend the fire season (Evans and Young 1970).

Desert Shrubland—Presettlement fire-free intervals were probably much longer in southwest desert shrubland than in northern sagebrush steppe because of lower fuel loads and less flammable canopy structure (Brooks and Minnich 2006). Paysen and others (2000) have suggested presettlement fire-return intervals of 35 to 100 years or longer in these deserts. Increased fire frequency has been reported in Mojave and Sonoran desert shrublands in recent years (Brooks and Esque 2002; Brooks and Minnich 2006; Brooks and Pyke 2001; Esque and Schwalbe 2002, review; McAuliffe 1995; Schmid and Rogers 1988).

Mediterranean grasses and red brome invade shrub interstices in southwestern desert shrublands (Beatley 1966; Brooks 1999a; Brooks and Berry 2006; Brooks and Pyke 2001; Salo 2005). In the Sonoran Desert, red brome is particularly invasive in saguaro (*Carnegiea gigantea*)-paloverde communities and Mediterranean grasses in creosotebush-white bursage (*Ambrosia dumosa*) desertscrub (Esque and Schwalbe 2002; McAuliffe 1995). Similarly, in the Mojave Desert red brome dominates elevations from 2,600 to about 3,300 feet (800 to 1,000 m), whereas Mediterranean grasses dominate below this range (Brooks and Berry 2006). A 1966 study concluded that red brome invasion was promoting fire in blackbrush communities in Nevada (Beatley 1966). Hunter (1991) continued the work at the Beatley sites in Nevada and suggested that increasing abundance of red brome, and cheatgrass to a lesser extent, were responsible for fueling large wildfires from 1978 to 1987. Reviews by Brooks and Pyke (2001), Esque and Schwalbe (2002), and Brooks and Esque (2002) support Beatley's pioneering research and add Mediterranean grasses to the list of dominant annual grasses that fuel large wildfires.

As with cheatgrass in Great Basin sagebrush habitats, senesced Mediterranean grasses and red brome increase the biomass and horizontal continuity of fine fuels that persist during the summer lightning season and allow fire to spread across landscapes where fire was previously restricted to isolated patches (Brooks 1999a; Brooks and Esque 2000; Brown and Minnich 1986; Esque and Schwalbe 2002). Thick layers of annual plant litter may accumulate rapidly and decompose very slowly in desert regions (Brooks 1999a; Brooks and others 2004). Following 2 or more years with above-average precipitation, annual grass litter may be sufficient to sustain a wildfire (Brooks and Berry 2006; Brown and Minnich 1986; Knapp 1998). Seed reserves of winter annual grasses such as Mediterranean grasses and red brome decline during normal dry years, and the growing season of the first wet year allows the winter annuals to build up their seed bank. If the next year is also relatively wet, these annuals establish and produce ample biomass to sustain a wildfire (McLaughlin and Bowers 1982). Red brome seeds, for example, exhibit uniform germination after just 1 cm of winter rainfall but do not have a persistent soil seed bank. Consequently, red brome abundance fluctuates greatly as a function of precipitation (Brooks 1999b; Brooks and Berry 2006; Brooks and Minnich 2006; Salo 2004). Swetnam and Betancourt (1998) suggested that wetter years and red brome invasion have resulted in fine fuel accumulations resulting in chronic large fires in the upper Sonoran Desert. An analysis of 29 years (1955 to 1983) of fire occurrences in Arizona Upland subdivision of the Sonoran Desert indicated increasing fire frequency over the study period, possibly caused in part by wetter than normal winters towards the end of the period and occurrence of nonnative annuals (Schmid and Rogers 1988). Esque and Schwalbe (2002) review studies that suggest the development of an annual grass/fire cycle in southwestern desert scrublands.

Invasion of salt desert shrub (*Atriplex* spp.) communities by cheatgrass was noted as early as 1947 (Hull and Pechanec 1947). Although the evidence was limited, West (1983, 1994) suggests that a combination of wet El Niño years and increased cheatgrass biomass were factors promoting several fires in these communities. Cheatgrass is also invasive in shadscale (*Atriplex confertifolia*) communities and may lead to conditions for creating an annual grass/fire cycle in this system (Meyer and others 2001).

Several fire ecologists speculate that nonnative perennial grass invasions may also lead to a nonnative grass/recurrent fire cycle in southern desert shrublands in the Interior West (Brooks and Esque 2002; Brooks and Pyke 2001; D'Antonio 2000; D'Antonio and Vitousek 1992). Native species in these communities are not adapted to frequent fire and thus would be threatened by these changes. Buffelgrass, Johnson grass, and fountain grass may increase fire frequency and spread in western deserts although the invasion process is at its early stages for these species (Brooks and Esque 2002; Williams and Baruch 2000). Buffelgrass is a warm-season, perennial, caespitose grass commonly used as a forage species in Texas, Oklahoma, and Mexico. It has established outside cultivation in desert shrublands of Arizona and New Mexico (fig. 8-6). Particularly in desert shrublands, it facilitates the spread of fire, is well-adapted for surviving fire, and resists control efforts (McPherson 1995; Williams and Baruch 2000). Burquez-Montijo and others (2002) examined newspaper accounts of wildfires in central Sonoran desertscrub and thornscrub. They concluded that fire frequency near suburban areas was increasing because buffelgrass invasion provides continuous fine fuel in the absence of intense grazing, particularly after high rainfall years. They acknowledge the lack of statistical data but suggest that most buffelgrass-invaded areas are associated with increased fire frequency.

Figure 8-6—Buffelgrass invading a desert plant community dominated by saguaro near Tucson, Arizona. (Photo by John M. Randall, The Nature Conservancy.)

Use of Fire to Manage Invasive Plants in Interior West Shrublands

Sagebrush Shrublands—Planning prescribed fires in sagebrush should include specific objectives and consider species and subspecies of sagebrush, dominant native grass species, soil types, fuel conditions, and climate (Paysen and others 2000).

Although prescribed fire may be used to suppress cheatgrass in the short term, sagebrush steppe is very susceptible to reestablishment and increases in abundance of annual nonnative grasses after burning unless the site has a sufficient component of native perennial grasses (Blaisdell and others 1982). According to Young and Evans (1974) and Evans and Young (1975), perennial bunchgrass density must be at least 2.5 plants per m^2 to prevent annual grass and/or shrub dominance after burning. Humphrey and Schupp (2001) have documented that native perennial seeds are almost absent from cheatgrass-dominated sagebrush steppe in the Great Basin, so even if a prescribed burn reduces the cheatgrass seed bank the site cannot return to native perennials without reseeding.

Burning cheatgrass, particularly in cheatgrass-dominated sagebrush steppe, is most useful as a seedbed preparation technique followed by immediate seeding of desirable species (Evans and Young 1987; Rasmussen 1994; Stewart and Hull 1949). Cheatgrass seed under the shrub canopy can be destroyed by the heat generated by woody fuels (Hassan and West 1986; Young and Evans 1978), but these areas must be planted with desirable species the year of the fire or they will be reinvaded by cheatgrass from the shrub interspaces (Evans and Young 1987).

Cheatgrass is not the only nonnative invasive species of concern in sagebrush shrublands. Both successes and failures have been reported with use of prescribed fire to reduce other nonnatives. An abstract by Dewey and others (2000) summarizes research on control of squarrose knapweed (*Centaurea triumfettii*) on a Utah site that had burned in an August wildfire. Picloram and 2,4-D were applied in the fall after fire.

Nearly 3 years later, the treatment was described as 98 to 100 percent effective in suppressing squarrose knapweed, whereas control in unburned areas was 7 to 20 percent. In a second study, fall applied herbicides were also more effective after wildfire than in unburned plots. Spring applied herbicides did not perform better in wildfire areas. The abstract does not include data or study designs (Dewey and others 2000).

In a sagebrush/fescue bunchgrass habitat type in western Montana, Kentucky bluegrass frequency was reported as reduced by 27.5 percent after a single late May (May 24) burn intended to suppress Douglas-fir encroachment. Kentucky bluegrass had not recovered in the second postfire growing season. However, the frequency of natives Idaho fescue and rough fescue also decreased. The reduction in litter (18.7 percent) indicates that this was a patchy, low-severity burn, so the decrease in Kentucky bluegrass may not be attributable entirely to the fire (Bushey 1985).

Mid-March prescribed fires used to reduce woody species in a big sagebrush-bluebunch wheatgrass habitat type in southwestern Montana did not reduce Dalmatian toadflax. Burns were severe enough to remove shrubs and encroaching trees but did not affect toadflax density or cover in the first postfire growing season. In fact, toadflax seed production and biomass per plant increased significantly (compared to unburned controls) during this period (Jacobs and Sheley 2003a). In the second and third postfire growing seasons, toadflax cover and biomass were significantly higher than during the first postfire growing season. Where treatments combined herbicide with prescribed fire, Dalmatian toadflax cover did not increase in postfire years 2 and 3 (Jacobs and Sheley 2005).

Desert Shrublands—Prescribed fire may be used to enhance grass production at the expense of woody plants in some desert shrubland communities. However, fire was uncommon in many desert shrublands, so contemporary fires also may contribute to the loss of desirable fire-intolerant species and their replacement by less desirable fire-tolerant species, such as invasive annual grasses (Paysen and others 2000).

Vertical and horizontal canopy structure and fuel load strongly affect heating patterns and seed mortality in shrublands. Differences in peak temperature caused spatial variation in control and succession of annual nonnatives (for example, red brome, Mediterranean grasses, and cutleaf filaree) in creosotebush communities in the Mojave desert (Brooks 2002). Sites were burned in spring and summer. Flame length varied from 4 inches (10 cm) at shrub interspace positions to 102 inches (260 cm) within creosotebush canopies. Burn temperatures under creosotebush canopies were high enough to kill nonnative annual seeds at the soil surface and those buried to a 0.8-inch (2-cm) depth. The nonnatives remained suppressed in these shrub microsites for 4 years following fire. At the shrub canopy drip line, where lethal temperatures were reached only above ground, annual nonnatives were suppressed for only 1 year, and by the third year Mediterranean grasses and cutleaf filaree had increased compared to unburned controls. The burns had little effect on seeds and response of nonnative species in the shrub interspaces, where soil heating was negligible. The investigator concluded that fire can temporarily reduce red brome and may allow managers to establish natives by postfire seeding, but Mediterranean grasses may quickly increase after fire (Brooks 2002).

Response of Lehmann lovegrass to prescribed fire in Arizona chaparral and Chihuahuan desert scrub may depend on season of fire and associated treatments. Lehmann lovegrass stands in southeastern Arizona Chihuahuan desert scrub were burned to remove the canopy and kill lovegrass prior to seeding native warm-season perennial grasses (Biedenbender and others 1995). An August burn followed by seeding allowed the best establishment of native green spangletop (*Leptochloa dubia*). Juvenile Lehmann lovegrass was also most abundant on these burn plots. The investigators suggest a two-step process for replacement of Lehmann lovegrass: first burning to stimulate germination and deplete the seed bank, then a follow-up herbicide treatment to kill juvenile and surviving adult Lehmann lovegrass plants prior to seeding the desired natives. Winter burns in Arizona chaparral, when soil moisture was high, had no significant effect on Lehmann lovegrass density or productivity (Pase 1971).

Piñon-Juniper Woodlands

The piñon-juniper ecosystem occupies areas of the Basin and Range province in Utah, Nevada, eastern California, southern Idaho and southeastern Oregon, and much of the Colorado Plateau in Arizona, New Mexico, Colorado and Utah. These woodlands occur in foothills, low mountains, mesas and plateaus at higher elevations and generally rougher terrain than semi-desert shrublands and grasslands (Garrison and others 1977; West 1988). A diverse assemblage of tree, shrub, and herbaceous species is associated with piñon-juniper woodlands (Evans 1988; West 1988). Typical dominant species include oneseed juniper, Utah juniper (*Juniperus osteosperma*), Colorado piñon (*Pinus edulis*), and singleleaf piñon (*P. monophylla*) (Küchler 1964).

Although people occupied and used piñon-juniper woodlands before Euro-American settlement (Samuels and Betancourt 1982), intensive use of these woodlands began with Spanish colonization of the Southwest in the 1600s (Evans 1988) and continued with European settlement in the 1800s (Evans 1988; Miller and Wigand

1994). Woodlands were used mainly for livestock grazing and fuelwood gathering (Evans 1988), and timber was harvested in some areas for use in mining (Evans 1988; Young and Budy 1987). The current utilization of piñon-juniper woodlands varies by region. However, livestock grazing continues to be a major use throughout the range (Evans 1988).

Piñon-juniper woodlands are characterized as having a mixed fire regime (Paysen and others 2000), although the exact nature of piñon-juniper fire regimes is uncertain (Bunting 1994; Evangelista and others 2004; Romme and others 2003). Fire histories are difficult to determine for piñon and juniper because these trees do not consistently form fire scars (but see Baker and Shinneman 2004; Floyd and others 2000). According to a review by Paysen and others (2000), presettlement fire regime characteristics in piñon-juniper woodlands depend largely on site productivity and plant community structure. On less productive sites with discontinuous grass cover, fires were probably infrequent, small, and patchy. Fire-return intervals were probably greater than 100 years in these areas, with larger, more severe fires occurring under extreme conditions. On productive sites where grass cover was more continuous, surface fires may have occurred at intervals of 10 years or less, maintaining open stands (Gottfried and others 1995). Similarly, in piñon-juniper communities in Great Basin National Park, fires occurred at intervals of 15 to 20 years in areas that had high cover of fine fuels and at intervals of 50 to 100 years on rocky slopes without fine fuels (Gruell 1999). Presettlement fire regimes in dense stands were probably a mixture of surface and crown fires, with surface fires at more frequent intervals (10 to 50 years) and crown fires at longer intervals (200 to 300+ years). Fire behavior and effects in piñon-juniper communities also depend on the stage of stand development. In young, open stands, shrubs and herbaceous cover may be sufficient to carry a surface fire, but as the stand approaches crown closure, herbaceous cover declines and eventually becomes too sparse to carry fire (Paysen and others 2000).

Prolonged livestock grazing and fire exclusion have contributed to a decline of perennial grasses and an increase in shrubs and trees at many piñon-juniper sites (Laycock 1991; Ott and others 2001). As piñon-juniper stands increase in density and approach crown closure, native herbaceous cover (Tausch and West 1995), seed production, and seed bank density decline (Everett and Sharrow 1983; Koniak and Everett 1982). Goodrich (1999) describes crown cover, stand structure, plant composition, and ground cover attributes representative of piñon-juniper seral stages. Nonnative annual grasses, especially cheatgrass, are common on many piñon-juniper sites. When cheatgrass is present in the understory with little or no perennial vegetation, removing piñon and juniper trees usually leads to cheatgrass dominance (Plummer 1959). Dominance of cheatgrass, in turn, may lead to increases in fire size and frequency, thus initiating an annual grass/recurrent fire cycle (Brooks and others 2004; D'Antonio and Vitousek 1992; Evangelista and others 2004). Successional trajectories in piñon-juniper woodlands are currently uncertain because of recent widespread tree mortality, approaching 100 percent in some locations, caused by extended, severe drought interacting with insects, root fungi, and piñon dwarf mistletoe (*Arceuthobium diviricatum*) (Breshears and others 2005; Shaw and others 2005).

Effects of Fire on Nonnative Plant Invasions in Interior West Woodlands

Postfire succession in piñon-juniper woodlands varies with prefire woodland condition. If burned before crown closure has eliminated the understory, the onset of precipitation and warm temperatures encourages native woody and herbaceous species to sprout and native seeds to germinate (Goodrich 1999). Conversely, in late-seral piñon-juniper stands native species are lacking in the understory and seed bank, and fire at this stage is likely to favor nonnative invasive species in the early successional stages if sites are not artificially seeded with native species (Barney and Frischknecht 1974; Bunting 1990; Erdman 1970; Goodrich 1999; Goodrich and Rooks 1999; Koniak 1985; Ott and others 2001; Young and Evans 1973; Young and Evans 1978). Nonnative species may establish from off-site seed sources or from the seed bank, even if they are not present in the community as mature plants (Koniak and Everett 1982).

Cheatgrass is the most commonly recorded nonnative annual grass in the early postfire environment in piñon-juniper woodlands (for example, see Barney and Frischknecht 1974; Erdman 1970; Evangelista and others 2004; Koniak 1985; Ott and others 2001). Japanese brome and red brome are also reported (Erdman 1970; Haskins and Gehring 2004; Ott and others 2001), and medusahead may be invasive on some sites (Archer 2001).

Vegetation response to wildfire was evaluated in piñon-juniper and sagebrush vegetation in west-central Utah following fires in 1996 (Ott and others 2001). Burned vegetation was compared to unburned vegetation at four piñon-juniper sites for 3 years following the fires. Cheatgrass frequency was the same before and after fire, while density decreased 1 year after fire and increased 2 and 3 years after fire. Cheatgrass frequency and density did not change in unburned plots. It was suggested that the cool, wet conditions that occurred 1 year after fire contributed to the subsequent increase of cheatgrass in burned plots. Cheatgrass canopy cover increased in the interspaces between piñon-juniper stumps 2 years after fire and in subcanopy microsites 3 years after fire. While cheatgrass was increasing, Japanese brome showed a decreasing trend. In burned subcanopy microsites, nonnative annual forbs declined over time. However, most species showed little difference in mean percent cover (less than 1 percent) between burned and unburned plots.

Chronosequence studies from piñon-juniper woodlands in Mesa Verde, Colorado (Erdman 1970), Nevada and California (Koniak 1985), and west-central Utah (Barney and Frischknecht 1974) suggest that nonnative plant species are most abundant in early postfire years and decline in later successional stages. These studies suggest that cheatgrass frequency and cover peak 2 to 8 years after fire, begin to decline around 20 years after fire, and continue to decline as succession proceeds (60 to 100+ years) (Barney and Frischknecht 1974; Erdman 1970; Koniak 1985). Similar trends are reported for nonnative annual forbs, although clear patterns are not evident (Barney 1972; Barney and Frischknecht 1974; Erdman 1970; Koniak 1985). The most common annual forbs recorded after fire in these studies include Russian-thistle, prickly lettuce (*Lactuca serriolata*), tumble mustard, flixweed tansy mustard, cutleaf filaree, and pale madwort (*Alyssum alyssoides*).

Goodrich and Gale (1999) indicate that postfire dominance by annual species (often dominated by cheatgrass) can achieve canopy coverage of 60 to 80 percent in 5 to 10 years. This stage can persist for 20 years or longer; it may persist until piñon and juniper return as dominants, or it may be perpetuated by frequent fires fueled by cheatgrass. For example, on some Colorado piñon-Utah juniper sites with south aspects in the Green River corridor, cheatgrass has dominated for 80 years or more. Cooler and more mesic woodlands may be less susceptible to invasion and complete dominance by introduced annuals, but more information is needed regarding factors that influence the susceptibility of piñon-juniper woodlands to invasion.

Five years after slash-pile burning in a piñon-juniper community of northern Arizona, Haskins and Gehring (2004) found that cover of nonnative plants was four times greater in burned than unburned sites. Japanese brome and the annual herbaceous dicot London rocket contributed the majority of the nonnative biomass. Red brome was the only other nonnative grass reported in this study. Minor amounts of the following herbaceous dicots were found: common dandelion (*Taraxacum officinale*), Dalmatian toadflax, yellow sweetclover (*Melilotus officinalis*), prickly lettuce, and common mullein (*Verbascum thapsus*).

Nonnative perennial grasses such as smooth brome and intermediate and crested wheatgrass are sometimes seeded for postfire rehabilitation in piñon-juniper communities (Erdman 1970; Ott and others 2001). These species may spread to unseeded areas (Erdman 1970).

Effects of Nonnative Plant Invasions on Fuels and Fire Regimes in Interior West Woodlands

The limited understanding of reference fire regimes in piñon-juniper woodlands and apparent large variation of fire regime attributes in these communities hamper assessment of changes in fuels and fire regimes. The distribution of late-seral piñon-juniper woodlands in the Interior West has increased over the last few centuries since European settlement; this change is attributed to several factors including livestock grazing, fire exclusion, and climate change (Bunting 1990; Monsen and Stevens 1999). Additionally, cheatgrass and medusahead invasions are now common in piñon-juniper woodlands (Billings 1990; Tausch 1999). It is suggested that both woody and fine fuel loads have increased with the absence of fire coupled with invasion of nonnative annual grasses (Gruell 1999).

Several researchers attribute increased wildfire size and frequency in piñon-juniper woodlands to increased fine fuel loads associated with annual grass invasion (Billings 1990; Tausch 1999; Young and Evans 1973). Researchers in Utah (Bradley and others 1991, review) and California (Stephenson and Calcarone 1999) associate increased probability of ignition and spread, thus higher fire frequency and larger fires in piñon-juniper communities, with invasion by cheatgrass. Postfire successional trajectories in contemporary piñon-juniper habitats may be altered such that native understory species are unable to establish in woodland stands with a cheatgrass understory that engenders frequent fire (Everett and Clary 1985). Cheatgrass dominance is thus perpetuated by the frequent fires that it fuels. On some Colorado piñon-Utah juniper sites with south aspects in the Green River corridor, cheatgrass has dominated for 80 years or more (Goodrich and Gale 1999).

Use of Fire to Manage Invasive Plants in Interior West Woodlands

The prevalence of cheatgrass in the early successional stages of piñon-juniper habitat types following fire or other disturbances can prevent the establishment of more desirable species. A severe fire can destroy much of the cheatgrass seed reserve in the litter layer and at the soil surface and thus facilitate successful seeding if planting is done within the first year after the burn (Evans 1988). Goodrich and Rooks (1999) tested this strategy on a Colorado piñon-Utah juniper site in Utah that formed a cheatgrass-dominated herbaceous layer following a 1976 wildfire. Part of the area was burned under prescription in June 1990, before cheatgrass seed had shattered. Nonnative perennial grasses (intermediate wheatgrass (*Thinopyrum intermedium*), orchard grass (*Dactylis glomerata*), and smooth brome) were seeded in the fall. Cheatgrass reestablished in seeded areas but was still suppressed in density and size when the measurements were made six years after the prescribed burn and seeding (Goodrich and Rooks 1999).

Open-Canopy Forest

Open-canopy forests in low mountains and foothills of the Interior West bioregion are dominated by one of three varieties of ponderosa pine (*Pinus ponderosa* var. *ponderosa*, *P. p.* var. *scopulorum*, and *P. p.* var. *arizonica*) singly and in mixtures, commonly with Douglas-fir and other conifers at moderate elevations (Garrison and others 1977). Ponderosa pine may occur as an early-seral or climax dominant species (Little 1971). The ponderosa pine ecosystem is the largest western forest type in the United States, occurring in parts of 14 western states, from Nebraska to the Pacific coast and from Arizona to Canada. The frost-free period ranges from 120 days in the north to 240 days in parts of the Southwest. Mean annual precipitation ranges from 15 to 20 inches (380 to 500 mm) where ponderosa pine dominates, to about 30 inches (760 mm) where it occurs with other conifers (Garrison and others 1977).

The presettlement fire regime in ponderosa pine forests was characterized in the northern part of its range by frequent (every 5 to 30 years) understory fires on most sites (Arno 1980; Cooper 1960), with less frequent (35 to 200 year), mixed-severity fire regimes on other sites (Arno and others 2000). In the southern part of the range of ponderosa pine, understory fires occurred at frequencies of 2 to 12 years (Cooper 1960; Heinlein and others 2005; Swetnam and Betancourt 1998; Weaver 1951), and mixed-severity fires occurred at frequencies of less than 35 years (Paysen and others 2000). Frequent fires shaped many ponderosa pine stands, and presettlement ponderosa pine stands were often described as open and parklike, with a ground cover of graminoids, forbs, and occasional shrubs. See Arno (2000) and Paysen and others (2000) for more details on ponderosa pine fire regimes and postfire succession.

Over the past century, stand structure and species composition of ponderosa pine forests have changed in character due to timber harvest, fire exclusion, and succession to other tree species. Stands with presettlement, old-growth characteristics are uncommon. Increasing densities of generally smaller, younger trees and changes in species composition (Cooper 1960; Covington and Moore 1994a,b) make these forests increasingly susceptible to crown fires, as well as insect and disease outbreaks (Covington and others 1994). The National Biological Service (Noss and others 1995) recently categorized old growth ponderosa pine forests as endangered ecosystems in the Interior West bioregion. Research in northern Mexico and at little-altered "reference sites" in the Western United States supports the idea that stand structure has been altered profoundly in most North American ponderosa pine stands as a result of anthropogenic activities; further, this research supports the notion that alterations in forest structure have been accompanied by changes in function, especially fire regime (Stephens and Fulé 2005).

Effects of Fire on Nonnative Plant Invasions in Interior West Open-Canopy Forest

Restoration of presettlement fire regimes using prescribed fire in ponderosa pine forests provides opportunities for the introduction and establishment of nonnative plants (Keeley 2004; Sieg and others 2003, review), possibly "trading one undesirable condition for another," according to Wolfson and others (2005). Stand-replacing wildfires that may occur without fuel reduction and fire regime restoration also risk enhancing nonnative plant invasions when uncharacteristically severe fires damage the native vegetation that is adapted to an understory fire regime (Hunter and others 2006).

Few studies report postfire responses of nonnative plant species in ponderosa pine communities, and the research available does not extend beyond 5 years after fire. Canada thistle, bull thistle, and knapweeds are the most frequently recorded nonnative forbs during the early postfire years (Cooper and Jean 2001; Phillips and Crisp 2001; Sackett and Haase 1998; Sieg and others 2003). Cheatgrass and Japanese brome are the most commonly reported nonnative grasses during this time (Cooper and Jean 2001; Crawford and others 2001; Merrill and others 1980). Some studies report greater increases of nonnative forbs than nonnative grasses in the early stages of post-wildfire succession in southwestern ponderosa pine forests (Crawford and others 2001; Griffis and others 2001). A review by Sieg and others (2003) lists several nonnative forbs targeted for herbicidal control in southwestern ponderosa pine, especially after fire, including two biennials (diffuse knapweed and Scotch thistle (*Onopordum acanthium*)) and five perennials (Canada thistle, leafy spurge,

Dalmatian toadflax, spotted knapweed, and Russian knapweed). Other authors (Phillips and Crisp 2001; Sackett and Haase 1998) also note that Dalmatian toadflax and Canada thistle commonly invade and persist in the wake of fire.

Research on fuel reduction treatments indicates that nonnative invasive plants are more abundant in areas with open forest canopies and in severely burned areas. In a ponderosa pine/Douglas-fir forest in western Montana, nonnative invasive species were consistently more abundant in thinned and burned plots than in untreated, thinned-only, or burned-only plots. Nonnative species richness was correlated with native species richness. The author notes that the ecological impacts of nonnative species on the study site may not be severe, since the total cover of nonnative species (less than 2 percent in the thin/burn treatment) was considerably less than that of native species (25.2 percent) (Dodson 2004).

Overall plant diversity increased after both prescribed fire and wildfire in open-canopy forests in Arizona, and nonnative plant diversity increased after wildfire. Griffis and others (2001) sampled several ponderosa pine stands that had been subjected to a range of density-reduction treatments. All sampling was conducted in the same year, but time since treatment varied. Unmanaged stands were untreated for 30 years. Thinned stands were sampled between 6 and 12 years after treatment. Stands that were thinned and burned under prescription were sampled 3 to 4 years after treatment. Stands burned by stand-replacing wildfire were sampled within 5 years after fire. The timing of the sampling should be considered when reflecting on the following results. The species richness of nonnative forbs was lowest in unmanaged and thinned stands and highest in stands burned by wildfire. Nonnative forb ranked abundance (in categories of percent cover) was significantly higher in the wildfire plots than in the other treatments. Nonnative grass responses were not significantly different between treatments.

Annual forbs accounted for most of the overall increase in nonnative species cover 2 years after wildfire at three ponderosa pine sites on the Mogollon and Kaibab Plateaus in central and northern Arizona (Crawford and others 2001) (table 8-3). Similar to results presented by Griffis and others (2001), species richness and abundance of all vascular plant species, including nonnatives, were higher in burned than in nearby unburned areas. Within each burned area, subplots were delineated by fire severity classes based on tree crown scorch and mortality (unburned, moderate severity, and high severity). Nonnative species, particularly annuals, generally had higher cover where fire severity was highest. On moderate-severity plots, nonnative annual forbs and perennial forbs averaged 42 and 11 percent cover, respectively, while nonnative annual and perennial grasses had 3 percent cover each. Crawford and others (2001) account

Table 8-3—Plant species not native to study area with at least 0.5% mean cover in unburned, moderate-severity, or high-severity burn plots from 2 years after fires in northern Arizona ponderosa pine forests (Crawford and others 2001).

Nonnative plant species	Unburned	Moderate severity	High severity
Forbs—annual		% cover	
Lambsquarter *(Chenopodium album)*[a]	<0.5	11	39
Canadian horseweed *(Conyza canadensis)*[b]	-	27	18
Prickly lettuce *(Lactuca serriola)*	-	1	2
Flatspine stickseed *(Lappula occidentalis)*[b]	-	<0.5	4
Yellow sweetclover *(Melilotus officinalis)*	<0.5	<0.5	1
Russian-thistle *(Salsola kali)*	-	<0.5	24
Fetid goosefoot *(Teloxys graveolens)*[b]	<0.5	1	<0.5
Common mullein *(Verbascum thapsus)*	<0.5	1	5
Forb—perennial			
Common dandelion *(Taraxacum officinale)*	<0.5	11	1
Grass—annual			
Cheatgrass *(Bromus tectorum)*	<0.5	3	19
Grasses—perennial			
Smooth brome *(Bromus inermis)*	<0.5	2	<0.5
Orchard grass *(Dactylis glomerata)*	-	<0.5	1
Kentucky bluegrass *(Poa pratensis)*	-	1	2

[a] Species listed as native and nonnative to U.S. depending on variety.

[b] Species considered nonnative to the specific study area but considered native to the United States by the USDA PLANTS database.

for canopy cover of nonnative species on high-severity plots as follows: annual forbs, 93 percent; perennial forbs, 1 percent; annual grasses, 19 percent; perennial grasses, 3 percent.

Average cover and frequency of cheatgrass was higher in burned than unburned sites 1 to 3 years following a wildfire in the Selway-Bitterroot Wilderness, Idaho in August 1973, although there was a great deal of variability among the seven sample sites (table 8-4). Individual cheatgrass plants were significantly taller on burned than unburned plots for the first 2 postfire seasons, after which there was no difference between burned and unburned plots. Yield (g/m^2) of cheatgrass was significantly higher in burned than unburned plots in 3 out of 4 postfire seasons (Merrill and others 1980).

In western Montana ponderosa pine communities, nonnative species richness more than doubled and nonnative species cover increased nearly 5-fold on burned plots between 1 and 3 years after fire. Over this period, cheatgrass frequency increased from 11 to 39 percent and knapweed cover increased from 0.4 to 2.6 percent on burned plots. On unburned plots, nonnative species richness, nonnative species cover, cheatgrass frequency, and spotted knapweed cover did not change over the same time period (Sutherland, unpublished data, 2007, 2008).

Effects of Nonnative Plant Invasions on Fuels and Fire Regimes in Interior West Open-Canopy Forest

There is speculation that cheatgrass may be increasing fire frequency in open-canopy forest, but there is little direct evidence of fire regime alteration. Considering the high frequency of understory fire in these forests during presettlement times, a significant increase due to cheatgrass would be hard to detect. Dense cheatgrass growth in 1938 was claimed to be the precursor of a fire that exceeded 100,000 acres in a ponderosa pine forest in north-central Oregon (Weaver 1959). Wildfires that originate in lower elevation cheatgrass-dominated big sagebrush sites sometimes spread into adjacent, upper elevation communities.

The many other species of nonnative plants that have invaded ponderosa pine and Douglas-fir forests have had no demonstrable impact on fire regimes. For example, the relatively common postfire invaders, Dalmatian toadflax and Canada thistle, do not measurably influence fire regimes (Sieg and others 2003).

Use of Fire to Manage Invasive Plants in Interior West Open-Canopy Forests

Although numerous nonnative plants are invasive in open-canopy forests in the Interior West, we did not find any published studies on the use of fire to control nonnative invasive species in these forests.

Closed-Canopy Forests

Closed-canopy forest types in the Interior West bioregion include Rocky Mountain Douglas-fir, western white pine, fir-spruce, western larch, and lodgepole pine ecosystems described by Garrison and others (1977). Elevations in these forests range from 500 feet (150 m) for Douglas-fir to over 11,000 feet (3,400 m) for lodgepole pine and southwestern fir-spruce forests. Mean annual precipitation ranges from 20 to 50 inches (510 to 1,300 mm), and the frost-free season lasts 80 to 120 days. See Garrison and others (1977) and Arno (2000) for more information on distribution, associated species, and fire regimes in each of these forest types.

Presettlement fire regimes in Douglas-fir forests are characterized as mixed-severity with return intervals of 25 to 100 years and understory fire regimes with more frequent return intervals. Each of the other forest types has both mixed and stand-replacement presettlement fire severity with relatively long fire intervals (Arno 2000).

Fire regimes in closed-canopy forests have generally not been affected by fire exclusion as much as fire regimes in open communities with historically more frequent fire-return intervals, such as ponderosa pine forests. This is due to fire-return intervals that typically exceed 100 years, longer than the period of effective fire exclusion (Barrett and others 1991; Romme 1982).

Table 8-4—Mean percent canopy coverage (± SD) of the three most frequently encountered nonnative plant species the first (1974) and third (1976) year following fire in the Selway-Bitterroot Wilderness, Idaho (Merrill and others 1980).

Nonnative plant species	Plant group	Postfire year 1 (% cover)		Postfire year 3 (% cover)	
		Burned	Unburned	Burned	Unburned
Cheatgrass *(Bromus tectorum)*	Annual grass	24 ± 12	9 ± 12	10 ± 6	6 ± 9
Common sheep sorrel *(Rumex acetosella)*	Perennial forb	Trace ± 1	Trace	2 ± 3	1 ± 2
Salsify (*Tragopogon* spp.)	Perennial forb	2 ± 2	1 ± 1	1 ± 1	2 ± 1

Nonnative species are not common in undisturbed, closed-canopy forests. Establishment has been recorded in these forests after disturbances such as logging, fire, and volcanic activity. Nonnative species are established either from wind-dispersed seed (Canada thistle, bull thistle, hawkweeds) or from seed in the soil seed bank (St. Johnswort). Research has not documented whether these species persist in the aboveground vegetation as succession proceeds and the canopy closes; some may persist in the seed bank and reestablish following subsequent disturbances.

Effects of Fire on Nonnative Plant Invasions in Interior West Closed-Canopy Forests

Most of the information available on nonnative plant species responses to wildfires in western closed-canopy forests is from studies on lodgepole pine and spruce-subalpine fir forests. Several studies were conducted in the Greater Yellowstone Area (Anderson and Romme 1991; Doyle and others 1998; Turner and others 1997), where lodgepole pine occurs on the broad plateaus and subalpine fir and spruce forests occur in the more mesic areas such as ravines (Romme and Knight 1981). These sites are thought to have a history of infrequent (300 to 400 year fire-return interval) stand-replacing fires (Romme 1982). Several studies focus on early postfire vegetation and patterns of succession (Anderson and Romme 1991; Doyle and others 1998; Lyon and Stickney 1976; Romme and others 1995; Stickney 1980, 1990; Turner and others 1997). Only one study aimed to document nonnative plant species responses to wildfire (Benson and Kurth 1995). Species reported in these studies include Canada thistle, bull thistle, redtop (*Agrostis gigantea*), fowl bluegrass (*Poa palustris*) (Benson and Kurth 1995), prickly lettuce (Turner and others 1997), common dandelion (Doyle and others 1998), and timothy (*Phleum pratense*) (Anderson and Romme 1991). Reports on these species do not show consistent responses to wildfire, but patterns for a few species are summarized below, followed by discussion of seed dispersal mechanisms, rates of establishment of nonnative species, the relationship of invasion to fire severity, and fire suppression activities in closed-canopy forests.

Canada thistle is the most commonly recorded nonnative species after fire in closed-canopy forests (Benson and Kurth 1995; Doyle and others 1998; Romme and others 1995; Turner and others 1997). It is described mainly as an opportunistic or transient species that is not present or is uncommon in mature forest or on unburned sites (Doyle and others 1998; Turner and others 1997). For example, Canada thistle was first detected 2 years after a wildfire in Grand Teton National Park, Wyoming. It increased to 5 percent cover by postfire year 9 on both moderately and severely burned sites, and it decreased to ≤1 percent cover by postfire year 17 as cover of tree saplings increased (Doyle and others 1998).

Bull thistle often establishes after disturbance (after logging with and without burning) in closed-canopy forests in Idaho, Oregon, and Montana and is occasionally reported in closed-canopy forests after wildfire (Shearer and Stickney 1991; Stickney 1980), but populations of this species tend to be short lived (Zouhar 2002b, FEIS review). Bull thistle probably establishes following disturbance either via long-distance seed dispersal or soil-stored seed. Bull thistle seeds are equipped with a feathery pappus that is suited to wind dispersal, and evidence suggests that buried bull thistle seeds experience an induced dormancy and decay more slowly with increasing depth (Zouhar 2002b). A buried seed bank will not maintain a bull thistle population from year to year, but it could provide seeds that would establish a new population after major physical disturbance of the soil (Doucet and Cavers 1996).

The combination of a high rate of production and persistence of seed in St. Johnswort suggests that any site that has supported a population of this species for even a few years has high potential for re-establishment from seed for several years after mature plants are removed, whether by fire or other means. Research in closed-canopy forests in other bioregions indicates that viable St. Johnswort seeds occur in the soil seed bank in areas where mature plants do not occur or occur only at some distance from the sampled sites (Halpern and others 1999; Leckie and others 2000; Thysell and Carey 2001a). Several references from the Interior West indicate that St. Johnswort often occurs in previously burned areas, especially forests (Cholewa 1977; Lavender 1958; Monleon and others 1999; Tisdale and others 1959). If St. Johnswort does become established after fire, it may not persist through stand maturity. It established several years after fire in several forested habitats in Idaho but declined as the tree canopy developed (Habeck 1985; Stickney 1986; Stickney and Campbell 2000). However, it has persisted in coast Douglas-fir (*Pseudotsuga menziesii* var. *menziesii*) forests in Washington, Oregon and California (Ruggiero and others 1991).

Managers should be alert to the possibility that species with easily dispersed seed may establish after fire in a closed-canopy forest even if they were not present in the aboveground vegetation or in the vicinity before fire. Establishment of species with either wind-dispersed seed (for example, bull thistle and Canada thistle) or a long-lived seed bank (for example, St. Johnswort, spotted knapweed, and possibly bull thistle) may be possible. In addition, species with animal-dispersed seed (including cheatgrass, spotted knapweed, and St. Johnswort) may establish within a few years after fire. Sutherland (unpublished data, 2008) examined the impact of wildfire on nonnative invasive plants in burned and unburned Douglas-fir, grand fir (*Abies grandis*), and lodgepole pine communities in western Montana following wildfire in 2000. Cheatgrass did not occur on any burned or unburned Douglas-fir plots 1 year after fire. By 2005, it had established in 7 of 26 burned Douglas-fir plots but did not occur on unburned plots; it did not occur on any burned or unburned grand fir or lodgepole pine plots. In Douglas-fir communities, spotted knapweed occurred on one burned plot and one unburned plot 1 year after fire. By 2005, spotted knapweed had established on 15 additional burned plots. All burned and unburned grand fir and lodgepole pine plots were free of nonnative invasive plants in 2001. Unburned plots in these forest types were still free of nonnative invasives in 2005, but spotted knapweed had established in four burned grand fir plots and three burned lodgepole pine plots (Sutherland, unpublished data, 2007, 2008). It is unknown how long cheatgrass or spotted knapweed may persist on these burned sites in the absence of further disturbance.

The influence of fire severity on nonnative plant establishment in closed-canopy forests after fire is not well understood. Doyle and others (1998) compared plant species composition on moderately burned (<60 percent canopy mortality) and severely burned (complete canopy mortality and aboveground portion of understory consumed) plots to plant composition on unburned plots for 17 years after fire in sites dominated by subalpine fir, spruce, and lodgepole pine in Grand Teton National Park. Canada thistle and common dandelion established 2 to 3 years after fire on both moderate and severe burn plots, while neither occurred in unburned forest. The authors inferred that both species were introduced by wind-dispersed seed from off-site seed sources. Common dandelion cover increased slightly over time in the severe burn and remained at low cover in the moderate burn during the 17-year period, while Canada thistle cover declined after the initial increase on both moderate and severe burn plots (Doyle and others 1998). In Yellowstone National Park, Canada thistle densities increased 2 to 5 years after wildfire; density was higher in sites burned by crown fire than in sites with severe surface burns, which in turn had higher densities than low-severity burn sites (Turner and others 1997). Density of prickly lettuce remained low on low-severity burns throughout the study, but densities of around 100 stems/ha were observed on severe surface burns and crown fire sites 3 years after fire (1991), which then decreased to less than 50 stems/ha by 5 years after fire (the end of the study).

Fire suppression activities may have a greater impact on nonnative species establishment in the postfire environment than the fire itself (chapter 14). After a 1988 wildfire in old-growth Douglas-fir, dog-hair lodgepole pine and Englemann spruce (*Picea engelmannii*) sites in the North Fork Valley of the Flathead National Forest in Montana, 23 nonnative plant species were observed in bulldozed plots, 5 in burned plots, and 3 in undisturbed plots (Benson and Kurth 1995). Sutherland (unpublished data, 2008) found similar results in dense ponderosa pine/Douglas-fir forest in western Montana and northern Idaho. One year after wildfire, nonnative species richness was 7 on dozer line plots, and 1.7 on adjacent burned and unburned plots.

Effects of Nonnative Plant Invasions on Fuels and Fire Regimes in Interior West Closed-Canopy Forests

There is no evidence to support fire regime change due to invasion of nonnative species in closed-canopy forests in the Interior West bioregion. Change in fire regimes due to nonnative plant invasions is unlikely, based on the relatively short-term persistence and/or low abundance of nonnative plants as overstory canopy develops.

Use of Fire to Manage Invasive Plants in Interior West Closed-Canopy Forests

No published studies are available on the use of fire to control nonnative invasive plants in closed-canopy forests. Compared to grasslands, shrublands, and open-canopy forests, nonnative plant invasion of these higher elevation, more mesic, cooler habitats has been limited.

Riparian Communities

Just as there are diverse upland communities in the Interior West bioregion, characteristics of riparian zones within these landscapes vary substantially. Nonetheless, plant communities of riparian zones tend to be distinctly different from those of adjacent uplands in this bioregion and so are treated separately here. Limited research has been conducted on how riparian characteristics influence fire properties; however, a review by Dwire and Kauffman (2003) provides some tentative speculations, based on available research, about how differences in topography, microclimate, geomorphology, and vegetation between riparian areas and surrounding uplands may lead to differences in fire behavior and fire effects.

Riparian areas may act as a fire barrier or a fire corridor, depending on topography, weather, and fuel characteristics. Different wind speeds (lower or higher, depending on topography) in riparian areas may affect the quantity of downed woody fuels as well as fire behavior. Similarly, periodic flooding can lead to a patchy distribution of both vegetation and surface fuels. Low topographic position, proximity to surface water, presence of saturated soils, and shade from riparian vegetation collectively contribute to riparian microclimates that are generally characterized by cooler air temperature, lower daily maximum air temperature, and higher relative humidity than those of adjacent uplands. These conditions likely contribute to higher moisture content of live and dead fuels and riparian soils relative to uplands, suggesting that fire intensity, severity, and frequency may be lower in riparian areas. Additionally, deciduous trees and shrubs often dominate riparian areas, further contributing to differences in fuel characteristics (chemistry, fuel composition, and moisture content) from conifer-, shrub-, or grassland-dominated uplands. Many riparian species have adaptations that contribute to their recovery and survival following flooding, which also facilitate their recovery following fire. These include thick bark, sprouting from stems and roots, enhanced flowering and fruit production after disturbance, and wind and water dispersal of seed (Dwire and Kauffman 2003).

The few studies on fire regime characteristics of riparian areas relative to adjacent uplands in the Western United States indicate that fire frequency and severity vary among regions and plant communities. Studies in the northwestern United States (western Cascade and Klamath mountains) suggest that fires were generally less frequent and fire severity was more moderate than in adjacent uplands, while studies in the southern Cascade and Blue mountains suggest that fire-return intervals were generally similar in upslope (ponderosa pine, grand fir, and Douglas-fir communities) and riparian areas. Fewer studies have investigated these relationships for semi-arid shrublands and grasslands, where deciduous hardwoods dominate riparian plant communities. More information is needed to understand the ecological role and importance of fire in western riparian ecosystems (see Dwire and Kauffman 2003).

Cumulative effects of human disturbance may strongly influence fire behavior and fire regimes in riparian areas, as they do in other communities. Activities such as forest cutting, road building, channel simplification, elimination of wildlife populations (for example, beaver), damming, flow regulation, livestock grazing, urbanization, agricultural and recreational development, and introduction of nonnative invasive species all alter various components of riparian ecosystems, can contribute to changes in riparian species composition, and affect the structure and function of riparian ecosystems (Dwire and Kauffman 2003). Increases in fire size or frequency have been reported in riparian areas along some rivers in the southern Interior West bioregion in recent decades. These increases are attributed to a number of factors including an increase in ignition sources, increased fire frequency in surrounding uplands, increased abundance of fuels, and changes in fuel characteristics brought about by invasion of nonnative plant species (Zouhar 2003c, FEIS review).

The literature on nonnative invasive species and fire in riparian areas is sparse and represented primarily by a handful of studies on tamarisk and Russian-olive in the southern part of the Interior West bioregion. These shrubs or small trees are among the most threatening nonnative invasive plants in Interior West riparian areas (table 8-1). They may be more likely to alter plant community structure, function, and fuel characteristics than the many nonnative perennial and biennial forbs that also commonly occur in riparian areas.

Effects of Fire on Nonnative Plant Invasions in Interior West Riparian Communities

Tamarisk species invaded and dominated many riparian zones in the southwestern United States during the last century, and in recent decades they have spread northward as far as Montana in the river corridors of the western portions of the Great Plains (Ohmart and others 1988). Tamarisk invasion and spread appear to be facilitated by fire, although experimental evidence is limited. Tamarisk produces abundant, wind-dispersed seed throughout the growing season that can germinate and establish on bare, saturated soil (Zouhar 2003c). Burned tamarisk plants that are not killed may show an enhanced flowering response (Stevens 1989) and thus produce abundant seed following fire. Tamarisk is also capable of vigorous, abundant sprouting from roots and especially the root crown following fire or other disturbance that kills or injures aboveground portions of the plant (Zouhar 2003c). Even when completely top-killed, it may produce sprouts 6 to 9 feet (2 to 3 m) tall in the first postfire year (Fox and others 1999). Postfire site conditions may favor tamarisk recovery over native woody species (Busch and Smith 1993; Smith and others 1998). Response of tamarisk to fire depends on season of burning, fire frequency, fire severity, and postfire plant community composition (also see chapter 7).

There is no literature on Russian-olive regeneration after fire; however, observational evidence indicates that Russian-olive sprouts from the trunk, root crown, and/or roots after top-kill or damage (Zouhar 2005b, FEIS review). Observations by Caplan (2002) suggest that mixed-species stands along the Rio Grande can become monospecific stands of Russian-olive due to vigorous root sprouting following fire. Managers should be prepared to manage sprouts and possibly seedlings of Russian-olive following fire in areas where it occurs.

Preliminary results from postfire surveys of nonnative invasives in western Montana suggest that several species may be more likely to establish in riparian areas after fire than in the absence of disturbance. Sutherland (unpublished data, 2008) examined seven burned and three unburned riparian communities in western Montana following wildfires in 2000. By 2005, cheatgrass had established on one, Canada thistle three, spotted knapweed six, and bull thistle all seven burned plots. None of these species occurred on unburned riparian plots in 2005.

Effects of Nonnative Plant Invasions on Fuels and Fire Regimes in Interior West Riparian Communities

Historic fire regimes for western riparian communities have not been well characterized (Dwire and Kauffman 2003), so it is difficult to assess the effects of nonnative species invasions on fire regimes in these communities. For example, no reports of riparian zone fires occur in historical fire accounts of the Southwest. However, it has been suggested that presettlement wildfires were infrequent in southwestern floodplain communities (Busch and Smith 1993) due to high moisture content of fuels, rapid litter decay, removal of surface litter and standing dead woody fuels by periodic flooding, and low wildfire frequency in many of the surrounding arid habitats (Busch 1995; Ellis and others 1998).

Tamarisk invasions may alter fuel characteristics of invaded communities. High water and salt content of green tamarisk foliage make it difficult to burn except under extreme conditions (Busch 1995; Racher and others 2002). However, the buildup of dry flammable litter under the tamarisk canopy and accumulation of senescent woody material within individual crowns increase flammability as plants age (Busch 1995; Ohmart and Anderson 1982; Racher and others 2002) (also see chapter 7). Additionally, both tamarisk and Russian-olive invasions may alter the structure of invaded communities and thereby increase horizontal and vertical fuel continuity. This aspect of invasion deserves further research.

Increases in fire size or frequency have been reported for riparian communities along several river systems in recent decades—the lower Colorado and Bill Williams (Busch 1994), Gila (Turner 1974), and Rio Grande (Stuever 1997; Stuever and others 1997). Fire generally appears to be less common in riparian ecosystems where tamarisk has not invaded (Busch 1994; Busch and Smith 1993; Hansen and others 1995; Turner 1974). Analysis of the spatial distribution of riparian wildfires from 1981 through 1992 on the lower Colorado River floodplain indicates that fires were more frequent and larger in tamarisk-dominated communities than in communities dominated by native woody species such as cottonwood, willow, and mesquite (Busch 1995). Tamarisk may promote more frequent severe wildfires in these ecosystems, but the role of fire is still not well understood (Ellis 2001; Stuever and others 1997); increases in fire size and frequency may be attributed to other factors, including an increase in ignition sources, increased fire frequency in surrounding uplands, increased fuel abundance, and altered fuel distribution.

Several interrelated factors have contributed to changes in southwestern riparian ecosystems, such that the effects of tamarisk invasion are difficult to disentangle from the impacts of modern development that have both enhanced the spread of nonnative species and otherwise altered these ecosystems (Glenn and Nagler 2005; Zouhar 2003c). Disturbance regimes in many southwestern riparian communities have been altered by factors including dams and diversions, groundwater pumping, agriculture, and urban development; these measures have reduced base flows, lowered water tables, reduced the frequency of inundation, and changed the frequency, timing and severity of flooding (Anderson 1998; Everitt 1998). The result is a drier floodplain environment where much of the native broad-leaved vegetation is unable to establish new cohorts, becomes senescent or dies, and is replaced by more drought-tolerant vegetation such as tamarisk (Anderson 1998; Everitt 1998; Glenn and Nagler 2005; Shafroth and others 2002; Smith and others 1997; Stromberg and Chew 2002) and Russian-olive (Campbell and Dick-Peddie 1964). Native cottonwood and willow, for example, release seeds in spring and depend on spring flooding for seedling establishment, while tamarisk disperses seed throughout the summer and early fall (Zouhar 2003c) and thus does not depend on spring flooding.

Natural flood regimes also served to clear away live and dead vegetation and redistribute it in a patchy nature on the floodplain. When flooding is suppressed, biomass and continuity of fuels increase (Busch and Smith 1993; Ellis 2001; Ellis and others 1998; Ohmart and Anderson 1982). Typical stands on the Middle Rio Grande, for example, are now characterized by mature and over-mature Rio Grande cottonwood trees, with accumulations of dead wood and litter on the forest floor (Molles and others 1998; Stuever and others 1997).

Fire may be replacing flooding as the most important disturbance in dammed riparian ecosystems. In the absence of flooding, regeneration of native trees is impeded and organic matter accumulates, thus increasing chances for fires that may further alter the species composition and structure of southwestern riparian forests and promote the spread of tamarisk and other fire tolerant species (Ellis 2001; Ellis and others 1998).

Use of Fire to Manage Invasive Plants in Interior West Riparian Communities

Fire is rarely used for controlling nonnative species in riparian communities. Tamarisk is the only species for which any literature is available on the use of fire for control. Fire is used both alone and in combination with other treatments in an effort to control tamarisk. Several authors suggest that an "ecosystem-based" approach may be most effective for controlling nonnative species and promoting native species in riparian ecosystems.

Burn-only treatments cause some tamarisk mortality, opening dense tamarisk canopies to allow establishment and growth of herbaceous species. High live fuel moisture content of tamarisk foliage (over 275 percent during the summer) and typical lack of surface fuels in dense tamarisk thickets make it difficult to damage tamarisk with fire (Deuser 1996; Inglis and others 1996). Direct mortality from prescribed burning of tamarisk stands along the Pecos River in New Mexico has averaged 30 percent, sufficient to prevent development of impenetrable thickets of tamarisk (Racher and Britton 2003). In the Lake Mead area, crown fires are prescribed with the objective of consuming as much aboveground tamarisk biomass and surface litter as possible. Summer burns are more effective for killing tamarisk and consuming biomass, but burning is typically conducted in the early fall (September-October) to avoid killing nesting birds. Air temperature and relative humidity typically allow crown fires during this season. Direct mortality of tamarisk is usually 10 percent or less, and plants sprout from the root crown after fire, necessitating further treatment. Sustaining heat at ground level maximizes root crown injury and tamarisk mortality (Deuser, personal communication, 2004). Piling and burning cut and dried tamarisk slash around larger tamarisk trees can burn out stumps and suppress root crown sprouting (Coffey 1990; Deuser 1996).

Broadcast prescribed burning or pile burning is often incorporated with other treatments in the management of tamarisk (Taylor and McDaniel 1998a). Burning is sometimes used to reduce initial stand density as preparation for mechanical or herbicide treatments (Racher and Britton 2003). In the Lake Mead area, postfire tamarisk sprouts are treated with low volume basal herbicide sprays within 6 to 12 months following fire, resulting in tamarisk mortality of over 95 percent (Deuser, personal communication, 2004). However, Sprenger and others (2002) advise against broadcast spraying tamarisk too soon after fire. They treated tamarisk stands in the Bosque del Apache National Wildlife Refuge with aerial applications of imazapyr in the fall 1 year after wildfire. The following summer, 2,500 tamarisk sprouts per hectare occurred on burned, sprayed plots, more than on mechanically treated plots (26/ha) and nearly as many as in burned parts of the study area outside the treatment plots (2,834/ha). The investigators suggest that 1 year after fire tamarisk top growth was insufficient to absorb and translocate enough imazapyr to kill the entire plant.

Prescribed broadcast fire can be used after herbicide spraying to remove biomass of dead vegetation (McDaniel and Taylor 1999; Taylor and McDaniel 1998b). Using broadcast burning to remove post-spray biomass can lower the overall cost of tamarisk suppression compared to just mechanical clearing, which requires blading topgrowth, stacking and burning debris, then removing root crowns by plowing, stacking, and burning the piles. A study project in New Mexico where aerial herbicide spraying was followed 3 years later by broadcast burning resulted in 93 percent reduction in tamarisk density 6 years after the initial herbicide spraying. The two-stage mechanical clearing gave 70 percent control (McDaniel and Taylor 2003).

Control of tamarisk and other nonnative species in Interior West riparian areas may be improved by manipulation of fire, flooding, and other environmental factors in a way that favors desired native species over invasive species (Hobbs and Humphries 1995). There is evidence, for example, that native cottonwood trees increased in abundance at tamarisk-dominated sites in response to appropriately timed flood pulses, high groundwater levels and soil moisture, and exclusion of livestock grazing (Zouhar 2003c). Ecosystem-based management of this type—incorporating all spatial and temporal patterns of affected ecosystems—emphasizes removing the ecological stressors that may underlie the invasion, rather than on direct control of invasive species (Bisson and others 2003; Ellis and others 1998; Hobbs and Humphries 1995). In applying this approach to management of tamarisk, Levine and Stromberg (2001) examine several studies that contrast response of tamarisk and native riparian trees and shrubs to particular environmental factors. These studies provide the basis for identifying environmental factors that could be manipulated to restore conditions under which the natives are most competitive.

Conclusions

The interactions of nonnative species with fire depend on many factors: (1) fire regime (season, frequency, (2) abiotic factors (pre- and postfire precipitation, temperature regimes, and edaphic attributes); (3) biotic factors (morphology, phenology, and physiology of species on and near the burned area, plus availability of propagules); and (4) site specific management practices (grazing, logging, etc.). In most cases, effective management practices are specific to a particular site, native biotic community, and nonnative invasive(s), and are best determined by people who are knowledgeable about these particulars and have monitored the effects of fire and other disturbances on invasive species in the area.

An ecosystem's invasibility may increase as its structure, functions, and disturbance processes diverge from historical patterns (see chapter 2). Such changes are likely to reduce the vigor and abundance of native plants, creating opportunities for nonnative species to establish, persist, and spread. Alterations in the Interior West bioregion that may have increased invasibility include

- Extensive removal of dominant cover (woody species from desert shrublands, sagebrush from the Great Basin, old-growth trees in open-canopy forests, and complete canopy removal by clearcutting closed-canopy forests);
- Exclusion of fire from ecosystems where it had been frequent (mountain grasslands, open-canopy forests);
- Increased fire severity in ecosystems where most fires had been low- or moderate-severity (piñon-juniper, open-canopy forests);
- Extensive grazing in dry ecosystems (desert and mountain grasslands, piñon-juniper); and
- Removal of flooding from riparian ecosystems.

Alterations of basic ecological functions and establishment of nonnative species populations have changed the mix of organisms and wildland fuels in many ecosystems of the Interior West. Fires in these "new" systems may favor nonnative species to a greater extent than native species and may also affect biological control agents (mainly insects) that have been introduced to reduce nonnative invasives (see chapter 4). Attempts to reduce or eliminate nonnative species are likely to be ineffective if changes in historic ecosystem processes are not addressed. Instead of restoring vigor to the native plant community, removal of nonnative invasives under these circumstances is likely to be short-lived or lead to invasion by another undesired species.

Closed-canopy forests in this bioregion currently seem somewhat resistant to invasion after fire, and research has thus far not documented widespread persistence of nonnative species after fire. However, closed-canopy forest types are susceptible to invasives after activities commonly associated with soil disturbance, such as clearcutting. If prescribed fire, wildfire, fire suppression activities, or postfire rehabilitation disturbs the soil, invasibility of closed-canopy forest may increase but is likely to decline as the canopy closes.

The potential for nonnative invasive plants to establish or spread after fire appears to be greater in hot, dry plant communities than in moist, cool ones. The fact that these systems have generally been more altered by anthropogenic activities could also be an important factor affecting their level of invasibility.

Desert shrublands and grasslands seem particularly susceptible to postfire increases in nonnative grasses. The lovegrasses, which were originally introduced to reduce soil erosion and enhance range productivity, are problematic in desert communities, especially grasslands. Two annual grasses—red brome and cheatgrass—establish and increase readily in burned desert shrublands. Nonnative grasses have increased the continuity of fine fuels in many of these ecosystems, a change that can increase fire frequency and size.

Mountain grasslands are vulnerable to establishment and spread of herbaceous nonnative species after fire, especially where the nonnatives were prevalent before fire. Where the native herbaceous community was vigorous before fire, increases in nonnatives are less likely.

Among the shrublands of the Interior West, the only ones where a relationship between fire and nonnative species seems clear are sagebrush and desert shrublands. In these types, annual grasses, particularly those from the Mediterranean region, establish and spread readily after fire and contribute to a grass/fire cycle that further reduces shrubs and often the quality of wildlife habitat. Other shrublands have a more complex relationship with nonnative invasives and fire, especially where woody species have been removed mechanically, nonnative grasses have been introduced to improve grazing, or presettlement fire regimes have been altered.

Research on fire-invasive relationships in piñon-juniper woodlands and open-canopy forests has mostly been limited to the first few years after fire. Annual grasses occur in these ecosystems, but it is not yet known whether they will contribute to significantly shorter fire intervals than in presettlement times, especially in open-canopy forests where presettlement fires were very frequent. Likewise, research is lacking on persistence of the many nonnative forbs that occur in these vegetation types after fire. Fire exclusion from open-canopy forests has led to increased surface and ladder fuels, a change that is followed by increases in fire severity in some areas. Where this has occurred, native plant communities are likely to be adversely impacted and nonnative species may be favored.

Research is sparse on fire-invasive interactions for riparian ecosystems, with the exception of tamarisk and Russian-olive. Both of these species may be favored by fire, especially in areas where flooding has been eliminated, diminishing opportunities for establishment of native species. The influence of tamarisk on fire regimes is equivocal: It has been reported to increase fire frequency due to accumulation of litter and standing dead fuels, but it may also be difficult to burn under some circumstances.

Few studies in the Interior West bioregion indicate that fire alone can be used to reduce or eliminate nonnative plant species. Prescribed fires must be timed carefully or combined with other treatments to reduce nonnative species. For some species or sites, fire may need to be avoided altogether. This situation produces a dilemma for managers of ecosystems where fire is needed to reduce fuels or restore native plant communities and ecosystem functions.

Anthropogenic activities have fragmented landscapes throughout the Interior West, blocking routes of dispersal for many native species. Former metapopulations no longer exist as propagule sources to replenish small, isolated populations. This factor may amplify the negative effects of even small fires on isolated populations of desired native species.

If a nonnative species is prevalent or potentially prevalent on a particular site, managers should exercise caution after wildfire and when attempting to reintroduce fire to benefit the ecosystem; after treatment, managers should monitor results and adapt treatments according to what has been learned. Given the number of nonnative species and their ubiquity, caution is warranted on all sites in the Interior West. Practical actions that managers can consider in handling the complex interactions of fire and nonnative invasive species include

- Identify and eradicate small populations of nonnative invasive species as quickly as possible;
- Limit postfire seeding to native species only;
- Monitor nonnative species' responses to wildfire and prescribed fire in terms of invasiveness and impact of on surrounding vegetation;
- Share knowledge about nonnative species and fire in specific ecosystems; and
- Consider the impacts of other disturbances and management activities, in addition to fire, on nonnative species.

Notes

Rob Klinger
Robin Wills
Matthew L. Brooks

Chapter 9:

Fire and Nonnative Invasive Plants in the Southwest Coastal Bioregion

Introduction

The Southwest Coastal bioregion is closely aligned with the geographic boundaries of the California Floristic Province. Excluding Great Basin and Mojave Desert plant communities, the bioregion is defined by the Transverse Ranges of Southern California, the eastern edge of the Sierra Nevada and southern Cascade Ranges, and the northern edge of the Siskiyou Mountains of southern Oregon. The California Flora represents a unique and diverse mix of species from temperate northern and xeric southern biomes. Climatic conditions in the late Tertiary and Quaternary are thought to have played a major role in the origin of the large number of species and high degree of endemism (47 percent) in the region (Axelrod 1966; Hickman 1993). High topographic relief, persistence of relict species, and periodic pulses of speciation have resulted in the most diverse assemblage of native plants in temperate North America (Raven and Axelrod 1978). At approximately 125,000 sq. miles (323,750 km^2), the California Floristic Province contains approximately 1,222 genera and 4,839 species of native vascular plants (Hickman 1993). Important factors contributing to the maintenance of this diversity include the region's isolation and tremendous variation in climate and geology. The diversity of discrete environments within California is greater than in any geographic unit of similar size within the continuous 48 states (Raven and Axelrod 1978). This has resulted in more than 300 natural communities classified within the region (Sawyer and Keeler-Wolf 1995).

The same climatic and topographic complexity that has generated high native plant diversity may also have helped facilitate the establishment and spread of many nonnative species. For example, Raven and Axelrod (1978) proposed that changes in climate since the end of the Pleistocene created an environment more favorable to annuals. The California Flora is generally considered annual-poor and may have offered unoccupied niches easily exploited by introduced annual plant species.

The earliest evidence of plant invasions into the bioregion associated with anthropogenic factors stems from three nonnative species that were found in construction materials used to build the Spanish missions between 1769 and 1824 (Hendry 1931). These species were curley dock (*Rumex crispus*), cutleaf filaree (*Erodium cicutarium*) and spiny sowthistle (*Sonchus asper*). The apparent abundance of these three nonnatives by the mid to late 1700s suggests that plant invasions occurred prior to widespread European settlement of the region, possibly as early as the 1730s (Mensing and Byrne 1999).

The number of introduced species steadily increased throughout the period of California's settlement. Sixteen species first appeared during the period of Spanish colonization (1769 to 1824), 63 during Mexican occupation (1825 to 1848), and 55 more while Euro-American pioneers arrived (1849 to 1860) (Frenkel 1970). Today as many as 1,360 nonnative plant species are known to occur within California (Hrusa and others 2002; Randall and others 1998). Nearly 60 percent of these species are annuals originating from the Mediterranean Basin including southern Europe, North Africa, and Eurasia (Raven 1977). Although the number of nonnative species established in California is large, fewer than 10 percent are recognized as serious threats (Randall and others 1998). Nonnative species continue to establish in the bioregion; however, the rate of new introductions may have slowed recently (Rejmánek and Randall 1994).

Invasion of nonnative plants into the bioregion was likely due to several interacting factors including agriculture, ranching, transportation of seed during the settlement period, climate trends, and availability of diverse microenvironments. Intensive farming and ranching were major land-use activities during the settlement period and were widespread by the mid 1800s. Grazing of domestic livestock is often cited as a significant factor in facilitating the dispersal and establishment of nonnative propagules in the region (Aschmann 1991). The overgrazing of native prairies, aligned with prolonged drought in the mid 1800s, may have initiated the widespread replacement of perennial bunchgrasses with nonnative annuals (Heady and others 1977). Cattle ranching alone does not, however, adequately explain the rapid rate of early plant invasion, because native perennials persisted on a number of heavily grazed sites.

Nonnative species are not evenly distributed throughout the bioregion. Very few occur throughout the region, and 33 percent are restricted to limited geographic subregions (Randall and others 1998). The establishment of many nonnative species was initially associated with coastal areas, with a subsequent spread inland. Nonnatives also tend to occur more frequently at lower elevations (Keeley 2001; Klinger and others 2006b; Mooney and others 1986; Randall and others 1998; Underwood and others 2004). Most of California's invasive plants originated from the Mediterranean Basin and, therefore, encountered an environment favorable to establishment in the large areas where California's climatic conditions are similar to those of the Mediterranean Basin (Blumler 1984). The Mediterranean Basin also has a long history of intensive human disturbance. Regions like the Great Central Valley in California, where agriculture and ranching have been common since the mid 1800s, provided favorable sites for species adapted to high levels of disturbance (Keeley 2001).

While the rate of new introductions of nonnative species into the bioregion may have slowed, previously uninvaded wildlands continue to be invaded as established species expand their range. This has been facilitated by widespread development and intensive land use, and continues to occur throughout the bioregion (Rejmánek and Randall 1994). Many nonnative species exhibit lag effects (Crooks and Soule 1996), occurring in small, often unnoticed populations for decades to a century or more, and suddenly spreading at alarming rates. For example, yellow starthistle (*Centaurea solstitialis*) has occurred in California for over a century. Beginning in the late 1950s, it expanded its range within the state at an exponential rate, increasing from 1.2 to 7.9 million acres (485,000 to 3,197,000 ha) by the early 1990s (Maddox and others 1985; Thomsen and others 1996). Fennel (*Foeniculum vulgare*) was present on Santa Cruz Island for 130 years in small localized areas but then doubled its range and increased in cover more than 250 percent in 1 year following removal of nonnative grazing animals (Brenton and Klinger 1994).

Most ecologists agree that fire and its management have influenced the patterns of nonnative plant invasions in the bioregion, and the occurrence of frequent fire events in the presettlement landscape may have facilitated the establishment and spread of invasive annual plants. Several authors have presented evidence supporting the idea that Native Americans actively converted coastal shrublands to herbaceous plant communities with frequent burning (Anderson, M. 2006; Keeley 2002b; Lewis 1993; Timbrook and others 1982). This frequent burning regime may have weakened the resistance of many sites to invasion. Keeley (2001) proposed that disturbance-adapted plants arrived on a landscape in which the competitive balance had been shifted towards annual-dominated communities. This same cycle of disturbance and nonnative plant establishment continued as early Euro-American settlers converted extensive areas of California into rangeland.

There was a gradual cessation of burning by Native Americans after Euro-Americans began to settle, eventually leading to a long era of fire exclusion. Federal Forest Reserves were established in 1891, and programs were established soon afterwards to systematically suppress wildland fires. Effectiveness of early suppression efforts varied. In many forest ecosystems (for example, the Sierra Nevada mountains), the combined effect of reduced aboriginal burning and active fire suppression resulted in a greatly reduced amount of area burned and fewer large wildfires resulting from human ignitions, as well as profound changes in forest structure (McKelvey and others 1996). In contrast, fire exclusion policies were largely ineffective in changing fire size and frequency in shrubland ecosystems such as chaparral (Keeley and Fotheringham 2001b).

In general, the current landscape now represents a wide range of conditions with various degrees of departure from reference fire regimes (Sugihara and others 2006b). For example, many mid- and high-elevation conifer forests have seen a significant reduction in fire frequency and a corresponding increase in both woody stem densities and surface fuels. Closed-canopy conifer forests are generally not favorable sites for plant invasion (Rejmánek 1989); thus, the "unnatural" conditions of some California forest types may actually be helping to limit nonnative species invasion. Therefore, while the reintroduction of fire is considered crucially important for managing many California forest communities (Biswell 1999; Minnich and others 1995; Parsons and DeBenedetti 1979; Parsons and Swetnam 1989; Pyne and others 1984), initial fire treatments may also facilitate nonnative establishment in these uninvaded sites (Keeley 2001). However, if fuels are not treated, fires in these forest types are typically larger and more severe, relative to presettlement fire regimes, and may increase establishment and spread of nonnative plants.

In contrast to conifer forests, lower elevation foothill woodlands, shrublands, and grasslands appear to burn more frequently than in the past (Keeley and others 1999). Their close proximity to development and highly receptive fuel beds may contribute to shortened fire-return intervals. Increased fire frequency in plant communities that tend to be resistant to invasion (for example, chaparral) may be exacerbating the problem of nonnative invasion into these communities (Keeley and Fotheringham 2001a; Keeley and others 1999). Complicating this are land management practices that result in the purposeful introduction of nonnative plants, such as postfire rehabilitation. For example, for many decades nonnative grass species were commonly seeded following wildfires in an effort to stabilize soils and minimize erosion (Beyers and others 1998).

Invasive species impacts range from displacement of native species to fundamental alteration of ecosystem processes, including fire intensity, severity, and frequency (Brooks and others 2004; D'Antonio and Vitousek 1992; Vitousek and others 1987, 1996; Vitousek and Walker 1989; Whisenant 1990a).

Fire has contributed to the establishment and spread of nonnative plant species in many plant communities in North America (Brooks and Pyke 2001; Grace and others 2001; Harrod and Reichard 2001; Keeley 2001; Richburg and others 2001). Still, appropriately targeted fire events can also be an effective tool for managing some invasive plants (DiTomaso and others 1999; Lonsdale and Miller 1993; Myers and others 2001; Nuzzo 1991). Relative to other control methods, prescribed fire has many unique properties (for example, generally broad social acceptance, defined policies, extensive planning and implementation infrastructure) that allow it to be implemented at a wide range of scales for minimal costs. Many weed control efforts have found prescribed fire is most effective when used in combination with other treatments (for example, herbicides or grazing). Fire is effective at killing individual plants and/or reducing aboveground biomass, and may also stimulate a sprouting response that increases the effectiveness of herbicide applications (fig. 9-1). However, responses of plants to fire are complex and species-specific, and can vary both within and between sites. Clearly, many different factors must be considered and tested before a reasonable likelihood of success can be achieved when managing weeds with prescribed fire.

In the following sections we describe the patterns and processes associated with fire and nonnative invasive plants. Where information is available, we discuss the role of fire in promoting plant invasions, the effects of plant invasions on fire regimes, and the role of fire in the management of nonnative invasive plants. We present this information separately for each of five major vegetation types in the bioregion where the management of both invasive plants and fire are primary concerns for land managers: grasslands, chaparral and coastal scrub, mixed evergreen forests, coniferous forests, and wetland/riparian communities (fig. 9-2) (table 9-1).

Grasslands

This vegetation type includes what is typically classified as grassland, savanna, and open woodland. Historically, there were three main regions in the bioregion where grasslands were well developed: south of the Transverse Ranges, the Central Valley, and the coastal prairie. Although records are sparse, enough historical information exists to make reasonable inferences about general characteristics of the structure and species composition of these ecosystems (Busch 1983; Conise 1868; Cook 1960, 1962; Jepson 1910; Thurber 1880). The grasslands in all three of these regions were characterized by native perennial bunchgrasses and a diversity of native annual and perennial forbs. There was considerable variation in grassland structure among the three regions, while species composition varied both among and within the regions.

The most extensive and continuous grasslands in the state occur below 3,900 feet (1,200 m) in the Central

Figure 9-1—(A) The rapid and unexpected expansion of fennel on Santa Cruz Island in the 1990s led to a 10-year management program (1991 to 2001) that examined the effects of different control methods on changes in community composition and structure, as well as fennel abundance. (B) A combination of prescribed burning and herbicide application decreased fennel cover 95 to 100 percent, (C) but fennel was replaced by cover of other nonnative herbaceous species. This illustrates that active planting or seeding will often have to follow control of undesirable nonnative species if a long-term shift to dominance by native species is a goal. (Photos by R. Klinger.)

Valley and the interior foothills of the Sierra Nevada and Central Coast Range (Heady 1977; Wills 2006). There are strong gradients of precipitation and soil type from the southern to the northern ends of the Central Valley, and from the valley floor to the surrounding foothills. These gradients were likely the most important factors producing differences in species composition in the valley grasslands (Holstein 2001). Purple needlegrass (*Nassella pulchra*) was believed to be the dominant grass throughout the Central Valley, and native forbs comprised a substantial component of the cover in some areas (Hamilton 1997). Coastal prairie occurs discontinuously below 3,300 feet (1,000 m) from Point Concepción northward into southern Oregon. Native vegetation in grasslands of the coastal prairie was dominated by the bunchgrasses Idaho fescue (*Festuca idahoensis*) and California oatgrass (*Danthonia californica*), but as in the Central Valley, species composition of other native grasses and forbs likely varied with elevation and from north to south (Heady and others 1977). Grasslands south of the Transverse Ranges probably occurred primarily as relatively small patches intermixed with extensive stands of chaparral and coastal scrub. They probably were most developed in areas where soils had high water-holding capacity and were characterized by native perennial bunchgrasses such as purple needlegrass and Sandberg bluegrass (*Poa secunda*) (Keeley 2006a).

Over the past 200 years many grassland areas have either been destroyed for urban and suburban

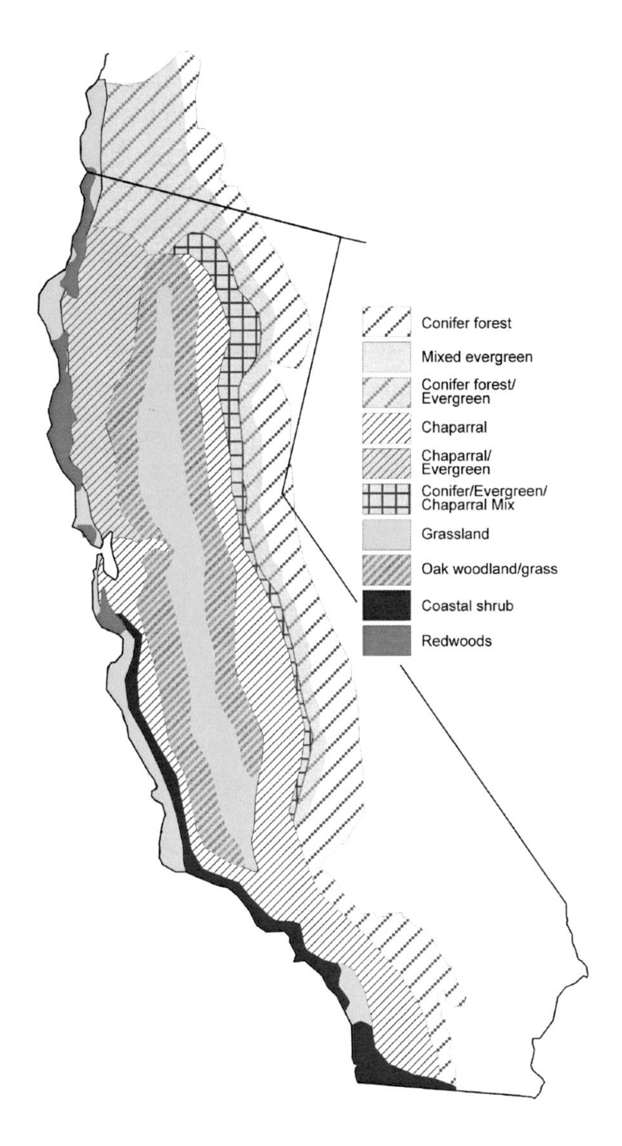

Figure 9-2—Approximate distribution of major vegetation types in the Southwest Coastal bioregion. Areas of overlap in vegetation types are included to show that ecotones are broad and often composed of mixtures or mosaics of vegetation types. Areas in white include parts of the Great Basin and Mojave deserts, which are covered in chapter 8.

Table 9-1—Major vegetation types in the Southwest Coastal bioregion and estimated threat potential of several nonnative plants in each vegetation type (L= Low threat, H = High threat, P = Potentially high threat). Shaded boxes indicate that information is either not applicable or not available.

Scientific name	Common name	Wetland/Riparian communities	Grasslands	Coastal scrub	Chaparral	Mixed evergreen forests	Coniferous forests
Aegilops spp.	Goatgrass		H	H	P		
Ailanthus altissima	Tree-of-heaven	P			P	P	P
Arundo donax	Giant reed	H					
Avena spp.	Oat		H	H	H		
Brassica spp.	Mustard		H	H	H		
Bromus diandrus	Ripgut brome		H	H	H		
Bromus rubens	Red brome		H	H	H		
Bromus tectorum	Cheatgrass	P					P
Carpobrotus edulis	Hottentot fig	L		L	L		
Centaurea solstitialis	Yellow starthistle		L	L	L		
Cortaderia spp.	Pampas grass	P	P	P			
Cytisus scoparius	Scotch broom		L	L	L	P	P
Eucalyptus globulus	Tasmanian bluegum	H	L	H	H	P	
Foeniculum vulgare	Fennel	L	L	L	L		
Genista monspessulana	French broom	L	L	L	L	P	P
Hordeum spp.	Barley		H	H	H		
Lepidium latifolium	Perennial pepperweed	H					
Lolium multiflorum	Italian ryegrass		H	H	H		
Taeniatherum caput-medusae	Medusahead		H	H	H		
Tamarix spp.	Tamarisk or saltcedar	H					
Vulpia myuros	Foxtail fescue		H	H	H		

development, or fragmented and converted into agricultural production. Nevertheless, grasslands still comprise a significant proportion of the area in California, especially in the central and northern parts of the state (Sawyer and Keeler-Wolf 1995). However, they have undergone severe alterations since the period of Euro-American settlement and now bear little resemblance to those that existed prior to the 1800s. Some grassland areas were type-converted from perennial grasslands, while many others were type-converted from shrublands (see below and page 185). Most have been heavily invaded and are now dominated by nonnative annual species (Bossard and others 2000). Many native plant species still occur in these areas, but they comprise a low proportion of the total cover.

Fire and Nonnative Invasives in Grasslands

Role of Fire in Promoting Invasions—The conversion of California's grasslands into ecosystems dominated by nonnative annual grasses and forbs was probably due to several interacting factors, including human disturbance, climate change, the introduction of large numbers of nonnative grazing mammals by Euro-American settlers, and high propagule pressure of nonnative species from seed transport on the hides and in the feed of the grazers (Hamilton 1997; Jackson 1985; Mack 1989; Mooney and others 1986). Fire likely played a role in this process by creating conditions conducive for invasion (Keeley 1995, 2001), but it is virtually impossible to determine the degree to which altered fire regimes alone were responsible for increasing invasion by nonnative plants. Fire does not necessarily promote invasion or dominance of nonnative species in grassland ecosystems; a pool of nonnative species and adequate propagule pressure are required to begin the invasion process (Rejmánek and others 2005b). Where the pool of nonnative species and/or propagule pressure is small, the invasion rate into burned areas is likely to be low (Klinger and others 2006b).

Although we do not know the degree to which fire promoted invasions in California grasslands, there are a number of nonnatives that consistently have high cover values in burned grassland areas (Klinger and Messer 2001; Parsons and Stohlgren 1989) (table 9-1). These include Eurasian annual grasses such as barbed goatgrass (*Aegilops triuncialis*), slender oat (*Avena barbata*), wild oat (*A. fatua*), ripgut brome (*Bromus diandrus*), soft brome (*B. hordeaceus*), red brome (*B. rubens*), seaside barley (*Hordeum marinum*), mouse barley (*H. murinum*), Italian ryegrass (*Lolium perenne* ssp. *multiflorum*), medusahead (*Taeniatherum caput-medusae*), brome fescue (*Vulpia bromoides*), and foxtail fescue (*V. myuros*). Many nonnative forbs also have high cover values in burned grasslands, especially filaree (*Erodium* spp.) and yellow starthistle.

Fire Regimes Changed by Plant Invasions—It is not clear how the shift to dominance by nonnative species in California's grasslands has affected fire regimes (Heady 1977; Simms and Risser 2000). The size of fires in these grasslands has likely decreased because of reduced continuity of fuels due to agriculture and urban development, as well as fire exclusion. However, it is unknown if the frequency, intensity, or seasonality of fire has changed since the region was settled in the 1700s and 1800s. Human-caused ignitions rather than changes in fuel characteristics are responsible for frequent fires in grasslands, but the overwhelming majority of fires are contained at less than 10 acres (4 ha). This pattern has not changed substantially in the last 50 years (Wills 2006).

Although many of the nonnative species in California grasslands are highly invasive (Heady 1977; Wills 2006), this has not resulted in any individual species having a particularly strong effect on fuel characteristics or fire behavior. For example, yellow starthistle is one of the most invasive species in California grasslands (DiTomaso and others 1999; Kyser and DiTomaso 2002), but it has no documented effect on fire regimes. While seasonality and frequency of fire play a role in the relative abundance of particular nonnative species in grasslands (Foin and Hektner 1986; Meyer and Schiffman 1999), fire alone typically does not change dominance of the assemblage of nonnative species in these systems (Klinger and Messer 2001; Parsons and Stohlgren 1989). Fire behavior in California's grasslands is generally determined by the assemblage of herbaceous species, and nonnative annual grasses comprise the most important fuel type in the system.

Role of Fire in Managing Nonnative Invasives—Fire has been used in attempts to control or manage invasive species in grasslands of the bioregion for over 50 years (DiTomaso and others 1999; Furbush 1953; Hervey 1949). Management goals were initially focused on range improvement (Furbush 1953; Hervey 1949) but have more recently shifted to more general, conservation-oriented objectives (Menke 1992; Randall 1996). The fundamental assumption has been that by decreasing abundance of nonnative species, there should be a concomitant increase of native species. Various projects have focused on reducing populations of specific invaders (for example, DiTomaso and others 1999, 2001; Hopkinson and others 1999; Pollak and Kan 1998), increasing populations of particular native herbaceous species (Dyer and Rice 1997, 1999; Gillespie and Allen 2004; Hatch, D. and others 1999), or inducing community-level changes by changing the relative abundance of entire assemblages of nonnative species (Klinger and Messer 2001; Marty 2002; Meyer and Schiffman 1999; Parsons and Stohlgren 1989; Wills 2001).

In terms of fire regime components, timing and frequency of burning are important factors for reducing the abundance of individual nonnative species. Barbed goatgrass can form dense stands and reduce richness and abundance of native herbaceous species. One study found that barbed goatgrass could be controlled with two late-spring or early-summer burns in consecutive years (DiTomaso and others 2001). However, another study (Hopkinson and others 1999) concluded that, while single summer burns were effective at reducing cover of barbed goatgrass, the burns did not impede its spread. Hopkinson and others (1999) also speculated that repeated burning in different seasons could be an effective strategy for reducing both cover and spread of barbed goatgrass; however, this has not been tested. Range ecologists initially thought efforts to control medusahead with fire would not be effective (Heady 1973), but Pollak and Kan (1998) reported that medusahead in a vernal pool/grassland system was almost completely eliminated the first growing season after a spring burn. Cover of yellow starthistle was reduced 91 percent after three consecutive summer burns (DiTomaso and others 1999); however, 4 years after burning ceased, yellow starthistle seed bank, seedling density, and cover had increased and were approaching pre-burn levels (Kyser and DiTomaso 2002). A simple conclusion would be that managing yellow starthistle with fire would require burning every 1 to 2 years. However, this approach may actually promote dominance by other nonnative species, so it is important to consider community-wide effects when implementing management programs focused on single species.

Season of burning appears to have variable community-level effects. Some studies have found that spring burns are more effective than fall and winter burns for reducing nonnative annual grasses (for example, medusahead and foxtail fescue) and increasing native herbaceous species (Meyer and Schiffman 1999; Pollak and Kan 1998; Wills 2001). However, other studies using fall burns documented significant decreases in nonnative species cover and significant increases in native species richness and cover, at least over short time periods (≤2 years) (Klinger and Messer 2001; Parsons and Stohlgren 1989).

Variation in site characteristics (for example, topography, climate, land use patterns and history, soils) and weather can have strong influences on burn effects. In burned grasslands on Santa Cruz Island, Klinger and Messer (2001) found that species composition for both natives and nonnatives was correlated with an interaction between rainfall, topography (aspect and elevation), and fire (fig. 9-3). Topography had a significant influence on responses of native and nonnative species to burning in a vernal pool-grassland complex (Pollak and Kan 1998), and species richness of

Figure 9-3—(A) Fire has been used extensively in California's grasslands to alter species composition away from dominance by nonnative grasses and forbs. Results from studies of grassland burns on Santa Cruz Island are representative of general patterns throughout the state. (B) Responses by both native and nonnative plants are species-specific, and shifts in relative abundance tend to be related more to factors such as rainfall and topographic position than to fire. (Photos by R. Klinger.)

nonnative plants in burned grasslands of the Central Coast Range was greater on more productive soils than on nutrient-poor soils (Harrison and others 2003).

Variation in spatial and temporal scales of prescribed fire studies likely has a strong influence on interpreting fire effects on nonnative species in grasslands. The most important of these may be study duration. Meyer and Schiffman (1999) and Pollak and Kan (1998) noted increases in native annual forbs, especially after spring burns. However, these burns were only monitored for 1 year after burning. Studies of longer duration have found that the decrease in nonnative species and increase in native species may be temporary (Klinger and Messer 2001; Kyser and DiTomaso 2002; Parsons and Stohlgren 1989). Both burn area and sampling unit size may introduce a spatial bias. Edge effects in studies using small sampling plots and/or small burn plots are probably substantial, and interpretations of postfire responses can vary at different scales (Klinger and Messer 2001, Marty 2002). A pilot study to determine an appropriate plot size and collecting data at least 1 year before and 3 to 4 years after fire treatments are useful strategies for reducing spatial and temporal bias and improving the evaluation of burn responses (Klinger and Messer 2001; Kyser and DiTomaso 2002).

In prescribed fire studies focused on community-level changes, a short term (≤2 years) decrease in cover (or biomass) of annual grasses and a short-term increase in annual forbs usually occur beginning in the first, and sometimes the second, growing season after fire (but see Marty 2002). However, nonnative grasses and forbs still dominate the burned sites, and within 2 to 3 years burn effects are largely gone (Klinger and Messer 2001; Parsons and Stohlgren 1989), a pattern that has been recognized for many decades (Heady 1973; Hervey 1949). Environmental variability and factors interacting with fire can have more important effects on both nonnative and native species than fire alone. Coordinated studies across multiple sites and conducted during the same period of time are needed (Klinger and others 2006a).

Generalizing from the studies that were conducted on managing invasive species with fire in grasslands of the bioregion is somewhat tenuous, because the studies have been site-specific and varied in objectives and study design. Despite these differences, three limited generalities can be drawn from them. First, fire can be effective at reducing abundance of some invasive species (for example, barbed goatgrass, yellow starthistle, medusahead). Second, timing and frequency of burning will be important to the success of these programs, but effects will likely be transient unless the burn program is sustained over long periods of time (for example, Kyser and DiTomaso 2002). Third, fire may be effective at altering population parameters such as abundance or biomass, but not other parameters such as dispersal and spread (Hopkinson and others 1999). This implies that a management program may need to vary the timing and frequency of burns and not rely on one prescription.

Chaparral and Coastal Scrub

Chaparral

Chaparral is characterized by dense stands of woody evergreen shrubs 3 to 10 feet (1 to 3 m) in height. It is the most extensive vegetation type in the bioregion, extending from Baja, California into southern Oregon, and typically from 660 to 9,800 feet (200 to 3,000 m) elevation. Chaparral stands are highly developed in lower mountains and foothills but rarely occur east of the major north-south trending mountain ranges in the bioregion or above 6,600 feet (2,000 m). Chaparral occurs on a variety of soil types and often occurs on coarse textured, shallow, and nutrient poor soils (Christensen 1973; Crawford 1962; Wells 1962). The long, hot dry season and relatively poor water holding capacity of many chaparral soils have resulted in a number of water-retention adaptations in chaparral shrubs, the most characteristic being small, thick, evergreen leaves (Mooney and Parsons 1973).

The variation in climate, topography, and soils throughout the range of chaparral has resulted in considerable geographic variation in species composition and physiognomy. Major subdivisions of chaparral have been based on dominant species (Hanes 1977), although this approach has been questioned (Rundel and Vankat 1989). These subdivisions include chamise (*Adenostoma fasciculatum*), ceanothus (*Ceanothus* spp.), scrub oak (*Quercus* spp.), manzanita (*Arctostaphylos* spp.), montane (chamise-ceanothus), and woodland (chamise-toyon (*Heteromeles arbutifolia*)-birchleaf mountain-mahogany (*Cercocarpus montanus* var. *glaber*)-California coffeeberry (*Rhamnus californica*)) (Hanes 1977). Some dominant woody species occur throughout most of the range of chaparral (for example, chamise, Nuttall's scrub oak (*Quercus dumosa*), toyon) as do some genera (for example, manzanita, ceanothus), but beta diversity throughout the range of chaparral tends to be high (Richerson and Lum 1980). Chaparral in the bioregion is thought to have evolved *in situ* (Axelrod 1958; Cooper 1922; Stebbins and Major 1965), and as a result a relatively high proportion of species in chaparral are endemic to the bioregion (Raven 1977).

Certain types of chaparral have developed under relatively restricted edaphic and/or climatic conditions. Serpentine chaparral has a unique physiognomy and species composition due to the nutrient-poor characteristics of these soils (Kruckeberg 1954; Whittaker

1954). Maritime chaparral occurs on sandy soils within 12 miles (20 km) of the coast and is characterized by a relatively diverse component of herbaceous species and a number of shrubs endemic to this type (Davis and others 1988; Van Dyke and Holl 2001).

Fire has had profound evolutionary and ecological influences on chaparral (Axelrod 1989; Rundel and Vankat 1989). At the evolutionary scale, woody chaparral species exhibit adaptations that enhance their ability to both survive fire and reproduce in postfire landscapes (Schwilk 2003; Zedler and Zammit 1989). These include sprouting from root crowns following top-kill, seeds that germinate after fire or have a fire-resistant seed coat, and heavy postfire seed production (Keeley 1987; Keeley and Fotheringham 1998a, b; Keeley and Keeley 1981; Keeley and Zedler 1978). Postfire regeneration of shrubs varies by species and occurs by both sprouting and seeding, with virtually all of the seed germination taking place in the first 2 years after burning (Keeley and Keeley 1981; Keeley and Zedler 1978). Herbaceous species that occur in chaparral communities also have dormant, fire resistant seeds in a persistent seed bank that germinate after burning, apparently as a result of stimulation by heat and/or chemical cues in smoke and ash (Keeley and Fotheringham 1998a, b; Keeley and others 1985; Keeley and Nitzberg 1984; Keeley and Pizzorno 1986). Germination tends to be from seeds in the soil seed bank and not from seeds that disperse from surrounding unburned areas (Vogl and Schorr 1972). Most germination of herbaceous species seed occurs in the first postfire year (Sweeney 1956).

Postfire succession follows a relatively predictable but highly dynamic pattern in chaparral communities (Keeley and others 2005; Keeley and Keeley 1981). Herbaceous species dominate total cover 1 to 3 years after fire, then gradually diminish as succession proceeds, shrub cover increases, and the canopy closes (Keeley and others 2005). Alpha diversity is initially high, but over several decades it drops substantially as the community becomes dominated by just a few woody species. The pulse pattern appears to result from an interaction between physical factors and competition (Safford and Harrison 2004). Nutrients are initially abundant and result in a flush of germination of both woody and herbaceous species, and the increase in light is especially important for herbaceous species (Christensen and Muller 1975). Species richness patterns are dependent on fire severity, resource availability (especially water), and plant life-form. A small group of forb species that establish the first year after burning persist throughout succession, but in general, turnover of herbaceous species is high and relatively rapid (Keeley and others 2005; Keeley and others 2006b). Sprouting by some woody species, such as chamise, is vigorous, as is seed production (Keeley and others 2006b). Diversity of pioneer shrubs is high 1 to 10 years after fire, and within several years they have excluded many of the herbaceous species (Sweeney 1956). As the canopy layer develops, many of the pioneer shrubs die off, resulting in a community dominated by a few large shrubs (Keeley and Keeley 1981).

Some aspects of the presettlement or reference fire regime in chaparral are controversial (Chou and others 1993; Keeley and Fotheringham 2001a, b; Minnich 1989, 2001; Minnich and Chou 1997), but typically it was characterized by stand-replacement crown fires of very high intensity where temperatures can exceed 1,290 °F (700 °C) (Borchert and Odion 1995; Sampson 1944). It is difficult to reconstruct fire history in shrublands, but there probably was a mix of smaller, moderate-sized burns ignited by summer lightning storms and infrequent (several times/century), very large, high-intensity fires in the fall that were driven by dry Santa Ana winds in southern California (Keeley and Fotheringham 2001a) or north winds in northern California. Inferences about the presettlement fire regime are complicated by anthropogenic burning practices (Davis 1992; Keeley 2002b; Timbrook and others 1982), but it is likely that fire-return intervals were typically on the order of 20 to 50 years (Biswell 1999; Caprio and Lineback 2002; Wright and Bailey 1982). Fire-return intervals have likely decreased during the 1900s, primarily because of increased ignition due to the growing human population. Modal fire size has decreased due to fire exclusion policies; however, fire exclusion has not altered the likelihood of large fires (Keeley and Fotheringham 2003; Moritz 1997).

Coastal Scrub

Coastal scrub forms dense but relatively low statured stands (<5 feet (1.5 m)) primarily in the maritime-influenced coastal band from Baja California to southern Oregon. Although coastal scrub occupies a broad latitudinal range, it is confined to a relatively narrow longitudinal range along coastal California, and is found almost exclusively below 3,900 feet (1,200 m) (Westman 1981a). Relative to chaparral shrubs, coastal scrub species tend to be shorter; have lower biomass and leaf area, shallower root systems, and a more restricted growth period; and be drought-deciduous (Gray and Schlesinger 1981; Hellmers and others 1955; Mooney 1977). Coastal scrub and chaparral are often in juxtaposition at the lower extent of chaparral's elevation range, with coastal scrub restricted to more xeric lower-elevation sites with shallow soils (Westman 1982, 1983). Coastal scrub species often co-occur with chaparral species on xeric sites that were recently burned or heavily grazed (Westman 1981a). Ecotones between chaparral, coastal scrub, and grassland can be dynamic (Axelrod 1978; Callaway and Davis 1993; O'Leary 1995; White 1995). Coastal scrub may be a transition state in mesic sites following disturbance such as grazing and burning, but stands are structurally stable on more xeric sites (Axelrod 1978; Callaway and Davis 1993; Westman 1981a).

There are two major coastal scrub types. Coastal sage scrub occurs from Baja California Norte north to Point Sur, and northern coastal scrub occurs north of Point Sur (Heady and others 1977; Mooney 1977; Munz and Keck 1959). There are approximately eight subdivisions within these two major types (Axelrod 1978; Westman 1981b). All of the subdivisions are dominated by shrubs, although relative abundance of the dominant species and overall species composition differ substantially among them (Westman 1981a,b).

Chaparral and coastal scrub have a number of species in common, but alpha diversity is generally lower in coastal scrub than chaparral (Kirkpatrick and Hutchinson 1977; Westman 1981a). Light penetration is higher in coastal scrub stands than chaparral, and as a result there is a greater proportion of herbaceous species in coastal scrub (Westman 1979, 1982). Successional changes in alpha diversity in coastal scrub are due primarily to changes in herbaceous species (Westman 1981a).

Fire has a major influence on vegetation structure and succession patterns in coastal scrub (Malanson 1984; Malanson and Westman 1985; O'Leary 1990). Light, soil nitrogen, and soil potassium increase during the first 2 years after fire (Westman and O'Leary 1986). These abrupt changes in the physical environment are likely the reason annual herbaceous species dominate cover in the first postfire years (Keeley and Keeley 1984; Westman 1981a). Within 5 to 7 years herbaceous cover substantially declines and shrub cover increases, but 15 to 25 years after fire there is often a second pulse of annual herbaceous species despite shrub cover increasing to over 70 percent. However, by 40 years after fire the herbaceous component has all but disappeared and cover at the site is dominated by a few shrub species (Westman 1981a). In coastal areas most regeneration of woody species is by resprouting (Keeley and Keeley 1984; Malanson and Westman 1985; Westman and O'Leary 1986), but in drier inland sites a greater proportion of species regenerate from seed (Westman 1981a, 1982). The amount of resprouting among shrubs appears to be variable but generally inversely related to fire intensity, at least for dominant species (Westman 1981a).

In summary, the major differences in postfire succession between coastal scrub and chaparral are (1) the persistence of herbaceous species in later successional stages of coastal scrub stands, (2) the persistence of seeding by shrub dominants in coastal scrub, (3) a higher proportion of sprouting versus seeding in coastal scrub species, and (4) a tendency for lower tolerance of sprouting coastal scrub species to high intensity fires (Westman 1982).

There is little quantitative data on fire regimes in coastal scrub, but there is evidence that they differ from chaparral fire regimes in important respects (Keeley and Keeley 1984; Malanson 1984; Minnich 1983; O'Leary 1995; Westman and O'Leary 1986; White 1995). Branches of coastal scrub shrubs tend to be smaller in diameter than those of chaparral shrubs (1-hour versus 10-hour fuels, respectively), and coastal scrub stands have a relatively high proportion of dry herbaceous fuels (O'Leary 1988). Litter accumulation under coastal scrub canopies can be high, and many of the shrubs have a higher content of volatile oils than chaparral shrubs (Westman and others 1981). These fuel structure characteristics, combined with higher rates of ignition and an extended season, result in coastal scrub having higher fire frequency than chaparral (Minnich 1983). When fires occur, they tend to be stand-replacing crown fires of high intensity, as in chaparral. Fire-return intervals have probably decreased in coastal scrub as population growth and attendant development in California surged in the last 50 years (Davis and others 1994; Goodenough 1992; Keeley and Fotheringham 2003; O'Leary 1995; White 1995). In general, succession in postfire interior coastal scrub stands occurs more slowly than in coastal stands (Keeley and others 2005).

Fire and Nonnative Invasives in Chaparral and Coastal Scrub

Many of the nonnative grass species that dominate the bioregion's grasslands are also species that invade chaparral and coastal scrub. These include the annual grasses slender oat, wild oat, ripgut brome, red brome, cheatgrass (*Bromus tectorum*), Mediterranean barley (*Hordeum geniculatum*), mouse barley, common barley (*Hordeum vulgare*), Italian ryegrass, and foxtail fescue (table 9-1). A suite of annual nonnative forbs is often associated with the nonnative grasses, including yellow starthistle, longbeak stork's bill (*Erodium botrys*), cutleaf filaree, burclover (*Medicago polymorpha*), and various species of mustards (for example, *Brassica* spp., *Sisymbrium* spp., *Descurainia sophia*). When these annual species dry, they become an integral part of the thatch (dried grasses and forbs) that comprises the light, flashy, 1-hour fuels that ignite easily and carry fire.

Perennial grasses such as pampas grass (*Cortaderia jubata*), Uruguayan pampas grass (*Cortaderia selloana*), and fountain grass (*Pennisetum setaceum*) are highly invasive in many areas of coastal scrub (fig. 9-4), and there is speculation that they contribute to stands being more prone to ignition (DiTomaso 2000a,b).

Figure 9-4. Dense stand of pampas grass on the central California coast near Big Sur. (Photo by Mandy Tu, The Nature Conservancy.)

While it has been shown that disturbance facilitates invasion by these species, it is important to note that they readily invade both unburned and burned sites, and there are no data to support the contention they have altered fire regimes (Lambrinos 2000).

Despite the differences in community structure, species composition, and fire regimes between chaparral and coastal scrub, the impact of nonnative invasive species on these two shrubland vegetation types has been similar (Keeley 2001; Keeley and Fotheringham 2003; O'Leary 1995; White 1995). However, it is important to emphasize that increased invasion rates and alteration of fire regimes in the bioregion have not occurred independently of each other (Keeley 2001). Rather, it has been their interaction that has resulted in the type-conversion of native-dominated chaparral and coastal scrub into grasslands dominated by nonnative species. The particular process for type-conversion of shrublands into grasslands has been documented in many ecosystems worldwide and is known as the grass/fire cycle (Brooks and others 2004; D'Antonio and Vitousek 1992). The general model for this is that nonnative invasive grass species establish in an area dominated by woody vegetation, and as the invasive grasses increase in abundance a continuous layer of highly combustible fine fuel develops. These fine fuels have a higher probability of ignition than the woody vegetation, resulting in an increase in rates of fire frequency and spread. In fire-type vegetation such as chaparral and coastal scrub, fuel structure changes when the fire-return interval is shortened and fire intensity decreases. Fires of lower intensity are crucial for nonnative seed survivorship (Keeley 2001), so if fires occur frequently, shrublands and forests are converted to grasslands dominated by nonnative species (Keeley 2006b; Zedler and others 1983).

The grass/fire cycle has occurred throughout the Southwest Coastal bioregion but is especially widespread in southern California. Beginning in the 1950s, vast areas of coastal scrub and, to a somewhat lesser degree, chaparral have either been fragmented or converted to urban and agricultural development (Davis and others 1994; O'Leary 1995; Minnich and Dezzani 1998). Fragmentation results in a landscape mosaic of annual grasslands, chaparral, and coastal scrub, leading to an interaction between landscape architecture, the species pool of nonnative invaders, propagule pressure from the nonnatives, and fire behavior (Bailey 1991; Keeley 2006b; Zedler 1995). Historically, most of the cover in the first few years after a chaparral or coastal scrub fire was dominated by native annual forbs (Sweeney 1956). But in the contemporary landscape, grassland areas dominated by nonnative grasses and forbs are typically intermixed with shrublands and serve as propagule sources harboring a relatively large pool of potentially invasive species (Allen 1998). The close proximity of grasslands to the stands of coastal scrub and chaparral, and the abundance of human activities that enhance dispersal of seeds into shrub stands and increase rates of ignition, all contribute to the conversion of native shrublands to nonnative annual grasslands.

In fragmented areas, burns are often small and have a high perimeter:area ratio, and they are often in close proximity to degraded plant communities with a high proportion of invasive species. There is evidence of a negative relationship between the distance of a burned area from source areas of nonnative species and rates of invasion into the burn (Giessow and Zedler 1996), so when a shrub stand burns, invasion rates from surrounding degraded vegetation into the burn edge (10 to 20 m) can be very high (Allen 1998; Minnich and Dezzani 1998).

Stands of chaparral and coastal scrub with intact canopies are relatively resistant to invasion by nonnative plants (Keeley 2002a ; Knops and others 1995). If the shrub stands burn but fire-return intervals remain within the range of 20 to 50 years, nonnative species may establish in the burned area, but their dominance typically declines as shrub cover re-establishes. This is because most nonnative species that invade burned areas are herbaceous and not shade-tolerant, so as the canopy closes these species are typically shaded out (Klinger and others 2006b). However, when fire occurs at more frequent intervals of 1 to 15 years, the dominance of shrubs, especially those regenerating from seed, declines rapidly (Haidinger and Keeley 1993). Nonnative annual grasses and forbs from surrounding grasslands establish in the first years after burning, but more importantly can regenerate, persist, and dominate cover in a fire regime characterized by short fire-return intervals (Parsons and Stohlgren 1989). Once dense stands of annual grasses and forbs form, it becomes extremely difficult for woody and herbaceous native species to establish and regenerate (Eliason and Allen 1997; Gordon and Rice 2000; Schultz and others 1955). The result is that the shrub canopy cannot re-establish, and nonnative herbaceous species (mainly annual grasses and forbs) become the dominants in what is now a grassland community.

The nonnative grass/fire cycle in the bioregion has been exacerbated by postfire management programs where seeds of nonnative grasses and forbs (for example, Italian ryegrass) were spread across burned sites to prevent erosion following a fire. This practice added to the already heavy propagule pressure by nonnative herbaceous species and essentially swamped out most native species attempting to regenerate in the burned areas. The practice has been shown to be of little benefit (Beyers and others 1998), and it is now uncommon. Still, previously seeded areas continue to be dominated by nonnatives, and these areas act as propagule sources of nonnative species (Allen 1998; Turner and others 1997; Zedler 1995).

Fire suppression activities such as fuel break construction and heavy equipment and vehicle use possibly serve as vectors of transport for invasive species (Giessow and Zedler 1996; Merriam and others 2006; Stylinski and Allen 1999). In a comprehensive study of 24 fuel breaks across the bioregion, Merriam and others (2006) found that nonnative plant cover was 200 percent greater along fuel breaks than in adjacent wildland areas and decreased rapidly with distance from the breaks (fig. 9-5). Relative cover of nonnative plants was 21 percent higher on breaks constructed with heavy equipment than on those constructed with hand tools.

Nonnative shrubs such as Scotch broom (*Cytisus scoparius*) and French broom (*Genista monspessulana*) can also establish in disturbed and degraded chaparral and coastal scrub stands (table 9-1). Although both of these shrubs can be highly flammable and carry fire under some circumstances (Bossard 2000a, b), they likely have little if any role altering fire regimes in chaparral and coastal scrub communities and are probably just as susceptible to elimination by frequent fire as native shrubs are (Alexander and D'Antonio 2003; Odion and Hausenback 2002).

Figure 9-5. Shepard's Saddle near the southwestern boundary of Sequoia National Park. (A) Wildland chaparral dominated by chamise and California yerba santa (*Eriodictyon californicum*). (B) A fire line constructed by bulldozers in 1960 and not maintained since. The most abundant nonnatives in the fire line are soft brome, ripgut brome, and wild oat. (Photos by Kyle Merriam, Plumas National Forest.)

Fire frequency may decrease as a result of invasion by some nonnative species (Brooks and others 2004). For example, the succulent perennial hottentot fig (*Carpobrotus edulis*) can be highly invasive in coastal scrub stands, but because of the high moisture content in its leaves it is difficult for fires to ignite or carry through it (Albert 2000). The perennial forb fennel is a common invader in many disturbed grasslands and coastal scrub areas of the bioregion and often forms extensive stands that can reach 10 feet (3 m) in height. Although there are many dead stems from previous year's growth and annual grasses grow in the understory of the stands, the tall canopy and high moisture content of the live stems creates conditions that reduce ignitions except in periods of extremely dry weather (for example, Santa Ana foehn winds) (Klinger 2000).

Use of fire to control nonnative invasive species in chaparral and coastal scrub—The controversy over presettlement or reference fire regimes in chaparral and coastal scrub has spilled over into the management arena. One hypothesis is that the fuel buildup in the bioregion's shrublands is an artifact of the fire exclusion era and that prescribed burning would reduce the number of large, catastrophic wildfires (Minnich 1989; Minnich and Chou 1997; Philpot 1974). Another hypothesis is that while fire exclusion, human population growth, and landscape fragmentation have reduced the average size and increased the frequency of fire, fuel loads and the frequency of large fires are within the normal range of variation, and prescribed burning would only worsen the rates of stand degradation and destruction (Keeley 2002a; Keeley and Fotheringham 2001a, 2003; Keeley and others 1999). The notion of older chaparral stands becoming "senescent" and needing to burn also contributed to this controversy, despite cautions that senescence was an inappropriate term to apply to stands that had not burned for a number of decades (Zedler and Zammit 1989).

From the standpoint of invasive species management, there is virtually no rationale for promoting management burning of stands of chaparral or coastal scrub. The reduction in fire-return interval more than anything else has exacerbated problems with invasive species in shrublands, and programs using fire to control nonnative plants in shrub communities may contribute to increased invasion rates (Alexander and D'Antonio 2003; Odion and Hausenback 2002).

Chaparral and coastal scrub are fire-prone environments, and stands are going to burn whether ignitions are caused by humans or not (Keeley 1995). Where management of invasive species is a priority, emphasis should be placed on not allowing any given stand to burn more than 1 to 2 times per century. Fuel breaks should be monitored and managed so that dense populations of nonnative plants do not establish along them. The same is true for fuel removal to create defensible spaces around structures in wildland urban interface (WUI) areas. Burned areas should be monitored for 5 to 6 years to ensure that fire suppression activities do not contribute to increased levels of invasion. The most effective way to manage invasions in burned shrublands is to allow the canopy to develop to its full extent.

Mixed Evergreen Forests

Mixed evergreen forests are characterized by stands with well developed tree and shrub layers dominated by evergreen species. The herbaceous understory is typically sparse, but there may be a substantial surface layer of dead leaves and branches. These forest types occur mainly west of the Coast Ranges from the Klamath Mountains in the northwestern part of the bioregion to the San Bernardino Mountains in the southwest, and in parts of the foothills of the Sierra Nevada (Holland and Keil 1995; Sawyer and others 1977). The extensive distribution of mixed evergreen forests spans a wide range of soil, topographic and climatic types, which in turn results in substantial variation in species composition. Although they do not necessarily co-occur throughout the range of mixed evergreen forest, the dominant species typically include Pacific madrone (*Arbutus menziesii*), giant chinquapin (*Chrysolepis chrysophylla*), tanoak (*Lithocarpus densiflorus*), California laurel (*Umbellularia californica*), Coulter pine (*Pinus coulteri*), coast Douglas-fir (*Pseudotsuga menziesii* var. *menziesii*; hereafter, Douglas-fir), and several species of oaks (for example, coast live oak (*Quercus agrifolia*), canyon live oak (*Q. chrysolepis*), and California black oak (*Q. kelloggii*)) (Sawyer and others 1977). Mixed evergreen forests generally occur as mosaics among other vegetation types, such as coastal pine forest, coastal prairie, and chaparral (Campbell 1980; Wainwright and Barbour 1984).

Compared to many of the other major vegetation types in the bioregion, there has been relatively little quantitative study of fire in mixed evergreen forests. Studies of fire history suggest that, prior to the fire exclusion era, surface fires were common and return intervals were in the range of 10 to 16 years (Rice 1985; Wills and Stuart 1994). Many mixed evergreen forests occur in areas with frequent lightning strikes and/or where burning by Native Americans was common, which likely resulted in frequent fires of low intensity and severity (Keeley 1981; Lewis 1993). In areas where fire was less frequent, there was considerable variation in fire behavior, size, and severity, resulting in a mosaic of sites of varied age and species composition

(Lewis 1993; Rice 1985; Wills and Stuart 1994). It is likely the spatial arrangement of the stands has also influenced fire behavior. Surface fires of moderate intensity are probably common in larger, more homogeneous stands, but in areas where stands are contiguous with vegetation types where fire intensity can be high, spread of fire into mixed evergreen forests can often result in both passive and active crown fire (Wills and Stuart 1994).

Exclusion of fire from mixed evergreen forests is generally regarded to have resulted in a build-up of fuels and a shift in species composition. Shrub density in the understory has increased during the fire exclusion era, leading to increased fuel loading and the development of ladder fuels that can carry fire into the crown. Shifts in species composition result in increased dominance by shade-tolerant species such as tanoak (Hunter and others 1999; Hunter and Price 1992). Dominance by tanoak can either be maintained or shifted depending upon fire severity and the time since burning (Stuart and others 1993). A severe fire will result in high mortality of all age classes of non-sprouting species, such as Douglas-fir, and prolific sprouting by tanoaks. Tanoak will dominate the early succession period, but in later periods of succession seedlings of shade-intolerant species establish in gaps created by self-thinning tanoak. In contrast, low- and moderate-severity burns will kill seedlings and saplings of Douglas-fir, but adults will survive and seeds will be shed into sites with favorable growing conditions. This will result in reduced dominance of tanoak throughout most stages of succession. It is important to note that these patterns likely vary along a moisture gradient, with greater fire intensity in stands in the drier foothills of the Sierra Nevada and the southern part of the bioregion (Skinner and Taylor 2006).

Beginning in the mid-1990s an epidemic disease known as Sudden Oak Death (SOD) has resulted in the death of great numbers of oaks and tanoaks in California forests (Meentemeyer and others 2004). The large number of dead trees could potentially result in larger, more intense fires. However, preliminary evidence indicates the spread of the SOD pathogen (*Phytophthora ramorum*) is extremely limited in areas that have burned in the last 50 years (Moritz and Odion 2005). Although there appears to be a fire-disease relationship, a better understanding of the mechanisms underlying the relationship is needed before fire is used in managing the epidemic.

Fire and Nonnative Invasives in Mixed Evergreen Forests

In contrast with other vegetation types such as chaparral and coastal scrub, invasive plant species have rarely been identified as a threat in mixed evergreen forests. Nonnative species that are of potential concern in mixed evergreen forests are predominantly woody, such as Scotch broom, French broom, tree-of-heaven (*Ailanthus altissima*), and eucalyptus (*Eucalyptus* spp.). For example, French broom is of concern in mixed evergreen forests in the San Francisco Bay region, but prescribed fire has also been recommended as a useful control method under certain conditions (Swezy and Odion 1997). In general, these species are found most frequently along roads and highways and seldom in intact forest. Nevertheless, populations and individuals along roads could act as a propagule source and the roads as dispersal corridors in the event of disturbances such as fire.

The lack of quantitative data in mixed evergreen forests makes it difficult to draw general conclusions about the relationship between fire and invasive species in these communities. Nevertheless, it is useful to make some preliminary statements about potential relationships, bearing in mind that extending these is tenuous. The apparent weak interactions between fire and invasive plants in mixed evergreen forests may be due to several factors. These factors are not necessarily independent and may vary in importance from location to location. The first is that the rapid development of a canopy layer by sprouting trees and shrubs results in low light levels that impede establishment by herbaceous nonnative species. Related to this is that effective fire suppression operations may preserve stand structure in all but the largest fires, so conditions are not suitable for most invading species. Another factor may be that many stands are in relatively inaccessible areas or where anthropogenic influences are of low to moderate magnitude. This may reduce propagule pressure, maintain stand structure, or both. Finally, because of the proximity of mixed evergreen forests to other vegetation types that do burn readily (for example, chaparral), it is possible that invasions into these forests have been underestimated or lumped with effects in the other vegetation types. It is likely that, if fire-return intervals were reduced to the same extent as they have been in many shrublands, mixed evergreen forests could be converted to nonnative-dominated communities.

Conifer Forests

Conifer forests are one of the most widespread and variable vegetation types in the bioregion. They occur throughout all parts of the bioregion, ranging in elevation from sea level to 9,800 feet (3,000 m). Aside from being dominated (at least in height) by cone-bearing trees, the various conifer forest types differ vastly in structure, species composition, soils, climatic conditions, and fire regimes.

Coastal Conifer Forests

The northwest coastal conifer forest zone extends from Monterey Bay northward into coastal Alaska. The climate where these coastal conifer forests are found differs from the mediterranean climate that typifies most of the rest of the bioregion in having cool, moist summers as a result of proximity to the north-coastal maritime environment (Stuart and Stephens 2006). Coastal conifer forests in the bioregion include relatively low-statured stands of closed-cone pines such as knobcone pine (*Pinus attenuata*), bishop pine (*Pinus muricata*), and Monterey pine (*Pinus radiata*); stands of redwood (*Sequoia sempervirens*) forest; Douglas-fir forests; Sitka spruce (*Picea sitchensis*) forests; and an edaphic type sometimes known as pygmy forest, which is characterized by Bolander beach pine (*Pinus contorta* var. *bolanderi*). With the exception of redwood stands, the horizontal and vertical arrangement of the understory of these forest types is complex and dominated primarily by woody shrubs. Because of the immense height and continuous canopy of redwood forests, the understory in these stands is shady and the shrub layer is less well developed. The herbaceous layer is not well developed in many of these forest types because of reduced light levels beneath their canopies.

Fire regimes in the coastal forests vary from intense, extensive, stand-replacing burns in closed-cone and pygmy forests to understory burns of moderate severity in redwood forests. Adaptations of species to fire show a similar range of variability, from species inhibited by fire, such as Sitka spruce, to those substantially enhanced by it, such as Bishop pine. Stuart and Stephens (2006) provide a comprehensive review of fire in coastal forests.

Klamath and Cascade Range Conifer Forests

Inland of the coastal conifer forests is the Cascade/Klamath conifer forest zone. The climate is a more typical mediterranean type than the coastal zone, but precipitation varies substantially because of steep elevation gradients, complex topography, and variability of soils (Skinner and Taylor 2006; Skinner and others 2006). Vegetation species composition and fire regimes reflect this complexity. Forests on the west side of the mountain ranges tend to be dominated in the lower to mid elevations by mixed stands of white fir (*Abies concolor*), Douglas-fir, incense cedar (*Calocedrus decurrens*), sugar pine (*Pinus lambertiana*), and ponderosa pine (*Pinus ponderosa*). Higher elevation stands are comprised of California red fir (*Abies magnifica*), mountain hemlock (*Tsuga mertensiana*), western white pine (*Pinus monticola*), Jeffrey pine (*Pinus jeffreyi*), whitebark pine (*Pinus albicaulis*), Sierra lodgepole pine (*Pinus contorta* var. *murrayana*), and foxtail pine (*Pinus balfouriana*). The well developed understory of shrubs in low-, mid-, and high-elevation stands includes Pacific dogwood (*Cornus nuttallii*), bigleaf maple (*Acer macrophyllum*), Pacific madrone, California black oak, canyon live oak, giant chinquapin, and tanoak. The herbaceous layer can be well developed in areas where the canopy is open but otherwise is sparse and patchily distributed.

The rain shadow effect on the east side of the crest of the Klamath and Cascade ranges results in a more open forest structure. Although many of the species that occur on the west side are also found on the east side, stands are typically dominated just by ponderosa pine at lower elevations and Jeffrey pine at higher elevations. The shrub layer is more open and comprised of rabbitbrush (*Chrysothamnus* spp.), big sagebrush (*Artemisia tridentata*), manzanita, whitethorn ceanothus (*Ceanothus cordulatus*), bush chinquapin (*Chrysolepis sempervirens*), snowbrush ceanothus (*Ceanothus velutinus*), and huckleberry oak (*Quercus vaccinifolia*). The patchier distribution of trees and shrubs on the east side of the range results in greater light levels and a better developed (but still relatively sparse) herbaceous layer than the west side.

Fire regimes in the Klamath and Cascade ranges vary by climate, topographic position, and vegetation structure. Lightning is a major source of ignition, but human-set fires were also common prior to the 1900s. In general, fires were probably relatively frequent and of low to moderate severity prior to the onset of the fire exclusion era in the early 1900s. From that point forward, fires have become less frequent but of greater severity and likely more continuous. Skinner and Taylor (2006) and Skinner and others (2006) have done full reviews of fire regimes in the Klamath and Cascade ranges.

Sierra Nevada Conifer Forests

The Sierra Nevada range is the most dominant landscape feature in the bioregion. It extends northward 430 miles (700 km) from the Transverse Ranges until it meets the Cascade Range in the general vicinity of Mount Lassen. Mount Whitney exceeds 14,000 feet (4,400 m) elevation, and there are numerous peaks over 12,000 feet (3,700 m). The climate is mediterranean, with most winter precipitation in the higher elevations occurring as snow. There are north-south and east-west gradients in precipitation, with greater amounts occurring in the northern and western parts of the range. Summer thunderstorms are common from July to September, and lightning occurs year-round but is most frequent during the thunderstorm season and at higher elevations.

Vegetation on the west side of the Sierra Nevada is patterned in belts that relate primarily to elevation and includes lower montane forest, upper montane forest, and subalpine forest zones (Stephenson 1998). The lower montane forest is the most widely distributed zone and typically occurs from 4,900 to 6,600 feet (1,500 to 2,000 m). It is characterized by mixed forests consisting of ponderosa pine, white fir, Douglas-fir, incense cedar, and California black oak. The understory is typically shrubby (for example, manzanita and ceanothus), with herbaceous species scattered sparsely in areas with canopy openings. The upper montane forest zone occurs from 6,600 feet to 9,800 feet (2,000 m to 3,000 m) and is characterized by stands of California red fir and Jeffrey pine, with occasional stands of western white pine, western juniper (*Juniperus occidentalis*), and quaking aspen (*Populus tremuloides*). The understory consists of patches of shrubs, including manzanita, bush chinquapin, whitethorn ceanothus, and huckleberry oak. Western needlegrass (*Achnatherum occidentale*) is a low bunchgrass that occurs in openings with adequate light. Subalpine forest occurs most commonly from 9,800 to 11,500 feet (3,000 m to 3,500 m) and is characterized by lodgepole pine, mountain hemlock (*Tsuga mertensiana*), limber pine (*Pinus flexilis*), foxtail pine, and whitebark pine.

The east side of the range is notably steeper and more arid than the west side. The vegetation zones dominated by conifers (3,400 feet to 9,400 feet (1,050 m to 2,850 m)) are somewhat less distinct than on the west side. Jeffrey pine and white fir typically occur in the upper elevations, with stands of mixed conifers in the middle elevations and singleleaf piñon (*Pinus monophylla*) in the lower elevations. The understory is characterized by sagebrush, rabbitbrush, and antelope bitterbrush (*Purshia tridentata*), with sparse herbaceous cover between shrubs.

Fire has been one of the dominant forces shaping vegetation structure and species composition in the Sierra Nevada (Anderson and Smith 1997; Smith and Anderson 1992). Fire regimes have varied through time as a result of climate change and burning by Native Americans, but most evidence indicates that there were many fires of low to moderate intensity and severity and variable size in lower montane forests, with fire-return intervals typically in the range of 5 to 15 years (Kilgore 1973; Kilgore and Taylor 1979; Skinner and Chang 1996; Swetnam 1993; Wagener 1961). Although large, high-severity fires did occur, they were relatively infrequent (McKelvey and Busse 1996). Burning by indigenous tribes beginning about 9,000 years ago undoubtedly decreased fire-return intervals, especially near areas of habitation (Anderson 1999; Vale 1998).

Fires were less frequent in upper montane and subalpine forests, with fire-return intervals in upper montane forests in the range of 60 to 70 years. This was due to the prevalence of lightning as the primary ignition source in upper elevations. Since lightning is usually accompanied by rain in the Sierra Nevada, ignitions are less frequent. Fuels are also more compact and difficult to ignite in upper montane forests, and the numerous rocky slabs and outcrops make it difficult for large fires to develop under anything but extremely hot, dry conditions (van Wagtendonk 1995).

During the era of fire exclusion, fire regimes and vegetation communities changed greatly (Skinner and Chang 1996). Several extensive reviews of fire regimes in the Sierra Nevada were done in the 1990s (Chang 1996; Erman and Jones 1996; McKelvey and Busse 1996; McKelvey and others 1996; Skinner and Chang 1996). These reviews emphasized the considerable variation in fire regimes before the fire exclusion era, primarily due to weather, topography, elevation, fuel type, and anthropogenic burning. Contemporary fire regimes also vary (Skinner and Chang 1996), but far less than historically. In general, exclusion of fire has increased fuel loads as a result of increase in density of white fir and incense cedar and the development of ladder fuels that extend into the canopy. Low- and moderate-severity fires have been effectively excluded, but this has resulted in large, stand-replacing fires of high intensity and severity. These large fires cannot be controlled quickly like ones of lower severity, resulting in less spatial heterogeneity in forest structure across the landscape.

Fire and Nonnative Invasives in Conifer Forests

Rates of invasion by nonnative plants into conifer forests have been far slower than in many other ecosystems in this bioregion (Mooney and others 1986; Randall and others 1998; Schwartz and others 1996). Although nonnative plants occur in conifer forests in the Klamath, Cascade, and Sierra Nevada ranges (Gerlach and others 2003), they are less abundant and, to date, their impacts have been relatively minor (Klinger and others 2006b). Factors that may reduce the likelihood of establishment by invasive species in these higher elevation conifer forests include less activity by humans, relatively intact shrub and tree canopies, harsh climates, and other physical conditions. Elevation *per se* probably has little to do with the decreasing incidence of invasive species in conifer forests. Instead this reduced incidence is more directly a result of physical factors affecting establishment and growth of plants (reduced moisture, low temperatures, short growing seasons, etc.) that are correlated with elevation. Rejmánek (1989) pointed out that, as a general rule, more extreme environments may be less susceptible to invasion; montane conifer forests are likely a good example. Presumably, less human activity results in lower propagule pressure (Randall and others 1998; Schwartz and others 1996). It is telling that sites in conifer forests where invasive species have established and are relatively abundant are usually lower elevation ponderosa pine forests that have been disturbed by human activities (Keeley and others 2003; Moore and Gerlach 2001).

Besides temperature and moisture, light may be a factor reducing invasion rates into conifer forests. Preliminary data suggest that invasive species are more abundant in conifer forests where the canopy is broken (Keeley 2001). Since most of the nonnative species that do invade conifer forests are shade intolerant (for example, cheatgrass, common velvet grass (*Holcus lanatus*), and Kentucky bluegrass (*Poa pratensis*)) (Klinger and others 2006b; Moore and Gerlach 2001), reduced light availability under intact canopies may generally prevent or minimize their establishment.

Some woody nonnative species may have the potential to invade conifer forests. Species of concern in the Sierra Nevada, Klamath, and Cascade ranges include Scotch broom, tamarisk (*Tamarix* spp.), Russian-olive (*Elaeagnus angustifolia*), and tree-of-heaven (Schwartz and others 1996). In coastal forests, Scotch broom and French broom are potentially problematic (Bossard 2000a,b).

Although rates of invasion into conifer forests are relatively low, there is concern that this situation could change, primarily because of disturbance resulting from land use and fire suppression activities (Harrod and Reichard 2001; Merriam and others 2006). Despite this concern over **potential** effects, there is relatively little published data on the relationship between fire and invasive species in montane conifer ecosystems in the bioregion (Bossard and Rejmánek 1994; Keeley 2001; Randall and Rejmánek 1993; Schoennagel and Waller 1999). There is evidence from ponderosa pine forests in Arizona that nonnative invasive species are altering postfire succession patterns (Crawford and others 2001), but the limited data available from California are inconclusive (Keeley 2001; Keeley and others 2003; Klinger and others 2006b). In Sequoia and Kings Canyon National Parks, Keeley and others (2003) found that frequency and cover values of nonnative invasive species were low in virtually all burned conifer sites, but that cover was greater in areas with higher-severity fires. Based on an analysis of two different data sets from Yosemite National Park, Klinger and others (2006b) found that species richness and cover of nonnative species were relatively low in conifer forests and did not differ significantly between burned and unburned sites. Of perhaps greater significance, Klinger and others (2006b) found a negative relationship between richness and cover of nonnative species and time since burn, suggesting that even though nonnative plants can establish in burned conifer forests they are shaded out as the canopy closes (Keeley and others 2003). Nevertheless, management agencies are taking an increasingly cautious approach when applying prescribed burns in some conifer forest systems, especially those where cheatgrass could invade (Keeley 2006b).

Once established, several factors influence the persistence of cheatgrass in these forests (Keeley and McGinnis 2007). A research project was initiated after substantial increases in cheatgrass cover were noticed in low-elevation ponderosa pine forests in Kings Canyon National Park in the late 1990s (Caprio and others 1999). Results showed that postfire cheatgrass abundance was affected positively by the cheatgrass seed bank, growing season precipitation, soil nitrogen, and number of hours of sunlight during the fall; and negatively affected by canopy coverage, summer sunlight hours, and fire intensity. Fire intensity was characterized by the maximum temperature at three vertical profile levels, total duration of elevated temperatures, and fireline intensity, and was most strongly affected by fuel load. Based on cheatgrass germination tests following heat exposure and field thermocouple measurements during experimental fires, over 40 percent of cheatgrass seeds are predicted to survive a fire with natural fuel loads. Where fuel loads were increased with additional pine needles, cheatgrass seed survival was predicted to decrease (fig. 9-6). Altering burning season to coincide with seed maturation is not likely to control cheatgrass in these forests because fuel loads and fire intensity were similar among seasons. The only treatment that showed potential for inhibiting cheatgrass persistence was needle accumulation, which directly affected cheatgrass establishment and, when burned, reduced cheatgrass seed banks. Longer fire-return intervals may result in this type of needle accumulation, though the relationship between fire-return intervals and litter accumulation sufficient to inhibit cheatgrass persistence is unclear (Keeley and McGinnis 2007).

Wetland and Riparian Communities

Wetland vegetation includes marshes, vernal pools, and riparian corridors. Fire effects in these vegetation types in the bioregion are poorly understood. More research has been initiated in riparian systems of the Interior West bioregion, though understanding of fire and invasive species interactions in riparian communities is still weak. Most fire research in western wetland systems has focused on forested riparian

Figure 9-6. Ignition of cheatgrass plot located in mixed ponderosa pine/black oak/white fir at Kings Canyon National Park by NPS crews. Fuel load on plot in foreground was augmented with additional litter, while adjacent, already burned plot had unmanipulated fuels. (Photo by Tony Caprio, National Park Service.)

communities and not on herbaceous-dominated systems like freshwater marsh. The limited spatial extent of these vegetation types and their less frequent burning regimes have contributed to the general lack of investigation.

Fire and Nonnative Invasives in Wetland and Riparian Communities

The lack of information on fire effects in wetland communities makes it difficult to generalize about the relationship between fire and invasive species in these systems. It is apparent that wetlands in the bioregion are highly susceptible to invasion regardless of whether they are burned (Bossard and others 2000), and increased fuel accumulation may be independent of invasion by nonnative plants. Several investigators have suggested that fire may facilitate invasion into riparian plant communities, but it is also a somewhat novel event (Bahre 1985; Busch 1995; Dobyns 1981; Kirby and others 1988). After invasion, nonnative plants such as giant reed (*Arundo donax*) (fig. 9-7) and tamarisk can increase the amount and continuity (especially vertical ladder fuels) of flammable, nonnative vegetation which may alter fire behavior (Bell 1997). Studies from other bioregions suggest this can shift fire regimes in riparian zones towards more frequent, intense, and severe fires (Busch and Smith 1993; Ohmart and Anderson 1982). Fire events on these sites can be of extremely high severity, with all overstory trees being killed (Stuever 1997). The lack of flooding on regulated rivers may also contribute to excessive fuel accumulation and more severe fire events (Ellis 2001).

Conclusions

In general, studies on fire in the bioregion have focused almost exclusively on two questions:

1. To what degree does fire promote invasion by nonnative plants?
2. To what degree can invasive plants be controlled with fire?

Although we have a much better understanding of the relationship between fire and invasive plants in the bioregion than 15 to 20 years ago (Brooks and others 2004), we are still far from being able to provide simple answers to either question that are useful for management. Much of this can be attributed to "ecological complexity" (Maurer 1999), but at least as much is due to other issues that are related to how research on fire and invasive plants has been conducted. These include the recent interest in the relationships between fire and invasive plants, the emerging view of fire in its proper ecological context when designing and interpreting studies, and the need to broaden the questions we ask to include studies of how higher-order interactions influence the relationship between fire and invasive plants.

Complexity is a feature of virtually all ecological patterns and processes. But rather than acting as an impediment to generalizing research findings, complexity stresses the necessity of designing research projects that, at a minimum, are conducted over large enough spatial and temporal scales to capture the variability inherent in ecological systems. This, of course, increases the logistical difficulty of conducting the studies, but the long-term payoff will almost certainly

Figure 9-7—(A) Tall, dense patches of giant reed at an experimental control site near the Santa Ana River in southern California. (B) Light green giant reed encroaching on native riparian vegetation above a small dam on the Santa Ana River near Ojai, California. (Photos by Mandy Tu, The Nature Conservancy.)

justify the effort. Virtually all studies of fire and invasive species in the bioregion are very limited in either spatial and/or temporal extent. An exception is an ongoing, relatively long-term experimental study of fire effects on composition of plant assemblages that is being conducted in vernal pool/grassland systems at three sites in northern California (Marty, personal communication, 2003). An even larger scale example includes the Fire and Fire Surrogate Study that has been implemented in forested ecosystems across the United States (Weatherspoon 2000). These and other examples of long-term, multi-site, and multi-variable studies in other areas of ecology and conservation serve as models for future research programs on fire/invasive plant interactions.

Although the first comprehensive review of biological invasions was published over 50 years ago (Elton 1958), the overwhelming majority of studies has been relatively recent (Kolar and Lodge 2001). Consistent with this pattern is that most of our understanding of the relationship between invasive plant species and fire has come in the last 12 years (Brooks and others 2004; D'Antonio 2000; D'Antonio and Vitousek 1992; Galley and Wilson 2001; Mack and D'Antonio 1998). Because it is still one of the most recent areas of study, consistent patterns and mechanistic explanations have not yet emerged, and variations to some of the patterns that have emerged in the last decade will likely be found. Therefore, while management actions can certainly be based on our current state of knowledge, these actions must be re-evaluated frequently. It is critical that agencies and organizations implement programs within an explicitly defined adaptive management framework (Holling 1978) and be prepared to modify programs if necessary.

Terms such as "fire ecology" may produce a subtle bias that results in fire not being placed in an appropriate ecological context. While "fire ecology" is a convenient descriptive term, fire is a disturbance event and has no more inherent ecology than a flood, a hurricane, or a freezing event. Fire certainly affects the ecology of species and communities and, in turn, is itself affected by the ecology of organisms, but it is only one of a number of processes interacting with the biotic components of an ecosystem. A holistic approach that considers all of the potential effects of a particular management action is always recommended. In studies of fire, there is often an unstated assumption that fire is the most important factor influencing species distribution and abundance patterns. In many instances this assumption is justified, but in others it may not be (Klinger and Messer 2001; Klinger and others 2006b). This subtle bias can have consequences for evaluating management burns and interpreting research findings. For example, in systems with multiple invaders or those adjacent to such areas, the species pool, propagule pressure, relative abundance, and seed bank of the nonnatives may have a far greater influence on postfire species response than the fire itself (Keeley and others 2003; Klinger and Messer 2001; Parsons and Stohlgren 1989). In contrast, a burn program targeted at controlling a specific invasive species may be effective if that is the primary objective and other community issues are secondary (DiTomaso and others 1999, 2001). Even then, fire may be effective at reducing abundance but not spread of some species, implying that dispersal limitation of the species will be at least as great a consideration as fire (Hopkinson and others 1999). In addition, control of one invasive species may provide opportunities for others to increase in dominance (Brooks 2000).

Most research on fire and invasive species in the last decade has focused on the direct effect of invasive species on fire regimes and the direct effect of fire on invasive species. But as we have seen, fire interacts with species and other processes in a variety of ways and at different scales, and studies of higher order interactions within this context are needed. For instance, our understanding of how the interaction between fire and large scale geophysical processes such as climate change and nitrogen deposition influences biological invasions is largely speculative (Fenn and others 2003). We know little about the relationship between fire, invasive species, and many functional components of communities and ecosystems such as trophic structure, predator-prey relationships, seed dispersal, granivory, and herbivory. There is now greater effort to understand these interactions (Bock and Bock 1992a; Collins and others 2002; D'Antonio and others 1993; Espeland and others 2005; Howe 1995; Howe and Brown 2001), but far more of these types of studies are needed in the bioregion.

Given our current state of knowledge, what can be said about managing fire and invasive species in the bioregion? In terms of fire enhancing invasion by nonnative species, it appears clear that frequent fires in chaparral and coastal scrub communities facilitate invasions and should be avoided. Fires in conifer forests often have good conservation and management justification. But it needs to be acknowledged in the planning stage that preburn site preparation, fire suppression activities, and holding activities as well as the fire itself may increase rates of invasion by nonnative plants into burn areas. The understory of most grassland and oak woodland communities is already dominated by nonnative plants, and it is unlikely that fire alone will change this. In general, the conceptual model proposed by Brooks and others (2004) is useful as a framework for developing fire management or research programs where postfire invasion by nonnative plants is an important consideration.

As discussed in other sections of this chapter, there are certain situations where fire can be an effective tool for reducing populations of some species (DiTomaso and others 1999; Gann and Gordon 1998; Lonsdale and Miller 1993; Myers and others 2001; Nuzzo 1991). However, this will not be the case for all species (Jacobs and Sheley 2003a; Ruyle and others 1988), and cessation of fire treatments may result in reinvasion by nonnatives (Kyser and DiTomaso 2002). In systems with multiple invaders, the use of fire alone is unlikely to alter community composition for any meaningful period of time. Responses of nonnative plants to fire will be species-specific and will vary between sites as well as over time within a site. Consequently, management programs should not be framed within the concept of "restoration" in these situations.

Despite lingering hopes (Gillespie and Allen 2004), there is little evidence other than Wills' (1999) study in grasslands on the Santa Rosa Plateau that burning alone will result in a long-term decrease in abundance of nonnative species and an increase in native species. Burning as a sole management technique will likely be most effective as a means of maintaining areas that already have a significant component of native species. In areas where nonnative species are not particularly abundant, fire will probably have to be integrated with other techniques, such as seeding and planting, if the goal is enhancement of native species diversity. In some instances, burning can be integrated with other types of management such as herbicides, grazing, plowing, and cutting. For example, fire was found to be effective at stimulating a sprouting response in fennel that enhanced the effectiveness of herbicide applications (Gillespie and Allen 2004). Establishment of native perennial bunchgrasses (for example, purple needlegrass) can be enhanced by plug planting in burned areas following herbicide application to reduce re-establishing nonnative grasses and forbs (Anderson, J. personal communication, 2006).

In conclusion, the variation observed in the relationships between fire and invasive plants in the bioregion reflects the historical, physiographic, climatic, and biological diversity of the region. Patterns are emerging, but the lack of systematic research on fire and invasive plants means that many relevant questions will remain unanswered for years to come. It is our hope that this chapter has captured the biological and physical diversity inherent in this region and the complexity of the interaction between fire and invasive nonnative plants, and will stimulate additional thoughts and efforts towards more effective fire and plant community management.

Notes

Dawn Anzinger
Steven R. Radosevich

Chapter 10:

Fire and Nonnative Invasive Plants in the Northwest Coastal Bioregion

Introduction

This chapter discusses the relationship between fire (natural and prescribed) and nonnative plant species within major vegetation communities of the Northwest Coastal bioregion, and specifically addresses the role of fire in promoting nonnative species invasions, the effects of nonnative species on fire regimes, and usefulness of fire as a management tool for controlling nonnative species. Four plant communities of western Washington and Oregon will be covered: coastal Douglas-fir forests, montane forests and meadows, riparian forests, and Oregon oak woodlands. Three plant communities of Alaska will also be examined: coastal hemlock-spruce forests, boreal forests, and tundra (fig. 10-1). Table 10-1 provides a list of important nonnative species in the Northwest Coastal bioregion and their estimated impact within these plant communities.

Conifer forests dominate much of the landscape of the Northwest Coastal bioregion. In this densely forested environment, fire promotes nonnative species establishment by creating open-canopy conditions for these predominantly shade-intolerant plants and exposing mineral soils for ruderal seedling establishment.

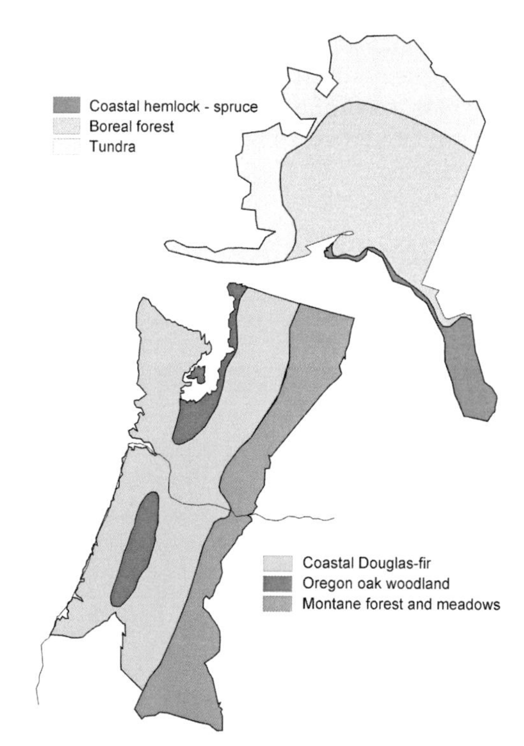

Figure 10-1—Approximate distribution of major plant communities in the Northwest Coastal bioregion. Riparian forests are not shown.

Table 10-1—Nonnative invasive plant species of concern in the Northwest Coastal bioregion and their approximate threat potential in each of the major plant communities covered in this chapter. L= low threat, H = high threat, P = potentially high threat, U = unknown threat, N= not invasive

		Plant communities						
		Pacific Northwest				Alaska		
Scientific name	Common name	Coastal Douglas-fir forests	Montane forests and meadows	Riparian forests	Oak woodlands	Hemlock-spruce forests	Boreal forests	Tundra
Anthoxanthum aristatum	Annual vernal grass	N	N	U	H	N	N	N
Arrhenatherum elatius	Tall oatgrass	N	U	U	H	N	N	N
Brachypodium sylvaticum	False brome	P	U	U	P	N	N	N
Cytisus scoparius	Scotch broom	H	N	L	H	L	N	N
Hedera helix	English ivy	L	N	H	U	N	N	N
Holcus lanatus	Common velvetgrass	P	U	U	P	N	N	N
Hypericum perforatum	Common St. Johnswort	L	U	L	P	L	N	N
Hypochaeris radicata	Hairy catsear	U	L	U	H	L	N	N
Linaria vulgaris	Yellow toadflax	N	N	N	N	P	L	N
Melilotus alba, M. officinale	Sweetclover	N	N	L	L	H	H	N
Phalaris arundinacea	Reed canarygrass	N	U	H	P	L	N	N
Phleum pratense	Timothy	U	U	L	U	L	H	N
Polygonum cuspidatum, P. sachalinense	Knotweed species	N	N	H	N	H	N	N
Rubus discolor, R. lacinatus	Blackberry species	P	U	H	H	N	N	N
Rumex acetosella	Common sheep sorrel	U	U	L	P	L	N	N
Ulex europeus	Gorse	P	N	U	N	L	N	N
Vicia cracca	Bird vetch	N	N	N	P	H	H	N

Throughout this region, nonnative plant species are largely restricted to early seral environments that follow disturbance and do not persist into later stages of forest succession. Persistent populations of nonnative plants tend to be restricted to naturally open environments, such as woodlands and meadows, or locations subject to repeated disturbance, such as riparian corridors or roadsides.

Coastal Douglas-fir Forests

At low to mid elevations west of the Cascade Range in Washington and Oregon, humid, maritime forests composed of evergreen conifers are the dominant vegetation type. Two distinct forest zones are recognized within the greater coastal Douglas-fir region: Sitka spruce (*Picea sitchensis*) and western hemlock (*Tsuga heterophylla*) (Franklin and Dyrness 1973). Sitka spruce zone forests dominate the western edge of this region, forming a narrow band along the coast from southern Oregon north to southeast Alaska. Inland from the coastal margin, western hemlock zone forests are the most extensive vegetation type west of the Cascade Crest in Washington and Oregon. Western hemlock zone forests are dominated by coast Douglas-fir (*Pseudotsuga menziesii* var. *menziesii*) during the predominant subclimax periods, with succession to western hemlock and western redcedar (*Thuja plicata*) only at later-seral (climax) stages. Broadleaf trees are of lesser importance than conifers in both forest types, though red alder (*Alnus rubra*) is abundant in disturbed locations and several hardwood species are common in riparian areas. Together, Sitka spruce and western hemlock zone forests make up the coastal Douglas-fir region (Garrison and others 1977); both forest zones are discussed in this section.

Humid forests of the coastal Douglas-fir region are characterized by infrequent fire (Arno 2000, table 5-1, pg. 98) and remain largely unaffected by fire exclusion policies. In Sitka spruce zone forests, where precipitation averages 80 to 120 inches (200 to 300 cm) per year and low clouds lead to high rates of fog drip, climate limits the ability of wildfires to burn through forest stands. In this zone, wind is the primary disturbance type, and fire intervals are on the order of hundreds to thousands of years (Agee 1993; Arno 2000, pg. 98). However, when fires do occur in coastal forests, they spread over large areas and are typically stand-replacing.

Presettlement fire regimes are highly variable for western hemlock zone forests. In drier regions of the Cascade foothills, mixed-severity fires occurred roughly every 100 years, while high-severity fires occurred about every 130 to 150 years or more; in mesic forests of the Olympic Peninsula, fire-return intervals greater than 750 years are common (Agee 1993). Large, intense, high-severity fires in both Sitka spruce and western hemlock forest zones may be associated with severe fire weather, regardless of fuel conditions (Agee 1997). Low- and moderate-severity fires in western hemlock zone forests, however, are related to a complex combination of weather, seral vegetation stage, and fuel characteristics (Agee 1997; Wetzel and Fonda 2000).

Forests of the coastal Douglas-fir region were extensively harvested during the 20th century. This practice continues to be the case, particularly on private industrial forestlands. As a result, coastal Douglas-fir forests are generally immature with scattered pockets of mature and old-growth stands. Logging has become the dominant disturbance agent, and prescribed fire has been used by forest managers for disposing of logging slash and preparing sites for reforestation.

Role of Fire in Promoting Nonnative Plant Invasions in Coastal Douglas-fir Forests

Nonnative species in the coastal Douglas-fir region are associated with high-light environments and disturbance (DeFerrari and Naiman 1994; Heckman 1999; Parendes and Jones 2000). Therefore, opportunities for nonnative species establishment are created by disturbances that open the forest canopy, such as fire, forest thinning and harvest, and road building and maintenance. Gradually, intense competition from residual native species and regenerating conifers and eventual shading by developing forest stands can eliminate some plants, including nonnative species, from understory plant communities (Halpern and Spies 1995; Oliver and Larson 1996), especially when recently harvested sites are densely planted with Douglas-fir (Schoonmaker and McKee 1988) or other conifers. For example, abundance and cover of nonnative plant species are negatively correlated with time since canopy opening disturbance (successional age) ($r = -0.61$, $P < 0.001$) and canopy density ($r = -0.37$, $P < 0.05$) in clearcut timber harvest units located in Sitka spruce and western hemlock zone forests of the Olympic Peninsula (DeFerrari and Naiman 1994). Nonnative species that establish after fire or other canopy disturbances must originate from the soil seed bank, be transported to the site by logging or fire-suppression equipment, or be dispersed from populations located along nearby roads, in riparian corridors, or in open habitats.

Few studies have been conducted on the effects of fire on nonnative species within the coastal Douglas-fir region, perhaps due to relatively long fire-return intervals. Neiland (1958) conducted a vegetation analysis in the Tillamook Burn more than a decade after a huge stand-replacing fire (1933) and subsequent smaller fires (1939, 1945) burned on the western slope of the Oregon Coast Range. Several nonnative herbaceous species were found in quadrats within the burned area but not in adjacent unburned forest (table 10-2). Since Neiland's study took place several years after the fires, it is not clear whether nonnative species established and spread in response to fire or were introduced by fire suppression and reforestation efforts and persisted and spread due to open-canopy conditions.

The ecological effects of slash and broadcast burns after clearcut timber harvests have been well studied (Dyrness 1973; Halpern 1989; Halpern and Spies 1995; Halpern and others 1997; Kraemer 1977; Lehmkuhl 2002; Morris 1958; Schoonmaker and McKee 1988; Stein 1995; Stewart 1978; West and Chilcote 1968). However, several critical topics remain unaddressed. Although slash fires are typically set in both spring and fall, the effects of season of burn on nonnative species' responses has not been examined. In addition, nonnative plant species' responses to natural fire and slash burning have not been compared, though similarities are assumed. Finally, the effects of mechanical disturbance associated with harvest have not been separated from the effects of burning.

Slash burning on clearcuts promotes the establishment and temporary dominance of both native and nonnative herbaceous species in the coastal Douglas-fir forest region (Kraemer 1977). Though disposing of

Table 10-2—Frequency of nonnative plant species observed in 200, 1-m^2 quadrats located within the perimeter of the Tillamook Burn, Oregon (Neiland 1958).

Scientific name	Common name	Frequency (%)
Hypochaeris radicata	Hairy catsear	25
Senecio vulgaris	Common groundsel	5
Cirsium vulgare	Bull thistle	4
Poa trivialis	Rough bluegrass	1
Elymus repens	Quackgrass	1
Cirsium arvense	Canada thistle	[a]

[a] Species present in area but not in quadrat.

logging slash without the use of fire may reduce the establishment and spread of nonnative invasive species (Lehmkuhl 2002), untreated logging slash increases overall fire danger (Graham and others 1999). Native and, to a lesser extent, nonnative ruderal herbs are the dominant vegetation during the first 4 to 5 years after slash burning; native residual species typically regain dominance after this point (Dyrness 1973; Halpern 1989; Halpern and Spies 1995; Schoonmaker and McKee 1988). Severe slash fires may kill residual native species (Schoonmaker and McKee 1988), thereby increasing the temporal window available to nonnative species to establish, reproduce, develop soil seed banks, and disperse seeds to neighboring locations.

The nonnative winter annual woodland groundsel (*Senecio sylvaticus*) is particularly prominent in slash burns following clearcuts in western hemlock zone forests of the Cascade and Coast Ranges (Dyrness 1973; Halpern 1989; Halpern and others 1997; Kraemer 1977; Morris 1958, 1969; Schoonmaker and McKee 1988; Stewart 1978; West and Chilcote 1968). Woodland groundsel rapidly increases in abundance and cover 1 to 2 years after broadcast burning (fig. 10-2) and then, just as quickly, declines to negligible amounts (Dyrness 1973; Halpern and others 1997; Kraemer 1977; Morris 1958; Schoonmaker and McKee 1988; Stewart 1978; West and Chilcote 1968). Woodland groundsel seed does not survive broadcast burning (Clark 1991); therefore, fast increases in woodland groundsel populations are related to its abundant production of wind-dispersed seed along with its life history traits of fall germination, rapid early growth, and annual lifecycle (Halpern and others 1997). The transient nature of woodland groundsel has been attributed to a high soil fertility requirement (West and Chilcote 1968) that is met by soil conditions associated with recent burns; and to poor competition with perennials for soil nutrients (van Andel and Vera 1977). However, Halpern and others (1997) demonstrated that interspecific competition was not responsible for this pattern and questioned the hypothesis that woodland groundsel populations decline after soil nutrients have been depleted. The timing of harvest and slash burning has a strong effect on the timing of population booms of woodland groundsel (Halpern 1989; Halpern and others 1997), indicating the importance of timing of disturbance to its establishment and initial growth. Fall broadcast burns consume wind-dispersed and buried seed and result in low population densities the following year, though dramatic increases in population densities typically occur during the second growing season (Halpern and others 1997). Similarly, logging activities that occur after seed dispersal in late summer and early fall prevent seedling establishment and also result in low initial population densities the following year (Halpern 1989).

In addition to woodland groundsel, the invasion and short-term abundance of several other nonnative herbs are associated with clearcut timber harvest followed by broadcast burning in coastal Douglas-fir forests. For example, in the Oregon Coast Range, the frequency and abundance of tansy ragwort (*Senecio jacobaea*) is related to the amount of site disturbance after clearcutting of western hemlock and Sitka spruce zone forests, being greatest with site preparation treatments such as broadcast burning and spraying of herbicides (Stein 1995). In areas with persistent seed banks of tansy ragwort—a function of the historic abundance of mature plants—invasions are triggered by localized disturbances that remove existing vegetation (McEvoy and Rudd 1993; McEvoy and others 1993).

Broadcast-burned clearcuts within western hemlock zone forests of the Cascade Range may also be invaded by nonnative bull and Canada thistles (*Cirsium vulgare*, *C. arvense*) (Dyrness 1973; Halpern 1989; Schoonmaker and McKee 1988), St. Johnswort (*Hypericum perforatum*) (Schoonmaker and McKee 1988), prickly lettuce (*Lactuca serriola*) and wall-lettuce (*Mycelis muralis*) (Dyrness 1973; Schoonmaker and McKee 1988). In a study of secondary succession in the western Oregon Cascades, severely burned (surface litter completely consumed) microsites located within broadcast burned harvest units were colonized and dominated by native fireweed (*Chamerion* spp.), though nonnative bull thistle (Dyrness 1973) and Canada thistle (Halpern 1989) also established in severely burned as well as lightly burned (surface litter charred but not completely removed) microsites. These two thistle species peaked

Figure 10-2—Woodland groundsel dominating a clearcut and broadcast burned site 1 year after disturbance, as is typical for this species in western hemlock forests. (Photo by Vegetation Management Research Cooperative, Oregon State University.)

in relative abundance 3 to 5 years after broadcast burning and then declined (Dyrness 1973; Halpern 1989). Similarly, in a survey of understory vegetation in broadcast burned clearcuts in the western Oregon Cascades (Schoonmaker and McKee 1988), bull and Canada thistles were several times more abundant in stands that had been harvested and burned 5 years previous than in younger or older stands. No statistical analysis was presented. Fire Effects Information System (FEIS) literature reviews indicate that bull and Canada thistles establish after fire in other regions of the Northwest (Zouhar 2001d, 2002b); however, it has not been determined whether establishment occurs from soil seed banks or from wind-dispersed seed. In an experiment examining the effects of heat and soil moisture treatments on bull thistle seeds collected from an old-growth Douglas-fir forest seed bank (Clark 1991; Clark and Wilson 1994), the authors determined that soil temperatures typical of even a low-severity fire could kill bull thistle seed in the soil seed bank. This observation suggests that bull thistle establishment in broadcast burned harvest units is achieved with seeds dispersed from mature plants located in nearby unburned locations.

St. Johnswort also establishes in clearcut and broadcast-burned harvest units of the western Cascade Range, Oregon (Schoonmaker and McKee 1988). Similar to the thistle species previously described, St. Johnswort cover was several times greater in stands that had been harvested and broadcast burned 5 years prior to Schoonmaker and McKee's study than in younger or older harvest units examined. Statistical analysis was not provided. Fire may stimulate sprouting and seed germination of St. Johnswort (Zouhar 2004, FEIS review) and, therefore, may have contributed to its establishment in the broadcast-burned harvest units examined. Alternatively, establishment and spread of St. Johnswort may have been related to soil disturbances associated with mechanical timber harvest and open-canopy conditions.

Finally, broadcast-burned clearcuts in western hemlock zone forests of the Cascade Range are frequently invaded by prickly lettuce (Dyrness 1973) and wall-lettuce (Schoonmaker and McKee 1988). Within two broadcast-burned clearcuts in the western Oregon Cascades, microsites with low-severity burns were invaded by prickly lettuce, while severely burned microsites were not (Dyrness 1973). In another western Oregon Cascade Range study examining the understory composition of clearcut and broadcast burned Douglas-fir stands, Schoonmaker and McKee (1988) observed that wall-lettuce was several hundred times more abundant in stands that had been clearcut and broadcast burned 5 years prior to the study than in younger or older stands. Again, statistical analysis was not presented. The influence of broadcast burning, mechanical disturbance, or open-canopy conditions on the establishment and spread of these species was not explored.

Seeding of nonnative grasses and legumes in slash-burned clearcuts (Lehmkuhl 2002), after wildfires (Agee 1993; Beyers 2004), and along forest roads (Dyrness 1975) and firelines (Beyers 2004) has been widely practiced in coastal Douglas-fir forests in order to increase herbaceous forage for ungulate populations, reduce browsing on conifer seedlings, suppress undesirable species, and reduce soil erosion (Beyers 2004; Lehmkuhl 2002). Seeding of nonnative herbaceous species may influence the establishment and growth of nonnative invasive species and/or alter successional development of native plant communities. In an experiment examining the ecological effects of spring broadcast burning coupled with seeding of common nonnative forage species (orchard grass (*Dactylis glomerata*), white clover (*Trifolium repens*), perennial ryegrass (*Lolium perenne*), annual ryegrass (*L. perenne* ssp. *multiflorum*), and birdsfoot trefoil (*Lotus corniculatus*)) after clearcut timber harvest in the coastal forests of western Washington, the total number and cover of nonnative species were significantly greater in burned plots than in unburned plots ($P = 0.018$ and $P = 0.094$, respectively), suggesting that broadcast burning contributed to the establishment of nonnative species (Lehmkuhl 2002). However, since both seeded forage species and invasive species were included in nonnative species counts and assessments, it is not clear whether burning encouraged the establishment of forage species, invasive species, or both. In addition, the number of nonnative species in burned plots, both forage and invasive, increased during the first 3 years after site treatment ($P = 0.054$), while the number of nonnatives in unburned plots did not change significantly over the same time period. Seeding of nonnative forage species had no effect on the total number of nonnative species observed in burned and unburned plots even though introduced forage species were included in counts, suggesting that introduced forage species may have displaced nonnative invasive species. In contrast, the cover of forage and invasive nonnative species was significantly greater in seeded versus unseeded plots ($P = 0.038$) and broadcast burning more than doubled the annual production of forage grasses for 3 years after treatment. Though the author concluded that seeding of nonnative forage species had "little long-term apparent effect on native plant communities" (Lehmkuhl 2002, pg. 57), the initially high cover of introduced forage species may have reduced the cover of both nonnative invasive species and native ruderal species that establish after fires (Beyers 2004). Therefore, seeding of forage species for invasive nonnative species control must be weighed against potential impacts to native early-seral plant communities.

Silvicultural thinning of dense, young forest stands is frequently used in the coastal Douglas-fir region to encourage structural and species diversity in the understory community (Halpern and others 1999; Thysell and Carey 2001a) and, to a lesser extent, to reduce the severity of future wildfires (Graham and others 1999). Forest thinning in the coastal Douglas-fir forest region stimulates germination of seeds in the soil seed bank, including seeds of nonnative species (Bailey and others 1998; Thysell and Carey 2001a). To assess potential nonnative species response to silvicultural thinning in the coastal Douglas-fir region, Halpern and others (1999) examined the composition of soil seed banks in 40- to 60-year-old Douglas-fir and Sitka spruce stands on the Olympic Peninsula, Washington. All plots were located in stands that originated after clearcut logging. Almost 30 percent of all species represented in soil seed banks and 50 percent of all germinants from litter and soil samples were nonnative species. These species may have originally invaded the stand after clearcut timber harvest. Nonnative species observed in this study represented two basic life histories: short-lived herbaceous species that establish after clearcut logging and common weeds of agricultural areas, waste places, and roadsides. The authors concluded that silvicultural thinning of young stands may provide a temporary window for the re-establishment of nonnative species (Halpern and others 1999). The maintenance of open stand conditions in order to decrease the threat or severity of wildfire may allow the persistence of nonnative plant species in forest understories.

Common roadside weeds of the coastal Douglas-fir region may spread into burns and timber harvest units located adjacent to road networks and along firelines. At the landscape scale, Parendes (1997) tracked the invasion status of woody nonnative Scotch broom (*Cytisus scoparius*) and Himalayan blackberry (*Rubus discolor*) as well as herbaceous nonnative species such as Canada thistle, bull thistle, tansy ragwort, and St. Johnswort on the H.J. Andrews Experimental Forest in the western Oregon Cascades. Of these species, Scotch broom was closely associated with disturbance, as it was more frequent (no statistical analysis presented) on roads adjacent to timber harvest units. St. Johnswort and bull thistle were present on more than 70 percent of the road network; Canada thistle, Scotch broom, and tansy ragwort were present on 10 percent to 30 percent of the road network; Himalayan blackberry was present in only a few isolated locations (Parendes 1997). The fire ecology of Canada thistle, bull thistle, tansy ragwort, and St. Johnswort were discussed previously. A review of Scotch broom and Himalayan blackberry fire ecology follows.

In a literature review published by The Nature Conservancy, the author notes that Scotch broom does not grow well in forest understories but rapidly invades after fire or logging disturbance throughout the coastal Douglas-fir region where it forms dense thickets, spreads into native vegetation, and prevents or slows reforestation (Hoshovsky 1986, TNC review). Seed germination of Scotch broom is increased by soil heating (Regan 2001) and broadcast burning (Parker 1996), suggesting that fire may facilitate invasion of this species in the coastal Douglas-fir region.

Himalayan blackberry is an aggressive invader within the coastal Douglas-fir region, invading wet sites that have been disturbed and abandoned by humans and forming impenetrable thickets in young forest plantations and riparian areas (fig. 10-3). Similar to Scotch broom, Himalayan blackberry grows poorly in forest understories, requiring high light levels for seedling survival and fruit production. However, rapid invasion of cleared forestland suggests that Himalayan blackberry seed may remain viable in the soil for many years (Hoshovsky 1989, TNC review). Himalayan blackberry probably sprouts from root crowns after fire. Its seeds may also survive fire, explaining observations of rapid seedling establishment of many blackberry species after fire (Tirmenstein 1989a, FEIS review). Fire in the coastal Douglas-fir region may encourage invasion by this species.

Similarly, in a multi-scale assessment of nonnative plants on the Olympic Peninsula, Washington, DeFerrari and Naiman (1994) identified several nonnative plant species in clearcuts and, to a much lesser extent, young forest understories. Nonnative species recorded

Figure 10-3—Himalayan blackberry establishing in a disturbed site within the coastal Douglas-fir region. (Photo by Jed Colquhoun, Extension Weed Specialist, Oregon State University.)

in this study include species that have already been discussed: Scotch broom, Himalayan blackberry, Canada thistle, bull thistle, woodland groundsel, and tansy ragwort. In addition, cutleaf blackberry (*Rubus laciniatus*), perennial ryegrass, and common dandelion (*Taraxacum officinale*) were observed. Cutleaf blackberry shares many of the ecological and life history traits of Himalayan blackberry (Tirmenstein 1989b, FEIS review) and invades after logging and slash burning in the coastal Douglas-fir region (Steen 1966). Fire may encourage the production of reproductive tillers in perennial ryegrass (Sullivan 1992b, FEIS review) and facilitate a short-term increase in common dandelion abundance (Esser 1993b, FEIS review).

Effects of Nonnative Plant Invasions on Fuels and Fire Regimes in Coastal Douglas-fir Forests

Two nitrogen-fixing shrubs, Scotch broom (fig. 10-4) and, in coastal environments, gorse (*Ulex europaeus*), may influence fire behavior and/or fire regimes in western hemlock and Sitka spruce zone forests. According to literature reviews, both species develop seed banks that may remain viable for up to 30 years (Zielke and others 1992) and are stimulated by fire to germinate (Parker 1996; Washington State Noxious Weed Control Board 2005; Zielke and others 1992). Both species rapidly invade disturbed areas and can prevent or slow reforestation by forming dense populations (Hoshovsky 1986; Huckins 2004; Washington State Noxious Weed Control Board 2005; Zielke and others 1992). Literature reviews note that, without further disturbance, populations of Scotch broom and gorse degrade after 6 to 8 years and senesce at 10 to 15 years, allowing later-seral plant species to gradually reoccupy the site (Huckins 2004; Zielke and others 1992). As populations of gorse and Scotch broom age, they have been observed to create large amounts of litter (Hoshovsky 1986; Zielke and others 1992). Gorse leaves a center of dead vegetation as it grows outward and its stems have high oil content (Washington State Noxious Weed Control Board 2005), increasing its flammability.

Though it has not been demonstrated that the dominance of either species can increase fire spread or intensity in the coastal Douglas-fir region, the characteristics of both species indicate that such a relationship may exist. In literature reviews, Scotch broom (Hoshovsky 1986; Huckins 2004; Zielke and others 1992) and gorse (Washington State Noxious Weed Control Board 2005; Zielke and others 1992) are described as fire hazards. Early-seral stages of forest succession in coastal Douglas-fir forests are the most flammable (Agee 1997). By slowing reforestation, Scotch broom and gorse prolong these flammable early-seral stand

Figure 10-4—Scotch broom (the shrub with yellow flowers) invades disturbed areas and forms dense populations. (Photo by Jed Colquhoun, Extension Weed Specialist, Oregon State University.)

conditions. Furthermore, dense populations of senescing Scotch broom or gorse provide continuous fuels that may increase the spread of surface fires. For example, Zielke and others (1992) describe a fire that spread rapidly through a gorse understory across 2,500 acres (1,000 ha) of New Zealand forestland. Similarly, a fire that quickly overran the coastal town of Bandon, Oregon in 1936 was attributed to dense populations of gorse found in neighborhood yards (Huber 2005). The density of litter associated with Scotch broom and gorse populations and the flammable oils found in gorse stems may increase fire intensity, though comparisons with native shrubs have not been made to support this assumption. Finally, fire stimulates seed germination of Scotch broom and gorse; therefore, if these shrubs are capable of increasing the frequency of wildfire, they create the conditions necessary for their continued recruitment and dominance.

False brome (*Brachypodium sylvaticum*), a nonnative perennial bunchgrass, is rapidly invading low- to mid-elevation western hemlock zone forests of Oregon's Coast and western Cascade Ranges (fig. 10-5); it is not

Figure 10-5—False brome, a nonnative bunchgrass that invades forest understories and roadsides in western Oregon, and forms dense, continuous populations. (Photo by Tom Kaye, Institute for Applied Ecology, Corvallis, Oregon.)

currently found in Washington. False brome invades roads, clearcuts, open habitats, and understories of both young and mature undisturbed mixed conifer stands, where it dominates the herbaceous layer and forms dense, continuous populations that exclude most native species (False Brome Working Group 2002, 2004; Kaye 2001). Members of the False Brome Working Group have speculated that false brome may alter fire regimes (False Brome Working Group 2002). False brome may increase biomass of fine fuels capable of carrying late-season fires, particularly in well established stands of false brome that have accumulated a heavy build-up of thatch. Alternatively, populations of false brome may decrease understory fire spread of early- and mid-season fires because it has been observed to stay green until late fall (False Brome Working Group 2002). False brome reproduces rapidly from seed but is not rhizomatous (Kaye 2001). Observations indicate that false brome is not controlled by burning (False Brome Working Group 2003).

Use of Fire to Manage Invasive Plants in Coastal Douglas-fir Forests

Though broadcast burning is regularly used to prepare timber harvest units for reforestation, fire is not commonly used for nonnative species control within the coastal Douglas-fir region.

Upper Montane Conifer Forests and Meadows

Above the extensive low- to mid-elevation western hemlock zone forests of Washington and Oregon, upper montane slopes of the Olympic Range and the western Cascade Range are dominated by cold, wet conifer forests. Two forest zones are represented in this region: Pacific silver fir (*Abies amabilis*) at middle to high elevations and mountain hemlock (*Tsuga mertensiana*) in subalpine environments. Wet meadows are found in both forest zones but are more common within the mountain hemlock zone, while dry grassy balds occur along high-elevation ridges of the Olympic Range (Franklin and Dyrness 1973).

Infrequent stand-replacing fires with return intervals of 200 or more years typically characterize the fire regimes of upper montane environments of the western Cascade and Olympic Ranges (Arno 2000, pg. 99). In the western Cascade Range, extensive areas of mountain hemlock-zone forest burned during the latter half of the 19th and early 20th centuries (Franklin and Dyrness 1973). Past fires probably contributed to the establishment and maintenance of high elevation meadows in both the Cascade and Olympic Ranges (Henderson 1973, as cited by Franklin and Dyrness 1973). Fire exclusion in the latter part of the 20th century may be allowing the gradual succession of mountain meadows to forest. Consequently, land managers are experimenting with the use of prescribed fire in order to prevent forest encroachment and maintain or restore meadow composition and structure (Halpern 1999).

Upper montane environments in the Northwest Coastal bioregion support fewer nonnative species than lower elevation sites (DeFerrari 1993; Parendes 1997; Sarr and others 2003). For instance, in the Olympic Mountains, nonnative plant diversity is lower in mature forests located in high-elevation protected wilderness areas than in lower elevation forests (DeFerrari and Naiman 1994). This trend also holds true along roadsides; in the western Oregon Cascades, many roadside nonnative species decrease in abundance with increased elevation (Parendes 1997). Microclimatic conditions found at high elevations may limit the spread of nonnative species; however, several other factors that influence nonnative species establishment and spread, such as precipitation, density and ages of roads, and proximity to source populations, are confounded with elevation and may play an even larger role in limiting invasions.

Though nonnative plant abundance is generally low in upper montane forests of the Northwest Coastal bioregion, nonnative species are common invaders along roadsides and after disturbances, such as landslides and debris flows. In a decommissioned parking lot located in the northeast corner of Olympic National

Park and within a subalpine meadow dominated by Idaho fescue (*Festuca idahoensis*), established populations of dandelion and timothy (*Phleum pratense*) maintained cover over an 8-year study period, while Kentucky bluegrass (*Poa pratensis*) continued to spread into heavily impacted areas (Schreiner 1982). On the debris flow that followed the eruption of Mt. St. Helens in 1980, several wind-dispersed nonnative species established, including Canada thistle (Dale 1989, 1991), woodland groundsel (Dale 1989; Dale and Adams 2003), and hairy catsear (*Hypochaeris radicata*) (Dale and Adams 2003; del Moral and others 1995). Hairy catsear and woodland groundsel were abundant in the post-eruption seed rain (Wood and del Moral 2000), along with lesser quantities of bull thistle, Canada thistle, common dandelion, common groundsel (*Senecio vulgaris*), and wall-lettuce. All of these species have wind-dispersed seed and are frequently observed invading recently disturbed environments. The presence of nonnative plants in roaded and disturbed environments may increase the likelihood of establishment in burned areas.

Though these nonnative species may invade burns in upper montane environments of the Northwest Coastal bioregion, their ability to do so has not been demonstrated. However, populations of Canada thistle (Zouhar 2001d, FEIS review), bull thistle (Zouhar 2002b, FEIS review), and common dandelion (Esser 1993b) sometimes increase after fire in mountain environments of the Interior Northwest. Fire has also been observed to stimulate tiller production and growth in timothy (Esser 1993a, FEIS review). In addition, woodland groundsel is closely associated with slash burns in coastal Douglas-fir forests of the Northwest Coastal bioregion.

Role of Fire in Promoting Invasions of Nonnative Plant Species in Upper Montane Communities

Past research suggests that burned areas in upper montane environments of the Cascade Range may be largely free of nonnative species. Fire effects studies conducted in wilderness areas and National Parks located along the crest of the Washington and Oregon Cascades have found no evidence of nonnative species in burned or unburned plots (Douglas and Ballard 1971; Fahnestock 1977; Hemstrom and Franklin 1982; Miller and Miller 1976). However, these studies were conducted more than 2 decades ago. Follow-up research is needed to determine whether nonnative plants have invaded these areas or other burned areas more recently. Whether this conspicuous absence of nonnative species is due to a lack of local seed source or to environmental barriers to establishment has also not been explored.

Agee and Huff (1980) examined the effects of the Hoh fire, which burned through both lower and upper montane forests of the western slope of the Olympic Range. One year after the blaze, burned plots supported three wind-dispersed nonnative species–hairy catsear, wall-lettuce, and tansy ragwort. These species were absent from adjacent undisturbed forest. The authors did not indicate whether these species were observed in lower montane environments, upper montane environments, or both.

There is some concern that reintroduction of fire into alpine and subalpine meadow communities to prevent forest encroachment and restore meadow community composition and structure may inadvertently facilitate invasion of nonnative species (Halpern 1999). Prior to the application of prescribed fire in a subalpine bunchgrass meadow in the western Oregon Cascades, Halpern (1999) noted that nonnative plant species contributed very little to the diversity of species or the vegetative cover of herbaceous plants in the meadow. Five nonnative species were observed in the meadow community prior to the prescribed burn: quackgrass (*Elymus repens*), creeping bentgrass (*Agrostis stolonifera*), Kentucky bluegrass, common sheep sorrel (*Rumex acetosella*), and yellow salsify (*Tragopogon dubius*). Although these investigators did not assess postfire vegetation responses, low-severity, late fall burns such as the one conducted in this study would be unlikely to affect the spread of Kentucky bluegrass (Uchytil 1993, FEIS review) or common sheep sorrel (Esser 1995, FEIS review), but might reduce the spread of quackgrass (Snyder 1992a, FEIS review).

Effects of Nonnative Plant Invasions on Fuels and Fire Regimes in Upper Montane Communities

There is no indication that nonnative species are changing the fire regimes of upper montane environments of the Pacific Northwest. Nonnative plant abundance is low throughout this region and, therefore, unlikely to cause significant changes to fuel characteristics. Furthermore, the wet, cold climate of upper montane environments prevents the ignition and limits the spread of fire. Fires in this region are closely associated with periods of extreme fire weather, regardless of fuel characteristics that may be influenced by the presence and abundance of nonnative species.

Use of Fire to Manage Invasive Plants in Upper Montane Communities

Nonnative plant species are not presently posing a serious threat to upper montane plant communities of the Northwest Coastal bioregion. Prescribed fire is not considered to be a useful management tool in subalpine forests as fires do not spread under controllable conditions, and native forest dominants tend to be fire avoiders (Agee 1993). However, there is growing interest in using prescribed fire to maintain or restore subalpine meadows that are gradually being lost to forest succession (Halpern 1999). Management activities in these areas should be designed to prevent the accidental introduction of nonnative species' propagules and the promotion of nonnative species establishment and spread.

Riparian Forests

Low- and mid-elevation riparian forests of the coastal region of Washington and Oregon are covered in this section. Dominant trees of riparian corridors which dissect coastal Douglas-fir forests include red alder, bigleaf maple (*Acer macrophyllum*), Sitka spruce, western redcedar, and western hemlock. Riparian forests of low-elevation valleys, such as the Willamette Valley of Oregon and the lower Columbia River Valley of western Oregon and Washington, are dominated by black cottonwood (*Populus trichocarpa*), Oregon ash (*Fraxinus latifolia*), bigleaf maple, red alder, white alder (*Alnus rhombifolia*), and Oregon white oak (*Quercus garryana*). Riparian corridors of the Coast Range and western Cascades are frequently located in steep ravines, while those of the western Olympic Peninsula are typically located on river terraces. Riparian forests in low-elevation valleys have been extensively cleared for agriculture and urban development.

Riparian communities are highly susceptible to invasion by nonnative plants because of frequent natural (floods and landslides) and anthropogenic disturbances. Because of the down gradient movement of soils and plant propagules, riparian communities are often subject to the cumulative ecological damage sustained by entire watersheds (Naiman and others 2000), including the effects of road building, timber harvesting, slash burning, grazing, mining, fire suppression activities, and water withdrawals. These activities have led to a loss of native species (Naiman and others 2000) and increased vulnerability to invasion of nonnative plants. Nonnative plants typically contribute 20 to 30 percent, and can contribute up to 75 percent, of total richness in riparian communities in the Northwest Coastal bioregion (DeFerrari and Naiman 1994; Naiman and others 2000; Planty-Tabacchi and others 1996). In general, the proportion of total richness composed of nonnative species increases from headwaters to piedmont (the transition zone between mountains and lowlands) and remains high along the lower reaches of streams and rivers where human impacts and development are concentrated (Planty-Tabacchi and others 1996). Young, disturbed patches of riparian vegetation are considerably more invaded by nonnative species than more stable, mature patches (DeFerrari and Naiman 1994; Parendes and Jones 2000; Planty-Tabacchi and others 1996). Similarly, riparian communities that dissect coastal Douglas-fir forests are more invaded than adjacent conifer forest (DeFerrari and Naiman 1994; Planty-Tabacchi and others 1996). High abundance of nonnative plants in riparian communities of the Northwest Coastal bioregion highlights the importance of riparian habitats as source populations and dispersal corridors for these species (DeFerrari and Naiman 1994; Naiman and others 2000; Parendes and Jones 2000).

The ecological role of fire in riparian forests of the Northwest Coastal bioregion is only beginning to be explored. Riparian environments tend to be cooler and moister than adjacent uplands; therefore, the flammability of riparian vegetation may be lower than that of vegetation in upland forests (Olson and Agee 2005). High moisture levels coupled with fewer ignitions due to lower slope positions suggest lower fire frequency and higher fire severity in riparian forests than in upland forests (Minnich 1977, as cited by Olson and Agee 2005). Thus, riparian corridors can serve as fire refugia. However, deviations from this trend have also been observed. In a study of historic fires in riparian and upland forests in the Umpqua River drainage of the southern Oregon Cascades, Olson and Agee (2005) found no difference in fire-return intervals between riparian and upland forests. Furthermore, when severe fire weather affects Northwest Coastal bioregion forests (Agee 1997), riparian forests may be as likely to burn as adjacent upland forests.

Riparian vegetation tends to recover rapidly after fire disturbance (Beschta and others 2004). However, fire exclusion in riparian habitats may alter the successional development of plant communities (Gregory 1997). After years of fire exclusion and timber harvest in both upland and riparian locations, the ecosystem impacts of fire may be severe and recovery may be slow or incomplete (Beschta and others 2004).

Role of Fire in Promoting Nonnative Plant Invasions in Riparian Forests

Fire in riparian forests of the Northwest Coastal bioregion may encourage the spread of nonnative plants (DeFerrari and Naiman 1994; Parendes and Jones 2000; Planty-Tabacchi and others 1996; Thompson 2001). Himalayan blackberry (Tirmenstein 1989a), cutleaf blackberry (Termenstein 1989b, FEIS review), St. Johnswort (Zouhar 2004), Canada thistle (Zouhar 2001d), bull thistle (Zouhar 2002b), common sheep sorrel (Esser 1995), perennial ryegrass, Scotch broom, white sweetclover (*Melilotus alba*), and common dandelion (Esser 1993b) have all been observed to establish or increase abundance after fire. The mechanism of response to fire varies by species; for example, fire stimulates the production of reproductive tillers in perennial ryegrass (Sullivan 1992b) and seed germination of Scotch broom (Regan 2001) and white sweetclover (Uchytil 1992a, FEIS review).

Fires in upslope forests may indirectly affect nonnative plant invasions in riparian forests by affecting the timing and magnitude of soil erosion (Bisson and others 2003). Forested slopes of the Cascade, Olympic, and Coast Ranges are steep, and landslides and debris flows are common on both disturbed (fire, logging, road building) and undisturbed slopes in these areas (Miles and Swanson 1986). Landslides and storm runoff deliver soil, wood, and plant propagules from upland sources to riparian communities. For example, the invasion of Scotch broom and foxglove into western Cascade riparian habitats of Oregon may be limited by upland seed sources, such as roads and clearcuts, and the distributions of both species in this area are consistent with down gradient movement of seed from upslope locations to riparian communities (Watterson 2004). In contrast, riparian corridors may not be important sources of nonnative propagules for upland forested locations (DeFerrari and Naiman 1994). Nonnative species are more common on landslides caused by road building and clearcutting than on landslides located on undisturbed slopes. Common nonnatives found on landslides located near such disturbances include species that are frequently sown on road embankments for erosion control (Miles and Swanson 1986).

Seeding of nonnative plant species in postfire recovery projects alters successional pathways and compounds fire-related impacts (Beschta and others 2004). Postfire seeding of nonnative grasses and forbs is often done to reduce soil erosion, prevent landslides, and protect riparian and aquatic habitats (Beyers 2004). However, the establishment of a dense cover of nonnative vegetation can inhibit the regeneration of native woody species and eliminate native ruderal herbs from postfire ecosystems (Beschta and others 2004). Furthermore, several of the nonnative species commonly sown to slow soil erosion, such as timothy, orchard grass, and tall fescue (*Lolium arundinaceum*), have invaded riparian habitats (DeFerrari and Naiman 1994; Planty-Tabacchi and others 1996). Postfire recovery projects should ideally be aimed at enhancing reestablishment of native vegetation. Unfortunately, it is often difficult to get native seed for postfire recovery projects when and where it is needed (Beyers 2004).

Effects of Nonnative Plant Invasions on Fuels and Fire Regimes in Riparian Forests

Despite increasing concern over the effects of nonnative plants in riparian communities, their effect on fire regimes has received limited attention. Nonnative species that may influence riparian fire regimes include forage grasses used in postfire seeding projects, English ivy (*Hedera helix*), and knotweed (*Polygonum* spp.). However, the magnitude and direction of these species' effects on fuel characteristics and fire regimes are unknown.

Postfire seeding of nonnative grasses to slow soil erosion and protect riparian and aquatic habitats may increase the flammability of burned sites. Seeded grasses can form continuous fuel beds with high surface to volume ratios that encourage rapid rates of fire spread. Furthermore, many nonnative grasses are dry and flammable during the late summer and early fall fire seasons (Beschta and others 2004). More research is needed to assess the relationship between fire recovery activities and fuels.

Given its growth habit, the invasion of English ivy may influence fire behavior. English ivy is an aggressive nonnative vine that poses a threat to nearly all forest habitats in the Northwest Coastal bioregion below 3,000 feet (900 m), but is especially problematic in moist and riparian forests and urban and suburban areas. English ivy grows both along the ground, where it covers over native vegetation, and up trees, where it attaches to tree bark with root-like structures and rapidly climbs into the canopy. Soll (2004c, TNC review) found that host trees have low vigor and, within a few years, are killed and/or vulnerable to tip-over and blow down. After tree canopies are destroyed by the invasion of English ivy, shade-intolerant nonnative species, such as Himalayan blackberry, may become established (Soll 2004c). Though dense populations of English ivy clearly affect the structure of surface and crown fuels, their impact on fire behavior is unknown. In the moist forests where English ivy occurs, extreme fire weather may be a more important driving force of fire intensity and severity than fuel characteristics (Agee 1997); therefore, even if English ivy causes marked changes in fuel characteristics it may have little or no influence on local fire regimes. However, given English ivy's abundance near populated areas, further research may be warranted.

Another group of invasive nonnative species that has spread rapidly in Pacific Northwest riparian communities and may affect fire regimes includes Japanese, giant, and cultivated knotweeds (*Polygonum cuspidatum*, *P. sachalinense*, *P. polystachyum*). Knotweeds are fast growing and invade recently disturbed soils where they quickly outgrow and suppress native vegetation (Soll 2004a, TNC review). It has been suggested that Japanese knotweed (*Polygonum cuspidatum*) populations pose a fire hazard during the dormant season due to dense accumulations of dead plant material (Ahrens 1975). However, tissues of Japanese knotweed have relatively low heat content (Dibble and others 2004), so fires in these populations may be of relatively low intensity and severity. More research is needed to determine whether knotweed populations may influence fire behavior, severity, or frequency.

Use of Fire to Manage Invasive Plants in Riparian Forests

There is little information available regarding the use of fire in Northwest riparian communities to control invasive plants, with the exception of reed canarygrass (*Phalaris arundinacea*). Reed canarygrass invades and dominates valley wetland and riparian communities in the Northwest Coastal bioregion. It typically forms dense monotypic stands in wetlands, moist meadows, and riparian communities, excluding native vegetation. Altered water levels and human-caused disturbances appear to facilitate the invasion of reed canarygrass (Lyons 1998, TNC review).

Reed canarygrass's response to prescribed burning is mixed. Hutchison (1992b) found that prescribed burning is an effective method of controlling reed canarygrass in productive sites containing seed banks of native, fire-adapted species, such as wet prairie habitats. In fact, some native wetland species may be unable to compete with reed canarygrass without prescribed fire (Hutchison 1992b). However, Lyons (1998) states that fire does not always kill mature reed canarygrass and may even stimulate stem production unless the fire burns through the sod layer to mineral soil. The high temperature required at the soil surface may be difficult to achieve as reed canarygrass stays green late into the fire season and "so does not burn very hot" (Tu 2004, pg. 6). According to a TNC regional management report (Tu 2004), prescribed fires are typically applied in the fall in the Pacific Northwest and, therefore, may not be severe enough to kill mature reed canarygrass when used in this region. Herbicide treatment prior to prescribed fire may help increase fuel loads and reed canarygrass mortality. Successive burn treatments may not be a control option, as it is frequently impossible to burn stands of reed canarygrass several consecutive years in a row due to a lack of fine fuels after only one burn. Overall, in Northwest Coastal bioregion riparian areas, prescribed fire may be more effective as a pretreatment before other types of control efforts, such as tillage, shade cloth, or herbicide application (Tu 2004).

Oregon Oak Woodlands and Prairies

Oregon oak woodlands and prairies comprised the historic vegetation of the Willamette Valley of Oregon and the Puget Lowlands of Washington. Prairie vegetation is also found on the San Juan Islands, in locations on the western Olympic Peninsula, on coastal headlands in Oregon, and along the shores of Puget Sound and the Straits of Georgia. Oregon oak woodlands were historically composed of native grasses and forbs with Oregon white oak either in open-grown stands or as solitary trees, and other low-stature broadleaved trees and shrubs. Prior to Euro-American settlement and widespread conversion of Oregon oak woodlands and prairie habitats to agriculture and urban development, American Indians burned Oregon oak woodland habitat every year to every several years in order to increase the production of desired plants and herd game (Boyd 1986; Johannessen and others 1971; Norton 1979; Wray and Anderson 2003). Presettlement fire regimes of Oregon oak woodlands were characterized by low-severity understory fires occurring every 35 years or less (Arno 2000, pg. 98). The dominant tree species, Oregon white oak, persisted due to its resistance to low-severity fire (Agee 1996b).

Today, due to extensive livestock grazing and agricultural and urban development, Oregon oak woodlands and prairies are severely fragmented and degraded. Less than 1 percent of presettlement condition Oregon oak woodland habitat remains (Crawford and Hall 1997; Kaye and others 2001; Pendergrass 1996). Nonnative species abound, competing with native plants and often dominating the herbaceous vegetation. Fire exclusion and cessation of burning by Native Americans allowed the invasion and establishment of woody species, such as native Douglas-fir and Oregon ash, as well as nonnative Scotch broom and Himalayan blackberry in previously open woodlands and prairies (Johannessen and others 1971; Pendergrass 1996; Thilenius 1968; Thysell and Carey 2001b; Towle 1982). Increasingly dense and widespread shrub layers are associated with a decreasing abundance of native forbs (Parker and others 1997; Thilenius 1968). The loss of Oregon oak woodland habitat endangers many species such as Bradshaw's lomatium (*Lomatium bradshawii*) (Kaye and others 2001; Pendergrass and others 1999), Curtis aster (*Symphyotrichum retroflexum*) (Clampitt 1993; Giblin 1997), Kincaid's lupine (*Lupinus organus* var. *kincaidii*), Fender's blue butterfly (*Icaricia icarioides fenderi*) (Schultz and Crone 1998), and Oregon silverspot butterfly (*Speyeria zerene hippolyta*) (Pickering and others 2000).

Reserves and wildlife refuges in this region are experimenting with the reintroduction of low-severity fire to maintain and restore Oregon oak woodland habitats and species. Prescribed fires are most often set in the fall, in keeping with the presettlement fire regime. However, with agricultural development and widespread livestock grazing, nonnative plant species now dominate the herbaceous and shrub layers of these plant communities, complicating restoration efforts.

Competition between native and nonnative species may also alter community structure and composition in the postfire environment, even when the reintroduced fire regime is similar to the presettlement regime (Agee 1996a).

Role of Fire in Promoting Nonnative Plant Invasions in Woodlands and Prairies

Though fire is reintroduced into Oregon oak woodlands and prairies by land managers with the dual goals of decreasing nonnative plants and increasing native plants, the diversity of nonnative species established in Oregon oak woodland habitats ensures that at least some will respond favorably to fire. Fire increases the reproduction, germination, establishment, and/or growth of a number of nonnative species that invade these communities (table 10-3).

Herbaceous species—Prescribed fires in Oregon oak woodland habitats have increased the establishment, frequency, and cover of a number of nonnative herbaceous species (table 10-3). Annual fires, in particular, appear to strongly favor nonnative ruderal herbaceous species over native grasses. For instance, in remnant Puget Lowland prairies located at Fort Lewis, Washington, 50 years of annual burning have resulted in the complete replacement of the dominant native species, Idaho fescue, with nonnative forbs and annual grasses, such as hairy catsear and annual vernal grass (*Anthoxanthum aristatum*) (Tveten 1997; Tveten and Fonda 1999). Researchers concluded that native prairie communities, while adapted to frequent, low-severity fires, are not adapted to prolonged annual burning. If prescribed fire is introduced too frequently, land managers may inadvertently encourage the invasion and dominance of nonnative plant species in these communities.

When fire is introduced less frequently or introduced after an extended period of fire-free conditions, impacts on the composition of Oregon oak woodland communities are less obvious and more complicated. In general, prescribed fire encourages the establishment and spread of nonnative ruderal herbaceous species in Oregon oak woodlands. Herbaceous species observed to establish or increase in frequency and/or cover after prescribed fire include annual vernal grass (Clark and Wilson 2001), colonial bentgrass (*Agrostis capillaris*) (Parker 1996), little quakinggrass (*Briza minor*) (Pendergrass 1996), garden cornflower (*Centaurea cyanus*) (Maret 1997), common velvetgrass (*Holcus lanatus*) (Agee 1996a; Dunwiddie 2002; Pickering

Table 10-3—Nonnative plant species' responses to prescribed fire in lowland prairies of the Northwest Coastal bioregion. Statistical significance provided where available.

Species	Burn season	Effect	Direction	Significance	Authors	Notes
Anthoxanthum odoratum sweet vernal grass	Fall	Flowering	+	$P = 0.02$	Clark and Wilson 2001	
Agrostis capillaris colonial bentgrass	Fall	Cover	+		Parker 1996	Correlated with burn temperature
	Spring	Frequency	–	Not significant	Tveten 1997 Tveten and Fonda 1999	
Briza minor little quakinggrass	Fall	Cover	+		Pendergrass 1996	
Carex pensylvanica Penn sedge	Fall	Frequency	–	$P < 0.05$	Tveten 1997 Tveten and Fonda 1999	
Centaurea cyanus garden cornflower	Fall	Establishment	+	$P < 0.005$	Maret 1997	
Cytisus scoparius scotch broom		Germination	+		Parker 1996	
			+		Regan 2001	Greenhouse study
			+		Agee 1996a	
		Cover	–		Agee 1996a	
	Fall		–	$P < 0.05$	Tveten 1997 Tveten and Fonda 1999	
	Fall	Density	–	$P < 0.05$	Tveten 1997 Tveten and Fonda 1999	
	Spring	Seedling density	–	$P < 0.05$	Tveten 1997 Tveten and Fonda 1999	

Table 10-3—(Continued)

Species	Burn season	Effect	Direction	Significance	Authors	Notes
Holcus lanatus common velvetgrass	Fall	Cover	+		Dunwiddie 2002	1st and 2nd burns in 15 yrs
	Fall		–		Dunwiddie 2002	3rd burn in 15 yrs
	Fall		–		Schuller 1997	
	Fall	Flowering	–	$P = 0.01$	Clark and Wilson 2001	
	Summer	Frequency	+		Schuller 1997	
	Fall		+	$P < 0.1$	Pickering and others 2000	
		Establishment	+		Agee 1996a, b	Positively associated with severely burned microsites
Hypericum perforatum St.Johnswort	Fall	Cover	+		Pendergrass 1996	
	Fall		–	$P < 0.01$	Clark and Wilson 2001	
	Fall	Frequency	+	$r > 0.50$	Streatfeild and Frenkel 1997	Correlated with time since burn
Hypochaeris radicata hairy catsear	Fall	Cover	+		Pendergrass 1996	
	Spring, Fall		–	$P < 0.05$	Tveten 1997	
					Tveten and Fonda 1999	
	Summer	Frequency	+		Schuller 1997	
	Fall	Establishment	+	$P < 0.005$	Maret 1997	
	Fall		+	$P < 0.05$	Tveten 1997 Tveten and Fonda 1999	
Luzula congesta heath woodrush	Spring, Fall	Frequency	+	$P < 0.05$	Tveten 1997 Tveten and Fonda 1999	
Leontodon hirtus rough hawkbit	Fall	Cover	+	Pendergrass 1996		
Parentucellia viscosa yellow glandweed	Fall	Frequency	+	$r > 0.5$	Streatfeild and Frenkel 1997	Time since burn
Poa pratensis Kentucky bluegrass	Fall	Cover	–	$P < 0.05$	Tveten 1997 Tveten and Fonda 1999	
		Frequency	+	$r > 0.50$	Streatfeild and Frenkel 1997	Burn history
Pyrus communis common pear	Fall	Sprouting	+	Not significant	Pendergrass and others 1998	
	Fall	Stem height	–		Pendergrass 1996	
Rosa eglanteria sweetbriar rose	Fall	Sprouting	+	Not significant	Pendergrass and others 1998	
	Fall	Stem height	–	Not significant	Pendergrass 1996	
	Fall		–		Streatfeild and Frenkel 1997	
Rubus discolor Himalayan blackberry	Fall	Establishment	+	Not significant	Pendergrass 1996	
Rubus laciniatus evergreen blackberry	Fall	Establishment	+	Not significant	Pendergrass 1996	
Rumex acetosella common sheep sorrel	Fall	Cover	+		Dunwiddie 2002	1st and 2nd burns in 15 yrs
	Fall	Cover	–		Dunwiddie 2002	3rd burn in 15 yrs
	Spring, Fall	Frequency	+		Tveten 1997	
	Spring, Fall		+	$P < 0.1$	Pickering and others 2000	
Senecio jacobaea tansy ragwort		Establishment	+		Agee 1996a	Positively associated with severely burned microsites

and others 2000; Schuller 1997), St. Johnswort (Pendergrass 1996; Streatfeild and Frenkel 1997), hairy catsear (Maret 1997; Pendergrass 1996; Schuller 1997; Tveten 1997; Tveten and Fonda 1999), heath woodrush (*Luzula congesta*) (Tveten 1997; Tveten and Fonda 1999), yellow glandweed (*Parentucellia viscosa*) (Streatfeild and Frenkel 1997), Kentucky bluegrass (*Poa pratensis*) (Tveten 1997; Tveten and Fonda 1999), common sheep sorrel (Dunwiddie 2002; Pickering and others 2000; Tveten 1997), and tansy ragwort (Agee 1996a). Results from regional fire effects studies are summarized in table 10-3.

Despite this trend, nonnative plants rarely respond predictably and consistently to prescribed fire in this region. Many species have been observed to respond both positively and negatively to fire, perhaps related to differences in environment, community composition, season of burn, or fire frequency. Short-term responses may not be predictive of long-term trends. For example, cover of velvetgrass is usually reduced by fire (Clark and Wilson 2001; Dunwiddie 2002; Schuller 1997), while its frequency is increased (Pickering and others 2000; Schuller 1997). Though fire damages mature plants, it strongly favors seedling establishment. A short-term reduction in cover may, therefore, give way to a long-term increase in population density and cover.

Due to the extensive invasion of nonnative plant species that has occurred over the last century or more, the environmental impact of fire has also been fundamentally altered in Oregon oak woodland and prairie communities. Many prairie and woodland habitats are no longer dominated by native bunchgrasses and instead support nonnative thatch-forming grasses. Because of this transition, prescribed fires in habitats extensively invaded by nonnative grasses may create microsites favorable for seedling establishment of herbaceous species, both nonnative and native, primarily through removal of accumulations of litter (Maret 1997). For example, in a study conducted in the Willamette Valley, fall broadcast burning prior to sowing of common nonnative species significantly increased establishment of garden cornflower ($P < 0.005$) in a prairie dominated by nonnative annual grasses, and significantly increased establishment of hairy catsear ($P < 0.005$) in a prairie dominated by nonnative tall oatgrass (*Arrhenatherum elatius*). The author speculated that burning may not create favorable microsites for seedling establishment in communities dominated by native bunchgrasses because litter accumulations are much less (Maret 1997). This observation points to a fundamental change that has occurred to the composition, structure, and dynamics of Oregon oak woodland communities since Euro-American settlement, one that affects nonnative and native species' responses to fire and cannot be simply undone by reintroducing presettlement fire regimes.

Woody species—Prescribed fire increases the abundance of several nonnative woody species in Oregon oak woodland habitats (table 10-3), primarily through sprouting of underground parts or seed scarification. For example, prescribed fires increase the stem density of several nonnative woody species that sprout from underground parts in response to disturbance, including sweetbriar rose (*Rosa eglanteria*) and common pear (*Pyrus communis*) (Pendergrass and others 1998), though stem heights are reduced (Pendergrass 1996; Streatfeild and Frenkel 1997). Burned prairie sites have also been associated with the establishment and increase of nonnative blackberries (*Rubus discolor*, *R. laciniatus*) (Pendergrass 1996).

Scotch broom is another woody species that often increases after prescribed fire. Though fire reduces the cover and density of mature plants (Agee 1996a; Tveten 1997; Tveten and Fonda 1999), burning and soil heating increase germination of Scotch broom by scarifying seed coats (Parker 1996; Regan 2001). In a greenhouse study conducted with soils from Fort Lewis, Washington, Regan (2001) found that Scotch broom germination greatly increased with soil heating (140 °F (60 °C) for 10 minutes), leading him to conclude that prescribed burns would increase the germination of Scotch broom. Field studies have confirmed this hypothesis (Agee 1996a; Parker 1996).

Scotch broom has extensively invaded prairies of the Puget Lowlands, forming monotypic stands in some locations (Parker 1996; Tveten 1997). Though Scotch broom invasion is typically associated with fire or other disturbance, a unique situation exists in some prairie communities of the Puget Lowlands of Washington. Unlike other prairie vegetation in the Northwest Coastal bioregion, prairies of the Puget Lowlands are usually found on glacial outwash and characterized by the presence of a biological soil crust composed of algae, lichens, and liverworts. Parker (1996, 2001) conducted a seeding experiment to assess the importance of seedbeds to the invasion of Scotch broom in these prairies. Significantly more Scotch broom seedlings emerged in untreated control plots than in any other treatment ($P = 0.01$), including broadcast burning before and after seeding. The author suggests that, in prairies located on nutrient-poor glacial outwash, biological soil crusts may be facilitating the invasion and establishment of Scotch broom. Though Parker (1996) noted that broadcast burning after seeding increased Scotch broom seed germination, she concluded that prescribed fire does not necessarily increase the success of Scotch broom seedling establishment in some Puget Lowland prairies. Rather, Scotch broom is more likely to establish in undisturbed prairies than ones that are regularly burned (Parker 1996).

Effects of Nonnative Plant Invasions on Fuels and Fire Regimes in Woodlands and Prairies

Due in part to invasion by native and nonnative woody plants, the fire regime of lowland prairies has shifted from a low-severity regime, maintained by frequent, anthropogenic ignitions and fueled by grasses and herbaceous vegetation, to a mixed-severity regime with lengthened fire-return intervals and accumulations of woody fuels. Over the last century, forest succession in Oregon oak woodlands and prairies has been associated with the spread of Douglas-fir and broadleaved trees into previously open habitats. Fire-resistant, open-grown Oregon white oak stands have grown increasingly dense with small, clustered stems. Native trees such as Douglas-fir, bigleaf maple, and Oregon ash, and nonnative trees such as sweet cherry (*Prunus avium*), common pear, paradise apple (*Malus pumila*), and oneseed hawthorn (*Crataegus monogyna*) have established in open prairies and oak understories and given oak stands a distinctly two layered appearance (Thilenius 1968). Canopy closure of Douglas-fir, which grows considerably taller than Oregon white oak, eventually kills overtopped oaks and may contribute to an accumulation of large woody fuel.

Shrub layers composed of native and nonnative plants have also thickened and spread due to fire exclusion (Chappell and Crawford 1997; Thilenius 1968). Invading nonnative shrubs include Himalayan and cutleaf blackberry, Scotch broom, and sweetbriar rose, all of which form dense, impenetrable thickets (Hoshovsky 1986, 1989; Pendergrass 1996; Soll 2004b). Himalayan blackberry (Soll 2004b) and Scotch broom (Hoshovsky 1986) populations are associated with accumulations of dead plant material. Dense stands of small oaks and other trees, and thick understories of Scotch broom and other nonnative shrubs provide fuels for intense, high-severity fires that sweep into the crowns of large oaks (Thysell and Carey 2001b). While resistant to low-severity fire, Oregon white oaks are vulnerable to high-severity fires fueled by native and nonnative woody plants.

The widespread replacement of native bunchgrasses with thatch-forming nonnative grasses such as tall oatgrass and false brome may change the behavior and severity of surface fires from historic conditions. No research is currently available on this topic.

Use of Fire to Manage Invasive Plants in Woodlands and Prairies

Several studies of lowland prairie restoration have examined the effectiveness of prescribed fire to both control nonnative invasive plants and encourage the establishment and growth of native plants (Clark and Wilson 2001; Ewing 2002; Maret and Wilson 2000; Parker 1996; Pendergrass 1996; Pickering and others 2000; Streatfeild and Frenkel 1997; Wilson and others 2004). In plant communities adapted to low-severity fire regimes, native plants are usually not killed by fire unless fuel buildup is excessively high or native plants have low vigor prior to burning (Agee 1996a). Both conditions are prevalent in lowland prairie communities of the Pacific Northwest (Dunwiddie 2002). Several native plants of lowland prairies respond favorably to prescribed fire (Agee 1996a; Clark and Wilson 2001; Kaye and others 2001). Others, such as Idaho fescue (Agee 1996a; Ewing 2002), Roemer's fescue (*Festuca roemeri*) (Dunwiddie 2002), and tufted hairgrass (*Deschampsia caespitosa*) (Clark and Wilson 2001) are more sensitive to fire but can be replanted after fires aimed at eradicating nonnative species (Agee 1996a) or creating impoverished soil conditions. Reducing soil nutrients and organic matter through application of fire may give some native species a competitive advantage over nonnative invasive species (Ewing 2002).

Restoration projects in lowland prairies are frequently based on the assumption that prescribed fires promote or maintain native herbaceous species, and that fire inhibits nonnative herbaceous and woody species because they are not fire-adapted. Studies of prairie restoration have either weakly supported or weakly refuted this assumption. For example, in a short-term replicated experiment conducted in the southern Willamette Valley, the effects of 2 years (1994, 1996) of prescribed burns (conducted in September-October) were compared with other restoration treatments in a remnant patch of wetland prairie extensively invaded by nonnative grasses and forbs and native and nonnative woody plants. Burning significantly reduced the cover of nonnative forbs as a group ($P = 0.03$) and significantly increased the cover of native forbs ($P = 0.04$), supporting the hypothesis that prescribed fires tend to favor native species over nonnative species. In particular, St. Johnswort cover and the flowering of common velvetgrass were reduced after fire (Clark and Wilson 2001); however, both of these species also responded favorably to fire in other regional studies (table 10-3).

In contrast, another study of Willamette Valley wet prairie restoration indicated that prescribed burns were effective at increasing native forbs but ineffective at controlling nonnative plant species. Prior to the burns, community species richness was dominated by native forbs, while nonnative perennial grasses dominated vegetation cover. Broadcast burns were conducted in fall of 1988 and 1989 with strip-head burning techniques and were reported to reach lethal temperatures at the soil surface, with short residence times. After the burns, the frequency of native annual and perennial forbs increased in most of communities sampled (four out of five, and three out of five communities sampled, respectively); however, cover of native annual forbs decreased in two communities as well. Though the frequency of nonnative perennial graminoids decreased in most communities sampled, the total cover of nonnative species increased in all but one community (Pendergrass 1996).

Similarly, at Oak Patch Natural Area Preserve, Washington, Oregon white oak regeneration, as well as establishment of nonnative herbaceous species, increased after prescribed burning (Agee 1996a, b). The site had been extensively invaded by Douglas-fir, and most of the mid-sized Douglas-fir were cut and removed prior to the burn. Oregon white oak regeneration was associated with high-severity burn patches where small logs had burned and most soil organic matter had been consumed. However, these same high-severity burn patches were also associated with establishment of nonnative invasive species such as tansy ragwort and common velvetgrass. No statistical analysis was presented (Agee 1996a, b).

In other situations, the reintroduction of fire appears to have little immediate impact, positive or negative, on the composition of degraded prairie communities. For example, 1 year after implementation of a prescribed fire program at the W.L. Finley National Wildlife Refuge, Willamette Valley, Oregon, Streatfeild and Frenkel (1997) found little difference in the relative proportion of native and nonnative plant species in treatment areas, regardless of fire history. In the study area, 20 plots were burned in September of the previous year; six plots were burned 4 to 6 years prior to the study; and ten plots were unburned controls. The authors concluded that prescribed fires were not (yet) achieving the management goals of reducing the cover of nonnative plant species or increasing the cover of native perennial herbaceous species (Streatfeild and Frenkel 1997). Whether the continued application of prescribed fire at the refuge will eventually favor native or nonnative species remains uncertain.

Though the initial reintroduction of fire may have a pronounced effect on community composition, individual applications of a frequent fire program may have few observable impacts and serve to maintain established community characteristics. For instance, a program of spring-applied prescribed fire was begun in 1978 and applied on a 3 to 5 year rotation on 7,500 acres (3,035 ha) of fescue grasslands and oak woodlands at Fort Lewis, Washington. The effects of 1 year (1994-95) of prescribed fire were examined within this management area. Prescribed fires were set in spring or fall under the following conditions: 50 to 68 °F (10 to 20 °C) ambient temperatures, 20 to 50 percent relative humidity, and wind speeds <3 miles/hour (4.8 km/hour). Flame heights were <3 feet (0.9 m). In the fescue grassland community, the majority of native and nonnative herbaceous species had no significant response to fire treatments. Nevertheless, the cover of one nonnative species, hairy catsear, decreased after both spring and fall burns, though its frequency did not change significantly due to dense postfire germination. Likewise, prescribed fire in the oak woodland community examined had little effect on herbaceous species. Though spring burns significantly decreased the frequency of colonial bentgrass ($P < 0.05$) and fall burns significantly decreased the frequency of Penn sedge ($P < 0.05$) (*Carex pensylvanica*), neither treatment reduced the cover of either nonnative graminoid (Tveten 1997; Tveten and Fonda 1999).

Similarly inconclusive results were obtained from a study of herbaceous species response to summer and fall burns conducted from 1985 to 1992 on the Mima Mounds Natural Area, Washington. Prescribed fires had, for the most part, mixed results with few significant effects on the frequency of nonnative or native species. The only lasting effect observed was a 3-year increase in the frequency of hairy catsear after a single summer burn. Though fall burns reduced the frequency of common velvetgrass, declines were also observed in the unburned control area, limiting interpretation of the results (Schuller 1997).

Prescribed fire is also applied to lowland prairies to control invading native and nonnative woody plants (Parker 1996; Thysell and Carey 2001b; Tveten and Fonda 1999). Though prescribed fire can reduce the spread of these species, it is not always an effective method of control. Fire exclusion allows Scotch broom to invade lowland prairies and oak woodlands of southern Washington (Tveten and Fonda 1999), and prescribed fire is commonly used to control this species in this region (Parker 2001; Tveten and Fonda 1999). Though fire has proven useful for this purpose, it must be applied frequently enough to prevent the buildup of fuels, which threaten oak overstories, but not so frequently that nonnative herbaceous species are favored over native ones (Thysell and Carey 2001b; Tveten and Fonda 1999). Scotch broom is least likely to sprout if treatments are applied during mid-summer, though care must be taken to avoid spreading mature seeds (Ussery and Krannitz 1998). Care must also be taken during dry conditions due to the volatile oils in Scotch broom foliage, which are capable of producing high-intensity fires (Huckins 2004). Though a single, severe fire can greatly reduce the cover of Scotch broom, it may also stimulate seed germination from the soil seed bank. A second, less intense fire roughly 2 or 3 years later, before Scotch broom seedlings begin to flower, is required to achieve long-term control (Agee 1996a). Spot treatment, such as using a flamethrower during winter months, can remove remaining Scotch broom seedlings that are not killed by prescribed fires (Agee 1996b).

In the prescribed fire program at Fort Lewis previously described, the effects of prescribed fires (spring and fall treatments) were examined in Scotch broom communities. Due to fuel conditions, fires were patchy, and flame heights were <6.5 feet (2 m). Fall burns caused more mortality and resulted in less sprouting of Scotch broom than spring burns. Fall fires caused a reduction in Scotch broom density and cover, but had little effect on seedling density. In contrast, spring burns reduced seedling density but had no effect on mature plants. Fall burns also reduced total fuels in Scotch broom thickets while spring burns did not. The authors concluded that several cycles of prescribed fire will be required to "restore the balanced fire regime" to Fort Lewis prairies (Tveten and Fonda 1999, p. 156).

Prescribed fire can help control the spread of Himalayan blackberry in lowland prairies but may not eliminate it from an area (Agee 1996a). In fact, fire may promote the spread of Himalayan blackberry in some prairie communities (Pendergrass and others 1998). Several burn treatments are necessary to control this species. According to a TNC management report (Soll 2004b), fire is not completely effective on its own but may be used to remove mature plants over large areas. The use of herbicides prior to burning may desiccate aboveground vegetation so that fires can take place during safe weather conditions. In addition, to ensure that fires carry, aboveground vegetation may need to be chopped or mown prior to ignition. For long-term control, burning may need to be followed by herbicide treatment, repeated burning or mowing to exhaust the soil seed banks and rhizome carbohydrate reserves, and/or planting of fast-growing or shade-tolerant native species. Prescribed fire may be most effective on slopes and in locations where grasses can help carry the fire (Soll 2004b).

Similar to the variable response of nonnative herbaceous species to prescribed fire, studies examining the effectiveness of prescribed fire for controlling nonnative woody plant species in lowland prairies have yielded mixed results. For example, prescribed fire has been shown to be largely ineffective at controlling sweetbriar rose, common pear, and nonnative blackberries in some locations. After 2 consecutive years of experimental prescribed burns in a wet Willamette Valley prairie, burned plots were associated with increased sprouting of sweetbriar rose and common pear, indicating that prescribed burns were not severe enough to kill belowground meristematic tissues of these species. Furthermore, burned plots were more invaded by Himalayan and cutleaf blackberry than unburned plots (Pendergrass and others 1998). Similarly, a program of frequent prescribed fire in a Willamette Valley refuge was deemed ineffective at controlling sweetbriar rose (Streatfeild and Frenkel 1997). In another prairie restoration experiment conducted in the Willamette Valley, the effects of 2 years of fall burns were compared with other restoration treatments in a remnant patch of wetland prairie extensively invaded by native and nonnative woody species, including Scotch broom and Himalayan blackberry. Though burning did significantly reduce the number of surviving native and nonnative shrubs ($P = 0.03$), other results were inconclusive, as the response was variable, perhaps due to variable fire severity and species' abilities to sprout after fire (Clark and Wilson 2001). Results from these studies suggest that the ecological changes caused by a century of fire exclusion, forest succession, and other human impacts are unlikely to be reversed by one or two low-severity broadcast burns (Pendergrass and others 1998).

Burning prior to the direct seeding of native plants may improve their establishment by removing accumulations of litter and destroying competing vegetation. However, burning may also increase the establishment of nonnative plants from the soil seed bank. In a seeding experiment conducted in the Willamette Valley, seeds of common native and nonnative grasses and forbs were planted in fall broadcast burned and unburned plots. Plots were located in three prairies distinguished by the relative dominance of different herbaceous vegetation types: annual nonnative grasses, perennial nonnative grasses, and native bunchgrasses. Burned seedbeds located in communities dominated by nonnative grasses had greater seedling establishment of native species (100 percent and 75 percent of species sown) than nonnative species (13 percent and 33 percent of species sown). In contrast, a greater proportion of nonnative species (50 percent) was favored by burned seedbeds than native species (25 percent) in the relatively pristine site dominated by native bunchgrasses. In general, broadcast burning prior to direct seeding of native species improved seedling establishment in low-quality, highly-invaded prairie habitats. However, broadcast burning in relatively pristine prairie communities may have created conditions that favored establishment of nonnative ruderal species present in the soil seed bank (Maret and Wilson 2000).

In conclusion, the utilization of fall burns for lowland prairie restoration is a "mixed bag" in terms of native and nonnative plant species response (Pickering and others 2000). Fire stimulates many nonnative species while controlling others and can have both negative and positive effects on native vegetation. Many nonnative species respond both favorably and unfavorably to fire, making community responses difficult if not impossible to predict. In addition, many studies have only examined the immediate impacts of recently reintroduced fire on plant communities. Extrapolation of short-term species response from one or two burns to predict long-term trends should be done with caution (Dunwiddie 2002).

A century of fire exclusion has greatly increased woody fuels in Oregon oak woodland communities. Prescribed fire may have limited usefulness to control woody vegetation, and short-term applications of fire are unlikely to reduce accumulations of woody fuels or to control species that sprout after aboveground disturbance (Pendergrass and others 1998). On some sites, repeated burning can reduce fuel loads without harming mature oaks. Alternatively, mechanical destruction and removal of woody fuels prior to burning may be necessary to reduce the risk of damaging overstory oaks (Thysell and Carey 2001b).

Careful evaluation of species composition and life history traits may help managers select between different management options for restoring some of the ecological functions of lowland prairies. The timing, frequency, and season of burning must be selected carefully to avoid damaging native species, particularly sensitive or endangered species, or promoting establishment and spread of nonnative species.

Alaska

The following three ecoregions will be discussed in this subsection: coastal hemlock-spruce forest, interior boreal forest, and tundra (classification after Küchler 1967). Currently, fire does little to contribute to the invasion of nonnative species into these plant communities. However, rapidly warming temperatures in the northern latitudes associated with global climate change, may increase fire frequency in these plant communities while simultaneously disrupting environmental barriers that currently limit nonnative plant species invasion.

Coastal Hemlock-Spruce Forests

Three distinct vegetation types occur in the coastal hemlock-spruce region of Alaska: western hemlock-Sitka spruce forests, deciduous brush thickets, and muskeg bogs. Fire regimes in this region may be characterized by major stand-replacing fires occurring every 200 years or more (Arno 2000, table 5-1, pg. 98).

Role of Fire in Promoting Nonnative Plant Invasions in Coastal Hemlock-Spruce Forests

Due to the moist conditions in the coastal hemlock-spruce region of Alaska, fires are typically rare, though more frequent in forest stands of the Kenai Peninsula which include black spruce (*Picea mariana*), white spruce (*Picea glauca*), and paper birch (*Betula papyrifera*). When fires in the coastal hemlock-spruce region do occur, they are relatively small (less than 10 acres) and tend to be located along road systems and near populated areas (Noste 1969). Nonnative plant species are largely restricted to locations that have been recently or frequently disturbed by humans (Densmore and others 2001; DeVelice, no date; Duffy 2003). These locations are also associated, coincidentally, with fire occurrence; however, fires have not been observed to promote the invasion of nonnative species into coastal hemlock-spruce communities of Alaska. Unfortunately, few fire effects studies have been conducted within coastal hemlock-spruce forest and none have been carried out in deciduous brush thickets or muskeg bogs.

In coastal hemlock-spruce forests on the Kenai Peninsula, spruce bark beetle outbreaks have killed mature spruce trees over large areas. Despite moist conditions, wildfires have burned after these outbreaks, further opening the forest canopy. As part of an inventory of nonnative plant species on the Chugach National Forest, Kenai Peninsula, Duffy (2003) examined two burns located within beetle outbreak areas and found that native herbaceous species dominated both sites. Nevertheless, two nonnative plant species that were present in unburned areas, field foxtail (*Alopecurus pratensis*) and timothy, a widespread nonnative species in the coastal hemlock-spruce region (Heutte and Bella 2003), also established within burned areas. Fall fires may encourage the spread of established populations of timothy in beetle outbreak areas, as late season fire stimulates growth, production of reproductive tillers, and increased seed production in this plant species in other geographical regions (Esser 1993a). Fire response information is unavailable for field foxtail.

Twentieth-century wildfires created favorable moose habitat in coastal hemlock-spruce forests on the Kenai Peninsula by eliminating conifer overstories and stimulating shoot production of willow, aspen, and birch (Miner 2000). Boucher (2001) evaluated the relative effectiveness of prescribed burns for creating moose habitat by examining 17 prescribed burns conducted in coastal hemlock-spruce stands on the Chugach National Forest, Kenai Peninsula. The "probably introduced" (Hultén 1968) Dewey's sedge (*Carex deweyana*) developed minor cover in 3 of the burns but was absent from paired transects located in adjacent unburned forest. However, the reverse was true in a fourth set of paired transects. This study found no statistically or ecologically significant relationship between prescribed fire and invasion by Dewey's sedge. Common dandelion, a species noted to increase in frequency after fire in the lower 48 states due to its abundant production of wind-dispersed seed (Esser 1993b), was also observed in this study but was no more abundant in prescribed burns than in adjacent unburned forest.

Nonnative plant inventories conducted within the coastal hemlock-spruce region of Alaska indicate that populations of nonnative species that may have the ecological potential to invade after fire are present in the region. Though these species have been observed to invade after fire in lower latitudes, there is no evidence to indicate whether these species will invade or increase after fire in the coastal hemlock-spruce region of Alaska. Currently, these nonnative plants are largely restricted to roadsides and populated areas within this region. For example, in an observational study of roadside vegetation along the coastal slope of the Haines Road in southeast Alaska (Lausi and Nimis 1985), four nonnative species that have been observed to tolerate or spread after fire in lower latitudes were found: yellow toadflax (*Linaria vulgaris*), common dandelion, and orchard grass (*Dactylis glomerata*) in all roadside locations regardless of vegetation type, and western wheatgrass (*Pascopyron smithii*) in deciduous thickets and muskeg bogs. Both yellow toadflax and common dandelion are established in coastal hemlock-spruce forest and boreal forest regions of Alaska (Lapina and Carlson 2005). Though yellow toadflax is more common in southcentral Alaska, it has also been identified in Juneau and Skagway (Heutte and Bella 2003). Due to its deep taproot, yellow toadflax typically survives even severe fires, and postfire environments are favorable to seedling establishment (Zouhar 2003b, FEIS review). Yellow toadflax may require an initial disturbance, such as fire, for establishment, but once a population is established yellow toadflax can spread into adjacent undisturbed locations within the coastal hemlock-spruce region (Lapina and Carlson 2005). Orchard grass has been observed throughout the coastal hemlock-spruce region of Alaska (Heutte and Bella 2003) and is reported to be somewhat tolerant of fire disturbance, perhaps even facilitating the spread of low-intensity fires when dormant (Sullivan 1992a, FEIS review). Western wheatgrass is noted to maintain or slightly increase in response to fire (Tirmenstein 1999, FEIS review). In conclusion, on some sites roadside fires within the coastal hemlock-spruce region may contribute to the invasion and establishment of nonnative species within natural areas located adjacent to roads.

In a compilation of inventory and field guide data, nonnative species observed in the coastal hemlock-spruce region of Alaska were identified and described (Heutte and Bella 2003), including several species that invade or increase after fire. Several species observed in the coastal hemlock-spruce region invade broadcast-burned clearcuts of the coastal Douglas-fir region of Washington and Oregon. For example, Canada thistle is located around human settlements in the region; this species is known to survive fire and establish in postfire environments (Zouhar 2001d). Bull thistle has been observed in Ketchikan, Haines, Gustavus, Juneau, and Prince of Wales Islands and, similar to Canada thistle, postfire environments are favorable to its establishment (Zouhar 2002b). Tansy ragwort, a species that responds favorably to a variety of disturbances including slash fires (Stein 1995), has been observed in Ketchikan and Juneau. Spotted knapweed (*Centaurea biebersteinii*), a species observed invading burned and logged land in British Columbia (Zouhar 2001c, FEIS review), has been located in Skagway, Valdez, and Prince of Wales Island. This species produces a taproot capable of surviving low-severity fire and large amounts of fire-tolerant seed (Zouhar 2001c). Scotch broom has also made inroads into the coastal hemlock-spruce region where it is found in Ketchikan, in private yards in Sitka, Hoonah, and Petersburg (Heutte and Bella 2003) and Prince of Wales Island (Lapina and Carlson 2005). A similar shrub species, gorse, has also made its way to the Queen Charlotte Islands (Heutte and Bella 2003).

Two species that increase after prescribed fire in Oregon oak woodlands of Washington and Oregon have also been identified in the coastal hemlock-spruce region. Hairy catsear has been observed along logging roads on northern Zarembo Island, in the upper Lynn Canal on Queen Charlotte Island, and in Juneau. St. Johnswort has been observed in Hoonah and Sitka (Heutte and Bella 2003).

In addition, white sweetclover is found in both the boreal forest and coastal hemlock-spruce regions of Alaska (Lapina and Carlson 2005). In the coastal hemlock-spruce region, white sweetclover has infested gravel bars and sand dunes along the Stikine River in Tongass National Forest (Heutte and Bella 2003).

In an extensive survey of nonnative plant species in Alaskan National Parks (Densmore and others 2001), three nonnative species, which literature reviews indicate are fire-adapted, were observed in both the Kenai Fjords National Park (KFNP) and Sitka National Historical Park (SNHP): common dandelion, yellow toadflax, and yellow sweetclover (*Melilotus officinalis*). As with the first two species described previously, yellow sweetclover invades disturbed areas in Alaska (Lapina and Carlson 2005). In a review published by the Alaska Natural Heritage Program (2004), the author suggests that yellow sweetclover easily invades open habitats such as those created by fire, although primary literature was not cited to support this observation. All three species are currently limited to roadsides and trails, but there is concern that future construction projects may encourage their spread (Densmore and others 2001).

Two inventories of nonnative species were conducted in Chugach National Forest, one in the mountains of the Copper River area and Kenai Peninsula (Duffy 2003) and the other restricted to trails of the Kenai Peninsula (DeVelice, no date). Nonnative plant species were confined to areas subject to human disturbance such as roads, boat ramps, trailheads (Duffy 2003), and trails (DeVelice, no date) and were rare in densely forested and alpine areas (DeVelice, no date; Duffy 2003).

FEIS literature reviews were consulted for fire effects information on nonnative species identified in the Chugach National Forest inventories. Six species are noted to either establish or increase after fire in other geographical regions (table 10-4). The remaining species in the inventories either do not respond favorably to fire, or fire effects information is unavailable. Fire response information has previously been described for three of the six species: common dandelion, timothy, and yellow sweetclover. In addition, fire stimulates the production of reproductive tillers in perennial ryegrass (Sullivan 1992b) and encourages the establishment or increase of common sheep sorrel in regions outside of Alaska (Dunwiddie 2002; Esser 1995; Pickering and others 2000; Tveten 1997). Kentucky bluegrass was among the most commonly encountered nonnative plant species in the Chugach National Forest inventories (DeVelice, no date; Duffy 2003). Kentucky bluegrass is a rhizomatous, mat-forming perennial that has been used for soil stabilization along Alaskan highways. Though established populations have been observed to displace native species and alter succession in plant communities located in other regions (Uchytil 1993), a review published by the Alaska Natural Heritage Program states that Kentucky bluegrass does not seriously alter successional development in Alaska (Alaska Natural Heritage Program 2004). The source of this information was not given.

Effects of Nonnative Plant Invasions on Fuels and Fire Regimes in Coastal Hemlock-Spruce Forests

There is no indication that fire regimes in the coastal spruce-hemlock region of Alaska have been altered by nonnative plant invasions. The climate is wet, and fire frequency and severity are probably more closely associated with rare periods of dry weather than with fuel conditions. Nevertheless, there are a few nonnative species that may have the potential to influence fire regimes in this region.

- Japanese, giant, and bohemian knotweed (*Polygonum* x *bohemicum*) are highly invasive nonnative plant species that are becoming increasingly common along streams and rivers, utility rights-of-way, and gardens in Alaska and the coastal northwest. Populations of Japanese knotweed are established in the Tongass National Forest (Alaska Natural Heritage Program 2004) and in Juneau, Sitka, and Port Alexander (Lapina and Carlson 2005). Knotweeds are well established in the Anchorage area, and there is concern that they could spread into adjacent forestland (Duffy 2003). In a review published by the Alaska Natural Heritage Program (Lapina and Carlson 2005), it is noted that Japanese knotweed may pose a fire hazard during the dormant season due to an abundance of dried leaves and stems, suggesting that dense populations of dormant plants may encourage fire spread during rare periods of abnormally dry winter weather.
- Orchard grass, a roadside weed in southeastern Alaska (Lausi and Nimis 1985), develops a dense thatch (Sullivan 1992a) that may aid fire spread.

Table 10-4—Nonnative plant species that were observed in Chugach National Forest inventories and are reported to establish or increase in response to fire in other locations.

Scientific name	Common name	Kenai Peninsula-Trails (DeVelice n.d.)	Kenai Peninsula-Mountains (Duffy 2003)	Seward area-Fjordland (Duffy 2003)	Cordova area-Foreland/Fjordland (Duffy 2003)
Lolium perenne	Perennial ryegrass	X		X	
Melilotus officinalis	Yellow sweetclover	X		X	
Phleum pratensis	Timothy	X	X	X	
Poa pratensis	Kentucky bluegrass	X	X		X
Rumex acetosella	Common sheep sorrel	X			
Taraxacum officinale	Common dandelion	X	X	X	X

- Bird vetch (*Vicia cracca*) populations occur in the Seward area of coastal Alaska (Duffy 2003) as well as Ketchikan and Unalaska (Lapina and Carlson 2005). Its potential impact on fire regimes is discussed in the boreal forest subsection that follows.

Use of Fire to Manage Invasive Plants in Coastal Hemlock-Spruce Forests

There is no indication that prescribed fire is currently being used to manage invasive nonnative plants in the coastal hemlock-spruce region of southeast Alaska. This may be due to the wet climate or the infeasibility of controlling the widely-dispersed, relatively small populations of nonnative plants found in this region. However, Heutte and Bella (2003) mention that wetland invasions of reed canarygrass, a plant of questionable nativity, may be effectively controlled with fire. Whether the authors are referring to control efforts within the coastal hemlock-spruce region is not clear, though fire is used to control this species in western Washington and Oregon.

Boreal Forests

Two general boreal forest types are represented in Alaska: black spruce (*Picea mariana*) and spruce (*P. mariana, P. glauca*)-birch (*Betula papyrifera*) forests. The most widespread boreal forest type in Alaska, black spruce forests are dense to open lowland forests composed of mixed hardwoods and black spruce or pure stands of black spruce. The fire regime of black spruce forests is characterized by major stand-replacing fires occurring approximately every 35 to 200 years (Duchesne and Hawkes 2000, table 3-1, pg. 36), and most of the plant species that occupy black spruce forest communities are adapted to fire disturbance. Interior black spruce forests of Alaska burn relatively frequently for several reasons: climatic conditions, fire-prone lichen-covered trees, and flammable organic ground cover (Lutz 1956; Viereck 1983).

Permafrost is common in black spruce forests. One of the most important ecological impacts of fire in black spruce forests is the long-term effect that fire has on soil temperatures (Dyrness and others 1986; Swanson 1996; Viereck 1973; Viereck and Dyrness 1979). Fire leads to considerable soil warming and a deepening of the biologically active layer in the soil profile that persists for several years subsequent to fire disturbance. These changes, along with the accelerated processes of decomposition and mineralization associated with fire, lead to enhanced productive capacity in postfire plant communities. In general, early-seral native species in black spruce forests have high nutrient requirements and fast growth rates that allow them to dominate the early stages of succession (Dyrness and others 1986).

In contrast, spruce-birch forests are dense interior forests composed of white spruce, paper birch, quaking aspen (*Populus tremuloides*), and poplar (*P. balsamifera*). Fire regimes in this forest type are characterized by minor mixed-severity and major stand-replacing fires with fire frequencies between <35 and 200 years (Duchesne and Hawkes 2000, table 3-1, pg. 36). Permafrost is rare in this forest type (Foote 1983).

Role of Fire in Promoting Nonnative Plant Invasions in Boreal Forests

Black spruce—Ecological studies conducted in black spruce forests do not mention the presence of nonnative plant species in postfire plant communities (Cater and Chapin 2000; Dyrness and Norum 1983; Swanson 1996; Van Cleve and others 1987; Viereck and Dyrness 1979; Viereck and others 1979). Instead, severely burned areas are quickly colonized by black spruce and several other native species with wind-borne propagules, such as fireweed (*Epilobium angustifolium*), bluejoint reedgrass (*Calamagrostis canadensis*), willow (*Salix* spp.), fire-adapted bryophytes, fire moss (*Ceratodon purpureus*) and liverwort (*Marchantia polymorpha*). Areas that burn at a lower severity are rapidly reoccupied by sprouts originating from underground parts of surviving vegetation (Dyrness and Norum 1983).

There has been a recent increase in nonnative plant species along road systems in Interior Alaska (Burned Area Response National-Interagency Team 2004; Lapina and Carlson 2005). Interior Alaskan road systems traverse a landscape of boreal forest composed of both black spruce and spruce-birch forest. Since over 100 miles (160 km) of road corridor were burned in the 2004 fire season alone, there is concern that fire and the use of bulldozers to create firelines may promote the invasion and spread of roadside nonnative plant species in black spruce and spruce-birch forest (Burned Area Response National-Interagency Team 2004). Invasive species can be transported into fire areas when bulldozers and other suppression equipment are not cleaned of soil and plant material prior to being moved, or when equipment is driven through populations of invasive species located adjacent to fire areas. In the Burned Area Emergency Stabilization and Rehabilitation Plan for the 2004 Alaska fire season, a list of priority nonnative plant species that occur either within or adjacent to burned areas was provided (table 10-5) to assist with postfire monitoring, assessment, and control (Burned Area Response National-Interagency Team 2004). Several

Table 10-5—Nonnative plant species of Interior Alaska that occurred within or adjacent to areas burned during the 2004 fire season (Burned Area Response National-Interagency Team 2004).

Scientific name	Common name
Avena fatua	Wild oats
Bromus inermis	Smooth brome
Capsella bursa-pastoris	Shepherd's purse
Centaurea cyanus	Garden cornflower
Cirsium arvense	Canada thistle
Convolvulus arvensis	Field bindweed
Crepis tectorum	Narrowleaf hawksbeard
Descurainia sophia	Flixweed tansymustard
Galeopsis tetrahit	Brittlestem hempnettle
Hieracium caespitosum	Yellow hawkweed
Lappula squarrosa	Bristly sheepburr
Lepidium densiflorum	Common pepperweed
Leucanthemum vulgare	Oxeye daisy
Linaria vulgaris	Yellow toadflax
Matricaria discoidea	Pineapple weed
Melilotus alba	White sweetclover
Melilotus officinalis	Yellow sweetclover
Phalaris arundinacea	Reed canarygrass
Plantago major	Common plantain
Rorippa sylvestris	Creeping yellowcress
Sisymbrium altissimum	Tumble mustard
Sonchus arvensis	Perennial sowthistle
Tanacetum vulgare	Common tansy
Taraxacum officinale	Common dandelion
Trifolium hybridum	Alsike clover
Vicia cracca	Bird vetch

of these species are observed to increase after fire in lower latitudes. Species previously mentioned in this chapter include Canada thistle, yellow toadflax, white and yellow sweetclovers, and common dandelion. In addition, flixweed tansy mustard (*Descurainia sophia*) can form dense populations from soil seed banks in the early stages of secondary succession after fire (Howard 2003a, FEIS review).

Fireline creation and revegetation impact both soils and native plant communities. When bulldozed firelines are constructed, organic matter is removed, leaving mineral soil exposed. Deep thawing of exposed mineral soil in firelines has resulted in extensive soil erosion in black spruce forests (DeLeonardis 1971; Dyrness and others 1986; Viereck 1973; Viereck and Dyrness 1979). While native grasses and forbs establish rapidly in burned areas without soil disturbance, vegetation does not establish readily on previously frozen, severely disturbed soils (DeLeonardis 1971). In the past, firelines have been fertilized and seeded with nonnative grasses, such as 'Manchar' smooth brome (*Bromus inermis*), creeping red fescue (*Festuca rubra*), and 'Rodney' oats (*Avena fatua*) (Bolstad 1971; Viereck and Dyrness 1979). Experimental plantings of nonnative grasses and legumes in boreal forests resulted in rapid initial growth followed by decreased cover in following years (Johnson and Van Cleve 1976).

The use of nonnative species in boreal forest and tundra revegetation projects has been controversial (Johnson and Van Cleve 1976). So far, the intentional introduction of nonnative species for fireline revegetation has not led to their long-term establishment or spread into burned areas. However, since native species such as bluejoint grass and fireweed are widely adapted and competitive in early seral environments, it may be preferable to revegetate firelines with these species. The Burned Area Emergency Stabilization and Rehabilitation Plan for the 2004 Alaska fire season specifies that only certified weed free native seed mixes will be used during postfire revegetation projects in boreal forests and the use of straw mulch for soil stabilization will be discouraged (Burned Area Response National-Interagency Team 2004).

Fire effects studies conducted in the boreal forest region of Alaska have not reported nonnative plant species establishment in postfire communities. However, plant inventories conducted in the boreal forest region record the presence of nonnative plant species observed to increase after fire disturbance in other regions. For example, in an extensive study of nonnative plant species in Alaska National Parks (Densmore and others 2001), common dandelion and white sweetclover were observed in boreal environments in Denali and Wrangell-St. Elias National Parks.

White sweetclover has invaded roadsides and river bars in the boreal forest region of Alaska. A cold-hardy cultivar of white sweetclover was seeded on highway cutbanks outside of Denali National Park and has repeatedly established in the park. It is probably transported on vehicle tires from highway plantings (Densmore and others 2001). White sweetclover has also invaded early-successional river bars in interior and south-central Alaska. In particular, it has invaded extensive acreage along the Nenena and Matanuska rivers. There is concern that white sweetclover may invade native boreal forest communities; it has been observed spreading into open areas and forest clearings in other regions (Lapina and Carlson 2005).

In a survey of roadside vegetation in the Susitna, Matanuska, and Copper River drainages, several nonnative species were identified that respond favorably to fire disturbance in other regions. Among the most frequent species observed were common dandelion and timothy (73 percent and 53 percent, respectively). In addition, white sweetclover was observed in 29 percent of survey sites. White sweetclover, timothy, and bird vetch were among the five species noted for the worst infestations observed. All three species form large, dense populations and have been observed invading native plant communities within the boreal forest region (Lapina and Carlson 2005).

In addition, several species recently introduced to the boreal forest region of Alaska were noted in this survey, including two species that invade after fire in other regions. Canada thistle was located at only one site but was recommended for immediate control by the survey's authors. Two small populations of cheatgrass (*Bromus tectorum*), a notorious fire-adapted grass that invades fire-disturbed areas and has altered fire regimes in arid habitats of the western United States (Zouhar 2003a, FEIS review), were located in southern Wasilla and Houston. Though cheatgrass often responds favorably to fire in arid environments, how this species will respond to fire in the boreal forest region is unknown.

Spruce-birch—While wildfire is less common in spruce-birch forest than in black spruce forest (Foote 1983), timber harvesting in the Alaskan interior has concentrated on the spruce-birch forest community. In an observational study that compared sites burned by wildfire with logged sites in the Tanana and Yukon River drainages of central Alaska (Rees and Juday 2002), 17 plant species were found only in burned sites in the early stages of postfire succession. One of these species was nonnative narrowleaf hawksbeard (*Crepis tectorum*), a roadside plant that produces abundant wind-dispersed seed (Lapina and Carlson 2005). However, this study neither demonstrates a statistically significant association between burned sites and the presence of narrowleaf hawksbeard nor provides evidence that fire is contributing to the invasion of this species. No other nonnative species were observed in burned plots of this study.

Effects of Nonnative Plant Invasions on Fuels and Fire Regimes in Boreal Forests

The invasion of bird vetch in both the boreal forest and coastal hemlock-spruce regions of Alaska may have the potential to alter fire regimes. A noxious weed in Alaska, bird vetch has rapidly invaded Alaska's right-of-ways since its initial establishment in the Matanuska Valley and Fairbanks area more than 20 years ago (Klebesadel 1980). It is most common in these areas but has also established in the Anchorage area, and there is concern that it could spread to adjacent forestland (Duffy 2003). It has also been observed near Denali National Park and in Seward, Kenai Peninsula (Nolen 2002). Bird vetch, a nitrogen-fixing perennial forage crop, thrives in areas of soil disturbance and is now abundant along roadsides, railroads, field edges, and abandoned fields where it climbs over bushes and small trees, such as alder and willow, and up fences to a height of 4 to 6 feet (1.2 to 1.8 meters). Bird vetch produces abundant seed that may be carried in tangled vegetation by maintenance and suppression equipment (Alaska Natural Heritage Program 2004). Observers note that fences overgrown with bird vetch alter winter wind flow, causing snowdrift accumulations (Klebesadel 1980). Though unsupported by citations from primary literature, a literature review published by the Alaska Natural Heritage Program states that bird vetch is fire tolerant (Alaska Natural Heritage Program 2004). The density and continuity of bird vetch in high use areas, coupled with its climbing habit, suggest that it might carry fire both along the ground and into shrub and tree canopies. There is also concern that forest fires could allow the movement of bird vetch into new areas (Burned Area Response National-Interagency Team 2004) where it may suppress the growth of native species (Nolen 2002).

Use of Fire to Manage Invasive Plants in Boreal Forests

Thus far, prescribed fire has only been used in control trials to manage invasive nonnative plants in the boreal forest region of Alaska (Conn, personal communication, 2005).

Tundra

Lightning-ignited fires are common in tundra vegetation, but they occur irregularly. Even though there is usually little standing fuel and organic soils tend to be moist year around, cottongrass (*Eriophorum vaginatum*) tussock communities are fire-prone (Racine 1979). Tundra communities experience major stand-replacement fires occurring at frequencies of about 35 to 200 years (Duchesne and Hawkes 2000, table 3-1, pg. 36). Tundra fires tend to be low-severity, with no vascular plant species completely eliminated by fire (Wein 1971).

Role of Fire in Promoting Nonnative Plant Invasions in Tundra

Though recovery of total primary production is usually quite rapid after fire in tundra communities (Wein and Bliss 1973), changes in relative species abundance can be long lasting (Fetcher and others 1984). Nonnative species have not been reported in postfire tundra communities (Landhausser and Wein 1993; Racine 1979, 1981; Racine and others 1987; Vavrek and others 1999; Wein and Bliss 1973).

Similar to black spruce forests, an ecologically important impact of fire in tundra communities is increased soil thaw depth, a condition that can last for more than 23 years after fire (Vavrek and others 1999). Extreme thawing results in exposure of mineral soil, which is then available to establishing species, especially those with wind-dispersed propagules. Firelines are of concern in tundra ecosystems, as they increase soil thaw depths. One year after firelines were constructed to help suppress lightning-caused fires in tundra communities on the Seward Peninsula, firelines had thaw depths of more than 20 inches (50 cm) on the burned side and more than 14 inches (35 cm) on the unburned side (Racine 1981).

The revegetation of disturbed tundra communities after fire and fireline creation has usually been accomplished by seeding northern varieties of commercially available nonnative grasses. While these nonnative grasses may reduce erosion, they do not establish permanent cover in tundra ecosystems without frequent fertilization (Chapin and Chapin 1980). In study plots that examined the effects of bulldozing in the Eagle Creek area of interior Alaska, nonnative grasses commonly used in tundra restoration (table 10-6) were seeded with and without fertilizer. Three years after sowing, the grasses did not interfere with the establishment of native sedges. After 10 years, plot vegetation was composed entirely of native species (Chapin and Chapin 1980).

While these results indicate that native graminoids are superior competitors in tundra ecosystems, environmental changes anticipated with global warming (Chapin and others 1995) may disrupt this advantage. Therefore, it is possible that "...the use of exotic species which have been selected for their performance under arctic conditions maximizes the possibility that an introduced grass or weed will establish in the community..." (Chapin and Chapin 1980, p. 454). Revegetation of disturbed sites, such as firelines, may be accomplished effectively with native sedges and forbs while simultaneously avoiding the introduction of potentially invasive nonnative species into the tundra ecosystem.

Table 10-6—Nonnative grasses used in tundra restoration and examined by Chapin and Chapin (1980).

Scientific name	Common name
Alopecurus pratensis	Meadow foxtail
Festuca rubra	Red fescue
Lolium perenne ssp. *perenne*	Perennial ryegrass
Phalaris arundinacea	Reed canarygrass
Phleum pratense	Timothy
Poa pratensis	Kentucky bluegrass

Effects of Nonnative Plant Invasions on Fuels and Fire Regimes in Tundra

While there is no indication that nonnative species are changing tundra fire regimes, fire regime and climatic changes initiated by global climate warming could influence the susceptibility of tundra plant communities to invasion by nonnative species. The effects of global warming are expected to be particularly pronounced in northern latitudes (Chapin and others 1995). In a spatially explicit model of vegetation response to warming climate on Seward Peninsula (Rupp and others 2000), a 3.6 °F (2 °C) temperature increase was associated with increased flammability of tundra vegetation, increased fire frequency, fires of greater spatial extent, and gradual expansion of spruce-birch forest into previously treeless tundra communities and/or conversion of tundra vegetation to grassland steppe. In an 11-year manipulated experiment conducted in a moist tundra community near Toolik Lake, located in the northern foothills of the Brooks Range of Alaska, environmental manipulations simulating global warming (increased air and soil temperatures, decreased light availability, increased nutrient mineralization and decomposition) resulted in decreases in species richness and shifts in community composition, including increased dominance of birch (*Betula* spp.) and decreased dominance of evergreen shrubs and understory forbs (Chapin and others 1995). Since tundra communities have few species to begin with, species loss coupled with increasing temperatures, soil fertility, and fire frequency may have dramatic ecosystem consequences, including increased vulnerability to invasion by nonnative species. It would be prudent under such uncertain climatic conditions to take steps to prevent the introduction of nonnative plant species to this region.

Use of Fire to Manage Invasive Plants in Tundra

Nonnative species are not well established in the tundra region; therefore, there is no need for control efforts such as prescribed fire at this time.

Summary of Fire-Invasive Plant Relationships in Alaska

In Alaska today, nonnative plants are largely restricted to areas heavily impacted by human activities. With the construction of roads, trails, and firelines in pristine native plant communities, the threat of nonnative plant species establishment and spread increases. Ecological barriers to nonnative species establishment may weaken with future climatic changes and should not be relied upon to slow the invasion of nonnative plants into areas disturbed by fire or human activities. To improve the knowledge base about which species are likely to invade after fires in Alaska, fires that intersect anthropogenic disturbances such as roads and firelines should be monitored for several years after burning. In order to ensure that nonnative species do not establish in Alaskan plant communities, development and fire suppression activities must be conducted in a careful manner that precludes the inadvertent introduction of nonnative plant propagules. Because of the frequency and size of fires in Alaska, existing infestations of fire-adapted species should be controlled to decrease the dispersal opportunities along fire perimeters. Revegetation with competitive, early-seral native species after fire and soil disturbance is also important, both for reducing soil thaw and erosion and for the prevention of nonnative species establishment.

Conclusions

Throughout much of the Northwest Coastal bioregion, fires are more closely tied to varying weather conditions than to fuel conditions (Agee 1997). Therefore, changes to local climate are likely to modify regional fire regimes. Paleoecological data, as reviewed by McKenzie and others (2004), indicate that periods of warmer temperatures and decreased precipitation have been associated with increased fire frequency and decreased fire severity in this region. Though future climatic conditions are difficult, if not impossible, to predict, global circulation models indicate that fire seasons (measured by degree-days, temperature, and drought indices) may lengthen throughout the Pacific Northwest over the next century. Fires may burn earlier and later in the year than they do now and total area burned may increase. In addition, carbon dioxide fertilization may contribute to fuel production, increasing potential fire severity. Warmer temperatures may reduce winter snow pack in the mountains, leading to increased moisture stress, insect and disease outbreaks, and fuel loading in montane forests, which could result in more frequent and severe fires in Northwest coastal forested communities (as reviewed by McKenzie and others 2004).

Given the uncertainties regarding future climatic conditions and fire regimes, fire management techniques should be developed that avoid transporting or facilitating the movement of nonnative plant propagules between different environments. Nonnative plants that are currently not invasive in particular local plant communities may become so in the future. Regionally organized programs of native seed collection, propagation, and storage for postfire restoration projects will help discourage the seeding of nonnative plants.

Montane communities of Washington and Oregon and coastal hemlock-spruce forests, boreal forests, and tundra communities of Alaska have relatively few established populations of nonnative plants, providing land managers with an opportunity to prevent the spread of these species into intact natural communities. Of these communities, boreal forests are the most vulnerable to the spread of fire-adapted nonnative species due to the frequency and scale of fire in this region. The best way to prevent future expansion of these species is through early detection and rapid control response. Whether nonnative species are absent from these ecosystems due to ecological barriers such as climate or to a lack of a local seed source is unclear, though the presence of increasing numbers of nonnative species in Alaska suggests the latter. If environmental conditions are indeed preventing nonnative plant invasions into high elevation and high latitude environments, global climate change could remove existing ecological barriers to species establishment. Therefore, preventing the introduction of nonnative plant propagules into these communities is also critically important. Human activities, such as fire suppression, that inadvertently or intentionally introduce nonnative plants into these communities may cause irreparable harm.

Few studies have examined the effects of natural fire on upland and riparian forests of the coastal Douglas-fir region of Washington and Oregon. Examination of short- and long-term changes in plant community composition and structure that follow natural fires should be a research priority, particularly with regard to invasion and establishment of nonnative plant species. In remote wildlands, the ecological impacts of natural fire need to be weighed against those of fire exclusion, particularly fire management activities that promote the invasion of nonnative species, such as fireline construction and postfire seeding.

Fire suppression activities may encourage the invasion of riparian communities by nonnative species. The building of firelines can increase runoff and soil erosion, facilitating the invasion of riparian communities through the delivery of soil and seed from upland communities. Firelines built through riparian forests and down the fall lines of steep slopes are especially damaging (Beschta and others 2004).

In the dense coastal Douglas-fir forests of the Northwest Coastal bioregion, the ecological impact of nonnative plant populations is currently restricted to the earliest stages of forest succession that follow logging and slash fires. Ruderal nonnative forbs, such as Canada thistle, are displacing native early seral vegetation in some locations and reducing tree regeneration in others. Though nonnative plants are typically eliminated from the plant community after a few years of forest stand development, nonnative shade-tolerant plant species are capable of persisting and/or invading forest understories if relatively open stand conditions are maintained through clearcutting, silvicultural thinning, or prescribed underburning. In particular, false brome poses a serious threat to forest understory communities and may affect fire behavior and spread. Research is needed to determine how invasive this species is in shaded forest understories, how it responds to natural and prescribed fire, and how its foliage and population characteristics might influence fuel characteristics and fire behavior.

In contrast, Oregon oak woodland communities of western Washington and Oregon are already extensively invaded by nonnative plant species. However, the populations of individual plant species, both native and nonnative, are distributed heterogeneously across the landscape. Due to the diversity in plant community composition, responses to prescribed fire are highly variable and site-specific. Disturbance regimes that effectively achieve management goals will need to be developed from long-term localized research and observation. In most situations, fire used alone will not be as effective as fire used in conjunction with other management techniques such as the seeding of native plants.

Notes

Anne Marie LaRosa
J. Timothy Tunison
Alison Ainsworth
J. Boone Kauffman
R. Flint Hughes

Chapter 11:

Fire and Nonnative Invasive Plants in the Hawaiian Islands Bioregion

Introduction

The Hawaiian Islands are national and global treasures of biological diversity. As the most isolated archipelago on earth, 90 percent of Hawai`i's 10,000 native species are endemic (Gagne and Cuddihy 1999). The broad range of elevation and climate found in the Hawaiian Islands supports a range of ecosystems encompassing deserts, rain forests and alpine communities often within the span of less than 30 miles. Recent analyses suggest that species diversity may not differ between island and continental ecosystems once habitat area is taken into account (Lonsdale 1999); however there are a disproportionate number of threats to Hawai`i's ecosystems.

Invasion of nonnative species is a leading cause of loss of biodiversity and species extinctions in Hawai`i (Loope 1998, 2004). On average, islands have about twice as many nonnative plant species as comparable mainland habitats (Sax and others 2002) and the number of plant invasions in Hawai`i is great. Over 8,000 plant species and cultivars have been introduced to Hawai`i and over 1,000 of these are reproducing on their own. Many watersheds in Hawai`i are now dominated by nonnative invasive species (Gagne and Cuddihy 1999).

Available evidence suggests that fire was infrequent in Hawaiian lowlands before human settlement, increasing in frequency with Polynesian and European colonization, the introduction and spread of invasive species, and most recently, the cessation of grazing by feral and domestic ungulates. Currently, most fires in the Hawaiian Islands occur in lowlands, communities that have been modified the most by fire.

Our understanding of fire ecology in Hawaiian ecosystems is limited by the relative scarcity of published information regarding fire and its effects on the native biota (but see reviews by Mueller-Dombois 1981; Smith and Tunison 1992; Tunison and others 2001). While anecdotal information is available, detailed studies are few and very little is known about the long term consequences of fire for Hawaiian ecosystems. But available evidence indicates that the altered fuel characteristics of many communities resulting from invasion by introduced grasses in Hawai`i and elsewhere in the tropics, coupled with an increase in ignitions, often results in frequent fires with severe consequences to the native biota

(D'Antonio and Vitousek 1992; Mueller-Dombois 2001; Mueller-Dombois and Goldhammer 1990).

In the following sections we review the interactions of fire and invasive species within grasslands, shrublands, woodlands, and forests of lowland and upland environments in Hawai`i (fig. 11-1; table 11-1). Where possible, distinctions are made between dry and mesic communities. Wet forests are discussed separately. For each environment we describe the affected vegetative communities and discuss how native fire regimes and native ecosystems have been affected by nonnative species invasions. Where information is available, we also discuss potential opportunities for the use of prescribed fire to manage invasive plants and restore native plant communities. We conclude with a summary of the current state of fire and invasive species in Hawai`i, the outlook for the future, and needs for further investigation.

Description of Dry and Mesic Grassland, Shrubland, Woodland, and Forest

Lowlands

Low-elevation forests, woodlands, shrublands, and grasslands generally occur above the salt spray zone and below 4,000 feet (1,200 m) in elevation and on a variety of soils ranging from relatively undeveloped, shallow soils to deeply weathered lava flows. The environment is dry to mesic, depending on soil type and precipitation, with annual rainfall from 4 to 50 inches (100 to 1,270 mm). Summer months are typically drier, but rainfall is variable and episodic both within and among years. Rainfall is less variable for mesic communities (Gagne and Cuddihy 1999).

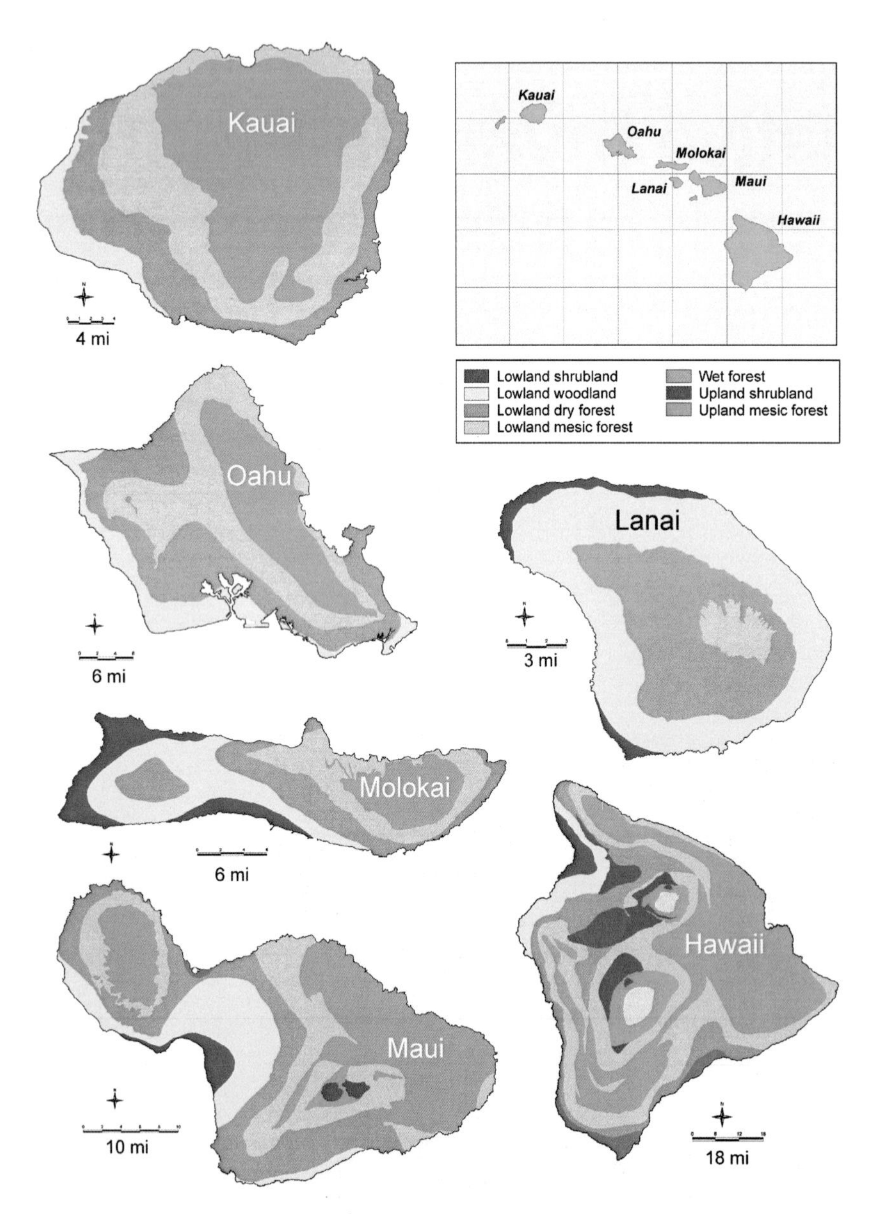

Figure 11-1—General distribution of shrublands, woodlands, and forests of lowland and upland environments in Hawai`i. No grasslands are shown on this map, although limited areas of native grasslands and extensive areas of nonnative grasslands are found in Hawai`i today. Submontane `ōhi`a woodlands are not differentiated and are lumped with mesic forest types. Map units shown here are based on the Bioclimatic Life Zone Maps (LZM) of Potential Natural Vegetation in Hawai`i (Tosi and others 2002), with modifications by Thomas Cole, Institute of Pacific Islands Forestry, USDA Forest Service, Hilo, Hawai`i.

Table 11-1—Principal nonnative species influenced by fire or influencing fire behavior, and their estimated threat potential in major plant communities in Hawai`i. L= low threat; H = high threat; PH = potentially high threat. Empty boxes reflect either a lack of information or no perceived threat.

		Community types								
		Lowland dry & mesic				Wet	Upland dry & mesic			
							Montane		Subalpine	
		Grassland	Shrubland	Forest	Woodland[a]	Forest	Shrubland	Forest	Shrubland	Forest
		Elevation range (feet)								
Scientific name[b]	**Common name**	- - - - - - - - up to 1,500 - - - - - - -			1,000-4,000	300-6,600	- - - 3,800-9,500 - - -		- - - 6,500-9,500 - - -	
Acacia farnesiana	Klu			PH						
Andropogon virginicus	Broomsedge	PH	H	H	H	PH	L	L		
Anthoxanthum odoratum	Sweet vernal grass							L	PH	PH
Ehrharta stipoides	Meadow ricegrass						PH	PH		
Festuca rubra	Red fescue								PH	
Holcus lanatus	Velvetgrass								H	H
Hyparrhenia rufa	Thatching grass	PH								
Leucaena leucocephala	Koa haole			PH	PH					
Melinis minutiflora	Molasses grass		PH		H					
Melinis repens	Natal redtop	H	H							
Morella faya	Firetree		PH	PH	PH	PH				
Nephrolepis multiflora	Asian sword fern				H	H				
Paspalum conjugatum	Hilo grass					H				
Pennisetum ciliare	Buffelgrass	PH	H	PH	PH					
Pennisetum clandestinum	Kikuyu grass				L			L		
Pennisetum setaceum	Fountain grass	H	H	H	H		PH	PH		PH
Prosopis pallida	Kiawe			PH	PH					
Schizachyrium condensatum	Bush beardgrass	PH	H	H	H		L	L		

[a] Includes submontane `ohi`a woodland and kiawe/koa haole woodlands.
[b] Scientific nomenclature follows Wagner and others (1999) and Palmer (2003).

Lowland dry and mesic forests consist of an array of communities with diverse canopy tree species, but many lowland forested areas are now dominated by introduced tree species, and remnants of native lowland forest communities remain primarily where human disturbances have been limited. Most native tree canopies are relatively open when compared to other tropical areas. Dominant native tree species across this broad spectrum of habitats include `ōhi`a (*Metrosideros polymorpha)*, wiliwili (*Erythrina sandwicensis*), lama (*Diospyros sandwicensis*), olopua (*Nestegis sandwicensis*), sandalwood (*Santalum freycinetianum*), koa (*Acacia koa*), and āulu (*Sapindus oahuensis*). Associated understory shrub species include many of the same species found in native shrublands described below (Gagne and Cuddihy 1999). `Ōhi`a forests are widespread and comprise 80 percent of all Hawai`i's native forests. `Ōhi`a occurs in nearly monotypic stands with closed to open canopies or as scattered trees (fig. 11-2), creating a gradient of communities that includes closed forests, woodlands, and savannas (D'Antonio and others 2000; Hughes and others 1991; Tunison and others 1995; Tunison and others 2001). Distinctions among types are often unclear, and most dry and mesic `ōhi`a forests have relatively open canopies; therefore, `ōhi`a community types will be referred to as "`ōhi`a woodlands." The composition of nonnative lowland forests is variable and dominated by such species as kiawe (*Prosopis pallida*), common guava (*Psidium guajava*), christmas berry (*Schinus terebinthifolius*) ironwood (*Casuarina equisetifolia*) and silk oak (*Grevillea robusta*) (Gagne and Cuddihy 1999). Kiawe and koa haole (*Leucaena leucocephala*) can form stands ranging from dense thickets to open woodlands, with an understory of introduced, drought resistant grasses.

Dry and mesic shrublands have been altered by grazing and fire and many have been replaced by nonnative communities. Today, native dry shrub communities in Hawai`i occur on a variety of soil types in leeward areas of most Hawaiian Islands. Shrub canopies are relatively open and consist of two or more dominant, widespread native species, including `a`ali`i (*Dodonaea viscosa*), `akia (*Wikstroemia* spp.), pūkiawe (*Styphelia tameiameiae*), and `ulei (*Osteomeles anthyllidifolia*). These shrub species often co-occur with an understory of nonnative grasses (Cuddihy and Stone 1990; Gagne and Cuddihy 1999) (fig. 11-3). The nonnative koa haole (*Leucaena leucocephala*) forms dense shrublands in coastal areas throughout the islands. Today native mesic shrublands occur in marginal habitats, such as on thin soils and exposed aspects, and consist of open- to closed-canopy communities with few understory herbs and shrubs up to 10 feet (3 m) tall. Dominants include: ohi'a pūkiawe, `a`ali`i, `ulei, and iliau (*Wilkesia* spp.) (Gagne and Cuddihy 1999).

Most dry and mesic lowland grasslands are floristically simple and likely anthropogenic. Gagne and Cuddihy (1999) recognize five dry, low-elevation grassland communities and three mesic lowland grasslands. Only two of these communities are dominated by native grasses: dry pili grasslands (*Heteropogon contortus*) and mesic kawelu grasslands (*Eragrostis variabilis*). Extensive pili grasslands are thought to have been maintained by burning in the lowlands prior to the 1700s (Daehler and Carino 1998; Kirch 1982) but they are now limited to small, scattered remnants (Gagne and Cuddihy 1999). The remaining grassland types are a result of past disturbances, including grazing, browsing, and fire, and are now dominated by one or two of

Figure 11-2—`Ōhi`a woodland with native shrub and uluhe (*Dicranopteris linearis*) understory invaded by nonnative grasses in Hawai`i Volcanoes National Park. (Photo by A. LaRosa.)

Figure 11-3—Low elevation dry shrublands dominated by `a`ali`i and nonnative grasses in Hawai`i Volcanoes National Park. (Photo by B. Kauffman.)

the many nonnative invasive grasses (fig. 11-4) that occur in these communities and fuel most of the fires occurring in Hawai`i today (table 11-2).

Montane and Subalpine

Montane and subalpine communities occur on the leeward slopes of Mauna Loa, Mauna Kea and Hualalai on Hawai`i Island and Haleakala, on Maui, between approximately 3,800 and 9,500 feet (1,500 to 3,000 m), as well as at higher elevations above the tradewind inversion layer (Gagne and Cuddihy 1999). Rainfall varies from 12 to 80 inches (300 to 1,970 mm) per year and is seasonal, coming largely during winter cyclonic storms. Summers are typically dry. Soils throughout the zone are either cinder, loams derived from volcanic ash, or thin soils over pahoehoe lava (Gagne and Cuddihy 1999).

Open- to closed-canopy shrubland communities of montane and subalpine areas are dominated by drought-tolerant native shrubs, including pūkiawe, `ōhelo (*Vaccinium calycinum*), and 'a`ali`i. The `ama`u fern (*Sadleria cyatheoides*) is also abundant at higher elevations on Maui. Shrubs often occur in a matrix of native grasses such as alpine hairgrass (*Deschampsia nubigena*) or hardstem lovegrass (*Eragrostis atropioides*). Invasive nonnative grasses, such as velvetgrass (*Holcus lanatus*), sweet vernal grass (*Anthoxanthum odoratum*), and dallis grass (*Paspalum dilatatum*) are locally abundant. Montane and subalpine shrublands on Mauna Loa are dominated by the native shrubs `a`ali`i or naio (*Myoporum sandwicense*) with lesser amounts of ko`oko`olau (*Bidens menziesii*), `aheahea (*Chenopodium oahuense*) and `akoko (*Chamaesyce multiformis*). Grasses include hardstem lovegrass and nonnative fountain grass (*Pennisetum setaceum*) (Gagne and Cuddihy 1999).

Figure 11-4—Dense fountain grass (*Pennisetum setaceum*) on the leeward slopes of Hualalai, Hawai`i. (Photo by C. Litton.)

Table 11-2—Nonnative species that are primary carriers of fire in Hawaiian ecosystems. All are drought tolerant, perennial, C4 grasses[a].

Common name	Origin	Habit	Date[b] of introduction	Elevational range (feet)
Broomsedge	Eastern North America	Bunchgrass	1924(c)	150 to 4,000
Molasses grass	Africa	Mat forming	1914(c)	400 to 4,000
Buffelgrass	Africa & Tropical Asia	Mat forming	1932(c)	0 to 400
Fountain grass	North Africa	Bunchgrass	1914	0 to 7,000
Bush beardgrass	Tropical/subtropical America	Bunchgrass	1932	650 to 4,500

[a] As reported in the literature.
[b] (c) = Date first collected, otherwise date first reported.

Many areas of montane open and closed forests on Hawai`i contain nearly monotypic stands of koa with a continuous understory of invasive nonnative pasture grasses including meadow ricegrass (*Ehrharta stipoides*) (fig. 11-5), dallis grass, and Kikuyu grass (*Pennisetum clandestinum*). Native brackenfern (*Pteridium acquilium)* is found in the understory of some forests. Subalpine forests are open-canopied, with either `ōhi`a or mamane (*Sophora chrysophylla*) as the dominant species. A mamane-naio closed forest subtype is also present in some areas (Gagne and Cuddihy 1999). Many upper-elevation koa forests were logged beginning in the mid 19th century and then grazed by domestic and wild cattle. Other forests were subject to uncontrolled grazing and browsing by feral animals (cattle, goats, sheep, and pigs) during this time period (Cuddihy and Stone 1990).

Figure 11-5—Montane koa forest on Mauna Loa invaded by meadow ricegrass. (Photo by P. Scowcroft.)

Fire History and the Role of Nonnative Invasives on Fire Regimes in Dry and Mesic Ecosystems

Fires of volcanic origin occurred in Hawaiian forests prior to human habitation and continue today, although volcanism is intermittent and highly localized (fig. 11-6). While the lack of annual growth rings in tropical trees precludes use of dendrochronology for detailed analysis of early fire history, some data from sediment cores and evidence from charcoals in soils exist. Palynological studies from a high elevation bog on Maui indicates that fires occurred infrequently in forested areas in both dry and wet periods, during the last 9,000 years, and were probably volcanic in origin (Burney and others 1995). Soils from mesic and wet forested areas on Mauna Loa show charcoal from 2,080, 1,040 and 340 years ago, consistent with a volcanic origin of most fires and a long fire-return interval (Mueller-Dombois 1981). Charcoal has also been found, but not dated, from montane forest soils on Mauna Kea (Wakida 1997), indicating the occurrence of fire but not the nature of the fire regime in that area. Some prehistoric fires may also have started from occasional lightning strikes (Tunison and Leialoha 1988; Vogl 1969).

Several authors suggest that presettlement fires in Hawai`i were so infrequent (in other words, long fire-return intervals) that they had little effect on the evolution of the Hawaiian flora (Mueller-Dombois 1981, 2001; Smith and Tunison 1992) and most native plant species do not exhibit specific fire adaptations (for example, thick bark or serotiny). However, Mueller-Dombois (1981) also suggests that fire was an important ecological and evolutionary factor in upland communities because of prevailing dry conditions in leeward and high-elevation environments, the abundance and continuity of fine fuels (native grasses), and the ability of the common native species in these communities, such as koa, mamane, and bracken fern, to regenerate after fire.

Prehistoric Polynesians dramatically altered vegetation in Hawai`i's lowlands when they burned to clear land for agriculture. Reviewing archeological and palynological studies, Kirch (1982) and Cuddihy and Stone (1990) outline the general impacts of early Polynesians, including those from burning, on Hawaiian ecosystems. Slash and burn agriculture is used commonly in the

Figure 11-6—Lava ignited fire in Hawai`i Volcanoes National Park, 2002. (Photo courtesy of Hawai`i Volcanoes National Park.)

tropics (Bartlett 1956), and Polynesians burned to clear vegetation since they first occupied Hawai`i around 400 AD (Cuddihy and Stone 1990; Kirch 1982). During the early years of settlement, population levels were low and fire was probably used occasionally to clear shrublands and forests for cultivation. A population spike between 1100 and 1650 AD led to large-scale expansion of agriculture throughout much of Hawai`i. Kirch (1982) presents observational and physical evidence from several sources that suggests widespread burning of lowland areas by early Polynesians to clear land for shifting agriculture, maintain grasslands to provide pili grass thatch for shelter and, according to McEldowney (1979), promote the growth of ferns and other plants used for famine food and pig fodder. Montane and subalpine areas were not cultivated by early Polynesians and early anthropogenic fire was improbable there (Kirch 1982).

Fire size and frequency increased in the lowlands following European contact (beginning ~200 YBP). Sediment cores from a high-elevation bog on Maui suggest an appreciable increase in fire frequency during the last 200 years (Burney and others 1995). The introduction and spread of ungulates and nonnative pasture grasses accompanied European settlement and transformed some Hawaiian native forest and shrubland ecosystems into mixed communities with a significant herbaceous component (Cuddihy and Stone 1990; Gagne and Cuddihy 1999), which increased fuel loading and continuity (D'Antonio and Vitousek 1992; Hughes and Vitousek 1993). Pollen records from the last 200 years contained a high proportion of grasses and charcoal derived from grasses. However, the charcoal may have come from burning sugar cane (a common practice) rather than wildfires (Burney and others 1995).

Changes in fuels, in conjunction with increases in human-caused ignitions, contributed to the sizeable increase in fire observed in the 20th century in many ecosystems. Throughout the state, the average acreage burned increased five-fold and the average number of fires increased six-fold from the early (1904 to 1939) to the mid (1940 to 1976) part of the 20th century (table 11-3) (Cuddihy and Stone 1990). A similar, but more pronounced, trend occurred in the lowlands of Hawai`i Volcanoes National Park (HVNP); fires were three times more frequent and 60 times larger, on average, from the late 1960s to 1995 when compared to the period 1934 to the late 1960s (Tunison and others 2001). The increase in fire frequency and size in HVNP coincides with a period of increased volcanic activity from 1969 to the present from Mauna Ulu and Pu`u O`o flows, the establishment and spread of nonnative invasive grasses, and the removal of nonnative goats from HVNP in the late 1960s (Hughes and Vitousek 1993; Tunison and others 2001). Over 90 percent of all fires in the lowlands of HVNP occurred after taller, fire-adapted, nonnative perennial grasses replaced nonnative short-stature annual and perennial grasses following goat removal (Tunison and others 1994; Williams 1990).

The existence of a grass/fire cycle fueled by nonnative invasive species has been well established in some of Hawai`i's `ōhi`a woodlands (fig. 11-7). There was a dramatic increase in fire size and frequency in `ōhi`a woodlands within HVNP when compared to earlier in the century: annual fire frequency increased more than 3-fold and annual fire size more than 100-fold (table 11-3) (Tunison and others 1995). Prior to the invasion of nonnative grasses, `ōhi`a woodlands consisted of open stands of shrubs and `ōhi`a with few native grasses (Mueller-Dombois 1976) and the discontinuous surface fine fuels would rarely have carried fire. Now, nonnative broomsedge (*Andropogon virginicus*), bush beardgrass (*Schizachryium condensatum*) and molasses grass (*Melinis minutiflora*) constitute over 30

Table 11-3—Increases in the size and number of fires in Hawai`i from the early to latter part of the 20th century.

Location	Time period	Average area burned[a] (acres)	Average number of fires/year[a]
State of Hawai`i[b]	1904 to 1939	1,044	4
	1940 to 1976	5,740	24
Hawai`i Volcanoes National Park, `ōhi`a woodlands[c]	1920 to ~1970	< 2.5	11
	~1970 to 1995	430	39

[a] Data are from historic fire records available for the period 1904 to 1995.
[b] Data from Cuddihy and Stone (1990).
[c] Data from Tunison and others (1995).

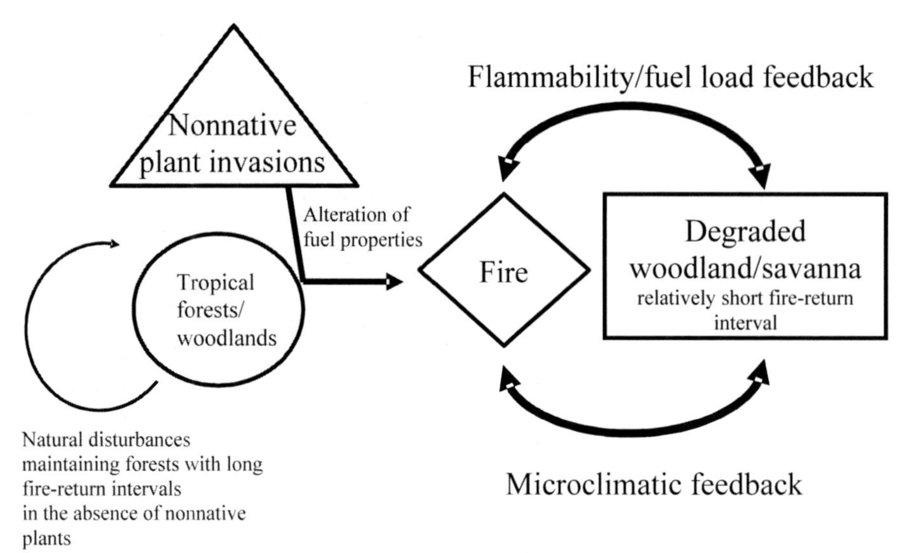

Hypothesized alterations of fuel properties due to nonnative invasions (grasses and ferns)

- Increased fine fuel loads
- Increased surface:volume ratio of surface fuels
- Decreases in fuel moisture content

Figure 11-7—The potential effects of grass and fern invasions into Hawaiian forest and woodland ecosystems. (Modified from D'Antonio and Vitousek 1992.)

percent of the understory biomass and 60 to 80 percent of the understory cover in HVNP's `ōhi`a woodlands (D'Antonio and others 1998), forming a continuous matrix of fine fuel between native shrubs (fig. 11-8) (D`Antonio and others 2000; Hughes and others 1991; Tunison and others 1995, 2001). Similarly, fountain grass initially invaded lama forests without the aid of fire, and it now forms a nearly continuous layer of surface fuels in many areas of former lama forest (Cordell and others 2004). Introduced Asian sword fern (*Nephrolepis multiflora*) also invades mesic `ōhi`a woodlands and forms a continuous understory of fine fuels (fig. 11-9) that carry fire (Ainsworth and others 2005).

These nonnative species possess characteristics that facilitate fire spread, including a high standing biomass and a high dead-to-live biomass ratio throughout most of the year. All recover rapidly after fire with increased vigor, by resprouting or seedling recruitment (Ainsworth and others 2005; D'Antonio and others 2000). Burned sites are predisposed to more severe and repeated fires due to increased fuel loads and higher wind speeds in the more open postfire savannas compared to unburned areas (D'Antonio and Vitousek 1992; Freifelder and others 1998; Tunison and others 1995). This invasive plant-fire cycle can be persistent (Hughes and others 1991; Tunison and

Figure 11-8—Principal nonnative grass fuels in `ōhi`a woodland: broomsedge (left), bush beardgrass (middle), and molasses grass (right). (Photos by A. LaRosa.)

Figure 11-9—Submontane `ōhi`a woodland with understory of nonnative Asian sword fern. (Photo courtesy of Hawai`i Volcanoes National Park.)

others 1995) and may represent a long lasting shift in plant community composition (fig. 11-7).

We have no information on the impacts of lowland nonnative woody species such as kiawe, koa haole, and klu (*Acacia farnesiana*) on fuel characteristics and fire regimes of other invaded lowland forest communities in Hawai`i. These species are legumes capable of symbiotic nitrogen-fixation. Kiawe stands ranging in height from 6 to 30 feet (2 to 10 m) are now prevalent in many dry lowland areas, forming grass and tree mosaics ranging from dense forest to open woodlands and savannas. Koa haole and klu often occur within or adjacent to kiawe stands. Increases in fire frequency, size, and severity are possible where invasions of these species result in increased aboveground biomass.

Historic fire records are available but inadequate to characterize changes in the fire regimes of montane and subalpine habitats, where historic disturbances have been fewer and native grasses are significant components of some plant communities (Loope and others 1990; Wakida 1997). Most documented high-elevation fires were small and of probable or known human origin; lightning fires occurred rarely (Loope and others 1990; Wakida 1997). Over 40 percent of fires in and around the Mauna Kea Forest Reserve (MKFR) on Hawai`i, were less than 10 acres (25 ha) in size and over 80 percent were less than 200 acres (80 ha) (Wakida 1997). On Haleakala most fires were less than 1 acre (0.4 ha) in size but several large fires, estimated at greater than 1,000 acres (400 ha), started in "pasture lands" (probably nonnative grasses) and moved into subalpine shrublands (Loope and others 1990). Although fires are reported from nearly every month of the year, most fires in high-elevation habitats occured during the dry summer months between May and August (52 percent for MKFR) (Wakida 1997), or during periods of drought (Loope and others 1990). Fuels in montane forests of MKFR are now composed of light to heavy accumulations of dead and downed wood and nonnative grasses. Within Haleakala National Park, Maui, nonnative grasses comprise a significant part of the biomass and standing fuel accumulation (Loope and others 1990). Although fire frequency might have increased in some upland areas with the removal of feral ungulates and the accompanying increase in nonnative grass cover, insufficient data are available to support this assumption.

Fire and Its Effects on Nonnative Invasives in Dry and Mesic Ecosystems

Most fires in Hawai`i's grasslands, shrublands and forests are caused by one or more species of introduced grasses (table 11-2). The dominant nonnative grasses recover rapidly after fire, often increasing in abundance, although dominance may shift from one nonnative species to another (Daehler and Carino 1998; Daehler and Goergen 2005; Tunison and others 1994, 2001). Detailed fire effects studies are few and limited to the last 20 years. Most have been conducted in HVNP on the island of Hawai`i in areas where fire is now frequent, detailed fire records have been kept, and nonnative invasive grasses are present (D'Antonio and others 2000, 2001a; Freifelder and others 1998; Hughes and others 1991; Hughes and Vitousek 1993: Ley and D'Antonio 1998; Mack and others 2001; Tunison and others 2001). Recent studies in low-elevation grasslands represent a wider geographic area, including several islands (Daehler and Carino 1998; Daehler and Goergen 2005; Goergen and Daehler 2001, 2002; Nonner 2006).

Fire potential is highest in certain dry and mesic ecosystems where substrate age, precipitation and plant productivity do not limit fuel continuity and biomass, and where dry periods create conditions of low fuel moisture suitable for combustion and fire spread (Asner and others 2005). On the Island of Hawai`i, Asner and others (2005) found that these areas generally receive between 30 and 60 inches (750 to 1,500 mm) of rainfall annually and often are dominated by nonnative grasses.

A majority of Hawai`i's grasslands are nonnative and often maintained by fire. Tunison and others (1994) studied the impacts of fire in nonnative grasslands in the central coastal lowland portion of HVNP. The canopy cover of Natal redtop (*Melinus repens*) generally decreased and that of thatching grass (*Hyparrhenia rufa*) increased within 2 years following fire. Cover of other nonnative species, such as molasses grass, changed little after 2 years from prefire values (Tunison and others 1994). On Oahu, Kartawinata and Mueller-Dombois (1972) noted the postfire encroachment of molasses grass into nonnative broomsedge communities.

Fountain grass is present in grasslands and as an understory species in forests over a wide elevational range. It is well adapted to persist after fire. Live fountain grass culms can sprout rapidly following top-kill and set seed within a few weeks (Goergen and Daehler 2001). Seeds are also blown in from neighboring unburned populations and rapidly germinate under favorable conditions (Nonner 2006). Unusually wet years result in rapid growth and expansion of fountain grass, which has a high net photosynthetic rate (Goergen and Daehler 2001, 2002). When dry conditions return, fires of high severity often result from the high fuel loads and low standing fuel moistures associated with fountain grass (Blackmore and Vitousek 2000; Nonner 2006).

Nonnative grasses can contribute to the decline of native pili grasslands where fire is excluded. Daehler and Carino (1998) compared the current (1998) and historic (1965 to 1968) cover at 41 sites on Oahu that were dominated by pili grass in the late 1960s and where fires had generally been suppressed. At these sites, nonnative grass monotypes, particularly buffelgrass, guineagrass (*Panicum maximum*) and fountain grass replaced pili grass in 2/3 of sites sampled, and pili was absent in the remaining 1/3 of the sites. In burned areas, however, the fire-adapted native pili grass competes well with some nonnative fire adapted grasses. In low-elevation grasslands within HVNP, pili grass cover was the same or higher 2 years after fire in areas where it was dominant or co-dominant (30 percent to 50 percent of total cover) prior to the fires and nonnative grass cover did not increase from preburn levels (Tunison and others 1994, 2001).

Establishment of nonnative invasive grasses, such as broomsedge and beardgrass, initiates a grass/fire cycle (fig. 11-7) that can increase the cover of nonnative grasses, inhibit shrub and tree colonization and growth, and result in changes in the composition of native woody plant communities (D'Antonio and Vitousek 1992; Hughes and Vitousek 1993). Although this cycle is best documented for `ōhi`a woodlands in Hawai`i (Tunison and others 1995) it is probable in other native plant communities here. Monocultures of molasses grass in `ōhi`a woodlands and fountain grass in montane shrublands have been documented following repeated fires (Hughes and others 1991; Shaw and others 1997).

Fire effects were monitored 2 to 5 years after fire on paired burned and unburned transects in open-canopy shrublands between 130 and 600 feet (40 to 190 m) in HVNP (D'Antonio and others 2000; Tunison and others 1994). Fires were fueled by nonnative broomsedge and bush beardgrass. Nonnative grasses maintained dominance of the sites following fire; composing a majority of the total cover on burned and unburned sites, although nonnative grass cover was slightly lower (5 percent to 15 percent) in four of the five burned sites. Total native plant cover was the same or slightly higher in burned areas (four out of five sites) but this was largely due to an increase in pili grass rather than woody species (D'Antonio and others 2001a; Tunison and others 2001).

The cover of fountain grass was reduced but recovered quickly following a low- to moderate-severity fire in montane shrublands in the saddle between Mauna Kea and Mauna Loa. The aerial cover of fountain grass was 50 percent of total herbaceous cover prior to the fire, but was less than half of preburn values 6 months after. Within 1 year, fountain grass had regained over half of its original aerial cover and one-third of its basal cover (Shaw and others 1997). At the same time, total shrub cover was much lower (4 percent total cover) than prefire shrub cover (20 percent) and stem densities were at half of preburn values. Shaw and others (1997) noted that fountain grass became a monoculture in nearby areas that had burned repeatedly.

Anecdotal accounts of historic fires in east Maui (Loope and others 1990) and observations from Haleakala and Hawai`i Volcanoes National Parks (Smith and Tunison 1992) suggest that severe fires may convert native subalpine shrublands into nonnative grasslands (fig. 11-10). Anderson and Welton (unpublished data) observed a four-fold increase in cover of

Figure 11-10—The result of wildfire in the Haleakala subalpine shrubland is a conversion of fuels from closed canopy of endemic shrubs (background) to nonnative grassland with sparse shrub recovery (foreground) as shown on the flank of the 1992 fire scar. (Photo by S. Anderson.)

nonnative grasses including velvetgrass, sweet vernal grass, and red fescue (*Festuca rubra*) in a burned area when compared to an adjacent unburned area 6 years after a 0.6 acre (0.2 ha), high-severity fire in February of 1992 at 8,000 feet (2,630 m) on Haleakala (Anderson, personal communication, 2005). The dominant native shrubs in the area, pūkiawe and `ohelo, do not tolerate fire well and are slow to reestablish (Anderson and Welton, unpublished data; Smith and Tunison 1992).

A fire fueled by fountain grass converted a lama woodland on the west side of Hawai`i Island into a nonnative grass-dominated savanna with few postfire sprouts and seedlings of lama after 3 years (Takeuchi 1991). Fountain grass had invaded the lama woodlands in previous decades to become a nearly continuous understory dominant (fig. 11-11). Fountain grass recovered rapidly after fire, returning to dominance. Little postfire recruitment of common native woody plants was observed after fire, but three native species, 'ilima (*Sida fallax*), kulu`i (*Nototrichium sandwicense*), and mamane, regenerated from seed or vegetative sprouts in scattered, localized patches within the vigorously recovering fountain grass matrix.

Nonnative grasses and Asian sword fern can carry fire in the ungulate-disturbed understory of mesic `ōhi`a forests (Tunsion, personal observation, May 2002, Kupukupu Fire, Holei Pali, Hawai`i Volcanoes National Park.). Short-term fire effects in mesic `ōhi`a forest with an understory of nonnative Asian sword fern were documented after two fires in HVNP (Ainsworth and others 2005; Tunison and others 1995). The response of sword fern was comparable to many nonnative grasses; shortly after fire it recovered

Figure 11-11—The highly flammable fountain grass forms a nearly continuous fuel bed in dry submontane lama forests on the island of Hawai`i. (Photo by S. Cordell.)

vegetatively and quickly dominated the understory. Survivorship of native woody plant species was generally high, with sprouting observed in most tree and shrub species, including `ōhi`a. However, a second fire occurring 1 year later dramatically increased `ōhi`a mortality (Ainsworth and others 2005).

Fire effects have been studied most frequently in `ōhi`a woodlands of HVNP. Wildland fire resulted in a rapid increase in cover and biomass of nonnative grasses following six wildfires on nine sites in submontane `ōhi`a woodlands of HVNP (D'Antonio and others 2000; Tunison and others 1995). All fires were fueled by the invasive grasses broomsedge, bush beardgrass and molasses grass. Total nonnative grass cover was about 30 percent higher and total native species cover lower for all burned transects when compared with unburned transects (nine out of nine sites). Molasses grass showed the greatest increase (D'Antonio and others 2000; Tunison and others 1995). Hughes and others (1991) also studied the effects of fire in `ōhi`a woodlands in HVNP, from 1 to 18 years later, with similar results. After 15 months, the total cover of nonnative grasses was the same or greater in burned areas than in unburned areas (table 11-4), indicating that nonnative grasses can quickly reclaim a site and increase in cover over time. The cover of individual grass species changed, however, as molasses grass replaced broomsedge and bush beardgrass as the dominant grass species (Hughes and others 1991). Cover of molasses grass was even higher in areas that had burned twice (table 11-4).

This cycle of repeated fires increases the potential that fires will burn into and through native woody plant communities further reducing cover of trees and shrubs, altering forest structure and composition, and depleting native seed banks (Blackmore and Vitousek 2000; D'Antonio and others 2000, 2001a; D'Antonio and Vitousek 1992; Hughes and Vitousek 1993; Tunison and others 2001). Changes in forest structure further alter the microclimate to warmer, drier, and windier conditions (Freifelder and others 1998), creating additional barriers to the recovery of native communities. Total cover of native woody species in `ōhi`a woodlands was lower by nearly two orders of magnitude in burned areas as fire top-killed three of the four common native shrubs (pūkiawe, `ulei, and `akia). An average of 55 percent of `ōhi`a was top-killed among all sites studied, although some surviving trees resprouted within 1 year after fire (Tunison and others 1995). More trees were killed with higher fire intensity (as estimated by char height) (D'Antonio and others 2000), and no `ōhi`a seedling recruitment was observed, resulting in conversion of open-canopied woodlands to savannas (Tunison and others 1995). Hughes and others (1991) noted that postfire sprouting of native shrubs occurred infrequently and the dominant shrub,

Table 11-4—Average cover of native and nonnative species and dominant nonnative grasses following fire in submontane `ōhi`a woodlands in Hawai`i Volcanoes National Park.*

Species	Unburned	Burned once and sampled 15 months after fire	Burned once and sampled 18 years after fire	Burned twice and sampled 1 year after second fire
	Fire history and sampling scheme			
Bush beardgrass	63.9[a]**	38.5[b]	39.1[b]	21.4[c]
Broomsedge	8.8[a]	3.6[b]	0.3[c]	0.1[c]
Molasses grass	7.2[a]	49.7[b]	62.1[b]	79.3[c]
Native species subtotal	117.0[a]	5.8[b]	31.6[c]	0.7[d]
Nonnative species subtotal	80.3[a]	92.4[a,b]	101.9[b]	101.2[b]

* Adapted from Hughes and others (1991).
** Numbers with different superscripts are significantly different (n = 5; $P > 0.05$).

pūkiawe, was absent from most burned areas 18 years after fire. The native shrub `ākia is highly intolerant of fire and was absent from burned sites of all ages in both shrublands and woodlands (D'Antonio and others 2000). A similar result was noted with a related species of `ākia (*W. oahuensis*) by Kartawinata and Mueller-Dombois (1972) on Oahu.

The vigorous postfire response of molasses grass, a mat-forming species, creates an environment that inhibits regeneration of many species, primarily due to low light levels under the thatch (fig. 11-12). In addition to reducing the cover of other nonnative grasses, total native understory shrub cover in `ōhi`a woodlands was lower by two orders of magnitude, from over 60 percent cover in unburned plots to 0.6 percent in twice-burned plots containing molasses grass (Hughes and others 1991). Some native shrubs, such as `a`ali`i, may persist because of rapid germination and growth, and tolerance of low light levels beneath grass canopies (Hughes and Vitousek 1993).

Molasses grass has also replaced other grasses, including nonnative broomsedge and bush beardgrass (Hughes and others 1991) and native pili grass. Pili grass was absent from the burned areas of native low-elevation shrublands where molasses grass occurred at very high postfire cover levels (96 percent) (D'Antonio and others 2000).

A long-term study of the effects of a single large wildfire in montane shrublands and forest on Mauna Loa (Haunss 2003) supports Mueller-Dombois' assertion that fire does not appreciably alter the composition of native communities that contain a high proportion of

Figure 11-12—Molasses grass (*Melinus minutiflora*) forms a dense mat preventing regeneration of other plants. (Photo by A. LaRosa.)

fire-tolerant species. There was no increase in cover of nonnative velvetgrass, dallis grass, and broomsedge in burned native shrublands 27 years after fire compared with unburned shrublands, although the native shrub, pukiawe, was replaced by the fire-tolerant native shrub `a`ali`i. Similarly, forest composition differed little in burned and unburned koa forests, and koa remained the dominant canopy tree species 27 years after fire (Haunss 2003). Koa recovered rapidly after the burn by sprouting from roots, and formed a canopy within a decade that was similar to that in unburned sites. The understories of both burned and unburned koa forests were dominated by nonnative meadow ricegrass, suggesting that disturbances other than fire may play a role in determining postfire community composition (Haunss 2003; Tunison and others 2001).

Banko and others (2004) studied the effects of fire on mortality and regeneration of mamane in MKFR and surrounding areas following a small (5.6 ha) fire in 1999. They reported higher fire severity (percent burned area) in open-canopy mamane forests with an understory of nonnative grasses (velvetgrass, sweet vernal grass) than in adjacent grazed pasture with scattered mamane. They attributed this difference largely to the lower fuel loading of grasses in heavily grazed pasture when compared to the forest. Relative grass cover and biomass were not reported, however, and the effects of fire on the nonnative grasses cannot be determined. Fire appeared to stimulate root suckering of mamane trees in both forest and pasture, but suckering was much higher in pasture areas where fire severity was reportedly lower. An increase in the number of mamane saplings in burned areas 10 months after fire may be attributed to the decrease in competition for light or moisture with the removal of the dense, nonnative grasses (velvetgrass, sweet vernal grass). This has been suggested for molasses grass in other systems and warrants closer study.

The potential for, and consequences of, the loss of threatened and endangered (T/E) species in Hawaiian forests is particularly serious. Hawai`i has nearly 280 listed T/E plant taxa (Bishop Museum 2004) and many of these occur in fire prone environments. Studies of the effects of fire on rare plants are few, and the documented effects are variable, depending upon species and, to a certain extent, life form. There was virtually no regeneration of rare small trees, such as koki`o (*Kokia drynarioides*), kauila (*Alphitonia ponderosa*), or uhiuhi (*Caesalpinia kavaiensis*) following a fire in lama forest with fountain grass (Takeuchi 1991). In high-elevation shrublands on Hawai`i, several endangered tree and shrub species showed no signs of regeneration 1 year after fire, while certain subshrubs and prostrate species with rhizomes or tuberous root systems (for example, the native mint *Stenogyne angustifolia*) survived and sprouted soon after fire (Shaw and others 1997).

Lowland and Montane Wet Forests

Wet forests are found on all the main Hawaiian Islands except Niihau and Kahoolawe, in lowland habitats ranging from 300 to 3,600 feet (100 to 1,200 m), and montane habitats ranging from 3,600 to 6,600 feet (1,200 to 2,200 m). Lowland wet forests occur across a wide range of substrate and soil conditions with annual rainfall from 60 to 200 inches (500 to 5,000 mm). Montane wet forests occur on windward aspects with more than 100 inches (2,500 mm) of rainfall per year. Fog and fog drip are common in montane wet forests (Gagne and Cuddihy 1999).

Wet forests have been altered by anthropogenic activities including logging, introductions of feral ungulates, and invasions by nonnative plant species. Despite widespread human-caused alterations, many wet forest habitats remain dominated by native species. Gagne and Cuddihy (1999) recognize seven lowland and four montane wet forest communities. `Ōhi`a is the dominant canopy species in all but two communities: the lowland nonnative wet forest community types, composed of java plum (*Syzygium cumuni*), rose apple (*S. jambos*), mountain apple (*S. malaccense*), common guava, strawberry guava (*Psidium cattleianum*), and others; and the montane nonnative firetree (*Morella faya*) forest community (Gagne and Cuddihy 1999).

Fire has been documented in `ōhi`a lowland wet forests and `ōhi`a / hapu`u (*Cibotium* spp.) tree/fern montane forests in and near HVNP (Tunison, personal observation, 30 September 1995, Volcano Dump Fire, Volcano, HI.; Tunison and others 2001). `Ōhi`a lowland wet forests vary widely in composition and structure, ranging from nearly monotypic `ōhi`a stands to older `ōhi`a forests with a diverse assemblage of trees, shrubs, ferns (including native uluhe, *Dicranopteris linearis*) and other herbs in the understory. Introduced shrubs, grasses and fern species often dominate disturbed areas. `Ōhi`a lowland wet forest grades into `ōhi`a/ hapu`u tree/fern montane forest as elevation increases. `ōhi`a trees are present as scattered emergents above a lower closed canopy of tree ferns and other native trees (Gagne and Cuddihy 1999).

Fire History and the Role of Nonnative Invasives in Wet Forest Fire Regimes

The historic role of fire in the Hawaiian wet forests is not well understood. Some scientists have speculated that disturbances such as fire may set wet forest communities back to an earlier stage of succession dominated by `ōhi`a and uluhe ferns and that the presence of these communities on the landscape is evidence of past fire disturbance (Gagne and Cuddihy 1999). Vogl (1969) suggested that the capacity of tree fern (hapu'u, *Cibotium glaucum*) to withstand disturbance may be an evolutionary adaptation to fire. Conversely, some scientists suggest that wildfires have had little influence on the development of tropical wet forests (Mueller-Dombois 1981). Studies have shown that fuel moisture content stays above the moisture of extinction in intact evergreen tropical forests with frequent, high precipitation coupled with constantly high relative humidity (Uhl and Kauffman 1990).

Few wildfires were recorded in Hawaiian wet forests prior to the 1990s (Tunison and others 2001). However, the record of fires in Hawai`i is relatively short (less than 100 years) compared to probable long fire-return intervals for forests with these fuel and weather patterns. Naturally ignited lava and lightning fires in wet forests are possible during extended droughts. Fuels are abundant and continuous in these wet forests. The pantropical climbing fern, uluhe, for example, can attain heights greater than 30 feet (9 m) (Kepler 1984), creating a laddered fuel bed (Holttum 1957) (fig. 11-2).

Since the early 1990s, at least 13 wildfires have occurred in relatively young (400- to 750-year-old) wet forests on the Island of Hawai`i. Following prolonged drought, five small fires (less than 30 acres (12 ha) each) occurred in wet montane forests near the Volcano community on the Island of Hawai`i. In HVNP, eight small to medium (less than 5,000 acres (2020 ha) each) wildfires were documented in lowland and montane wet forests from 1995 to 2004 (Loh, unpublished data, 2004). At least three of these wildfires occurred following severe droughts associated with El Niño events (Ainsworth and others 2005). Native and nonnative plant species fueled these wildfires (fig. 11-13). Disturbance of tropical wet forests can create in-stand weather and fuel conditions (altered loading and arrangement) that increase fire spread (D'Antonio and Vitousek 1992; Uhl and Kauffman 1990).

Fire and Its Effects on Nonnative Invasives in Wet Forests

Many nonnative species in wet forests of Hawai`i appear able to survive fire and/or recolonize after fire. Our information comes from a few studies conducted 1 year following the 2003 Luhi Fire in the southeastern region of the Island of Hawai`i on 400- to 750-year-old lava flows (Ainsworth, unpublished data, 2005). The nonnative trees, firetree and strawberry guava, sprouted after fire. On average, burned sites had four times as many nonnative species as neighboring unburned reference sites, and cover of many nonnative herbaceous species was significantly greater in burned sites than unburned reference sites. For example, cover of nonnative Hilo grass (*Paspalum conjugatum*), was 2 percent in unburned plots and 30 percent in burned plots (Ainsworth, unpublished data, 2005). Due to their capacity for invasion and rapid postfire site occupation, nonnative ferns and grasses may reduce the regeneration of native species following fire.

Many native species in wet forests possess traits that enable them to survive fire, but these may be adaptations to other disturbance factors such as volcanism and hurricanes. For example, following the recent lava-ignited fires in 2002 and 2003 at Hawai`i Volcanoes National Park, many native species including `ōhi`a

Figure 11-13—Lowland wet forest with native and nonnative dominated understory fuels. Photo on left shows native tree fern (*Cibotium glaucum*) in unburned area; on right is nonnative Hilo grass (Paspalum conjugatum) 2 years after fire. (Photo courtesy of Hawai`i Volcanoes National Park.)

(fig. 11-14), `ilima (*Sida fallax*), `iliahi (*Santalum paniculatum var. paniculatum*), and `a`ali`i (*Dodonaea viscosa*), resprouted and/or established from seed during the first 2 years after fire. Over 90 percent of the tree ferns, hapu`u, also survived wildfire by sprouting (Ainsworth and others 2005).

Nonnative species invasions may pose increasing threats to native wet forest ecosystems in Hawai`i because of their ability to dominate postfire environments. However, fire effects are site-specific and depend on microclimate, fire severity, and prefire species composition. Long-term research is needed to elucidate the interactions of climate, nonnative species, and fire on composition of Hawaiian wet forests.

Use of Fire to Control Invasive Species and Restore Native Ecosystems

While little is known about the long-term effects of fire on native species, many lowland native woody species appear to regenerate poorly after fire. There are two major areas of research on restoration of biologically and culturally important ecosystems degaded by fire: (1) researchers are experimenting with fire to enhance pili grass in formerly Polynesian-maintained grasslands, and (2) in native forest and woodlands, fire and other methods are being tested in an attempt to reduce nonnative grass dominance and maintain native woody plant communities.

Current research indicates that prescribed fire, in combination with other treatments, may be used to maintain or restore pili grasslands invaded by nonnative grasses. Managers at HVNP have used prescribed fire successfully to enhance pili grass in areas dominated by Natal redtop but have been less successful where thatching grass is present. Park managers continue to refine burn prescriptions and identify fire-adapted native shrubs useful for rehabilitation of shrublands with pili grass in the understory (Tunison and others 2001). Daehler and Goergen (2005) successfully reestablished pili grass in areas dominated by buffelgrass using biennial, low-intensity backing fires in winter followed by herbicides or mechanical removal of buffelgrass, or fire without other treatments. In areas of combined treatments, pili grass became dominant. The absolute cover of pili grass in burned areas averaged 35 percent in plots where buffelgrass had been removed and 10 percent in plots where buffelgrass remained (Goergen and Daehler 2002). These results indicate that fire alone may be insufficient to restore pili grasslands where competition from nonnative grasses is substantial. In areas where pili grass has been absent for some time, the addition of seed becomes necessary for restoration due to the absence of a soil seed bank (Daehler and Goergen 2005). Burning pili grasslands also increases pili grass seed production and could provide seed for large scale restoration (Daehler and Goergen 2005; Nonner 2006).

Figure 11-14—`Ōhi`a trees resprouting from base following fire in 2003 in Hawai`i Volcanoes National Park. (Photo courtesy of Hawai`i Volcanoes National Park.)

One of the most widespread and problematic invasive species in dry and mesic communities in Hawai`i is fountain grass (Cordell and others 2004; Smith, C. 1985; Smith and Tunison 1992). Scientists are working to restore highly altered native dry forests invaded by fountain grass using ungulate exclusion and grass removal followed by direct seeding of fast-growing, weedy native species. Some fast growing species are able to establish and persist on their own. These early successional species may then create suitable microsites for the establishment of other native species (Cabin and others 2002; Cordell and others 2002). Attempts at controlling fountain grass with prescribed fire, herbicides, and grazing have met with varying degrees of success. To date, herbicide use (glyphosate) appears to be the most successful technique (Cordell and others 2002).

Managers at Hawai`i Volcanoes National Park are attempting to reduce impacts of fires fueled by nonnative grasses by establishing self-sustaining, fire-tolerant "near native" communities in place of communities dominated by formerly common fire-intolerant native species found on those sites. Between 1993 and 2000 they tested the capacity of native species to survive and recolonize after fire with laboratory heat trials and seven research burns. Fourteen native plant species, among them mamane, `a`ali`i, and `iliahi, were identified as fire-tolerant (Tunison and others

2001). Intensive restoration efforts were initiated in 2001 on 2,000 acres (815 ha) of `ōhi`a woodlands burned by wildfires. Outplantings and direct seeding of shrubs and trees were completed in the 1,000-acre (405-ha) Broomsedge Fire area. Survival of outplantings after 1 to 3 years was greater than 80 percent, and direct seeding has resulted in establishment of some species, including koa, mamane, naio, `a`ali`i, naupaka (*Scaevola kilauea*), and ko`oko`olau (*Bidens hawaiensis*) (Loh and others 2004). Park managers intend to keep fire out of these rehabilitation sites for 15 to 20 years to allow maturation of fire-tolerant plantings and development of a soil seed bank. In other sites, small prescribed fires have been used to reduce biomass of nonnative grasses and promote the successful establishment of individuals via outplantings and direct seedings. Other methods to temporarily control grasses, including the use of herbicides, are under investigation (Loh and others 2007).

We could find no published descriptions of attempts to use prescribed fire for control of invasive species and restoration of upland communities. Results of the effects of wildfires on montane and subalpine shrublands with nonnative grasses suggest that fire can increase the cover of introduced grasses and decrease that of many native shrub species. The limited data also suggest that some shrubs may recover in the long term. Some species, notably koa and mamane, tolerate fire and resprout readily after fire in these upland ecosystems but the abundance of invasive grasses after wildfire may hamper recovery of native understory species. Experimental restoration of the understory in unburned montane seasonal koa forest at HVNP uses herbicidal control of grass with planting and direct seeding of understory species to develop a soil seed bank (McDaniel, unpublished, 2003).

Given the probable long fire-return intervals in wet forests and the facilitation of nonnative invasive species spread, few ecological benefits are expected to accrue from prescribed burning of these forests.

Conclusions

The historic fire regimes of many Hawaiian ecosystems have been altered from typically rare events to more frequent, and in some instances, more severe fires today. This is largely a result of the increase in continuous, fine fuels associated with the spread of nonnative fire-tolerant grasses into dry and mesic native plant communities that were once dominated by relatively open stands of shrub and trees (Gagne and Cuddihy 1999; Smith and Tunison 1992). Invasive grasses are altering fuel loading and continuity, microsite conditions, and the postfire recovery rate of native species.

Fire is now an important disturbance factor in Hawai`i (Smith and Tunison 1992). It contributes to the maintenance and spread of many nonnative grasses, even though some of these grasses invaded ecosystems following other human disturbances and can persist without fire (Tunison and others 2001; Williams 1990). A few species of nonnative grasses are the primary carriers of most fires in Hawai`i today (table 11-2). Many of these nonnative pyrophytic species alter community structure and ecosystem properties including resource availability, primary productivity, decomposition, and nutrient cycling (D'Antonio and others 2001a; Hughes and Vitousek 1993; Ley and D'Antonio 1998; Vitousek and others 1996).

Much of the available information on the effects of fire on invasive species in Hawai`i is anecdotal. Fire effects studies have been conducted primarily on the Island of Hawai`i and are few, narrowly focused, and generally of short duration (1 to 5 years after fire). In reviewing the available literature, it is clear that plant responses are complex, not well understood, and often site specific. For example, molasses grass cover increased very little in dry, low-elevation grasslands in HVNP while in higher and wetter grasslands and woodlands, molasses grass cover increased appreciably (Hughes and others 1991; Tunison and others 1994, 2001). Even within HVNP, where most fire research has occurred, many questions remain.

As nonnative grasses continue to spread, damage to and potential loss of native shrublands and forests in Hawai`i will increase. The best documented examples in Hawai`i come from lama forests and `ōhi`a woodlands, where several species of nonnative grasses readily carry fire and where relatively few native species are fire-tolerant. In the submontane `ōhi`a woodlands, nonnative grasses have set in motion a grass/fire cycle resulting in changes in plant community structure and composition, fuel properties, and microclimate. Many of these areas have been converted to savannas dominated by invasive grasses (D'Antonio and Vitousek 1992; D'Antonio and others 2001a; Hughes and others 1991; Hughes and Vitousek 1993). This invasive grass/fire cycle can be long lasting (Hughes and others 1991) and may represent a permanent shift in plant community composition (D'Antonio and others 2001a). A similar cycle is occurring with the nonnative Asian sword fern and may be found in other plant communities upon further study.

In ecosystems where the dominant native species exhibit some fire tolerance, for example koa and a`ali`i, the detrimental effects of fire at the community level may be limited as these species often recover from fire by sprouting or establishing from seed. In montane shrublands and koa forests on Mauna Loa, where fire-promoting grasses are fewer, native communities have changed little, even three decades after a fire (Haunss 2003; Tunison and others 2001).

One of the most problematic nonnative species in Hawai`i today is fountain grass, which is spreading in dry and mesic shrublands and forests, particularly in western Hawai`i (Island). Fountain grass has a wide ecological range in Hawai`i. It has invaded many site types, from bare lava flows to late successional forests, from near sea level to over 9,000 feet (2,800 m), in areas with rainfall ranging from 10 to 50 inches (25 to 125 cm) (Jacobi and Warshauer 1992). Fountain grass forms a nearly continuous bed of highly flammable fuel in many areas and has the potential to dramatically alter entire landscapes and contribute to the loss of remnant native dry forests (Cordell and others 2004; Shaw and others 1997). There is a critical need for more research on fire and fountain grass interactions.

The cycle of invasive grasses and fire must be broken to protect native plant communities and ecosystems. The potential for loss of threatened and endangered species in Hawaiian forests is of particular concern. With little history of fire in many ecosystems and limited knowledge of fire and its effects, managers in Hawai`i have been cautious with their use of fire to manage disturbed environments. However research into the use of fire to manage some lowland ecosystems, for example native grasslands, is increasing. Prescribed fire is not likely to be a useful tool in most Hawaiian environments where fire-return intervals were historically long and most native species are intolerant of fire. Attempts to reduce fuel loads and restore native ecosystems using a variety of methods other than fire are underway. More research is needed on the effects of fire in native and altered ecosystems to understand the mechanisms of invasion, the nature of competitive interactions among native and nonnative species after fire, and the use of fire and other techniques in restoration of degraded ecosystems. Longer term studies are needed to determine if the short-term competitive advantage observed for many fire-promoting nonnative grasses over native woody species persists over time.

Notes

Kristin Zouhar
Gregory T. Munger
Jane Kapler Smith

Chapter 12:

Gaps in Scientific Knowledge About Fire and Nonnative Invasive Plants

"The issue I am attempting to deal with... is not knowledge but ignorance. In ignorance I believe I may pronounce myself a fair expert."

Wendell Berry (2000), *Life is a Miracle*

Abstract—*The potential for nonnative, invasive plants to alter an ecosystem depends on species traits, ecosystem characteristics, and the effects of disturbances, including fire. This study identifies gaps in science-based knowledge about the relationships between fire and nonnative invasive plants in the United States. The literature was searched for information on 60 nonnative invasives. Information was synthesized and placed online in the Fire Effects Information System (FEIS, www.fs.fed.us/database/feis), and sources were tallied for topics considered crucial for understanding each species' relationship to fire. These tallies were analyzed to assess knowledge gaps. Fewer than half of the species examined had high-quality information on heat tolerance, postfire establishment, effects of varying fire regimes (severities, seasons, and intervals between burns), or long-term effects of fire. Information was generally available on biological and ecological characteristics relating to fire, although it was sometimes incomplete. Most information about species distribution used too coarse a scale or unsystematic observations, rendering it of little help in assessing invasiveness and invasibility of ecosystems, especially in regard to fire. Quantitative information on the impact of nonnative plants on native plant communities and long-term effects on ecosystems was sparse. Researchers can improve the knowledge available on nonnative invasive plants for managers by applying rigorous scientific methods and reporting the scope of the research, in both scientific papers and literature reviews. Managers can use this knowledge most effectively by applying scientific findings with caution appropriate to the scope of the research, monitoring treatment results over longer periods of time, and adapting management techniques as new information becomes available.*

Introduction

Wildland managers face challenges in obtaining and using information about nonnative invasive plant species (as defined in chapter 1) and fire. What they require is detailed knowledge about complex issues, including:

- The likelihood of establishment, persistence, and spread of nonnative invasives under various disturbance regimes;
- The probable interactions of invasive species with native plant species, and how these interactions influence community and ecosystem properties; and
- Quantitative descriptions of results of management actions, particularly fire exclusion, use of prescribed fire, and postfire rehabilitation.

What is usually available to managers is a smattering of knowledge about the biology of the nonnative plant itself (sometimes available only from the region of origin or not available in English); information from agricultural science that focuses on interactions of the nonnative species with crop plants and tillage systems; and some knowledge about North American ecosystems and fire, framed almost entirely in terms of native species. Relatively little information specifically addresses nonnative invasive species' interactions with fire in native North American plant communities. The scope of this problem is greater than a lack of knowledge about invasives themselves, because the nature and condition of a plant community strongly influence its susceptibility to invasion (chapter 2). Even where scientists have reported interactions between nonnative invasives and wildland fire in specific ecosystems, the knowledge may be anecdotal or incomplete (D'Antonio 2000; Grace and others 2001; McPherson 2001), applicable only to a specific ecosystem under a narrow range of conditions (Klinger and others 2006a), or limited to laboratory conditions (so applicability to field conditions is unknown).

To assess the quality of information on fire and invasive species that is available to managers, we identified information gaps on the basic biology, ecology, and relationship to fire for 60 nonnative invasive plant species. Our goal was to address two main questions:

1. How can research contribute most meaningfully to increasing and sharing knowledge about nonnative invasives and fire?
2. How can managers best apply current scientific knowledge about nonnative invasives and fire?

Methods

In spring 2001, we began a 4-year project to synthesize knowledge on fire and nonnative invasive plants for the Fire Effects Information System (FEIS, online at www.fs.fed.us/database/feis). Our task was to produce literature reviews covering 60 nonnative invasives. We selected the species to be covered by asking land managers from federal agencies throughout the continental United States (excluding Alaska) for nominations, resulting in a list of 162 species. We excluded species recently covered in FEIS (medusahead (*Taeniatherum caput-medusae*), leafy spurge (*Euphorbia esula*), smooth brome (*Bromus inermis*), and red brome (*B. rubens*)) and then excluded the species with the least scientific literature available on basic biology, ecology, and fire. Table 12-1 lists the species selected, their grouping into knowledge syntheses (called "species reviews" in FEIS), and the date that each was completed. Our list is neither a random nor a systematic sample of nonnative plant species in North America, but it represents many nonnative invasives about which managers are concerned and about which at least some scientific research has been published.

For each species or group of species on our list, we obtained, reviewed, and synthesized information from the scientific literature. We searched for information by scientific and common names using two main sources: (1) the Citation Retrieval System, which is the citation database for the Fire Effects Library at the USDA Forest Service Fire Sciences Laboratory, Missoula, Montana (available at http://feis-crs.org); and (2) WEBSPIRS from Silver Platter, provided by the USDA Forest Service, Rocky Mountain Research Station Library, Fort Collins, Colorado. These searches yielded peer-reviewed journal articles, literature reviews, proceedings from scientific meetings, theses and dissertations, book chapters, and technical papers from research groups in state and federal agencies. Our sources were not restricted to single-species studies; any literature that included the species of interest was reviewed and pertinent information was included in the review. We also conducted Internet searches for each species, which generally yielded information from non-peer-reviewed sources such as University Extension Services and natural history organizations. In the process of reviewing the literature, we frequently discovered and obtained additional pertinent articles. Finally, where knowledge gaps remained after the literature search, we occasionally obtained information in the form of personal communications from researchers and managers familiar with the species.

We used the FEIS species review template (table 12-2) to ensure completeness and consistency of information. While planning and writing species reviews, we kept

Table 12-1—Nonnative invasive plant species used for this analysis. Common names are from the Fire Effects Information System (FEIS, www.fs.fed.us/database/feis) or PLANTS database (plants.usda.gov).

Scientific name(s)	Common name(s)	Number species[a]	Date completed[b]
Acer platanoides	Norway maple	1	May-03
Acroptilon repens	Russian knapweed	1	Feb-02
Ailanthus altissima	Tree-of-heaven	1	Feb-04
Alliaria petiolata	Garlic mustard	1	Oct-01
Arundo donax	Giant reed	1	April-04
Bromus tectorum	Cheatgrass	1	Apr-03
Cardaria draba, C. pubescens, C. chalapensis	Hoary cress species	3	Feb-04
Carduus nutans	Musk thistle	1	Jun-02
Celastrus orbiculatus	Oriental bittersweet	1	March-05
Centaurea diffusa	Diffuse knapweed	1	Oct-01
Centaurea maculosa	Spotted knapweed	1	Sep-01
Centaurea solstitialis	Yellow starthistle	1	Jun-03
Chondrilla juncea	Rush skeletonweed	1	Mar-04
Cirsium arvense	Canada thistle	1	Nov-01
Cirsium vulgare	Bull thistle	1	Aug-02
Convolvulus arvensis	Field bindweed	1	Jul-04
Cynoglossum officinale	Houndstongue	1	Aug-02
Cytisus scoparius, C. striatus	Scotch and Portuguese broom	2	Dec-05
Elaeagnus angustifolia	Russian-olive	1	Aug-05
Elaeagnus umbellata	Autumn-olive	1	Oct-03
Genista monspessulana	French broom	1	Dec-05
Hypericum perforatum	Common St. Johnswort	1	Jan-05
Imperata cylindrica, I. brasiliensis	Cogongrass, Brazilian satintail	2	July-05
Lepidium latifolium	Perennial pepperweed	1	Oct-04
Lespedeza cuneata	Sericea lespedeza	1	Feb-04
Ligustrum vulgare, L. sinense, L. japonicum, L. amurense	Privet species	4	Jun-03
Linaria dalmatica, L. vulgaris	Dalmatian and yellow toadflax	2	Aug-03
Lonicera japonica	Japanese honeysuckle	1	Dec-02
Lonicera fragrantissima, L. maackii, L. morrowii, L. tatarica, L. xylosteum, L. x *bella*	Bush honeysuckles	6	Nov-04
Lygodium microphyllum, L. japonicum	Climbing fern species	2	Dec-05
Lythrum salicaria	Purple loosestrife	1	Jun-02
Melaleuca quinquenervia	Melaleuca	1	Sept-05
Microstegium vimineum	Japanese stiltgrass	1	Jan-05
Potentilla recta	Sulfur cinquefoil	1	Dec-03
Pueraria montana var. *lobata*	Kudzu	1	Jul-02
Rosa multiflora	Multiflora rose	1	Sept-02
Schinus terebinthifolius	Brazilian pepper	1	Jan-06
Sonchus arvensis	Perennial sowthistle	1	Aug-04
Sorghum halepense	Johnson grass	1	May-04
Spartium junceum	Spanish broom	1	Dec-05
Tamarix ramosissima, T. chinensis, T. parviflora, T. gallica	Saltcedar, small-flowered tamarisk, French tamarisk	4	Aug-03
Triadica sebifera	Chinese tallow	1	Sept-05
Ulex europaeus	Gorse	1	March-06

[a] Number of species included in the review

[b] Date the species review went online; literature that became available after that date is not included in the review or this analysis.

Table 12-2—Structure of FEIS plant species reviews and topics covered in each section of a review. Sections and topics highlighted in bold print were considered crucial for understanding relationship between plant species and fire. Sources providing information on these topics were tallied for this analysis.

Review section	Topics covered
Introductory information	Scientific and common names, abbreviations, synonyms, code names Taxonomy description **Life form** (tree, shrub, herb, etc.) Legal status (threatened, endangered, etc.) Authorship and citation
Distribution and occurrence[a]	
Botanical and ecological characteristics	General characteristics **Raunkiaer life form**[b] Reproduction (includes breeding system, pollination, **seed production, seed dispersal, seed banking, germination/establishment/ seedbed requirements, growth, and asexual reproduction and regeneration**[c] **Site characteristics**[a] (includes topography, elevation, climate, and soils) **Successional information** (includes **longevity,**[a] **response to disturbance**[c], and **competitive interactions**[a]) Seasonal patterns (**aboveground phenology, belowground phenology**)
Fire ecology	Fire adaptations (including **heat tolerance of tissues** and **seed**), fire regimes **Postfire regeneration strategies**
Fire effects	**Immediate fire effect on plant** Species response to fire (includes **postfire establishment** and postfire **vegetative response**) Fire management considerations (includes fire as a control agent)
Fire Research Project (fire experiment)[d]	
Management considerations	Importance to livestock and wildlife Other uses Impacts and control
Literature cited	

[a] Information on distribution, site characteristics, succession, longevity, and competitive interactions was combined for this paper to examine available information on where a nonnative species occurs and where it may become invasive.

[b] Raunkiaer (1934)

[c] Information on asexual regeneration and response to disturbance was combined for this paper to examine available information on post-injury regeneration potential.

[d] "Fire Research Project" is an optional category in a FEIS species review that describes research providing quantitative information on the prefire and postfire plant community, burning conditions, and fire behavior. It is included only if such research is available.

track of knowledge gaps as follows: We identified sections and topics crucial for understanding the plant's relationship to fire (shown in bold print, table 12-2). To this list of topics, we added questions related to fuels and fire regimes: Does any available research provide information about nonnative invasive plants' fuel characteristics or compare the fuel characteristics of invaded versus uninvaded sites? Does research provide evidence that nonnative plant invasions alter presettlement or reference fire regimes? We then added questions related to fire experiments: Does any research describe effects of fires of different severities, fires in different seasons, or fire treatments repeated at different intervals? Does research describe fire effects after the first postfire year?

As we wrote each species review, we used key phrases, such as "research is needed" and "incompletely understood" to identify knowledge gaps. A subsequent search of completed reviews for these key phrases highlighted topics with knowledge gaps.

Knowledge gaps were often attributable to lack of information, but some occurred because the available information covered only a narrow range of conditions or a small geographic area. Knowledge gaps also occurred when information was of uncertain quality, such as anecdotal evidence and assertions unsubstantiated by data. Such knowledge can be useful to managers, but it is important that readers recognize its limited scope of inference. To help readers apply published knowledge appropriately, Krueger and Kelley (2000) suggest identifying the nature of cited publications and classifying them as either professional resource knowledge, experimental research, case history, or scientific synthesis. In a similar vein, we developed a numerical scale to rank publications on fire and invasive species. This scale represents a continuum of information "quality," based on the study's evident rigor and clear scope of inference—from "high" (rank of 4) to "no information" (rank of 0):

- 4 Evidence from primary research published in a peer-reviewed journal
- 3 Evidence from primary research published in a technical paper from a research group in a state or federal agency, thesis or dissertation, book chapter, proceedings, or flora
- 2 Other substantial, published or unpublished experimental or observational data
- 1 Assertion with no experimental or observational data (that is, source of evidence for the assertion is unknown)
- 0 No information or assertions at all

The highest value in the information-quality scale (4) represented primary research published in peer-reviewed journals; for these articles, the population, variables, and scope of inference were usually well described, and blind peer review indicated the knowledge was probably reliable. An information quality value of 3 represented similar information published in an outlet that was reviewed by peers, but not anonymously. We classified publications ranked 3 and 4 as "high-quality" information. A value of 2 represented reports that had not been reviewed, such as reports of management or control activities, as well as information reported without a description of rigorous scientific procedure (that is, not containing hypotheses, controls, replication, or statistical analyses) and thus having unknown certainty and scope of inference. A value of 1 represented knowledge considered poor in quality or reliability, such as anecdotal knowledge and casual observations, for which the scope of inference was poorly defined or not described at all. Anecdotal information of this type was often found in literature or knowledge reviews.

Because the information-quality scale was subjective it had the potential to misrepresent the quality of information. We recognized, for example, that blind peer review does not guarantee accuracy even though we ranked its information quality as high (4), and a single peer-reviewed study may not have provided sufficient information to support widespread application. In contrast, a manager may possess bountiful, accurate, unpublished data (ranked 2) or accurate anecdotal information (ranked 1) that applies directly to management. Therefore we consider the information-quality scale a rough but useful indicator of information quality.

We examined the knowledge gaps in each species review using the information-quality scale. For botanical and ecological information, we recorded the highest quality of information available on identified topics. For fire-related information, we identified the highest quality of information available on each topic and also tallied the total number of citations, of any quality, available on each topic.

Results

No knowledge gaps were identified in any of the 43 species reviews for three of the highlighted topics in table 12-2: life form, seed production, and aboveground phenology. High-quality information was also available for every review on species distribution. However, much of this information comes from documents such as floras and reviews, whose main objective is not necessarily to gather and report distribution information, and thus has limited usefulness for estimating a species' potential to invade a particular plant community. Other distribution information comes from sources reporting coarse scale information. For example, low-resolution state and county distribution maps, such as those in the Plants (http://plants.usda.gov/) and Invaders (http://invader.dbs.umt.edu/) databases, are widely available but have insufficient detail for determining a species' ecological amplitude and potential to invade other sites and plant communities. Comprehensive inventory information was not available for any of the species that we reviewed.

High-quality information on site requirements was available for most species reviewed, but this information was usually not systematic or detailed enough to help managers assess invasiveness. For most species, the literature provided site information primarily for areas where the species is most problematic[1], where research has been conducted[2], or where it occurs in its native range[3]. For some species[4], this information was provided in reports from agricultural settings, suggesting that the species will spread into natural areas but not describing the sites or plant communities most likely to be invaded. Ironically, we sometimes inferred distribution of nonnative invasives from publications describing the geographic range where planting of those species has been recommended[5].

Knowledge gaps were identified for several species on the remaining topics highlighted in table 12-2. Table 12-3 shows the highest quality of information found for biological and ecological topics in each species review. For example, seed dispersal for Norway maple (*Acer platanoides*) is described by at least one article ranked 4, that is, containing primary research and published in a peer-reviewed journal. Only anecdotal information (ranked 1) is available regarding seed dispersal for sericea lespedeza (*Lespedeza cuneata*). The prevalence of high-quality citations (ranked 3 or 4) regarding biological and ecological topics is shown in figure 12-1.

Information on phenology of flowering and seed production was generally available for all species examined, although seed production information is typically limited to a particular set of conditions and rarely related to or available for postfire conditions. While high-quality information on post-injury regeneration was available for 88 percent of the species examined knowledge about belowground phenology and regeneration from underground tissue was often lacking (table 12-3, fig. 12-1). Information on seasonal changes in carbohydrate reserves of roots and other underground tissues was found for only about half (48 percent) of species reviews on biennial and perennial plants (fig. 12-1). Knowledge about depth of belowground perennating tissue was rarely available for the species reviewed.

High-quality information on seedbed requirements was available for 81 percent (table 12-3; fig. 12-1) of species reviews, but little of this information was specific to postfire situations. Most species reviews had high-quality information for seed dispersal (88 percent) and seed banking (77 percent) (table 12-3; fig. 12-1). However, seed banking information for several of these species reviews comes either from outside North America or from laboratory experiments, so they still lack information that is directly applicable to North American ecosystems. For example, the literature on rush skeletonweed (*Chondrilla juncea*) comes primarily from Australia, so its applicability to field conditions in North American wildlands is difficult to assess. Evidence of seed longevity has implications for seed banking, and for sulfur cinquefoil (*Potentilla recta*) and perennial pepperweed (*Lepidium latifolium*) is provided by only one laboratory study for each species (table 12-3), while only anecdotal observations from the field provide information on seed banking for these species. Information on the relationship between seed banking, field conditions, and disturbance is rarely available for nonnative invasive species.

We found high-quality information on longevity and/or succession for 77 percent of our species reviews (table 12-3; fig. 12-1); however, this information was often limited in scope. As with species distributions, information may be available where the species is most problematic and where studies have been conducted, but is lacking in other areas. Thus it was typically insufficient to provide a clear understanding of the potential for a particular invasive species to alter successional trajectories in newly invaded communities.

Table 12-4 describes both the quantity and quality of information found on fire-related topics for each species review. Quantity is expressed as the total number of sources cited in the species review, of any quality. Quality is expressed as in table 12-3, that is, the highest rank given any citation for that topic. For example, two sources provided information on postfire seedling establishment of bull thistle (*Cirsium vulgare*); the highest information-quality rank among these citations was 3, indicating that either one or both sources described primary research and was published in a technical paper, thesis, dissertation, book chapter, proceedings, or flora.

---

[1] Examples: *Bromus tectorum*, *Chondrilla juncea*, *Centaurea solstitialis*, *C. diffusa*, *C. maculosa*, *Elaeagnus angustifolia*, *E. umbellata*, *Hypericum perforatum*, *Lepidium latifolium*, *Lespedeza cuneata*, *Ligustrum* spp., *Lonicera japonica*, *Lythrum salicaria*, *Potentilla recta*, *Pueraria montana* var. lobata, and *Rosa multiflora*

[2] Examples: *Acer platanoides*, *Alliaria petiolata*, *Cardaria* spp. *Celastrus orbiculatus*, *Centaurea repens*, *Cytisus* spp., *Elaeagnus umbellata*, *Genista monspessulana*, *Lespedeza cuneata*, *Ligustrum* spp., *Linaria* spp., *Lonicera japonica*, *Lythrum salicaria*, *Pueraria montana* var. lobata, *Rosa multiflora*, and *Spartium junceum*

[3] Examples: *Acer platanoides*, *Centaurea repens*, *Cynoglossum officinale*, and *Lythrum salicaria*

[4] Examples: *Cardaria* spp., *Centaurea repens*, *Chondrilla juncea*, *Convolvulus arvensis*, *Imperata cylindrica*, and *Lespedeza cuneata*

[5] Examples: *Acer platanoides*, *Elaeagnus angustifolia*, *E. umbellata*, *Lespedeza cuneata*

Table 12-3—Highest quality ranking of information available on aspects of biology and ecology for nonnative invasive plant species reviews. See "Methods" for explanation of ranks 0 to 4 used in table.

Species review	Seed dispersal	Seed banking	Optimum seed bed	Post-injury regeneration	Succession /longevity	Below-ground phenology
Norway maple	4	0	4	4	4	0
Russian knapweed	4	0	4	4	4	4
Tree-of-heaven	4	3	4	3	4	1
Garlic mustard	4	4	4	4	0	0
Giant reed	3	0	0	4	3	2
Cheatgrass	4	4	4	4	4	n/a[a]
Hoary cress[b]	4	3	4	4	0	4
Musk thistle	4	4	4	4	4	4
Oriental bittersweet	4	4	4	4	4	0
Diffuse knapweed	4	0	4	3	3	0
Spotted knapweed	4	4	4	4	4	0
Yellow starthistle	4	4	4	4	4	n/a[a]
Rush skeletonweed	4	4	4	4	0	0
Canada thistle	4	4	4	3	3	4
Bull thistle	4	4	4	4	4	4
Field bindweed	4	4	4	4	4	4
Houndstongue	4	4	4	4	4	4
Brooms[b]	4	4	4	4	4	2
Russian-olive	4	4	4	3	4	0
Autumn-olive	4	0	0	1	0	0
French broom	4	4	3	3	2	2
Common St. Johnswort	4	4	4	4	4	0
Cogongrass[b]	4	4	4	4	4	0
Perennial pepperweed	4	4	0	4	4	4
Sericea lespedeza	1	1	1	1	1	0
Privet[b]	4	4	0	4	4	0
Toadflax[b]	4	4	4	4	3	0
Japanese honeysuckle	4	4	3	3	4	4
Bush honeysuckles[b]	4	4	4	4	4	0
Climbing ferns[b]	2	1	0	1	1	0
Purple loosestrife	4	4	4	4	4	0
Melaleuca	4	4	4	3	2	2
Japanese stiltgrass	4	4	4	3	4	n/a[a]
Sulfur cinquefoil	0	4	4	4	4	0
Kudzu	0	2	0	3	0	3
Multiflora rose	4	1	0	2	0	0
Brazilian pepper	3	4	4	3	3	0
Perennial sowthistle	4	4	4	4	4	4
Johnson grass	4	4	4	4	4	4
Spanish broom	2	2	4	2	4	0
Tamarisk[b]	4	4	4	4	4	4
Chinese tallow	4	4	4	4	4	4
Gorse	3	4	4	4	4	0

[a] Topic not applicable to annual species.

[b] Two or more species included in review; see table 12-1 for complete list of species included. Ranked information may not apply to all species in that review.

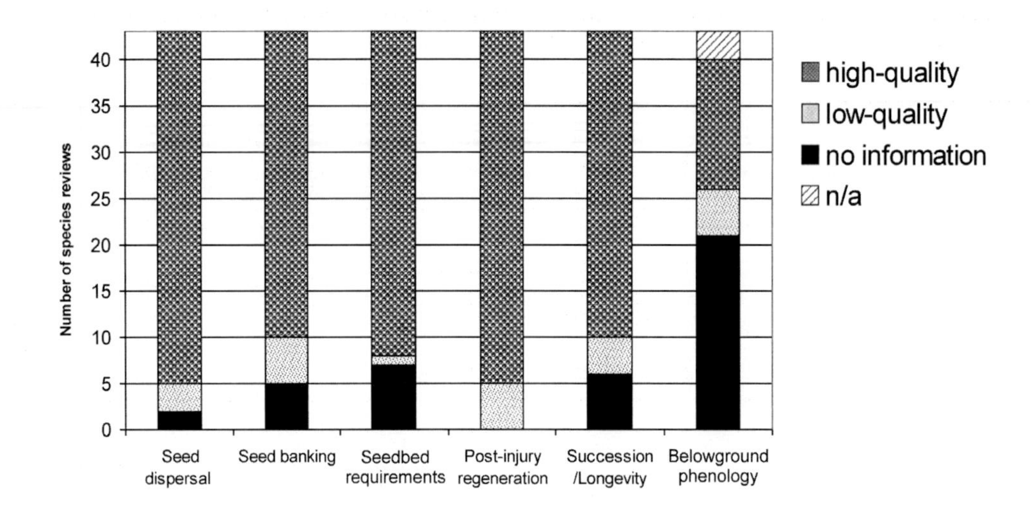

Figure 12-1—Highest quality of information available on botanical and ecological topics for 43 species reviews in FEIS. "Low quality" = rank of 1 or 2. "High quality" = rank 3 or 4. See "Methods" for explanation of information-quality ranking.

When quality of information is displayed for all species reviews, it is clear that less information is available on fire-related topics than on biological and ecological topics (figs. 12-1 and 12-2). For all fire-related topics except immediate fire effects and postfire vegetative response, more than half of reviews have no information at all (fig. 12-2).

For species that do have information on fire related topics, research results are still sparse and incomplete. While figure 12-2 breaks this information down according to quality, figure 12-3 breaks it down by the number of citations (of any quality) on each topic. As in figure 12-2, the dark portion of each bar indicates the number of reviews with no information (zero citations). If one considers the remaining reviews, those that have at least some information on a given fire-related topic, about half have only one or two citations on that topic (fig. 12-3).

Managers and members of the public often express concern about establishment and spread of nonnative invasives after fire, but even this topic shows a paucity of information. We found some information on postfire seedling establishment for 44 percent of species reviews (figs. 12-2 and 12-3). Examples include musk thistle (*Carduus nutans*) (Goodrich 1999; Grace and others 2001; Heidel 1987; Hulbert 1986), Canada thistle (*Cirsium arvense*) (Ahlgren 1979; Doyle and others 1998; Floyd-Hanna and others 1997; Goodrich and Rooks 1999; Hutchison 1992a; Rowe 1983; Smith, K. 1985; Thompson and Shay 1989; Turner and others 1997; Willard and others 1995), bull thistle (Messinger 1974; Shearer and Stickney 1991), and houndstongue (*Cynoglossium officinale*) (Johnson 1998). Ten of the 17 articles cited for these species rank high on the information-quality scale; however of the ten, only seven provide information on characteristics of prefire or unburned vegetation, six provide descriptions of fuels, fire behavior, burn conditions, or fire severity, and one gives detailed information on proximity and productivity of seed sources. For most species that we examined, the available information is not sufficient to conclude that, if the species occurs in a particular area and a fire occurs, it is likely to become invasive in the burned area.

We found high-quality information on vegetative response to fire for only 37 percent of species reviews (table 12-4; fig. 12-2), making predictions of postfire persistence and vegetative spread difficult for most species. While high-quality information was available for most species reviews on regeneration after mechanical disturbance (table 12-3; fig. 12-1), and this information can alert managers to the possibility of postfire regeneration, fires and mechanical disturbances cannot be assumed to evoke equivalent responses.

Fewer than half the species reviews (40 percent) had information on fuel characteristics of that species, or information on how fuel characteristics in invaded communities may be altered from uninvaded conditions. Where information is available, it is mostly anecdotal or speculative, as reflected by the number of reviews for which only low-quality information was available—8 of the 17 reviews provide only low-quality information on fuels. Similarly, only 30 percent of the species reviewed had information available on fire regime changes in invaded communities, and more than half of this information was anecdotal (table 12-4; fig. 12-2).

Table 12-4—Information available on relationships of nonnative invasive plants to fire. Total number of sources available on each topic for each species review is given in parentheses, followed by highest-ranked quality of information. See "Methods" section for explanation of ranks 0 to 4 used in table.

Species review	Heat tolerance, tissue	Heat tolerance, seed	Immediate fire effects on plant	Postfire seedling establishment	Postfire vegetative response	Postfire increase (source unknown)	Fuels	Fire regimes
Norway maple	0	0	0	0	(1)2	0	0	0
Russian knapweed	0	0	(1)2	0	0	0	0	0
Tree-of-heaven	0	0	0	0	(2)2	0	0	0
Garlic mustard	0	0	(4)4	0	(2)4	(2)4	(1)2	0
Giant reed	0	0	(2)2	0	(4)2	0	(1)1	(1)1
Cheatgrass[a]	0	(10)4	(17)4	(29)4	0	0	(32)4	(37)4
Hoary cress[b]	0	(1)1	(1)2	(1)1	(1)2	(1)1	0	0
Musk thistle	0	0	(3)1	(4)2	(3)2	(3)2	0	0
Oriental bittersweet	0	0	0	0	0	0	0	0
Diffuse knapweed	0	(1)1	(2)2	(2)2	(1)1	(4)2	0	0
Spotted knapweed	0	(1)4	0	0	0	(5)3	(3)3	0
Yellow starthistle	0	(1)2	(6)4	(4)4	(2)3	0	(2)2	0
Rush skeletonweed	0	0	0	0	0	(1)1	0	0
Canada thistle	0	0	(2)3	(10)4	(4)4	(8)4	0	(1)2
Bull thistle	0	(1)4	0	(2)3	0	(4)3	0	0
Field bindweed	0	(3)4	0	0	0	0	0	0
Houndstongue	0	0	0	(1)3	0	0	0	0
Brooms[b]	0	(4)4	(1)4	(4)4	(3)4	0	(3)3	(3)2
Russian-olive	0	0	(2)2	0	(4)2	(1)2	0	0
Autumn-olive	0	0	0	0	(3)1	0	0	0
French broom	0	0	(7)4	(8)4	(1)3	0	(4)3	0
Common St. Johnswort	0	(1)3	(1)3	(3)4	(1)4	(6)2	(1)1	0
Cogongrass[b]	0	0	(1)2	(2)4	(9)3	(2)4	(9)4	(15)4
Perennial pepperweed	0	0	(2)3	0	(1)3	0	0	0
Sericea lespedeza	0	(3)4	(3)2	(4)1	(1)1	(1)3	0	0
Privet[b]	0	0	(2)4	0	(2)4	0	0	0
Toadflax[b]	0	0	0	0	0	(6)4	0	0
Japanese honeysuckle	0	0	(5)4	0	(6)4	(1)4	0	(2)3
Bush honeysuckles[b]	0	(2)1	(3)1	0	(6)3	0	0	0
Climbing ferns[b]	0	0	0	0	0	0	(2)1	(2)1
Purple loosestrife	0	0	(4)3	0	0	0	(2)3	0
Melaleuca	0	0	(13)3	(6)4	(7)2	0	(10)4	(8)2
Japanese stiltgrass	0	0	(1)4	(4)4	(1)1	0	(2)2	0
Sulfur cinquefoil	0	0	(1)4	(1)2	0	0	0	(1)2
Kudzu	0	(3)4	(4)3	0	(2)3	0	0	0
Multiflora rose	0	0	0	0	0	0	0	0
Brazilian pepper	0	(1)4	(5)4	(1)1	(4)4	0	(2)2	(5)4
Perennial sowthistle	0	0	0	(2)4	(1)4	(1)4	0	0
Johnson grass	0	(1)4	(2)2	0	0	(4)4	0	0
Spanish broom	0	(1)1	(1)1	0	(1)1	0	0	0
Tamarisk[b]	0	(1)3	(6)4	0	(5)4	(3)4	(8)3	(7)4
Chinese tallow	0	0	(5)3	0	(2)2	0	(4)2	(3)2
Gorse	0	(4)4	(10)4	(9)4	(9)4	(5)4	(12)4	(6)4

[a] For all species except cheatgrass, we cited every source found on fire-related topics. A few cheatgrass studies were not cited because information was ample and they added no new information to the species review, so the total number of citations for topics under cheatgrass may be conservative. Similarly, cheatgrass studies reporting on a topic in a preliminary document and continued in a subsequent document were counted as one study.

[b] Two or more species included in review; see table 12-1 for complete list of species included. Ranked information may not apply to all species in that review.

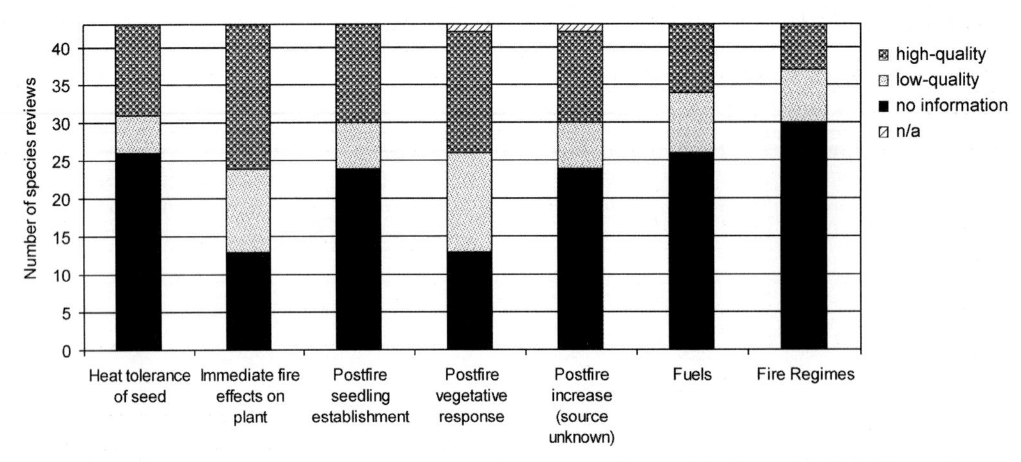

Figure 12-2—Highest quality of information available on fire and fuels topics for 43 species reviews in FEIS. "Low quality" = rank of 1 or 2. "High quality" = rank 3 or 4. See "Methods" for explanation of information-quality ranking.

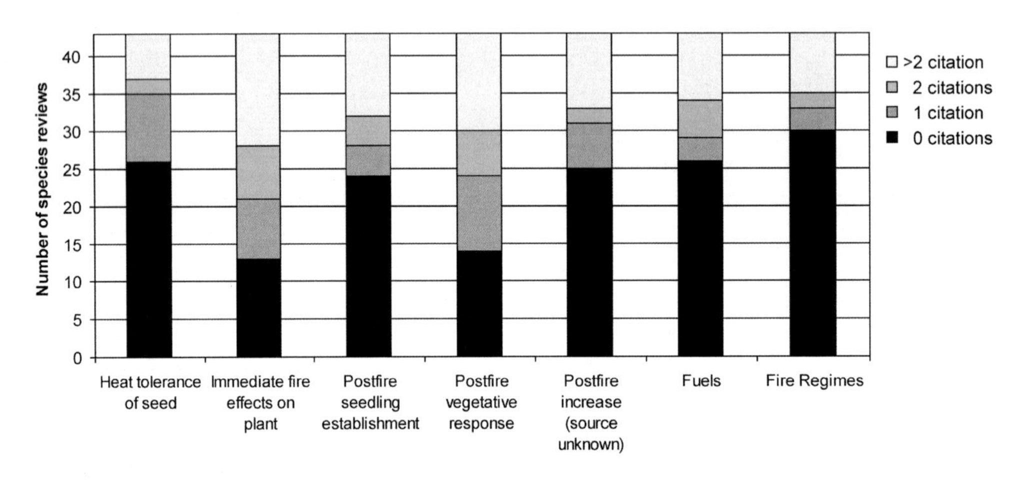

Figure 12-3—Frequency of zero, one, two, and more than two citations of any quality covering fire and fuels topics for 43 species reviews in FEIS. "0" in this graph corresponds to "no information" in figure 12-2.

Many species reviews refer to experiments on fire effects, and most of these references are of high quality. However, very few fire experiments report the effects of variation in fire severity, season, or burn interval; furthermore, most of these studies report results from only 1 postfire year so they are not useful for understanding postfire succession (table 12-5; fig. 12-4). Of our 43 species reviews, 37 percent cite no direct experimental evidence of the effects of fire. Where experimental evidence of fire effects is reported, it often lacks important information regardless of quality rating. In the case of Russian knapweed (*Acroptilon repens*), for example, an article published in a peer-reviewed journal (Bottoms and Whitson 1998) concludes that the species cannot be effectively controlled by burning. However, the article fails to provide any information on prefire plant community, fuels, fire behavior, or burn conditions.

Of the experiments cited, very few address the effects of varied fire regime characteristics and postfire succession. Less than a third of our species reviews contain any information on the differential effects of fire severity, season of burn, or interval between fires on nonnative invasives (table 12-5; fig. 12-4). Among the remaining reviews that contain some information on variation in a fire regime characteristic or succession, only a handful—one to three for each topic—have more than two citations (fig. 12-5). For most species, it is unclear how responses to fire might change over time. Among our species reviews, 30 percent had studies reporting plant responses over multiple years (fig. 12-4), and few offered insight about the potential effects of long-term maintenance of native fire regimes on nonnative invasive plants.

Table 12-5—Information available on fire experiments and addressing relationships between nonnative invasive species and aspects of the fire regime. Total number of sources available on each topic for each species review is given in parentheses, followed by highest-ranked quality of information. See "Methods" section for explanation of ranks 0 to 4 used in table.

Species review	Fire experiment	Varying fire severities	Varying burn seasons	Varying burn intervals	Multiple postfire years
Norway maple	0	0	0	0	0
Russian knapweed	(2)2	0	0	0	0
Tree-of-heaven	0	0	0	0	0
Garlic mustard	(5)4	(1)4	(3)4	0	(2)4
Giant reed	0	0	0	0	0
Cheatgrass	(19)4	(3)4	(2)4	0	(15)4
Hoary cress[a]	0	0	0	0	0
Musk thistle	(1)2	0	0	0	0
Oriental bittersweet	0	0	0	0	0
Diffuse knapweed	0	0	0	0	0
Spotted knapweed	(2)3	0	0	0	(1)3
Yellow starthistle	(3)4	0	0	0	0
Rush skeletonweed	0	0	0	0	0
Canada thistle	(13)4	(1)4	(2)4	(2)3	(2)4
Bull thistle	(1)3	0	0	0	0
Field bindweed	0	0	0	0	0
Houndstongue	0	0	0	0	0
Brooms[a]	(2)4	0	(2)4	0	0
Russian-olive	0	0	0	0	0
Autumn-olive	0	0	0	0	0
French broom	(4)4	0	0	(1)4	(1)3
Common St. Johnswort	(6)4	(1)4	(1)4	0	(2)4
Cogongrass[a]	(2)4	0	0	0	(1)3
Perennial pepperweed	(1)3	0	0	0	0
Sericea lespedeza	(1)3	0	(1)3	0	0
Privet[a]	0	0	0	0	0
Toadflax[a]	(6)4	0	(1)3	0	0
Japanese honeysuckle	(8)4	0	(2)4	0	(3)4
Bush honeysuckles[a]	(2)3	0	0	0	(2)3
Climbing ferns[a]	0	0	0	0	0
Purple loosestrife	0	0	0	0	0
Melaleuca	(3)4	0	(1)3	(1)3	(2)3
Japanese stiltgrass	(1)4	0	0	0	0
Sulfur cinquefoil	(1)4	0	(1)4	0	0
Kudzu	0	0	0	0	0
Multiflora rose	(1)4	(1)4	0	0	0
Brazilian pepper	(3)4	0	0	0	(2)4
Perennial sowthistle	(3)4	0	(1)4	0	0
Johnson grass	(2)4	0	(1)4	0	(1)4
Spanish broom	0	0	0	0	0
Tamarisk[a]	(8)4	(2)4	(1)3	(2)2	0
Chinese tallow	(2)3	0	(2)2	0	0
Gorse	(8)4	(3)4	(2)4	0	(7)4

[a] Two or more species included in review; see table 12-1 for complete list of species included. Ranked and tallied information may not apply to all species in that review.

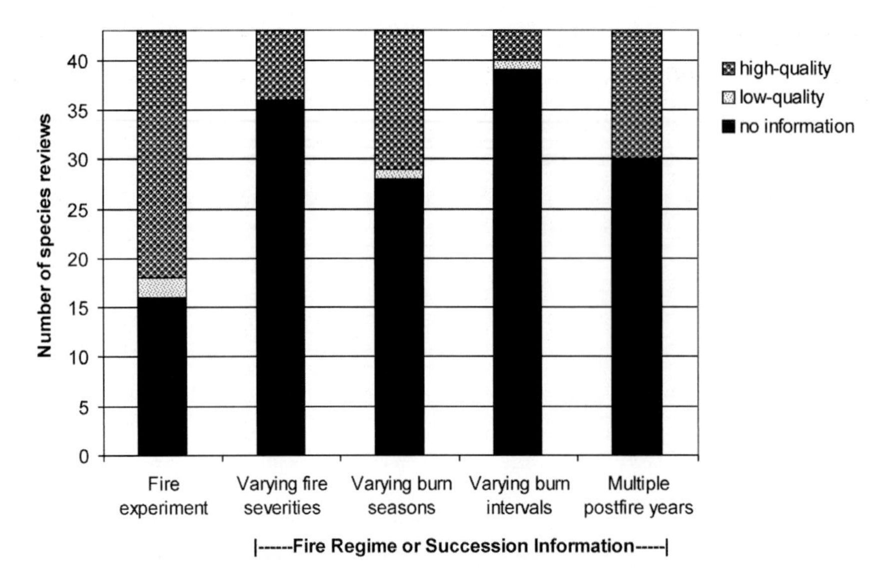

Figure 12-4—Highest quality of information available from fire experiments for 43 species reviews in FEIS. "Low quality" = rank of 1 or 2. "High quality" = rank 3 or 4. See "Methods" for explanation of information-quality ranking.

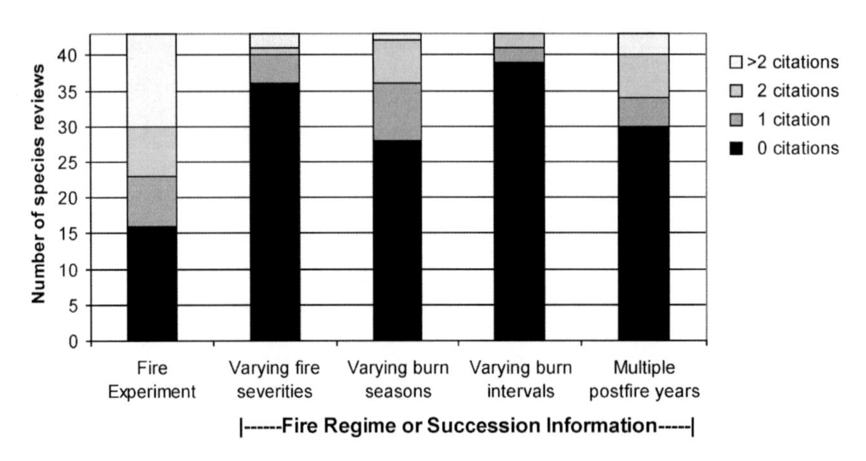

Figure 12-5—Frequency of zero, one, two, and more than two citations of any quality covering fire experiments for 43 species reviews in FEIS. "0" in this graph corresponds to "no information" in figure 12-4.

Discussion

Many articles describe fire's relationship with nonnative invasive species, but the quality and quantity of information are often inadequate for managers to use with confidence. A manager planning a prescribed burn, for example, needs to know which nonnative invasives are of concern and assess the potential for establishment, persistence, and/or spread of those species after fire in a particular area. At the very least, the manager needs information on basic biological traits of each species, such as vegetative reproduction and requirements for seedling establishment. Better would be information on how those traits are expressed in response to fire. Additional information on the ecology and invasiveness of each species under various environmental conditions would further improve the manager's basis for decisions. If no information is available for the particular ecosystem under consideration, a literature review synthesizing research from other ecosystems or a model predicting species response based on basic biological traits and ecological relationships could be helpful. The best information that a manager could hope for would describe long-term outcomes from fire research that has a scope of inference covering that ecosystem under various burning conditions, at varying times of year, with varying fire severities and intervals between burns. Publications with such a comprehensive scope and content are few when considered in light of the number of nonnative, invasive plants in the United States and the probability that invasiveness varies from one plant community to another. Here we discuss the ways in which information on basic biology, invasiveness and invasibility, distribution, ecology, and responses to fire, heat, and fire regimes pertains to managing invasives and fire. We compare the knowledge available for FEIS species reviews with the knowledge needed to manage with confidence, and we offer suggestions on how to deal with the fact that managers frequently need more knowledge than is available.

Basic Biology

The ability of a plant to establish, persist, and/or spread in a postfire community depends partly on its resistance to heat injury (chapter 2), and may be inferred from experimental evidence or, with less certainty, from information on reproductive strategies. Responses to fire vary with plant phenology relative to timing of the fire, the location of perennating buds and seeds relative to lethal heat loads, seed production, seed dispersal, seed longevity, and requirements for successful seedling establishment (chapter 2). Information on seasonal changes in carbohydrate reserves of roots and other underground tissues may help managers understand when fire will have the greatest impact on perennial species. Our results indicate that while information on phenology of flowering and seed production is generally available for the species reviewed, information on seasonal changes in carbohydrate reserves of roots and other underground parts is less abundant (fig. 12-1). Descriptions of depth of underground perennating tissues, crucial for understanding the varying effects of different fire severities, are rarely available.

In the absence of information on postfire regeneration, knowledge of a plant's ability to regenerate after mechanical injury or removal of top growth may help managers assess the likelihood that a plant will sprout after fire. High-quality information on post-injury regeneration was available for the majority of the species examined in this study; however, fires and mechanical disturbances alter a site in different ways, so biological responses cannot be assumed to be equivalent. Where high-quality information is lacking on this topic, it does not always indicate scant or poor information. For species such as melaleuca, for example, post-injury regeneration is so prolific and so obvious that there is no need for peer-reviewed literature to demonstrate this response.

The ability of a plant to establish from seed in a postfire environment depends on seed production and dispersal, requirements for germination and seedling establishment, and seed bank dynamics. We found high-quality information on seed production, dispersal, and seedbed requirements for germination for most species reviews, although this was rarely available for postfire conditions. Similarly, most species reviews had high-quality information on seed banking (fig. 12-1); however, the scope of applicability of this information was usually limited to laboratory experiments or field studies on other continents. A description of seed bank dynamics, including seed longevity, temporal and spatial variation in the number of viable seeds stored in the soil, and the seed bank's relationship to disturbances, can help managers assess the potential role of invasive species in a postfire environment (Pyke 1994). In many communities, nonnative species are common in soil seed banks, and there are differences between the species growing on the site and those present in the soil seed bank (for example, Halpern and others 1999; Kramer and Johnson 1987; Laughlin 2003; Leckie and others 2000; Livingston and Allessio 1968; Pratt and others 1984; Rice 1989). These differences may lead to substantial changes in community composition following fire, including establishment of nonnatives. Seed longevity also influences fire's effectiveness in controlling annual plants (Brooks and Pyke 2001). More research is needed on the relationship between seed banking, germination requirements, field conditions, and fire.

Impacts, Invasiveness, and Invasibility

To make informed decisions about nonnative invasive species and fire, managers need to know when a nonnative species threatens a native ecosystem. For example, species that alter fuel characteristics of invaded communities may alter fire regimes such that an invasive plant/fire cycle is established (chapter 3). Assertions regarding impacts of particular nonnative species on native ecosystems are abundant in the literature; however, quantitative evaluations of these impacts are not common. In fact, little formal attention has been given to defining what is meant by "impact" or to connecting ecological theory with particular measures of impact (Parker and others 1999). For example, reviews by Hager and McCoy (1998) and Anderson (1995) describe purported negative impacts caused by purple loosestrife (*Lythrum salicaria*) in North America. Both papers express concern that claims of ecological harm caused by purple loosestrife (for example, Thompson and others 1987) are not supported by quantitative assessments, so some management activities aimed at controlling the species could be inappropriate. Parker and others (1999) point out that disagreements on the impact of historical invasions reflect the fact that ecologists have no common framework for quantifying or comparing the impacts of invaders. Managers are therefore cautioned to read generalizations regarding the impacts of nonnative, invasive species with care.

Distribution and Site Information

The likelihood that a nonnative, invasive species will establish, persist and spread in an area is determined not only by properties of the species, but also by the structure, composition, and successional status of the native plant community, site factors and conditions, landscape structure (Rejmánek and others 2005a; Sakai and others 2001; Simberloff 2003), and the species' current distribution. In this study, the information available on distribution of nonnative invasive species had limited usefulness for estimating the potential for a particular species to invade after fire. Comprehensive information on distribution and site requirements was not available for any of the species reviewed. The information currently available is often based on county records rather than systematic surveys. Because of this, it may reflect the density of botanists in a particular area more than the density of invasive plants (Moerman and Estabrook 2006; Schwartz 1997).

The most consistent predictor of invasiveness may be a species' success in previous invasions (for example, Kolar and Lodge 2001; Williamson 1999). Based on this idea, Williamson (1999) emphasizes the need for more studies on the population dynamics of invaders and better definitions of their demographic parameters. Similarly, in a review of literature on sulfur cinquefoil, Powell (1996) suggests that surveys including geographic location, plant community type, seral stage, site characteristics (including disturbance and management history), and size, density, and canopy cover of infestations can help establish ecological limits of nonnative plants, define potential North American distributions, and identify other areas where a nonnative species is likely to be invasive. Such surveys could also provide a baseline for monitoring populations, direction for management activities, and a means for evaluating management effectiveness (Powell 1996). Mack and others (2000) agree, adding that such information would be useful for calculating an invasive's rate of spread. While surveys such as these may seem unrealistic given the resources needed to survey large areas, it may be possible to detect occurrence and spread of invasives using satellite remote sensing, aerial photography, hyperspectral imagery, or other spatial information technologies (review by Byers and others 2002).

Ecological Information

A nonnative species' invasiveness in a postfire environment depends not only on the species' location and response to fire, but also on the response of other plants in the community (chapter 2). Like Grace and others (2001), we found that information is very incomplete with regard to fire effects on competitive interactions between nonnative invasives and the native plant community. It is routinely asserted that a nonnative, invasive species "outcompetes" native species, but rarely are these assertions supported by quantitative data. A review by Vilà and Weiner (2004) of published pair-wise experiments between invading and native plant species[6] suggests that the effect of nonnative invasives on native species is usually stronger than vice versa. However, because the selection of invaders and natives for study is not random (that is, the plants most frequently chosen for study are those that cause the most trouble), the data could be biased towards highly competitive invaders and natives that may be weaker than average competitors. Furthermore, the reviewers point out, methods that have been used to investigate competition between invasive and native species are often limited in scope and applicability (Vilà and Weiner 2004).

Information on persistence of a nonnative, invasive plant species on a particular type of site, and how persistence of this species may change successional trajectories, is important for assessing potential impacts of invasion, but is available only for a limited number of species and locations. Long-term research in a variety of locations or plant communities with contrasting characteristics might help managers assess potential persistence, spread, and successional trajectories after a species has become established in an area, and understand what changes may occur after fire. Additionally, control plots maintained without intervention or attempts at reducing invasives are essential for long-term research. Results from a study on Illinois prairie vegetation illustrate this point. Anderson and Schwegman (1991) studied 20 years of change in a prairie plant community, which included substantial cover of Japanese honeysuckle (*Lonicera japonica*), in response to four prescribed burns. The study compared different burn treatments, but it did not compare burned with unburned plots. Control plots were established and measured in the first 2 years of the study, and no changes were observed during these

[6] Nonnative, invasive plant species included in the Vilà and Weiner (2004) review and also addressed in our project include *Ailanthus altissima*, *Bromus tectorum*, *Centaurea diffusa*, *C. maculosa*, *C. solstitialis*, *Hypericum perforatum*, and *Lonicera japonica*.

2 years (Anderson and Schwegman 1971). No data or results from control plots were reported for subsequent years (Anderson 1972; Anderson and Schwegman 1991; Schwegman and Anderson 1986). Without long-term data from controls, however, the researchers could not compare long-term variation within burns to variation over the same time in unburned areas. They could have attributed long-term changes to fire alone when variation may have been caused by other factors, such as weather. Where information on competition and long-term successional patterns is unavailable—and this would be in most ecosystems—sustained monitoring, analysis of local patterns of change, and flexible, adaptive approaches to management can provide guidance.

Responses to Fire, Heat, and Postfire Conditions

While information on heat tolerance of perennating tissue and seed would be helpful to managers, measuring heat transfer into plant tissues is complex. Observations and models describing heat tolerance currently focus mainly on damage to trees during fires with relatively low fireline intensities (Dickinson and Johnson 2001, 2004; Jones and others 2004). We found no sources describing heat tolerance of perennating tissues for the nonnative invasives examined in this study, and only 28 percent of reviews include high-quality information on heat tolerance of seed (table 12-4; fig. 12-2)[7]. Most of this research is based on laboratory observations, which may not replicate field conditions. Exposure to smoke or chemicals leached from charred material contributes to breaking seed dormancy for some species (Keeley and Fotheringham 1998a). Information on smoke and char effects was not found for our species reviews, although one study is available on how exposure of cheatgrass (*Bromus tectorum*) seeds to smoke affects its seedling development (Blank and Young 1998).

Field Experiments Addressing Fire Effects

Information on basic biology, ecological interactions, and responses to heat may not apply directly to fire responses under field conditions (Harrod and Reichard 2001). Information concerning the effects of specific fire behavior on nonnative invasives in a specific plant community under specific fuel and weather conditions may be essential for unraveling the effects of fire from those of other variables. Reports from comprehensive fire research studies and well-documented prescribed fires are sparse in the literature, as are reports of experiments describing the use of fire to control nonnative invasive plants (but see chapter 4). Where such information is available, it is sometimes too limited in scope (one or two sites) to support application on other sites, in different ecosystems, under different burning conditions.

We found relatively few papers on fire effects that distinguished between postfire seedling establishment and postfire vegetative recovery. If research does not differentiate between seedlings and stems of vegetative origin, managers will have limited ability to predict postfire population dynamics. In the first year after fire, it is often relatively simple for field observers to determine whether a plant originated from seed or from underground parts; researchers should record and report this information as a routine part of fire effects studies.

Even when the relationship between a nonnative invasive species and fire is described by high-quality research, the information may not be widely applicable to management. Causes of limited scope of inference and suggestions for addressing these limitations are presented in table 12-6.

Responses to varying fire regime characteristics—When using fire as a tool to change or maintain floristic composition in a plant community, one must consider not only the effects of individual fires, but also the effects of the imposed fire regime (chapter 1) over a long time. In some cases, fire managers aim to promote native species by introducing fire at seasons and intervals that approximate presettlement or reference fire regimes, but little information is available regarding the effects of these fire regimes on nonnative invasive species. In other cases, the presettlement fire regime of the invaded ecosystems is unknown; examples include ecosystems where yellow starthistle (*Centaurea solstitialis*) and tamarisk (*Tamarix* spp.) are most problematic (chapters 8 and 9). Likewise, little is known about the differential effects of fire severity, season of burn, or interval between fires on nonnative invasives. In some cases, comparative studies on the effects of burning in different seasons may be lacking because management constraints require that burning be conducted during a particular season; wherever possible, research should measure and report variation in fire severity and fire season relative to plant phenology.

[7] Reviews for *Bromus tectorum* and *Genista monspessulana* do include experimental evidence describing fire effects on seed (Alexander and D'Antonio 2003; Keeley and others 1981; Odion and Haubensak 2002; Young and Evans 1978; Young and others 1976), though the research does not directly address heat tolerance. Several references describe germination of *Hypericum perforatum* seed after fire (for example, Briese 1996; Sampson and Parker 1930; Walker 2000), and one study examines heat tolerance of *H. perforatum* seed in the laboratory (Sampson and Parker 1930).

Table 12-6—Reasons for limited scope of inference from high quality research on fire effects.

Cause of limited scope	Example, explanation	Ways to address
Research not specifically designed to assess interactions between fire and nonnative invasive plants	Fire effects on *Lonicera japonica* are described by 11 studies, 9 of high quality, but fire effects are incidental to the study and not thoroughly covered for 7 of these.	Incidental data on fire might improve the usefulness of such research for fire managers; better would be research designed for understanding nonnatives and fire.
Fire effects described by only one study. A single study rarely covers multiple ecosystems, seasons, and burning conditions, so the information is usually not sufficient for generalizing.	Occurred for 7 of the 27 FEIS species reviews with experimental information, while another 7 reviews cite only 2 studies (fig. 12-5).	Monitor results of management actions based on results from a single study, and use adaptive management as new information becomes available.
Fire effects described by several studies but only one vegetation type	Several studies have been conducted on prescribed burning to control *Centaurea solstitialis* (DiTomaso and others 1999; Hastings and DiTomaso 1996; Martin and Martin 1999), but all were within a single ecosystem type, so the information is insufficient to generalize to other ecosystems.	Monitor results of management actions based on results from studies in other vegetation types, and use adaptive management as new information becomes available.
Fire effects described primarily outside North America	Some fire effects experiments on *Hypericum perforatum* (for example, Briese 1996) and *Ulex europaeus* (for example, Johnson 2001; Soto and others 1997) were conducted in Australia, New Zealand, or Europe. It is difficult to have confidence that this experimental evidence applies to North American plant communities.	Monitor results of management actions based on results from studies on other continents, and use adaptive management as new information becomes available.
Complex patterns of fire severity that cannot be correlated with postfire vegetation	Faulkner and others (1989) conducted research on invasives including *Ligustrum sinense*. Fire behavior varied from plot to plot, apparently confounding detection of immediate effects on aboveground plant parts.	Design burn studies to account for as much variation as possible; avoid burning when conditions for fire spread are marginal.
Incomplete or ineffective burn treatments	Rawinski (1982) attempted to compare the effects of burning after cutting with cutting alone on *Lythrum salicaria*. Attempts to burn *Lythrum salicaria* stems that had been cut were generally ineffective, so treatments could not be compared.	Avoid burning when conditions for fire spread are marginal.

The lack of information on plant responses over multiple years and the potential effects of long-term maintenance of native fire regimes on nonnative invasive plants impedes long-term planning and restoration of ecosystem processes. Anderson and Schwegman's (1991) study illustrates both the value of long-term research and the need to use control plots for the duration of a study. They examined effects of burning on southern Illinois prairie vegetation, including invasive Japanese honeysuckle, over the course of 20 years. Japanese honeysuckle decreased with frequent fire and increased after burning treatments ceased. However, this study lacked long-term control plots and so failed to compare changes in burned plots with changes in the surrounding unburned plant community.

Representing Information Quality in Literature Reviews: Potential for Illusions of Knowledge

While managers need the knowledge produced by science to make decisions, they generally rely on scientists to search the scientific literature and synthesize information. When scientists write literature reviews (including reviews within articles presenting primary research), book chapters, and agricultural extension literature, they need to frame and qualify information so managers will understand the kind of knowledge being reported (for example, Krueger and Kelley's (2000) categorization of natural resources literature). In our study, unsubstantiated assertions (quality ranking of 1) were found in one or more species reviews for every fire-related topic (table 12-4). Without context and hedging, such assertions create an illusion of certainty about a subject for which no empirical evidence is available, and the reader cannot determine how well the results apply to a particular management question. Readers should note when information is provided on a study's scope of inference and apply unsubstantiated assertions with caution. The 2003 Data Quality Act (Public Law 106-554, Section 515) requires that Federal land managers base decisions on high-quality information; literature reviews that fail to identify what kind of information is cited or misquote original research do a disservice to their readers.

Monitoring, Data Sharing, and Adaptive Management

Managers need not depend completely on published reports to form a useful body of knowledge. Records of management treatments, especially those preceded by measurement and followed by monitoring and data analysis, can inform and contribute to local management (Christian 2003) when a flexible, adaptive approach is used. When supplemented by complete site descriptions and shared across sites, landscapes, and regions, monitoring data could provide substantial guidance for management of invasives and fire. Well-designed, long-term monitoring programs can provide valuable ecological information about the invasion process and how individual ecosystems are affected (Blossey 1999); the ways in which data will be analyzed and presented must be addressed in the design phase for monitoring to be useful in assessing treatment success (Christian 2003). Suggestions for monitoring interactions between fire and nonnative species are discussed more thoroughly in chapter 15.

Conclusions

Current scientific knowledge about the relationship between invasive plants, ecosystem characteristics, and fire regimes is limited in quantity and quality. Scientists have the responsibility and require the necessary resources to study interactions between invasives, native communities, and fire, and variation in all of these factors. In addition, timely reporting of research is critical, including careful descriptions of the population studied and variables controlled, in both primary research and reviews of the literature. It is very difficult for managers to access information in unpublished reports; even if they can obtain such data, it may not be provided with contextual information that enables managers to assess its applicability to the ecosystems they are managing. Furthermore, the Data Quality Act (Public Law 106-554, Section 515) obligates managers to rely mainly on results published in peer-reviewed literature.

McPherson (2001) suggests that the enormity, complexity, and importance of management make the creative application of existing knowledge as important, and as difficult, as the development of new knowledge. High-quality information on the relationships between nonnative, invasive plants and fire is sparse when compared with the need for knowledge. More information is continually becoming available, but research cannot possibly investigate every possible combination of nonnative species and plant community in the United States, especially since nonnative plants continue to be introduced. Where research specific to a species and community is lacking, managers often rely on the synthesis provided by literature reviews, so it is important that reviews describe not only general ecological patterns but also the scope and limitations of the knowledge presented. When managers apply science to management in a specific plant community, they have the responsibility to recognize the limitations of current knowledge, apply generalizations cautiously, identify needs for site-specific knowledge, monitor results over many years, and use results adaptively, improving the management of nonnative invasive species in impacted plant communities over time.

Notes

Erik J. Martinson
Molly E. Hunter
Jonathan P. Freeman
Philip N. Omi

Chapter 13:

Effects of Fuel and Vegetation Management Activities on Nonnative Invasive Plants

Introduction

Twentieth century land use and management practices have increased the vertical and horizontal continuity of fuels over expansive landscapes. Thus the likelihood of large, severe wildfires has increased, especially in forest types that previously experienced more frequent, less severe fire (Allen and others 2002). Disturbances such as fire may promote nonnative plant invasions by increasing available light and nutrients, as well as by decreasing competition from native plants for these resources (Fox 1979; Melgoza and others 1990). Once established, nonnative species may further alter fuel bed characteristics and increase the likelihood of future wildfires (Whisenant 1990a). Land managers increasingly rely on prefire fuel manipulations to reduce wildfire potential, and these efforts have expanded significantly under the current National Fire Plan (USDI and USFS 2001).

However, fuel treatments themselves are disturbances that may promote invasion by nonnative plant species. Depending on the intensity, severity, size, and seasonality of a fuel treatment, increased availability of light, water, and nutrients may result (Covington and others 1997; Gundale and others 2005; Kaye and Hart 1998); these conditions can favor spread of nonnative species (Brooks 2003; Hobbs and Huenneke 1992; Stohlgren and others 1999b). Response of nonnative species to fuel treatments may also vary by treatment type, whether accomplished by means of prescribed fire, heavy equipment, hand tools, or chemicals. For example, unseen nonnative seeds may be carried by humans and mechanical equipment used in some types of fuel reduction treatments. This can increase propagule pressure of nonnative species in an area, which is an important factor in predicting success of nonnative plant invasions (D'Antonio and others 2001b; Lockwood and others 2005). Use of mechanical equipment may also result in soil disturbances that favor nonnative plant establishment (Hobbs and Huenneke 1992; Kotanen 1997). Based on the potential effects of fuel treatments on ecosystem structure and function, it is possible for nonnative species to thrive in areas treated for fuels reduction (Sieg and others 2003).

Potential treatment outcomes are myriad, from interactions among the wide variety of possible fuel treatment characteristics discussed above, and site factors such as topography, soils, climate, and proximity to existing nonnative seed sources. As with most topics related to fire ecology, confusing and conflicting literature that addresses fuel treatment effects on nonnative species has begun to accumulate. A fire regime construct often provides the most useful organization for such literature (Kilgore and Heinselman 1990). For example, ecosystems where surface fire has become less frequent due to changes in land use practices are the most common locations for fuel treatments that target hazardous accumulations of native fuels. But these ecosystems may also be the most resilient to post-treatment invasion by nonnative species due to adaptations of the native species that evolved under a regime of frequent disturbance in the past. Contrastingly, in ecosystems where fire has become more frequent due to the establishment of a positive feedback cycle between fire and nonnative grasses, fuel treatments often focus on the nonnative species themselves, and subsequent treatments may be necessary to prevent their reestablishment (chapters 3 and 4).

Here we review the extant research that addresses effects of fuel modification on nonnative invasive plant species. The discussion is organized by broad ecosystem types that are based on historic fire regimes: (1) those characterized by frequent, low-severity fires; (2) those characterized by mixed fire regimes; and (3) those characterized by infrequent, high severity fire regimes. The third group is subdivided into ecosystems where fires have become more frequent due to a type conversion to more flammable nonnative species and ecosystems where the fire regime has remained essentially unchanged. The discussion is limited to studies conducted in the United States and focuses mainly on nonfire vegetation management activities to avoid overlap with the other chapters of this volume, though some of the reviewed studies address treatment combinations that include prescribed fire. A discussion of the literature on combining fire with other fuel treatments is also provided in chapter 2 (see "Question 3. Does Additional Disturbance Favor Invasions?" page 22).

High-Frequency, Low-Severity Historic Fire Regime

Ecosystems where frequent fires historically limited fuel accumulation and favored fire resistant plant species have been most affected by the fire exclusion practices of the 20th Century (Allen and others 2002). Fires in these ecosystems have become less frequent and tend to be more severe when they do occur. Management activities in these ecosystems are thus most likely to include fire hazard reduction and/or ecological restoration as a primary objective. This ecosystem group includes vegetation dominated by long-needled conifers in the West and Southeast, as well as oak savannas in the Midwest. While a few publications from eastern forests with a high-frequency and low-severity historic fire regime discuss the effects of silvicultural treatments on understory vegetation, none mention nonnative species (Dolan and Parker 2004; Gilliam and others 1995; Ruben and others 1999; Stransky and others 1986). This lack of mention does not necessarily imply that nonnative species were absent from treated areas, only that these studies did not include an assessment of nonnative species as a research objective. Investigators who are currently monitoring the effects of fuel treatments as part of the national Fire and Fire Surrogates study in the piedmont of South Carolina (Waldrop, personal communication, 2006) and the coastal plain of Florida (Brockway, personal communication, 2006) indicate that nonnatives are generally not a concern in their study areas. Thus most of the information on the effects of fuel treatments on nonnative plant species in ecosystems with a frequent, low-severity fire regime comes from ponderosa pine (*Pinus ponderosa*) forests, particularly those in the Southwest.

Prescribed fire and mechanical thinning of trees, alone and in combination, are common treatments used to reduce the potential for spread of hazardous wildfire in ponderosa pine forests throughout the West (Agee and Skinner 2005; Arno and Fiedler 2005; Covington and others 1997; Kaufmann and others 2003). Such treatments have been shown to be effective in reducing the severity of wildfires in ponderosa pine forests (Cram and others 2006; Finney and others 2005; Martinson and Omi 2003; Martinson and others 2003; Pollet and Omi 2002). However, an undesirable consequence of such treatments may be the creation of environments that are conducive to nonnative species establishment and spread.

Investigations of nonnative species abundance following fuel treatments in ponderosa pine forests have shown mixed results. Higher nonnative species cover was found following thinning and burning treatments in ponderosa pine forests in Montana (Metlen and Fiedler 2006). In northern Arizona and New Mexico, higher cover of nonnative plants has been found following some thinning and burning treatments but not others (Abella and Covington 2004; Fulé and others 2005; Griffis and others 2001; Hunter and others 2006; Moore and others 2006; Speer and Baily in review). Results have also been mixed in the Black Hills of South Dakota (Thompson and Gartner 1971; Uresk and Severson 1998; Wienk and others 2004). However, research to date has not found significant increases in nonnative species following thinning and burning treatments in ponderosa pine forests of the Pacific Northwest or in the Front Range of Colorado. Several studies found no evidence of nonnative species following fuel treatments in ponderosa pine forests of the Pacific

Northwest (Busse and others 2000; McConnell and Smith 1965, 1970; Metlen and others 2004), while one study found post-treatment nonnative cover to be very low (Page and others 2005). Neither Fornwalt and others (2003) nor Hunter and others (2006) found any difference in nonnative species cover between treated and untreated areas of the Colorado Front Range, and both studies found overall nonnative cover to be very low.

Researchers have proposed several hypotheses for the observed variability in the response of nonnative species to fuel treatments. Nonnative species cover tends to increase following intense treatments that result in severe site disturbances, such as extensive canopy removal or soil alteration. Higher nonnative species cover has been found in areas where large portions of the overstory were removed (Fulé and others 2005; Uresk and Severson 1998; Wienk and others 2004). Burned slash piles, which produce intense soil heating on a small scale, are also prone to invasion by nonnative species (Korb and others 2004). Several studies have found that a combination of thinning and burning resulted in higher cover of nonnatives than either thinning or burning alone (Fulé and others 2005; Metlen and Fiedler 2006; Moore and others 2006; Wienk and others 2004).

Treatments involving high-severity disturbance do not always result in invasion by nonnative species (Laughlin and others 2004); variations are sometimes related to propagule availability. For example, nonnative species were abundant after fuel treatments in northern Arizona in areas that have a history of human caused disturbance and numerous nonnative seeds in the seed bank; however, they were not found following restoration treatments in more remote areas without substantial nonnative plant populations (Korb and others 2005). Correspondingly, Fornwalt and others (2003) found little difference in nonnative species richness between adjacent heavily managed and historically protected ponderosa pine landscapes in the Colorado Front Range that likely had similar pressure from nonnative seed sources. However, site factors such as topography, soil characteristics, and weather conditions may also influence invasion success (Hunter and others 2006).

Most of the nonnative species found following fuel treatments in ponderosa pine forests (such as common mullein (*Verbasucum thapsus*), lambsquarters (*Chenopodium album*), and prickly lettuce (*Lactuca serriola*)) are not considered a threat to native plant communities. Weink and others (2004) suggest that observed increases in such species after fuel treatments may not be problematic in the long term, since the species are usually transient members of recently disturbed communities. However, only one study has reported data to support this presumption of no long-term effect from management activities on the abundance of nonnative plants in ponderosa pine forests (Fornwalt and others 2003). The lack of long-term studies is conspicuous and requires attention before conclusions can be made regarding the effects of fuel treatments on nonnative species and potential mitigation strategies.

If nonnative invasive species are present in an area, they are likely to be favored by fuel treatments (Wolfson and others 2005) and may create problems for long-term native species diversity and ecosystem function. Seeding treatments may provide a deterrent to nonnative establishment (Korb and others 2004), although contaminated seed mixes can render such treatments counterproductive (Hunter and others 2006; Springer and others 2001). Some authors question the benefits of seeding entirely (Keeley and others 2006a). Careful monitoring and control actions should take place following fuel treatments, particularly if known nonnative invasive populations are in the vicinity of treatment areas.

Mixed-Severity Historic Fire Regime

Mid-elevation mixed conifer forests in the western states exhibit historical evidence of surface fires as well as less frequent crown fires (Agee 1993). The mesic environmental conditions in these ecosystems are such that fuels are less frequently available for combustion than in the fire regime discussed above (Martin 1982). Canopies tend to be denser and ladder fuels more abundant, facilitating crown fire ignition when surface combustion becomes possible, such as following a prolonged drought. Fuel treatment efficacy is less well established in these ecosystems. Most of the evidence is from commercial harvests rather than from treatments to reduce fire hazard *per se* (for example, see Weatherspoon and Skinner 1995). Likewise, most of the research on how fuel manipulation affects invasion by nonnative species in these ecosystem types is from commercial harvests.

The studies that represent mixed fire regime ecosystems are all from mixed conifer forests in the Cascade and Sierra Nevada ranges of California, Oregon, and Washington. All but one (Dyrness 1973) found significant increases in nonnative species following timber harvest, though severity of invasion varied substantially. Much of this variation may be attributable to the intensity and age of treatments.

Several short-term (1 to 6 years following treatment) studies have compared nonnative species cover in different types of harvest that varied in intensity. Nelson and Halpern (2005) found the frequency of nonnative species to range from 0 percent in intact mixed conifer forest to 1 percent in aggregated retention harvests, to as high as 31 percent in clearcuts. North and others (1996) compared the effects of clearcutting and green tree retention (27 live trees/ha) and found that both treatments resulted in significantly higher nonnative species cover than found in intact forest, though the clearcuts had slightly higher nonnative cover than the green tree retention harvests. Thysell and Carey (2001a) found significantly higher nonnative cover 3 years after second-growth stands were treated with variable-density thinning than in untreated second-growth stands. They hypothesize that variable-density thinning may result in less persistence of nonnatives than conventional thinning, since variable density thinning produces greater within-stand heterogeneity. However, no data were presented to support this hypothesis.

Longer term studies have found nonnative species to persist for several decades after intense fuel treatments. Thomas and others (1999) surveyed plots from three levels of thinning intensity (retention of 1,236, 865, and 494 trees/ha, respectively) and found that the more intense treatments resulted in increased levels of three nonnative species for as long as 27 years after treatment. Battles and others (2001) found nonnative species richness to be slightly higher in selection harvests than in control plots (two species versus one), but substantially higher in shelterwood cuts and clearcut plantations (seven nonnative species). The selection cuts were surveyed 1 year after the last entry, while the shelterwood and clearcut harvests were 20 years old. Several other studies from commercial harvests report increases in nonnative species richness or cover for as long as 30 years after treatment (Bailey and others 1998; Gray 2005; Isaac 1940; Thysell and Carey 2000), but these do not include treatment type or intensity as an explanatory variable.

Several chronosequence studies of harvests that were similar in type but spanned a range of ages have found a negative correlation between treatment age and nonnative abundance, but all found a nonnative presence for at least 20 years after treatment. DeFerrari and Naiman (1994) sampled clearcuts on the east side of the Olympic peninsula that ranged from 2 to 30 years old and found that the richness and cover of nonnative species began to decline after about 8 years. Similarly, Schoonmaker and McKee (1988) found nonnative abundance to peak at about 5 years in the Cascades of Washington and Oregon after sampling clearcuts up to 40 years old where the slash was subsequently broadcast burned. Halpern and Spies (1995) analyzed 27 years of repeated samples following two clearcuts that were subsequently burned in Oregon and found that nonnative cover peaked in year 2.

Few studies have investigated the effects of post-harvest slash treatments on nonnative species in ecosystems with mixed-severity fire regimes. Slash treatments, such as burning, yarding, or crushing unmerchantable material, is necessary for timber harvest to be an effective fuel treatment in these systems (Weatherspoon and Skinner 1995), but adding surface disturbance to canopy removal might result in even greater potential for invasion by nonnative plants. Scherer and others (2000) compared nonnative plant cover among various methods of post-harvest slash treatment at several sites in eastern Washington, but their findings were inconclusive. At their lowest elevation site, nonnative cover was about four times greater where slash was burned or mechanically chopped than in stands that were not harvested or where slash was left untreated or simply piled. However, at other sites there was either no difference among treatments or nonnatives were most abundant in the stands that were not harvested or where the slash was untreated.

Additional research is needed to clarify the benefits and undesirable consequences of treatments for reducing fire hazard in ecosystems with a mixed-severity fire regime. However, a recent study suggests that areas burned by wildfire may be at greater risk of nonnative invasion than are treated areas in these types of ecosystems. Freeman and others (2007) studied the effects of several forestry practices (shelterwood cutting, commercial and precommercial thinning, underburning, broadcast burning, and pile burning) on nonnative invasive species in western forests, including mixed conifer forests of northern California, Oregon, and Washington. Treatments did not produce clear increases in nonnative species, whereas wildfire was associated with significant increases in both richness and cover of nonnative invasives.

Low-Frequency, High-Severity Historic Fire Regime

Fire ignition and spread in ecosystems where environmental conditions limit the availability of fuel for combustion require extreme weather that typically produces rare, but severe, stand-destroying events (Martin 1982). Such ecosystems include high elevation and boreal forests dominated by spruce (*Picea* spp.) and fir (*Abies* spp.) (Bessie and Johnson 1995) and low elevation ecosystems dominated by shrub and woodland (Moritz 2003). Human activities have altered the historic fire regime in many shrublands (Keeley and Fotheringham 2003), while the fire regime of high elevation and boreal forests remains essentially unchanged (Gutsell and others 2001). The threat posed by nonnative plants is likewise very different between these two types of ecosystems, and they are thus considered separately.

Altered Fire Regime

Many western shrublands, such as Intermountain sagebrush (dominated by *Artemisia* species), Hawaiian dry tropical woodland, California chaparral, and California coastal scrub, have been converted into grasslands dominated by highly flammable nonnative species (Keeley 2006b). This, in addition to increased ignition sources from encroaching human habitation, has resulted in a fire regime that has become much more frequent but somewhat less severe than in the past due to decreased woody fuel loads. Frequent fire and nonnative grasses seem to have become symbiotic in these systems, each supporting the persistence of the other.

Fuel treatments in shrublands, particularly in southern California, have traditionally focused on the woody fuels, disrupting continuous shrub canopies with fuel breaks to provide ingress and egress for firefighters as well as anchor points for prescribed fire operations (Agee and others 2000). Prescribed fire is advocated by some as a fuel treatment for chaparral in southern California to create a mosaic of stand ages, since younger stands tend to be less flammable (Minnich and Chou 1997). However, others have suggested that heterogeneous fuel conditions are overwhelmed by the extreme weather conditions that accompany most wildfires, making fuel treatments in chaparral largely ineffective (Moritz 2003). Further, management ignitions may strengthen the positive feedback between fire and nonnative grasses (chapter 3; Keeley and Fotheringham 2003).

Mechanically constructed fuel breaks have also been found to promote invasion by nonnative plants in California shrubland ecosystems. Merriam and others (2006) inventoried numerous fuel breaks in various vegetation types that ranged in age from 1 to 67 years. Nonnatives were found on 65 percent of the fuel break plots and 43 percent of plots in adjacent untreated areas. Relative nonnative cover was greatest in coastal scrub (68 percent), followed by chaparral (39 percent), oak (*Quercus* spp.) woodland (25 percent), and coniferous forests (4 percent). Cheatgrass (*Bromus tectorum*) and red brome (*Bromus rubens*) were the most frequently encountered nonnative species and were most highly correlated to disturbance severity as indicated by method of fuel break construction; those created with bulldozers had significantly greater nonnative cover than those constructed by hand crews. Time since treatment was a marginally significant predictor but indicated a disturbing trend toward greater abundance of nonnatives as treatments age. Giessow and Zedler (1996) also sampled numerous fuel breaks in California coastal scrub and found them to promote invasion of nonnative species into these systems.

There is some evidence that mechanical treatments promote nonnative invasion into shrubland and woodland ecosystems in regions outside California, as well. Haskins and Gehring (2004) found percent cover of nonnative species to be nearly 8 times greater in Arizona piñon-juniper (*Pinus* spp.-*Juniperus* spp.) woodlands 5 years following thinning and slash burning than in adjacent control areas.

The most effective fuel treatments in shrubland systems that have become dominated by flammable nonnative grasses are likely to be those that focus on eradication of the nonnative species and reestablishment of less flammable native species. Cione and others (2002) found that both herbicide application and hand removal effectively reduced the cover of wild oats (*Avena fatua*) in a California coastal sage scrub habitat. Seeded native shrubs established only in areas where the nonnative annual grass was successfully removed. A subsequent accidental fire burned through areas that remained grass-dominated but did not enter areas where native shrub cover had been restored, suggesting that restoration of the scrub habitat may have also been an effective fuel treatment.

Nonnative eradication efforts have been successful in other shrubland systems, at least on small scales in the short term. Herbicide applications successfully removed cheatgrass from sagebrush systems in Wyoming (Whitson and Koch 1998) and Nevada (Evans and Young 1977), medusahead (*Taeniatherum caput-medusae*) from Utah sagebrush (Monaco and others 2005), and fountain grass (*Pennisetum setaceum*) from tropical dry forest in Hawai`i (Cabin and others 2000, 2002). Older studies also found medusahead to be effectively controlled in California with herbicides that are no longer available (Kay 1963; Kay and McKell 1963). In Hawai`i, D'Antonio and others (1998) found native shrub biomass increased where nonnative grasses were removed by hand weeding, and Cabin and others (2002) found that bulldozing was more effective than herbicides for removing fountain grass and establishing seeded native species.

Seeding with native species over expansive landscapes may not always be possible due to budget or resource limitations. Thus noninvasive nonnative species have been used as a strategy to reduce reestablishment by more invasive nonnative species, particularly after large wildfires (chapter 14). For example, tilling followed by seeding with introduced wheatgrasses (Triticaceae) reduced reestablishment of cheatgrass in Idaho (Klemp and Hull 1971) and Wyoming (Whitson and Koch 1998). Ott and others (2003) found that wheatgrass seeding after wildfire in a Utah piñon-juniper woodland reduced cheatgrass proliferation if the area was subsequently chained to bury the seeds. Aerial seeding alone was found to be ineffective, which is commonly the case (Robichaud and others 2000).

Postfire rehabilitation seeding may also inhibit establishment of native species (Beyers 2004). The chained areas sampled by Ott and others (2003) had lower cover of some native species as well as cheatgrass. This may have been due to the effects of the chaining treatment itself or competition from the seeded species. Use of native species for seeding treatments would, therefore, seem to be imperative (Richards and others 1998), though we are unaware of any research that compares the cost-effectiveness of the three alternatives for rehabilitation seeding: native species versus noninvasive nonnative species versus no seeding treatment.

Unaltered Fire Regime

In contrast to many shrubland ecosystems, the infrequent crown fire regime of forests at high elevations and latitudes remains essentially unchanged (Gutsell and others 2001). These ecosystems tend to have low priority for fuel treatments, though silvicultural treatments may affect the distribution of nonnative invasive species and may also alter the severity of subsequent wildfires. Omi and Kalabokidis (1991) found that the effects of a wildfire were less severe in lodgepole pine (*Pinus contorta*) stands that originated from clearcuts on land managed by the U.S. Forest Service than in adjacent stands in a National Park that had received no active management; nonnative species were not investigated. Caution is warranted in applying treatments with the intent of reducing fire hazard in ecosystems with historically infrequent, severe fire. Observations from the International Crown Fire Modeling Experiment in the Northwest Territories of Canada suggest that fuel treatments can create environmental conditions, such as lower fuel moisture and higher wind speed on the forest floor, that will exacerbate fire behavior in boreal forests (Alexander and others 2001).

Remoteness and unfavorable environmental conditions have thus far limited invasion of nonnative species into treatments that have been implemented in most boreal and high-elevation areas. Deal (2001) inventoried understory vegetation in Alaskan spruce stands that had been thinned in the past 12 to 96 years but noted no nonnative species. Dandelion (*Taraxacum officinale*) occurred on 1 out of 72 plots established 6 months after thinning in a spruce-hemlock (*Tsuga* spp.) forest in coastal Oregon, but no nonnative plants were found when the plots were resampled 17 years later (Alaback and Herman 1988).

Nonnative invasive species have invaded some harvested sites in forests with long-interval, severe fire regimes. Selmants and Knight (2003) found at least one nonnative species present in 26 of 30 old (30 to 50 years) harvest treatments in Wyoming montane and subalpine forests, though nonnative cover was insignificant on all plots. DeFerrari and Naiman (1994) found up to seven species of nonnative plants on clearcuts of various ages in spruce-hemlock forest on the west side of the Olympic Peninsula. Disturbingly, no relationship was found between nonnative cover and time, with cover of four unspecified nonnative species remaining as high as 40 percent in the oldest clearcut sampled (24 years old).

Summary

Much research remains to be conducted on the subject of fuel treatment effects on nonnative species. Many ecosystems are not represented in the literature and longer-term monitoring is needed in most of the ecosystems that have been studied. Several critical issues have received little scientific attention, including (1) the challenge of balancing invasion potential in fuel treatments versus large wildfires, (2) the cost effectiveness of alternative reseeding treatments to deter invasions after wildfires, and (3) the influence of fuel treatment size and seasonality on nonnative invasion.

Nonetheless, the extant literature does suggest some trends. Broad ecosystem groups distinguished by historic fire regime exhibit different responses by nonnative plant species to disturbances created by fuel treatment activities, at least in the western United States. Potential for invasion by nonnative species appears to increase with treatment intensity in most of the ecosystems that have been studied, though persistence seems to vary by fire regime type. Undesirable impacts from fuel treatments should be weighed against the possible benefits (such as reduced wildfire hazard), which also vary by fire regime.

Continued application of fuel treatments in ecosystems with high-frequency, low-severity fire regimes is probably justified in most cases, though most of the supporting research is from ponderosa pine systems. Fuel treatments have been effective in reducing wildfire severity in some of these systems, and several studies suggest that wildfire may pose a greater threat than fuel treatments with respect to establishment and spread of nonnative species (Freeman and others 2007; Griffis and others 2001; Hunter and others 2006). Also, nonnatives appear to be more ephemeral in these systems than others, at least in the short term. Long-term studies are needed. The short- and long-term effects of fuel treatments should be carefully monitored, as should treatments to reduce nonnative species cover and richness.

The benefits of fuel treatments for fire hazard reduction are not well established in ecosystems where a mixed-severity fire regime was the historical norm.

There are theoretical reasons to suspect that fuel treatments in these systems may increase fire hazard by creating greater exposure to wind and solar radiation, as well as encouraging production of fine surface fuels by herbaceous species (Agee 1996c). Nonetheless, fuel treatments probably have a role to play in mixed-severity fire regimes to forestall a shift to more frequent crown fires. Based on studies that have been conducted primarily in Douglas-fir forests of the Pacific Northwest, nonnative species appear to decline with time since treatment in systems with mixed-severity fire regimes, though the process may take several decades.

Nonnative species that establish after disturbances in low frequency crown fire regimes may become persistent members of the vegetation community. While opportunities for establishment of nonnative species may be currently rare in the coldest and most remote of these ecosystems, such as boreal and subalpine forests, nonnative species have already become dominant in more favorable environments, such as some shrublands in the Great Basin and southern California. Nonnative species are favored by traditional fuel treatments that focus on removal of woody vegetation from these shrubland systems, and the undesirable influence of nonnative grasses on fire hazard is evident. Thus the only effective fuel treatments in crown-fire regime ecosystems that have been altered by a grass/fire cycle may be those that focus on eradication of nonnative species themselves.

Notes

Matthew L. Brooks

Chapter 14:
Effects of Fire Suppression and Postfire Management Activities on Plant Invasions

This chapter explains how various fire suppression and postfire management activities can increase or decrease the potential for plant invasions following fire. A conceptual model is used to summarize the basic processes associated with plant invasions and show how specific fire management activities can be designed to minimize the potential for invasion. The recommendations provided are focused specifically on invasive plant management, although other considerations can take precedence under certain situations. Every fire presents a unique combination of site history and management goals, and the approaches adopted for management always involve tradeoffs between alternative combinations of management actions. The information in this chapter is designed to help land managers make more informed decisions on integrating invasive plant management into fire suppression and postfire management operations.

Challenges of Identifying Postfire Plant Invasions

Invasion means the establishment, persistence, and spread of a species outside of its native range into a region that it did not historically occupy, with the demonstrated or potential ability to cause significant ecological consequences (chapter 1). However, it is often difficult to know whether or not a species was present prior to a discrete event (for example, before a fire or fire management action), because comprehensive plant surveys do not exist for most areas. Even where prior plant surveys do exist, they are not typically designed to detect nonnative invasive plants. In addition, these surveys may not include species that were present but not detected or species that dispersed into the region subsequent to the most recent sampling date (Brooks and Klinger, in press). Although regional invasive databases are becoming more available (for example, for

Hawai`i, www.hear.org), they typically do not possess the spatial resolution necessary for planning postfire management actions at local scales.

Most cases of postfire invasions reported in the literature involve plant species that were present prior to fire and then expanded their distribution and dominance following fire (D'Antonio 2000, review). Although fire may not be necessary for a species to establish within a region, it may trigger an increase in dominance to the point that it begins to cause ecological harm—for example, by altering fire regimes (chapter 3).

From an ecological standpoint and from the perspective of land managers, the potential for plant invasions to cause ecological harm is the reason why fire operation guidelines are needed to help minimize the chances of plant invasions and mitigate their negative effects. Thus, the details of whether a species was not previously present or was only present in low numbers prior to a fire may not be critical to the development of management actions designed to prevent the species from becoming a management problem. In this chapter, the term invasion and all derivatives thereof (including invasibility and invasion potential) are used in the context of both the (1) establishment of new species in an area they did not previously occupy, and (2) increase in abundance or dominance of species previously present but relatively less common.

The Invasion Process

Plant invasions have been associated with many factors including disturbances, proximity to previously invaded sites, pathways and vectors of spread, characteristics of potential invaders, altered resource availability, and disruption of ecological processes (Brooks and Klinger, in press; D'Antonio 1993; Davis and others 2000; Hobbs and Huenneke 1992; Lonsdale 1999; Maron and Connors 1996). In the current chapter, these factors are combined into two primary groups: (1) resource availability, and (2) propagule pressure (modified from Brooks 2007a). The concept of "propagule pressure" as used in this chapter includes both the rates of dispersal (numbers per dispersal event and frequency of dispersal events (Williamson and Fitter 1996)) and the characteristics of those species, including their ability to survive and reproduce. This two-part model predicts that landscapes are more invasible if the availability of limiting resources is high than if resource availability is low, but only if propagule pressure is sufficiently high and comprised of species with characteristics that allow them to establish new populations under prevailing environmental conditions (fig. 14-1). This approach to characterizing plant invasions differs from that of chapter 2 and other publications (for example, Davis and others 2000; Lonsdale 1999) only in the sense that it distills the major causative factors affecting invasions down into two primary factors for the purposes of developing and explaining management recommendations.

Plant resource availability is a function of the supply of light, water, and mineral nutrients and the proportions of these resources that are unused by vegetation or other organisms, such as soil microbes. Using mineral nutrients as an example, resource availability can increase due to direct additions to the landscape (fertilization), increased rates of production within

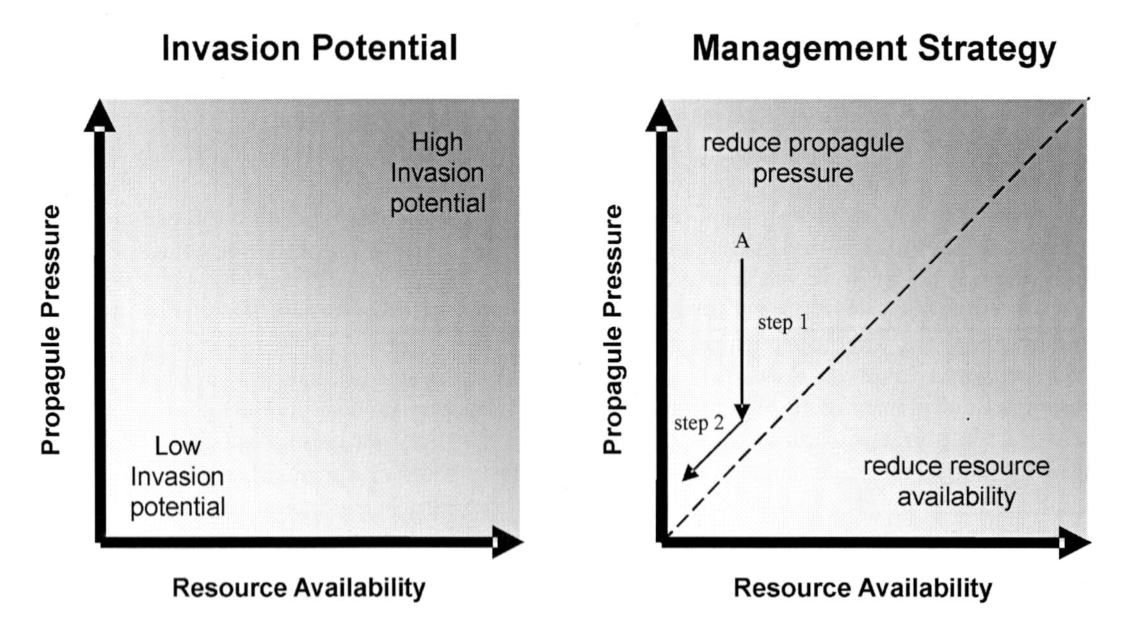

Figure 14-1—Main factors influencing invasion potential and a recommended management strategy to most efficiently minimize invasion potential. (Adapted from Brooks 2007a.)

the landscape (increased nutrient cycling following fire), or reduced rates of uptake following declines in resource use from extant plants after they are thinned or removed (biomass consumed by fire). Alternatively, mineral nutrient availability can decrease by volatilization during fire, rapid recovery of vegetation following fire, or success of revegetation efforts (for example, seeding).

Propagule pressure is typically used to mean the number of viable propagules available to establish and increase populations, and traditional definitions have focused on long-distance dispersal of individuals into regions to which they are not native (for example, Blackburn and Duncan 2001). This term has also been applied to the spread of nonnative species within regions where they have already established (for example, see Colautti and MacIsaac 2004). Percent cover of invading species has been found to decrease with increasing distance from initial points of invasion (Rouget and Richardson 2003), suggesting that dispersal rates are highest near established populations. These findings suggest that the concept of propagule pressure can be applied to different parts of the invasion process (Colautti and MacIsaac 2004; Lockwood and others 2005) and, in particular, to the stages of initial introduction and subsequent spread. By this more inclusive definition, propagule pressure can increase as a result of long distance dispersal from offsite populations (for species not previously present), local dispersal from onsite populations (for species previously present), or a combination of both. Propagule pressure can also be negatively affected by predators or diseases that reduce the reproductive rates of invading populations. This broad definition of propagule pressure is adopted here because it coincides with the definition presented earlier in this chapter that postfire invasions include both the establishment of new species in an area they did not previously occupy and the increase in dominance of species previously present.

Propagule pressure as used in the current chapter is also affected by the suitability of the component species to reproduce under prevailing environmental conditions. This approach places resource availability and propagule pressure on even par related to their theoretical scope. Just as the importance of resource availability varies among potentially limiting resource—such as light, water, and mineral nutrients—so too does the importance of propagule pressure vary among species, which can range from those likely to establish and cause undesirable effects to those not likely to establish. Phrased another way, it is not the increase in resource availability that necessarily matters, but rather the increase in resources that would otherwise be limiting to plant growth. Similarly, it is not the increase in propagule dispersal rates that matters, but rather the increase in propagules that can establish and reproduce under prevailing environmental conditions and ultimately cause undesirable ecological effects.

Since resource availability and propagule pressure of nonnative species are positively related to landscape invasibility, minimizing these two factors should be a significant consideration in land management activities (Brooks 2007a). Prioritizing which of the two factors to focus management actions on will depend on their relative importance on the landscape (fig. 14-1). For example, if propagule pressure is high but resource availability is moderately low (point A in fig. 14-1), then management actions should focus on reducing propagule pressure as a first step, which alone can significantly reduce invasion potential. If a further reduction of invasion potential is needed, then a management strategy focused on reducing both propagule pressure and resource supply is a potentially efficient and effective second step. In the sections that follow, these concepts are used to explain ways in which fire suppression and postfire management activities can influence plant invasions, both positively and negatively.

Effects of Fire Suppression Activities on Plant Invasions

Resource Availability

Fire suppression activities rarely lead to increased resource availability, although there are a few possible exceptions (table 14-1). For example, the use of fire retardants composed of ammonium phosphate adds a source of nitrogen and phosphorus that can lead to increased productivity of invasive plants in landscapes where these nutrients limit plant growth. Ripgut brome (*Bromus diandrus*), a highly flammable nonnative annual grass of significant management concern in western North America, increased by a factor of five in response to fire retardant added to burned areas, and by a factor of eight in response to the same retardant added to unburned areas during the first post-treatment year (Larson and Duncan 1982). Responses may depend on the effects of other factors limiting plant growth, such as soil moisture. This variable response seems to be exhibited by the nonnative Kentucky bluegrass (*Poa pratensis*), which increased significantly in growth following fire retardant application in a mesic northern prairie ecosystem (Larson and Newton 1996) but not in a more arid Great Basin ecosystem where soil moisture was assumed to be more limiting to plant growth than mineral nutrients (Larson and others 1999). Even if fire retardant increases growth rates of nonnative plants for a few postfire years, these increases may be less over the long term than those caused by fireline construction

Table 14-1—Recommendations for minimizing the potential of plant invasions during fire suppression activities. (Adapted from Asher and others 2001; Goodwin and Sheley 2001; and the U.S. Department of Agriculture, Forest Service 2001.)

Resource Availability

Minimizing Resource Input

Minimize the use of fire retardants containing nitrogen and/or phosphorus, except potentially where their use reduces the need for vegetation removal.

Maximizing Resource Uptake

Minimize vegetation removal in the construction of control lines.

- Use wet lines and foam lines as much as possible.
- Use narrow handlines in preference to broad dozer lines or blacklines.

Tie control lines into pre-existing fuel breaks (for example, bare rock and managed fuel zones) to minimize the amount of new vegetation removal.

Cover exposed soil with an organic mulch (for example, chipped fuels) where control lines were established to promote microbial activity that will use nitrogen and phosphorus, thus reducing their availability to invading plants.

Propagule Pressure

Preventing Deliberate Dispersal

There are no fire suppression activities with the potential to deliberately introduce nonnative propagules.

Minimizing Accidental Dispersal

Implement a postfire monitoring and control plan for invasive plants, focusing on populations of high priority invasive plants known to exist before the fire and on areas of significant fire management activity during the fire (for example, fire camps and dozer lines).

Ensure that vehicles, equipment, and personnel do not disperse propagules into burned areas.

- Coordinate with local personnel who know the locations of high priority invasive plants or who can quickly survey sites for their presence.
- Include warnings to avoid known areas infested with invasive plants during briefings at the beginning of each shift.
- Avoid establishing staging areas (for example, fire camps and helibases) in areas dominated by high priority invasive plants.
- If populations of high priority invasive plants occur within or near staging areas, flag their perimeters so that vehicle and foot traffic can avoid them.
- Inspect vehicles and equipment and wash them if they have propagules or materials that may contain propagules (such as mud) on them. Inspections should be done when vehicles first arrive at the fire and periodically during the fire as they return from the field.
- Avoid using water from impoundments infested with invasive plants.

If fire management options include prescribed fire or wildland fire use for resource benefits, address invasive plants in the environmental assessment. The assessment should document the distribution of high priority invasive plants and evaluate the potential for the burn to increase their dominance. If this potential is high, either remove those areas from the burn unit or develop and implement a postfire mitigation plan.

Identify populations of high priority invasive plants within areas burned by wildfire and focus postfire control efforts in those areas.

---

or the increased acreage burned if retardant is not used. Research on this topic is currently lacking, but enough evidence exists to consider fire retardants a potential contributor to plant invasions (table 14-1).

The construction of fuelbreaks and some firelines, both by hand crews and by heavy equipment, could lead to increased nutrient availability due to reduced consumption because plants have been removed (fig. 14-2; table 14-1). Merriam and others (2006) found that nonnative plants were often more abundant within fuelbreaks than in the surrounding landscape in California shrublands. In another example, a 16-fold increase in spotted knapweed (*Centaurea biebersteinii*) density was found on dozer lines between postfire years 1 and 3 in ponderosa pine forests in western Montana (Sutherland, unpublished data, 2008). Adjacent burned plots were free of spotted knapweed the first year after fire but had been invaded by knapweed by the third year after fire; propagules within the dozer lines were the apparent source. Over many decades, nonnative species may increase in dominance both within fuelbreaks and in adjacent areas, up to about 10 to 20 m (Giessow 1997; Merriam and others 2006).

Pre-existing fuelbreaks that are planted with less flammable noninvasive vegetation (that is, greenstripping) may reduce the need for complete vegetation removal during a fire (Pellant 1990) and thus reduce the likelihood of invasion. In addition, less destructive control lines, such as wet lines or foam lines, may be less likely to increase plant invasions because extant vegetation is left in place. Mop-up activities that include raking organic material back over control lines may reduce subsequent increases in nutrient availability;

Figure 14-2—Fuelbreak construction in a sagebrush (*Artemisia* spp.) steppe/pinyon-juniper (*Pinus-Juniperus*) woodland ecotone on the Colorado Plateau in northwestern Arizona. (Photo by Tim Duck, BLM, Arizona Strip Field Office.)

organic mulch added in this process can increase microbial metabolism of available soil nutrients and reduce incident light, thereby suppressing germination of invasive plants. These recommendations follow from the plant invasion theory discussed earlier in this chapter and in chapter 2, but they have not been rigorously studied and should be evaluated in future studies.

Fire suppression activities may promote plant invasions, but their influence on the amount of area that is ultimately burned needs to be considered as a potential counter-balancing factor. For example, if 100 acres of control line reduce a wildfire's area by 1,000 acres, then there may be a net reduction in the invasion potential of the landscape compared to the situation if no control lines were established. This is a simplistic example, and in reality many factors need to be considered, including the potential effects of the fire on native vegetation and the fire's proximity to populations of invasive nonnative plants. In addition, invasion potential is only one of many considerations in fire planning, and the benefits of fire as an ecosystem process (for example, in a frequent surface fire regime) may be more ecologically valuable than the potential negative effects of fire as a promoter of plant invasions.

Propagule Pressure

Fire suppression activities seem more likely to influence propagule pressure than resource availability (table 14-1). Firefighting crews and their equipment may disperse invasive plant propagules as they travel from other regions. They may also be vectors for local dispersal within the area of the fire. For example, fire camps are typically set up where the terrain is hospitable and where their ecological impacts will be minimal. These areas are typically large, flat clearings that have been disturbed in the past (for example, campgrounds, pastures, clearcuts, old fields). In many respects, it makes sense to localize the impacts caused by fire camps in areas that are already significantly altered. However, these areas often support populations of invasive plants. Propagules of these plants may adhere to fire personnel and their equipment as they move about camp and thereby may be dispersed elsewhere into the management unit as crews leave camp for the fireline.

Fire crew equipment largely consists of personal belongings (boots, clothes, sleeping bag, tent), personal protective equipment (gloves, helmet, goggles, fire pack, fire shelter), and hand tools (shovels, pulaskis, axes, fire rakes, hoes). This equipment can serve as vectors for the dispersal of invasive plants unless it is cleaned prior to reuse at other locations. At the least, firefighters should be given instructions to clean these items prior to leaving and arriving at a fire site. This practice has become standard operating procedure for fire crews following recent increased awareness of invasive plant management issues (Roberts, personal communication, 2005). It should also be adopted by contractors who provide support services such as food, restrooms, and showers.

Bulldozers and other heavy equipment can potentially spread invasives since they often accumulate significant amounts of soil and vegetation debris in their undercarriages (Matt Brooks, personal observation, St. George Utah, summer 2005). When heavy equipment is used, it should be washed prior to transport, at a commercial washing station enroute, or on-site when it arrives. It is becoming increasingly common for heavy equipment to be inspected prior to entering a fire zone, and in some cases equipment has been turned away if it shows signs of mud and other debris of unknown origin (Anderson, personal communication, 2004).

Aircraft are often used to transport and disperse water, foam, or other fire retardant materials. There is concern that aircraft such as helicopters with buckets or airplanes with holding tanks could become vectors for the introduction and local dispersal of invasive aquatic or riparian species, especially into local waterways from other regions, but also into upland areas. Nonnative species could establish in or along springs and creeks occurring within upland areas, but the risk for establishment in the fire area is probably low because the water is typically deposited onto non-aquatic upland sites.

The probability of dispersing aquatic or riparian plant propagules into burned areas from long distances is also probably low because water is typically obtained from local sources near fires. However, propagules can remain on equipment after water is released, and they may be dispersed into new geographic regions if aircraft are not decontaminated before being assigned to a new fire. There is also probably more potential for aircraft to disperse propagules between water sources in the vicinity of a fire than for them to disperse propagules into burned areas. Propagules may adhere to water holding tanks or buckets after water drops and then fall off during the next filling event. Repeated use of the same water source can help reduce the chances of such cross contamination. In addition, to help reduce the chance for local dispersal of invasive nonnative propagules, resource advisors assigned to a fire should identify preferable sources of water based on where existing populations of invasives occur. This requires pre-existing information that typically comes from the personal knowledge of local land managers but could also be based on comprehensive surveys and mapping efforts. Inspection and decontamination of water holding equipment before or immediately after fire should also reduce the dispersal risk.

Effects of Postfire Management Activities on Plant Invasions

There are three primary stages of postfire management planning and treatment implementation in the United States: (1) emergency stabilization, (2) rehabilitation, and (3) restoration. These terms reflect the policies and funding sources associated with all federal, and some state, land management agencies. The first two stages are generally under the purview of emergency fire management funding authorities, whereas the third stage is typically associated with nonfire programs and funding authorities (for example, natural resource management).

Emergency stabilization is focused on mitigating the immediate effects of fire and fire suppression activities during the first postfire year. The specific objectives are "to determine the need for and to prescribe and implement emergency treatments to minimize threats to life or property or to stabilize and prevent further unacceptable degradation to natural and cultural resources resulting from the effects of a fire" (USDA and USDI 2006a). The time period for emergency stabilization begins with containment of a fire and continues for 1 year. Emergency stabilization plans can be developed by local land management units (for example, BLM field offices, NPS park units, FWS refuges, Forest Service districts) or by overhead crews that specialize in this task (for example, Burned Area Emergency Response teams). In either case, a Burned Area Emergency Response (BAER) plan needs to be developed outlining the specific treatments and other activities that are proposed.

The most common objective of emergency stabilization plans is the prevention of soil erosion, but treatments for this purpose may have unintended impacts on invasive species. For example, contour felling of ponderosa pines in western Montana trapped not only overland sediment but also spotted knapweed seeds. Spotted knapweed densities were four- to five-fold higher above felled logs than 3 meters below the logs. Application of straw mulch had the opposite impact in perennial bunchgrass communities in western Montana (Sutherland, unpublished data, 2008). Burned, unmulched grasslands had spotted knapweed densities 50 times higher than burned, mulched grasslands 1 year after wildfire and 5 times higher 3 years after wildfire. Although burned, mulched grasslands had lower forb and total vegetation cover 1 year after fire; by 3 years following fire there was little difference.

During recent years, the prevention of plant invasions has increasingly been identified as a goal of emergency stabilization. Recent interagency guidelines from the United States Departments of Agriculture and Interior provide an excellent summary of how invasive plant management can be integrated into BAER plans (table 14-2) (USDA and USDI 2006a). Activities may focus on managing resource availability (for example, revegetation with noninvasive vegetation to minimize the availability of soil nutrients or light) or managing propagule pressure of invasive nonnatives (for example, postfire detection and monitoring, chemical, biological, mechanical, cultural and/or physical control treatment methods). Under emergency stabilization, this work can be done only if the management unit has a pre-existing program and/or approved plan to treat invasive plants. Emergency stabilization for invasive plant management is hampered by the policy of setting targets for reducing invasive plant numbers at prefire levels and not at some more ecologically relevant level. Also, the effective management of invasive plants often requires an integrated pest management approach, which is extremely difficult to implement within the 1-year emergency stabilization timeframe. For example, a single BAER application of the herbicide picloram eliminated spotted knapweed 1 year following wildfire in Montana perennial bunch grasslands (Sutherland, unpublished data, 2008). However, 3 years after herbicide application, spotted knapweed had re-established on 90 percent of these sprayed plots and knapweed cover was approaching pretreatment levels.

Rehabilitation plans focus on mitigating the effects of fire and fire suppression activities during the first 3 postfire years (USDA and USDI 2006b). They often involve reconstruction of minor infrastructure damaged as the result of fire (for example, fences and outbuildings), but they are increasingly addressing invasive plant issues as well. The management of propagule pressure, via monitoring for and direct control of plants known to be invasive in the area, is the most common approach during the rehabilitation phase. Species that are known to be the greatest management problems are typically the focus of these monitoring and control efforts. If target species are not previously known, then prioritization systems may be applied

Table 14-2—Federal interagency guidelines for the management of invasive plants within Emergency Stabilization (ES) and Rehabilitation plans (R) (USDA and USDI 2006a,b).

Emergency Stabilization and Rehabilitation funds can be used to control nonnative invasive plants in burned areas only if an approved plan for their management is in place prior to the wildfire. Integrated pest management methods are preferred, and they can include chemical, biological, mechanical, cultural, and physical treatments for minimizing the establishment of invasive species used in conjunction with vegetative treatments, or for site preparation for other treatments. Pesticides must be previously approved for use on public lands, and all applicable label and environmental restrictions must be adhered to.

Allowable Actions

- Assessments to determine the need for treatment associated with:
    - Known infestations
    - Possibility of new infestation due to management actions
    - Suspected contaminated equipment use areas (ES, R)
- Treatments to prevent detrimental invasion (not present on the site) by nonnative invasive species (ES, R)
- Treatment of invasive plants introduced or increased by the wildfire. The treatment objective when the population is increased is to maintain the invasion at no more than pre-wildfire conditions. (ES, R)
- Treatments to prevent the permanent impairment of designated critical habitat for federal and state listed, proposed, or candidate threatened and endangered species (ES)

Prohibited Actions

- Systematic inventories of burned areas (ES, R)
- Treatments designed to achieve historic conditions or conditions described in an approved land management plan, but that did not exist before the fire (ES)
- Treatments beyond 1 year post wildfire containment (ES)

to help identify them. Herbicide treatments may be proposed as follow-ups to initial treatments applied during emergency stabilization actions, which together can be designed as two phases of an integrated pest management plan. Seeding treatments are not frequently included in rehabilitation plans because land managers generally believe that the window of opportunity for pre-empting resources for invasive plants is mostly confined to the first postfire year (Matt Brooks, personal observation during the Hackberry Fire Complex BAER team planning session, Primm Nevada, Summer 2005). Revegetation is frequently proposed in rehabilitation plans, though it is not necessarily to suppress the establishment and spread of invasives. It is typically proposed to help native vegetation recover following fire, especially if the fire was thought to be excessively severe or otherwise undesirable.

Restoration is focused on the management of vegetation beyond the first 3 postfire years. It has been defined as "the continuation of rehabilitation beyond the initial 3 years, or the repair or replacement of major facilities damaged by fire" (USDA and USDI 2006a,b). The restoration phase has a much more comprehensive and long-term perspective than either emergency response or rehabilitation. Because restoration is separated in time from the emergency responses elicited by fire, it is almost universally managed by nonfire programs and funding authorities such as natural resources. The one exception may be fuels management (chapter 13), which is funded through fire programs and often has objectives that align with long-term restoration plans. Such objectives include the manipulation of vegetation (fuels) to restore more natural conditions and desired fire regimes. An example of this would be the thinning of understory vegetation in ponderosa pine forests with the objectives of reducing the potential for severe crown fire and restoring a more historically natural fire regime of frequent, low- to moderate-intensity surface fires. This long-term perspective of restoration projects is often very helpful in developing comprehensive plans for managing nonnative invasive plants, because short-term dominance by these species may be acceptable if over time their dominance wanes as native species recover.

Resource Availability

The use of fertilizers that may be pelletized with seed prior to application is not generally recommended because it can increase levels of available nutrients (table 14-3). Pelletized seed is also very expensive, adding significantly to the cost of seeding treatments (Roberts, personal communication, 2005). Invasive plants often utilize these extra mineral resources to the detriment of native species (for example, Brooks 2003). Nonnative and potentially invasive nitrogen fixing plants such as dryland alfalfa (*Medicago* spp.) and sweetclover (*Melilotus* spp.) have historically been included in seed mixes because they can provide a relatively inexpensive way to increase available soil nitrogen. More recently, native nitrogen fixing plants such as lupines (*Lupinus* spp.) have been included in seeding mixes. Although these nitrogen fixers could increase levels of available soil nitrogen and thereby increase dominance of invasive plants, such causative links have not been established by research and are not readily observed in the field (Pellant, personal communication, 2005). Alternatively, the addition of recalcitrant carbon sources, such as hydromulch, hay, or chipped fuels, can reduce available soil nutrients and shade the soil, thus suppressing the germination of invasive plant seeds.

Seeding of plant species that can rapidly establish and grow has the potential to usurp soil resources and intercept light, thus potentially reducing postfire dominance of invasive nonnative species (Pellant and Monsen 1993). In large-scale applications, seed is typically applied aerially (fig. 14-3). In smaller-scale applications, seed can be applied using a rangeland drill or broadcast and integrated into the soil by discing, harrowing, chaining, or raking. Although the establishment rate of seeded species is generally improved by integrating the seed into the soil, the associated tilling may damage existing vegetation and increase invasibility (Lynch 2003). Research is needed to compare the net effects that seeding versus seeding plus tilling has on the short- and long-term dominance of invasive plants.

Figure 14-3—Aerial seeding operation as part of the Emergency Stabilization Plan following the 2004 Chrome fire in southern Nevada. (BLM, Ely Field Office file photo.)

In the past, mostly nonnative species such as the perennial crested wheatgrass (*Agropyron* spp.) and Russian wildrye (*Psathrostachys juncea*) have been used in postfire seeding mixes (Pellant, personal communication, 2005). Nonnatives have been used because they are relatively inexpensive and readily available for seeding compared to most native species, and observations over the years suggest that they can compete with and suppress undesirable invasive nonnative plants (Roberts, personal communication, 2005). The logic of establishing one nonnative plant to prevent increased dominance by another is based on the idea that some nonnative species can dominate without producing severe negative ecological effects. For example, many nonnative annual grasses (such as *Bromus* spp. *Avena* spp.) produce more continuous fuels with lower fuel moisture during the heat of summer than nonnative perennial grasses, which grow more discontinuously and remain green throughout the year (Brooks and others 2004). It is believed that replacement of the nonnative annuals with nonnative perennials may increase the fire-return interval to the point were native vegetation adapted to longer fire-return intervals can recover in the Intermountain West of North America. Observations of decades-old crested wheatgrass seedings suggest that this may be occurring naturally in the Great Basin desert (Pellant and Lysne 2005).

Although seedings of nonnative perennial grasses have often been used in postfire landscapes to compete with other less desirable nonnative plants, relatively little has been known about the effectiveness of these treatments until recently (Pellant 1990; Pellant and Lysne 2005; Pellant and Monsen 1993). Some older publications provide evidence that nonnative perennial grass seeding can suppress cheatgrass (*Bromus tectorum*) (Hull 1974; Hull and Holmgren 1964; Hull and Pechanec 1947; Hull and Stewart 1948; Robertson and Pearse 1945), which alters fire regimes in some ecosystems of western North America (Brooks and Pyke 2001; chapter 3). However, these studies were very limited and relied largely on observational data.

A recent publication (Chambers and others 2007) reports that establishment, growth, and reproduction of cheatgrass is much higher following fire where herbaceous perennial plants (mostly native and nonnative bunchgrasses) were removed than where they were left intact in Intermountain West shrublands. Herbaceous perennials typically have high survival rates after fires in semiarid shrublands (Wright and Bailey 1982), and their quick recovery results in high utilization rates of soil nutrients such as nitrate, reducing nutrient availability and the subsequent productivity of cheatgrass in postfire landscapes (Chambers and others 2007). In contrast, where herbaceous species are removed, postfire levels of soil nitrate are relatively higher, resulting in increased production of cheatgrass. These results suggest that the maintenance of herbaceous perennials as a major prefire vegetation component may reduce the need for postfire management actions to control fine fuels created by cheatgrass. This study also suggests that postfire seedings of herbaceous perennials may suppress the dominance of invasive plants such as cheatgrass. However, the suppressive effects of seedings on invasive plants may not be evident during the first few postfire years, while they are only established as seedlings. It may take a number of years until mature stands develop and reach levels that effectively suppress invasives.

There is often strong pressure to quickly re-establish prevailing land use activities following fires. If these activities affect resource availability, they may inadvertently increase the invasibility of the landscape. For example, livestock grazing is a common use of public lands, and one of its primary effects is the removal of plant biomass, mostly herbaceous perennials that are typical forage species (Vallentine 2001). Biomass removal generally reduces competition and increases the availability of soil nutrients, which in theory increases landscape invasibility. If it is possible to target grazing on undesirable invasive plants, then it may help counteract the effects of increased soil nutrients. Reduced biomass of invasives that alter fire regimes may also help mitigate the ecosystem impacts of those species. However, it is difficult to control what livestock eat. In addition, repeated grazing in focused areas over long periods can lead to other problems such as soil erosion, soil compaction, and loss of native species diversity; and even short periods of grazing may allow nonnatives to rise to dominance. Further research is clearly needed in this area.

Propagule Pressure—Management of propagule pressure of invasive nonnative plants often focuses on direct control of nascent populations in postfire landscapes (table 14-3). For maximum effectiveness, this approach should include the following steps: (1) initial monitoring to locate nascent populations that may spread across the postfire landscape, (2) prioritization to decide which species need to be actively managed and where they need to be managed, (3) implementation of control treatments, (4) evaluation of treatment effectiveness, and (5) determination of the need for retreatment.

The first three steps need to be implemented during the first postfire year if they are supported by emergency stabilization funds. This timeframe makes the most sense ecologically as well, because during this time there is minimal competition from the extant vegetation and invasive plants have the greatest potential for establishment and spread.

Monitoring for new invaders should focus on likely pathways of invasion. These include linear corridors

Table 14-3—Recommendations for minimizing the potential of plant invasions during emergency stabilization, rehabilitation, and restoration activities. (Adapted from Asher and others 2001; Goodwin and Sheley 2001; and U.S. Department of Agriculture Forest Service 2001.)

Resource Availability
Minimizing Resource Input
Do not use fertilizers to promote plant growth.
Consider not using nitrogen-fixing plants in landscapes where increased nitrogen may increase invasibility.
Maximizing Resource Uptake
Consider covering exposed soil with an organic mulch (hydromulch or chipped fuels) to promote microbial activity that will take up N and P and reduce its availability to invading plants.
Minimize land uses that may reduce vigor of resprouting or establishing native plants (for example, livestock grazing).
Consider revegetating with fast-growing but noninvasive species to increase the uptake of resources that would otherwise be utilized by invasive species.
Propagule Pressure
Preventing Deliberate Dispersal
Revegetate with native species or nonnatives that are not likely to become invasive.
Minimizing Accidental Dispersal
Consider temporary closure of public access to burned areas to minimize propagule pressure.
Survey burned areas to locate nascent populations of invasive nonnative plants and eradicate or contain them so they don't spread across the postfire landscape.
Ensure that vehicles, equipment, and personnel do not disperse propagules into the project site.
Test seed mixes or other types of revegetation materials to ensure that they do not contain invasive species as contaminants.
Implement a monitoring and retreatment plan for invasive plants after the initial treatments are applied.

along which invaders can spread, such as roadsides, railroads, and utility rights-of-way (Brooks and Berry 2006; Brooks and Pyke 2001). They also include focused areas of disturbance to which invaders may disperse over long distances, such as livestock corrals or watering sites, mines, camping areas, OHV and military staging areas, old townsites, firelines, and backcountry landing zones (Brooks and Pyke 2001; Brooks and others 2006). Because these areas are extensive, monitoring should also be extensive, necessitating rapid assessment techniques, such as visual surveys of a given area (for example, between mile markers along a roadside) for a given amount of time. Ideally, this process can be complemented by pre-existing invasive plant maps to get the most comprehensive distributional assessment upon which prioritization and control plans can be based.

Prioritization of species and site may not be completed in time to implement control efforts within 1 postfire year. It requires a pre-existing prioritization of species within or adjacent to the burned area (for example, Brooks and Klinger, in press). Prioritization systems typically consider (1) the relative ecological and/or economic threats that the species pose, (2) their potential to spread and establish populations quickly, (3) their potential geographic and/or ecological ranges, and (4) the feasibility of control (Fox, A. and others 2001; Hiebert and Stubbendieck 1993; Morse and others 2004; Timmins and Williams 1987; Warner and others 2003; Weiss and McLaren 1999). Effectiveness monitoring should continue for an additional 2 years beyond control treatments with emergency stabilization funds, but follow-up treatments typically require additional funding before they can be implemented as part of rehabilitation or restoration plans.

Any land-use activity increases the chance for accidental introduction of invasive plant propagules, so minimizing these activities in postfire landscapes can reduce the potential for plant invasions. Any person or thing traveling into a recently burned area should be considered a potential vector, and the temporary closure of postfire landscapes to people and livestock can help reduce the potential for dispersal of nonnative invasive plants. In addition, postfire treatments that include the addition of organic materials (for example, straw mulch) or seed mixes have the potential to inadvertently introduce propagules of invasive nonnative species. It is imperative that these materials be certified weed-free and tested before they are applied. This practice would have been very beneficial after the 2000 Cerro Grande fire in New Mexico, where over 1 billion cheatgrass seeds were estimated to have contaminated an aerial seed mix that was applied as part of postfire management treatments (Keeley and others 2006a).

The intentional introduction of nonnative species is another source of invasive plant propagules that is not often scrutinized. Any species included in a seed mix should be evaluated for its potential to become a management problem in the future. Native species appropriate for the local vegetation (that is, local genotypes) are generally not a concern. In contrast, species that are not native to the area have the potential to introduce new functional types to the local vegetation that may change plant community relationships and ecosystem dynamics in the future. Many species of nonnative plants have been used for years in revegetation applications and appear to have some positive effects on plant community diversity as a result of their ability to compete with other, less desirable nonnative plants (Pellant, personal communication 2005; Roberts, personal communication, 2005). However, the research supporting these assumptions is limited, and decisions to include nonnative species in seed mixes should only be made after careful consideration of potential positive and negative outcomes.

Summary

The adage that an ounce of prevention is worth a pound of cure surely applies to the management of invasive plants. Fire managers share the responsibility of managing pubic lands with other resource management professionals, and they can play a key role in the prevention of plant invasions associated with wildland fires that may otherwise become significant and often intractable problems in the future. Postfire invasions that are prevented by some relatively simple actions by fire management personnel can reap great future rewards in terms of managing invasive plants.

The recommendations presented in this chapter are not meant to be comprehensive lists of actions that land managers should take to reduce the potential for plant invasions following fires. Rather, they are designed to provide some examples of procedures that can be integrated into land management plans. References to resource availability and propagule pressure as the primary causative factors of plant invasions were made to demonstrate how any land management action can be evaluated for its potential to affect landscape invasibility.

Many steps can be taken to minimize postfire plant invasions. Some are relatively simple and should not significantly impede fire management activities, whereas others may impose significant new layers of procedures. As always, firefighter safety is paramount in fire operations, and protection of natural resources and property is secondary. However, a fair amount of discretion is involved in determining how fire and postfire operations are carried out. Within this discretionary range, actions to reduce the potential for plant invasions need to be weighed against other considerations to arrive at a successful strategy for managing nonnative invasive species associated with fire.

Notes

Steve Sutherland

Chapter 15: Monitoring the Effects of Fire on Nonnative Invasive Plant Species

Monitoring, as defined by Elzinga and others (1998), is "the collection and analysis of repeated observations or measurements to evaluate changes in condition and progress towards meeting a management objective." Analyses of monitoring data may indicate that a project is meeting land management goals, or it may indicate that goals are not being met and management methods need to be adapted to reach them. Monitoring is an essential step in adaptive management (Chong and others 2006). For many federal agencies, monitoring is required by Agency or Congressional mandates (Elzinga and others 1998).

Monitoring can be qualitative or quantitative and can be applied with varying levels of rigor (see Elzinga and others 1998 for a complete discussion). Monitoring with a high level of rigor is essential for change to be detected **and** for cause and effect relationships to be inferred. In other words, a high level of rigor is needed for monitoring to produce defensible results that are useful for management decisions. In the rest of this chapter, when I use the term "monitoring," I mean monitoring with a high level of rigor.

Monitoring is essential for understanding the relationship between fire and invasive species, whether documenting new invaders following fire, postfire changes in established nonnative populations, recovery of native plant communities after wildfire (fig. 15-1), or efficacy of prescribed fire in controlling nonnative plant species. Monitoring can be used to detect change between sampling periods (before and after fire) or between treatments (burned and unburned areas), and it provides quantitative data for statistically analyzing the probability that the observed differences are due to chance. Statistics provides an objective and defensible means of evaluating the relationship between invasives and fire; it provides "high quality information" as described in chapter 12. However, statistical significance does not necessarily imply biological significance (see "Statistical Analysis" page 285).

This chapter is not a guide on how to monitor nonnative invasive plants. Many such guides are available, some in textbooks and others in agency manuals, which are referenced throughout the chapter. Instead of providing how-to instructions, this chapter (1) identifies

Figure 15-1—Monitoring effect of fire on understory vegetation in the Selway-Bitterroot Wilderness Area, northern Idaho, one year after wildfire. (Photo by Steve Sutherland.)

common elements of effective monitoring programs and suggests sources for additional information on those elements, (2) provides examples of how decisions on monitoring design can affect results and interpretation of the results, and (3) stresses the importance of integrating vegetation, fire behavior, and fire effects monitoring to better understand the impact of fire on nonnative, invasive species.

Vegetation Monitoring

In the past 10 years, many Federal Agencies have produced vegetation monitoring manuals (table 15-1) with two goals: (1) help managers improve monitoring and (2) provide a standard methodology for measuring plant attributes. Three of the manuals have associated relational databases for storing, managing, and statistically analyzing monitoring data. Efforts are currently underway to integrate the Fire Monitoring Handbook (FEAT) and FIREMON databases to create a new monitoring tool (Lutes and others, in press). Although none of these manuals specifically address monitoring nonnative invasive species, the standard methodologies for vegetation monitoring apply to monitoring of invasives.

Wirth and Pyke (2007) reviewed these seven publications and two national assessment programs (National Resources Conservation Service's National Resource Inventory (Spaeth and others 2003) and U.S. Department of Agriculture, Forest Service's Forest Inventory and Analysis (U.S. Department of Agriculture, Forest Service 2003)) and found seven common elements of effective monitoring program design:

- Objectives,
- Stratification,
- Controls (untreated plots),
- Random sampling,
- Sample size (data quality),
- Statistical analysis, and
- Field techniques.

Objectives

Managers need to be strategic about monitoring vegetation (Elzinga and others 1998; Lutes and others 2006). Elzinga and others (1998) describe the process in detail. The steps, as related to invasive species and fire, include (1) gathering and reviewing background information on invasive species, fire, and native communities; (2) identifying the priority invasive species and native communities; and (3) determining resources available for monitoring (for example, funding and personnel with botanical and ecological expertise). Once priorities are established and the amount of available resources is determined, the scale (local or regional) and intensity of monitoring can be determined.

Because resources are limited, the number of priority species and communities to be monitored may be inversely related to the scale and intensity of monitoring. As the number of priority species and communities increases, the scale and/or intensity of monitoring these species often decreases to stay within limited resources. Often managers are faced with the decision to monitor a few invasive species intensively at a few locations and monitor other species either qualitatively or not at all.

Well defined management and monitoring objectives are essential for successful monitoring. Management objectives provide a standard for determining management success and indicate which variables should be measured to determine when success has been achieved (Wirth and Pyke 2007). Management objectives should be realistic, specific, and measurable (Elzinga and others 1998). Monitoring objectives should be paired with each management objective. Monitoring objectives should specify the desired level of precision (desired confidence level and confidence interval width) or the desired minimum detectable change, acceptable probability of detecting a change when there is none (alpha or type I error), and acceptable probability of not detecting a change when there is one (beta or type II error). These variables also influence adequate sample size (see Elzinga and others (1998) and Wirth and Pyke (2007) for more details).

Table 15-1—Selected monitoring publications by federal entities.

Title	Topics covered by monitoring protocols			
	Vegetation	Invasives	Fire	Relational database
Measuring and monitoring plant populations (Elzinga and others 1998). U.S. Department of the Interior, Bureau of Land Management	Y	N	N	N
Sampling vegetation attributes (Interagency Technical Reference 1999). U.S. Department of the Interior, Bureau of Land Management	Y	N	N	N
Fuel and fire effects monitoring guide (U.S. Department of the Interior, Fish and Wildlife Service 1999)	Y	N	Y	N
Fire monitoring handbook (U.S. Department of the Interior, National Park Service 2003)	Y	N	Y	Y
Monitoring manual for grassland, shrubland, and savanna ecosystems (Herrick and others 2005a,b). U.S. Department of Agriculture, Agricultural Research Service	Y	N	N	Y
FIREMON: fire effects monitoring and inventory protocol (Lutes and others 2006). U.S. Department of Agriculture, Forest Service	Y	N	Y	Y
Range and training land assessment (U.S. Army, Sustainable Range Program 2006)	Y	N	N	N

Stratification

Stratification divides the treatment unit into ecologically similar areas to reduce variation and increase precision (Wirth and Pyke 2007). Use of GIS with data layers for elevation, aspect, slope, fire perimeters, fire severity, treatment areas, soils, roads, and watersheds may be useful for stratification. (See U.S. Department of the Interior, National Park Service (2003), Lutes and others (2006), Herrick and others (2005b), and Wirth and Pyke (2007) for more details on stratification). Because of limited monitoring resources, there may be a trade-off between the number of strata that can be monitored and monitoring intensity. For example, changing a study from two burn severities in three plant communities (2 X 3 = 6 strata) to two burn severities in three plant communities on north vs. south slopes (2 X 3 X 2 = 12 strata) doubles the number of strata sampled. In this example, unless stratification reduces variance so it reduces the needed sample size by half, the sampling effort will have to increase to cover all strata adequately.

Figure 15-2 illustrates the advantage of stratification. When spotted knapweed (*Centaurea biebersteinii*) density was compared on burned vs. unburned plots, data indicated that knapweed density was 40 percent lower in burned areas. However, when the burned areas were stratified by fire severity, more information was available. Knapweed density on low-severity burned plots was virtually the same as on unburned plots (38 vs. 34 stems/m^2), but severely burned plots had only 6 percent the knapweed density of unburned plots in the first postfire year (Sutherland, unpublished data, 2008). If management decisions are based on

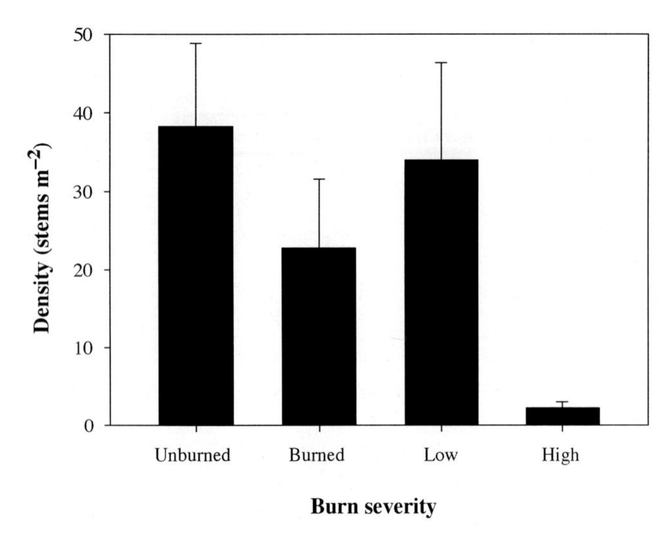

Figure 15-2—Spotted knapweed density (mean + standard error). "Burned" bar shows data from all burned plots, while "low" and "high" bars show the same data broken down (stratified) by fire severity classes. Data are from burned and unburned plots with spotted knapweed 1 year after wildfire in western Montana. (Data from Sutherland, unpublished, 2008.)

knapweed density, then stratifying the data according to fire severity could result in more efficient measures for controlling this nonnative invasive species.

Controls

While only two of the monitoring manuals in table 15-1 (Fire Monitoring Handbook, and FIREMON: Fire Effects Monitoring and Inventory Protocol) mention the use of controls (untreated plots), Wirth and Pyke (2007) stress that controls are essential for understanding treatment effects. Controls allow managers to determine whether or not a specific treatment caused the observed change, as opposed to other factors, such as year-to-year variation in weather conditions (Christian 2003). To avoid pseudoreplication (Hurlbert 1984), controls should be placed randomly within each monitoring unit rather than in an adjacent unburned (untreated) area, which may differ in some environmental variables from the treatment unit (see "Other Elements of an Effective Monitoring Program Design" page 288). While this may be possible for prescribed fires, it may not be possible for wildfires. If monitoring after wildfire, multiple wildfires and unburned areas adjacent to each can be treated as replicates.

The importance of controls can be illustrated using an example of postfire herbicide application to control leafy spurge (*Euphorbia esula*) in western Montana (fig. 15-3). There was a significant, 70 percent increase in leafy spurge cover 1 year after application of picloram (2 years after fire) (Sutherland, unpublished data, 2008). Because there was no control, the cause of the increase in leafy spurge density could not be determined. It was possible, but not likely, that picloram caused an increase in leafy spurge cover. A more likely explanation is that leafy spurge density increased because precipitation was 20 percent higher the year after herbicide application than the year before. An untreated control would have allowed these two hypotheses to be tested.

Random Sampling

Random sampling is a requirement for statistical analysis and the basis for defensible monitoring (Sokal and Rohlf 2000; Zar 1999). Simple random sampling requires that each member of a population has an equal chance of being sampled. It ensures that the data are unbiased and representative of the monitoring unit. There are several types of random sampling (for example, simple, stratified, systematic, restricted, cluster, two-stage, double, and individual plants). Elzinga and others (1998) review the advantages and disadvantages of these eight types of random sampling.

Sampling plots can be either permanently marked at random locations or randomly located in each sampling period. If plots are permanently marked and recorded

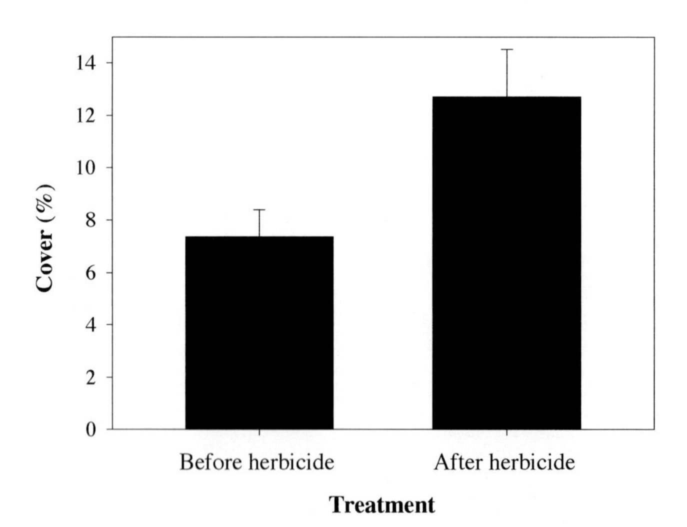

Figure 15-3—Cover of leafy spurge (*Euphorbia esula*) before and after herbicide application in a bunchgrass community in western Montana that burned in 2000. "Before herbicide" measurements were taken in summer 2001, herbicide was applied in fall 2001, and "after herbicide" measurements were taken in 2002. (Data from Sutherland, unpublished, 2008.)

in GIS, then the same plots can be remeasured each sampling period. Elzinga and others (1998) review the advantages and disadvantages of using permanent plots and conclude that permanent plots usually outperform temporary plots for detecting change. Statistical tests with permanent plots can be more powerful than tests with temporary plots, and the minimum sample size to detect a given change can be smaller with permanent plots. Locations of treatment areas including permanent plots or transects must, of course, be made known to managers to prevent management actions that would confound treatment effects and the usefulness of monitoring data.

Sample Size

Wirth and Pyke (2007) stress the importance of an adequate sample size to ensure that the manager can detect treatment-induced changes in populations. Adequate sample size depends on monitoring objectives, particularly on desired minimum detectable change and acceptable type I and type II errors (see "Objectives" page 282). Once monitoring objectives are chosen, the easiest way to determine adequate sample size is to collect preliminary data, calculate means and standard deviations, and use established equations to estimate minimum sample size. If multiple sampling techniques are used (for example, both cover and density will be measured), adequate sample size should be calculated independently for each technique. Many monitoring and statistical texts provide the needed equations (for example, Elzinga and others 1998; Lutes and others 2006; U.S. Army, Sustainable Range Program 2006; Wirth and Pyke 2007; Zar 1999).

Statistical Analysis

Statistical analysis provides an objective and defensible means of evaluating the relationship between vegetation change and treatment. Statistics assign a probability that the observed change is due to chance and allow the manager to infer that the treatment caused the change in vegetation. If the five preceding topics are properly addressed, data can be statistically analyzed. A very useful first step in this process is to examine graphic depictions of the data. Look for patterns. Do you see any trends between treated and untreated plots? Do you see differences between strata? Do you see any trends occurring over time? Did you expect to see differences not shown by the data? While statistical analysis produces defensible, quantitative results, simply *looking at* the data may provide insights as to what patterns are biologically meaningful and what trends may be occurring even if they are not captured by your data in a way that indicates statistical significance.

There are several options for statistical testing, including parametric and nonparametric methods. All of them assume random sampling; there are no statistical tests for data that are not randomly sampled. Standard parametric statistics (such as t-tests and analysis of variance) assume random sampling, normally distributed data, and equal variance among analysis groups. If the last two assumptions are not met, either the data should be transformed so that the assumptions of normality and variance homogeneity are met (see Elzinga and others 1998 for details) or other analytical methods should be used. Nonparametric statistics (including chi square, McNemar's test, Mann-Whitney U test, Wilcoxon's signed rank test, Kruskal-Wallis test, and Friedman's test) could be used to formulate tests when data are distributed similarly among groups but are non-normal (Hollander and Wolfe 1999). Generalized linear models could extend analysis of variance to data that are adequately approximated by binomial, Poisson, negative binomial, gamma or lognormal distributions (Dobson 2002), while mixed effects models (linear, generalized linear, or nonlinear) could account for correlation among observations (such as when observations are remeasured in time) or heterogeneous variance among groups (Littell and others 2006).

If the results are statistically significant, they should be examined to determine if they are biologically meaningful. Look again at graphic depictions of the data. Because all populations vary through time and space, a large sample size may make even small differences in mean values statistically significant. These differences, however, may not be biologically meaningful (Elzinga and others 1998). For example, Sutherland (unpublished data, 2008) measured total vegetation cover on burned and unburned bunchgrass plots 5 years after wildfires in 2000 in western Montana. Average total vegetation cover was 28.1 percent on burned plots and 23.5 percent on unburned plots. This 4.6 percent difference in total cover was statistically significant ($P < 0.05$) but less than the size of the cover classes (10 percent) used to record data. Thus the biological significance of the 5 percent difference in total cover on burned plots is questionable.

Do not necessarily ignore monitoring results that are not statistically significant. They may indicate trends that need further examination in the field or additional monitoring in the future using more precise sampling designs.

Field Techniques

Five monitoring methods are commonly used for estimating vegetation abundance: frequency, density, and cover measured in three ways—ocular, point-intercept, and line-intercept. Frequency is a measure of the proportion of sample plots that contain the target species.

Density is a measure of the number of individual plants or stems per given area. Cover measurements estimate the proportion of ground covered by target species. Each method has its strengths and weaknesses (see table 15-2), and the various methods may or may not give similar qualitative or quantitative results. Two examples of studies using a variety of measurement methods illustrate what can be learned from different methods.

Leafy spurge abundance in two unburned bunchgrass communities in western Montana was examined using frequency, ocular cover, point-intercept cover, and density measures (fig. 15-4). The methods produced similar qualitative results (leafy spurge abundance was lower in community B), but the absolute difference in spurge abundance varied with method. Spurge abundance was 22 percent lower in community B than in community A based on frequency measurements, 67 percent lower based on point intercept cover measurements, 70 percent lower based on ocular cover measurements, and 82 percent lower based on density measurements.

Sutherland (unpublished data, 2008) examined spotted knapweed abundance in 2001 and 2002 for a western Montana bunchgrass community using frequency, ocular cover, and density measures (fig. 15-5). He found that each method indicated a different pattern of change over time. Frequency did not change between the first and second postfire years, while cover decreased and density increased.

Why do these monitoring methods produce different results? As a measure of vegetation abundance, frequency is sensitive to the size, density, and distribution of the vegetation and also to the size and shape of

Table 15-2—Comparison of five monitoring methods.

Method		Advantage	Disadvantage
Frequency (How many plots contain the target species?)		• Sensitive to changes in spatial arrangement • Effective in monitoring establishment and spread of invasive species • Stable throughout growing season • Objective and repeatable • Easiest and fastest	• Affected by spatial arrangement, size, and density of vegetation • Difficult to interpret biologically • Sensitive to size and shape of sampling frame
Cover estimates (How much ground does the target species cover?)	Ocular (Visual estimate)	• Equalizes contribution of species • Most directly related to biomass • Does not require identification of individuals • Good for diversity and species richness • Easy and fast	• Changes during growing season • Sensitive to changes in number and vigor • Not sensitive to reproduction • Change detection is difficult with broad cover classes • Observer bias (but see discussion below)
	Point intercept (What proportion of the points hit the target species?)	• Least biased and most objective cover estimate	• Changes during growing season • Sensitive to wind • Sensitive to angle of point • Overestimation of cover with large pins • Not effective for species with low cover
	Line intercept (What proportion of a transect is occupied by the target species?)	• Effective for shrubs • Effective for mat-forming plants	• Less effective for single-stem plants • Less effective for grasses • Not effective for species with low cover
Density (How many stems per area of the target species?)		• Effective for recruitment and mortality • Independent of quadrat shape • Measurements are repeatable • Observer bias is low	• Less sensitive to changes in vigor • More time consuming • Difficult to use with bunchgrasses

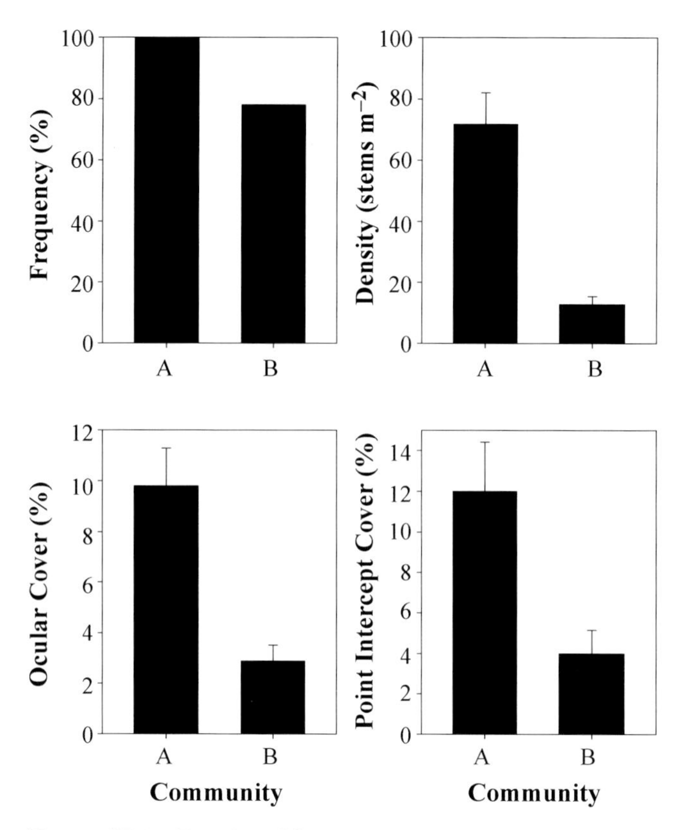

Figure 15-4—Results of four monitoring techniques used to characterize leafy spurge abundance in two unburned bunchgrass communities (A and B) in western Montana. (Unpublished data on file at USDA, Forest Service, Rocky Mountain Research Station, Fire Sciences Laboratory, Missoula, MT.)

the sampling frame (table 15-2). Therefore, the same size and shape sample frame must be used throughout monitoring, and only studies using the same sample frame should be compared. Because frequency varies with frame size, the absolute frequency has no biological meaning. It is the relative frequency and change in frequency that are important. To detect change, Elzinga and others (1998) recommend frequencies between 30 percent and 70 percent. Frequency can be adjusted by altering frame size; in general, a larger sampling frame will increase frequency and a smaller frame will decrease frequency. The optimal frame size for one species may not be optimal for other species. If you are monitoring several species, a series of nested quadrats may be useful—smaller frames for common species and larger frames for uncommon and rare species.

The 22 percent difference in spurge frequency shown in figure 15-4 resulted from using a 1-m^2 sampling frame. If a different sampling frame had been used, a different change in spurge frequency would have been observed. Sample-frame size also explains why Sutherland's spotted knapweed frequency (fig. 15-5) was not different in bunchgrass communities 1 and 2 years after fire. The 1-m^2 frame was so large that spotted knapweed frequency was 100 percent in both years. A smaller sample frame, that produced an initial spotted knapweed frequency of 30 percent to 70 percent, would have allowed an increase in knapweed to be detected.

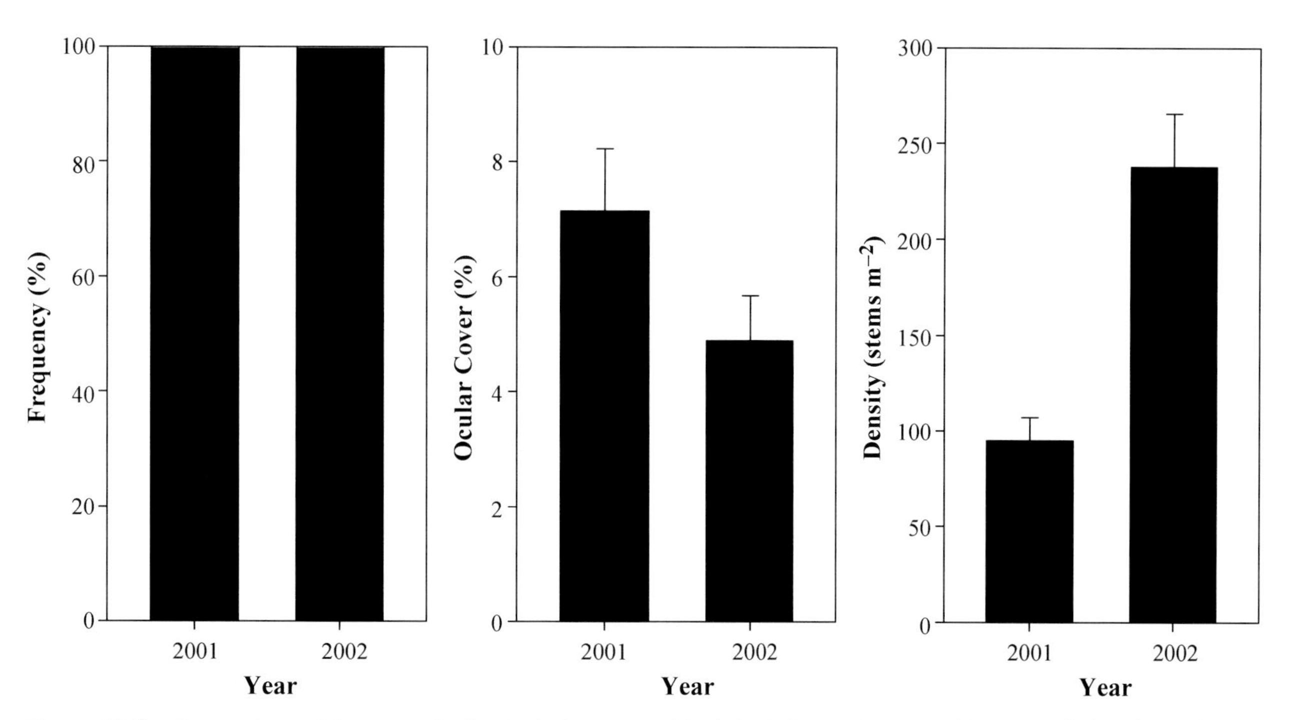

Figure 15-5—Comparison of three monitoring techniques used to detect change in spotted knapweed abundance between 2001 and 2002 in a western Montana grassland burned in 2000. (Data from Sutherland, unpublished, 2008.)

While Wirth and Pyke (2007) do not recommend using frequency for monitoring postfire vegetation treatments, they do acknowledge that frequency may be important in detecting invasion by nonnative species. Sutherland (unpublished data, 2008) used frequency data to document the establishment of new nonnative species on 24 burned plots in western Montana and northern Idaho between 1 and 3 years following wildfire. Sampling frequency at a large scale (30 m by 30 m macroplot) was more effective in detecting establishment than sampling at smaller scales (multiple 1 m by 1 m quadrats). Monitoring to detect new invasives should focus on invasion pathways and disturbed areas (see chapter 14 for more details).

Monitoring vegetation frequency is easy, fast, repeatable, and objective, but the results may be difficult to interpret. If there is a change in nonnative species frequency, it is unknown whether the change is caused by an increase in numbers, size, or distribution of nonnative species. Cover estimates or density measurements can be used to determine if the change in frequency is due to an increase in numbers, size, or distribution.

Where individual plants can be identified, density measures are repeatable with little observer bias. Density measures are independent of frame size and are effective for documenting mortality and reproduction. The counting unit must be identified (individual or stem), and adults and seedlings should be counted separately. Density monitoring may be time-consuming if the densities are high.

Cover estimates have the advantage of being fairly easy and fast, are related to biomass (Elzinga and others 1998), and are sensitive to changes in number and size of individuals. Disadvantages of cover estimates include (1) cover changes throughout the growing season so plots need to be resampled when the plants are in the same phenological condition each year, (2) cover is sensitive to changes in both numbers and vigor, and (3) cover is not effective for detecting changes in species with low cover (for example, rare species or species with seedlings).

The line-intercept method of cover estimation is usually used for shrubs, while ocular estimation and point-intercept methods are used for herbaceous vegetation. Ocular cover estimation has been considered more subjective than point- or line-intercept methods (Wirth and Pyke 2007). However, Booth and others (2006) found little difference between ocular and point-intercept estimations of known plant covers, whereas line-intercept consistently underestimated the known covers. Korb and others (2003), using unknown plant cover, concluded that ocular estimates of cover were more accurate than point-intercept because of violations of point-intercept assumptions (variable point size and non-vertical projection). Violations of these two assumptions can be avoided by training and consistent use of standard techniques.

Because it is difficult to make exact ocular estimation of plant cover, plants are usually placed in cover classes. Table 15-3 lists some common cover classes. The breadth of cover classes can impact monitoring results. For example, Sutherland (unpublished data, 2008) examined spotted knapweed after a BAER application of picloram in fall of 2001 in a bunchgrass community in western Montana, using the fine scale EcoData (Jensen and others 1993) cover classes in the initial data collection. If he had used Daubenmire's (1959) broader cover classes, then the eradication of spotted knapweed in 2002, its re-establishment in 2003, and the exponential increase in 2004 would have been undetected (see fig. 15-6). Daubenmire's (1959) and Braun-Blanquet's (1965) 25 percent cover classes are probably too coarse to detect change in many nonnative species populations.

The observed differences in Sutherland's data between estimates of spotted knapweed cover and density in non-herbicided plots (fig. 15-5) are caused by a postfire increase in spotted knapweed seedlings but no increase in adults. Total spotted knapweed density more than doubled, but the seedlings contributed little to cover.

Other Elements of an Effective Monitoring Program

There are two additional elements of an effective monitoring program: (2) avoiding pseudoreplication and (2) monitoring for multiple years. Hurlbert (1984)

Table 15-3—Percent cover classes recommended by Daubenmire (1959), Braun-Blanquet (1965), and EcoData (Jensen and others 1993).

Class	Daubenmire	Braun-Blanquet	EcoData
1	0-5	<1	0-1
2	6-25	1-5	1-5
3	26-50	6-25	6-15
4	51-75	26-50	16-25
5	76-95	51-75	26-35
6	96-100	>76	36-45
7			46-55
8			56-65
9			66-75
10			76-85
11			86-95
12			96-100

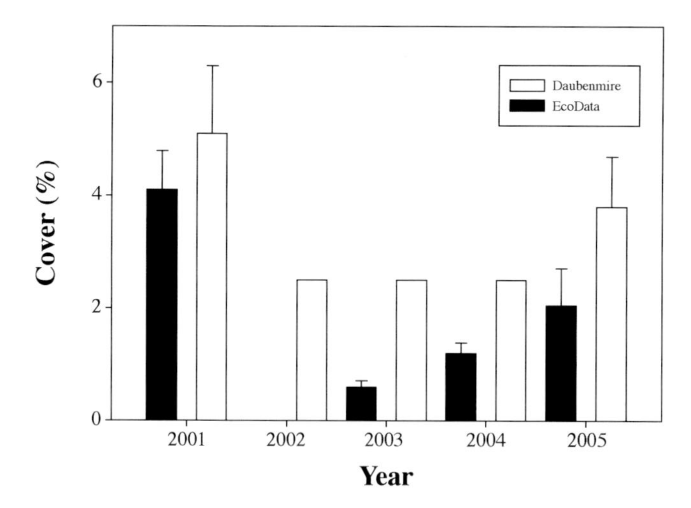

Figure 15-6—Comparison of percent cover as measured by EcoData and Daubenmire cover classes for documenting the efficacy of picloram in controlling spotted knapweed in a bunchgrass community in western Montana. Plots were burned by wildfire in 2000, sprayed with picloram in fall 2001, and monitored in summer 2001 to 2005. (Data from Sutherland, unpublished, 2008.)

defined pseudoreplication as the use of inferential statistics on data from treatments that are not replicated or replicates that are not statistically independent. For most nonnative monitoring programs, site heterogeneity and sampling efficiency require that data be collected from many sampling frames or transects per treatment unit. This is especially true for ocular cover and density measurements, where data can be collected efficiently only from relatively small sampling frames. These frames (see fig. 15-7) and transects are subsamples rather than experimental units; they should be treated as subsamples or averaged before statistical analyses. Defining a target population for a study and concurrently defining independent elements of the target population help distinguish between replicates and subsamples. Ideally, treatments and controls should be physically interspersed to minimize the effects of environmental gradients and random events. While this may be possible for prescribed fire, it is virtually impossible for wildfires.

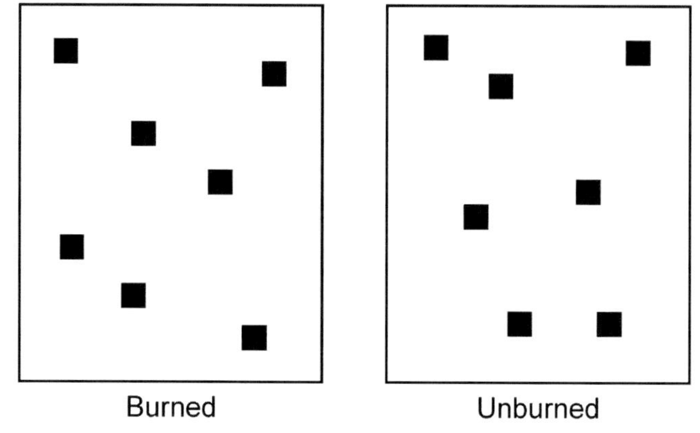

Figure 15-7—Pseudoreplication. The large rectangles are treatment units in a nonreplicated study, and the small squares (sample frames) within are subsamples. If statistics were performed using the subsamples as independent samples, that would be pseudoreplication.

Most studies of the impact of fire on nonnative invasive plants are of short duration. Seventy percent of the fire effects studies available for reviews of nonnative invasive plants written for the Fire Effects Information System from 2001 to 2006 were only followed for 1 year after fire (chapter 12). If postfire effects are ephemeral or delayed, then plant response the first year after fire may not reflect the true impact of fire on nonnative plants or the ecosystems where they reside. Managers and scientists agree that monitoring must cover periods long enough to provide results that can inform future management actions (White 2004). Two examples of studies using multiple-year sampling illustrate what can be learned from data sets that cover several years.

In the previous example of postfire application of picloram to control spotted knapweed in a Montana bunchgrass community (fig. 15-6), spotted knapweed cover was 0 percent 1 year after herbicide application. If the study had only lasted 1 year after treatment, the results would have suggested that BAER application of picloram was effective in controlling spotted knapweed after wildfire. The 5-year study indicated that the effect of picloram treatment was ephemeral; 4 years after herbicide treatment, spotted knapweed cover was half of pretreatment cover, and cover was doubling annually.

In the second example, Sutherland (unpublished data, 2008) found that spotted knapweed density was five times higher on unburned than burned ponderosa pine (*Pinus ponderosa*) plots 1 year after wildfire (fig. 15-8). By 3 years after fire, the pattern had reversed and spotted knapweed density was four times higher on burned than unburned plots. Two years later (5 years after fire), spotted knapweed density was five times higher on burned than unburned plots. This pattern was attributed to fire damage to seeds in 2000 and higher seed production and germination on burned plots in 2001 to 2005. If the study had only lasted 1 year (as do many fire effect studies), he would have concluded that wildfire reduces spotted knapweed; in reality, fire promoted spotted knapweed in ponderosa pine plots in western Montana.

While research and monitoring results are sparse regarding the relationship between fire and nonnative invasive species in the first decade after fire, there is almost no information on these relationships after the first postfire decade.

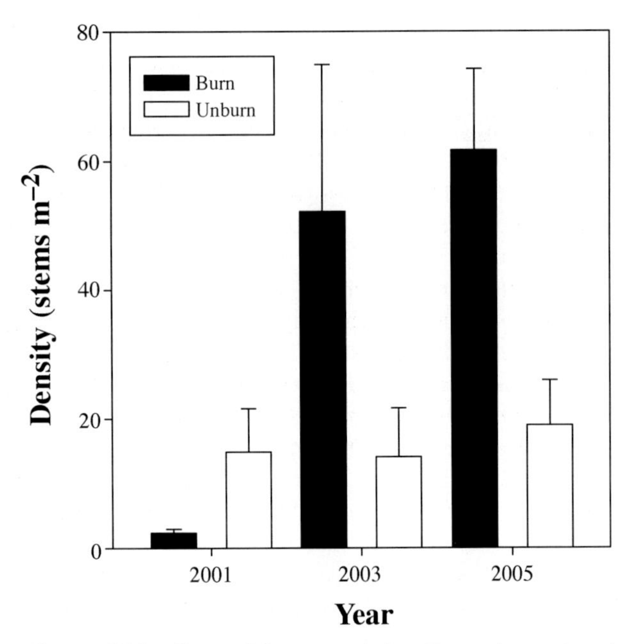

Figure 15-8—Spotted knapweed density on burned and unburned ponderosa pine (*Pinus ponderosa*) plots 1 to 5 years after wildfire in western Montana. (Data from Sutherland, unpublished, 2008.)

of the Interior, Fish and Wildlife Service 1999) and thus may be poor predictors of postfire responses of nonnative invasive species. Monitoring of changes in fuels and fire severity indicators may be more useful for understanding fire effects on vegetation.

Fuel Loading

Fuel load monitoring measures amount of litter, duff, and downed woody debris and is commonly based on Brown's (1974) planar intercept method (see U.S. Department of the Interior, Fish and Wildlife Service 1999; U.S. Department of the Interior, National Park Service 2003; or Lutes and others 2006 for further detail). Fuel consumption influences the magnitude and duration of soil heating, lethal soil temperatures, and consequent mortality of buds and seeds in the organic and mineral soil horizons. Fuel load monitoring is most useful when pre- and postfire data are collected and fuel consumption is calculated. Fuel consumption data can then be used to create relative fire severity classes or as input into models such as the First Order Fire Effects Model (FOFEM) (Reinhardt and others 1997) to estimate depth of lethal temperatures.

Fire Monitoring

To understand the impact of fire on nonnative invasive species, fire behavior and effects need to be monitored along with vegetation. Because invasion by nonnative species may be sensitive to severity of disturbance (chapter 2), fire monitoring can provide the context for understanding variations in postfire changes in invasive species as they relate to fuels and fire characteristics. Three of the monitoring manuals (Fuels and Fire Effects Monitoring Guide, Fire Monitoring Handbook, and FIREMON: Fire Effects Monitoring and Inventory Protocol) specifically address fire monitoring (tables 15-1 and 15-4).

Fire Behavior

Fire Monitoring Handbook (U.S. Department of the Interior, National Park Service 2003) and FIREMON: Fire Effects Monitoring and Inventory Protocol (Lutes and others 2006) suggest monitoring flame length, flame depth, and rate of spread. These fire behavior measurements are indicators of fire intensity (the rate of heat released by the flaming front during a fire). Other researchers have used instrumentation, temperature sensitive paint (Iverson and others 2004), and remote sensing (Hardy and others 2007) to measure fire intensity and maximum temperatures. But fire intensity and maximum temperature may be unrelated to fire severity (chapter 2; U.S. Department

Table 15-4—Contents of federal vegetation and fire monitoring manuals. "X" indicates topics covered in each manual.

Topic	FFEMG[a]	FMH[b]	FIREMON[c]
Fire behavior:			
Flame length		X	X
Flame depth		X	X
Spread rate		X	X
Fuel loading:			
Litter	X	X	X
Duff	X	X	X
1 hour	X	X	X
10 hour	X	X	X
100 hour	X	X	X
Fire severity:			
Crown scorch	X	X	X
Char height	X	X	X
Burn severity			
Substrate	X	X	X
Vegetation	X	X	X

[a] Fuel and Fire Effects Monitoring Guide (U.S. Department of the Interior, Fish and Wildlife Service 1999)
[b] Fire Monitoring Handbook (U.S. Department of the Interior, National Park Service 2003)
[c] FIREMON: Fire Effects Monitoring and Inventory Protocol (Lutes and others 2006)

Duff pins (Gundale and others 2005; U.S. Department of the Interior, Fish and Wildlife Service 1999) are an inexpensive way to measure changes in litter and duff. The pins are placed before fire with their tops or cross pieces at the top of the litter layer. After fire, the length of the exposed pin is measured to determine absolute litter-duff consumption, and residual duff depth is measured to determine percent litter-duff consumption. These data can also be used to create fire severity classes or estimate depth of lethal temperatures using FOFEM.

Monitoring fuel loads is most useful for prescribed fires, because pre- and postfire measurements can be taken. With wildfire, prefire measurements are uncommon, and other measurements (described below) must be taken to estimate fire severity.

Fire Severity

Crown scorch, char height, and vegetation burn severity classes are used to estimate aboveground fire severity (Lutes and others 2006; U.S. Department of the Interior, Fish and Wildlife Service 1999; U.S. Department of the Interior, National Park Service 2003). Crown scorch is expressed as maximum scorch height (from ground to highest point of foliar death) (U.S. Department of the Interior, Fish and Wildlife Service 1999; U.S. Department of the Interior, National Park Service 2003) or percent canopy scorch (Lutes and others 2006; U.S. Department of the Interior, Fish and Wildlife Service 1999; U.S. Department of the Interior, National Park Service 2003). Bole char height is measured as either maximum char height (ground to highest point of char) (U.S. Department of the Interior, Fish and Wildlife Service 1999; U.S. Department of the Interior, National Park Service 2003) or continuous char height (ground to lowest point of continuous char) (Lutes and others 2006). Crown scorch and char height are usually used to predict postfire tree mortality for species that do not sprout.

Vegetation burn severity monitoring is a visual estimate of fire damage to aboveground vegetation, and individual plants are assigned to a burn severity class based on amount of consumption: not burned; foliage scorched; foliage and small twigs partially consumed; foliage and small twigs consumed and branches partially consumed; or all plant parts consumed (U.S. Department of the Interior, National Park Service 2003). Vegetation burn severity may be an accurate predictor of plant mortality for species that do not sprout, but many perennial nonnative species sprout after being top-killed (chapter 2).

Although crown scorch, char height, and forest burn severity classes are not accurate predictors of mortality for sprouting species, they may predict changes in light levels that may impact invasive understory vegetation. Other methods to measure changes in light levels include measurements of photosynthetically active radiation and gap light analysis from digital photographs (Frazer and others 1999).

Substrate burn severity monitoring is a visual estimate of fire-caused changes in litter, duff, and soil (fig. 15-9). A sample point is assigned to a burn severity class based upon litter and duff consumption

Figure 15-9—Monitoring substrate burn severity in the Selway-Bitterroot Wilderness Area, northern Idaho, 1 year after wildfire. (Photo by Steve Sutherland.)

and changes in soil properties: not burned; litter blackened and duff unchanged; litter charred or partially consumed and upper duff charred; litter consumed and duff deeply charred; litter and duff consumed with mineral soil exposed and red (U.S. Department of the Interior, National Park Service 2003). Substrate severity may be correlated with soil heating, lethal temperature regimes, and bud and seed mortality.

Success and Failure in Monitoring Programs

Several common problems, both technical and institutional, determine the success of monitoring programs. Technical problems can be caused by poor monitoring design where the management and/or monitoring objectives are not well identified, where the minimum detectable change and type I and type II errors are inappropriate, where controls (untreated plots) are not included and sample size is too small or sampling period is too short, or where an inappropriate sampling method is selected. Success of a program is also influenced by personnel; failures can occur if staff have inadequate botanical skills, high observer error (especially ocular estimation of cover), or inadequate statistical skills. If the monitoring design and data collection are appropriate, the organization must be committed to documenting the information adequately, storing it appropriately (with backup copies), analyzing it with statistical skill, and protecting sampling sites from impacts that would confound treatment effects.

Monitoring may fail to detect change or differences between burned and unburned plots because the treatment impacts are smaller than natural variations in the plant populations over the time period measured. This outcome is a biologically meaningful result, not a failure of the monitoring program. It may indicate that additional variables should be measured in the future. Lack of statistically significant results should not necessarily be interpreted as a sign that a site is "safe" from invasion after fire and monitoring can be discontinued; the time period measured may simply be too short for significant differences to be detected.

Long-term agency commitment to monitoring determines its ultimate success and the usefulness of monitoring data for management planning and decisions. Failure of agency commitment at any stage in the monitoring process will degrade the results and can render them useless. A monitoring program can be well designed but never implemented; pretreatment data can be collected but post-treatment data not collected; pre- and post-treatment data may be collected but not analyzed; data may be collected and analyzed but the results never presented; and data may be collected, analyzed, and presented but ignored. In many cases these failures are due to inadequate resources (personnel and equipment) over an extended time and can be related to changing budgets, priorities, or politics.

Monitoring may be difficult to plan, implement, analyze, interpret, and integrate into the adaptive management process, but long term monitoring on permanent plots often provides the best way, and sometimes the only way, to evaluate the impact of fire on nonnative invasive plant species and make defensible decisions regarding management of fire and invasives.

Jane Kapler Smith
Kristin Zouhar
Steve Sutherland
Matthew L. Brooks

Chapter 16:

Fire and Nonnative Plants—Summary and Conclusions

This volume synthesizes scientific information about interactions between fire and nonnative invasive plants in wildlands of the United States. If the subject were clear and simple, this volume would be short; obviously, it is not. Relationships between fire and nonnative species are variable and difficult to interpret for many reasons:

- Fire and invasions are both inherently complex, responding to site and climate factors and the condition of the plant community. In addition, the nationwide scope of this volume incorporates great variation across ecosystems, climates, and regions.
- Fire effects and invasions interact with other ecosystem processes and land use history and patterns, and these interactions and effects can vary over time.
- Research tends to focus on highly successful invasions even though comparisons to failed or marginally successful invasions could be instructive (Beyers and others 2002).
- To date, research on fire and nonnative invasives has been limited, with few studies covering more than 1 year after fire (chapter 12).

The complexity of this subject makes it difficult to identify trends and implications for management. In this chapter we summarize the patterns (and lack of patterns) currently demonstrated by research regarding fire effects on nonnatives and the use of prescribed fire to reduce invasions, and we suggest some management implications. We also present some of today's burning questions about relationships between plant invasions and fire. Background for the assertions in this chapter can be found in Parts I and III of this volume. Readers interested in specific regional problems and issues should refer to Part II.

Nonnative Invasive Species and Wildland Fire

The literature shows that fire in many cases favors nonnative species over natives and thereby may lead to postfire invasions—that is, cases where ecosystems, habitats, or species are threatened because fire has promoted the establishment and spread of invasive plants (chapter 1). In some cases, nonnative species alter the native plant community and fuel characteristics to the extent that the fire regime is altered, and the altered fire regime favors further dominance of the plant community by the nonnative invader. This positive feedback loop is sometimes referred to as a grass/fire cycle or invasive plant/fire regime cycle (chapter 3).

The potential for nonnatives to negatively impact wildland ecosystems after fire suggests that managers should give priority to (1) controlling nonnative species known to be invasive after fire in the area burned or similar areas, especially if they are likely to alter the fire regime; (2) preventing new invasions through early detection and eradication of likely invaders; and (3) long-term monitoring and adaptive management after fire to control or reduce invasions. Tables provided at the beginning of each bioregional chapter in Part II may be helpful for identifying potential invaders. However, postfire invasion cannot be assumed for every ecosystem or for every nonnative species—even those mentioned in the bioregional tables. Invasion potential varies with prefire plant community condition, fire characteristics, and climate, and it depends on which plants and propagules (native and nonnative) are present within and near the burn. Invasions can be exacerbated by other disturbances and management activities, and they can be transient or persistent (chapter 2). While postfire invasion cannot be assumed, neither can a burn that is not invaded immediately after fire be assumed "safe" from invasion. Nonnative species may persist at low density for years before becoming invasive. Postfire disturbances, such as grazing or logging, may "tip the scales" toward invasion by altering resource availability, increasing nonnative propagules, and stressing native plants. In addition, wildlands are constantly exposed to new nonnative species with unknown invasive potential.

Generalizations commonly made about fire and nonnative species are supported by the literature under some circumstances but not others (chapter 2). For example, nonnative species establishment may increase with increasing fire severity, but this pattern can also be influenced by condition of the prefire plant community, postfire response of onsite species (both native and nonnative), propagule pressure, and the uniformity and size of high-severity burn patches. Plant communities dominated by native species that sprout after fire may be more resistant to invasion than communities where desired natives must regenerate from seed. Invasions are more likely in some plant communities when the baseline fire regime is disrupted, including locations where a nonnative grass/fire cycle has developed and native grasslands from which fire has been excluded for periods exceeding the baseline fire-return interval. Unfortunately, current conditions can diverge from presettlement conditions in so many ways that this generalization may not be helpful for predicting postfire responses. Postfire invasions tend to become less severe with increasing time since fire in closed-canopy forests and chaparral, but there are exceptions, and few long-term studies have investigated this pattern. Postfire invasions tend to be more severe with time since fire in shrub/grass ecosystems invaded by trees where native understory species (often sprouters) have been reduced. Postfire invasions are less likely in high-elevation than low-elevation ecosystems. Where human-caused disturbance occurs in high-elevation ecosystems, however, postfire invasions are more likely than in similar undisturbed systems.

Management Implications

Scientific study of the relationship between fire and nonnative invasive species is a relatively young field of investigation (Klinger and others 2006a) with many uncertainties. Many studies *describe* postfire invasions, but our scientific knowledge base is not yet extensive enough in space and time to *explain* or *predict* patterns of invasion across a range of ecosystems (Rejmánek and others 2005a), with or without fire. Information about fire effects on specific plant communities with specific invasive species provides the best knowledge base for management decisions regarding those communities and species. Knowledge of nonnative species biology, ecology, and responses in similar environments may also be useful for directing management decisions, although the effect of the particular environment must be considered (Rejmánek and others 2005a). One of the few consistent predictors of a nonnative species' potential to invade is its success in previous invasions (Daehler and Carino 2000; Kolar and Lodge 2001; Reichard and Hamilton 1997; Williamson 1999). Yet, as Williamson (1999) comments, "...we know that that can fail badly." This approach may be useful for predicting which species are "risky" but not for predicting which species are "safe". The most useful predictions likely require the integration of several approaches (Rejmánek and others 2005a).

While more knowledge is needed on fire and invasive species, research will never eliminate uncertainty, so scientists and managers must integrate many kinds of knowledge while remaining aware of their applications and limitations. The more the prefire plant community and conditions in a burned area diverge from conditions described in published research, the less reliable predictions based on that research will be. This is why the location and scope of research projects should be presented clearly in publications and read with care by managers. Management actions in general, including those based on extrapolation of research results, should be implemented with caution, monitored, and adapted as new knowledge develops. Partnerships between scientists and managers are likely to increase the pace and effectiveness of adaptive management (Beyers and others 2002). Hobbs and Mooney's (2005) comment on nonnative species invasions in light of global change applies well to fire/invasive interactions: "Scientists need to become smarter at considering potential scenarios based on multiple levels of uncertainty. Even qualitative analyses of likely outcomes can provide useful input to decision-making processes."

Many strategies for evaluating and addressing potential postfire invasions by nonnative species are described in this volume and the supporting literature. A brief summary is provided here.

During Fire Suppression—Wildfire managers should include training for crews on identifying nonnative invasive plants and preventing their spread. Firelines and fire suppression facilities should be located away from known invasions whenever possible. Equipment should be washed before being used on a fire so it will not introduce propagules of invasive species. Camps, staging areas, and helibases should be monitored during and after the fire to prevent establishment and spread of nonnative species. Additional guidelines and specific recommendations and requirements are available in chapter 14 and guides and manuals cited therein.

During Postfire Mitigation—Preventing invasive plants from establishing in burned areas is the most effective and least costly management approach. This can be accomplished through early detection and eradication, careful monitoring, and limiting invasive plant seed dispersal into burned areas. Opportunities for postfire establishment can be minimized by following the guidelines presented in chapter 14 and in the agency and extension publications referenced there; see also Goodwin and others (2002).

Successful postfire mitigation is likely to require that managers prioritize species and sites for exclusion, containment, control, or eradication based on their current distribution and potential to cause ecological harm. This process requires information on the distribution, dominance, and ecological effects of invasive and potentially invasive plants on and near the burn, especially species likely to alter fuel characteristics and fire regimes.

Use of Fire in Suppression Activities and for Other Management Goals—The potential for introducing or increasing the abundance of nonnative invasive species must be addressed in *all* land management planning and activities, including choice of appropriate management response to wildfire and use of prescribed fire for goals such as wildlife habitat improvement and site preparation. Minimize or avoid use of fire in areas at high risk for establishment or spread of invasives due to high propagule pressure or fire tolerance (for example, persistent seed bank and vegetative sprouting) (chapters 2 and 14). Where fire is used, incorporate precautions mentioned above ("During Fire Suppression").

Addressing Invasive-Caused Changes in Fire Regimes—If a species has already changed one or more fire regime characteristics, evaluate the altered regime and prioritize species control based on potential for negative effects on native species diversity, ecosystem processes, natural resources, public safety, property, and local economies. In some cases, it may not be possible to restore communities to their pre-invasion state, and managers may need to establish communities of native species that can coexist with the nonnatives (chapters 3 and 11).

Sharing Information—More high-quality research on relationships between fire and nonnative species is sorely needed (chapter 12). Managers will benefit from such studies only if they are published in a timely fashion. Managers could also benefit by sharing information from postfire monitoring of invasive species patterns after wildland and prescribed burns.

Use of Fire to Control Nonnative Invasive Species

Reduction of nonnative species abundance is usually just one facet of management to improve the condition of a native plant community. The use of prescribed fire to reduce nonnatives and contribute to overall management objectives is generally complex, and little research is available on the successes and failures of such efforts beyond the first few years after treatment (chapter 4). Therefore, integration of prescribed fire with other management techniques and long term monitoring of results are crucial to ensure that management objectives are being met.

Management Implications

To achieve long-term control of a nonnative invasive population and/or to favor native species with fire, managers must consider the regeneration strategies, phenology, and site requirements of all species in the management area. If invasive species are generally promoted by fire, fire alone is not likely to reduce them, although it may be effective during certain seasons or in combination with other treatments. Mechanical and chemical treatments may be useful to prepare for prescribed burning, especially on sites with sparse fuels. In addition, fire may be useful to prepare a site for introduction of desired native species, increase herbicide efficacy, or even promote population expansion of some biocontrol organisms.

In planning prescribed fire to control invasive species, managers can consider manipulating any or all aspects of the fire regime. Where the phenology of native plants differs from that of nonnatives, it is sometimes possible to schedule prescribed fires and manage their behavior to reduce nonnatives without damaging the native species. In addition, managers can manipulate the type, intensity, and severity of fire, fire size and uniformity, and fire frequency to damage nonnatives, consume nonnative seed, or favor native species. In the mixtures of species that characterize many wildland ecosystems, however, it is difficult to develop a strategy that reduces all nonnatives and favors all native species. Combinations or sequences of treatments may be needed. As mentioned often in this volume, monitoring is essential so managers can adjust their techniques to meet objectives (chapter 15).

Questions

Considerable information is needed to better understand and manage the relationships between nonnative invasive species and fire in wildlands of the United States. Long-term experimental studies and monitoring are essential for their descriptive value and also for their potential contribution to the development of tools for predicting postfire invasions. The following issues require consideration by scientists and managers in the near future:

- Nonnative species can negatively impact wildland ecosystems, but in field situations it is often difficult to distinguish the impacts of nonnatives from the impacts of other factors. How can scientists isolate and measure the impacts of invaders? How can managers distinguish minor, possibly transitory, effects of invaders from major impacts that are likely to persist? Do the effects of invaders change over time, and if so, how?
- Wildland conditions will continue to change in the face of continuing urbanization and accompanying ecosystem fragmentation, including increasing global trade and introduction of new nonnative species, changing atmospheric composition and climate, and interactions of these factors (Hobbs and Mooney 2005; Mooney and Hobbs 2000). Some of these changes are likely to facilitate invasions that alter fuels and fire regimes. In light of these problems, what are the most useful indicators that a nonnative species is likely to become invasive and alter ecosystem processes? Do these indicators vary by ecosystem?
- What tools are available, and at what scales, to help managers assess the invasibility or resistance of a particular plant community? What tools help assess the potential for establishment and spread of nonnatives after wildfire? How can managers prevent unintended consequences from prescribed fire? How can spatial information technology be used to obtain information on the presence and abundance of invasive species?
- Nonnative species exert selective pressures in wildland plant communities (Parker and others 1999), and the genetic structure of nonnative species may influence their potential to become invasive (Lee 2002). How does fire affect the gene pools of nonnative and native species interacting within a plant community?
- How do nonfire management activities affect undesirable nonnative species and desirable native species and communities? What combinations of management approaches will yield the most desirable results?

In discussing genetic influences of invasions, Barrett (2000) comments, "One of the most remarkable aspects of biological invasions is how unpredictable they are. Because of this, we should not be surprised if totally unexpected plant invaders appear, aided by new environmental conditions arising from global change." This idea can also be applied to the interactions of fire with nonnative invasive species. Scientists continue to seek explanations for invasions, develop predictive models, and seek ways to assess and address uncertainty in predictions (see, for example, Caley and others 2006; Colautti and others 2006; Cuddington and Hastings 2004; Daehler and Carino 2000; Drake and Lodge 2006; Lee 2002; Lockwood and others 2005; Sutherland 2004). However, the potential for surprise is practically limitless. It is crucial that we consider relationships between fire and nonnative invasive species with inquiring, open minds, paying careful attention to how fire and invasives interact in different situations and continually asking how they might be influenced by management practices.

References

Abella, S. R.; Covington, W. W. 2004. Monitoring an Arizona ponderosa pine restoration: sampling efficiency and multivariate analysis of understory vegetation. Restoration Ecology. 12: 359-367.

Abella, Scott. R.; MacDonald, Neil. W. 2000. Intense burns may reduce spotted knapweed germination. Ecological Restoration. 18(2): 203-205.

Abrahamson, W. G. 1984. Species responses to fire and the Florida Lake Wales ridge. American Journal of Botany. 71: 35-43.

Acker, Steven A. 1992. Wildfire and soil organic carbon in sagebrush-bunchgrass vegetation. The Great Basin Naturalist. 52(3): 284-287.

Adger, Neil; Aggarwal, Pramod; Agrawala, Shardul; [and others]. 2007. Climate Change 2007: impacts, adaptation and vulnerability. Contribution of Working Group II to the 4th assessment report of the Intergovernmental Panel on Climate Change, [Online]. Geneva, Switzerland: Intergovernmental Panel on Climate Change (IPCC) (Producer). Available: http://www.ipcc.ch/SPM13apr07.pdf [2007, July 12].

Agee, J. K. 1993. Fire ecology of Pacific Northwest forests. Washington, DC: Island Press. 493 p.

Agee, J. K. 1996a. Achieving conservation biology objectives with fire in the Pacific Northwest. Weed Technology. 10: 417-421.

Agee, J. K. 1996b. Fire in restoration of Oregon white oak woodlands. In: Hardy, Colin C.; Arno, Stephen F., eds. The use of fire in forest restoration: A general session of the Society for Ecological Restoration; 1995 September 14-16; Seattle, WA. Gen. Tech. Rep. INT-GTR-341. Ogden, UT: U.S. Department of Agriculture, Forest Service, Intermountain Research Station: 72-73.

Agee, J. K. 1996c. The influence of forest structure on fire behavior. In: Proceedings, 17th annual forest vegetation management conference; 1996 January 16-18; Redding, CA. Redding, CA: [Publisher unknown]: 52-68.

Agee, J. K. 1997. The severe weather wildfire—too hot to handle? Northwest Science. 71(1): 153-156.

Agee, J. K.; Bahro, B.; Finney, M. A.; Omi, P. N.; Sapsis, D. B.; Skinner, C. N.; van Wagtendonk, J. W.; Weatherspoon, C. P. 2000. The use of shaded fuelbreaks in landscape fire management. Forest Ecology and Management. 127: 55-66.

Agee, J. K.; Huff, M. H. 1980. First year ecological effects of the Hoh Fire, Olympic Mountains, Washington. In: Martin, Robert E.; Edmonds, Donald A.; Harrington, James B.; [and others], eds. Proceedings, 6th conference on fire and forest meteorology; 1980 April 22-24; Seattle, WA. Bethesda, MD: Society of American Foresters: 175-181.

Agee, J. K.; Skinner, C. N. 2005. Basic principles of fuel reduction treatments. Forest Ecology and Management. 211: 83-96.

Agee, James K.; Huff, Mark H. 1987. Fuel succession in a western hemlock/Douglas-fir forest. Canadian Journal of Forest Research. 17: 697-704.

Ahlgren, Clifford E. 1979. Emergent seedlings on soil from burned and unburned red pine forest. Minnesota Forestry Research Notes No. 273. St. Paul, MN: University of Minnesota, College of Forestry. 4 p.

Ahrens, J. F. 1975. Preliminary results with glyphosate for control of *Polygonum cuspidatum*. Proceedings of the Northeastern Weed Science Society. 29: 326.

Ahshapanek, C.C. 1962. Phenology of a tall-grass prairie in central Oklahoma. Ecology 43: 135-138.

Ainsworth, Alison. 2005. [Unpublished data]. On file at: Hawai`i Volcanoes National Park, Hawai`i National Park, HIi.

Ainsworth, Alison; Tetteh, Michel; Kaufmann, J. Boone. 2005. Relationships of an alien plant, fuel dynamics, fire weather, and unprecedented wildfires in Hawaiian rain forests: implications for fire management at Hawaii Volcanoes National Park. Unpublished report. On file at: Hawai`i Volcanoes National Park. Hawai`i National Park, HI. 12 p.

Alaback, P. B.; Herman, F. R. 1988. Long-term response of understory vegetation to stand density in Picea-Tsuga forests. Canadian Journal of Forest Research. 18: 1522-1530.

Alaska Natural Heritage Program. 2004. Weed ranking project, [Online]. Anchorage, AK: Alaska Natural Heritage Program, Environment and Natural Resources Institute, University of Alaska (Producer). Available: http://akweeds.uaa.alaska.edu/akweeds_ranking_page.htm [2005, January 15].

Albert, M. 2000. *Carpobrotus edulis*. In: Bossard, C. C.; Randall, J. M.; Hoshovsky, M. C., eds. Invasive plants of California's wildlands. Berkeley, CA: University of California Press: 90-94.

Albini F.; Amin, M. R.; Hungerford R. D.; Frandsen W. H.; Ryan, K. C. 1996. Models for fire-driven heat and moisture transport in soils. Gen. Tech. Rep. INT-GTR-335. Ogden, UT: U.S. Department of Agriculture, Forest Service, Intermountain Reasearch Station. 16 p.

Alexander, Janice M.; D'Antonio, Carla M. D. 2003. Seed bank dynamics of French broom in coastal California grasslands: effects of stand age and prescribed burning on control and restoration. Restoration Ecology. 11(2): 185-197.

Alexander, M.; Stefner, C.; Beck, J.; Lanoville, R. 2001. New insights into the effectiveness of fuel reduction treatments on crown fire potential at the stand level. In: Pearce, G.; Lester, L., tech. cords. Bushfire 2001 conference proceedings; [Meeting date unknown]; [Location unknown]. Christchurch, New Zealand: Forest Research: 318.

Alexander, Martin E. 1982. Calculating and interpreting forest fire intensities. Canadian Journal of Botany. 60: 349-357.

Allaby, Michael. 1992. The concise Oxford dictionary of botany. New York: Oxford University Press. 442 p.

Allain, L.; Billock A.; Grace, J.B. 2004. Plants of the USFWS's Midcoast Wildlife Refuge Complex. Unpublished report on file at: U.S. Geological Survey, National Wetland Research Center, Lafayette, LA.

Allen, C. D.; Savage, M.; Falk, D. A.; Suckling, K. F.; Swetnam, T. W.; Schulke, T.; Stacey, P. B.; Morgan, P.; Hoffman, M.; Klingel, J. T. 2002. Ecological restoration of southwestern ponderosa pine ecosystems: a broad perspective. Ecological Applications. 12: 1418–1433.

Allen, E. B. 1998. Restoring habitats to prevent exotics. In: Kelly, M.; Wagner, E.; Warner, P., eds. Proceedings: California exotic pest plant council symposium, Ontario, CA. Volume 4: 41-44.

Allen, Edith B. 1995. Restoration ecology: limits and possibilities in arid and semiarid lands. In: Roundy, Bruce A., McArthur, E. Durant, Haley, Jennifer S., Mann, David K., comps. Proceedings: Proceedings: wildland shrub and arid land restoration symposium. Gen. Tech. Rep. Las Vegas, NV. Ogden, UT: U.S. Department of Agriculture, Forest Service: 7-15.

Allen, L. H., Jr.; Sinclair, T. R.; Bennett, J. M. 1997. Evapotranspiration of vegetation in Florida: Perpetuated misconceptions versus mechanistic processes. Soil and Crop Science Society of Florida Proceedings. 56: 1-10.

Allen-Diaz, Barbara; Bartolome, James W.; McClaran, Mitchel P. 1999. California oak savanna. In: Anderson, Roger C.; Fralish, James S.; Baskin, Jerry M., eds. Savannas, barrens, and rock outcrop plant communities of North America. New York: Cambridge University Press: 322-339.

Alley, Richard; Berntsen, Terje; Bindoff, Nathaniel L.; [and others]. 2007. Climate change 2007: the physical science basis—summary for policymakers. Contribution of Working Group I to the 4th assessment report of the Intergovernmental Panel on Climate Change, [Online]. Geneva, Switzerland: Intergovernmental Panel on Climate Change (IPCC) (Producer). Available: http://www.ipcc.ch/SPM2feb07.pdf [2007, April 3].

Ambrose, J. P.; Bratton, S. P. 1990. Trends in landscape heterogeneity along the borders of Great Smoky Mountains National Park. Conservation Biology. 4(2): 135-143.

Amor, R. L.; Stevens, P. L. 1975. Spread of weeds from a roadside into sclerophyll forests at Dartmouth, Australia. Weed research. 16: 111-118.

Anable, Michael E.; McClaran, Mitchel P.; Ruyle, George B. 1992. Spread of introduced Lehmann lovegrass *Eragrostis lehmanniana* Nees. in southern Arizona, USA. Biological Conservation. 61(3): 181-188.

Anderson, Bertin. 1998. The case for salt cedar. Restoration and Management Notes. 16(2): 130-134.

Anderson, Bruce. 1994. Converting smooth brome pasture to warm-season grasses. In: Wickett, Robert G., Lewis, Patricia Dolan, Woodliffe, Allen, Pratt, Paul, eds. Proceedings: Proceedings of the thirteenth North American prairie conference: spirit of the land, our prairie legacy Windsor, ON. Windsor, ON: City of Windsor: 157-160.

Anderson, Darwin; Hamilton, Louis P.; Reynolds, Hudson G.; Humphrey, Robert R. 1953. Reseeding desert grassland ranges in southern Arizona. Bulletin 249. Tucson, AZ: University of Arizona, Agricultural Experiment Station. 32 p.
Anderson, Howard G.; Bailey, Arthur W. 1980. Effects of annual burning on grassland in the aspen parkland of east-central Alberta. Canadian Journal of Botany. 58: 985-996.
Anderson, J. 2006. [Personal communication]. October 2006. Hedgerow Farms, Chico, CA.
Anderson, J. E.; Romme, W.H. 1991. Initial floristics in lodgepole pine (*Pinus contorta*) forests following the 1988 Yellowstone fires. International Journal of Wildland Fire. 1(2): 119-124.
Anderson, K. L. 1965. Time of burning as it affects soil moisture in an ordinary upland bluestem prairie in the Flint Hills. Journal of Range Management. 18: 311-316.
Anderson, M. K. 1999. The fire, pruning, and coppice management of temperate ecosystems for basketry material by California Indian tribes. Human Ecology. 27: 79-113.
Anderson, M. K. 2006. The use of fire by Native Americans in California. In: Sugihara, Neil G.; van Wagtendonk, Jan W.; Shaffer, Kevin E.; Fites-Kaufman, Joann; Thode, Andrea E., eds. Fire in California's ecosystems. Berkeley, CA: University of California Press: 417-430.
Anderson, M.; Michelsen, A.; Jensen, M.; Kjoller, A. 2004. Tropical savannah woodland: effects of experimental fire on soil microorganisms and soil emissions of carbon dioxide. Soil Biology and Biochemistry. 36: 849-858.
Anderson, Mark G. 1995. Interactions between *Lythrum salicaria* and native organisms: a critical review. Environmental Management. 19(2): 225-231.
Anderson, R. S.; Smith, S. J. 1997. The sedimentary record of fire in montane meadows, Sierra Nevada, California, USA: a preliminary assessment. In: Clark, J. S.; Cachier, H.; Goldammer, J. G.; Stocks, B. J., eds. Sediment records of biomass burning and global change. NATO ASI Series. New York: Springer-Verlag: 313-327.
Anderson, Roger C. 1972. Prairie history, management and restoration in southern Illinois. In: Zimmerman, James H., ed. Proceedings of the second Midwest prairie conference; 1970 September 18-20; Madison, WI. Madison, WI: University of Wisconsin Arboretum: 15-21.
Anderson, Roger C.; Schwegman, John E. 1991. Twenty years of vegetational change on a southern Illinois barren. Natural Areas Journal. 11(2): 100-107.
Anderson, Roger C.; Schwegman, John E.; Anderson, M. Rebecca. 2000. Micro-scale restoration: a 25-year history of a southern Illinois barrens. Restoration Ecology. 8(3): 296-306.
Anderson, Roger C.; Schwegman, John. 1971. The response of southern Illinois barren vegetation to prescribed burning. Transactions, Illinois Academy of Science. 64(3): 287-291.
Anderson, Steve. 2004. [Personal communication]. August 18, Portland, OR. U.S. Department of the Interior, National Park Service, Haleakela National Park.
Anderson, Steve. 2005. [Personal Communication] April 25, Makawao, Hawai`i i. U.S. Department of the Interior, National Park Service, Haleakala National Park.
Anderson, Steve; Welton, Patricia. [Unpublished data.] On file at: Haleakala National Park, Makawao, HI.
Andreas, Barbara K.; Knoop, Jeffrey K. 1992. 100 years of changes in Ohio peatlands. Ohio Journal of Science. 92(5): 130-138.
Antenen, Susan. 1996. *Ampelopsis brevipedunculata*—porcelain berry. In: Randall, John M.; Marinelli, Janet, eds. Invasive plants: Weeds of the global garden. Handbook #149. Brooklyn, NY: Brooklyn Botanic Garden: 91.
Antenen, Susan; Yost, Susan; Hartvigsen, Gregg. 1989. Porcelainberry vine control methods explored (New York). Restoration and Management Notes. 7(1): 44.
Antos, Joeseph A.; McCune, Bruce; Bara, Cliff. 1983. The effect of fire on an ungrazed western Montana grassland. American Midland Naturalist. 110(2): 354-364.
Apfelbaum, S. I.; Sams, C. E. 1987. Ecology and control of reed canary grass (*Phalaris arundinacea* L.). Natural Areas Journal. 7(2): 69-74.
Apfelbaum, Steven I.; Haney, Alan W. 1990. Management of degraded oak savanna remnants in the upper Midwest: preliminary results from three years of study. In: Hughes, H. Glenn; Bonnicksen, Thomas M., eds. Restoration `89: the new management challenge: Proceedings, 1st annual meeting of the Society for Ecological Restoration; 1989 January 16-20; Oakland, CA. Madison, WI: The University of Wisconsin Arboretum, Society for Ecological Restoration: 280-291.
Arabas, Karen B. 2000. Spatial and temporal relationships among fire frequency, vegetation, and soil depth in an eastern North American serpentine barren. Journal of the Torrey Botanical Society. 127(1): 51-65.
Archer, Amy J. 2001. *Taeniatherum caput-medusae*. In: Fire Effects Information System, [Online]. U.S. Department of Agriculture, Forest Service, Rocky Mountain Research Station, Fire Sciences Laboratory (Producer). Available: http://www.fs.fed.us/database/feis/ [2005, December 14].
Archer, S.; Boutton, T. W.; Hibbard, K. A. 2001: Trees in grasslands: biogeochemical consequences of woody plant expansion. In: Schulze, E.-A.; Harrison, S. P.; Heimann, M.; Holland, E. A.; Lloyd, J.; Prentice, I. C.; Schimel, D. S., eds. Global biogeochemical cycles and their interrelationship with climate. San Diego, CA: Academic Press. 115-137.
Archibold, O. W.; Brooks, D.; Delanoy, L. 1997. An investigation of the invasive shrub European Buckthorn, *Rhamnus cathartica* L., near Saskatoon, Saskatchewan. Canadian Field-Naturalist. 111: 617-621.
Archibold, O. W.; Ripley, E. A.; Delanoy, L. 2003. Effects of season of burning on the microenvironment of fescue prairie in central Saskatchewan. Canadian Field Naturalist. 117(2): 257-266.
Armour, Charles D.; Bunting, Stephen C.; Neuenschwander, Leon F. 1984. Fire intensity effects on the understory in ponderosa pine forests. Journal of Range Management. 37(1): 44-48.
Arno, S. F. 1980. Forest fire history in the northern Rockies. Journal of Forestry. 78: 460-465.
Arno, S. F.; Fiedler, C. E. 2005. Mimicking nature's fire: Restoring fire prone forests in the west. Washington, DC: Island Press. 242 p.
Arno, Stephen F. 2000. Fire in western forest ecosystems. In: Brown, James K.; Smith, Jane Kapler, eds. Wildland fire in ecosystems: Effects of fire on flora. Gen. Tech. Rep. RMRS-GTR-42-vol. 2. Ogden, UT: U.S. Department of Agriculture, Forest Service, Rocky Mountain Research Station: 97-120.
Arno, Stephen F.; Gruell, George E. 1983. Fire history at the forest-grassland ecotone in southwestern Montana. Journal of Range Management. 36(3): 332-336.
Arno, Stephen F.; Parsons, David J.; Keane, Robert E. 2000. Mixed-severity fire regimes in the northern Rocky Mountains: consequences of fire exclusion and options for the future. In: Cole, David N.; McCool, Stephen F.; Borrie, William T.; O'Loughlin, Jennifer, comps. Wilderness science in a time of change conference—Volume 5: wilderness ecosystems, threats, and management; 1999 May 23-27; Missoula, MT. Proceedings RMRS-P-15-VOL-5. Ogden, UT: U.S. Department of Agriculture, Forest Service, Rocky Mountain Research Station: 225-232.
Artman, Vanessa L.; Hutchinson, Todd F.; Brawn, Jeffrey D. 2005. Fire ecology and bird populations in eastern deciduous forests. In: Saab, Victoria A.; Powell, Hugh D. W., eds. Fire and avian ecology in North America. Studies in Avian Biology No. 30. Ephrata, PA: Cooper Ornithological Society: 127-138.
Aschmann, H. 1991. Human impact on the biota of mediterranean-climate regions. In: Groves, R. H.; DiCastri, F., eds. Biogeography of Mediterranean Invasions. Cambridge, UK: Cambridge University Press: 33-42.
Asher, Jerry E.; Dewey, Steve; Johnson, Curt; Olivarez, Jim. 2001. Reducing the spread of invasive exotic plants following fire in western forests, deserts, and grasslands. In: Galley, Krista E. M.; Wilson, Tyrone P., eds. Proceedings of the invasive species workshop: The role of fire in the control and spread of invasive species; Fire conference 2000: the first national congress on fire ecology, prevention, and management; 2000 November 27-December 1; San Diego, CA. Misc. Publ. No. 11. Tallahassee, FL: Tall Timbers Research Station: 102-103. Abstract.
Asher, Jerry; Dewey, Steven; Olivarez, Jim; Johnson, Curt. 1998. Minimizing weed spread following wildland fires. Proceedings, Western Society of Weed Science. 51: 49.
Ashton, Isabel W.; Hyatt, Laura A.; Howe, Katherine M.; Gurevitch, Jessica; Lerdau, Manuel T. 2005. Invasive species accelerate

decomposition and litter nitrogen loss in a mixed deciduous forest. Ecological Applications. 15(4): 1263-1272.
Asner, Greg P.; Hughes, R. Flint; Elmore, Andrew J.; Warner, Amanda.S.; Vitousek, Peter, M. 2005. Ecosystem structure along bioclimatic gradients in Hawaii from imaging spectroscopy. Remote Sensing of Environment. 96: 497-508.
Auclair, Allan N.; Cottam, Grant. 1971. Dynamics of black cherry (*Prunus serotina* Erhr.) in southern Wisconsin oak forests. Ecological Monographs. 41(2): 153-177.
Axelrod, D. I. 1958. Evolution of the Madro-tertiary flora. Botanical Review. 24: 433-509.
Axelrod, D. I. 1966. The Pleistocene Soboba flora of southern California. University of California Publications in Geology. 60: 1-79.
Axelrod, D. I. 1978. The origin of coastal sage scrub, Alta and Baja California. American Journal of Botany. 65: 1117-1131.
Axelrod, D. I. 1989. Age and origin of chaparral. In: Keeley, S. C., ed. The California chaparral: paradigms reexamined. Los Angeles, CA: Natural History Museum of Los Angeles County: 7-20.
Backer, Dana M.; Jensen, Sara E.; McPherson, Guy R. 2004. Impacts of fire-suppression activities on natural communities. Conservation Biology. 18(4): 937-946.
Baez S.; Fargione, J.; Moore, D. I.; Collins, S. L.; Gosz, J. R. 2007. Nitrogen deposition in the northern Chihuahuan desert: Temporal trends and potential consequences. Journal of Arid Environments 68: 640-651.
Bahre, C. J. 1985. Wildfire in southeastern Arizona between 1859 and 1890. Desert Plants. 7: 190-194.
Bahre, C. J.; Shelton, M. L. 1993. Historic vegetation change, mesquite increases, and climate in southeastern Arizona. Journal of Biogeography. 20(5): 489-504.
Bailey, D. W. 1991. The wildland-urban interface: social and political implications in the 1990's. Fire Management Notes. 52: 11-18.
Bailey, John D.; Mayrsohn, Cheryl; Doescher, Paul S.; St. Pierre, Elizabeth; Tappeiner, John. 1998. Understory vegetation in old and young Douglas-fir forests of western Oregon. Forest Ecology and Management. 112: 289-302.
Bailey, Robert G. 1995. Description of the ecoregions of the United States. 2nd ed. Misc. Pub. 1391. Washington, DC: U.S. Department of Agriculture, Forest Service. 108 p.
Bakeman, M. E.; Nimlos, J. 1985. The genesis of mollisols under Douglas-fir. Soil Science. 140: 449-452.
Baker, W. L.; Shinneman, D. J. 2004. Fire and restoration of pinon-juniper woodlands in the western United States: a review. Forest Ecology and Management. 189(1-3): 1-21.
Banko, P.; Farmer, C.; Hess, S.; Brinck, K.; Beauprez, G.; Castner, J.; Crummer, J.; Danner, R.; Hsu, B.; Kozar, K.; Leialoha, J.; Lindo, A.; Muffler, B.; Murray, C.; Nelson, D.; Pollock, D.; Schwarzfeld, M.; Rapozo, K.; Severson, E. 2004. Palila restoration project, 2004. Summary of results, 1996-2004. Hawai`i National Park, HI: U.S. Geological Survey, Biological Resources Discipline, Pacific Islands Ecosystem Research Center Kilauea Field Station.
Barden, L. S.; Matthews, J. F. 1980. Change in abundance of honeysuckle (*Lonicera japonica*) and other ground flora after prescribed burning of a Piedmont pine forest. Castanea. 45: 257-260.
Barden, Lawrence S. 1987. Invasion of *Microstegium vimineum* (Poaceae), an exotic, annual, shade-tolerant, C4 grass, into a North Carolina floodplain. The American Midland Naturalist. 118(1): 40-45.
Barnes, Thomas, G. 2004. Strategies to convert exotic grass pastures to tall grass prairies communities. Proceedings of the IPINAMS Conference. Weed Technology. 18. 1364-1370.
Barnes, William J. 1972. The autecology of the *Lonicera X bella* complex. Madison, WI: University of Wisconsin. 169 p. Dissertation.
Barney, M. A.; Frischknecht, N. C. 1974. Vegetation changes following fire in the pinyon-juniper type of west-central Utah. Journal of Range Management. 27(2): 91-96.
Barney, Milo A. 1972. Vegetation changes following fire in the pinyon-juniper type of west central Utah. Provo, UT: Brigham Young University. 71 p. Thesis.
Barrett, S. W.; Arno, S. F.; Key, C. H. 1991. Fire regimes of western larch-lodgepole pine forests in Glacier National Park, Montana. Canadian Journal of Forest Research. 2: 1711.
Barrett, Spencer C. H. 2000. Microevolutionary influences of global changes on plant invasions. In: Mooney, Harold A.; Hobbs, Richard J., eds. Invasive species in a changing world. Washington, DC: Island Press. 115-139.
Barrilleaux, T. C.; Grace, J. B. 2000. Effects of soil type and moisture regime on the growth and invasive potential of Chinese tallow tree (*Sapium sebiferum*). American Journal of Botany. 87: 1099-1106.
Bartlett, H. H. 1956. Fire, primitive agriculture, and grazing in the tropics. In: Thomas, W.L., ed. Man's role in changing the face of the earth. Vol. 2. Chicago, IL: University of Chicago Press. 692-720
Barton, Andrew M.; Brewster, Lauri B.; Cox, Anne N.; Prentiss, Nancy K. 2004. Non-indigenous woody invasive plants in a rural New England town. Biological Invasions. 6: 205-211.
Bartuszevige, Anne M.; Gorchov, David L. 2006. Avian seed dispersal of an invasive shrub. Biological Invasions. 8: 1013-1022.
Bashkin, M.; Stohlgren, T. J.; Otsuki, Y.; Lee, M.; Evangelista, P.; Belnap, J. 2003. Soil characteristics and plant exotic species invasion in the Grand Staircase-Escalante National Monument, Utah, USA. Applied Soil Ecology. 22: 67-77.
Baskin, Carol C.; Baskin, Jerry M. 2001. Seeds: ecology, biogeography, and evolution of dormancy and germination. San Diego, CA: Academic Press. 666 p.
Baskin, J. M.; Baskin, C. C. 1981. Ecology of germination and flowering in the weedy winter annual grass *Bromus japonicus*. Journal of Range Management. 34: 369-372.
Batcher, M. S. 2000a. Element stewardship abstract: *Ligustrum* spp. (privet), [Online]. In: The Global Invasive Specie Initiative: The Nature Conservancy (Producer). Available: http://tncweeds.ucdavis.edu/esadocs/documnts/ligu_sp.pdf [2007, July 12].
Batcher, M. S. 2000b. Element stewardship abstract: *Melia azadarach* (chinaberry), [Online]. In: The global invasive specie initiative. Arlington, VA: The Nature Conservancy (Producer). Available: http://tncweeds.ucdavis.edu/esadocs/documnts/meliaze.pdf [2005, January 4].
Batcher, M. S. 2004. Element stewardship abstract: *Festuca arundinacea* (Schreb.) (tall fescue), [Online]. In: The global invasive specie initiative. Arlington, VA: The Nature Conservancy (Producer). Available: http://tncweeds.ucdavis.edu/esadocs/documnts/festaru.pdf [2004, December 12].
Batcher, M. S.; Stiles, S. A. 2000. Element stewardship abstract: *Lonicera maackii* (Rupr.) Maxim (Amur's honeysuckle), *Lonicera morrowii* A. Gray (Morrow's honeysuckle), *Lonicera tatarica* L. (Tatarian honeysuckle), *Lonicera x bella* Zabel (Bell's honeysuckle), the bush honeysuckles, [Online]. In: The global invasive specie initiative. Arlington, VA: The Nature Conservancy (Producer). Available: http://tncweeds.ucdavis.edu/esadocs/documnts/loni_sp.pdf [2007, July 12].
Battles, J. J.; Shlisky, A. J.; Barrett, R. H.; Heald, R. C.; Allen-Diaz, B. H. 2001. The effects of forest management on plant species diversity in a Sierran conifer forest. Forest Ecology and Management. 146: 211-222.
Bauhus, J.; Khanna, P. K.; Raison, R. J. 1993. The effects of fire on carbon and nitrogen mineralization and nitrification in an Australian forest soil. Australian Journal of Soil Research. 31: 621-639.
Bazzaz, F. A. 1986. Life history of colonizing plants: some demographic, genetic, and physiological features. In: Mooney, Harold A.; Drake, James A., eds. Ecology of biological invasions of North America and Hawaii. Ecological Studies 58. New York: Springer-Verlag: 96-110.
Beadle, N. C. W. 1940. Soil temperatures during forest fires and their effect on the survival of vegetation. Journal of Ecology. 28(2): 180-192.
Bean, Ellen; McClellan, Linnea, eds. 1997. Oriental bittersweet—*Celatrus orbiculata* Thunb., [Online]. In: Tennessee exotic plant management manual. Tennessee Exotic Pest Plant Council (Producer). Available: http://www.tneppc.org/Manual/Oriental_Bittersweet.htm [2005, February 11].
Beatley, J. C. 1966. Ecological status of introduced brome grasses (*Bromus* spp.) in desert vegetation of southern Nevada. Ecology. 47(4): 548-554.
Becker, Donald A. 1989. Five years of annual prairie burns. In: Bragg, Thomas A.; Stubbendieck, James, eds. Prairie pioneers: ecology, history and culture: Proceedings, 11th North American prairie conference; 1988 August 7-11; Lincoln, NE. Lincoln, NE: University of Nebraska: 163-168.
Beever, J. W., III; Beever, L. B. 1993. The effects of annual burning on the understory of a hydric slash pine flatwoods in southwest

Florida. (Poster abstract). In: Cerulean, Susan I.; Engstrom, R. Todd, eds. Fire in wetlands: A management perspective: 19th Tall Timbers Fire Ecology Conference; 1993 November 3-6; Tallahassee, FL. Tall Timbers Research Station, The Nature Conservancy, U.S. Fish and Wildlife Service: 163.

Belcher, J. W.; Wilson, S. D. 1989. Leafy spurge and the species composition of a mixed-grass prairie. Journal of Range Management. 42: 172-175.

Bell, Gary P. 1997. Ecology and management of *Arundo donax*, and approaches to riparian habitat restoration in southern California. In: Brock, J. H.; Wade, M.; Pyšek, P.; Green, D., eds. Plant invasions: studies from North America and Europe. Leiden, The Netherlands: Backhuys Publishers: 103-113.

Bellemare, Jesse; Motzkin, Glenn; Foster, David R. 2002. Legacies of the agricultural past in the forested present: an assessment of historical land-use effects on rich mesic forests. Journal of Biogeography. 29(10/11): 1401-1420.

Belles, H. A.; Myers, R. L.; Thayer, D. D. 1999. Physical control. In: Laroche, F. ed. Melaleuca management plan, ten years of successful melaleuca management in Florida 1988-1998. Florida Exotic Pest Plant Council. 29-31.

Bennett, Karen; Dibble, Alison C.; Patterson III, William A. 2003. Using fire to control invasive plants: What's new, what works in the Northeast? 2003 workshop proceedings; 2003 January 24; Portsmouth, NH. Durham, NH: University of New Hampshire Cooperative Extension. 53 p.

Benninger-Truax, Mary; Vankat, John L.; Schaefer, Robert L. 1992. Trail corridors as habitat and conduits for movement of plant species in Rocky Mountain National Park, Colorado, USA. Landscape Ecology. 6(4): 269-278.

Benson, Emily J. 2001. Effects of fire on tallgrass prairie plant population dynamics. Manhattan, KS: Kansas State University. 59 p. Thesis.

Benson, Nathan C.; Kurth, Laurie L. 1995. Vegetation establishment on rehabilitated bulldozer lines after the 1988 Red Bench Fire in Glacier National Park. In: Brown, James K.; Mutch, Robert W.; Spoon, Charles W.; Wakimoto, Ronald H., technical coordinators. Proceedings: symposium on fire in wilderness and park management; 1993 March 30-April 1; Missoula, MT. Gen. Tech. Rep. INT-GTR-320. Ogden, UT: U.S. Department of Agriculture, Forest Service, Intermountain Research Station: 164-167.

Bentley, J. R.; Fenner, R. L. 1958. Soil temperatures during burning related to postfire seedbeds on woodland range. Journal of Forestry. 56: 737-740.

Berg, William A. 1993. Old World bluestem response to fire and nitrogen fertilizers. Journal of Range Management. 46(5): 421-425.

Berry, Wendell. 2000. Life is a miracle, an essay against modern superstition. Washington, DC: Counterpoint. 153 p.

Beschta, Robert L.; Rhodes, Jonathan J.; Kauffman, J. Boone; Gresswell, Robert E.; Minshall, G. Wayne; Karr, James R.; Perry, David A.; Hauer, F. Richard; Frissell, Christopher A. 2004. Postfire management on forested public lands of the western United States. Conservation Biology. 18(4): 957-967.

Bessie, W. C.; Johnson, E. A. 1995. The relative importance of fuels and weather on fire behavior in subalpine forests. Ecology. 26: 747-762.

Beyers, J. L.; Wakeman, C. D.; Wohlgemuth, P. M.; Conard, S. G. 1998. Effects of postfire grass seeding on native vegetation in southern California chaparral. Proceedings of the Annual Forest Vegetation Management Conference. 19: 52-64.

Beyers, James E.; Reichard, Sarah; Randall, John M.; Parker, Ingrid M.; Smith, Carey S.; Lonsdale, W. M.; Atkinson, I. A. E.; Seastedt, T. R.; Williamson, Mark; Chornesky, E.; Hayes, D. 2002. Directing research to reduce the impacts of nonindigenous species. Conservation Biology. 16(3): 630-640.

Beyers, Jan L. 2004. Postfire seeding for erosion control: effectiveness and impacts on native communities. Conservation Biology. 18(4): 947-956.

Biedenbender, S. H.; Roundy, B. A.; Abbott, L. 1995. Replacing Lehmann lovegrass with native grasses. In: Roundy, Bruce A.; McArthur, E. Durant; Haley, Jennifer S.; Mann, David K., comps.. Proceedings: Wildland Shrub and Arid Land Restoration Symposium. Gen. Tech. Rep. INT-GTR-315. Ogden, UT: U.S. Department of Agriculture, Forest Service, Intermountain Research Station: 52-56.

Biesboer, D. D.; Eckardt, N. 1987. Element stewardship abstract: *Euphorbia esula* (leafy spurge), [Online]. In: The global invasive specie initiative. Arlington, VA: The Nature Conservancy (Producer). Available: http://tncweeds.ucdavis.edu/esadocs/documnts/euphesu.pdf [2007, July 10].

Billings, W. D. 1990. *Bromus tectorum*, a biotic cause of ecosystem impoverishment in the Great Basin. In: Woodell, George M., ed. Proceedings: The Earth in transition: patterns and processes of biotic impoverishment ;Woods Hole, MA. New York: Cambridge University Press: 301-322.

Billings, W. D. 1994. Ecological impacts of cheatgrass and resultant fire on ecosystems in the western Great Basin. In: Monsen, Stephen B.; Kitchen, Stanley G., comps. Proceedings: Ecology and management of annual rangelands; 1992 May 18-22; Boise, ID. Gen. Tech. Rep. INT-GTR-313. Ogden, UT: U.S. Department of Agriculture, Forest Service, Intermountain Research Station: 22-30.

Billock, Arlene. 2003. [Personal communication]. Center for Ecology and Environmental Technology, Lafayette, LA.

Bishop Museum. 2004. Hawaii's endangered plants, [Online]. In: Honolulu, HI: Bishop Museum (Producer). Available: http://hbs.bishopmuseum.org/endangered/endangered-new.html [2007, June 15].

Bisson, Peter A.; Rieman, Bruce E.; Luce, Charlie; Hessburg, Paul F.; Lee, Danny C.; Kershner, Jeffrey L.; Reeves, Gordon H.; Gresswell, Robert E. 2003. Fire and aquatic ecosystems of the western USA: current knowledge and key questions. In: Young, Michael K.; Gresswell, Robert E.; Luce, Charles H., guest eds. Selected papers from an international symposium on effects of wildland fire on aquatic ecosystems in the western USA; 2002 April 22-24; Boise, ID. In: Forest Ecology and Management. Special Issue: The effects of wildland fire on aquatic ecosystems in the western USA. New York: Elsevier Science B. V. 178(1-2): 213-229.

Biswell, H. 1999. Prescribed burning in California wildlands vegetation management, 2nd ed. University of California Press, Berkeley, California USA. 1085 p.

Blackburn, T. M.; Duncan, R. P. 2001. Determinants of establishment success in introduced birds. Nature. 414: 195-197.

Blackmore, Murray; Vitousek, Peter M. 2000. Cattle grazing, forest loss and fuel loading in a dry forest ecosystem at Pu`u Wa`a Wa`a Ranch, Hawaii. Biotropica. 32: 625-632.

Blaisdell, James P.; Murray, Robert B.; McArthur, E. Durant. 1982. Managing Intermountain rangelands—sagebrush-grass ranges. Gen. Tech. Rep. INT-134. Ogden, UT: U.S. Department of Agriculture, Forest Service, Intermountain Forest and Range Experiment Station. 41 p.

Blank, Robert R.; Allen, Fay L.; Young, James A. 1996. Influence of simulated burning of soil-litter from low sagebrush, squirreltail, cheatgrass, and medusahead on water-soluble anions and cations. International Journal of Wildland Fire. 6(3): 137-143.

Blank, Robert R.; Trent, James D.; Young, James A. 1992. Sagebrush communities on clayey soils of northeastern California: a fragile equilibrium. In: Clary, Warren P., McArthur, E. Durant, Bedunah, Don, Wambolt, Carl L. Proceedings: Symposium on ecology and management of riparian shrub communities. Gen. Tech. Rep. GTR- INT-289 Sun Valley, ID. Intermountain Research Station, Ogden, UT: U.S. Department of Agriculture, Forest Service: 198-202.

Blank, Robert R.; White, Robert H.; Ziska, Lewis H. 2006. Combustion properties of *Bromus tectorum* L.: influence of ecotype and growth under four CO_2 concentrations. International Journal of Wildland Fire. 15: 227-236.

Blank, Robert R.; Young, James A. 1998. Heated substrate and smoke: influence on seed emergence and plant growth. Journal of Range Management. 51(5): 577-583.

Blankespoor, Gilbert W.; Bich, Brian S. 1991. Kentucky bluegrass response to burning: interactions between fire and soil moisture. Prairie Naturalist. 23(4): 181-192.

Blankespoor, Gilbert W.; Larson, Eric A. 1994. Response of smooth brome (*Bromus inermis* Leyss.) to burning under varying soil moisture conditions. The American Midland Naturalist. 131: 266-272.

Blossey, Bernd. 1999. Before, during and after: the need for long-term monitoring in invasive plant species management. Biological Invasions. 1: 301-311.

Blossey, Bernd; Notzold, Rolf. 1995. Evolution of increased competitive ability in invasive nonindigenous plants: a hypothesis. Journal of Ecology. 83: 887-889.

Blumler, M. A. 1984. Climate and the annual habit. Berkeley, CA: The University of California. 119 p. Thesis.

Bock, C. E.; Bock, J. H. 1992a. Response of birds to wildfire in native versus exotic Arizona grassland. Southwestern Naturalist. 37: 73-81.

Bock, C. E.; Bock, J. H. 1998. Factors controlling the structure and function of desert grasslands: a case study from southeastern Arizona. In: Tellman, Barbara, Finch, Deborah M., Edminster, Carl, Hamre, Robert, eds. Proceedings: The future of arid grasslands: identifying issues, seeking solutions Tucson, AZ. Fort Collins, CO: U.S. Department of Agriculture, Forest Service, Rocky Mountain Research Station: 33-44.

Bock, J. H.; Bock, C. E. 1992b. Vegetation responses to wildfire on native versus exotic Arizona grassland. Journal of Vegetation Science. 3: 439-446.

Bodle, M.; Van, T. 1999. Biology of melaleuca, In: Laroche, F., ed. Melaleuca management plan, ten years of successful melaleuca management in Florida 1988-1998. Florida Exotic Pest Plant Council. 7-12.

Boerner, Ralph E. J.; Decker, Kelly L. M.; Sutherland, Elaine K. 2000a. Prescribed burning effects on soil enzyme activity in a southern Ohio hardwood forest: a landscape-scale analysis. Soil Biology and Biochemistry. 32: 899-908.

Boerner, Ralph E. J.; Morris, Sherri J.; Sutherland, Elaine K.; Hutchinson, Todd F. 2000b. Spatial variability in soil nitrogen dynamics after prescribed burning in Ohio mixed-oak forests. Landscape Ecology. 15: 425-239.

Boggs, K. W.; Story, J. M. 1987. The population age structure of spotted knapweed (*Centaurea maculosa*) in Montana. Weed Science. 35: 194-198.

Bolstad, Roger. 1971. Catline rehabilitation and restoration. In: Slaughter, C. W.; Barney, Richard J.; Hansen, G. M., eds. Fire in the northern environment—a symposium: Proceedings of a symposium; 1971 April 13-14; Fairbanks, AK. Portland, OR: U.S. Department of Agriculture, Forest Service, Pacific Northwest Range and Experiment Station: 107-116.

Bonnicksen, Thomas M.; Stone, Edward C. 1982. Reconstruction of a presettlement giant sequoia-mixed conifer forest community using the aggregation approach. Ecology. 63(4): 1134-1148.

Booth, D. T.; Cox, S. E.; Meikle, T. W.; Fitzgerald, C. 2006. The accuracy of ground-cover measurements. Rangeland Ecological Management. 59: 179–188.

Borchert, M. I.; Odion, D. C. 1995. Fire intensity and vegetation recovery in chaparral: a review. In: Keeley, J. E.; Scott, T., eds. Brushfires in California wildlands: ecology and resource management. Fairfield, WA: International Association of Wildfire: 91-100.

Borell, A. E. 1971. Russian-olive for wildlife and other conservation uses. Leaflet 292. Washington, DC: U.S. Department of Agriculture. 8 p.

Bossard, C.; Randall, J. M.; Hoshovsky, M. C. 2000. Invasive plants of California's wildlands. Berkeley, CA: University of California Press. 360 p.

Bossard, C.; Rejmánek, M. 1994. Herbivory, growth, seed production, and resprouting of an exotic invasive shrub. Biological Conservation. 67:193-200.

Bossard, Carla. 1993. Seed germination in the exotic shrub *Cytisus scoparius* (Scotch broom) in California. Madroño. 40(1): 47-61.

Bossard, Carla. 2000a. *Cytisus scoparius* (L.) Link. In: Bossard, Carla C.; Randall, John M.; Hoshovsky, Marc C., eds. Invasive plants of California's wildlands. Berkeley, CA: University of California Press: 145-150.

Bossard, Carla. 2000b. *Genista monspessulana*. In: Bossard, Carla C.; Randall, John M.; Hoshovsky, Marc C., eds. Invasive plants of California's wildlands. Berkeley, CA: University of California Press: 203-208.

Bottoms, Rick. 2002. [Email to Kristin Zouhar]. January 8. Columbia, MO: University of Missouri, Agronomy Specialist. On file at: U.S. Department of Agriculture, Forest Service, Rocky Mountain Research Station, Fire Sciences Laboratory, Missoula, MT.

Bottoms, Rick M.; Whitson, Tom D. 1998. A systems approach for the management of Russian knapweed (*Centaurea repens*). Weed Technology. 12(2): 363-366.

Boucher, Tina V. 2001. Vegetation response to prescribed fire in the Kenai Mountains, Alaska. Corvallis, OR: Oregon State University. 98 p. Thesis.

Boudreau, D.; Willson, G. 1992. Buckthorn research and control at Pipestone National Monument (Minnesota). Restoration and Management Notes. 10: 94-95.

Bovey, R. W. 1965. Control of Russian olive by aerial application of herbicides. Journal of Range Management. 18: 194-195.

Bowles, Marlin L.; Jones, Michael D.; McBride, Jenny L. 2003. Twenty-year changes in burned and unburned sand prairie remnants in Northwestern Illinois and implications for management. American Midland Naturalist. 149: 35-45.

Bowles, Marlin L.; McBride, Jenny L. 1998. Vegetation composition, structure, and chronological change in a decadent midwestern North American savanna remnant. Natural Areas Journal. 18(1): 14-27.

Boyd, David. 1995. Use of fire to control French broom. In: Lovich, Jeff; Randall, John; Kelly, Mike, eds. Proceedings, California Exotic Pest Plant Council: Symposium '95; 1995 October 6-8; Pacific Grove, CA. Berkeley, CA: California Exotic Pest Plant Council: 9-12.

Boyd, David. 1998. Use of fire to control French broom. Proceedings, California Weed Science Society. 50: 149-153.

Boyd, R. 1986. Strategies of Indian burning in the Willamette Valley. Canadian Journal of Anthropology. 5: 65-86.

Bradley, Anne F.; Noste, Nonan V.; Fischer, William C. 1991. Fire ecology of forests and woodlands in Utah. Ogden, UT: U.S. Department of Agriculture, Forest Service, Intermountain Research Station. 128 p.

Bram, Margot R.; McNair, James N. 2004. Seed germinability and its seasonal onset of Japanese knotweed (*Polygonum cuspidatum*). Weed Science. 52: 759-767.

Braun-Blanquet, J. 1965. Plant sociology: the study of plant communities. New York: Hafner Publishing. 437 p.

Brender, E. V. 1961. Control of honeysuckle and kudzu. Station Paper Number 120. Asheville, NC: U.S. Department of Agriculture, Forest Service, Southeastern Forest Experiment Station. 9 p.

Brender, E. V.; Merrick, E. 1950. Early settlement and land use in the present Toccoa Experimental Forest. Scientific Monthly. 71: 318-325.

Brenner, J.; Wade, D. 2003. Florida's revised prescribed fire law: protection for responsible burners. In: Gallery, K. E. M.; Klinger, R. C.; Sugihara, N. G., eds. Proceedings of the fire conference 2000: The 1st national congress on fire ecology, prevention, and management; 2000 November 27-December 1; San Diego, CA. Miscellaneous Publication No.13. Tallahassee, FL: Tall Timbers Research Station, pp.132-136.

Brenton, B.; Klinger, R. C. 1994. Modeling the expansion and control of fennel (*Foeniculum vulgare*) on the Channel Islands. In: Maender, G. J.; Halvorson, W. L., eds. The 4th California Islands Symposium: update on the status of resources. Santa Barbara, CA: Santa Barbara Museum of Natural History: 497-504.

Breshears, David D.; Cobb, Neil S.; Rich, Paul M.; Price, Kevin P.; Allen, Craig D.; Balice, Randy G.; Romme, William H.; Kastens, Jude H.; Floyd, M. Lisa; Belnap, Jayne; Anderson, Jesse J.; Myers, Orrin B.; Meyer, Clifton W. 2005. Regional vegetation die-off in response to global-change-type drought. Proceedings National Academy of Sciences. 102(42): 15144-15148.

Briese, D. T. 1996. Biological control of weeds and fire management in protected natural areas: Are they compatible strategies? Biological Conservation. 77(2-3): 135-141.

Brock, John H. 1998. Invasion, ecology and management of *Elaeagnus angustifolia* (Russian olive) in the southwestern United States. In: Starfinger, U.; Edwards, K.; Kowarik, I.; Williamson, M., eds. Plant invasions: ecological mechanisms and human responses. Leiden, The Netherlands: Backhuys Publishers: 123-136.

Brockway, Dale G. (Research Ecologist, USDA Forest Service Southern Research Station). 2006. [Letter to Molly E. Hunter]. March 17. 1 Leaf. On file at: Colorado State University, Warner College of Natural Resources, Western Forest Fire Research Center, Fort Collins, CO.

Brooks, M. L. 1999a. Alien annual grasses and fire in the Mojave desert. Madroño. 46: 13-19.

Brooks, M. L. 1999b. Habitat invasibility and dominance by alien annual plants in the western Mojave Desert. Biological Invasions. 1: 325-337.

Brooks, M. L. 2000. Competition between alien annual grasses and native annual plants in the Mojave Desert. The American Midland Naturalist. 144: 92-108.

Brooks, M. L. 2002. Peak fire temperatures and effects on annual plants in the Mojave Desert. Ecological Applications. 12(4): 1088-1102.

Brooks, M. L. 2003. Effects of increased soil nitrogen on the dominance of alien annual plants in the Mojave Desert. Journal of Applied Ecology. 40(2): 344-353.

Brooks, M. L. 2006. Effects of fire on plant communities. In: DiTomaso, J. M.; Johnson, D. W., eds. The use of fire as a tool for controlling invasive weeds. Bozeman, MT: Center for Invasive Plant Management: 29-32.

Brooks, M. L. 2007a. Effects of land management practices on plant invasions in wildland areas. In: Nentwig, W., ed. Biological Invasions: Ecological Studies 193. Heidelberg, Germany: Springer: 147-162.

Brooks, M. L. 2007b. [Review comments to J.B. Grace]. June 8, Henderson, NV, ID: US Geological Survey, Western Ecological Research Center.

Brooks, M. L.; Berry, K. H. 2006. Dominance and environmental correlates of alien annual plants in the Mojave Desert, USA. Journal of Arid Environments 67 (supplement 1): 100-124.

Brooks, M. L.; D'Antonio, C. M.; Richardson, D. M.; Grace, J. B.; Keeley, J. E.; DiTomaso, J. M.; Hobbs, R. J.; Pellant, M.; Pyke, D. 2004 . Effects of invasive alien plants on fire regimes. BioScience. 54(7): 677-688

Brooks, M. L.; Esque, T. C. 2000. Alien grasses in the Mojave and Sonoran deserts. In: Proceedings of the California Exotic Pest Plant Council symposium (eds unknown). Berkeley, CA: California Invasive Plant Council: 39-44.

Brooks, M. L.; Esque, T. C. 2002. Alien plants and fire in desert tortoise (*Gopherus agassizii*) habitat of the Mojave and Colorado deserts. Chelonian Conservation Biology. 4(2): 330-340.

Brooks, M. L.; Esque, T. C.; Duck, T. 2003. Fuels and fire regimes in creosotebush, blackbrush, and interior chaparral shrublands. Report for the Southern Utah Demonstration Fuels Project, U.S. Department of Agriculture, Forest Service, Rocky Mountain Research Station, Fire Science Lab, Missoula, Montana. 17 p.

Brooks, M. L.; Klinger, R. C. [In press]. Prioritizing species and sites for early detection programs. Chapter 5 In: Geissler, P.; Welch, B., eds. Early Detection of Invasive Plants: A Handbook.

Brooks, M. L.; Matchett, J. R. 2003. Plant community patterns in unburned and burned blackbrush (*Coleogyne ramosissima* Torr.) shrublands in the Mojave Desert. Western North American Naturalist. 63(3): 283-298.

Brooks, M. L.; Matchett, J. R. 2006. Spatial and Temporal Patterns of Wildfires in the Mojave Desert. 1980-2004. Journal of Arid Environments. 67: 148-164.

Brooks, M. L.; Matchett, J. R.; Berry, K. H. 2006. Effects of livestock watering sites on alien and native plants in the Mojave Desert, USA. Journal of Arid Environments. 67: 125-147.

Brooks, M. L.; Minnich, R. A. 2006. Southeastern deserts bioregion. In: Sugihara, Neil G.; van Wagtendonk, Jan W.; Shaffer, Kevin E.; Fites-Kaufman, Joann; Thode, Andrea E., eds. Fire in California's ecosystems. Berkeley, CA: University of California Press: 391-414.

Brooks, M. L.; Pyke, D. A. 2001. Invasive plants and fire in the deserts of North America. In: Galley, Krista E. M.; Wilson, Tyrone P., eds. Proceedings of the invasive species workshop: The role of fire in the control and spread of invasive species; Fire conference 2000: the first national congress on fire ecology, prevention, and management; 2000 November 27-December 1; San Diego, CA. Misc. Publ. No. 11. Tallahassee, FL: Tall Timbers Research Station: 1-14.

Brooks, Matthew Lamar. 1998. Ecology of a biological invasion: alien annual plants in the Mojave Desert. Riverside, CA: University of California. 186 p. Dissertation.

Brose, Patrick; Schuler, Thomas; Van Lear, David; Berst, John. 2001. Bring fire back: the changing regimes of the Appalachian mixed-oak forests. Journal of Forestry. 99(11): 30-35.

Brothers, Timothy S.; Spingarn, Arthur. 1992. Forest fragmentation and alien plant invasion of central Indiana old-growth forests. Conservation Biology. 6(1): 91-100.

Brown, D. E.; Minnich, R. A. 1986. Fire and changes in Creosote bush scrub of the western Sonoran Desert, California. The American Midland Naturalist. 116(2): 411-422.

Brown, David E., editor. 1982. Biotic communities of the American Southwest-United States and Mexico. Desert Plants. 4(1-4): 1-342.

Brown, J. K. 1974. Handbook for inventorying downed woody material. Gen. Tech. Rep. GTR-INT-16. Ogden, UT: U.S. Department of Agriculture, Forest Service, Intermountain Forest and Range Experiment Station. 34 p.

Brown, James K. 2000. Introduction and fire regimes. In: Brown, James K.; Smith, Jane Kapler, eds. Wildland fire in ecosystems: Effects of fire on flora. Gen. Tech. Rep. RMRS-GTR-42-vol. 2. Ogden, UT: U.S. Department of Agriculture, Forest Service, Rocky Mountain Research Station: 1-8.

Brown, James K.; Arno, Stephen F.; Barrett, Stephen W.; Menakis, James, P. 1994. Comparing the prescribed natural fire program with presettlement fires in the Selway-Bitterroot Wilderness. International Journal of Wildland Fire. 4(3): 157-168.

Brown, James K.; Oberheu, Rick D.; Johnston, Cameron M. 1982. Handbook for inventorying surface fuels and biomass in the Interior West. Gen. Tech. Rep. INT-129. Ogden, UT: U.S. Department of Agriculture, Forest Service, Intermountain Forest and Range Experiment Station. 48 p.

Brown, James K.; Smith, Jane Kapler, eds. 2000. Wildland fire in ecosystems: Effects of fire on flora. Gen. Tech Rep. RMRS-GRT-42-vol. 2. Ogden, UT: U.S. Department of Agriculture, Forest Service, Rocky Mountain Research Station. 257 p.

Brown, R. L.; Peet, R. K. 2003. Diversity and invasibility of southern Appalachian plant communities. Ecology. 84: 32-39.

Bruce, K.A.; Cameron, G. N.; Harcombe, P. A.; Jubinsky, G. 1997. Introduction, impact on native habitats, and management of a woody invader, the Chinese tallow tree, *Sapium sebiferum* (L.) Roxb. Natural Areas Journal. 17: 255-260.

Bruce, Katherine A.; Cameron, Guy N.; Harcombe, Paul A. 1995. Initiation of a new woodland type on the Texas Coastal Plain by the Chinese tallow tree (*Sapium sebiferum* (L.) Roxb.). Bulletin of the Torrey Botanical Society. 122(3): 215-225.

Brueckheimer, W. R. 1979. The quail plantations of the Thomasville-Tallahassee-Albany regions. Proceedings, Tall Timbers Fire Ecology Conference. 16: 141-165.

Bryan, Justin B.; Mitchell, Robert B.; Racher, Brent J.; Schmidt, Charles. 2001. Saltcedar response to prescribed burning in New Mexico. In: Zwank, Phillip J.; Smith, Loren M., eds. Research highlights—2001: range, wildlife, and fisheries management. Volume 32. Lubbock, TX: Texas Tech University, Department of Range, Wildlife, and Fisheries Management: 24.

Bryson, Charles T.; Carter, Richard. 2004. Biology of pathways for invasive weeds. Weed Technology. 18: 1216-1220.

Buckley, David S.; Crow, Thomas R.; Nauertz, Elizabeth A.; Schulz, Kurt E. 2003. Influence of skid trails and haul roads on understory plant richness and composition in managed forest landscapes in Upper Michigan, USA. Forest Ecology and Management. 175: 509-520.

Bunting, Stephen C. 1990. Prescribed fire effects in sagebrush-grasslands and pinyon-juniper woodlands. In: Alexander, M. E., and Bisgrove, G. F., technical coordinators. Proceedings: The art and science of fire management: Proceedings of the 1st Interior West Fire Council annual meeting and workshop. Information Rep. Kananaskis Village, AB. Edmonton, AB: Forestry Canada, Northwest Region, Northern Forestry Centre: 176-181.

Bunting, Stephen C. 1994. Effects of fire on juniper woodland ecosystems in the Great Basin. In: Monsen, Stephen B.; Kitchen, Stanley G., comps. Proceedings: ecology and management of annual rangelands; 1992 May 18-22; Boise, ID. Gen. Tech. Rep. INT-GTR-313. Ogden, UT: U.S. Department of Agriculture, Forest Service, Intermountain Research Station: 53-55.

Bunting, Stephen C.; Kilgore, Bruce M.; Bushey, Charles L. 1987. Guidelines for prescribed burning sagebrush-grass rangelands in the northern Great Basin. Gen. Tech. Rep. INT-231. Ogden, UT: U.S. Department of Agriculture, Forest Service, Intermountain Research Station. 33 p.

Bunting, Stephen C.; Peters, Erin F.; Sapsis, David B. 1994. Impact of fire management on rangelands of the Intermountain West.

Scientific Contract Report: Science Integration Team, Terrestrial Staff, Range Task Group. Walla Walla, WA: Interior Columbia Basin Ecosystem Management Project. 32 p.

Burke, M. J. W.; Grime, J. P. 1996. An experimental study of plant community invasibility. Ecology. 77(3): 776-790.

Burkett, V.; Grace, J.; Larson D.; Kallemeyn, L.; Stohlgren, T. 2000. Invasive species science strategy for DOI lands, central region. USGS Unnumbered Report. [Pages unknown].

Burkhardt, J. W.; Tisdale, E. W. 1976. Causes of juniper invasion in southwestern Idaho. Ecology. 57: 472-484.

Burned Area Response National-Interagency Team. 2004. Burned area emergency stabilization and rehabilitation plan: 2004 Alaskan fires, [Online]. [Place of publication unknown]: [Burned Area Response National-Interagency Team]. Available: http://www.ak.blm.gov/baer/ [2005, January 15].

Burney; D. A.; DeCandido, R. V.; Burney, L. P.; Kostel-Hughes, F. N.; Stafford, T. W., Jr.; James, H. F. 1995. A Holocene record of climate change, fire ecology, and human activity from montane Flat Top Bog, Maui. Journal of Paleolimnology. 13: 209-217.

Burns, Russell M.; Honkala, Barbara H., tech. coords. 1990a. Silvics of North America. Volume 1. Conifers. Agric. Handb. 654. Washington, DC: U.S. Department of Agriculture, Forest Service. 675 p.

Burns, Russell M.; Honkala, Barbara H., tech. coords. 1990b. Silvics of North America. Volume 2. Hardwoods. Agric. Handb. 654. Washington, DC: U.S. Department of Agriculture, Forest Service. 877 p.

Burquez-Montijo, Alberto; Miller, Mark E.; Martinez-Yrizar, Angelina. 2002. Mexican grasslands, thornscrub, and the transformation of the Sonoran Desert by invasive exotic bufffelgrass (Pennisetum ciliare). In: Tellman, Barbara, ed. Invasive exotic species in the Sonoran region. Tucson, AZ: The University of Arizona Press; The Arizona-Sonora Desert Museum: 126-146.

Busch, Briton Cooper, ed. 1983. Alta California, 1840-1842: The journal and observations of William Dane Phelps, master of the ship «Alert». Glendale, CA: Arthur H. Clark Co. 364 p.

Busch, D. E. 1994. Fire in southwestern riparian habitats: functional and community responses. In: Covington, W. W.; DeBano, L. F., tech. coords. Sustainable ecological systems: implementing an ecological approach to lands management. Gen. Tech. Rep. GTR-RM-247. Fort Collins, CO: U.S. Department of Agriculture, Forest Service, Rocky Mountain Forest and Range Experiment Station: 304-305.

Busch, D. E. 1995. Effects of fire on southwestern riparian plant community structure. The Southwestern Naturalist. 40: 259-267.

Busch, David E.; Smith, S. 1993. Effects of fire on water and salinity relations of riparian woody taxa. Oecologia. 94: 186-194.

Bushey, Charles L. 1985. Summary of results from the Galena Gulch 1982 spring burns (Units 1b). Missoula, MT: Systems for Environmental Management. 9 p.

Busse, M. D.; Simon, S. A.; Riegel, G. M. 2000. Tree-growth and understory responses to low-severity prescribed burning in thinned Pinus ponderosa forests of Central Oregon. Forest Science. 46: 258-268.

Busse, Matt; Shestak, Carol; Knapp, Eric; Hubbert, Ken; Fiddler, Gary. 2006. Lethal soil heating during burning of masticated fuels: effects of soil moisture and texture, [Online]. In: 3rd international fire ecology and management congress: Proceedings; 2006 November 13-17; San Diego, CA. [Davis, CA: The Association for Fire Ecology]. Pullman, WA: Washington State University (Producer). 4 p. Available: http://www.emmps.wsu.edu/2006firecongressproceedings/Extended%20Abstracts%20PDf%20Files/Contributed%20Papers/6%20Soils,%20Watershed,%20Aquatic/Busse.pdf [2007, June 27].

Butler, J. L.; Cogan, D. R. 2004. Leafy spurge effects on patterns of plant species richness. Journal of Range Management. 57: 305-311.

Butterfield, B. J.; Rogers, W. E.; Siemann, E. 2004. Growth of Chinese tallow tree (*Sapium sebiferum*) and four native trees under varying water regimes. Texas Journal of Science. 56: 335-346.

Byers, James E.; Reichard, Sarah; Randall, John M.; Parker, Ingrid M.; Smith, Carey S.; Lonsdale, W.M.; Atkinson, I.A.E.; Seastedt, T.R.; Williamson, M.; Chornesky, E.; Hayes, D. 2002. Directing research to reduce the impacts of nonindigenous species. Conservation Biology. 16(3): 630-640.

Cabin, R. J.; Weller, S. G.; Lorence, D. H.; Cordell, S.; Hadway, L. J. 2002. Effects of microsite, water, weeding, and direct seeding on the regeneration of native and alien species within a Hawaiian dry forest preserve. Biological Conservation. 104: 181-190.

Cabin, R. J.; Weller, S. G.; Lorence, D. H.; Flynn, T. W.; Sakai, A. K.; Sandquist, D.; Hadway, L. J. 2000. Effects of long-term ungulate exclusion and recent alien species control on the preservation and restoration of a Hawaiian tropical dry forest. Conservation Biology. 14: 439-453.

Cable, Dwight R. 1965. Damage to mesquite, Lehman lovegrass, and black grama by a hot June fire. Journal of Range Management. 18: 326-329.

Cable, Dwight R. 1971. Lehmann lovegrass on the Santa Rita Experimental Range, 1937-1968. Journal of Range Management. 24: 17-21.

Cable, Dwight R. 1973. Fire effects in southwestern semidesert grass-shrub communities. In: Proceedings, annual Tall Timbers fire ecology conference; 1972 June 8-9; Lubbock, TX. Number 12. Tallahassee, FL: Tall Timbers Research Station: 109-127.

Caley, Peter; Lonsdale, W. M.; Pheloung, P. C. 2006. Quantifying uncertainty in predictions of invasiveness, with emphasis on weed risk assessment. Biological Invasions. 8(8): 1595-1604.

Call, Lara J.; Nilsen, Erik T. 2003. Analysis of spatial patterns and spatial association between the invasive tree-of-heaven (*Ailanthus altissima*) and the native Black locust (Robinia *pseudoacacia*). The American Midland Naturalist. 150: 1-14.

Call, Lara J.; Nilsen, Erik T. 2005. Analysis of interactions between the invasive tree-of-heaven (*Ailanthus altissima*) and the native black locust (*Robinia pseudoacacia*). Plant Ecology. 176: 275-285.

Callaway, R. M.; Aschehoug, E. T. 2000. Invasive plants versus their new and old neighbors: a mechanism for exotic invasion. Science. 290: 521-523.

Callaway, R. M.; Davis, F. W. 1993. Vegetation dynamics, fire, and the physical environment in coastal central California. Ecology. 74: 1567-1578.

Callaway, R. M.; Ridenour, W. M. 2004. Novel weapons: invasive success and the evolution of increased competitive ability. Frontiers in Ecology and the Environment. 2: 436-443.

Callison, J.; Brotherson, J.D.; Bowns, J.E. 1985. The effects of fire on the blackbrush [*Coleogyne ramosissima*] community of southwestern Utah. Journal of Range Management. 38(6): 535-538.

Cameron, G. N.; Glumac, E. G.; Eshelman, B. D. 2000. Germination and dormancy in seeds of *Sapium sebiferum* (Chinese tallow tree). Journal of Coastal Research. 16: 391-395.

Campbell, B. 1980. Some mixed hardwood communities of the coastal ranges of southern California. Phytocoenologia. 8: 297-320.

Campbell, C. J.; Dick-Peddie, W. A. 1964. Comparison of phreatophyte communities on the Rio Grande in New Mexico. Ecology. 45(3): 492-502.

Campbell, G. S.; Jungbauer J. D.; Bidlake W. R.; Hungergord, R. D. 1994. Predicting the effect of temperature on soil thermal conductivity. Soil Science. 158: 307-313.

Caplan, Todd. 2002. Controlling Russian olives within cottonwood gallery forests along the Middle Rio Grande floodplain (New Mexico). Ecological Restoration. 20(2): 138-139.

Caplan, Todd. 2005. [Personal communication]. May 3. Albuquerque, NM: Parametrix, Inc. On file with: U.S. Department of Agriculture, Forest Service, Rocky Mountain Research Station, Missoula, MT; RWU 4403 files.

Cappuccino, Naomi; Mackay, Robin; Eisner, Candice. 2002. Spread of the invasive alien vine *Vincetoxicum rossicum*: tradeoffs between seed dispersability and seed quality. The American Midland Naturalist. 148(2): 263-270.

Caprio, Anthony C.; Lineback, Pat. 2002. Pre-twentieth century fire history of Sequoia and Kings Canyon National Park: A review and evaluation of our knowledge. In: Sugihara, Neil G.; Morales, Maria; Morales, Tony, eds. Fire in California ecosystems: integrating ecology, prevention and management: Proceedings of the symposium; 1997 November 17-20; San Diego, CA. Misc. Pub. No. 1. [Davis, CA]: Association for Fire Ecology: 180-199.

Caprio, Anthony; Haultain, Sylvia; Keifer, MaryBeth; Manley, Jeff. 1999. Problem evaluation and recommendations: invasive cheatgrass (*Bromus tectorum*) in Cedar Grove, Kings Canyon

National Park, [Online]. Three Rivers, CA: Sequoia and Kings Canyon National Parks, Science and Natural Resources Division (Producer). 19 p. Available: http://www.nps.gov/archive/seki/fire/pdf/cheatrecom2.pdf [2007, August 1].

Carey, Jennifer H. 1995. *Agrostis gigantea*. In: Fire Effects Information System, [Online]. U.S. Department of Agriculture, Forest Service, Rocky Mountain Research Station, Fire Sciences Laboratory (Producer). Available: http://www.fs.fed.us/database/feis/ [2007, February 1].

Carpenter, A. T.; Murray, T. A. 1999. Element stewardship abstract: *Bromus tectorum* L. (cheatgrass), [Online]. In: The global invasive species initiative. Arlington, VA: The Nature Conservancy (Producer). Available: http://tncweeds.ucdavis.edu/esadocs/documnts/bromtec.pdf [2007, July 10].

Carpenter, Jeffrey L. 1986. Responses of three plant communities to herbicide spraying and burning of spotted knapweed (*Centaurea maculosa*) in western Montana. Missoula, MT: University of Montana. 110 p. Thesis.

Carpinelli, Michael F. 2005. Effect of fire and imazapic application timing and rate on medusahead and desirable species, (abstract). Proceedings: 58th Annual Meeting Vancouver, British Columbia, Canada. Western Society of Weed Science. [Pages unknown].

Cater, Timothy C.; Chapin, F. Stuart, III. 2000. Differential effects of competition or microenvironment on boreal tree seedling establishment after fire. Ecology. 81(4): 1086-1099.

Catling, Paul M.; Brownell, Vivian R. 1998. Importance of fire in alvar ecosystems—evidence from the Burnt Lands, eastern Ontario. The Canadian Field-Naturalist. 112(4): 661-667.

Catling, Paul M.; Sinclair, Adrianne; Cuddy, Donald. 2001. Vascular plants pf a successional alvar burn 100 days after a severe fire and their mechanisms of re-establishment. The Canadian Field-Naturalist. 115: 214-222.

Catling, Paul M.; Sinclair, Adrianne; Cuddy, Donald. 2002. Plant community composition and relationships of disturbed and undisturbed alvar woodland. The Canadian Field-Naturalist. 116(4): 571-579.

Cave, George H.; Patten, Duncan T. 1984. Short-term vegetation responses to fire in the upper Sonoran Desert. Journal of Range Management. 37(6): 491-496.

Chambers, J. C.; Roundy, B. A.; Blank, R. R.; Meyer, S. E.; Whittaker, A. 2007. What makes Great Basin sagebrush ecosystems invasible by *Bromus tectorum*? Ecological Monographs. 77: 117-145.

Chang, Chi-ru. 1996. Ecosystem responses to fire and variations in fire regimes. In: Status of the Sierra Nevada. Sierra Nevada Ecosystem Project: Final report to Congress. Volume II: Assessments and scientific basis for management options. Wildland Resources Center Report No. 37. Davis, CA: University of California, Centers for Water and Wildland Resources: 1071-1099.

Chapin, F. S., III; Reynolds, H. L.; D'Antonio, C. M.; Eckhart, V. M. 1996. The functional role of species in terrestrial ecosystems. In: Walker, B.; Steffen, W., eds. Global change and terrestrial ecosystems. Cambridge, UK: Cambridge University Press: 403-428

Chapin, F. Stuart, III; Chapin, Melissa C. 1980. Revegetation of an arctic disturbed site by native tundra species. The Journal of Applied Ecology. 17(2): 449-456.

Chapin, F. Stuart, III; Shaver, Gaius; Giblin, Anne E.; Nadelhoffer, Knute J.; Laundre, James A. 1995. Responses of arctic tundra to experimental and observed changes in climate. Ecology. 76(3): 694-711.

Chapman, H. H. 1932. Is the longleaf type a climax? Ecology. 13: 328-334.

Chapman, R. R.; Crow, G. E. 1981. Applications of Raunkiaer's life form system to plant species survival after fire. Bulletin of the Torrey Botanical Club. 108: 472-478.

Chappell, Christopher B.; Crawford, Rex C. 1997. Native vegetation of the south Puget Sound prairie landscape. In: Dunn, Patrick; Ewing, Kern, eds. Ecology and conservation of south Puget Sound prairie landscape. Seattle, WA: The Nature Conservancy. 107-122.

Child, L. E.; Wade, P. M. 2000. The Japanese knotweed manual: the management and control of an invasive alien weed. Chichester, UK: Packard Publishing Ltd. 152 p.

Cholewa, Anita F. 1977. Successional relationships of vegetational composition to logging, burning, and grazing in the Douglas-fir/Physocarpus habitat type of northern Idaho. Moscow, ID: University of Idaho. 65 p. [+ appendices]. Thesis.

Chong, Geneva W.; Otsuki, Yuka; Stohlgren, Thomas J.; Guenther, Debra; Evangelista, Paul; Villa, Cynthia; Waters, Alycia. 2006. Evaluating plant invasions from both habitat and species perspectives. Western North American Naturalist. 66(1): 92-105.

Chou, Y. H.; Minnich, R. A.; Dezzani, R. J. 1993. Do fire sizes differ between southern California and Baja California? Forest Science. 39: 835-844.

Christensen, N. L. 1973. Fire and the nitrogen cycle in California chaparral. Science. 181: 66-68.

Christensen, N. L.; Muller, C. H. 1975. Effects of fire on factors controlling plant growth in Adenostema chaparral. Ecological Monographs. 45: 29-55.

Christian, Caroline E. 2003. A plan for monitoring and evaluating the effects of fire on plant communities at the Lassen Foothills Project. [Prepared for the Nature Conservancy]. Unpublished report on file at: U.S. Department of Agriculture, Forest Service, Rocky Mountain Research Station, Fire Sciences Laboratory, Missoula, MT. Variously paginated.

Cincotta, Richard P.; Uresk, Daniel W.; Hansen, Richard M. 1989. Plant compositional change in a colony of black-tailed prairie dogs in South Dakota. In: Bjugstad, Ardell J.; Uresk, Daniel W.; Hamre, R. H., tech. coords. 9th Great Plains wildlife damage control workshop proceedings; 1989 April 17-20; Fort Collins, CO. Gen. Tech. Rep. RM-171. Fort Collins, CO: U.S. Department of Agriculture, Forest Service, Rocky Mountain Forest and Range Experiment Station: 171-177.

Cione, N. K.; Padgett, P. E.; Allen, E. B. 2002. Restoration of a native shrubland impacted by exotic grasses, frequent fire, and nitrogen deposition in southern California. Restoration Ecology. 10: 376-384.

Clampitt, Christopher A. 1993. Effects of human disturbance on prairies and the regional endemic Aster curtis in western Washington. Northwest Science. 67: 163-169.

Clark, Deborah L. 1991. Factors determining species composition of post-disturbance vegetation following logging and burning of an old-growth Douglas-fir forest. Corvallis, OR: Oregon State University. 73 p. Thesis.

Clark, Deborah L.; Wilson, Mark V. 1994. Heat-treatment effects on seed bank species of an old-growth Douglas-fir forest. Northwest Science. 68(1): 1-5.

Clark, Deborah L.; Wilson, Mark V. 2001. Fire, mowing, and hand-removal of woody species in restoring a native wetland prairie in the Willamette Valley of Oregon. Wetlands. 21(1): 135-144.

Clark, James S.; Royall, P. Daniel. 1996. Local and regional sediment charcoal evidence for fire regimes in presettlement north-eastern North America. The Journal of Ecology. 84: 365-382.

Clark, Kennedy H. 1998. Use of prescribed fire to supplement control of an invasive plant, *Phragmites australis*, in marshes of southeast Virginia. In: Pruden, Teresa L.; Brennan, Leonard A., eds. Fire in ecosystem management: shifting the paradigm from suppression to prescription: Proceedings, Tall Timbers fire ecology conference; 1996 May 7-10; Boise, ID. No. 20. Tallahassee, FL: Tall Timbers Research Station: 140.

Coffey, Jenness. 1990. Summary report on tamarisk control: Joshua Tree National Monument. In: Kunzmann, Michael R., Johnson, R. Roy, Bennett, Peter S., eds. Proceedings: Tamarisk control in southwestern United States—proceedings of tamarisk conference Tucson, AZ. Tucson, AZ: Cooperative National Park Resources Studies Unit, University of Arizona: 25-27.

Colautti, R. I.; MacIsaac, H. J. 2004. A neutral terminology to define 'invasive' species. Diversity and Distributions. 10: 135-141.

Colautti, Robert I.; Grigorovich, Igor A.; MacIsaac, Hugh J. 2006. Propagule pressure: a null model for biological invasions. Biological Invasions. 8(5): 1023-1037.

Cole, Margaret A. R. 1991. Vegetation management guideline: white and yellow sweet clover [*Melilotus alba* Desr. and *Melilotus officialis* (L.) Lam.]. Natural Areas Journal. 11(4): 214-215.

Cole, Patrice G.; Weltzin, Jake F. 2004. Environmental correlates of the distribution and abundance of *Microstegium vimineum*, in east Tennessee. Southeastern Naturalist. 3(3): 545-562.

Cole, Patrice G.; Weltzin, Jake F. 2005. Light limitation creates patchy distribution of an invasive grass in eastern deciduous forests. Biological Invasions. 7: 477-488.

Collins, S. L.; Glenn, S. M.; Briggs, J. M. 2002. Effects of local and regional processes on plant species richness in tallgrass prairie. Oikos. 99: 571-592.
Collins, S. L.; Glenn, S. M.; Gibson, D. J. 1995. Experimental analysis of intermediate disturbance and initial floristic composition: decoupling cause and effect. Ecology. 76: 486-492.
Collins, S. L.; Knapp. A. K.; Briggs, J. M.; Blair, J. M.; Steinauer, E. M. 1998. Modulation of diversity by grazing and mowing in native tallgrass prairie. Science. 280: 745-747.
Conise, T. F. 1868. The natural wealth of California. San Francisco, CA: H. H. Bancroft & Co. 512 p.
Conn, Jeffrey. 2005. [Personal communication]. November 2005. Fairbanks, AK: U.S. Department of Agriculture, Agricultural Research Service.
Converse, Carmen K. 1984a. Element stewardship abstract: *Rhamnus cathartica*, *Rhamnus frangula* (syn. *Frangula alnus*) (buckthorns), [Online]. In: The global invasive species initiative. Arlington, VA: The Nature Conservancy (Producer). Available: http://tncweeds.ucdavis.edu/esadocs/documnts/franaln.pdf[April 06, 2006].
Converse, Carmen K. 1984b. Element stewardship abstract: *Robinia pseudoacacia* (black locust), [Online]. In: The global invasive species initiative. Arlington, VA: The Nature Conservancy (Producer). Available: http://tncweeds.ucdavis.edu/esadocs/documnts/robipse.pdf [2007, July 10].
Cook, J. G.; Hershey, T. J.; Irwin, L. L. 1994. Vegetative response to burning on Wyoming mountain-shrub big game ranges. Journal of Range Management. 47(4): 296-302.
Cook, S. F. 1960. Colonial expeditions to the interior of California: Central Valley, 1800-1820. University of California Antrhropological Records. 16: 239-292.
Cook, S. F. 1962. Colonial expeditions to the interior of California: Central Valley, 1820-1840. University of California Anthropological Records. 20: 151-214.
Cooper, C. R. 1960. Changes in vegetation, structure, and growth of a southwestern pine forests since white settlement. Ecological Monographs. 30: 129-164.
Cooper, S. V.; Jean, C. 2001. Wildfire succession in plant communities natural to the Alkali Creek vicinity, Charles M. Russell National Wildlife Refuge, Montana. MT Natural Heritage Program. Helena. Unpublished report to the US Fish and Wildlife Service. 32 p. [+ appendices].
Cooper, W. S. 1922. The broad-sclerophyll vegetation of California-an ecological study of the chaparral and its related communities. Publication No. 319. Washington, DC: Carnegie Institute of Washington. 122 p.
Cooperative Quail Study Association. 1961. Third annual report—1933-34. In: The Cooperative Quail Study Association: May 1, 1931-May 1, 1943. Misc. Publ. No. 1. Tallahassee, FL: Tall Timbers Research Station: 29-45.
Coppedge, B. R.; Engle, D. M.; Toepfer, C. S.; Shaw, J. H. 1998. Effects of seasonal fire, bison grazing and climatic variation on tallgrass prairie vegetation. Plant Ecology. 139(2): 235-246.
Cordell, S.; Cabin, R. J.; Weller, S. G.; Lorence, D. H. 2002. Simple and cost-effective methods control fountain grass in dry forests (Hawaii). Ecological Restoration. 20: 139-140.
Cordell, S.; Sandquist, D. R.; Litton, C.; Cabin, R. J.; Thaxton, J.; Hadway, L.; Castillo, J. M.; Bishaw. 2004. An invasive grass has significant impacts on tropical dry forest ecosystems in Hawaii: The role of science in landscape level resource management and native forest restoration in West Hawaii. Proceedings; 16th International Conference of the Society for Ecological Restoration, Victoria, BC: Society for Ecological Restoration. 4 p.
Cost, N. D.; Carver, G. C. 1981. Distribution of melaleuca in south Florida measured from the air. In: Geiger, R. K., ed. Proceedings of the melaleuca symposium; 1980 September 23-24; [Ft. Meyers, FL]. Tallahassee, FL: Florida Department of Agriculture and Consumer Services, Division of Forestry: 1-8
Coupland, R. T. 1992. Overview of the grasslands of North America. In: Coupland, R. T., ed. Natural grasslands: Introduction and western hemisphere. Ecosystems of the World 8A. Amsterdam, Netherlands: Elsevier Science Publishers B. V.: 147-149.
Covington, W. W.; Everett, R. L.; Steele, R.; Irvin, L. L.; Daer, T. A.; Auclair, A. N. D. 1994. Historical and anticipated changes in forest ecosystems of the Inland West of the United States. Journal of Sustainable Forestry. 2(1/2): 13-63.
Covington, W. W.; Fulé, P. Z.; Moore, M. M.; Hart, S. C.; Kolb, T. E.; Mast, J. N.; Sackett, S. S.; Wagner, M. R. 1997. Restoring ecosystem health in ponderosa pine forests of the southwest. Journal of Forestry. 95: 23-29.
Covington, W. W.; Moore, M. M. 1994a. Postsettlement changes in natural fire regimes and forest structure: ecological restoration of old-growth ponderosa pine forests. Journal of Sustainable Forestry. 2(1/2): 153-181.
Covington, W. W.; Moore, M. M. 1994b. Southwestern ponderosa forest structure resource conditions: changes since Euro-American settlement. Journal of Forestry. 92(1): 39-47.
Cox, J. R.; Morton, H. L.; Johnsen, T. N. Jr.; Jordan, G. L.; Martin, S. C.; Fierro, L. C. 1984. Vegetation restoration in the Chihuahuan and Sonoran deserts of North America. Rangelands. 6: 112-115.
Cox, J. R.; Ruyle, G. B.; Roundy, B. A. 1990. Lehmann lovegrass in southeastern Arizona: biomass production and disappearance. Journal of Range Management. 43: 367-372.
Cox, Jerry R. 1992. Lehmann lovegrass live component biomass and chemical composition. Journal of Range Management. 45: 523-527.
Cox, Jerry R.; DeAlba-Avila, Abraham; Rice, Richard W.; Cox, Justin N. 1993. Biological and physical factors influencing Acacia constricta and Prosopis velutina establishment in the Sonoran Desert. Journal of Range Management. 46(1): 43-48.
Cox, R. D.; Anderson, V. J. 2004. Increasing native diversity of cheatgrass-dominated rangeland through assisted succession. Rangeland Ecology and Management. 57: 203–210.
Cram, D. S.; Barker, T. T.; Boren, J. C. 2006. Wildland fire effects in silviculturally treated vs. untreated stands of New Mexico and Arizona. Res. Pap. RMRS-55. Fort Collins, CO: U.S. Department of Agriculture, Forest Service, Rocky Mountain Research Station. 28 p.
Crawford, J. A.; Wahren, C. H. A.; Kyle, S.; Moir, W. H. 2001. Responses of exotic plant species to fires in *Pinus ponderosa* forests in northern Arizona. Journal of Vegetation Science. 12(2): 261-268.
Crawford, J. M. 1962. Soils of the San Dimas Experimental Forest. Miscellaneous Paper PSW-76, USDA Forest Service, Pacific Southwest Forest and Range Experimental Station, Berkeley, California USA.
Crawford, Rex C.; Hall, Heidi. 1997. Changes in the south Puget prairie landscape. In: Dunn, Patrick; Ewing, Kern, eds. Ecology and conservation of south Puget Sound prairie landscape. Seattle, WA: The Nature Conservancy. 11-15.
Crider, Franklin J. 1945. Three introduced lovegrasses for soil conservation. Circular number 730. U.S. Department of Agriculture, Washington, DC.
Cronon, William. 1983. Changes in the land: Indians, colonists, and the ecology of New England. New York: Hill and Wang. 242 p.
Crooks, J.; Soule, M. E. 1996. Lag times in population explosions of invasive species: causes and implications. In: Sandlund, O. T.; Schei, P. J.; Viken, A., eds. Proceedings Norway/UN Conference on alien species. Trondheim, Norway: Directorate for Nature Management and Norwegian Institute for Nature Research: 39-46.
Cuddihy, Linda H.; Stone, Charles P. 1990. Alteration of native Hawaiian vegetation: effects of humans and their introductions. Honolulu, HI: University of Hawai`i Press, University of Hawai`i Cooperative National Park Studies Unit. 138 p.
Cuddington, Kim; Hastings, Alan. 2004. Invasive engineers. Ecological Modeling. 178: 335-347.
Curtis, J. T.; Partch, M. L. 1948. Effect of fire on the competition between blue grass and certain prairie plants. The American Midland Naturalist. 39: 437-443.
Curtis, John T. 1959. Weed communities. In: Curtis, John T. The vegetation of Wisconsin. Madison, WI: The University of Wisconsin Press: 412-434.
D'Antonio, C. M. 1993. Mechanisms controlling invasions of coastal plant communities by the alien succulent, *Carpobrotus edulis*. Ecology. 74: 83-95.
D'Antonio, C. M. 2000. Fire, plant invasions, and global changes. In: Mooney, Harold A.; Hobbs, Richard J., eds. Invasive species in a changing world. Washington, DC: Island Press: 65-93.

D'Antonio, C. M.; Dudley, T.; Mack, M. 1999. Disturbance and biological invasions. In: L. Walker (ed), Ecosystems of disturbed ground. Elsevier: 429-468.

D'Antonio, C. M.; Hughes, R. Flint; Mack, Michelle; Hitchcock, Doug; Vitousek, Peter M. 1998. The response of native species to removal of invasive exotic grasses in a seasonally dry Hawaiian woodland. Journal of Vegetation Science. 9: 699-712.

D'Antonio, C. M.; Hughes, R. Flint; Vitousek, Peter M. 2001a. Factors influencing dynamics of two invasive C4 grasses in seasonally dry Hawaiian woodlands. Ecology. 82: 89-104.

D'Antonio, C. M.; Levine, J.; Thomsen, M. 2001b. Ecosystem resistance to invasion and the role of propagules supply: a California perspective. Journal of Mediterranean Ecology. 2: 233-45.

D'Antonio, C. M.; Meyerson, Laura A. 2002. Exotic plant species as problems and solutions in ecological restoration: a synthesis. Restoration Ecology. 10(4): 703-713.

D'Antonio, C. M.; Odion, D. C.; Tyler, C. M. 1993. Invasion of maritime chaparral by the introduced succulent *Carpobrotus edulis*: the roles of fire and herbivory. Oecologia. 95: 14-21.

D'Antonio, C. M.; Vitousek, P. M. 1992. Biological invasions by exotic grasses, the grass/fire cycle, and global change. Annual Review of Ecology and Systematics. 23: 63-87.

D'Antonio, Carla M.; Tunison, J. Timothy; Loh, Rhonda K. 2000. Variations in impact of exotic grasses and fire on native plant communities in Hawaii. Journal of Australian Ecology. 25: 507-22.

D'Appollonio, Jennifer. 2006. Regeneration strategies of Japanese barberry (*Berberis thunbergii* DC.) in coastal forests of Maine. Machias, ME: University of Maine. 93 p. Thesis.

Daehler, Curtis C.; Carino, Debbie A. 1998. Recent replacement of native pili grass (*Heteropogon contortus*) by invasive African grasses in the Hawaiian Islands. Pacific Science. 52: 220-227.

Daehler, Curtis C.; Carino, Debbie A. 2000. Predicting invasive plants: prospects for a general screening system based on current regional models. Biological Invasions. 2: 93-102.

Daehler, Curtis C.; Goergen, Erin M. 2005. Experimental restoration of an indigenous Hawaiian grassland after invasion by *Cenchrus ciliaris* (buffelgrass). Restoration Ecology. 13: 380-389.

Dailey, R. 2001. [Email to Kris Zouhar]. Sioux Falls, SD: The Nature Conservancy of the Dakotas, South Dakota. On file at: U.S. Department of Agriculture, Forest Service, Rocky Mountain Research Station, Fire Sciences Laboratory, Missoula, MT.

Dale, V. H.; Gardner, R. H.; DeAngelis D. L.; Eagar, C. C.; Webb, J. W. 1991. Elevation mediated effects of balsam wooly adelgid on southern Appalachian spruce-fir forests. Canadian Journal of Forest Research. 21: 1639–1648.

Dale, Virginia H. 1989. Wind dispersed seeds and plant recovery on the Mt. St. Helens debris avalanche. Canadian Journal of Botany. 67: 1434-1441.

Dale, Virginia H. 1991. Mount St. Helens: revegetation of Mount St. Helens debris avalanche 10 years post eruption. National Geographic Research and Exploration. 7(3): 328-341.

Dale, Virginia H.; Adams, Wendy M. 2003. Plant establishment 15 years after the debris avalanche at Mount St. Helens, Washington. The Science of the Total Environment. 313: 101-113.

Dale, Virginia H.; Joyce, Linda A. McNulty, Steve; Neilson, Ronald P.; Ayres, Matthew P.; Flannigan, Michael D.; Hanson, Paul J.; Irland, Lloyd C.; Lugo, Ariel E.; Peterson, Chris J.; Simberloff, Daniel; Swanson, Frederick J.; Stocks, Brian J.; Wotton, Michael B. 2001. Climate change and forest disturbances. BioScience. 51(9): 723-734.

Dark, S. J. 2004. The biogeography of invasive alien plants in California: an application of GIS and spatial regression analysis. Diversity and Distributions. 10: 1-9.

Daubenmire, R. 1968a. Ecology of fire in grasslands. Advances in Ecological Research. 5: 209-266.

Daubenmire, R. 1968b. Plant communities: a textbook of plant synecology. New York, NY: Harper & Row Publishers, Inc. 300 p.

Daubenmire, R. 1978. Plant geography—with special reference to North America. Physiological Ecology. New York: Academic Press. 338 p.

Daubenmire, R. F. 1959. A canopy-coverage method. Northwest Science. 33: 43-64.

Davis, Frank W.; Hickson, Diana E.; Odion, Dennis C. 1988. Composition of maritime chaparral related to fire history and soil, Burton Mesa, Santa Barbara County, California. Madroño. 35(3): 169-195.

Davis, Frank W.; Stine, Peter A.; Stoms, David M. 1994. Distribution and conservation status of coastal sage scrub in southwestern California. Journal of Vegetation Science. 5: 743-756.

Davis, Mark A.; Grime, J. Philip; Thompson, Ken. 2000. Fluctuating resources in plant communities: a general theory of invasibility. Journal of Ecology. 88(3): 528-534.

Davis, O. K. 1992. Rapid climatic change in coastal southern California inferred from pollen analysis of San Joaquin Marsh. Quartenary Research. 37: 89-100.

Day, Gordon M. 1953. The Indian as an ecological factor in the northeastern forest. Ecology. 34(2): 329-346.

Deal, R. 2001. The effects of partial cutting on forest plant communities of western hemlock: Sitka spruce stands in southeast Alaska. Canadian Journal of Forest Research. 31: 2067-2079.

DeBano, Leonard F.; Neary, Daniel G. 2005. Part A-The soil resource: its importance, characteristics, and general response to fire. In: Neary, Daniel G.; Ryan, Kevin C.; DeBano, Leonard F., eds. 2005. Wildland fire in ecosystems: effects of fire on soils and water. Gen. Tech. Rep. RMRS-GTR-42-vol.4. Ogden, UT: U.S. Department of Agriculture, Forest Service, Rocky Mountain Research Station. 250 p.

DeBano, Leonard F.; Neary, Daniel G.; Ffolliott, Peter F. 1998. Fire's effects on ecosystems. New York: John Wiley & Sons, Inc. 333 p.

DeBano, Leonard F.; Neary, Daniel G.; Ffolliott, Peter F. 2005. Soil physical properties. In: Neary, Daniel G.; Ryan, Kevin C.; DeBano, Leonard F., eds. Wildland fire in ecosystems: effects of fire on soils and water. Gen. Tech. Rep. RMRS-GTR-42-vol. 4. Ogden, UT: U.S. Department of Agriculture, Forest Service, Rocky Mountain Research Station: 29-52.

DeBano, Leonard F.; Rice, Raymond M.; Conrad, C. Eugene. 1979. Soil heating in chaparral fires: effects on soil properties, plant nutrients, erosion, and runoff. Res. Pap. PSW-145. Berkeley, CA: U.S. Department of Agriculture, Forest Service, Pacific Southwest Forest and Range Experiment Station. 21 p.

DeCoster, James K.; Platt, William J.; Riley, Sarah A. 1999. Pine savannas of Everglades National Park—an endangered ecosystem. In: Jones, David T.; Gamble, Brandon W., eds. Florida's garden of good and evil: Proceedings of the 1998 joint symposium of the Florida Exotic Pest Plant Council and the Florida Native Plant Society; 1998 June 3-7; Palm Beach Gardens, FL. West Palm Beach, FL: South Florida Water Management District: 81-88.

Deeming, J. E.; Burgan, R. E.; Cohn, J. D. 1977. The National Fire Danger Rating System—1978. Gen. Tech. Rep. INT-39. Ogden, UT: U.S. Department of Agriculture, Forest Service, Intermountain Forest and Range Experiment Station. 63 p.

DeFerrari, Collette M. 1993. Exotic plant invasions across landscape patch types on the Olympic Peninsula. Seattle, WA: University of Washington. [Number of pages unknown]: Thesis.

DeFerrari, Collette M.; Naiman, Robert J. 1994. A multi-scale assessment of the occurrence of exotic plants on the Olympic Peninsula, Washington. Journal of Vegetation Science. 5: 247-258.

Deiter, Laurie. 2000. *Elaeagnus angustifolia* L. In: Bossard, Carla C.; Randall, John M.; Hoshovsky, Marc C., eds. Invasive plants of California's wildlands. Berkeley, CA: University of California Press: 175-178.

del Moral, R.; Titus, J. H.; Cook, A. M. 1995. Early primary succession on Mount St. Helens, Washington, USA. Journal of Vegetation Science. 6: 107-120.

Delcourt, H. R.; Delcourt, P. A. 1997. Pre-Columbian Native American use of fire on southern Appalachian landscapes. Conservation Biology. 11: 1010-1014.

DeLeonardis, Salvatore. 1971. Effects of fire and fire control methods in interior Alaska. In: Slaughter, C. W.; Barney, Richard J.; Hansen, G. M., eds. Fire in the northern environment—a symposium: Proceedings of a symposium; 1971 April 13-14; Fairbanks, AK. Portland, OR: U.S. Department of Agriculture, Forest Service, Pacific Northwest Range and Experiment Station: 101-105.

Densmore, Roseann V.; McKee, Paul C.; Roland, Carl. 2001. Exotic plants in Alaskan National Park units, [Online]. Anchorage, AK: U.S. Geological Survey, Alaska Biological Science Center and Denali National Park and Preserve (Producer). Available: http://akweeds.uaa.alaska.edu/pdfs/literature/non-native_NPS_Densmore.pdf [2005, January 15].

Deuser, Curt. 1996. Appendix A: Sacatone prescribed burn report and evaluation: U.S. Department of the Interior, National Park Service. Technical Report NPS/NRWRD/NRTR-96/93. vi + 58 p.
Deuser, Curt. 2004. [Personal communication]. March 29. ID: US Department of the Interior, National Park Service, Lake Mead EPM Team. Boulder City, NV.
DeVelice, R. L. [n.d.]. Non-native plant inventory: Kenai trails. R10-TP-124, [Online]. Anchorage, AK: U.S. Department of Agriculture, Forest Service, Chugach National Forest (Producer). Available: http://akweeds.uaa.alaska.edu/pdfs/literature/Kenai_non-native_DeVelice.pdf [2005, January 15].
Dewey, Steven A.; Mace, R. W.; Buhler, Lillian A.; Andersen, Kimberly. 2000. The interaction of fire and herbicides in the control of squarrose knapweed [Abstract]. Proceedings of the Western Society of Weed Science. 53: 8-9.
Dey, Daniel C.; Guyette, Richard P. 2000. Anthropogenic fire history and red oak forests in south-central Ontario. The Forestry Chronicle. 76(2): 339-347.
Dey, Daniel. 2002a. Fire history and postsettlement disturbance. In: McShea, William J.; Healy, William M., eds. Oak forest ecosystems: ecology and management for wildlife. Baltimore, MD: The Johns Hopkins University Press: 46-59.
Dey, Daniel. 2002b. The ecological basis for oak silviculture in eastern North America. In: McShea, William J.; Healy, William M., eds. Oak forest ecosystems: ecology and management for wildlife. Baltimore, MD: The Johns Hopkins University Press: 60-79.
Dibbern, J. C. 1947. Vegetative response of *Bromus inermis* to certain variations in environment. Botanical Gazette. 109: 44-58.
Dibble, Alison C. 2005. Manassas National Battlefield Park, Manassas, VA. Unpublished data on file at: U.S. Department of Agriculture, Forest Service, Northern Research Station, Penobscot Experimental Forest, 686 Government Rd, Bradley, ME.
Dibble, Alison C.; Patterson, William A., III; White, Robert H. 2004. Fire and invasive plants: combustibility of native and invasive exotic plants, [Online]. Durham, NH: U.S. Department of Agriculture, Forest Service, Joint Fire Science Program, Northeastern Research Station (Producer). Available: http://www.fs.fed.us/ne/durham/4155/fire/dibble3_jfsp.html [2005, January 15].
Dibble, Alison C.; Rees, Catherine A. 2005. Does the lack of reference ecosystems limit our science? A case study in nonnative invasive plants as forest fuels. Journal of Forestry. 103(7): 1-10.
Dibble, Alison C.; Rees, Catherine A.; Ducey, Mark J.; Patterson III, William A. 2003. Fuel bed characteristics of invaded forest stands. In: Using fire to control invasive plants: What's new, what works in the Northeast. 2003 workshop proceedings, [Online]. Durham, NH: University of New Hampshire Cooperative Extension (Producer): 26-29. Available: http://invasives.eeb.uconn.edu/ipane/summit05/plantsummithome.htm [2007, July 12].
Dibble, Alison C.; White, Robert H.; Lebow, Patricia K. 2007. Combustion characteristics of north-eastern USA vegetation tested in the cone calorimeter: invasive versus non-invasive plants. International Journal of Wildland Fire 16: 426-443.
Dickinson, M. B.; Johnson, E. A. 2001. Fire effects on trees. In: Johnson, E. A.; Miyanishi, K., eds.. Forest fires: behavior and ecological effects. New York: Academic Press: 477-525.
Dickinson, Matthew B.; Johnson, Edward A. 2004. Temperature-dependent rate models of vascular cambium cell mortality. Canadian Journal of Forest Research. 34: 546-559.
Dick-Peddie, William A. 1993. New Mexico vegetation: past, present, and future. Albuquerque, NM: University of New Mexico Press. 244 p.
Dillon, Stephen P.; Forcella, Frank. 1984. Germination, emergence, vegetative growth and flowering of two silvergrasses, *Vulpia bromoides* (L.) S. F. Gray and *V. myuros* (L.) C. C. Gmel. Australian Journal of Botany. 32(2): 165-175.
DiTomaso, J. M. 1997. Risk analysis of various weed control methods. Proceedings, California exotic pest plant council symposium 3: 34-39.
DiTomaso, J. M. 2000a. *Cortaderia jubata*. In: Bossard. C. C.; Randall, J. M.; Hoshovsky, M. C., eds. Invasive plants of California's wildlands. Berkeley, CA: University of California Press: 124-128.
DiTomaso, J. M. 2000b. *Cortaderia selloana*. In: Bossard. C. C.; Randall, J. M.; Hoshovsky, M. C., eds. Invasive plants of California's wildlands. Berkeley, CA: University of California Press: 128.133.
DiTomaso, J. M. 2004. [Personal communication]. March 30. University of California-Davis, Weed Science Program, Davis, CA.
DiTomaso, J. M.; Brooks, M. L.; Allen, E. B.; Minnich, R.; Rice, P. M.; Kyser, G. B. 2006a. Control of invasive weeds with prescribed burning. Weed Technology. 20: 535-548.
DiTomaso, Joseph M. 2006a. Control of invasive plants with prescribed fire. In: DiTomaso, J. M.; Johnson, D. W., eds. The use of fire as a tool for controlling invasive weeds. Bozeman, MT: Center for Invasive Plant Management: 6-18.
DiTomaso, Joseph M. 2006b. Using prescribed burning in integrated strategies. In: DiTomaso, J. M.; Johnson, D. W., eds. The use of fire as a tool for controlling invasive weeds. Bozeman, MT: Center for Invasive Plant Management: 19-27.
DiTomaso, Joseph M.; Healy, Evelyn A. 2003. Aquatic and riparian weeds of the West. Publication 3421. Davis, CA: University of California, Agriculture and Natural Resources. 442 p.
DiTomaso, Joseph M.; Heise, Kerry L.; Kyser, Guy B.; Merenlender, Adina M.; Keiffer, Robert J. 2001. Carefully timed burning can control barb goatgrass. California Agriculture. 55(6): 47-53.
DiTomaso, Joseph M.; Kyser, Guy B.; Hastings, Marla S. 1999. Prescribed burning for control of yellow starthistle (*Centaurea solstitialis*) and enhanced native plant diversity. Weed Science. 47: 233-242.
DiTomaso, Joseph M.; Kyser, Guy B.; Miller, Jessica R.; Garcia, Sergio; Smith, Richard F.; Nader, Glenn; Connor, J. Michael; Orloff, Steve B. 2006b. Integrating prescribed burning and clopyralid for the management of yellow starthistle (*Centaurea solstitialis*). Weed Science. 54: 757-767.
DiTommaso, Antonio; Lawlor, Frances M.; Darbyshire, Stephen J. 2005. The biology of invasive alien plants in Canada. 2. *Cynanchum rossicum* (Kleopow) Borhidi [=*Vincetoxicum rossicum* (Kleopow) Barbar.] and *Cynanchum louiseae* (L.) Kartesz & Gandhi [=*Vincetoxicum nigrum* (L.) Moench]. Canadian Journal of Plant Science. 85: 243-263.
Dix, R. L. 1960. The effects of burning on the mulch structure and species composition of grasslands in Western North Dakota. Ecology. 41: 49-56.
Dobberpuhl, J. 1980. Seed banks of forest soils in east Tennessee. Knoxville, TN: University of Tennessee. 219 p. Thesis.
Dobson, Annette J. 2002. An introduction to generalized linear models, 2nd ed. Boca Raton, FL: Chapman and Hall/CRC. 225p.
Dobyns, H. F. 1981. From fire to flood: historic human destruction of Sonoran Desert riverine oases. Socorro, NM: Ballena Press. 222 p.
Dodson, Erich K.; Fiedler, Carl E. 2006. Impacts of restoration treatments on alien plant invasion in *Pinus ponderosa* forests, Montana, USA. Journal of Applied Ecology. 43(5): 887-897.
Dodson, Erich Kyle. 2004. Monitoring change in exotic plant abundance after fuel reduction/restoration treatments in ponderosa pine forests of western Montana. Missoula, MT: The University of Montana. 95 p. Thesis.
Dolan, Benjamin J.; Parker, George R. 2004. Understory response to disturbance: an investigation of prescribed burning and understory removal treatments. In: Spetich, Martin A., ed. Upland oak ecology symposium: history, current conditions, and sustainability: Proceedings; 2002 October 7-10; Fayetteville, AR. Gen. Tech. Rep. SRS-73. Asheville, NC: U.S. Department of Agriculture, Forest Service, Southern Research Station: 285-291.
Donahue, C.; Rogers, W. E.; Siemann, E. 2004. Effects of temperature and mulch depth on Chinese tallow tree (*Sapium sebiferum*) seed germination. Texas Journal of Science. 56: 347-356.
Dooley, T. 2003. Lesson learned from eleven years of prescribed fire at the Albany Pine Bush Preserve. In: Using fire to control invasive plants: what's new, what works in the Northeast. 2003 workshop proceedings. Durham, NH: University of New Hampshire Cooperative Extension: 7-11.
Doren, R. F.; Platt, W. J.; Whiteaker, L. D. 1993. Density and size structure of slash pine stands in the Everglades region of south Florida. Forest Ecology and Management. 59: 295-311.
Doren, R. F.; Whiteaker, L. D.; LaRosa, A. M. 1991. Evaluation of fire as a management tool for controlling *Schinus terebinthifolius* as secondary successional growth on abandoned agricultural land. Environmental Management 15: 121-129.
Doren, Robert F.; Whiteaker, Louis D. 1990. Effects of fire on different size individuals of *Schinus terebinthifolius*. Natural Areas Journal. 10(3): 107-113.

Doucet, Colleen; Cavers, Paul B. 1996. A persistent seed bank of the bull thistle *Cirsium vulgare*. Canadian Journal of Botany. 74: 1386-1391.

Douglas, George W.; Ballard, T. M. 1971. Effects of fire in alpine plant communities in the North Cascades, Washington. Ecology. 52(6): 1058-1064.

Downey, Paul O. 2000. Broom (*Cytisus scoparius* (L.) Link) and fire: management implications. Plant Protection Quarterly. 15(4): 178-183.

Doyle, Kathleen M.; Knight, Dennis H.; Taylor, Dale L.; Barmore, William J., Jr.; Benedict, James M. 1998. Seventeen years of forest succession following the Waterfalls Canyon Fire in Grand Teton National Park, Wyoming. International Journal of Wildland Fire. 8(1): 45-55.

Drake, J. A. 1990. The mechanics of community assembly and succession. Journal of Theoretical Biology. 147: 213-233.

Drake, John M.; Lodge, David M. 2006. Allee effects, propagule pressure and the probability of establishment: risk analysis for biological invasions. Biological Invasions. 8: 365-375.

Drewa, P. B.; Platt, W. J.; Moser, E. B. 2002. Fire effects on resprouting of shrubs in southeastern longleaf pine savannas. Ecology. 83: 755-767.

Drewa, Paul B. 2003. Effects of fire season and intensity on *Prosopis glandulosa* Torr. var. *glandulosa*. International Journal of Wildland Fire. 12: 147-157.

Drewa, Paul B.; Havstad, Kris M. 2001. Effects of fire, grazing, and the presence of shrubs on Chihuahuan Desert grasslands. Journal of Arid Environments. 48: 429-443.

Drewa, Paul B.; Peters, Debra P. C.; Havstad, Kris M. 2001. Fire, grazing, and honey mesquite invasion in black grama-dominated grasslands of the Chihuahuan Desert: a synthesis. In: Galley, Krista E. M.; Wilson, Tyrone P., eds. Proceedings of the invasive species workshop: The role of fire in the control and spread of invasive species; Fire conference 2000: the first national congress on fire ecology, prevention, and management; 2000 November 27-December 1; San Diego, CA. Misc. Publ. No. 11. Tallahassee, FL: Tall Timbers Research Station: 31-39.

Ducey, Mark J. 2003. Modifying the BEHAVE Fuel Model for northeastern conditions: research needs for managing invasives. In: Using fire to control invasive plants: what's new, what works in the Northeast. 2003 workshop proceedings. Durham, NH: University of New Hampshire Cooperative Extension: 30-33.

Duchesne, Luc C.; Hawkes, Brad C. 2000. Fire in northern ecosystems. In: Brown, James K.; Smith, Jane Kapler, eds. Wildland fire in ecosystems: Effects of fire on flora. Gen. Tech. Rep. RMRS-GTR-42-vol. 2. Ogden, UT: U.S. Department of Agriculture, Forest Service, Rocky Mountain Research Station: 35-51.

Dudley, T.; Brooks M. L. 2006. Saltcedar invasions can change riparian fire regimes. In: Sugihara, N. G.; van Wagtendonk, J. W.; Fites-Kaufman, J.; Shaffer, K. E.; Thode, A. E., eds. Fire in California's ecosystems. Berkeley, CA: University of California Press: 9.

Duffy, Michael. 2003. Non-native plants of Chugach National Forest: a preliminary inventory. R10-TP-111, [Online]. Anchorage, AK. U.S. Department of Agriculture (Producer). Available: http://akweeds.uaa.alaska.edu/pdfs/literature/non-native_MDuffChugachNF.pdf [2005, January 15].

Duke, Sara E.; Caldwell, Martyn M. 2001. Nitrogen acquisition from different spatial distributions by six Great Basin plant species. Western North American Naturalist. 61(1): 93-102.

Dukes, Jeffrey S. 2000. Will the increasing atmospheric CO_2 concentration affect the success of invasive species? In: Mooney, Harold A.; Hobbs, Richard J., eds. 2000. Invasive species in a changing world. Washington, DC: Island Press: 95-113.

Duncan, C. A.; Clark, J. K., eds. 2005. Invasive plants of range and wildlands and their environmental, economic, and societal impacts. Lawrence, KS: Weed Science Society of America.

Duncan, K. W. 1994. Saltcedar: establishment, effects, and management. Wetland Journal. 6: 10-13.

Dunwiddie, Peter W. 1997. Long-term effects of sheep grazing on coastal sandplain vegetation. Natural Areas Journal. 17(3): 261-264.

Dunwiddie, Peter W. 1998. Ecological management of sandplain grasslands and coastal heathlands in southeastern Massachusetts. In: Pruden, Teresa L.; Brennan, Leonard A., eds. Fire in ecosystem management: shifting the paradigm from suppression to prescription. Tall Timbers Fire Ecology Conference Proceedings, No. 20. Tallahassee, FL: Tall Timbers Research Station: 83-93.

Dunwiddie, Peter W. 2002. Management and restoration of grasslands on Yellow Island, San Juan Islands, Washington, USA. In: Garry oak ecosystem restoration: Progress and prognosis—Proceedings, 3rd Annual Meeting of the British Columbia Chapter of the Society for Ecological Restoration, [Online]; 2002 April 27-28; Victoria, B.C. Victoria, B.C.: British Columbia Chapter of the Society for Ecological Restoration. Available: http:conserveonline.org/TNC [2005, March 10].

Dunwiddie, Peter W.; Zaremba, Robert E.; Harper, Karen A. 1996. A classification of coastal heathlands and sandplain grasslands in Massachusetts. Rhodora. 98(894): 117-145.

Dwire, Kathleen A.; Kauffman, J. B. 2003. Fire and riparian ecosystems in landscapes of the western USA. Forest Ecology and Management. 178(1-2): 61-74.

Dyer, A. R.; Rice, K. J. 1997. Intraspecific and diffuse competition: the response of *Nassella pulchra* in a California grassland. Ecological Applications. 7: 484-492.

Dyer, A. R.; Rice, K. J. 1999. Effects of competition on resource availability and growth of a California bunchgrass. Ecology. 80: 2697-2710.

Dyrness, C. T. 1973. Early stages of plant succession following logging and burning in the western Cascades of Oregon. Ecology. 54(1): 57-69.

Dyrness, C. T. 1975. Grass-legume mixtures for erosion control along forest roads in western Oregon. Journal of Soil and Water Conservation. 30(4): 169-173.

Dyrness, C. T.; Norum, Rodney A. 1983. The effects of experimental fires on black spruce forest floors in interior Alaska. Canadian Journal of Forest Research. 13: 879-893.

Dyrness, C. T.; Viereck, L. A.; Van Cleve, K. 1986. Fire in taiga communities of interior Alaska. Ecological Studies. 57: 74-86.

Ebinger, J. E. 1983. Exotic shrubs: A potential problem in natural area management in Illinois. Natural Areas Journal. 3(1): 3-6.

Ebinger, John E.; McClain, William. 1996. Recent exotic woody plant introductions into the Illinois flora. In: Warwick, Charles, ed. 15th North American prairie conference: Proceedings; 1996 October 23-26; St. Charles, IL. Bend, OR: The Natural Areas Association: 55-58.

Eckardt, N. 1987. Element stewardship abstract: *Melilotus officianalis* (sweetclover), [Online]. In: The global invasive specie initiative. Arlington, VA: The Nature Conservancy (Producer). Available: http://tncweeds.ucdavis.edu/esadocs/documnts/euphesu.pdf [2007, July 10].

Ehrenfeld, J. G. 1997. Invasion of deciduous forest preserves in the New York metropolitan region by Japanese barberry (Berberis thunbergii DC.). Journal of the Torrey Botanical Society. 124(2): 210-215.

Ehrenfeld, J. G. 1999. Structure and dynamics of populations of Japanese barberry (*Berberis thunbergii* DC.) in deciduous forests of New Jersey. Biological Invasions. 1: 203-213.

Ehrenfeld, Joan G. 2003. Soil properties and exotic plant invasions: a two-way street. In: Fosbroke, Sandra L. C.; Gottschalk, Kurt W., eds. Proceedings: U.S. Department of Agriculture interagency research forum on gypsy moth and other invasive species: 13th annual meeting; 2002 January 15-18; Annapolis, MD. Gen. Tech. Rep. NE-300. Newtown Square, PA: U.S. Department of Agriculture, Forest Service, Northeastern Research Station: 18-19.

Ehrenreich, John H. 1959. Effect of burning and clipping on growth of native prairie in Iowa. Journal of Range Management. 12: 133-137.

Eidson, James A. 1997. *Festuca arundinacea* (F. elatior)—tall fescue. In: Randall, John M.; Marinelli, Janet, eds. Invasive plants: Weeds of the global garden. Handbook #149. Brooklyn, NY: Brooklyn Botanic Garden: 87.

Eliason, S. A.; Allen, E. B. 1997. Exotic grass competition in suppressing native shrubland re-establishment. Restoration Ecology. 5: 245-255.

Ellis, Lisa M. 2001. Short-term response of woody plants to fire in a Rio Grande riparian forest, central New Mexico, USA. Biological Conservation. 97(2): 159-170.

Ellis, Lisa M.; Crawford, Clifford S.; Molles, Manuel C. Jr. 1998. Comparison of litter dynamics in native and exotic riparian

vegetation along the middle Rio Grande of central New Mexico, U.S.A. Journal of Arid Environments. 38(2): 283-296.
Elton, C. S. 1958. The ecology of invasions by plants and animals. London, UK: Methuen and Company. 181 p.
Elzinga, C. L.; Salzer, D. W.; Willoughby, J. W. 1998. Measuring and monitoring plant populations. Technical Reference 1730-1. Denver, CO: U.S. Department of the Interior, Bureau of Land Management National Business Center. 492 p.
Emery, Sarah M.; Gross, Katherine L. 2005. Effects of timing of prescribed fire on the demography of an invasive plant, spotted knapweed *Centaurea maculosa*. Journal of Applied Ecology. 40: 60-69.
Erdman, J. A. 1970. Pinyon-juniper succession after natural fires on residual soils of Mesa Verde, Colorado. Science Bulletin of Biology Series. 11: 11-24.
Erman, D. C.; Jones, R. 1996. Fire frequency analysis of Sierra forests. In: Sierra Nevada Ecosystem Project: final report to Congress, Volume II. Assessments and scientific basis for management options. Davis, CA: Centers for Water and Wildland Resources: 1139-1154.
Ervin, G.; Smothers, M.; Holly, C.; Anderson, C.; Linville, J. 2006. Relative importance of wetland type versus anthropogenic activities in determining site invasibility. Biological Invasions. 8: 1425-1432.
Espeland, E. K.; Carlsen, T. M.; Macqueen, D. 2005. Fire and dynamics of granivory on a California grassland forb. Biodiversity and Conservation. 14: 267-280.
Esque, Todd C.; Schwalbe, Cecil R. 2002. Alien annual grasses and their relationships to fire and biotic change in Sonoran desertscrub. In: Tellman, Barbara, ed. Invasive exotic species in the Sonoran region. Tucson, AZ: The University of Arizona Press; The Arizona-Sonora Desert Museum: 165-194.
Esser, Lora L. 1993a. *Phleum pratense*. In: Fire Effects Information System, [Online]. U.S. Department of Agriculture, Forest Service, Rocky Mountain Research Station, Fire Sciences Laboratory (Producer). Available: http://www.fs.fed.us/database/feis/ [2005, March 28].
Esser, Lora L. 1993b. *Taraxacum officinale*. In: Fire Effects Information System, [Online]. U.S. Department of Agriculture, Forest Service, Rocky Mountain Research Station, Fire Sciences Laboratory (Producer). Available: http://www.fs.fed.us/database/feis/ [2005, March 29].
Esser, Lora L. 1995. *Rumex acetosella*. In: Fire Effects Information System, [Online]. U.S. Department of Agriculture, Forest Service, Rocky Mountain Research Station, Fire Sciences Laboratory (Producer). Available: http://www.fs.fed.us/database/feis/ [2005, March 29].
Etter, A. G. 1951. How Kentucky bluegrass grows. Annals of the Missouri Botanical Garden. 38: 293-375.
Evangelista, Paul; Stohlgren, Thomas J.; Guenther, Debra; Stewart, Sean. 2004. Vegetation response to fire and postburn seeding treatments in juniper woodlands of the Grand Staircase-Escalante National Monument, Utah. Western North American Naturalist. 64(3): 293-305.
Evans, R. A. 1988. Management of pinyon-juniper woodlands. Gen. Tech. Rep. INT-249. Ogden, UT: U.S. Department of Agriculture, Forest Service, Intermountain Research Station. 34 p.
Evans, R. A.; Young, J. A. 1970. Plant litter and establishment of alien annual weed species in rangeland communities. Weed Science. 18(6): 697-703.
Evans, R. A.; Young, J. A. 1972. Microsite requirements for establishment of annual rangeland weeds. Weed Science. 20(4): 350-356.
Evans, R. A.; Young, J. A. 1975. Aerial application of 2-4,D plus picloram for green rabbitbrush control. Journal of Range Management. 28: 315-318.
Evans, R. A.; Young, J. A. 1977. Weed control-revegetation systems for big sagebrush-downy brome rangelands. Journal of Range Management. 30: 331-336.
Evans, Raymond A.; Young, James A. 1987. Seedbed microenvironment, seedling recruitment, and plant establishment on rangelands. In: Frasier, Gary W.; Evans, Raymond A., eds. Seed and seedbed ecology of rangeland plants: proceedings of symposium; 1987 April 21-23; Tucson, AZ. Washington, DC: U.S. Department of Agriculture, Agricultural Research Service: 212-220.
Everett, Richard L.; Clary, Warren P. 1985. Fire effects and revegetation on juniper-pinyon woodlands. In: Sanders, Ken, and Durham, Jack, eds. Proceedings: Rangeland fire effects: a symposium. Boise, ID. Boise, ID: U.S. Department of the Interior, Bureau of Land Management, Idaho State Office: 33-37.
Everett, Richard L.; Sharrow, Steven H. 1983. Understory seed rain on tree-harvested and unharvested pinyon-juniper sites. Journal of Environmental Management. 17(4): 349-358.
Everitt, Benjamin L. 1998. Chronology of the spread of tamarisk in the central Rio Grande. Wetlands. 18(4): 658-668.
Ewel, J. J. 1986. Invasibility: Lessons from south Florida. In: Mooney, H. A.; Drake, J. A., eds. Ecology of biological invasions of North America and Hawaii. New York: Springer-Verlag. 214-230.
Ewel, John J.; Ojima, Dennis S.; Karl, Dori A.; DeBusk, William F. 1982. *Schinus* in successional ecosystems of Everglades National Park. Report T-676. Homestead, FL: Everglades National Park, South Florida Research Center. 141 p.
Ewing, Kern. 2002. Effects of initial site treatments on early growth and three-year survival of Idaho fescue. Restoration Ecology. 10(2): 282-288.
Ewing, Kern; Windhager, Steve; McCaw, Matt. 2005. Effects of summer burning and mowing on central Texas juniper-oak savanna plant communities during drought conditions. Ecological Restoration. 23(4): 255-260.
Fahey, Timothy J.; Reiners, William A. 1981. Fire in the forests of Maine and New Hampshire. Bulletin of the Torrey Botanical Club. 108: 362-373.
Fahnestock, George Reeder. 1977. Interactions of forest fires, flora, and fuels in two Cascade Range wildernesses. Seattle, WA: University of Washington. 179 p. Dissertation.
False Brome Working Group. 2002. [Meeting notes]. November 19, [Online]. Available: http://www.appliedeco.org/BRSYweb/False-Bromemtgnotes11_19_02. htm [2005, February 20].
False Brome Working Group. 2003. [Meeting notes]. January 15, [Online]. Available: http://www.appliedeco.org/BRSYweb/False-Bromemtgnotes1_15_03.htm [2005, February 20].
False Brome Working Group. 2004. [Newsletter]. January, [Online]. Available: http://www.appliedeco.org/BRSYweb/finalJan-2004newsletter.pdf [2005, February 20].
Faulkner, Jerry L.; Clebsch, Edward E. C.; Sanders, William L. 1989. Use of prescribed burning for managing natural and historic resources in Chickamauga and Chattanooga National Military Park, U.S.A. Environmental Management. 13(5): 603-612.
Feeney, Shelly R.; Kolb, Thomas E.; Covington, W. Wallace; Wagner, Michael R. 1998. Influence of thinning and burning restoration treatments on presettlement ponderosa pines at the Gus Pearson Natural Area. Canadian Journal of Forest Research. 28(9): 1295-1306.
Felger, Richard S. 1990. Non-native plants of Organ Pipe Cactus National Monument, Arizona. Tech. Rep. No. 31. Tucson, AZ: University of Arizona, School of Renewable Natural Resources, Cooperative National Park Resources Studies Unit. 93 p.
Fellows, David P.; Newton, Wesley E. 1999. Prescribed fire effects on biological control of leafy spurge. Journal of Range Management. 52: 489-493.
Fenn, M. E.; Haeuber, R.; Tonnesen, G. S.; Baron, J. S.; Grossman-Clarke, S.; Hope, D.; Jaffe, D. A.; Copeland, S.; Geiser, L.; Rueth, H. M.; Sickman, J. O. 2003. Nitrogen emissions, deposition, and monitoring in the western United States. BioScience. 53: 391-403.
Fernald, Merritt Lyndon. 1950. Gray's Manual of Botany, 8th Ed. New York, NY: American Book Company. 1632 p.
Fernandez, R. J.; Reynolds, J. F. 2000. Potential growth and drought tolerance of eight desert grasses: lack of a trade-off? Oecologia. 123: 90-98.
Ferriter, A. P. 1999. Extent of melaleuca infestation in Florida. In: Laroche, F., ed. Melaleuca management plan, ten years of successful melaleuca management in Florida 1988-1998. [Orlando, FL]: Florida Exotic Pest Plant Council. 12-16.
Ferriter, Amy, ed. 1997. Brazilian pepper management plan for Florida: Recommendations from the Florida Exotic Pest Plant Council's Brazilian Pepper Task Force. [Orlando, FL]: Florida Exotic Pest Plant Council. 26 p.
Ferriter, Amy, ed. 2001. Lygodium management plan for Florida: A report from the Florida Exotic Pest Plant Council's Lygodium Task Force. [Orlando, FL]: Florida Exotic Pest Plant Council. 51 p.

Ferriter, Amy. 2007. [Email to Kristin Zouhar]. July 31. Boise State University, Boise, ID. On file at U.S. Department of Agriculture, Forest Service, Rocky Mountain Research Station, Fire Sciences Laboratory, Missoula, MT.

Fetcher, Ned; Beatty, Thomas F.; Mullinax, Ben; Winkler, Daniel S. 1984. Changes in arctic tussock tundra thirteen years after fire. Ecology. 65(4): 1332-1333.

Fike, Jean; Niering, William A. 1999. Four decades of old field vegetation development and the role of *Celastrus orbiculatus* in the northeastern United States. Journal of Vegetation Science. 10(4): 483-492.

Finney, M. A.; McHugh, C. W.; Grenfell, I. C. 2005. Stand- and landscape-level effects of prescribed burning on two Arizona wildfires. Canadian Journal of Forest Research. 35: 1714-1722.

Fire Effects Information System (FEIS), [Online]. U.S. Department of Agriculture, Forest Service, Rocky Mountain Research Station, Fire Sciences Laboratory (Producer). Available: http://www.fs.fed.us/database/feis.

Fisher, M. A.; Fulé, P. Z. 2004. Changes in forest vegetation and arbuscular mycorrhizae along a steep elevation gradient in Arizona. Forest Ecology and Management. 200: 293-311.

Fisser, Herbert G.; Johnson, Kendall L.; Moore, Kellie S.; Plumb, Glenn E. 1989. 51-year change in the shortgrass prairie of eastern Wyoming. In: Bragg, Thomas B.; Stubbendieck, James, eds. Prairie pioneers: ecology, history and culture: Proceedings, 11th North American prairie conference; 1988 August 7-11; Lincoln, NE. Lincoln, NE: University of Nebraska: 29-31.

Fitch, Henry S.; Kettle, W. Dean. 1983. Ecological succession in vegetation and small mammal populations on a natural area of northeastern Kansas. In: Kucera, Clair L., ed. Proceedings, 7th North American prairie conference; 1980 August 4-6; Springfield, MO. Columbia, MO: University of Missouri: 117-121.

FLEPPC. 1996. Florida Exotic Pest Plant Council occurrence database. Available at: http://www.fleppc.org/. [2006, October 18].

Flory, S. L.; Clay, K. 2006. Invasive shrub distribution varies with distance to roads and stand age in eastern deciduous forests in Indiana, USA. Plant Ecology. 184: 131-141.

Flory, S. Luke; Rudgers, Jennifer A.; Clay, Keith. 2007. Experimental light treatments affect invasion success and the impact of Microstegium vimineum on the resident community. Natural Areas Journal. 27: 124-132.

Flowers, John D., II. 1991. Subtropical fire suppression in *Melaleuca quinquenervia*. In: Center, Ted D.; Doren, Robert F.; Hofstetter, Ronald L.; [and others], eds. Proceedings of the symposium on exotic pest plants; 1988 November 2-4; Miami, FL. Tech. Rep. NPS/NREVER/NRTR-91/06. Washington, DC: U.S. Department of the Interior, National Park Service: 151-158.

Floyd, M. L.; Romme, W. H.; Hanna, D. D. 2000. Fire history and vegetation pattern in Mesa Verde National Park, Colorado, USA. Ecological Applications. 10(6): 1666-1680.

Floyd, M. Lisa; Hanna, David; Romme, William H.; Crews, Timothy E. 2006. Predicting and mitigating weed invasions to restore natural post-fire succession in Mesa Verde National Park, Colorado, USA. International Journal of Wildland Fire. 15: 247-259.

Floyd-Hanna, Lisa; DaVega, Anne; Hanna, David; Romme, William H. 1997. Chapin 5 Fire vegetation monitoring and mitigation: First year report. [Mesa Verde, CO]: [U.S. Department of the Interior, National Park Service, Mesa Verde National Park]. Unpublished report on file at: U.S. Department of Agriculture, Forest Service, Rocky Mountain Research Station, Fire Sciences Laboratory, Missoula, MT. 7 p. [+ appendices].

Foin, T. C.; Hektner, M. M. 1986. Secondary succession and the fate of native species in a California coastal prairie. Madroño. 33: 189-206.

Foote, M. Joan. 1983. Classification, description, and dynamics of plant communities after fire in the taiga of interior Alaska. Res. Pap. PNW-307. Portland, OR: U.S. Department of Agriculture, Forest Service, Pacific Northwest Forest and Range Experiment Station. 104 p.

Forcella, Frank; Harvey, Stephen J. 1983. Eurasian weed infestation in western Montana in relation to vegetation and disturbance. Madroño. 30(2): 102-109.

Forman, J.; Kesseli, R. V. 2003. Sexual reproduction in the invasive species *Fallopia japonica* (Polygonaceae). American Journal of Botany. 90(4): 586-592.

Fornwalt, Paula J.; Kaufmann, Merrill R.; Huckaby, Laurie S.; Stoker, Jason M.; Stohlgren, Thomas J. 2003. Non-native plant invasions in managed and protected ponderosa pine/Douglas-fir forests of the Colorado Front Range. Forest Ecology and Management. 177: 515-527.

Fox, A. M.; Gordon, D. R.; Dusky, J. A.; Tyson, L.; Stocker, R. K. 2001. IFAS assessment of non-native plants in Florida's natural areas. SS-AGR-79, [Online]. University of Florida, Institute of Food and Agricultural Sciences, Agronomy Department, Florida Cooperative Extension Service (Producer). Available: http://agronomy.ifas.ufl.edu/docs/IFASAssessment2001.pdf [2007, August 1].

Fox, J. F. 1979. Intermediate-disturbance hypothesis. Science. 204: 1344-1345.

Fox, Russell B.; Mitchell, Robert B.; Davin, Michael. 2000. Saltcedar management at Lake Meredith National Recreation Area. In: Zwank, Phillip J.; Smith, Loren M., eds. Research highlights-2000: range, wildlife, & fisheries management. Volume 31. Lubbock, TX: Texas Tech University, College of Agricultural Sciences and Natural Resources: 27-28.

Fox, Russell; Mitchell, Rob; Davin, Mike. 2001. Managing saltcedar after a summer wildfire in the Texas rolling plains. In: McArthur, E. Durant; Fairbanks, Daniel J., comps.. Shrubland ecosystem genetics and biodiversity: proceedings; 2000 June 13-15; Provo, UT. Proc. RMRS-P-21. Ogden, UT: U.S. Department of Agriculture, Forest Service, Rocky Mountain Research Station: 236-237.

Fox, Russell; Mitchell, Robert B.; Davin, Mike. 1999. Managing saltcedar after a summer wildfire in the Texas Rolling Plains. In: Wester, David B.; Britton, Carlton M., eds. Research highlights-1999: Noxious brush and weed control: Range, wildlife, and fisheries management. Volume 30. Lubbock, TX: Texas Tech University, College of Agricultural Sciences and Natural Resources: 15.

Foxx, Teralene S. 1996. Vegetation succession after the La Mesa Fire at Bandelier National Monument. In: Allen, Craig D., ed. Fire effects in southwestern forests: Proceedings, 2nd La Mesa fire symposium; 1994 March 29-31; Los Alamos, NM. RM-GTR-286. Fort Collins, CO: U.S. Department of Agriculture, Forest Service, Rocky Mountain Forest and Range Experiment Station: 47-69.

Frandsen W. H.; Ryan, K. C. 1986. Soil moisture reduces belowground heat flux and soil temperature under a burning fuel pile. Canadian Journal of Forest Research. 16: 244-248.

Franklin, Jerry F.; Dyrness, C. T. 1973. Natural vegetation of Oregon and Washington. Gen. Tech. Rep. PNW-8. Portland, OR: U.S. Department of Agriculture, Forest Service, Pacific Northwest Forest and Range Experiment Station. 417 p.

Frappier, Brian; Eckert, Robert T.; Lee, Thomas D. 2003. Potential impacts of the invasive exotic shrub *Rhamnus frangula* L. (glossy buckthorn) on forests of southern New Hampshire. Northeastern Naturalist. 10(3): 277-296.

Fraver, Shawn. 1994. Vegetation responses along edge-to-interior gradients in the mixed hardwood forests of the Roanoke River Basin, North Carolina. Conservation Biology. 8(3): 822-832.

Frazer, G. W.; Canham, C. D.; Lertzman, K. P. 1999. Gap Light Analyzer (GLA), Version 2.0: Imaging software to extract canopy structure and gap light transmission indices from true-colour fisheye photographs, users manual and program documentation. Burnaby, BC: Simon Fraser University. In cooperation with: The Institute of Ecosystem Studies.

Freckleton, R. P. 2004. The problem of prediction and scale in applied ecology: the example of fire as a management tool. Journal of Applied Ecology. 41: 599-603.

Freckleton, R. P.; Watkinson, A. R. 2001. Non-manipulative determination of plant community dynamics. Trends in Ecology and Evolution. 16: 301-307.

Freeman, J. P.; Stohlgren, T. J.; Hunter, M. E.; Omi, P. N.; Martinson, E. J.; Chong, G. W.; Brown, C. S. 2007. Rapid assessment of postfire plant invasions in coniferous forests of the western United States. Ecological Applications 17(6): 1656-1665.

Freifelder, Rachel; Vitousek, Peter M.; D'Antonio, Carla M. 1998. Microclimate effects of fire-induced forest/grassland conversion in a seasonally dry Hawaiian woodlands. Biotropica. 30: 286-297.

Frenkel, R. E. 1970. Ruderal vegetation along some California roadsides. University of California Publications in Geography Vol. 20. Berkeley, CA: University of California Press. 163 p.

Friederici, Peter. 1995. The alien saltcedar. American Forests. 101(1-2): 45-47.

Frolik, A. L. 1941. Vegetation on the peat lands of Dane County, Wisconsin. Ecological Monographs. 11(1): 117-140.
Frost, C. 1993. Four centuries of changing landscape patterns in the longleaf pine ecosystem. In: Proceedings, 18th Tall Timbers Fire Ecology Conference; The longleaf pine ecosystem: ecology, restoration and management; 1991 May 30-June 2; Tallahassee, FL. Tallahassee, FL: Tall Timbers Research Station. 17-43.
Frost, Cecil C. 1995. Presettlement fire regimes in southeastern marshes, peatlands, and swamps. In: Cerulean, Susan I.; Engstrom, R. Todd, eds. Fire in wetlands: a management perspective: Proceedings, 19th Tall Timbers fire ecology conference; 1993 November 3-6; Tallahassee, FL. No. 19. Tallahassee, FL: Tall Timbers Research Station: 39-60.
Frost, Cecil C. 1998. Presettlement fire frequency regimes of the United States: a first approximation. In: Pruden, Teresa L.; Brennan, Leonard A., eds. Fire in ecosystem management: shifting the paradigm from suppression to prescription: Proceedings, Tall Timbers fire ecology conference; 1996 May 7-10; Boise, ID. No. 20. Tallahassee, FL: Tall Timbers Research Station: 70-81.
Fuhlendorf, S. D.; Engle, D. M. 2004. Application of the fire-grazing interaction to restore a shifting mosaic on tallgrass prairie. Journal of Applied Ecology. 41(4): 604-614.
Fulé, P. Z.; Laughlin, D. C.; Covington, W. W. 2005. Pine-oak forest dynamics five years after ecological restoration treatments, Arizona, USA. Forest Ecology and Management. 218: 129-145.
Furbush, Paul. 1953. Control of medusa-head on California ranges. Journal of Forestry. 51: 118-121.
Gabbard, Bethany Lynn. 2003. The population dynamics and distribution of the exotic grass, *Bothriochloa ischaemum*. Austin, TX: University of Texas. 156 p. Dissertation.
Gagne, Wayne; Cuddihy, Linda. 1999. Vegetation. In: Wagner, Warren L.; Herbst, Darrell H.; Sohmer, Sy H. Manual of the flowering plants of Hawaii. Revised Edition. Honolulu, HI: University of Hawai`i Press: 45-114.
Galley, Krista E. M.; Wilson, Tyrone P., eds. 2001. Proceedings of the invasive species workshop: The role of fire in the control and spread of invasive species; Fire conference 2000: the first national congress on fire ecology, prevention, and management; 2000 November 27-December 1; San Diego, CA. Misc. Publ. No. 11. Tallahassee, FL: Tall Timbers Research Station. 146 p.
Gann, G.; Gordon, D. R. 1998. *Paederia foetida* (skunk vine) and *P. cruddasiana* (sewer vine): threats and management strategies. Natural Areas Journal. 18: 169-174.
GAO 2003. Wildland fires: Better information needed on effectiveness of emergency stabilization and rehabilitation treatments. [Washington, DC]: U.S. General Accounting Office, GAO-03-430.
Garrison, George A.; Bjugstad, Ardell J.; Duncan, Don A.; Lewis, Mont E.; Smith, Dixie R. 1977. Vegetation and environmental features of forest and range ecosystems. Agric. Handb. 475. Washington, DC: U.S. Department of Agriculture, Forest Service. 68 p.
Gartner, F. R.; Lindsey, J. R.; White, E. M. 1986. Vegetation responses to spring burning in western South Dakota. In: Clambey, Gary K.; Pemble, Richard H., eds. The prairie: past, present and future: Proceedings of the 9th North American Prairie Conference; 1984 July 29-August 1; Moorhead, MN. Fargo, ND: Tri-College University Center for Environmental Studies: 143-146.
Geary, T. F.; Woodall, S. L. 1990. *Melaleuca quinquenervia* (Cav.) S. T. Blake melaleuca. In: Burns, Russell M.; Honkala, Barbara H., tech. coords. Silvics of North America. Vol. 2. Hardwoods. Agric. Handb. 654. Washington, DC: U.S. Department of Agriculture, Forest Service: 461-465.
Geiger, Erika L.; McPherson, Guy R. 2005. No positive feedback between fire and a nonnative perennial grass. In: Gottfried, Gerald J.; Gebow, Brooke S.; Eskew, Lane G.; Edminster, Carleton B., comps. Connecting mountain islands and desert seas: biodiversity and management of the Madrean Archipelago II; 2004 May 11-15; Tucson, AZ. Proceedings RMRS-P-36. Fort Collins, CO: U.S. Department of Agriculture, Forest Service, Rocky Mountain Research Station: 465-468.
Gelbard, Jonathan L.; Belnap, Jayne. 2003. Roads as conduits for exotic plant invasions in a semiarid landscape. Conservation Biology. 17(2): 420-432.
Gelbard, Jonathan L.; Harrison, Susan. 2003. Roadless habitats as refuges for native grasslands: interactions with soil, aspect, and grazing. Ecological Applications. 13(2): 404-415.
George, Melvin R. 1992. Ecology and management of medusahead. Range Science Report, University of California Dept. of Agronomy and Range, Davis, CA. 32 p.
Gerlach, John D., Jr.; Moore, Peggy E.; Johnson, Brent; Roy, D. Graham; Whitmarsh, Patrick; Lubin, Daniel M.; Graber, David M.; Haultain, Sylvia; Pfaff, Anne; Keeley, Jon E. 2003. Alien plant species threat assessment and management prioritization for Sequoia-Kings Canyon and Yosemite National Parks. Open File Report 02-170. Carson City, NV: U.S. Department of the Interior, Geological Survey, Western Ecological Research Center. 149 p.
Giblin, David. 1997. *Aster curtis*: current knowledge of its biology and threats to its survival. In: Dunn, Patrick; Ewing, Kern, eds. Ecology and Conservation of South Puget Sound Prairie Landscape. Seattle, WA: The Nature Conservancy. 93-100.
Gibson, Carly; Sproul, Fred; Rudy, Susan; Davis, Linh. 2003. Vegetation Resource Assessment: Cedar and Paradise Fires. Ramona, CA: Cleveland National Forest, Palomar Ranger District. 18 p. Unpublished report on file with: U.S. Department of Agriculture, Forest Service, Rocky Mountain Research Station, Fire Sciences Laboratory, Missoula, MT; RWU 4403 files.
Gibson, David J. 1988. Regeneration and fluctuation of tallgrass prairie vegetation in response to burning frequency. Bulletin of the Torrey Botanical Club. 115(1): 1-12.
Gibson, David J. 1989. Hulbert's study of factors effecting botanical composition of tallgrass prairie. In: Bragg, Thomas B.; Stubbendieck, James, eds. Prairie pioneers: ecology, history and culture: Proceedings, 11th North American prairie conference; 1988 August 7-11; Lincoln, NE. Lincoln, NE: University of Nebraska: 115-133.
Giessow, Jason H. 1997. Effects of fire frequency and proximity to firebreaks on the distribution and abundance of non-native herbs in coastal sage scrub. San Diego, CA: San Diego State University. 76 p. Thesis.
Giessow, Jason; Zedler, Paul. 1996. The effects of fire frequency and firebreaks on the abundance and species richness of exotic plant species in coastal sage scrub. In: Lovich, Jeff; Randall, John; Kelly, Mike, eds. Proceedings, California Exotic Pest Plant Council symposium; 1996 October 4-6; San Diego, CA. Volume 2. Berkeley, CA: California Exotic Pest Plant Council: 86-94.
Gill, A. M. 1995. Stems and fire. In: Gartner, N.G., ed. Plant stems physiology and functional morphology. San Diego, CA: Academic Press: 323-342.
Gillespie, I. G.; Allen, E. B. 2004. Fire and competition in a southern California grassland: impacts on the rare forb *Erodium macrophyllum*. Journal of Applied Ecology. 41: 643-652.
Gilliam, F. S.; Turrill, N. L.; Adams, M. B. 1995. Herbaceous and overstory species in clear-cut and mature Central Appalachian hardwood forests. Ecological Applications. 5: 947-955.
Glasgow, Lance S.; Matlack, Glenn R. 2007. The effects of prescribed burning and canopy openness on establishment of two non-native plant species in a deciduous forest, southeast Ohio, USA. Forest Ecology and Management. 238: 319-329.
Glass, William D. 1991. Vegetation management guideline: cut-leaved teasel (*Dipsacus laciniatus* L.) and common teasel (*Dipsacus sylvestris* L.). Natural Areas Journal. 11(4): 213-214.
Gleason, Henry A.; Cronquist, Arthur. 1991. Manual of vascular plants of northeastern United States and adjacent Canada. 2nd ed. New York: New York Botanical Garden. 910 p.
Glendening, George E.; Paulsen, Harold A., Jr. 1955. Reproduction and establishment of velvet mesquite as related to invasion of semidesert grasslands. Tech. Bull. 1127. Washington, DC: U.S. Department of Agriculture, Forest Service. 50 p.
Glenn, Edward P.; Nagler, Pamela L. 2005. Comparative ecophysiology of *Tamarix ramosissima* and native trees in western U.S. riparian zones. Journal of Arid Environments. 61(3): 419-446.
Godwin, H. 1936. Studies in the ecology of Wicken Fen. III. The establishment and development of fen scrub (carr). Journal of Ecology. 24: 82-116.
Goergen Erin; Daehler, Curtis C. 2001. Reproductive ecology of a native Hawaiian grass (*Heteropogon contortus*; Poaceae) versus its invasive alien competitor (*Pennisetum setaceum*; Poaceae). International Journal of Plant Science. 162: 317-326.
Goergen Erin; Daehler, Curtis C. 2002. Factors affecting seedling recruitment in an invasive grass (*Pennisetum setaceum*) and a

native grass (*Heteropogon contortus*) in the Hawaiian Islands. Plant Ecology. 161: 147-156.

Gogue, G. J.; Emino, E. R. 1979. Seed coat scarification of *Albizia julibrissin* Durazz. by natural mechanisms. Journal of the American Society of Horticultural Science. 104(3): 421-423.

Good, R. B. 1981. The role of fire in conservation reserves. In: Gill, A. M.; Groves, R. H.; Noble, I. R., eds. Fire and the Australian biota. Canberra, Australia: Australian Academy of Science. 529-550.

Goodenough, R. 1992. The nature and implications of recent population growth in California. Geography. 77: 123-133.

Goodrich, Sherel. 1999. Multiple use management based on diversity of capabilities and values within pinyon-juniper woodlands. In: Monsen, Stephen B.; Stevens, Richard, comps. Proceedings: Ecology and management of pinyon-juniper communities within the Interior West: Sustaining and restoring a diverse ecosystem; 1997 September 15-18; Provo, UT. Proceedings RMRS-P-9. Ogden, UT: U.S. Department of Agriculture, Forest Service, Rocky Mountain Research Station: 164-171.

Goodrich, Sherel; Gale, Natalie. 1999. Cheatgrass frequency at two relic sites within the pinyon-juniper belt of Red Canyon. In: Monsen, Stephen B.; Stevens, Richard, comps. Proceedings: Ecology and management of pinyon-juniper communities within the Interior West: Sustaining and restoring a diverse ecosystem; 1997 September 15-18; Provo, UT. Proceedings RMRS-P-9. Ogden, UT: U.S. Department of Agriculture, Forest Service, Rocky Mountain Research Station: 69-71.

Goodrich, Sherel; Rooks, Dustin. 1999. Control of weeds at a pinyon-juniper site by seeding grasses. In: Monsen, Stephen B.; Stevens, Richard, comps.. Proceedings: Ecology and management of pinyon-juniper communities within the Interior West: Sustaining and restoring a diverse ecosystem; 1997 September 15-18; Provo, UT. Proceedings RMRS-P-9. Ogden, UT: U.S. Department of Agriculture, Forest Service, Rocky Mountain Research Station: 403-407.

Goodwin, K. M.; Sheley, R. L. 2001. What to do when fires fuel weeds: A step-by-step guide for managing invasive plants after a wildfire. Rangelands. 23: 15-21.

Goodwin, Kim; Sheley, Roger; Clark, Janet. 2002. Integrated noxious weed management after wildfires. EB-160. Bozeman, MT: Montana State University, Extension Service. 46 p. Available: http://www.montana.edu/wwwpb/pubs/eb160.html [2003, October 1].

Gordon, D. R. 1998. Effects of invasive, non-indigenous plant species on ecosystem processes: lessons from Florida. Ecological Applications. 8: 975-989.

Gordon, D. R.; Rice, K. J. 2000. Competitive suppression of *Quercus douglasii* (Fagaceae) seedling emergence and growth. American Journal of Botany. 87: 986-994.

Gordon, D. R.; Thomas, K. P. 1997. Florida's invasion by nonindigenous plants: history, screening, and regulations. In: Simberloff, Daniel; Schmitz, Don C.; Brown, Tom C., eds. Strangers in paradise, impact and management of nonindigenous species in Florida. Washington, DC: Island Press. 21-37.

Gordon, R. A.; Scifres, C. J. 1977. Burning for improvement of Macartney rose-infested coastal prairie. Texas Agricultural Experiment Station Bulletin. 1183: 4-15.

Gordon, R. A.; Scifres, C. J.; Mutz, J. L. 1982. Integration of burning and picloram pellets for Macartney rose control. Journal of Range Management. 35(4): 427-430.

Gori, David; Enquist, Carolyn. 2003. An Assessment of the spatial extent and condition of grasslands in central and southern Arizona, southwestern New Mexico, and Northern Mexico. Tucson, Arizona: The Nature Conservancy.

Gorman, KellyAnn. 2005. [Email to Alison Dibble]. January 13. Luray, VA: USDI, National Park Service, Shenandoah National Park. On file at U.S. Department of Agriculture, Forest Service, Rocky Mountain Research Station, Fire Sciences Laboratory, Missoula, MT.

Gottfried, Gerald J.; Swetnam, Thomas W.; Allen, Craig D.; Betancourt, Julio L.; Chung-MacCoubrey, Alice L. 1995. Pinyon-juniper woodlands. In: Finch, Deborah M.; Tainter, Joseph A., eds. Ecology, diversity, and sustainability of the Middle Rio Grande Basin. Gen. Tech. Rep. RM-GTR-268. Fort Collins, CO: U.S. Department of Agriculture, Forest Service, Rocky Mountain Forest and Range Experiment Station: 95-132.

Grace, J. B. 1998. Can prescribed fire save the endangered coastal prairie ecosystem from Chinese tallow invasion? Endangered Species Update. 15(5): 70-76.

Grace, James B. 1999. Can prescribed fire save the endangered coastal prairie ecosystem from Chinese tallow invasion? Wildland Weeds. 2(2): 9-14.

Grace, James B. 2006. [Letter to Matthew Brooks]. August 15. On file at: U.S. Department of the Interior, United States Geological Survey, Yosemite Field Station, El Portal, CA.

Grace, James B.; Allain, Larry K.; Baldwin, Heather Q.; Billock, Arlene G.; Eddleman, William R.; Given, Aaron M.; Jeske, Clint W.; Moss, Rebecca. 2005. Effects of prescribed fire in the coastal prairies of Texas. USGS Open-File Report 2005-1287. Reston, VA: U.S. Department of the Interior, Fish and Wildlife Service, Region 2; U.S. Geological Survey. 46 p.

Grace, James B.; Smith, Melinda D.; Grace, Susan L.; Collins, Scott L.; Stohlgren, Thomas J. 2001. Interactions between fire and invasive plants in temperate grasslands of North America. In: Galley, Krista E. M, and Wilson, Tyrone P., eds. Proceedings of the invasive species workshop: the role of fire in the control and spread of invasive species San Diego, CA. Tallahassee, FL: Tall Timbers Research Station: 40-65.

Graham, Russell T.; Harvey, Alan E.; Jain, Theresa B.; Tonn, Jonalea R. 1999. The effects of thinning and similar stand treatments on fire behavior in western forests. Gen. Tech. Rep. PNW-GTR-463. Portland, OR: U.S. Department of Agriculture, Forest Service, Pacific Northwest Research Station. 27 p.

Gray, A. N. 2005. Eight nonnative plants in western Oregon forests: associations with environment and management. Environmental Monitoring and Assessment. 100: 109-127.

Gray, J. T.; Schlesginger, W. H. 1981. Biomass, production, and litterfall in the coastal sage scrub of southern California. American Journal of Botany. 68: 24-33.

Great Plains Flora Association. 1986. Flora of the Great Plains. Lawrence, KS: University Press of Kansas. 1392 p.

Greenberg, Cathryn H.; Smith, Lindsay M.; Levey, Douglas J. 2001. Fruit fate, seed germination and growth of an invasive vine-an experimental test of 'sit and wait' strategy. Biological Invasions. 3: 363-372.

Gregory, Stanley V. 1997. Riparian management in the 21st century. In: Kohm, Kathryn A.; Franklin, Jerry F., eds. Creating a Forestry for the 21st century: the science of ecosystem management. Covelo, CA: Island Press. 69-85.

Grese, R. 1992. The landscape architect and problem exotic plants. In: Burley, J. B., ed. Proceedings of the American Society of Landscape Architects' open committee on reclamation: reclamation diversity; 1991 October 29; San Diego, CA. [Place of publication unknown]: [American Society of Landscape Architects]: 7-15. On file with: U.S. Department of Agriculture, Forest Service, Rocky Mountain Research Station, Fire Sciences Laboratory, Missoula, MT.

Griffis, K. L.; Crawford, J. A.; Wagner, M. R.; Moir, W. H. 2001. Understory response to management treatments in northern Arizona ponderosa pine forests. Forest Ecology and Management. 146: 239-245.

Griffith, C. 1996. Sericea lespedeza—a friend or foe? Ag News and Views. 14(10): 4.

Grilz, P. L.; Romo, J. T. 1994. Water relations and growth of *Bromus inermis* Leyss (smooth brome) following spring or autumn burning in a fescue prairie. The American Midland Naturalist. 132: 340-348.

Grilz, P. L.; Romo, J. T. 1995. Management considerations for controlling smooth brome in fescue prairie. Natural Areas Journal. 15(2): 148-156.

Gruell, George E. 1985. Fire on the early western landscape: an annotated record of wildland fire. Northwest Science. 59(2): 97-107.

Gruell, George E. 1999. Historical and modern roles of fire in pinyon-juniper. In: Monsen, Stephen B.; Stevens, Richard, comps.. Proceedings: Ecology and management of pinyon-juniper communities within the Interior West: Sustaining and restoring a diverse ecosystem; 1997 September 15-18; Provo, UT. Proc. RMRS-P-9. Ogden, UT: U.S. Department of Agriculture, Forest Service, Rocky Mountain Research Station: 24-28.

Guala, G. F. 1995. Element stewardship abstract: *Neyraudia reynaudiana* (silk reed), [Online]. In: The global invasive species

initiative. Arlington, VA: The Nature Conservancy (Producer). Available: http://tncweeds.ucdavis.edu/esadocs/documnts/neyrrey.pdf [2006, July 18].
Gucker, Corey. 2004. Canyon grassland vegetation changes following the Maloney Creek wildfire. Moscow, ID: University of Idaho. 80 p. Thesis.
Gundale, Michael J.; DeLuca, Thomas H.; Fiedler, Carl E.; Ramsey, Philip W.; Harrington, Michael G.; Gannon, James E. 2005. Restoration treatments in a Montana ponderosa pine forest: effects on soil physical, chemical and biological properties. Forest Ecology and Management. 213(1-3): 25-38.
Guo, Qinfeng. 2001. Early post-fire succession in California chaparral: changes in diversity, density, cover and biomass. Ecological Research. 16: 471-485.
Gupta, J. N.; Trivedi, B. K. 2001. Impact of fire on rangeland species. Range Management and Agroforestry. 22(2): 237-240.
Gutsell, S. L.; Johnson, E. A.; Miyanishi, K.; Keeley, J. E.; Dickingson, M.; Bridges, S. R. J. 2001. Varied ecosystems need different fire protection. Nature. 409: 977.
Guyette, R. P.; Muzika, R. M.; Dey, D. C. 2002. Dynamics of an anthropogenic fire regime. Ecosystems. 5(5): 472-486.
Guyette, Richard P.; Dey, Daniel C.; Stambaugh, Michael C. 2003. Fire and human history of a barren-forest mosaic in southern Indiana. American Midland Naturalist. 149: 21-34.
Habeck, James R. 1985. Impact of fire suppression on forest succession and fuel accumulations in long-fire-interval wilderness habitat types. In: Lotan, James E.; Kilgore, Bruce M.; Fisher, William C.; Mutch, Robert W., technical coordinators. Proceedings, symposium and workshop on wilderness fire; 1983 November 15-18; Missoula, MT. Gen. Tech. Rep. INT-182. Ogden, UT: U.S. Department of Agriculture, Forest Service, Intermountain Forest and Range Experiment Station: 110-118.
Haferkamp, M. R.; Ganskopp, D. C.; Miller, R. F.; [and others]. 1987. Establishing grasses by imprinting in the northwestern United States. In: Frasier, Gary W.; Evans, Raymond A. Proceedings: Seed and seedbed ecology of rangeland plants: proceedings of symposium Tucson, AZ. Washington, DC: U.S. Department of Agriculture, Agricultural Research Service: 299-308.
Hager, Heather A.; McCoy, Karen D. 1998. The implications of accepting untested hypotheses: a review of the effects of purple loosestrife (*Lythrum salicaria*) in North America. Biodiversity and Conservation. 7(8): 1069-1079.
Haidinger, T. L.; Keeley, J. E. 1993. Role of high fire frequency in destruction of mixed chaparral. Madroño. 40: 141-147.
Hall, I. V. 1955. Floristic changes following the cutting and burning of a woodlot for blueberry production. Canadian Journal of Agricultural Science. 35: 143-152.
Halpern, Charles B. 1989. Early successional patterns of forest species: Interactions of life history traits and disturbance. Ecology. 70(3): 704-720.
Halpern, Charles B. 1999. Effects of prescribed burning at bunchgrass meadow: an establishment report and baseline data. An unpublished report submitted to the McKenzie Ranger District, Willamette National Forest, McKenzie Bridge, OR. [Number of pages unknown].
Halpern, Charles B.; Antos, Joseph A.; Geyer, Melora A.; Olson, Annette M. 1997. Species replacement during early secondary succession: the abrupt decline of a winter annual. Ecology. 78(2): 621-631.
Halpern, Charles B.; Evans, Shelley A.; Nielson, Sarah. 1999. Soil seed banks in young, closed-canopy forests of the Olympic Peninsula, Washington: potential contributions to understory reinitiation. Canadian Journal of Botany. 77(7): 922-935.
Halpern, Charles B.; Spies, Thomas A. 1995. Plant species diversity in natural and managed forests of the Pacific Northwest. Ecological Applications. 5(4): 913-934.
Hamilton, J. G. 1997. Changing perceptions of pre-European grasslands in California. Madroño. 44: 311-333.
Hamilton, W. T.; Scifres, C. J. 1982. Prescribed burning during winter for maintenance of buffelgrass. Journal of Range Management. 35(1): 9-12.
Hanes, T. L. 1977. California chaparral. In: Barbour, M. G.; Major, J., eds. Terrestrial vegetation of California. Sacramento, CA: Wiley-Interscience, reprinted by the California Native Plant Society 1988: 417-469.
Hanselka, C. Wayne. 1988. Buffelgrass—south Texas wonder grass. Rangelands. 10(6): 279-281.
Hansen, Allison K.; Six, Diana L.; Ortega, Yvette K. 2006. Effects of spotted knapweed invasion on ground beetle (Coleoptera: Carabidae) assemblages in Rocky Mountain savannas. Unpublished manuscript. University Montana School of Forestry, Missoula, MT.
Hansen, Paul L.; Pfister, Robert D.; Boggs, Keith; Cook, Bradley J.; Joy, John; Hinckley, Dan K. 1995. Classification and management of Montana's riparian and wetland sites. Miscellaneous Publication No. 54. Missoula, MT: The University of Montana, School of Forestry, Montana Forest and Conservation Experiment Station. 646 p.
Hardy, C. C.; Riggan, P. J.; Butler, B. W.; Freeborn, P. H.; Hood, S. M.; Macholz, L.; Kremens, R. 2007. Demonstration and integration of systems for fire remote sensing, ground-based fire measurement, and fire modeling; final report; Joint Fire Science Program agreement JFSP-03-S-01. Available: JFSP Program Office, Boise, Idaho. [Pages unknown].
Hare, Robert C. 1961. Heat effects on living plants. U.S. Department of Agriculture, Forest Service, Southern Forest Experiment Station. New Orleans, LA: Occasional Paper 183. 32 p.
Harper R. M. 1927. Natural resources of southern Florida. 18th Annual Report. Tallahassee, FL: State of Florida, Florida Geological Survey. 206 pp.
Harrison, S.; Hohn, C.; Ratay, S. 2002. Distribution of exotic plants along roads in a peninsular nature reserve. Biological Invasions. 4: 425-430.
Harrison, S.; Inouye, B. D.; Safford, H. D. 2003. Ecological heterogeneity in the effects of grazing and fire on grassland diversity. Conservation Biology. 17(3): 837-845.
Harrod, Richy J.; Reichard, Sarah. 2001. Fire and invasive species within the temperate and boreal coniferous forests of western North America. In: Galley, Krista E. M.; Wilson, Tyrone P., eds. Proceedings of the invasive species workshop: The role of fire in the control and spread of invasive species; Fire conference 2000: the first national congress on fire ecology, prevention, and management; 2000 November 27-December 1; San Diego, CA. Misc. Publ. No. 11. Tallahassee, FL: Tall Timbers Research Station: 95-101.
Hartford, Roberta A.; Frandsen, William H. 1992. When it's hot, it's hot...or maybe it's not! (Surface flaming may not portend extensive soil heating). International Journal of Wildland Fire. 2(3): 139-144.
Hartley, M.; Rogers, W.E.; Grace, J.B.; Siemann, E. 2007. Responses of prairie arthropod communities to fire and fertilizer: balancing plant and arthropod conservation. American Midland Naturalist 157:92-105.
Hartnett, David C.; Hickman, Karen R.; Walter, Laura E. Fischer. 1996. Effects of bison grazing, fire, and topography on floristic diversity in tallgrass prairie. Journal of Range Management. 49(5): 413-420.
Haskins, Kristin E.; Gehring, Catherine A. 2004. Long-term effects of burning slash on plant communities and arbuscular mycorrhizae in a semi-arid woodland. Journal of Applied Ecology. 41: 379-388.
Hassan, M. A.; West, N. E. 1986. Dynamics of soil seed pools on burned and unburned sagebrush semi-deserts. Ecology. 67: 292-272.
Hastings, J. R.; Turner, R. M. 1965. The changing mile: An ecological study of vegetation change in time in the lower mile of an arid and semiarid region. Tucson, AZ: University of Arizona Press.
Hastings, Marla S.; DiTomaso, Joseph M. 1996. Fire controls yellow starthistle in California grasslands. Restoration and Management Notes. 14(2): 124-128.
Hatch, Daphne A.; Bartolome, James W.; Fehmi, Jeffrey S.; Hillyard, Deborah S. 1999a. Effects of burning and grazing on a coastal California grassland. Restoration Ecology. 7(4): 376-381.
Hatch, S. L.; Schuster, J. L.; Drawe, D. L. 1999b. Grasses of the Texas gulf prairies and marshes. College Station, TX: Texas A&M University Press. [Pages unknown].
Haubensak, Karen A.; D'Antonio, Carla; Alexander, Janice. 2004. Effects of nitrogen-fixing shrubs in Washington and coastal California. Weed Technology. 18: 1475-1479.
Haunss, Verena. 2003. Long term fire effects in the montane seasonal zone, Hawai`i Volcanoes National Park. Unpublished report on file at: Hawai`i Volcanoes National Park. Volcano, HI . 65 p.

Heady, Harold F. 1956. Changes in a California annual plant community induced by manipulation of natural mulch. Ecology. 37(4): 798-811.

Heady, Harold F. 1973. Burning and the grasslands in California. In: Komarek, Edwin V., Sr., technical coordinator. Proceedings, 12th annual Tall timbers fire ecology conference; 1972 June 8-9; Lubbock, TX. Number 12. Tallahassee, FL: Tall Timbers Research Station: 97-107.

Heady, Harold F. 1977. Valley grassland. In: Barbour, Michael G.; Major, Jack, eds. Terrestrial vegetation of California. New York: John Wiley and Sons: 491-514.

Heady, Harold F.; Foin, Theodore C.; Hektner, Mary M.; Taylor, Dean W.; Barbour, Michael G.; Barry, W. James. 1977. Coastal prairie and northern coastal scrub. In: Barbour, Michael G.; Major, Jack, eds. Terrestrial vegetation of California. New York: John Wiley and Sons: 733-760.

Healy, William M.; McShea, William J. 2002. Goals and guidelines for managing oak ecosystems for wildlife. In: McShea, William J.; Healy, William M., eds. Oak forest ecosystems: Ecology and management for wildlife. Baltimore, MD: The Johns Hopkins University Press: 333-341.

Heckman, Charles W. 1999. The encroachment of exotic herbaceous plants into the Olympic National Forest. Northwest Science. 73(4): 264-276.

Heidel, B. 1987. Element stewardship abstract: *Carduus nutans* (musk thistle), [Online]. In: The global invasive species initiative. Arlington, VA: The Nature Conservancy (Producer). Available: http://tncweeds.ucdavis.edu/esadocs/documnts/cardnut.pdf [2002, March 11].

Heidorn, Randy. 1991. Vegetation management guideline: Exotic buckthorns—common buckthorn (*Rhamnus cathartica* L.), glossy buckthorn (*Rhamnus frangula* L.), Dahurian buckthorn (*Rhamnus davurica* Pall.). Natural Areas Journal. 11(4): 216-217.

Heinlein, Thomas A.; Moore, Margaret M.; Fulé, Peter Z.; Covington, W. Wallace. 2005. Fire history and stand structure of two ponderosa pine-mixed conifer sites: San Francisco Peaks, Arizona, USA. International Journal of Wildland Fire. 14: 307-320.

Heitlinger, Mark E. 1975. Burning a protected tallgrass prairie to suppress sweetclover, *Melilotus alba* Desr. In: Wali, Mohan K, ed. Prairie: a multiple view. Grand Forks, ND: University of North Dakota Press: 123-132.

Heitschmidt, R. K.; Ansley, R. J.; Dowhower, S. L.; Jacoby, P. W.; Price, D. L. 1988. Some observations from the excavation of honey mesquite root systems. Journal of Range Management. 41: 227-231.

Hellmers, H.; Horton, J.; Juhren, G.; O'Keefe, J. 1955. Root systems of some plants in southern California. Ecology. 36: 667-678.

Helms, John A., ed. 1998. The dictionary of forestry. Bethesda, MD: The Society of American Foresters. 210 p.

Hemstrom, Miles A.; Franklin, Jerry F. 1982. Fire and other disturbances of the forests in Mount Rainier National Park. Quaternary Research. 18: 32-51.

Henderson, Richard A. 1990. Controlling reed canary grass in a degraded oak savanna (Wisconsin). Restoration and Management Notes. 8(2): 123-124.

Henderson, Richard A. 1992. Ten-year response of a Wisconsin prairie remnant to seasonal timing of fire. In: Smith, Daryl D., and Jacobs, Carol A., eds. Proceedings: Proceedings of the twelfth North American prairie conference: recapturing a vanishing heritage Cedar Falls, IA. Cedar Falls, IA: University of Northern Iowa: 121-125.

Hendry, G. W. 1931. The adobe brick as an historical source. Agricultural History. 5: 110-127.

Hensel, R. L. 1923. Recent studies on the effect of burning on grassland vegetation. Ecology. 4: 183-188.

Herrick, J. E.; Van Zee, J. W.; Havstad, K. M.; Burkett, L. M.; Whitford, W. G. 2005a. Monitoring manual for grassland, shrubland, and savanna ecosystems. Volume 1: Quick start. Las Cruces, NM: U.S. Department of Agriculture, Agriculture Research Service, Jornada Experimental Range. 36 p.

Herrick, J. E.; Van Zee, J. W.; Havstad, K. M.; Burkett, L. M.; Whitford, W. G. 2005b. Monitoring manual for grassland, shrubland, and savanna ecosystems. Volume 2: Design, supplementary methods and interpretation. Las Cruces, NM: U.S. Department of Agriculture, Agriculture Research Service, Jornada Experimental Range. 200 p.

Hervey, Donald F. 1949. Reaction of a California annual-plant community to fire. Journal of Range Management. 2: 116-121.

Hessl, A.; Spackman, S. 1995. Effects of fire on threatened and endangered plants: an annotated bibliography. Washington, DC: U.S. Fish and Wildlife Service. [Pages unknown].

Heutte, Tom; Bella, Elizabeth. 2003. Invasive plants and exotic weeds of southeast Alaska. Anchorage, AK: U.S. Department of Agriculture, Forest Service, State and Private Forestry, and Chugach National Forest. 79 p.

Heyward, F. 1939. The relation of fire to stand composition of longleaf pine forests. Ecology. 20: 287-304.

Hickman, James C., ed. 1993. The Jepson manual: Higher plants of California. Berkeley, CA: University of California Press. 1400 p.

Hiebert, R. D.; Stubbendieck, J. 1993. Handbook for ranking exotic plants for management and control. Resources Report NPS/NRM-WRO/NRR-93/08. Denver, CA: U. S. Department of the Interior, Natural National Park Service, Natural Resources Publication Office. [Pages unknown].

Hilty, Julie H.; Eldridge, David J.; Rosentreter, Roger; Wicklow-Howard, Marcia C.; Pellant, Mike. 2004. Recovery of biological soil crusts following wildfire in Idaho. Journal of Range Management. 57(1): 89-96.

Hintz, Tom. 1996. Tillium cernuum, the rediscovery of the species and the ecological restoration of its surrounding habitat. In: Warwick, Charles, ed. 15th North American prairie conference: Proceedings; 1996 October 23-26; St. Charles, IL. Bend, OR: The Natural Areas Association: 124-126.

Hobbs, E. R. 1988. Using ordination to analyze the composition and structure of urban forest islands. Forest Ecology and Management. 23: 139-158.

Hobbs, R. J.; Gulmon, S. L.; Hobbs, V. J.; Mooney, H. W. 1988. Effects of fertilizer addition and subsequent gopher disturbance on a serpentine annual grassland community. Oecologia. 75: 291-295.

Hobbs, R. J.; Huenneke, L. F. 1992. Disturbance, diversity, and invasion: implications for conservation. Conservation Biology. 6: 324-337.

Hobbs, Richard J.; Humphries, Stella E. 1995. An integrated approach to the ecology and management of plant invasions. Conservation Biology. 9(4): 761-770.

Hobbs, Richard J.; Mooney, Harold A. 2005. Invasive species in a changing world: the interactions between global change and invasives. Chapter 12. In: H. A. Mooney, R. N. Mack, J. A. McNeely, L. E. Neville, P. J. Schei, and J. K. Waage, eds.. Invasive Alien Species: A New Synthesis. SCOPE 63. Washington, DC: Island Press: 310-331.

Hoffman, R.; Kearns, K. 2004. Wisconsin manual of control recommendations for ecologically invasive plants, [Online]. Available: http://www.dnr.state.wi.us/invasives/pubs/manual_TOC.htm [2007, July 12].

Hofstetter, Ronald H. 1991. The current status of *Melaleuca quinquenervia* in southern Florida. In: Center, Ted D.; Doren, Robert F.; Hofstetter, Ronald L.; [and others], eds. Proceedings of the symposium on exotic pest plants; 1988 November 2-4; Miami, FL. Tech. Rep. NPS/NREVER/NRTR-91/06. Washington, DC: U.S. Department of the Interior, National Park Service: 159-176.

Hogenbirk, John C.; Wein, Ross W. 1991. Fire and drought experiments in northern wetlands: a climate change analogue. Canadian Journal of Botany. 69: 1991-1997.

Hogenbirk, John C.; Wein, Ross W. 1995. Fire in boreal wet-meadows: implications for climate change. In: Cerulean, Susan I.; Engstrom, R. Todd, eds. Fire in wetlands: a management perspective: Proceedings, 19th Tall Timbers fire ecology conference; 1993 November 3-6; Tallahassee, FL. No. 19. Tallahassee, FL: Tall Timbers Research Station: 21-29.

Hohlt, Jason C.; Racher, Brent J.; Bryan, Justin B.; Mitchell, Robert B.; Britton, Carlton. 2002. Saltcedar response to prescribed burning in New Mexico. In: Wilde, Gene R.; Smith, Loren M., eds. Research highlights—2002: range, wildlife, and fisheries management. Volume 33. Lubbock, TX: Texas Tech University, College of Agricultural Sciences and Natural Resources: 25.

Holechek, Jerry L.; Pieper, Rex D.; Herbel Carlton H. 1998. Range management history. In: Range management: principles and practices. 3rd ed. Upper Saddle River, NJ: Prentice Hall: 29-39.

Holland, V. L.; Keil, D. J. 1995. California Vegetation. Dubuque, IA: Kendall/Hunt Publishing. 516 p.
Hollander, Myles; Wolfe, Douglas A. 1999. Nonparameteric statistical methods, 2nd ed. New York: John Wiley and Sons, Inc. 787p.
Holling, C. S. 1978. Adaptive environmental management and assessment. New York: John Wiley & Sons. 377 p.
Holm, LeRoy G.; Plocknett, Donald L.; Pancho, Juan V.; Herberger, James P. 1977. The world's worst weeds: distribution and biology. Honolulu, HI: University Press of Hawai`i. 609 p.
Holstein, G. 2001. Pre-agricultural grassland in central California. Madroño. 48: 253-264.
Holttum, Richard. E. 1957. Morphology, growth-habitat and classification in the family Gleicheniaceae. Phytomorphology. 7: 168-184.
Hopkinson, P. J.; Fehmi, S.; Bartolome, J. W. 1999. Summer burns reduce cover, but not spread, of barbed goatgrass in California grassland. Ecological Restoration. 17: 168-169.
Horton, J. S.; Kraebel, C. J. 1955. Development of vegetation after fire in the chamise chaparral of southern California. Ecology. 36(2): 244-262.
Hoshovsky, M. C. 1988. Element stewardship abstract: *Ailanthus altissima* (tree-of-heaven), [Online]. The global invasive species initiative. Arlington, VA: The Nature Conservancy (Producer). Available: http://tncweeds.ucdavis.edu/esadocs/documnts/ailaalt.pdf [2007, July 13].
Hoshovsky, Marc. 1986. Element stewardship abstract: *Cytisus scoparius* and *Genista monspessulanus* (Scotch broom, French broom), [Online]. In: The global invasive species initiative. Arlington, VA: The Nature Conservancy (Producer). Available:http://tncweeds.ucdavis.edu/esadocs/documnts/cytisco.pdf [2005, January 5].
Hoshovsky, Marc. 1989. Element stewardship abstract: *Rubus discolor*, (*Rubus procerus*), [Online]. In: The global invasive species initiative. Arlington, VA: The Nature Conservancy (Producer). Available: http://tncweeds.ucdavis.edu/esadocs/documnts/rubudis.pdf [2005, March 20].
Hosten, P. E.; West, N. E. 1994. Cheatgrass dynamics following wildfire on a sagebrush semidesert site in central Utah. In: Monsen, Stephen B.; Kitchen, Stanley G., comps. Proceedings: ecology and management of annual rangelands; 1992 May 18-22; Boise, ID. Gen. Tech. Rep. INT-GTR-313. Ogden, UT: U.S. Department of Agriculture, Forest Service, Intermountain Research Station: 56-62.
Houghton, J. T.; Jenkins, G. J.; Ephraume, J. J., eds. 1990. Climate Change: The IPCC Scientific Assessment. Cambridge, UK: Cambridge University Press. 365 p.
Howard, Janet L. 1992a. *Erodium cicutarium*. In: Fire Effects Information System, [Online]. U.S. Department of Agriculture, Forest Service, Rocky Mountain Research Station, Fire Sciences Laboratory (Producer). Available: http://www.fs.fed.us/database/feis/ [2006, September 21].
Howard, Janet L. 1992b. *Salsola kali*. In: Fire Effects Information System, [Online]. U.S. Department of Agriculture, Forest Service, Rocky Mountain Research Station, Fire Sciences Laboratory (Producer). Available: http://www.fs.fed.us/database/feis/ [2006, September 21].
Howard, Janet L. 1994. *Bromus japonicus*. In: Fire Effects Information System, [Online]. U.S. Department of Agriculture, Forest Service, Rocky Mountain Research Station, Fire Sciences Laboratory (Producer). Available: http://www.fs.fed.us/database/feis/ [2006, January 17].
Howard, Janet L. 1996. *Bromus inermis*. In: Fire Effects Information System, [Online]. U.S. Department of Agriculture, Forest Service, Rocky Mountain Research Station, Fire Sciences Laboratory (Producer). Available: http://www.fs.fed.us/database/feis/ [2006, January 15].
Howard, Janet L. 2003a. *Descurainia sophia*. In: Fire Effects Information System, [Online]. U.S. Department of Agriculture, Forest Service, Rocky Mountain Research Station, Fire Sciences Laboratory (Producer). Available: http://www.fs.fed.us/database/feis/ [2005, March 29].
Howard, Janet L. 2003b. *Sisymbrium altissimum*. In: Fire Effects Information System, [Online]. U.S. Department of Agriculture, Forest Service, Rocky Mountain Research Station, Fire Sciences Laboratory (Producer). Available: http://www.fs.fed.us/database/feis/ [2005, December 14].
Howard, Janet L. 2004a. *Ailanthus altissima*. In: Fire Effects Information System, [Online]. U.S. Department of Agriculture, Forest Service, Rocky Mountain Research Station, Fire Sciences Laboratory (Producer). Available: http://www.fs.fed.us/database/feis/ [2005, May 13].
Howard, Janet L. 2004b. *Sorghum halepense*. In: Fire Effects Information System, [Online]. U.S. Department of Agriculture, Forest Service, Rocky Mountain Research Station, Fire Sciences Laboratory (Producer). Available: http://www.fs.fed.us/database/feis/ [2006, January 17].
Howard, Janet L. 2005a. *Celastrus orbiculatus*. In: Fire Effects Information System, [Online]. U.S. Department of Agriculture, Forest Service, Rocky Mountain Research Station, Fire Sciences Laboratory (Producer). Available: http://www.fs.fed.us/database/feis/ [2005, September 21].
Howard, Janet L. 2005b. *Imperata brasiliensis*, I. cylindrical. In: Fire Effects Information System, [Online]. U.S. Department of Agriculture, Forest Service, Rocky Mountain Research Station, Fire Sciences Laboratory (Producer). Available: http://www.fs.fed.us/database/feis/ [2005, November 10].
Howard, Janet L. 2005c. *Microstegium vimineum*. In: Fire Effects Information System, [Online]. U.S. Department of Agriculture, Forest Service, Rocky Mountain Research Station, Fire Sciences Laboratory (Producer). Available: http://www.fs.fed.us/database/feis/ [2005, September 16]
Howard, S. W.; Dirar, A. E.; Evens, J. O.; Provenza, R. D. 1983. The use of herbicides and/or fire to control saltcedar (*Tamarix*). Proceedings: Western Society of Weed Science. 36: 65-72.
Howard, Timothy G.; Gurevitch, Jessica; Hyatt, Laura; Carreiro, Margaret; Lerdau, Manuel. 2004. Forest invasibility in communities in southeastern New York. Biological Invasions. 6: 393-410.
Howe, H. F.; Brown, J. S. 2001. The ghost of granivory past. Ecology Letters. 4: 371-378.
Howe, Henry F. 1994a. Managing species diversity in tallgrass prairie: assumptions and implications. Conservation Biology. 8(3): 691-704.
Howe, Henry F. 1994b. Response of early- and late-flowering plants to fire season in experimental prairies. Ecological Applications. 4(1): 121-133.
Howe, Henry F. 1995. Succession and fire season in experimental prairie plantings. Ecology. 76(6): 1917-1925.
Howe, W. H.; Knopf, F. L. 1991. On the imminent decline of Rio Grande cottonwoods in central New Mexico. Southwestern Naturalist. 36: 218-224.
Hrusa, F.; Ertter, B.; Sanders, A.; Leppig, G.; Dean, E. 2002. Catalogue of nonnative vascular plants occurring spotaneously in California beyond those addressed in The Jepson Manual-Part 1. Madroño. 49: 61-98.
Hruska, Mary C.; Ebinger, John E. 1995. Monitoring a savanna restoration in east-central Illinois. Transactions of the Illinois State Academy of Science. 88: 109-117.
Huber, Diane. 2005. Invincible gorse needs to be tamed. The News-Review, [Online]. Available: http://www.newsreview.info/article/20050220/FEATURES02/102210048/0/FEATURES [2005, June 29].
Huckins, Eddie. 2004. Controlling Scotch (Scots) broom (*Cytisus scoparius*) in the Pacific Northwest, [Online]. In: The global invasive species initiative. Arlington, VA: The Nature Conservancy (Producer). Available: http://tncweeds.ucdavis.edu/moredocs/cytsco01.pdf [2005, January 31].
Huenneke L. F.; Hamburg, S. P.; Koide, R.; Mooney, H. A.; Vitousek, P. M. 1990. Effects of soil resources on plant invasion and community structure in California serpentine grassland. Ecology. 71: 478-491.
Huenneke, Laura Foster. 1997. Outlook for plant invasions: Interactions with other agents of global change. In: Luken, J. O.; Thieret, J. W., eds. Assessment and management of plant invasions. Springer Series on Environmental Management; New York: Springer-Verlag: 95-103.
Hughes, R. Flint; Vitousek, Peter M. 1993. Barriers to shrub reestablishment following fire in the seasonal submontane zone of Hawaii. Oecologia. 93: 557-563.
Hughes, R. Flint; Vitousek, Peter M.; Tunison, J. Timothy. 1991. Alien grass invasion and fire in the seasonal submontane zone of Hawaii. Ecology. 72: 743-746.

Huisinga, Kristin D.; Laughlin, Daniel C.; Fulé, Peter Z.; Springer, Judith D.; McGlone, Christopher M. 2005. Effects of an intense prescribed fire on understory vegetation in a mixed conifer forest. Journal of the Torrey Botanical Society. 132(4): 590-601.

Hulbert, L. C. 1955. Ecological studies of *Bromus tectorum* and other annual bromegrasses. Ecological Monographs. 25: 181-213.

Hulbert, Lloyd C. 1986. Fire effects on tallgrass prairie. In: Clambey, Gary K.; Pemble, Richard H., eds. The prairie: past, present and future: Proceedings, 9th North American prairie conference; 1984 July 29-August 1; Moorhead, MN. Fargo, ND: Tri-College University Center for Environmental Studies: 138-142.

Hull, A. C. 1974. Species for seeding arid rangeland in southern Idaho. Journal of Range Management. 27: 216-218.

Hull, A. C., Jr.; Holmgren, R. C. 1964. Seeding southern Idaho rangelands. Res. Pap. INT-10. Ogden, UT: U.S. Department of Agriculture, Forest Service, Intermountain Forest and Range Experiment Station. 31 p.

Hull, A. C.; Pechanec, J. F. 1947. Cheatgrass—a challenge to range research. Journal of Forestry. 45: 555-564.

Hull, A.C.; Stewart, G. 1948. Replacing cheatgrass by reseeding with perennial grass on southern Idaho ranges. American Society of Agronomy Journal. 40: 694-703.

Hull, James C.; Scott, Ralph C. 1982. Plant succession on debris avalanches of Nelson County, Virginia. Castanea. 47(2): 158-176.

Hultén, Eric. 1968. Flora of Alaska and neighboring territories. Stanford, CA: Stanford University Press. 1008 p.

Humphrey, L. D. 1984. Patterns and mechanisms of plant succession after fire on Artemisia-grass sites in southeastern Idaho. Vegetatio. 57: 91-101.

Humphrey, L. D.; Schupp, Eugene W. 2001. Seed banks of *Bromus tectorum*-dominated communities in the Great Basin. Western North American Naturalist. 61(1): 85-92.

Humphrey, R. R. 1958. The desert grassland. Botanical Review. 24: 193-253.

Humphrey, R. R. 1974. Fire in the deserts and desert grassland of North America. In: Kozlowski, T. T.; Ahlgren, C. E., eds. Fire and ecosystems: 366-400.

Humphrey, R. R.; Mehrhoff, L. A. 1958. Vegetation changes on a southern Arizona grassland range. Ecology. 39(4): 720-726.

Hungerford, R. D.; Harrington, M. G.; Frandsen, W. H.; Ryan, K. C.; Niehoff, G. J. 1991. Influence of fire on factors that affect site productivity. In: Harvey, Alan E.; Neuenschwander, Leon F., comps.. Proceedings—Management and Productivity of Western-Montane Forest Soils; 1990 April 10-April 12; Boise. Gen. Tech. Rep. INT-280. Ogden, UT: U.S. Department of Agriculture, Forest Service, Intermountain Research Station: 32-50.

Hunter, J. C.; Parker, V. T.; Barbour, M. G. 1999. Understory light and gap dynamics in an old-growth forested watershed in coastal California. Madroño. 46: 1-6.

Hunter, John C.; Mattice, Jennifer A. 2002. The spread of woody exotics into the forests of a northeastern landscape, 1938-1999. Journal of the Torrey Botanical Society. 129(3): 220-227.

Hunter, M. D.; Price, P. W. 1992. Playing chutes and ladders: heterogeneity and the relative roles of bottom-up and top-down forces in natural communities. Ecology. 73: 724-732.

Hunter, Molly E.; Omi, Philip N.; Martinson, Erik J.; Chong, Geneva W. 2006. Establishment of non-native plant species after wildfires: effects of fuel treatments, abiotic and biotic factors, and post-fire grass seeding treatments. International Journal of Wildland Fire. 15: 271-281.

Hunter, Richard. 1991. *Bromus* invasions on the Nevada Test Site: present status of *B. rubens* and *B. tectorum* with notes on their relationship to disturbance and altitude. Great Basin Naturalist. 51(2): 176-182.

Hurlbert, S. H. 1984. Pseudoreplication and the design of ecological field experiments. Ecological Monographs. 54: 187–211.

Hutchings, M. J.; Price, E. A. C. 1999. *Glechoma hederacea* L. (Nepeta glechoma Benth., N. hederacea (L.) Trev.). The Journal of Ecology. 87(2): 347-364.

Hutchinson, Todd; Rebbeck, Joanne; Long, Robert. 2004. Abundant establishment of *Ailanthus altissima* (tree-of-heaven) after restoration treatments in an upland oak forest. In: Yaussy, Daniel; Hix, David M.; Goebel, P. Charles; Long, Robert P., eds. Proceedings, 14th central hardwood forest conference; 2004 March 16-19; Wooster, OH. Gen. Tech. Rep. NE-316. Newton Square, PA: U.S. Department of Agriculture, Forest Service, Northeastern Research Station: 514.

Hutchison, Max. 1992a. Vegetation management guideline: Canada thistle (*Cirsium arvense* (L) Scop.). Natural Areas Journal. 12(3): 160-161.

Hutchison, Max. 1992b. Vegetation management guideline: Reed canary grass (*Phalaris arundinacea* L.). Natural Areas Journal. 12(3): 159.

Hyatt, Laura A.; Casper, Brenda B. 2000. Seed bank formation during early secondary succession in a temperate deciduous forest. Journal of Ecology. 88(3): 516-527.

Ibarra-F., Fernando A.; Cox, Jerry R.; Martin-R., Martha H.; Croel, Todd A.; Call, Christopher A. 1995. Predicting buffelgrass survival across a geographical and environmental gradient. Journal of Range Management. 48(1): 53-59.

Inglis, Richard; Deuser, Curt; Wagner, Joel. 1996. The effects of tamarisk removal on diurnal ground water fluctuations: United States Department of the Interior, National Park Service. NPS/NRWRD/NRTR-96/93. 64 p.

Interagency Technical Reference. 1999. Sampling vegetation attributes. BLM Technical Reference 1734-4. Denver, CO: U.S. Department of the Interior, Bureau of Land Management National Business Center. 158 p.

Intergovernmental Panel on Climate Change. 2001. Climate Change 2001: The Scientific Basis, [Online]. Houghton, J. T.; Ding, Y.; Griggs, D.J.; Noguer, M.; van der Linden, P. J.; Xiaosu, D. (eds.). Cambridge, UK: University Press (Producer). Available: http://www.grida.no/climate/ipcc_tar/wg2/554.htm [2007, July 12].

Invasive Species Specialist Group. 2006. Ecology of *Urochloa maxiuma* (grass), [Online]. In: Global Invasive Species Database. The World Conservation Union, Invasive Species Specialist Group (Producer). Available: http://www.issg.org/database/species/search.asp?sts=sss&st=sss&fr=1&sn= Urochloa+maxima&rn=&hci=-1&ei=-1 [2007, August 23].

Irving, Robert S.; Brenholts, Susan; Foti, Thomas. 1980. Composition and net primary production of native prairies in eastern Arkansas. The American Midland Naturalist. 103(2): 298-309.

Isaac, L. A. 1940. Vegetative succession following logging in the Douglas-fir region with special reference to fire. Journal of Forestry. 38: 716-721.

ITIS Database. 2004. Integrated Taxonomic Information System, [Online]. Available: http://www.itis.usda.gov/index.html.

Iverson, L. R.; Yaussy, D. A.; Rebbeck, J.; Hutchinson, T. F.; Long, R. P.; Prasad, A. M. 2004. A comparison of thermocouples and temperature paint to monitor spatial and temporal characteristics of landscape-scale prescribed fire. International Journal of Wildland Fire. 13: 311-322.

Jackson, L. E. 1985. Ecological origins of California's Mediterranean grasses. Journal of Biogeography. 12: 349-361.

Jacobi, J. D.; Warshauer, F. R. 1992. Distribution of six alien plant species in upland habitats on the Island of Hawaii. In: Stone, Charles P.; Smith, Clifford W.; Tunison, Timothy, eds. Symposium on alien plant invasions in native ecosystems of Hawaii: management and research. Honolulu, HI: University of Hawai`i, Cooperative Parks Resources Studies: 155-188.

Jacobs, James S.; Sheley, Roger L. 2003a. Prescribed fire effects on Dalmatian toadflax. Journal of Range Management. 56(2): 193-197.

Jacobs, James S.; Sheley, Roger L. 2003b. Testing the effects of herbicides and prescribed burning on Dalmatian toadflax (Montana). Ecological Restoration. 21(2): 138-139.

Jacobs, James S.; Sheley, Roger L. 2005. The effect of season of picloram and chlorsulfuron application on Dalmatian toadflax (*Linaria genistifolia*) on prescribed burns. Weed Technology. 19: 319-324.

Jensen, M. E., Hann, W. J.; Keane, R. E.; Caratti, J.; Bougeron, P. S. 1993. ECODATA—a multiresource database and analysis system for ecosystem description and evaluation. In: Jensen, M. E.; Bourgeron, P. S., tech. eds. Eastside forest ecosystem health assessment Volume II—ecosystem management: principles and applications. Gen. Tech. Rep. PNW-GTR-318. Portland, OR: U.S. Department of Agriculture, Forest Service. 203-217.

Jepson, W. L. 1910. The silva of California. University of California Memoirs. 2: 1-480.

Johannessen, Carl L.; Davenport, William A.; Millet, Artimus; McWilliams, Steven. 1971. The vegetation of the Willamette Valley. Annals of the Association of American Geographers. 61: 286-302.

Johnson, A. S.; Hale, P. E. 2000. The historical foundations of prescribed burning for wildlife: a southern perspective. In: Ford, W. Mark; Russell, Kevin R.; Moorman, Christopher E., eds. The role of fire in nongame wildlife management and community restoration: traditional uses and new directions: Proceedings of a special workshop; 2000 December 15; Nashville, TN; Gen. Tech. Rep. NE-288. Newtown Square, PA: U.S. Department of Agriculture, Forest Service, Northeastern Research Station. [Pages unknown].

Johnson, Charles Grier, Jr. 1998. Vegetation response after wildfires in national forests of northeastern Oregon. R6-NR-ECOL-TP-06-98. Portland, OR: U.S. Department of Agriculture, Forest Service, Pacific Northwest Region. 128 p. (+ appendices).

Johnson, E. 1996. *Berberis thunbergii*, Japanese barberry. In: Randall, John M.; Marinelli, Janet, eds. 1996. Invasive plants: Weeds of the global garden. Handbook #149. Brooklyn, NY: Brooklyn Botanic Garden. 47.

Johnson, Edward A. 1992. Fire and vegetation dynamics: studies from the North American boreal forest. Cambridge Studies in Ecology. Cambridge: Cambridge University Press. 129 p.

Johnson, K. 1996. *Paulownia tomentosa*, princess tree. In: Randall, John M.; Marinelli, Janet, eds. 1996. Invasive plants: Weeds of the global garden. Handbook #149. Brooklyn, NY: Brooklyn Botanic Garden. 38.

Johnson, Larry; Van Cleve, Keith. 1976. Revegetation in arctic and subarctic North America-a literature review. CRREL Report 76-15. Hanover, U.S. Army, Corp of Engineers, Cold Regions Research and Engineering Laboratory. 32 p.

Johnson, P. N. 2001. Vegetation recovery after fire on a southern New Zealand peatland. New Zealand Journal of Botany. 39(2): 251-267.

Johnson, Vanessa S.; Litvaitis, John A.; Lee, Thomas D.; Frey, Serita D. 2006. The role of spatial and temporal scale in colonization and spread of invasive shrubs in early successional habitats. Forest Ecology and Management. 228(1-3): 124-134.

Jones, Joshua L.; Webb, Brent W.; Jimenez, Dan; Reardon, James; Butler, Bret. 2004. Development of an advanced one-dimensional stem heating model for application in surface fires. Canadian Journal of Forest Research. 34: 20-30.

Kan, Tamara; Pollak, Oren. 2000. *Taeniatherum caput-medusae*. In: Bossard, Carla C.; Randall, John M.; Hoshovsky, Marc C., eds. Invasive plants of California's wildlands. Berkeley, CA: University of California Press: 309-312.

Kartawinata, Kuswata; Mueller-Dombois, Dieter. 1972. Phytosociology and ecology of the natural dry-grass communities on Oahu, Hawaii. Reinwartdia. 8: 369-472.

Kartesz, J. T.; Meacham, C. A. 1999. Synthesis of the North American flora, Version 1.0. North Carolina Botanical Garden, Chapel Hill, NC.

Kartesz, J.T. 1999. A synonymized checklist and atlas with biological attributes for the vascular flora of the United States, Canada, and Greenland. First Edition. In: Kartesz, J. T.; Meacham, C. A. 1999. Synthesis of the North American flora, Version 1.0. North Carolina Botanical Garden, Chapel Hill, NC.

Katz, Gabrielle L.; Shafroth, Patrick B. 2003. Biology, ecology, and management of *Elaeagnus angustifolia* L. (Russian olive) in western North America. Wetlands. 23(4): 763-777.

Kaufmann, M. R.; Huckaby, L. S.; Fornwalt, P. J.; Stoker, J. M.; Romme, W. H. 2003. Using tree recruitment patterns and fire history to guide restoration of an unlogged ponderosa pine/Douglas fir landscape in the southern Rocky Mountains after a century of fire suppression. Forestry. 76: 231-241.

Kay, B. L. 1963. Effects of dalapon on a medusahead community. Weeds. 11: 207-209.

Kay, B. L.; McKell, C. M. 1963. Pre-emergence herbicides as an aid in seeding annual rangelands. Weeds. 11: 260-264.

Kaye, J. P.; Hart, S. C. 1998. Ecological restoration alters nitrogen transformations in ponderosa pine bunchgrass ecosystem. Ecological Applications. 8: 1052-1060.

Kaye, T. N.; Pendergrass, K. L.; Finley, K.; Kauffman, J. B. 2001. The effect of fire on the population viability of an endangered prairie plant. Ecological Applications. 11: 1366-1380.

Kaye, Thomas. 2001. *Brachypodium sylvaticum* (Poaceae) in the Pacific Northwest. Botanical Electronic News Number 277, [Online]. Available: http://www.ou.edu/cas/botany-micro/ben/ben277.html [2005, February 20].

Kearney, T. H.; Briggs, L. J.; Shantz, H. L.; McLane, J. W.; Piemeisel, R. L. 1914. Indicator significance of vegetation in Tooele Valley, Utah. Journal of Agric. Research. 1: 365-417.

Keeley, J. E. 1981. Distribution of lightning and man caused wildfires in California. In: Conard, C. E.; Oechel, W. C., eds. Proceedings of the symposium on dynamics and management of Mediterranean-type ecosystems. Gen. Tech. Rep. GTR-PSW-58. Berkeley, CA: U.S. Department of Agriculture, Forest Service, Pacific Southwest Forest and Range Experiment Station: 431-437.

Keeley, J. E. 1987. Role of fire in the germination of woody taxa in California chaparral. Ecology. 68: 434-443.

Keeley, J. E. 1995. Future of California floristics and systematics: wildfire threats to the California flora. Madroño. 42: 175-179.

Keeley, J. E. 2001. Fire and invasive species in Mediterranean-climate ecosystems in California. In: Galley, Krista E. M.; Wilson, Tyrone P., eds. Proceedings of the invasive species workshop: The role of fire in the control and spread of invasive species; Fire conference 2000: the first national congress on fire ecology, prevention, and management; 2000 November 27-December 1; San Diego, CA. Misc. Publ. No. 11. Tallahassee, FL: Tall Timbers Research Station: 81-94.

Keeley, J. E. 2002a. Fire management of California shrubland landscapes. Environmental Management. 29: 395-408.

Keeley, J. E. 2002b. Native American impacts on fire regimes of the California coastal ranges. Journal of Biogeography. 29: 303-320.

Keeley, J. E. 2004. Ecological impacts of wheat seeding after a Sierra Nevada wildfire. International Journal of Wildland Fire. 13(1): 73-78.

Keeley, J. E. 2006a. Fire in the South Coast Bioregion. In: Sugihara, N.; van Wagtendonk, J. W.; Fites, J.; Thode, A., eds. Fire in California ecosystems. Berkeley, CA: University of California Press: 350-390.

Keeley, J. E. 2006b. Fire management impacts on invasive plants in the western United States. Conservation Biology. 20(2): 375-384.

Keeley, J. E.; Allen, C. D.; Betancourt, J.; Chong, G. W.; Fotheringham, C. J.; Safford, H. D. 2006a. A 21st century perspective on postfire seeding. Journal of Forestry. 104: 103-104.

Keeley, J. E.; Fotheringham, C. J. 1998a. Mechanism of smoke-induced seed germination in a post-fire chaparral annual. Journal of Ecology. 86: 27-36.

Keeley, J. E.; Fotheringham, C. J. 1998b. Smoke-induced seed germination in California chaparral. Ecology. 79: 2320-2336.

Keeley, J. E.; Fotheringham, C. J. 2001a. Historic fire regime in southern California shrublands. Conservation Biology. 15: 1536-1548.

Keeley, J. E.; Fotheringham, C. J. 2001b. History and management of crown-fire ecosystems: a summary and response. Conservation Biology. 15: 1561-1567.

Keeley, J. E.; Fotheringham, C. J. 2003. Impact of past, present, and future fire regimes on North American Mediterranean shrublands. In: Veblen, Thomas T.; Baker, William L.; Montenegro, Gloria; Swetnam, Thomas W., eds. Fire and climatic change in temperate ecosystems of the western Americas. Ecological Studies, Vol. 160. New York: Springer: 218-262.

Keeley, J. E.; Fotheringham, C. J.; Baer-Keeley, M. 2005. Determinants of post-fire recovery and succession in Mediterranean-climate shrublands in California. Ecological Applications. 15: 1515-1534.

Keeley, J. E.; Fotheringham, C. J.; Baer-Keeley, Melanie. 2006b. Demographic patterns of postfire regeneration in mediterranean-climate shrublands of California. Ecological Monographs. 76(2): 235-255.

Keeley, J. E.; Fotheringham, C. J.; Morais, Marco. 1999. Reexamining fire suppression impacts on brushland fire regimes. Science. 284(5421): 1829-1831.

Keeley, J. E.; Keeley, S. C. 1981. Post-fire regeneration of southern California chaparral. American Journal of Botany. 68(4): 524-530.

Keeley, J. E.; Keeley, S. C. 1984. Post-fire recovery of California coastal sage scrub. The American Midland Naturalist. 111: 105-117.

Keeley, J. E.; Lubin D.; Fotheringham, C. J. 2003. Fire and grazing impacts on plant diversity and alien plant invasions in the southern Sierra Nevada. Ecological Applications. 13(5): 1355–1374.

Keeley, J. E.; McGinnis, T. W. 2007. Impact of prescribed fire and other factors on cheatgrass persistence in a Sierra Nevada ponderosa pine forest. International Journal of Wildland Fire. 16: 96-106.

Keeley, J. E.; Morton, B. A.; Pedrosa, A.; Trotter, P. 1985. Role of allelopathy, heat, and charred wood in the germination of chaparral herbs and suffrutescents. Journal of Ecology. 73: 445-458.

Keeley, J. E.; Nitzberg, N. E. 1984. The role of charred wood in the germination of the chaparral herbs *Emmenanthe penduliflora* (Hydrophyllaceae) and *Eriophyllum confertifolium* (Asteraceae). Madroño. 31: 208-218.

Keeley, J. E.; Pizzorno, M. 1986. Charred wood stimulated germination of two fire-following herbs of the California chaparral and the role of hemicellulose. American Journal of Botany. 73: 1289-1297.

Keeley, J. E.; Zedler, P. H. 1978. Reproduction of chaparral shrubs after fire: a comparison of sprouting and seeding strategies. The American Midland Naturalist. 99: 142-161.

Keeley, Sterling C.; Keeley, Jon E.; Hutchinson, Steve M.; Johnson, Albert W. 1981. Postfire succession of the herbaceous flora in southern California chaparral. Ecology. 62(6): 1608-1621.

Kepler, Angela Kay. 1984. Hawaiian heritage plants. Honolulu, HI: Oriental Publication Company. 240 p.

Keyser, Tara; Smith, Frederick; Lentile, Leigh. 2006. Monitoring fire effects and vegetation recovery on the Jasper Fire, Black Hills National Forest, SD. Final report RJV-01-CS-11221616-074 Mod. 2. Fort Collins, CO: Colorado State University, Department of Forest, Rangeland, and Watershed Stewardship. 56 p.

Kilgore, B. M. 1973. The ecological role of fire in Sierran conifer forests. Journal of Quaternary Research. 3: 496-513.

Kilgore, B. M.; Taylor, D. 1979. Fire history of a sequoia mixed-conifer forest. Ecology. 60: 129-142.

Kilgore, Bruce M.; Heinselman, Miron L. 1990. Fire in wilderness ecosystems. In: Hendee, John C.; Stankey, George H.; Lucas, Robert C., eds. Wilderness management. 2d ed. (Revised). Golden, CO: North American Press: 297-335. In cooperation with U.S. Department of Agriculture, Forest Service.

Kindschy, Robert R. 1994. Pristine vegetation of the Jordan Crater kipukas: 1978-91. In: Monsen, Stephen B.; Kitchen, Stanley G., comps. Proceedings: ecology and management of annual rangelands; 1992 May 18-22; Boise, ID. Gen. Tech. Rep. INT-GTR-313. Ogden, UT: U.S. Department of Agriculture, Forest Service, Intermountain Research Station: 85-88.

King, Richard S. 2003. Habitat management for the Karner blue butterfly (*Lycaeides Melissa samuelis*): Evaluating the short-term consequences. Ecological Restoration. 21(2): 101-106.

King, S. E.; Grace, J. B. 2000. The effects of gap size and disturbance type on invasion of wet pine savanna by cogongrass *Imperata cylindrica* (Poaceae). American Journal of Botany. 87: 1279-1286.

Kirby, Ronald E.; Lewis, Stephen J.; Sexson, Terry N. 1988. Fire in North American wetland ecosystems and fire-wildlife relations: an annotated bibliography. Biological Report 88(1). Washington, DC: U.S. Department of the Interior, Fish and Wildlife Service. 146 p.

Kirch, Patrick V. 1982. The impact of the prehistoric Polynesians on the Hawaiian ecosystem. Pacific Science. 36: 1-14.

Kirkpatrick, J. B.; Hutchinson, C. F. 1977. The community composition of Californian coastal sage scrub. Vegetatio. 35(1): 21-33.

Kirsch, Leo M.; Kruse, Arnold D. 1973. Prairie fires and wildlife. In: Proceedings, annual Tall Timbers fire ecology conference; 1972 June 8-9; Lubbock, TX. Number 12. Tallahassee, FL: Tall Timbers Research Station: 289-303.

Kitajima, Kaoru; Tilman, David. 1996. Seed banks and seedling establishment on an experimental productivity gradient. Oikos. 72(2): 381-391.

Kiviat, Eric. 2004. Occurrence of *Ailanthus altissima* in a Maryland freshwater tidal estuary. Castanea. 69: 139-142.

Klebesadel, L. J. 1980. Birdvetch: Forage crop, ground cover, ornamental, or weed? Agroborealis. 12(1):46-49.

Kleintjes, Paula K.; Sporrong, Jill M.; Raebel, Christopher A.; Thon, Stephen F. 2003. Habitat type conservation and restoration for the Karner blue butterfly. Ecological Restoration. 21(2): 107-115.

Klemp, G. J.; Hull, A. C. 1971. Methods for seeding three perennial wheatgrass on cheatgrass ranges in southern Idaho. Journal of Range Management. 25: 266-268.

Kline, Virginia M. 1983. Control of sweet clover in a restored prairie (Wisconsin). Restoration and Management Notes. 1(4): 30-31.

Kline, Virginia M. 1986. Response of sweet clover (*Melilotus alba* Desr.) and associated prairie vegetation to seven experimental burning and mowing treatments. In: Clambey, Gary K.; Pemble, Richard H., eds. The prairie: past, present and future: Proceedings, 9th North American prairie conference; 1984 July 29-August 1; Moorhead, MN. Fargo, ND: Tri-College University Center for Environmental Studies: 149-152.

Kline, Virginia M.; McClintock, Tom. 1994. Effect of burning on a dry oak forest infested with woody exotics. In: Wickett, Robert G.; Lewis, Patricia Dolan; Woodliffe, Allen; Pratt, Paul, eds. Spirit of the land, our prairie legacy: Proceedings, 13th North American prairie conference; 1992 August 6-9; Windsor, ON. Windsor, ON: Department of Parks and Recreation: 207-213.

Klinger, R. C. 2000. *Foeniculum vulgare*. In: Bossard, C. C.; Randall, J. M.; Hoshovsky, M. C., eds. Pages. Invasive plants of California's wildlands. Berkeley, CA: University of California Press: 198-202.

Klinger, Robert. 2006. [Personal communication]. April 10, Davis, CA: University of California, Davis (BINGO).

Klinger, R. C.; Brooks, M. L.; Randall, J. A. 2006a. Fire and invasive plant species. In: Sugihara, N.; van Wagtendonk, J. W.; Fites, J.; Thode, A., eds. Fire in California ecosystems. Berkeley, CA: University of California Press: 499-519.

Klinger, R. C.; Messer, Ishmael. 2001. The interaction of prescribed burning and site characteristics on the diversity and composition of a grassland community on Santa Cruz Island, California. In: Galley, Krista E. M.; Wilson, Tyrone P., eds. Proceedings of the invasive species workshop: The role of fire in the control and spread of invasive species; Fire conference 2000: the first national congress on fire ecology, prevention, and management; 2000 November 27-December 1; San Diego, CA. Misc. Publ. No. 11. Tallahassee, FL: Tall Timbers Research Station: 66-80.

Klinger, R. C.; Underwood, E. C.; Moore, P. E. 2006b. The role of environmental gradients in non-native plant invasions into burnt areas of Yosemite National Park, California. Diversity and distributions. 12: 139-156.

Knapp, P. A. 1995. Intermountain West lightning-caused fires: climatic predictors of area burned. Journal of Range Management. 48(1): 85-91.

Knapp, P. A. 1996. Cheatgrass (*Bromus tectorum* L.) dominance in the Great Basin desert. Global Environmental Change. 6: 37-52.

Knapp, Paul A. 1998. Spatio-temporal patterns of large grassland fires in the Intermountain West, U.S.A. Global Ecology and Biogeography Letters. 7(4): 259-273.

Knick, Steven T. 1999. Requiem for a sagebrush ecosystem? Northwest Science. 73(1): 53-57.

Knick, Steven T.; Rotenberry, John T. 1997. Landscape characteristics of disturbed shrubsteppe habitats in southwestern Idaho (U.S.A.). Landscape Ecology. 12: 287-297.

Knoepp, J. D.; DeBano, L. F.; Neary, D. G. 2005. Soil chemistry. In: Neary, Daniel G.; Ryan, Kevin C.; DeBano, Leonard F., eds. 2005. Wildland fire in ecosystems: effects of fire on soils and water. Gen. Tech. Rep. RMRS-GTR-42-vol.4. Ogden, UT: U.S. Department of Agriculture, Forest Service, Rocky Mountain Research Station: 53-72.

Knopf, F. L.; Olson, T. E. 1984. Naturalization of Russian-olive: Implications to Rocky Mountain Wildlife. Wildlife Society Bulletin. 12: 289-298.

Knopf, Fritz L.; Johnson, R. Roy; Rich, Terrell; Samson, Fred B.; Szaro, Robert C. 1988. Conservation of riparian ecosystems in the United States. Wilson Bulletin. 100(2): 272-284.

Knops, J. M. H.; Griffin, J. R.; Royalty, A. C. 1995. Introduced and native plants of the Hastings Reservation, central coastal California: a comparison. Biological Conservation. 71: 115-123.

Kolar, Cynthia S.; Lodge, David M. 2001. Progress in invasion biology: predicting invaders. Trends in Ecology and Evolution. 16(4): 199-204.

Komarek, E. V. 1964. The natural history of lightning. In: Proceedings, 3rd annual Tall Timbers fire ecology conference. Tallahassee, FL: Tall Timbers Research Station. 139-183.

Komarek, E. V. 1974. Effects of fire on temperate forests and related ecosystems: Southeastern United States. In Kozlowski, T. T.; Ahlgren, C. E., eds. Fire and ecosystems. New York: Academic Press: 251-277.

Koniak, S. 1985. Succession in pinyon-juniper woodlands following wildfire in the Great Basin. Great Basin Naturalist. 45(3): 556-566.

Koniak, Susan; Everett, Richard L. 1982. Seed reserves in soils of successional stages of pinyon woodlands. The American Midland Naturalist. 108(2): 295-303.

Korb, J. E.; Springer, J. D.; Powers, S. R.; Moore, M. M. 2005. Soil seed banks in *Pinus ponderosa* forests in Arizona: clues to site history and restoration potential. Applied Vegetation Science. 8: 103-112.

Korb, Julie E.; Covington, W. W.; Fulé, Peter Z. 2003. Sampling techniques influence understory plant trajectories after restoration: an example from ponderosa pine restoration. Restoration Ecology. 11(4): 504-515.

Korb, Julie E.; Johnson, Nancy C.; Covington, W. W. 2004. Slash pile burning effects on soil biotic and chemical properties and plant establishment: recommendations for amelioration. Restoration Ecology. 12(1): 52-62.

Kotanen, P. M. 1997. Effects of experimental soil disturbance on revegetation by natives and exotics in coastal California meadows. Journal of Applied Ecology. 34: 631-644.

Kotanen, Peter M.; Bergelson, Joy; Hazlett, Donald L. 1998. Habitats of native and exotic plants in Colorado shortgrass steppe: a comparative approach. Canadian Journal of Botany. 76: 664-672.

Kourtev, P. S.; Ehrenfeld, J. G.; Huang, W. Z. 1998. Effects of exotic plant species on soil properties in hardwood forests of New Jersey. Water, Air, and Soil Pollution. 105: 493-501.

Kourtev, P. S.; Huang, W. Z.; Ehrenfield, J. G. 1999. Differences in earthworm densities and nitrogen dynamics in soils under exotic and native plant species. Biological Invasions. 1: 237-245.

Kraemer, James Fred. 1977. The long term effects of burning on plant succession. The long term effects of burning on plant succession. Corvallis, OR: Oregon State University. 123 p. Thesis.

Kramer, Neal B.; Johnson, Frederic D. 1987. Mature forest seed banks of three habitat types in central Idaho. Canadian Journal of Botany. 65: 1961-1966.

Kruckeberg, A. R. 1954. The ecology of serpentine soils. III. Plants species in relation to serpentine soils. Ecology. 35: 267-274.

Krueger, William C.; Kelley, Claudia E. 2000. Describing and categorizing natural resources literature. Rangelands. 22(4): 37-39.

Kucera, C. L. 1992. Tall-grass prairie. In: Coupland, R. T., ed. Natural grasslands: Introduction and western hemisphere. Ecosystems of the World 8A. Amsterdam, Netherlands: Elsevier Science Publishers B. V.: 227-268.

Kucera, Clair L. 1981. Grasslands and fire. In: Mooney, H. A.; Bonnicksen, T. M.; Christensen, N. L.; Lotan, J. E.; Reiners, W. A., tech. coords. Fire regimes and ecosystem properties: Proceedings of the conference; 1978 December 11-15; Honolulu, HI. Gen. Tech. Rep. WO-26. Washington, DC: U.S. Department of Agriculture, Forest Service: 90-111.

Küchler, A. W. 1964. Potential natural vegetation of the conterminous United States. (Manual and map). Am. Geogr. Soc. Spec. Public. 36, 1965 rev. New York: American Geographical Society. 116 p.

Küchler, A.W. 1967. Potential natural vegetation of Alaska. Washington, DC: United States Geological Survey.

Kuddes-Fischer, Linda M.; Arthur, Mary A. 2002. Response of understory vegetation and tree regeneration to a single prescribed fire in oak-pine forests. Natural Areas Journal. 22 (1): 43-52.

Kumar, V.; DiTommaso, A. 2005. Mile-a minute (*Polygonum perfoliatum* L.): An increasingly problematic invasive species. Weed Technology. 19(4): 1071-1077.

Kyser, Guy B.; DiTomaso, Joseph M. 2002. Instability in a grassland community after the control of yellow starthistle (*Centaurea solstitialis*) with prescribed burning. Weed Science. 50(5): 648-657.

Lacey, J. R.; Marlow, C. B.; Lane J. R. 1989. Influence of spotted knapweed (*Centaurea maculosa*) on surface runoff and sediment yield. Weed Technology. 3: 627-631.

Lambrinos, J. G. 2000. The impact of the invasive alien grass *Cortaderia jubata* (Lemoine) Stapf on an endangered mediterranean-type shrubland in California. Diversity and Distributions. 6: 217-231.

Lamp, H. F. 1952. Reproductive activity in *Bromus inermis* Leyss. in relation to phases of tiller development. Botanical Gazette. 113: 413-438.

LANDFIRE Rapid Assessment. 2005a. Potential Natural Vegetation Group R4OASA—Oak savanna: Description, [Online]. In: Rapid assessment reference condition models. In: LANDFIRE. Washington, DC: U.S. Department of Agriculture, Forest Service, Rocky Mountain Research Station, Fire Sciences Lab; U.S Geological Survey; The Nature Conservancy (Producers). Available: http://www.landfire.gov/models_EW.php http://www.landfire.gov/models_EW.php [2007, July 12].

LANDFIRE Rapid Assessment. 2005b. Potential Natural Vegetation Group R7NMAR—Northern coastal marsh: Description, [Online]. In: Rapid assessment reference condition models. In: LANDFIRE. Washington, DC: U.S. Department of Agriculture, Forest Service, Rocky Mountain Research Station, Fire Sciences Lab; U.S Geological Survey; The Nature Conservancy (Producers). Available: http://www.landfire.gov/ModelsPage2.html [2006, February 9].

LANDFIRE Rapid Assessment. 2005c. Rapid assessment reference condition models, [Online]. In: LANDFIRE Washington, DC: U.S. Department of Agriculture, Forest Service, Rocky Mountain Research Station, Fire Sciences Lab; U.S. Geological Survey; The Nature Conservancy (Producers). Available: http://www.landfire.gov/models_EW.php [2007, July 12].

Landgraf, B. K.; Fay, P. K.; Havstad, K. M. 1984. Utilization of leafy spurge (*Euphorbia esula*) by sheep. Weed Science. 32: 348-352.

Landhausser, Simon M.; Wein, Ross W. 1993. Postfire vegetation and tree establishment at the arctic treeline: climate-change-vegetation-response hypotheses. The Journal of Ecology. 81(4): 665-672.

Langdon, Keith R.; Johnson, K. D. 1994. Additional notes on invasiveness of Paulownia tomentosa in natural areas. Natural Areas Journal. 14(2): 139-140.

Langeland, K. 2006. Management techniques—fire. In: Hutchinson, J.; Ferriter, A.; Serbesoff-King, K.; Langeland, K.; Rodgers, L., eds. 2006. Lygodium Management Plan for Florida, 2nd Edition, [Online]. Florida Exotic Pest Plant Council Lygodium Task Force (Producer). 51 p. Available at: http://www.fleppc.org/Manage_Plans/lymo_mgt.pdf

Lapina, Irina; Carlson, Matthew L. 2005. Weed risk assessment reports. Plant Invasiveness Assessment System for Alaska, Alaska Natural Heritage Program, [Online]. Anchorage, AK: University of Alaska (Producer). Available: http://akweeds.uaa.alaska.edu/pdfs/weed_risk_assess_pdfs/Weed_Risk_Asses_CRTE.pdf [2005, December 15].

Larson, Diane. 2006. [Review comments to Kristin Zouhar]. August 15. St. Paul, MN: USGS, Northern Prairie Wildlife Research Center. On file at: U.S. Department of Agriculture, Forest Service, Rocky Mountain Research Station, Fire Sciences Laboratory, Missoula, MT.

Larson, D. L. 2003. Native weeds and exotic plants: relationships to disturbance in mixed-grass prairie. Plant Ecology. 169: 317-333.

Larson, D. L.; Anderson, P. J.; Newton, W. 2001. Alien plant invasion in mixed-grass prairie: effects of vegetation type and anthropogenic disturbance. Ecological Applications. 11: 128-141.

Larson, D. L.; Grace, J. B. 2004. Temporal dynamics of leafy spurge (*Euphorbia esula*) and two species of flea beetles (*Aphthona* spp.) used as biological control agents. Biological Control. 29: 207-214.

Larson, D. L.; Grace, J. B.; Rabie, P. A.; Andersen, P. 2007. Short-term disruption of a leafy spurge (*Euphorbia esula*) biocontrol program following herbicide application. Biological Control. 40: 1-8.

Larson, D. L.; Newton, W. E. 1996. Effects of fire retardant chemical and fire suppressant foam on North Dakota prairie vegetation. Proceedings of the North Dakota Academy of Science. 50: 137-144.

Larson, D. L.; Newton, W. E.; Anderson, P. J.; Stein, S. 1999. Effects of fire retardant chemical and fire suppressant foam on shrub steppe vegetation in northern Nevada. International Journal of Wildland Fire. 9: 115-127.

Larson, J. R.; Duncan, D. A. 1982. Annual grassland response to fire retardant and wildfire. Journal of Range Management. 35: 700-703.

Larson, John L.; Stearns, Forest W. 1990. Effects of mowing on a woolgrass (*Scirpus cyperinus* (L.) Kunth) dominated sedge meadow in southeastern Wisconsin. In: Hughes, H. Glenn; Bonnicksen, Thomas M., eds. Restoration`89: the new management challenge: Proceedings, 1st annual meeting of the Society for Ecological Restoration; 1989 January 16-20; Oakland, CA. Madison, WI: The University of Wisconsin Arboretum, Society for Ecological Restoration: 549-560.

Larson, K. C. 2000. Circumnutation behavior of an exotic honeysuckle vine and its native congener: influence on clonal mobility. American Journal of Botany. 87: 533-538.

Lauenroth, W. K.; Milchunas, D. G. 1992. Short-grass steppe. In: Coupland, R. T., ed. Natural grasslands: Introduction and western hemisphere. Ecosystems of the World 8A. Amsterdam, Netherlands: Elsevier Science Publishers B. V.: 183-226.

Laughlin, D. C.; Bakker, J. D.; Stoddard, M. T.; Daniels, M. L.; Springer, J. D.; Gildar, C. N.; Green, A. M.; Covington, W. W. 2004. Toward reference conditions: wildfire effects on flora in an old-growth ponderosa pine forests. Forest Ecology and Management. 199: 137-152.

Laughlin, Daniel C. 2003. Lack of native propagules in a Pennsylvania, USA, limestone prairie seed bank: futile hopes for a role in ecological restoration. Natural Areas Journal. 23(2): 158-164.

Lausi, D.; Nimis, P. L. 1985. Roadside vegetation in boreal South Yukon and adjacent Alaska. Phytocoenologia.13(1): 103-138.

Lavender, Denis P. 1958. Effect of ground cover on seedling germination and survival. Research Note No. 38. Corvallis, OR: State of Oregon, Forest Lands Research Center. 32 p.

Lavergne, S.; Molofsky, J. 2006. Control strategies for the invasive reed canarygrass (*Phalaris arundinacea* L.) in North American wetlands: the need for and integrated management plan. Natural Areas Journal. 26(2): 208-214.

Lawlor, Fran. 2002. Element stewardship abstract: *Vincetoxicum nigrum* (L.) Moench. & *Vincetoxicum rossicum* (Kleopov) Barbarich (swallow-wort), [Online]. In: The global invasive species initiative. Arlington, VA: The Nature Conservancy (Producer). Available: http://tncweeds.ucdavis.edu/esadocs/documnts/vinc_sp.html [2007, July 12].

Lawlor, Frances M. 2000. Herbicidal treatment of the invasive plant *Cynachum rossicum* and experimental post control restoration of infested sites. Syracuse, NY: State University of New York, College of Environmental Science and Forestry. 78 p. Thesis.

Laycock, W. A. 1991. Stable states and thresholds of range condition on North American rangelands: a viewpoint. Journal of Range Management. 44(5): 427-433.

le Maitre D. C.; Midgley, J. J. 1992. Plant reproductive ecology. In: Cowling, R. M., ed. The ecology of fynbos (nutrients, fire, and diversity). Cape Town, South Africa: Oxford University Press. 135-173.

Leck, Mary Allessio; Leck, Charles F. 2005. Vascular plants of a Delaware River tidal freshwater wetland and adjacent terrestrial areas: seed bank and vegetation comparisons of reference and constructed marshes and annotated species list. Journal of the Torrey Botanical Society. 132(2): 323-354.

Leckie, Sara; Vellend, Mark; Bell, Graham; [and others]. 2000. The seed bank in an old-growth, temperate deciduous forest. Canadian Journal of Botany. 78(2): 181-192.

Lee, Carol Eunmi. 2002. Evolutionary genetics of invasive species. Trends in Ecology and Evolution. 17(8): 386-391.

Leenhouts, W. 1982. Maintaining a fire subclimax salt marsh community with prescribed fire (Florida). Restoration and Management Notes. 1: 20.

Leete, B. E. 1938. Forest fires in Ohio 1923 to 1935. Ohio Agricultural Experiment Station, Bulletin 598, Wooster, OH. Department of Forestry, Ohio Agricultural Experiment Station: [Pages unknown].

Lehmkuhl, John F. 2002. The effects of spring burning and grass seeding in forest clearcuts on native plants and conifer seedlings in coastal Washington. Northwest Science. 76(1): 46-60.

Leopold, Aldo. 1924. Grass, brush, timber and fire in southern Arizona. Journal of Forestry. 22: 1-10.

Lesica, P. 1996. Effects of fire on the demography of the endangered, geophytic herb Silene spaldinghii (Caryophyllaceae). American Journal of Botany. 86: 996-1003.

Lesica, Peter; Martin, Brian. 2003. Effects of prescribed fire and season of burn on recruitment of the invasive exotic plant, *Potentilla recta*, in a semiarid grassland. Restoration Ecology. 11(4): 516-523.

Lesica, Peter; Miles, Scott. 2001. Natural history and invasion of Russian olive along eastern Montana Rivers. Western North American Naturalist. 61(1): 1-10.

Levine, C. M.; Stromberg, J. C. 2001. Effects of flooding on native and exotic plant seedlings: implications for restoring southwestern riparian forests by manipulating water and sediment flows. Journal of Arid Environments. 49(1): 111-131.

Levitt, J. 1980. Responses of plants to environmental stresses, volume I. Chilling, freezing, and high temperature stresses. New York: Academic Press. 497 p.

Lewis, Henry T. 1993. Patterns of Indian burning in California: ecology and ethnohistory. In: Blackburn, Thomas C.; Anderson, Kat, eds. Before the wilderness: environmental management by native Californians. Menlo Park, CA: Ballena Press: 55-58.

Ley, Ruth E.; D'Antonio, Carla M. 1998. Exotic grass invasion alters potential rates of N fixation in Hawaiian woodlands. Oecologia. 113: 179-187.

Li, Hiram W. 1995. Non-native species. In: Laroe, Edward T.; Farris, Gaye S.; Puckett, Catherine E.; [and others], eds. Our living resources: a report to the nation on the distribution, abundance, and health of U.S. plants, animals, and ecosystems. Washington, DC: U.S. Department of the Interior, National Biological Service: 427-428.

Lincoln, Roger; Boxshall, Geoff; Clark, Paul. 1998. A dictionary of ecology, evolution and systematics. 2nd ed. Cambridge, UK: Cambridge University Press. 361 p.

Lippincott, C. L. 2000. Effects of *Imperata cylindrical* (L.) Beauv. (cogongrass) invasion on fire regime in Florida sandhill (USA). Natural Areas Journal. 20: 140-149.

Littell, Ramon C.; Milliken, George A.; Stroup, Walter W.; Wolfinger, Russell D.; Schabenberger, Oliver. 2006. SAS for mixed models, 2nd ed. Cary, NC: SAS Institute Inc. 813p.

Little, E. L. 1971. Atlas of United States Trees: Volume 1. Conifers and Important Hardwoods. Division of Timber Management Research, USDA Forest Service. Washington, DC. Miscellaneous Publication 1146. 25 p.

Livingston, R. B.; Allessio, Mary L. 1968. Buried viable seed in successional field and forest stands, Harvard Forest, Massachusetts. Bulletin of the Torrey Botanical Club. 95(1): 58-69.

Lockwood, J. L.; Cassay, P.; Blackburn, T. 2005. The role of propagule pressure in explaining species invasions. Trends in Ecology and Evolution. 20: 223-228.

Lodge, David M.; Williams, Susan; MacIsaac, Hugh J.; Hayes, Keith R.; Leung, Brian; Reichard, Sarah; Mack, Richard N.; Moyle, Peter B.; Smith, Maggie; Andow, David A.; Carlton, James T.; McMichael, Anthony. 2006. Biological invasions: Recommendations for U.S. Policy and Management. Ecological Applications. 16(6): 2035-2054.

Loh, Rhonda. 2004. Fire History of Hawaii Volcanoes National Park. [Unpublished database]. On file at: US Department of the Interior, National Park Service, Hawai`i Volcanoes National Park, Resources Management Division. Volcano, Hawai`i. 4 pages.

Loh, Rhonda; McDaniel, Sierra; Benitez, David; Schultz, Matthew; Palumbo, David; Ainsworth, Alison; Smith, Kimberly; Tunison, Tim; Vaidya, Maya. 2007. Rehabilitation of seasonally dry 'ohi'a woodlands and mesic koa forest following the Broomsedge Fire, Hawai`i Volcanoes National Park. Technical Report No. 147. Pacific Cooperative Studies Unit, University of Hawai'i, Honolulu, HI. 21 pg.

Loh, Rhonda; Tunison, J. Timothy; Benitez, David; McDaniel, Sierra.; Schultz, Matthew; Smith, Kimberly; Vaidya, Maya. 2004. Broomsedge Burn: Hawaii Volcanoes National Park, Burned Area Emergency Rehabilitation. Final Accomplishment Report prepared for the National Park Service. Unpublished Report on file at Hawai`i Volcanoes National Park, Hawai`i Volcanoes National Park, HI. 18 p.

Long, R. W. 1974. Vegetation of southern Florida. Florida Scientist. 37: 33-45.

Lonsdale, W. M. 1999. Global patterns of plant invasions and the concept of invasibility. Ecology. 80(5): 1522-1536.

Lonsdale, W. M.; Miller, I. L. 1993. Fire as a management tool for a tropical woody weed: Mimosa pigra in North Australia. Journal of Environmental Management. 33: 7-87.

Loope, Lloyd L. 1998. Hawaii and Pacific islands. In: Mac, M. J.; Opler, P. A.; Haecker, C. E. Puckett; Doran, P. D., eds. Status and trends of the nation's biological resources. Volume 2. Reston, VA: U.S. Department of the Interior, U.S. Geological Survey: 747–774.

Loope, Lloyd L. 2004. The challenge of effectively addressing the threat of invasive species to the National Park System. Park Science. 22: 14-20.

Loope, Lloyd L.; Dunevitz, V. L. 1981. Impact of fire exclusion and invasion of Schinus terebinthifolius on limestone rockland pine forests of southeastern Florida. Report T-645. Homestead, FL: U.S. Department of the Interior, Everglades National Park, South Florida Research Center. 30 p.

Loope, Lloyd L.; Medeiros, Art; Nagata, Ron; LaRosa, Anne Marie; Sanchez, Pete; Newton, Karen; Hill, Greg. 1990. Haleakala National Park Fire Management Plan. Unpublished report on file at Haleakala National Park, Makawao, HI. 75 p.

Lorimer, Craig G. 1977. The presettlement forest and natural disturbance cycle of northeastern Maine. Ecology. 58: 139-148.

Lorimer, Craig G. 1993. Causes of the oak regeneration problem. In: Oak Regeneration: Serious Problems, Practical Recommendations. Symposium Proceedings Knoxville, TN 8-10 Sept. 1992. Loftis, D.; McGee, C. E., eds. Asheville, NC: U. S. Department of Agriculture, Forest Service, Southeastern Forest Experiment Station: 14-39.

Lotan, James E.; Alexander, Martin E.; Arno, Stephen F.; [and others]. 1981. Effects of fire on flora: A state-of-knowledge review: Proceedings of the national fire effects workshop; 1978 April 10-14; Denver, CO. Gen. Tech. Rep. WO-16. Washington, DC: U.S. Department of Agriculture, Forest Service. 71 p.

Louisiana State University. 2001. *Ligustrum sinense* Lour In: Louisiana invasive plants, [Online]. AgCenter Research and Extension (Producer). Available: http://www.lsuagcenter.com/invasive/chineseprivet.asp [2003, March 14].

Luken, J. O. 2003. Invasions of forests in the eastern United States. In: Gilliam, Frank S. and Roberts Mark R. (eds.). The herbaceous layer in forests of eastern North America. Oxford University Press, New York.

Luken, James O.; Shea, Margaret. 2000. Repeated prescribed burning at Dinsmore Woods State Nature Preserve (Kentucky, USA): responses of the understory community. Natural Areas Journal. 20(2): 150-158.

Lumer, Cecile; Yost, Susan E. 1995. The reproductive biology of *Vincetoxicum nigrum* (L.) Moench (Asclepiadaceae), a Mediterranean weed in New York state. Bulletin of the Torrey Botanical Club. 122(1): 15-23.

Lundgren, Marjorie R.; Small, Christine J.; Dreyer, Glenn D. 2004. Influence of land use and site characteristics on invasive plant abundance in the Quinebaug Highlands of southern New England. Northeastern Naturalist. 11(3): 313-332.

Lunt, I. D. 1990. Impact of an autumn fire on a long-grazed *Themeda trian*dra (kangaroo grass) grassland: implications for management of invaded, remnant vegetations. Victorian Naturalist. 107: 45-51.

Lutes, D.C.: Benson, N. C.; Keifer, M.; Caratti, J.F.; Johnson, K.A. [In press]. FFI: An interagency fire ecology monitoring tool. In: EastFIRE conference proceedings. June 5-8, 2007; Fairfax, VA: George Mason University.

Lutes, Duncan C.; Keane, Robert, E.; Caratti, John. F.; Key, Carl H.; Benson, Nathan C.; Sutherland, Steve; Gangi, Larry J. 2006. FIREMON: Fire Effects Monitoring and Inventory System. Gen. Tech. Rep. RMRS-GTR-164-CD. Fort Collins, CO: U.S. Department of Agriculture, Forest Service, Rocky Mountain Research Station. 400 p.

Lutz, H. J. 1956. Ecological effects of forest fires in the interior of Alaska. Tech. Bull. No. 1133. Washington, DC: U.S. Department of Agriculture, Forest Service. 121 p.

Lym, R. G.; Tober, D. A. 1997. Competitive grasses for leafy spurge reduction. Weed Technology. 11: 787-792.

Lynch, N. 2003. An evaluation of post-fire rehabilitation of Nevada rangelands. Reno, NV: The University of Nevada. 70 p. Thesis.

Lyon, L. Jack; Brown, James K.; Smith, Jane Kapler. 2000a. Introduction. In: Smith, Jane Kapler, ed. Wildland fire in ecosystems: Effects of fire on fauna. RMRS-GTR-42-vol. 1. Ogden, UT: U.S. Department of Agriculture, Forest Service, Rocky Mountain Research Station: 1-7.

Lyon, L. Jack; Crawford, Hewlette S.; Czuhai, Eugene; Fredriksen, Richard L.; Harlow, Richard F.; Metz, Louis J.; Pearson, Henry A. 1978. Effects of fire on fauna: A state-of-knowledge review: Proceedings of the national fire effects workshop; 1978 April 10-14; Denver, CO. Gen. Tech. Rep. WO-6. Washington, DC: U.S. Department of Agriculture, Forest Service. 41 p.

Lyon, L. Jack; Hooper, Robert G.; Telfer, Edmund S.; Schreiner, David Scott. 2000b. Fire effects on wildlife foods. In: Smith, Jane Kapler, ed. Wildland fire in ecosystems: Effects of fire on fauna. RMRS-GTR-42-vol. 1. Ogden, UT: U.S. Department of Agriculture, Forest Service, Rocky Mountain Research Station: 51-58.

Lyon, L. Jack; Stickney, Peter F. 1976. Early vegetal succession following large northern Rocky Mountain wildfires. In: Proceedings, Tall Timbers fire ecology conference and Intermountain Fire Research Council fire and land management symposium; 1974 October 8-10; Missoula, MT. No. 14. Tallahassee, FL: Tall Timbers Research Station: 355-373

Lyons, K. E. 1998. Element stewardship abstract: Phalaris arundinacea L. (reed canarygrass), [Online]. In: The global invasive species initiative. Arlington, VA: The Nature Conservancy (Producer). Available: http://tncweeds.ucdavis.edu/esadocs/documnts/phalaru.html [2005, November 26].

Ma, Jinshuang; Moore, Gerry. 2004. Euonymus alatus. In: Francis, John K., ed. Wildland shrubs of the United States and its territories: thamnic descriptions: volume 1. Gen. Tech. Rep. IITF-GTR-26. San Juan, PR: U.S. Department of Agriculture, Forest Service, International Institute of Tropical Forestry, and Fort Collins, CO: U.S. Department of Agriculture, Forest Service, Rocky Mountain Research Station: 331-332.

MacDonald, Neil W.; Bosscher, Peter J.; Mieczkowski, Christopher A.; Sauter, Emily M.; Tinsley, Brenda J. 2001. Pre- and post-germination burning reduces establishment of spotted knapweed seedlings. Ecological Restoration. 19(4): 262-263.

MacDougall, Andrew S. 2005. Responses of diversity and invasibility to burning in a northern oak savanna. Ecology. 86(12): 3354-3363.

MacDougall, Andrew S.; Turkington, Roy. 2004. Relative importance of suppression-based and tolerance-based competition in an invaded oak savanna. Journal of Ecology. 92(3): 422-434.

Mack, M. C.; D'Antonio, C. M. 1998. Impacts of biological invasions on disturbance regimes. Trends in Ecology and Evolution. 13: 195-198.

Mack, Michelle C.; D'Antonio, Carla M. 2003. Exotic grasses alter controls over soil nitrogen dynamics in a Hawaiian woodland. Ecological Applications. 13(1): 154-166.

Mack, Michelle; D'Antonio, Carla; Ley, Ruth. 2001. Pathways through which exotic grasses alter ecosystem N cycling in a Hawaiian forest. Ecological Applications. 3(1): 69-73.

Mack, R. N. 1989. Temperate grasslands vulnerable to invasions: characteristics and consequences. In: Drake, J. A.; Mooney, H. A.; DiCastri, F.; Groves, R. H.; Kruger, F. J.; Rejmánek, M.; Williamson, M., eds. Biological invasions: a global perspective. New York: Wiley and Sons: 155-179.

Mack, Richard N. 1996. Predicting the identity and fate of plant invaders; emergent and emerging approaches. Biological Conservation. 78: 107-121.

Mack, Richard N.; Simberloff, Daniel; Lonsdale, W. Mark; Evans, Harry; Clout, Michael; Bazzaz, Fakhri A. 2000. Biotic invasions: causes, epidemiology, global consequences, and control. Ecological applications. 10(3): 689-710.

Maddox, D. M.; Mayfield, A.; Portiz, N. H. 1985. Distribution of yellow starthistle (*Centaurea solstitialis*) and Russian knapweed (*Centaurea repens*). Weed Science. 33: 315-327.

Maffei, M. D. 1991. Melaleuca control on Arthur R. Marshall Loxahatchee National Wildlife Refuge. In: Center, T. D.; Doren, R. F.; Hofstetter, R. L.; Myers, R. L.; Whiteaker, L. D., eds. Proceedings of the symposium on exotic pest plants. U.S. National. Park Service Tech. Rep. NPS/NREVER/NRTR-91/06. 197-207.

Maguire, J. 1995. Restoration plan for Dade County's pine rocklands following Hurricane Andrew. Miami, FL: Dade County Department of Environmental Resources Management Publication. 32 p.
Maine Department of Conservation, Natural Areas Program. 2004. Maine invasive plant fact sheet: Shrubby honeysuckles. Augusta, ME: Maine Natural Areas Program. 2 p.
Major, J.; McKell, C. M.; Berry, L. J. 1960. Improvement of medusahead-infested rangeland. California Agricultural Experiment Station, Davis, CA: University of California Cooperative Extension. Leaflet No. 123. 6 p.
Malanson, G. P. 1984. Fire history and patterns of Venturan subassociations of Californian coastal sage scrub. Vegetatio. 57: 121-128.
Malanson, G. P.; Westman, W. E. 1985. Postfire succession in coastal sage scrub: the role of continual basal sprouting. The American Midland Naturalist. 113: 309-318.
Maret, Mary P. 1997. Effects of fire on seedling establishment in upland prairies of the Willamette Valley, Oregon: Oregon State University. 159 p. Thesis.
Maret, Mary P.; Wilson, Mark V. 2000. Fire and seedling population dynamics in western Oregon prairies. Journal of Vegetation Science. 11: 307-314.
Marietta, K. L.; Britton, C. M. 1989. Establishment of seven high yielding grasses on the Texas High Plains. Journal of Range Management. 42(2): 289-294.
Marks, Marianne; Lapin, Beth; Randall, John. 1993. Element stewardship abstract: *Phragmites australis* (common reed), [Online]. In: The global invasive species initiative. Arlington, VA: The Nature Conservancy (Producer). Available: http://tncweeds.ucdavis.edu/esadocs/documnts/phraaus.html [2005, November 26].
Marks, P. L. 1974. The role of pin cherry (*Prunus pensylvanica* L.) in the maintenance of stability in northern hardwood ecosystems. Ecological Monographs. 44: 73-88.
Marks, P. L. 1983. On the origin of the field plants of the northeastern United States. The American Naturalist. 122(2): 210-228.
Maron J. L.; Connors, P. G. 1996. A native nitrogen-fixing shrub facilitates weed invasion. Oecologia. 105: 302-312.
Marsh, Michael A.; Fajvan, Mary Ann; Huebner, Cynthia D.; Schuler, Thomas M. 2005. The effects of timber harvesting and prescribed fire on invasive plant dynamics in the central Appalachians. In: Gottschalk, Kurt W., ed. Proceedings, 16th U.S. Department of Agriculture interagency research forum on gypsy moth and other invasive species 2005; 2005 January 18-21; Annapolis, MD. Gen. Tech. Rep. NE-337. Newton Square, PA: U.S. Department of Agriculture, Forest Service, Northeastern Research Station: 64.
Martin, Eveline L.; Martin, Ian D. 1999. Prescribed burning and competitive reseeding as tools for reducing yellow starthistle (*Centaurea solstitialis*) at Pinnacles National Monument. In: Kelly, Mike; Howe, Melanie; Neill, Bill, eds. Proceedings, California Exotic Pest Plant Council; 1999 October 15-17; Sacramento, CA. Volume 5. [Davis, CA]: California Exotic Pest Plant Council: 51-56.
Martin, J. Lynton. 1956. An ecological survey of burned-over forest land in southwestern Nova Scotia. Forestry Chronicle. 31(2): 313-335.
Martin, Patrick H.; Marks, Peter L. 2006. Intact forests provide only weak resistance to a shade-tolerant invasive Norway maple (*Acer platanoides* L.). Journal of Ecology. 94: 1070-1079.
Martin, Robert E. 1982. Fire history and its role in succession. In: Means, Joseph E., ed. Forest succession and stand development research in the Northwest: Proceedings of a symposium; 1981 March 26; Corvallis, OR. Corvallis, OR: Oregon State University, Forest Research Laboratory: 92-99.
Martin, Robert E.; Anderson, Hal E.; Boyer, William D.; [and others]. 1979. Effects of fire on fuels: A state-of-knowledge review: Proceedings of the national fire effects workshop; 1978 April 10-14; Denver, CO. Gen. Tech. Rep. WO-13. Washington, DC: U.S. Department of Agriculture, Forest Service. 64 p.
Martin, S. Clark. 1983. Responses of semidesert grasses and shrubs to fall burning. Journal of Range Management. 36(5): 604-610.
Martin, Tunyalee. 2000. Weed alert!: *Euonymus alatus* (Thunb.) Siebold, [Online]. In: Weed alert archives. In: The global invasive species initiative. Arlington, VA: The Nature Conservancy (Producer). Available: http://tncweeds.ucdavis.edu/alert/alrteuon.html [2006, August 28].
Martin-Rivera, Martha; Cox, Jerry R.; Ibarra-F, F.; Alston, Diana G.; Banner, Roger E.; Malecheck, John C. 1999. Spittlebug and buffelgrass responses to summer fires in Mexico. Journal of Range Management. 52: 621-625.
Martinson, E. J.; Omi, P. N. 2003. Performance of fuel treatments subject to wildfires. In: Omi, Philip N.; Joyce, Linda A., tech. eds. Fire, fuel treatments, and ecological restoration: conference proceedings; 2002 April 16-18; Fort Collins, CO. Proceedings RMRS-P-29. Fort Collins, CO: U.S. Department of Agriculture, Forest Service, Rocky Mountain Research Station: 7-13.
Martinson, E. J.; Omi, P. N.; Shepperd, W. D. 2003. Fire behavior, fuel treatments, and fire suppression on the Hayman fire. Part 3: Effects of fuel treatments on fire severity. In: Graham, R.L., ed. Hayman fire case study. Gen. Tech. Rep. RMRS-GTR-114. Ogden, UT: U.S. Department of Agriculture, Forest Service, Rocky Mountain Research Station: 96-126.
Marty, J. 2003. [Personal communication]. June 2003. The Nature Conservancy. Jepson Prairie, CA.
Marty, Jaymee Theresa. 2002. Spatially-dependent effects of fire and grazing in a California annual grassland plant community (Chapter 3). Davis, CA: University of California. 116 p. Dissertation.
Masters, R. A.; Beran, D. D.; Gaussoin, R. E. 2001. Restoring tallgrass prairie species mixtures on leafy spurge-infested rangeland. Journal of Range Management. 54: 362-369.
Masters, Robert A. 1994. Response of leafy spurge to date of burning. In: Ohlsen, Nancy S., comp. Proceedings: leafy spurge strategic planning workshop; 1994 March 29-30; Dickinson, ND. [Publisher unknown]: 102-105.
Masters, Robert A.; Nissen, Scott J. 1998. Revegetating leafy spurge (*Euphorbia esula*)-infested rangeland with native tallgrasses. Weed Technology. 12(2): 381-390.
Masters, Robert A.; Nissen, Scott J.; Gaussoin, Roch E.; Beran, Daniel D.; Stougaard, Robert N. 1996. Imidazolinone herbicides improve restoration of Great Plains grasslands. Weed Technology. 10(2): 392-403.
Masters, Robert A.; Vogel, Kenneth P. 1989. Remnant and restored prairie response to fire, fertilization, and atrazine. In: Bragg, Thomas B.; Stubbendieck, James, eds. Prairie pioneers: ecology, history and culture: Proceedings, 11th North American prairie conference; 1988 August 7-11; Lincoln, NE. Lincoln, NE: University of Nebraska: 135-138.
Mata-Gonzalez, R.; Hunter, R. G.; Coldren, C. L.; McLendon, T.; Paschke, M. W. 2007. Modelling plant growth dynamics in sagebrush steppe communities affected by fire. Journal of Arid Environments. 69(1): 144-157.
Matlack, G. R. 2002. Exotic plant species in Mississippi, USA: critical issues in management and research. Natural Areas Journal. 22: 241-247.
Matlack, Glenn R.; Good, Ralph E. 1990. Spatial heterogeneity in the soil seed bank of a mature coastal plain forest. Bulletin of the Torrey Botanical Club. 117(2): 143-152.
Mau-Crimmins, Theresa M.; Schussman, Heather R.; Geiger, Erika L. 2005. Can the invaded range of a species be predicted sufficiently using only native-range data? Lehmann lovegrass (*Eragrostis lehmanniana*) in the southwestern United States. Ecological Modelling. 193(3-4): 736-746.
Maurer, B. A. 1999. Untangling ecological complexity: the macroscopic perspective. Chicago, IL: University of Chicago Press: 251 p.
Mayeux, H. S., Jr.; Hamilton, W. T. 1983. Response of common goldenweed (*Isocoma coronopifolia*) and buffelgrass (*Cenchrus ciliaris*) to fire and soil-applied herbicides. Weed Science. 31(3): 355-360.
McAuliffe, J. R. 1995. The aftermath of wildfire in the Sonoran Desert. Sonoran Quarterly. 49: 4-8.
McCament, Corinne L.; McCarthy, Brian C. 2005. Two-year response of American chestnut (*Castanea dentata*) seedlings to shelterwood harvesting and fire in a mixed-oak forest ecosystem. Canadian Journal of Forest Research. 35: 740-749.
McCarthy, Brian C. 1997. Response of a forest understory community to experimental removal of an invasive nonindigenous plant (*Alliaria petiolata*, Brassicaceae). In: Luken, James O.; Thieret, John W., eds. Assessment and management of plant invasions. New York: Springer-Verlag: 117-130.

McCleery, D. W. 1993. American forests–a history of resiliency and recovery. Durham, NC: U.S. Department of Agriculture, Forest Service. 58 p. In cooperation with: The Forest History Society.

McClendon, Terry; Redente, Edward F. 1992. Effects of nitrogen limitation on species replacement dynamics during early secondary succession on a semiarid sagebrush site. Oecologia. 91: 312-317.

McConnell, B. E.; Smith, J. G. 1965. Understory response three years after thinning pine. Journal of Range Management. 18: 129-132.

McConnell, B. E.; Smith, J. G. 1970. Response of understory vegetation to ponderosa pine thinning in eastern Washington. Journal of Range Management. 23: 208-212.

McCoy, S. D.; Mosley, J. C.; Engle, D. M. 1992. Old world bluestem seedlings in western Oklahoma. Rangelands. 14(1): 41-44.

McCune, Bruce. 1978. First-season fire effects on intact palouse prairie. Unpublished report on file with: U.S. Department of Agriculture, Forest Service, Rocky Mountain Research Station, Fire Sciences Laboratory, Missoula, MT. 12 p.

McDaniel, K. C.; Taylor, J. P. 1999. Steps for restoring bosque vegetation along the middle Rio Grande of New Mexico. In: Eldridge, D.; Freudenberger, D., eds. People and rangelands: building the future: Proceedings, 6th international rangeland congress; 1999 July 19-23; Queensland, Australia. Volume 1 & 2. Aitkenvale, Queensland: The Congress: 713-714.

McDaniel, Kirk C.; Taylor, John P. 2003. Saltcedar recovery and herbicide-burn and mechanical clearing practices. Journal of Range Management. 56(5): 439-445.

McDaniel, Sierra. 2003. Koa forest rehabilitation on lower Mauna Loa. [Unpublished paper]. On file at: U.S. Department of the Interior, National Park Service, Hawai`i Volcanoes National Park, Division of Natural Resources Management, Hawai`i Volcanoes National Park, HI.

McDonald, Robert I.; Urban, Dean L. 2006. Edge effects on species composition and exotic species abundance in the North Carolina piedmont. Biological Invasions. 8: 1049-1060.

McEldowney, Holly. 1979. Archeological and historical literature search and research design. Lava Flow Control Study, Hilo, Hawaii. Honolulu, HI: University of Hawai`i, Anthropology Dept. Manuscript # 050879.

McEvoy, Peter B.; Rudd, Nathan T. 1993. Effects of vegetation disturbances on insect biological control of tansy ragwort, *Senecio jacobaea*. Ecological Applications. 3(4): 682-698.

McEvoy, Peter; Rudd, Nathan T.; Cox, Caroline S.; Huso, Manuela. 1993. Disturbance, competition, and herbivory effects on ragwort *Senecio jacobaea* populations. Ecological Monographs. 63(1): 55-75.

McFarland, J. Brent; Mitchell, Rob. 2000. Fire effects on weeping lovegrass tiller density and demographics. Agronomy Journal. 92: 42-47.

McGlone, Christopher M.; Huenneke, Laura F. 2004. The impact of a prescribed burn on introduced Lehmann lovegrass versus native vegetation in the northern Chihuahuan Desert. Journal of Arid Environments. 57(3): 297-310.

McGowan-Stinski, Jack. 2001. [Email to Kristin Zouhar]. October 11. Lansing, MI: The Nature Conservancy, Michigan Chapter. On file at: U.S. Department of Agriculture, Forest Service, Rocky Mountain Research Station, Fire Sciences Lab, Missoula, MT; RWU 4403 files.

McGowan-Stinski, Jack. 2006. Removal method for seedling buckthorn (*Rhamnus* spp.), [Online]. In: The global invasive species initiative. Arlington, VA: The Nature Conservancy (Producer). Available: http://tncweeds.ucdavis.edu/moredocs/rhaspp01.doc [2007, July 12].

McKell, Cyrus M.; Wilson, Alma M.; Kay, B. L. 1962. Effective burning of rangelands infested with medusahead. Weeds. 10(2): 125-131.

McKelvey, Kevin S.; Busse, Kelly K. 1996. Twentieth-century fire patterns on Forest Service lands. In: Status of the Sierra Nevada. Sierra Nevada Ecosystem Project: Final report to Congress. Volume II: Assessments and scientific basis for management options. Wildland Resources Center Report No. 37. Davis, CA: University of California, Centers for Water and Wildland Resources: 1119-1138.

McKelvey, Kevin S.; Skinner, Carl N.; Chang, Chi-ru; [and others]. 1996. An overview of fire in the Sierra Nevada. In: Status of the Sierra Nevada. Sierra Nevada Ecosystem Project: Final report to Congress. Volume II: Assessments and scientific basis for management options. Wildland Resources Center Report No. 37. Davis, CA: University of California, Centers for Water and Wildland Resources: 1033-1040.

McKenzie, Donald; Gedalof, Ze-ev; Peterson, David L.; Mote, Philop. 2004. Climate change, wildfire, and conservation. Conservation Biology. 18(4): 890-902.

McKinney, M. L. 2002. Influence of settlement time, human population, park shape, and age, visitation and roads on the number of alien plant species in protected areas in the USA. Diversity and Distributions. 8: 311-318.

McLaughlin, S. P.; Bowers, J. 1982. Effects of wildfire on a Sonoran Desert plant community. Ecology. 63(1): 246-248.

McPherson, Guy R. 1995. The role of fire in the desert grasslands. In: McClaran, Mitchel P.; Van Devender, Thomas R., eds. The desert grassland. Tucson, AZ: The University of Arizona Press: 130-151.

McPherson, Guy R. 1997. Ecology and management of North American savannas. Tucson, AZ: University of Arizona Press.

McPherson, Guy R. 2001. Invasive plants and fire: integrating science and management. In: Galley, Krista E. M.; Wilson, Tyrone P., eds. Proceedings of the invasive species workshop: The role of fire in the control and spread of invasive species; Fire conference 2000: the 1st national congress on fire ecology, prevention, and management; 2000 November 27-December 1; San Diego, CA. Misc. Publ. No. 11. Tallahassee, FL: Tall Timbers Research Station: 141-146.

McPherson, Guy R.; Wade, Dale D.; Phillips, Clinton B., comps.. 1990. Glossary of wildland fire management terms used in the United States. SAF-90-05. Washington, DC: Society of American Foresters. 138 p.

McPherson, Guy R.; Weltzin, Jake F. 2000. Disturbance and climate change in United States/Mexico borderland plant communities: a state-of-knowledge review. Gen. Tech. Rep. RMRS-GTR-50. Fort Collins, CO: USDA, Forest Service, Rocky Mountain Research Station. 24 p.

McWhiney, G. 1988. Cracker culture: Celtic ways in the old south. Tuscaloosa, AL: University of Alabama Press. [Pages unknown].

McWilliams, John D. 2004. *Arundo donax*. In: Fire Effects Information System, [Online]. U.S. Department of Agriculture, Forest Service, Rocky Mountain Research Station, Fire Sciences Laboratory (Producer). Available: http://www.fs.fed.us/database/feis/ [2007, August 9].

Meentemeyer, R.; Rizzo, D. M.; Mark, W.; Lotz, E. 2004. Mapping the risk of establishment and spread of Sudden Oak Death in California. Forest Ecology and Management. 200: 195-214.

Mehrhoff, L. J.; Silander, Jr., J. A.; Leicht, S. A.; Mosher, E. S.; Tabak, N. M. 2003. IPANE: Invasive Plant Atlas of New England, [Online]. In: Storrs, CT: Department of Ecology and Evolutionary Biology, University of Connecticut (Producer). Available: http://invasives.eeb.uconn.edu/ipane/ [2007, July 12].

Meiners, Scott J.; Pickett, Steward T. A.; Cadenasso, Mary L. 2002. Exotic plant invasions over 40 years of old field successions: community patterns and associations. Ecography. 25(2): 215-233.

Melgoza, Graciela; Nowak, Robert S.; Tausch, Robin J. 1990. Soil water exploitation after fire: competition between *Bromus tectorum* (cheatgrass) and two native species. Oecologia. 83(1): 7-13.

Menakis, James P.; Osborne, Dianne; Miller, Melanie. 2003. Mapping the cheatgrass-caused departure from historical natural fire regimes in the Great Basin, USA. In: Omi, Philip N., and Joyce, Linda A., tech. eds. Proceedings: Fire, fuel treatments, and ecological restoration: Conference proceedings Fort Collins, CO. Fort Collins, CO: U.S. Department of Agriculture, Forest Service, Rocky Mountain Research Station: 281-287.

Menke, John W. 1992. Grazing and fire management for native perennial grass restoration in California grasslands. Fremontia. 20(2): 22-25.

Mensing, S.; Byrne, R. 1999. Invasion of Mediterranean weeds into California before 1769. Fremontia. 27: 6-9.

Menvielle M. F.; Scopel, A. L. 1999. Life history traits of an alien tree invading a national park in temperate South America. 84th

Annual Meeting, Ecological Society of America. Book of abstracts. [Publication location unknown]: [Publisher unknown] 282.

Merriam, Kyle E.; Keeley, Jon E.; Beyers, Jan L. 2006. Fuel breaks affect nonnative species abundance in Californian plant communities. Ecological Applications. 16(2): 515-527.

Merrill, E. H.; Maryland, H. F.; Peek, J. M. 1980. Effects of a fall wildfire on herbaceous vegetation on xeric sites in the Selway-Bitterroot Wilderness, Idaho. Journal of Range Management. 33(5): 363-367.

Meskimen, George F. 1962. A silvical study of the melaleuca tree in south Florida. Gainesville, FL: University of Florida. 178 p. Thesis.

Messier, Christian; Parent, Sylvian; Bergeron, Yves. 1998. Effects of overstory and understory vegetation on the understory light environment in mixed boreal forests. Journal of Vegetation Science. 9: 511-520.

Messinger, Richard Duane. 1974. Effects of controlled burning on waterfowl nesting habitat in northwest Iowa. Ames, IA: Iowa State University. 49 p. Thesis.

Metlen, K. L.; Fiedler, C. E.; Youngblood, A. 2004. Understory responses to fuel reduction treatments in the Blue Mountains of northeastern Oregon. Northwest Science. 78: 175-185.

Metlen, Kerry L.; Fiedler, Carl E. 2006. Restoration treatment effects on the understory of ponderosa pine/Douglas-fir forests in western Montana, USA. Forest Ecology and Management. 222: 355-369.

Meyer, Marc D.; Schiffman, Paula M. 1999. Fire season and mulch reduction in a California grassland: a comparison of restoration strategies. Madroño. 46(1): 25-37.

Meyer, Rachelle. 2005a. Schinus terebinthifolius. In: Fire Effects Information System, [Online]. U.S. Department of Agriculture, Forest Service, Rocky Mountain Research Station, Fire Sciences Laboratory (Producer). Available: http://www.fs.fed.us/database/feis/ [2006, September 18].

Meyer, Rachelle. 2005b. *Triadica sebifera*. In: Fire Effects Information System, [Online]. U.S. Department of Agriculture, Forest Service, Rocky Mountain Research Station, Fire Sciences Laboratory (Producer). Available: http://www.fs.fed.us/database/feis/ [2007, January 30].

Meyer, Susan E.; Garvin, Susan C.; Beckstead, Julie. 2001. Factors mediating cheatgrass invasion of intact salt desert shrubland. In: McArthur, E. Durant; Fairbanks, Daniel J., comps.. Shrubland ecosystem genetics and biodiversity: proceedings; 2000 June 13-15; Provo, UT. Proc. RMRS-P-21. Ogden, UT: U.S. Department of Agriculture, Forest Service, Rocky Mountain Research Station: 224-232.

Miles, D. W. R.; Swanson, F. J. 1986. Vegetation composition on recent landslides in the Cascade Mountains of western Oregon. Canadian Journal of Forestry. 16: 739-744.

Miller, Heather C.; Clausnitzer, David; Borman, Michael M. 1999. Medusahead. In: Sheley, Roger L.; Petroff, Janet K., eds.. Biology and Management of Noxious Rangeland Weeds. Corvallis, OR: Oregon State University Press: 271-281.

Miller, J. H. 1988. Guidelines for kudzu eradication treatments. In: Miller, J. H.; Mitchell, R. J., eds. A manual on ground applications of forestry herbicides. Atlanta, GA: U.S. Department of Agriculture, Forest Service, Southern Region: 6-1 to 6-7.

Miller, James H. 2003. Nonnative invasive plants of southern forests: A field guide for identification and control. Gen. Tech. Rep. SRS-62, [Online]. Asheville, NC: U.S. Department of Agriculture, Forest Service, Southern Research Station (Producer). 93 p. Available: http://www.srs.fs.usda.gov/pubs/gtr/gtr_srs062/ [2007, July 12].

Miller, Margaret M.; Miller, Joseph W. 1976. Succession after wildfire in the North Cascades National Park complex. In: Proceedings, annual Tall Timbers fire ecology conference: Pacific Northwest; 1974 October 16-17; Portland, OR. No. 15. Tallahassee, FL: Tall Timbers Research Station: 71-83.

Miller, Melanie. 2000. Fire autecology. In: Brown, James K.; Smith, Jane Kapler, eds. Wildland fire in ecosystems: Effects of fire on flora. Gen. Tech. Rep. RMRS-GTR-42-vol. 2. Ogden, UT: U.S. Department of Agriculture, Forest Service, Rocky Mountain Research Station: 9-34.

Miller, Melanie. 2007. [Review comments to Kristin Zouhar]. April 27. Missoula, MT: U.S. Department of Agriculture, Forest Service, Rocky Mountain Research Station, Fire Sciences Laboratory.

Miller, R. F.; Heyerdahl, E. K. 2008. Fine-scale variation of historical fire regimes in sagebrush-steppe and juniper woodland: An example from California, USA. International Journal of Wildland Fire. 17: 245-254.

Miller, R. F.; Wigand, P. E. 1994. Holocene changes in semiarid pinyon-juniper woodlands. Bioscience. 44(7): 465-474.

Miller, Richard F.; Rose, Jeffery A. 1995. Historic expansion of *Juniperus occidentalis* (western juniper) in southeastern Oregon. The Great Basin Naturalist. 55(1): 37-45.

Miller, Richard F.; Tausch, Robin J. 2001. The role of fire in juniper and pinyon woodlands: a descriptive analysis. In: Galley, Krista E. M.; Wilson, Tyrone P., eds. Proceedings of the invasive species workshop: The role of fire in the control and spread of invasive species; Fire conference 2000: the first national congress on fire ecology, prevention, and management; 2000 November 27-December 1; San Diego, CA. Misc. Publ. No. 11. Tallahassee, FL: Tall Timbers Research Station: 15-30.

Miner, Brandon. 2000. Moose habitat experiments evaluated, Refuge Notebook, [Online]. Kenai, AK: U.S. Fish and Wildlife Service, Kenai National Wildlife Refuge (Producer). Available: http://kenai.fws.gov/overview/notebook/2000/aug/4aug00.htm [2005, March 27].

Minnich, R. A. 1983. Fire mosaics in southern California and northern Baja California. Science. 219: 1287-1294.

Minnich, R. A. 1989. Chaparral fire history in San Diego County and adjacent northern Baja California: an evaluation of natural fire regimes and the effects of suppression management. In: Keeley, S. C., ed. The California chaparral: paradigms reexamined. Los Angeles, CA: Natural History Museum of Los Angeles County: 37-48.

Minnich, R. A. 2001. An integrated model of two fire regimes. Conservation Biology. 15: 1549-1553.

Minnich, R. A.; Barbour, M. G.; Burk, J. H.; Fernau, R. A. 1995. Sixty years of change in Californian conifer forests of the San Bernardino Mountains. Conservation Biology. 9: 902-914.

Minnich, R. A.; Chou, Y. H. 1997. Wildland fire patch dynamics in the chaparral of southern California and northern Baja California. International Journal of Wildland Fire. 7: 221-248.

Minnich, R. A.; Dezzani, R. J. 1998. Historical decline of coastal sage scrub in the Riverside-Perris Plain, California. Western Birds. 29: 366-391.

Minnich, Ralph. 2006. Planning and implementing prescribed burns. In: DiTomaso, J. M.; Johnson, D. W., eds. The use of fire as a tool for controlling invasive plants. Bozeman, MT: Center for Invasive Plant Management: 3-6.

Mishra, B. K.; Ramakrishnan, P. S. 1983. Secondary succession subsequent to slash and burn agriculture at higher elevations of north-east India. I.—Species diversity, biomass and litter production. Acta Ecologica. 4(2): 95-107.

Mitchell, Laura R.; Malecki, Richard A. 2003. Use of prescribed fire for management of old fields in the Northeast. In: Galley, Krista E. M.; Klinger, Robert C.; Sugihara, Neil G., eds. Proceedings of fire conference 2000: the 1st national congress on fire ecology, prevention, and management; 2000 November 27-December 1; San Diego, CA. Miscellaneous Publication No. 13. Tallahassee, FL: Tall Timbers Research Station: 60-71.

Mitchell, Rob; Dabbert, Brad. 2000. Potential fire effects on seed germination of four herbaceous species. Texas Journal of Agriculture and Natural Research. 13: 99-103.

Mitich, L.W. 1994. Ground ivy. Weed Technology. 8(2): 413-415.

Moerman, Daniel E.; Estabrook, George F. 2006. The botanist effect: counties with maximal species richness tend to be home to universities and botanists. Journal of Biogeography. 33: 1969-1974.

Moffatt, S. F.; McLachlan, S. M.; Kenkel, N. C. 2004. Impacts of land use on riparian forest along an urban-rural gradient in southern Manitoba. Plant Ecology. 174(1): 119-135.

Molles, Manuel C., Jr.; Crawford, Clifford S.; Ellis, Lisa M.; [and others]. 1998. Managed flooding for riparian ecosystem restoration. BioScience. 48(9): 749-756.

Molnar, George; Hofstetter, Ronald H.; Doren, Robert F.; Whiteaker, Louis D.; Brennan, Michael T. 1991. Chapter 17: Management of *Melaleuca quinquenervia* within East Everglades wetlands. In: Center, Ted D., Doren, Robert F., Hofstetter, Ronald L., Myers, Ronald L., Whiteaker, Louis D., eds. Proceedings: Proceedings of the symposium on nonnative pest plants. Technical Report

Miami, FL. Washington, DC: U.S. Department of the Interior, National Park Service: 237-249.
Monaco, T. A.; Osmond, T. M.; Dewey, S. A. 2005. Medusahead control with fall- and spring-applied herbicides on northern Utah foothills. Weed Technology. 19: 653-658.
Monleon, V. J.; Newton, M.; Hooper, C.; Tappeiner, J. C., II. 1999. Ten-year growth response of young Douglas-fir to variable density varnishleaf ceanothus and herb competition. Western Journal of Applied Forestry. 14(4): 208-213.
Monsen, Stephen B.; Stevens, Richard. 1999. Proceedings: Ecology and management of pinyon-juniper communities within the Interior West. RMRS-P-9. Ogden, UT: U.S. Department of Agriculture, Forest Service, Rocky Mountain Research Station.
Mooney, H. A. 1977. Southern coastal scrub. In: Barbour, Michael G.; Major, Jack, eds. Terrestrial vegetation of California. New York: John Wiley & Sons: 471-490.
Mooney, H. A. 2005. Invasive alien species: the nature of the problem. In: Mooney, H. A.; Mack, R. N.; McNeely, J. A.; Neville, L. E.; Schei, P. J.; Waage, J. K., eds. Invasive alien species: a new synthesis. SCOPE 63. Washington, DC: Island Press: 1-15.
Mooney, H. A.; Hamburg, S. P.; Drake, J.A. 1986. The invasion of plants and animals into California. In: Mooney, H. A.; Drake, J. A., eds. Ecology of biological invasions of North America and Hawaii. Ecological Studies 58. New York: Springer-Verlag: 250-272.
Mooney, H. A.; Hobbs, R. J., eds. 2000. Invasive species in a changing world. Washington, DC: Island Press. 457 p.
Mooney, H. A.; Parsons, D. J. 1973. Structure and function in the California chaparral—an example from San Dimas. In: DiCastri, F.; Mooney, H. A., eds. Mediterranean type ecosystems: origin and structure. Heidelberg, Germany: Springer-Verlag: 83-112.
Moore, M. M.; Casey, C. A.; Bakker, J. D.; Springer, J. D.; Fulé, P. Z.; Covington, W. W.; Laughlin, D. C. 2006. Herbaceous vegetation responses (1992-2004) to restoration treatments in a ponderosa pine forest. Rangeland Ecology and Management. 59(2): 135-144.
Moore, P. E.; Gerlach, J. D., Jr. 2001. Exotic Species Threat Assessment in Sequoia, Kings Canyon and Yosemite National Parks. In: Harmon, D., ed. Crossing boundaries in park management: proceedings of the 11th conference on research and resource management in parks and on public land; 2001 April 16-20; Denver, CA. Hancock, MI: The George Wright Society: 96-103.
Moorhead, David J.; Johnson, Kevin D. 2002. Controlling kudzu in CRP stands, [Online]. In: The Bugwood Network—Forest weed control. Athens, GA: University of Georgia, Warnell School of Forestry and Natural Resources, College of Agricultural and Environmental Sciences (Producer). Available: http://bugwood.org/crp/kudzu.html [2004, December 2].
Moran, Robbin C. 1981. Prairie fens in northeastern Illinois: floristic composition and disturbance. In: Stuckey, Ronald L.; Reese, Karen J., eds. The Prairie Peninsula—in the “shadow” of Transeau: Proceedings, 6th North American prairie conference; 1978 August 12-17; Columbus, OH. Ohio Biological Survey Biological Notes No. 15. Columbus, OH: Ohio State University, College of Biological Sciences: 164-171.
Morgan, P.; Neuenschwander, L. F. 1988. Shrub response to high and low severity burns following clearcutting in Northern Idaho. Western Journal of Applied Forestry. 3(1): 5-9.
Morghan, Kimberly J. Reever; Seastedt, Timothy R.; Sinton, Penelope J. 2000. Frequent fire slows invasion of ungrazed tallgrass prairie by Canada thistle. Ecological Restoration. 18(2): 194-195.
Morisawa, TunyaLee. 1999a. Weed notes: *Lespedeza bicolor*, [Online]. In: The global invasive species initiative. Arlington, VA: The Nature Conservancy (Producer). Available: http://tncweeds.ucdavis.edu/moredocs/lesbic01.pdf [2005, January 1].
Morisawa, TunyaLee. 1999b. Weed notes: *Dioscorea bulbifera, D. alata, D. sansibarensis*, [Online]. In: The global invasive species initiative. Arlington, VA: The Nature Conservancy (Producer). Available: http://tncweeds.ucdavis.edu/moredocs/diospp01.pdf [2004, December 29].
Moritz, M. A. 1997. Analyzing extreme disturbance events: fire in the Los Padres National Forest. Ecological Applications. 7: 1252-1262.
Moritz, M. A. 2003. Spatiotemporal analysis of controls on shrubland fire regimes: age dependency and fire hazard. Ecology. 84: 351-361.
Moritz, M. A.; Odion, D. C. 2005. Examining the strength and possible causes of the relationship between fire history and Sudden Oak Death. Oecologia. 144: 106-114.
Morris, William G. 1958. Influence of slash burning on regeneration, other plant cover, and fire hazard in the Douglas-fir region: A progress report. Res. Pap. PNW-29. Portland, OR: U.S. Department of Agriculture, Forest Service, Pacific Northwest Forest and Range Experiment Station. 49 p.
Morris, William G. 1969. Effects of slash burning in overmature stands of the Douglas-fir region. Forest Science. 16: 258-270.
Morrow, L. A.; Stahlman, P. W. 1984. The history and distribution of downy brome (*Bromus tectorum*) in North America. Weed Science. 32(Supplement 1): 2-6.
Morse, L. E.; Randall, J. R.; Benton, N.; Hiebert, R.; Lu, S. 2004. An invasive species assessment protocol: Evaluating non-native plants for their impact on biodiversity. Version 1, [Online]. Arlington, VA: NatureServe (Producer) Available: http://www.natureserve.org/getData/plantData.jsp [2007, August 1].
Morton, J. F. 1962. Ornamental plants with toxic and/or irritant properties. II. Proceedings of the Florida State Horticultural Society. 75: 484-491.
Motzkin, Glenn; Foster, David; Allen, Arthur; Harrod, Jonathan; Boone, Richard. 1996. Controlling site to evaluate history: vegetation patterns of a New England sand plain. Ecological Monographs. 66(3): 345-365.
Mueller-Dombois, Dieter. 1976. The major vegetation types and their ecological zones in Hawaii Volcanoes National Park and their application to park management and research, In: Smith, C. W., ed. First conference in natural science, Hawai`i Volcanoes National Park: Proceedings. Honolulu, HI: University of Hawai`i, Cooperative Parks Resources Studies Unit: 149-161.
Mueller-Dombois, Dieter. 1981. Fire in tropical ecosystems. In: Mooney, H. A.; Bonnicksen, T. M.; Christensen, N. L.; Lotan, J. E.; Reiners, W. A., technical coordinators. Fire regimes and ecosystem properties: Proceedings of the conference; 1978 December 11-15; Honolulu, HI. Gen. Tech. Rep. WO-26. Washington, DC: U.S. Department of Agriculture, Forest Service: 137-176.
Mueller-Dombois, Dieter. 2001. Biological invasion and fire in tropical biomes. In: Galley, Krista E. M.; Wilson, Tyrone P., eds. Proceedings of the invasive species workshop: The role of fire in the control and spread of invasive species; Fire conference 2000: the first national congress on fire ecology, prevention, and management; 2000 November 27-December 1; San Diego, CA. Misc. Publ. No. 11. Tallahassee, FL: Tall Timbers Research Station: 112-121.
Mueller-Dombois, Dieter; Goldhammer, J. G. 1990. Fire in tropical ecosystems and global environmental change: an introduction. In: Goldhammer, J.G., ed. Fire in the tropical biota: ecosystem processes and global challenges. Heidelberg, Germany: Springer-Verlag. 1-10.
Munda, Bruce. 2001. [Personal communication]. March 22, Tucson, Arizona: U.S. Department of Agriculture, Natural Resources Conservation Service, Tucson Plant Materials Center.
Munger, Gregory T. 2001. *Alliaria petiolata*. In: Fire Effects Information System, [Online]. U.S. Department of Agriculture, Forest Service, Rocky Mountain Research Station, Fire Sciences Laboratory (Producer). Available: http://www.fs.fed.us/database/feis/ [2006, February 9].
Munger, Gregory T. 2002a. *Lonicera japonica*. In: Fire Effects Information System, [Online]. U.S. Department of Agriculture, Forest Service, Rocky Mountain Research Station, Fire Sciences Laboratory (Producer). Available: http://www.fs.fed.us/database/feis/ [2005, May 13].
Munger, Gregory T. 2002b. *Pueraria montana* var. *lobata*. In: Fire Effects Information System, [Online]. U.S. Department of Agriculture, Forest Service, Rocky Mountain Research Station, Fire Sciences Laboratory (Producer). Available: http://www.fs.fed.us/database/feis/ [2005, May 13].
Munger, Gregory T. 2002c. *Rose multiflora*. In: Fire Effects Information System, [Online]. U.S. Department of Agriculture, Forest Service, Rocky Mountain Research Station, Fire Sciences Laboratory (Producer). Available: http://www.fs.fed.us/database/feis/ [2005, November 10].
Munger, Gregory T. 2002d. *Lythrum salicaria*. In: Fire Effects Information System, [Online]. U.S. Department of Agriculture,

Forest Service, Rocky Mountain Research Station, Fire Sciences Laboratory (Producer). Available: http://www.fs.fed.us/database/feis/ [2005, November 28].

Munger, Gregory T. 2003a. *Acer platanoides*. In: Fire Effects Information System, [Online]. U.S. Department of Agriculture, Forest Service, Rocky Mountain Research Station, Fire Sciences Laboratory (Producer). Available: http://www.fs.fed.us/database/feis/ [2005, Nov 10].

Munger, Gregory T. 2003b. *Elaeagnus umbellata*. In: Fire Effects Information System, [Online]. U.S. Department of Agriculture, Forest Service, Rocky Mountain Research Station, Fire Sciences Laboratory (Producer). Available: http://www.fs.fed.us/database/feis/ [2005, November 10].

Munger, Gregory T. 2003c. *Ligustrum* spp. In: Fire Effects Information System, [Online]. U.S. Department of Agriculture, Forest Service, Rocky Mountain Research Station, Fire Sciences Laboratory (Producer). Available: http://www.fs.fed.us/database/feis/ [2005, May 13].

Munger, Gregory T. 2004. *Lespedeza cuneata*. In: Fire Effects Information System, [Online]. U.S. Department of Agriculture, Forest Service, Rocky Mountain Research Station, Fire Sciences Laboratory (Producer). Available: http://www.fs.fed.us/database/feis/ [2005, November 4].

Munger, Gregory T. 2005a. *Lonicera* spp. In: Fire Effects Information System, [Online]. U.S. Department of Agriculture, Forest Service, Rocky Mountain Research Station, Fire Sciences Laboratory (Producer). Available: http://www.fs.fed.us/database/feis/ [2005, November 14].

Munger, Gregory T. 2005b. *Melaleuca quinquenervia*. In: Fire Effects Information System, [Online]. U.S. Department of Agriculture, Forest Service, Rocky Mountain Research Station, Fire Sciences Laboratory (Producer). Available: http://www.fs.fed.us/database/feis/ [2006, September 18].

Munz, Philip A.; Keck, David D. 1959. A California flora. Berkeley, CA: University of California Press. 1104 p.

Murphy, A. H.; Lusk, W. C. 1961. Timing medusahead burns. California Agriculture. 15: 6-7.

Murphy, Alfred H.; Turner, David. 1959. A study on the germination of Medusa-head seed. Bulletin. [Sacramento, CA: California Department of Agriculture]; 48: 6-10.

Mutch, R. W.; Philpot, C. W. 1970. Relation of silica content to flammability in grasses. Forest Science. 16(1): 64-65.

Mutch, Robert W.; Arno, Stephen F.; Brown, James K.; [and others]. 1993. Forest health in the Blue Mountains: a management strategy for fire-adapted ecosystems. Gen. Tech. Rep. PNW-GTR-310. Portland, OR: U.S. Department of Agriculture, Forest Service, Pacific Northwest Research Station. 14 p.

Myers, Brenda R.; Walck, Jeffrey L.; Plum, Kurt E. 2004. Vegetation change in a former chestnut stand on the Cumberland Plateau of Tennessee during an 80-year period (1921-2000). Castanea. 69(2): 81-91.

Myers, R. L. 1975. The relationship of site conditions to the invading capability of melaleuca in southern Florida. Gainesville, FL: University of Florida.151 p. Thesis.

Myers, R. L. 1983. Site susceptibility to invasion by the exotic tree *Melaleuca quinquenervia* in southern Florida. Journal of Applied Ecology. 20: 645-658.

Myers, R. L. 1984. Ecological compression of *Taxodium distichum* var. *nutans* by *Melaleuca quinquenervia* in southern Florida. In: Ewel, K. C.; Odum, H. T., eds. Cypress swamps. Gainesville, FL: University Presses of Florida. 358-364.

Myers, Ronald L. 2000. Fire in tropical and subtropical ecosystems. In: Brown, James K.; Smith, Jane Kapler, eds. Wildland fire in ecosystems: Effects of fire on flora. Gen. Tech. Rep. RMRS-GTR-42-vol. 2. Ogden, UT: U.S. Department of Agriculture, Forest Service, Rocky Mountain Research Station: 161-173.

Myers, Ronald L.; Belles, Holly A.; Snyder, James R. 2001. Prescribed fire in the management of *Melaleuca quinquenervia* in subtropical Florida. In: Galley, Krista E. M.; Wilson, Tyrone P., eds. Proceedings of invasive species workshop: Role of fire in the control and spread of invasive species; Fire conference 2000: 1st national congress on fire ecology, prevention, and management; 2000 November 27-December 1; San Diego, CA. Misc. Publ. No. 11. Tallahassee, FL: Tall Timbers Research Station: 132-140.

Nagel, Harold G. 1995. Vegetative changes during 17 years of succession on Willa Cather Prairie in Nebraska. In: Hartnett, David C., ed. Prairie biodiversity: Proceedings, 14th North American prairie conference; 1994 July 12-16; Manhattan, KS. Manhattan, KS: Kansas State University: 25-30.

Naiman, Robert J.; Bilby, Robert E.; Bisson, Peter A. 2000. Riparian ecology and management in the Pacific coastal rain forest. BioScience. 50(11): 996-1011.

National Assessment Synthesis Team (NAST). 2003. Climate change impacts on the United States—the potential consequences of climate variability and change, [Online] National Assessment Synthesis Team, U.S. Global Change Research Program (Producer). Available: http://www.usgcrp.gov/usgcrp/Library/nationalassessment/overviewforests.htm [2006, October 24].

National Planning Association (NPA). 1999. Regional economic projection series: economic databases three growth projections 1967-2020, [Online]. NPA Data Services Inc. (Producer). Available: http://www.ghcc.msfc.nasa.gov/regional/assessment_progress.html [Access date unknown].

National Wildfire Coordinating Group. 1996. Glossary of wildland fire terminology. Incident Operations Working Team, National Wildfire Coordinating Group NFES 1832, PMS 205. Boise, Idaho: National Interagency Fire Center. 162 p.

Neal, John L.; Wright, Ernest; Bollen, Walter B. 1965. Burning Douglas-fir slash: physical, chemical, and microbial effects in the soil. Research Paper. Corvallis, OR: Oregon State University, Forest Research Laboratory, Forest Management Research. 32 p.

Neary, D. G.; Ffolliot, P. F. 2005. The water resource: its importance, characteristics, and general response to fire. In: Neary, Daniel G.; Ryan, Kevin C.; DeBano, Leonard F., eds. 2005. Wildland fire in ecosystems: effects of fire on soils and water. Gen. Tech. Rep. RMRS-GTR-42-vol.4. Ogden, UT: U.S. Department of Agriculture, Forest Service, Rocky Mountain Research Station. 250 p.

Neary, Daniel G.; Ryan, Kevin C.; DeBano, Leonard F., eds. 2005a. Wildland fire in ecosystems: effects of fire on soil and water. Gen. Tech. Rep. RMRS-GTR-42-vol. 4. Ogden, UT: U.S. Department of Agriculture, Forest Service, Rocky Mountain Research Station. 250 p.

Neary, Daniel G.; Ryan, Kevin C.; DeBano, Leonard F.; Landsberg, Johanna D.; Brown, James K. 2005b. Introduction. In: Neary, Daniel G.; Ryan, Kevin C.; DeBano, Leonard F., eds. Wildland fire in ecosystems: effects of fire on soil and water. Gen. Tech. Rep. RMRS-GTR-42-vol. 4. Ogden, UT: U.S. Department of Agriculture, Forest Service, Rocky Mountain Research Station: 1-18.

Neiland, Bonita J. 1958. Forest and adjacent burn in the Tillamook burn area of northwestern Oregon. Ecology. 39(4): 660-671.

Nelson, C. R.; Halpern, C. B. 2005. Edge-related responses of understory plants to aggregated retention harvest in the Pacific Northwest. Ecological Applications. 15: 196-209.

Nelson, J. A.; Lym, R. G. 2003. Interactive effects of Aphthona nigriscutis and picloram plus 2,4-D in leafy spurge (*Euphorbia esula*). Weed Science. 51: 118-124.

Nelson, T. C. 1957. The original forests of the Georgia Piedmont. Ecology. 38: 390-397.

Neumann, David D.; Dickmann, Donald I. 2001. Surface burning in a mature stand of *Pinus resinosa* and *Pinus strobus* in Michigan: effects on understory vegetation. International Journal of Wildland Fire. 10: 91-101.

Niering, William A. 1992. The New England forests. Restoration & Management Notes. 10(1): 24-28.

Niering, William A.; Dreyer, Glenn D. 1987. Prairies and prairie restoration in the East (Connecticut). Restoration & Management Notes. 5(2): 83.

Niering, William A.; Dreyer, Glenn D. 1989. Effects of prescribed burning on *Andropogon scoparius* in postagricultural grasslands in Connecticut. The American Midland Naturalist. 122: 88-102.

Nijjer, S.; Lankau, R. A.; Rogers, W. E.; Siemann, E. 2002. Effects of temperature and light on Chinese tallow (*Sapium sebiferum*) and Texas sugarberry (*Celtis laevigata*) seed germination. Texas Journal of Science. 54: 63-68.

Nilsen, E.; Muller, W. 1980. A comparison of the relative naturalization ability of two *Schinus* species in southern California. I. Seed germination. Bulletin of the Torrey Botanical Club. 107: 51-56.

Nolen, Andrew. 2002. Vetch infestations in Alaska. Prepared for Alaskan Department of Transportation and Public Facilities, [Online]. Palmer, AK: Alaska Plant Materials Center, Division of Agriculture, Department of Natural Resources (Producer). Available: http://akweeds.uaa.alaska.edu/pdfs/literature/VetchInfestationsAK_Nolan.pdf [2005, January 15].

Nonner, Edith. 2006. Seed bank dynamics and germination ecology of fountain grass (*Pennisetum setaceum*). Honolulu, HI: University of Hawai`i. 56 p. Thesis.

North, M.; Chen, J.; Smith, G.; Krakowiak, L.; Franklin, J. 1996. Initial response of understory plant diversity and overstory tree diameter growth to a green tree retention harvest. Northwest Science. 70: 24-34.

Norton, Helen H. 1979. The association between anthropogenic prairies and important food plants in western Washington. Northwest Anthropological Research Notes. 13(2): 175-200.

Noss, Reed F.; LaRoe, Edward T., III; Scott, J. Michael. 1995. Endangered ecosystems of the United States: a preliminary assessment of loss and degradation. Biological Report 28. Washington, DC: U.S. Department of the Interior, National Biological Services. 58 p.

Noste, Nonan V. 1969. Analysis and summary of forest fires in coastal Alaska. Juneau, AK: U.S. Department of Agriculture, Forest Service, Pacific Northwest Forest and Range Experiment Station, Institute of Northern Forestry. 12 p.

Nuzzo, V. A. 1986. Extent and status of midwest oak savanna: presettlement and 1985. Natural Areas Journal. 6: 6-36.

Nuzzo, V. A. 1991. Experimental control of garlic mustard [*Alliaria petiolata* (Bieb.) Cavara & Grande] in northern Illinois using fire, herbicide, and cutting. Natural Areas Journal. 11(3): 158-167.

Nuzzo, V. A. 1996. Impact of dormant season herbicide treatment on the alien herb garlic mustard (*Alliaria petiolata* [Bieb.] Cavara and Grande) and groundlayer vegetation. Transactions, Illinois State Academy of Science. 89(1/2): 25-36.

Nuzzo, Victoria A.; McClain, William; Strole, Todd. 1996. Fire impact on groundlayer flora in a sand forest 1990-1994. American Midland Naturalist. 136(2): 207-221.

Nuzzo, Victoria. 1997. Element stewardship abstract: *Lonicera japonica* (Japanese honeysuckle), [Online]. In: The global invasive species initiative. Arlington, VA: The Nature Conservancy (Producer). Available: http://tncweeds.ucdavis.edu/esadocs.html [2004, December 29].

Nyboer, Randy. 1990. Vegetation management guideline—bush honeysuckles: Tartarian, Morrow's, Belle, and Amur honeysuckle (*Lonicera tatariva* L., L. *morrowii* Gray, L. x *bella Zabel*, and L. *maackii* (Rupr.) Maxim.), [Online]. In: Vegetation management guideline: Vol. 1. No. 6. Champaign, IL: The Illinois Natural History Survey (Producer). Available: www.inhs.uiuc.edu/chf/outreach/VMG/bhnysckl.html [2004, December 2].

Nyboer, Randy. 1992. Vegetation management guideline: bush honeysuckles—tatarian, Marrow's, belle, amur honeysuckle. Natural Areas Journal. 12(4): 218-219.

Nyman, J. A.; Chabreck, R. H. 1995. Fire in coastal marshes: history and recent concerns. In: Cerulean, S. I.; Engstrom, R. T., eds. Proceedings of the Tall Timbers Fire Ecology Conference, No. 19. Fire in wetlands: a management perspective. Tallahassee, FL: Tall Timbers Research Station. 134-141.

Odion, Dennis C.; Haubensak, Karen A. 2002. Response of French broom to fire. In: Sugihara, Neil G.; Morales, Maria; Morales, Tony, eds. Fire in California ecosystems: integrating ecology, prevention and management: Proceedings of the symposium; 1997 November 17-20; San Diego, CA. Misc. Pub. No. 1. [Davis, CA]: Association for Fire Ecology: 296-307.

Ogden, Jennifer A. Erskine; Rejmánek, Marcel. 2005. Recovery of native plant communities after the control of a dominant invasive plant species, *Foeniculum vulgare*: Implications for management. Biological Conservation. 125: 427-439.

Ohmart, R. D.; Anderson, B. W.; Hunter, W. C. 1988. The Ecology of the Lower Colorado River from Davis Dam to the Mexico—United States International Boundary: A Community Profile. (Biological Report 85[7.19]) U.S. Fish and Wildlife Service.

Ohmart, Robert D.; Anderson, Bertin W. 1982. North American desert riparian ecosystems. In: Bender, Gordon L., ed. Reference handbook on the deserts of North America. Westport, CT: Greenwood Press: 433-479.

Okay, Judith A. 2005. Mile-a-minute weed: *Polygonum perfoliatum* L. buckwheat family (Polygonaceae), [Online]. Washington, DC: U.S. National Park Service (Producer). Available: http://www.nps.gov/plants/alien/fact/pope1.htm [2006, June 13].

Old, Sylvia M. 1969. Microclimate, fire, and plant production in an Illinois prairie. Ecological Monographs. 39(4): 355-384.

O'Leary, J. F. 1988. Habitat differentiation among herbs in postburn California chaparral and coastal sage scrub. The American Midland Naturalist. 120: 41-49.

O'Leary, J. F. 1990. Post-fire diversity patterns in two subassociations of California coastal sage scrub. Journal of Vegetation Science. 1: 173-180.

O'Leary, J. F. 1995. Coastal sage scrub: threats and current status. Fremontia. 23: 27-31.

O'Leary, J. F. and Minnich, R. A. 1981. Postfire recovery of creosote bush scrub vegetation in the western Colorado desert. Madroño. 28(2): 61-66.

Oliver, Chadwick D.; Larson, Bruce C. 1996. Forest stand dynamics. Updated ed. New York: John Wiley & Sons, Inc. 520 p.

Olson, Diana L.; Agee, James K. 2005. Historical fires in Douglas-fir dominated riparian forests of the southern Cascades, Oregon. Fire Ecology. 1(1): 50-74.

Olson, T. E.; Knopf, F. L. 1986. Naturalization of Russian-olive in the western United States. Western Journal of Applied Forestry. 1: 65-69.

Olson, Wendell W. 1975. Effects of controlled burning on grassland within the Tewaukon National Wildlife Refuge. Fargo, ND: North Dakota University of Agriculture and Applied Science. 137 p. Thesis.

Omi, P. N.; Kalabokidis, K. D. 1991. Fire damage on extensively versus intensively managed forest stands within the North Fork Fire, 1988. Northwest Science. 65: 149-157.

Ortega, Yvette K.; Pearson, Dean E.; McKelvey, Kevin S. 2004. Effects of biological control agents and exotic plant invasion on deer mouse populations. Ecological Applications. 14(1): 241-253.

Ortega, Yvette Katina; McKelvey, Kevin Scot; Six, Diana Lee. 2006. Invasion of an exotic forb impacts reproductive success and site fidelity of a migratory songbird. Oecologia. 149(2): 340-351.

Orwig, David A.; Foster, David R. 1998. Forest response to the introduced hemlock woolly adelgid in southern New England, USA. Journal of the Torrey Botanical Club. 125(1): 60-73.

Oswalt, Christopher M.; Oswalt, Sonja N.; Clatterbuck, Wayne K. 2007. Effects of Microstegium vimineum (Trin.) A. Camus on native woody species density and diversity in a productive mixed-hardwood forest in Tennessee. Forest Ecology and Management. 242: 727-732.

Ott, J. E.; McArthur, E. D.; Roundy, B. A. 2003. Vegetation of chained and non-chained seedlings after wildfire in Utah. Journal of Range Management. 56: 81-91.

Ott, Jeffrey E.; McArthur, E. Durant; Sanderson, Stewart C. 2001. Plant community dynamics of burned and unburned sagebrush and pinyon-juniper vegetation in west-central Utah. In: McArthur, E. Durant; Fairbanks, Daniel J., comps.. Shrubland ecosystem genetics and biodiversity: proceedings; 2000 June 13-15; Provo, UT. Proc. RMRS-P-21. Ogden, UT: U.S. Department of Agriculture, Forest Service, Rocky Mountain Research Station: 177-191.

Owensby, Clenton E.; Smith, Ed F. 1973. Burning true prairie. In: Hulbert, Lloyd C., ed. 3rd Midwest prairie conference proceedings; 1972 September 22-23; Manhattan, KS. Manhattan, KS: Kansas State University, Division of Biology: 1-4.

Packard, Steve. 1988. Just a few oddball species: restoration and the rediscovery of the tallgrass savanna. Restoration & Management Notes. 6(1): 13-22.

Padgett, P. E.; Allen, E. B.; Bytnerowicz, A.; Minnich, R. A. 1999. Changes in soil inorganic nitrogen as related to atmospheric nitrogenous pollutants in southern California. Atmospheric Environment. 33: 769-781.

Page, H. N.; Bork, E. W.; Newman, R. F. 2005. Understory responses to mechanical restoration and drought within montane forests of British Columbia. British Columbia Journal of Ecosystems and Management. 6: 8-21.

Paisley, C. 1968. From cotton to quail: an agricultural chronicle of Leon County, Florida. Gainesville, FL: University of Florida Press. 162 p.

Palmer, Daniel. H. 2003. Hawaii's Ferns and Fern Allies. Honolulu, HI: University of Hawai`i Press. 324 p.

Panetta, F. D.; Dodd, J. 1987. The biology of Australian weeds. 16. *Chondrilla juncea* L. The Journal of the Australian Institute of Agricultural Science. 53(2): 83-95.

Parendes, Laurie A.; Jones, Julia A. 2000. Role of light availability and dispersal in exotic plant invasion along roads and streams in the H.J. Andrews Experimental Forest, Oregon. Conservation Biology. 14(1): 64-75.

Parendes, Laurie Anne. 1997. Spatial patterns of invasion by exotic plants in a forested landscape. Corvallis, OR: Oregon State University. 208 p. Thesis.

Parker, I. M. 2001. Safe site and seed limitation in *Cytisus scoparius* (Scotch broom): invasibility, disturbance, and the role of cryptogams in a glacial outwash prairie. Biological Invasions. 3: 323-332.

Parker, I. M.; Harpole, William; Dionne, Diana. 1997. Plant community diversity and invasion of the exotic shrub Cytisus scoparius: testing hypotheses of invasibility and impact. In: Dunn, Patrick; Ewing, Kern, eds.. Ecology and conservation of south Puget Sound prairie landscape. Seattle, WA: The Nature Conservancy. 149-161.

Parker, I. M.; Mertens, Shoshana K.; Schemske, Douglas W. 1993. Distribution of seven native and two exotic plants in a tallgrass prairie in southeastern Wisconsin: the importance of human disturbance. The American Midland Naturalist. 130(1): 43-55.

Parker, I. M.; Simberloff, D.; Lonsdale, W. M.; Goodell, K.; Wonham, M.; Kareiva, P. M.; Williamson, M. H.; Von Holle, B.; Moyle, P. B.; Byers, J. E.; Goldwasser, L. 1999. Impact: toward a framework for understanding the ecological effects of invaders. Biological Invasions. 1: 3-19.

Parker, Ingrid Marie. 1996. Ecological factors affecting rates of spread in *Cytisus scoparius*, an invasive exotic shrub. Seattle, WA: University of Washington. 175 p. Dissertation.

Parshall, R.; Foster, D. R. 2002. Fire on the New England landscape: regional and temporal variation, cultural and environmental controls. Journal of Biogeography. 29: 1305-1317.

Parsons, D. J.; DeBenedetti, S. 1979. Impact of fire suppression on a mixed-conifer forest. Forest Ecology and Management. 2: 21-33.

Parsons, D. J.; Swetnam, T. W. 1989. Restoring natural fire to the Sequoia-mixed conifer forest: should intense fire play a role. Tall Timbers Fire Ecology Conference. 20: 20-30.

Parsons, David J.; Stohlgren, Thomas J. 1989. Effects of varying fire regimes on annual grasslands in the southern Sierra Nevada of California. Madroño. 36(3): 154-168.

Parsons, W. T.; Cuthbertson, E. G. 1992. Noxious weeds of Australia. Melbourne: Indata Press. 692 p.

Pase, Charles P. 1971. Effect of a February burn on Lehmann lovegrass. Journal of Range Management. 24: 454-456.

Patterson, William A., III.; Clarke, Gretel L.; Haggerty, Sarah A.; Sievert, Paul R.; Kelty, Matthew J. 2005. Wildland fuel management options for the central plains of Martha's Vineyard: impacts on fuel loads, fire behavior and rare plant and insect species. Final report submitted to the Massachusetts Department of Conservation and Recreation. 140 p.

Patterson, William A., III.; Crary, David W., Jr. 2004. Managing fuels in northeastern barrens v-2.0. A field tour sponsored by the Joint Fire Sciences Program, June 14-18, 2004. Great River, NY: Pine Barrens Commission. 40 p.

Patterson, William. A., III; Sassaman, K. E. 1988. Indian fires in the prehistory of New England, In: Nichols, G. P., ed. Holocene human ecology in northeastern North America. Plenum, NY: Plenum Press: 107-135.

Pauchard, A.; Alaback, P. B. 2004. Influence of elevation, land use, and landscape context on patterns of alien plant invasions along roadsides in protected areas of south-central Chile. Conservation Biology. 18(1): 238-248.

Pauchard, Anibal; Alaback, Paul B. 2006. Edge type defines alien plant species invasions along Pinus contorta burned, highway and clearcut forest edges. Forest Ecology and Management. 223(1-3): 327-335.

Pavek, Diane S. 1992. Halogeton glomeratus. In: Fire Effects Information System, [Online]. U.S. Department of Agriculture, Forest Service, Rocky Mountain Research Station, Fire Sciences Laboratory (Producer). Available: http://www.fs.fed.us/database/feis/ [2005, December 15].

Paysen, Timothy E.; Ansley, R. James; Brown, James K.; Gottfried, Gerald J.; Haase, Sally M.; Harrington, Michael G.; Narog, Marcia G.; Sackett, Stephen S.; Wilson, Ruth C. 2000. Fire in western shrubland, woodland, and grassland ecosystems. In: Brown, James K.; Smith, Jane Kapler, eds. Wildland fire in ecosystems: Effects of fire on flora. Gen. Tech. Rep. RMRS-GTR-42-vol. 2. Ogden, UT: U.S. Department of Agriculture, Forest Service, Rocky Mountain Research Station: 121-159.

Pearson, D.E. 2006. Spotted knapweed structure favors Dictyna spiders such that it increases their populations 11-fold and doubles their web size and prey capture rates. Unpublished data. US Forest Service Rocky Mountain Research Station, Missoula MT.

Pearson, D. E.; McKelvey, K. S.; Ruggiero, L. F. 2000. Non-target effects of an introduced biological control agent on deer mouse ecology. Oecologia. 122: 121-128.

Pellant, M. 1990. The cheatgrass-wildfire cycle—are there any solutions? In: McArthur, E. D.; Romney, E. M.; Smith, S. D.; and Tueller, P. T., comps.. Cheatgrass invasion, shrub die-off, and other aspects of shrub biology and management symposium: proceedings. Gen. Tech. Rep. INT-GTR-276. Ogden, UT: U.S. Dept. of Agriculture, Forest Service: 11-18.

Pellant, Mike. 2005. [Personal communication]. August 10, Henderson, NV: U.S. Department of the Interior, Bureau of Land Management, Idaho State Office.

Pellant, Mike; Lysne, Cindy R. 2005. Strategies to enhance plant structure and diversity in crested wheatgrass seedings. In: Shaw, Nancy L.; Pellant, Mike; Monsen, Stephen B., eds. Sage-grouse habitat restoration symposium proceedings; 2001 June 4-7; Boise, ID. Proc. RMRS-P-38. Fort Collins, CO: U.S. Department of Agriculture, Forest Service, Rocky Mountain Research Station: 81-92.

Pellant, Mike; Monsen, Stephen B. 1993. Rehabilitation on public rangelands in Idaho, USA: a change in emphasis from grass monocultures. In: Proceedings, 17th international grassland congress; 1993 February 8-21; Palmerston North, New Zealand. [Place of publication unknown]: [Publisher unknown]: 778-779.

Peloquin, R. L.; Hiebert, R. D. 1999. The effects of black locust (*Robinia pseudoacacia*) on species diversity and composition of black oak savanna / woodland communities. Natural Areas Journal. 19: 121-131.

Pendergrass, K. L.; Miller, P. M.; Kauffman, J. B. 1998. Prescribed fire and the response of woody species in Willamette Valley wetland prairies. Restoration Ecology. 6(3): 303-311.

Pendergrass, K. L.; Miller, P. M.; Kauffman, J. B.; Kaye, T. N. 1999. The role of prescribed burning in maintenance of an endangered plant species, *Lomatium bradshawii*. Ecological Applications. 9(4): 1420-1429.

Pendergrass, Kathy L. 1996. Vegetation composition and response to fire of native Willamette Valley wetland prairies. Corvallis, OR: Oregon State University. 241 p. Thesis.

Pernas, Antonio J.; Snyder, William A. 1999. Status of melaleuca control at Big Cypress National Preserve. In: Jones, David T.; Gamble, Brandon W., eds. Florida's garden of good and evil: Proceedings of the 1998 joint symposium of the Florida Exotic Pest Plant Council and the Florida Native Plant Society; 1998 June 3-7; Palm Beach Gardens, FL. West Palm Beach, FL: South Florida Water Management District: 133-137.

Peters, Erin F.; Bunting, Stephen C. 1994. Fire conditions pre- and postoccurrence of annual grasses on the Snake River Plain. In: Monsen, Stephen B.; Kitchen, Stanley G., comps. Proceedings: ecology and management of annual rangelands; 1992 May 18-22; Boise, ID. Gen. Tech. Rep. INT-GTR-313. Ogden, UT: U.S. Department of Agriculture, Forest Service, Intermountain Research Station: 31-36.

Petranka, J. W.; Holland, R. 1980. A quantitative analysis of bottomland communities in south-central Oklahoma. Southwestern Naturalist. 25: 207-214.

Phillips, Barbara Goodrich; Crisp, Debra. 2001. Dalmatian toadflax, an invasive exotic noxious weed, threatens Flagstaff pennyroyal community following prescribed fire. In: Maschinski, Joyce; Holter, Louella, tech. coords. Southwestern rare and endangered plants: Proceedings of the 3rd conference; 2000 September 25-28; Flagstaff, AZ. Proceedings RMRS-P-23. Fort Collins, CO: U.S.

Department of Agriculture, Forest Service, Rocky Mountain Research Station: 200-205.

Phillips, Sherman A., Jr.; Brown, C. Mark; Cole, Charles L. 1991. Weeping lovegrass, *Eragrostis curvula* (Schrader) Nees Von Esenbeck, as a harborage of arthropods on the Texas high plains. The Southwestern Naturalist. 36(1): 49-53.

Phillips, W. A.; Coleman, S. W. 1995. Productivity and economic return of three warm season grass stocker systems for the southern Great Plains. Journal of Production Agriculture. 8(3): 334-339.

Philpot, Charles W. 1974. The changing role of fire on chaparral lands. In: Rosenthal, Murray, ed. Symposium on living with the chaparral: Proceedings; 1973 March 30-31; Riverside, CA. San Francisco, CA: The Sierra Club: 131-150.

Pickering, Debbie L.; Bierzychudek, Paulette; Salzer, Dan; Rudd, Nathan. 2000. Oregon silverspot butterfly dispersal patterns and habitat management. An unpublished RJ/KOSE Funded Research Project, Final Report, [Online]. Available: Invasive Species Subject, http://conserveonline.org/csd;internal&action=buildframes.action [2005, March 22].

Pickford, G. D. 1932. The influence of continued heavy grazing and of promiscuous burning on spring-fall ranges in Utah. Ecology. 13(2): 159-171.

Piemeisel, Robert L. 1951. Causes affecting change and rate of change in a vegetation of annuals in Idaho. Ecology. 32(1): 53-72.

Pimm, S. L. 1984. The complexity and stability of ecosystems. Nature. 307: 321-326.

Planty-Tabacchi, Anne-Marie; Tabacchi, Eric; Naiman, Robert J.; DeFerrari, Collette; Decamps, Henri. 1996. Invasibility of species-rich communities in riparian zones. Conservation Biology. 10(2): 598-607.

Platt, W. J. 1999. Southeastern pine savannas. In: Anderson, R. C.; Fralish, J. S.; Baskin, J., eds. The savanna, barren, and rock outcrop communities of North America. Cambridge, UK: Cambridge University Press. 23-51.

Platt, W. J.; Gottschalk, R. M. 2001. Effects of exotic grasses on potential fine fuel loads in the ground cover of south Florida slash pine savannas. International Journal of Wildland Fire. 10: 155-159.

Platt, W. J.; Peet, R. K. 1998. Ecological concepts in conservation biology: lessons from southeastern ecosystems. Ecological Applications. 8: 907-908.

Platt, W. J.; Stanton, L. 2003. Managing invasions of fire-frequented ecosystems: hardwoods and graminoids in southeastern savannas, prairies, and marshes. In: Abstracts—Invasive plants in natural and managed systems: Linking science and management: Proceedings, 7th international conference on the ecology and management of alien plant invasions; 2003 November 3-7; Ft. Lauderdale, FL. [Lawrence, KS]: [Weed Science Society of America]. Abstract.

Plummer, A. Perry. 1959. Restoration of juniper-pinyon ranges in Utah. In: Meyer, Arthur B.; Eyre, F. H., eds. Proceedings, Society of American Foresters meeting: multiple-use forestry in the changing West; 1958 September 28-October 2; Salt Lake City, UT. Washington, DC: Society of American Foresters: 207-211.

Plummer, G. L. 1975. 18th century forests in Georgia. Bulletin of the Georgia Academy of Science. 33: 1-19.

Pollak, O., and T. Kan. 1998. The use of prescribed fire to control invasive exotic weeds at Jepson Prairie Preserve. In: Witham, C. W.; Bauder, E. T.; Belk, D.; Ferren, W. R.; Ornduff, R., eds. Ecology, conservation, and management of vernal pool ecosystems. Sacramento, CA: California Native Plant Society: 241-249.

Pollet, J.; Omi, P. N. 2002. Effect of thinning and prescribed burning on crown fire severity in ponderosa pine forests. International Journal of Wildland Fire. 11: 1-10.

Poole, Thomas. 2005. [Email to Alison Dibble]. January 4. Fort Devens, Massachusetts: Devens Reserve Forces Training Area. On file at U.S. Department of Agriculture, Forest Service, Rocky Mountain Research Station, Fire Sciences Laboratory, Missoula, MT.

Post, Thomas W.; McCloskey, Elizabeth; Klick, Kenneth F. 1990. Glossy buckthorn resists control by burning (Indiana). Restoration and Management Notes. 8(1): 52-53.

Powell, George W. 1996. Analysis of sulphur cinquefoil in British Columbia. Working Paper 16. Victoria, BC: British Columbia Ministry of Forests Research Program. 36 p.

Powell, R. D. 1990. The role of spatial pattern in the population biology of Centaurea diffusa. Journal of Ecology. 78: 374-388.

Prater, Margaret R.; Obrist, Daniel; Arnone, John A., III; DeLucia, Evan H. 2006. Net carbon exchange and evapotranspiration in postfire and intact sagebrush communities in the Great Basin. Oecologia. 146(4): 595-607.

Pratt, David W.; Black, R. Alan; Zamora, B. A. 1984. Buried viable seed in a ponderosa pine community. Canadian Journal of Botany. 62: 44-52.

Preuninger, Jill S.; Umbanhowar, Charles, E., Jr. 1994. Effects of burning, cutting, and spraying on reed canary grass studied (Minnesota). Restoration & Management Notes. 12(2): 207.

Prosser, C. W.; Sedivec, K. K.; Barker, W. T. 1999. Effects of prescribed burning and herbicide treatments on leafy spurge (*Euphorbia esula*), [Online]. Proceedings of a symposium on leafy spurge, Medora, North Dakota. Available: (www.team.ars.usda.gov).

Pyke, D. A.; Brooks, M. L.; D'Antonio, C. M. [In review]. Fire as a restoration tool: a life form decision framework for predicting the control or enhancement of plants using fire. Restoration Ecology. [Volume unknown]: [Pages unknown].

Pyke, David A. 1994. Ecological significance of seed banks with special reference to alien annuals. In: Monsen, Stephen B.; Kitchen, Stanley G., comps. Proceedings: ecology and management of annual rangelands; 1992 May 18-22; Boise, ID. Gen. Tech. Rep. INT-GTR-313. Ogden, UT: U.S. Department of Agriculture, Forest Service, Intermountain Research Station: 197-201.

Pyle, Laura L. 1995. Effects of disturbance on herbaceous exotic plant species on the floodplain of the Potomac River. The American Midland Naturalist. 134: 244-253.

Pyne, S. J. 1982a. Fire in America: a cultural history of wildland and rural fire. Princeton University Press, Princeton, NJ. 653 pp.

Pyne, Stephen J. 1982b. Fire primeval. The Sciences. Aug/Sept: 13-20.

Pyne, Stephen J.; Andrews, Patricia L.; Laven, Richard D. 1984. Introduction to wildland fire. 2nd ed. New York: Wiley & Sons, Inc. 769 p.

Racher, Brent J.; Mitchell, Robert B. 1999. Management of saltcedar in eastern New Mexico and Texas. In: Wester, David B.; Britton, Carlton M., eds. Research highlights—1999: Noxious brush and weed control: Range, wildlife, and fisheries management. Volume 30. Lubbock, TX: Texas Tech University, College of Agricultural Sciences and Natural Resources: 14-15.

Racher, Brent J.; Mitchell, Robert B.; Britton, Carlton; Wimmer, S. Mark; Bryan, Justin B. 2002. Prescription development for burning two volatile fuels. In: Wilde, Gene R.; Smith, Loren M., eds. Research highlights—2002: Range, wildlife, and fisheries management. Volume 33. Lubbock, TX: Texas Tech University, College of Agricultural Sciences and Natural Resources: 25.

Racher, Brent J.; Mitchell, Robert B.; Schmidt, Charles; Bryan, Justin. 2001. Prescribed burning prescriptions for saltcedar in New Mexico. In: Zwank, Phillip J.; Smith, Loren M.; eds. Research highlights—2001: Range, wildlife, and fisheries management. Volume 32. Lubbock, TX: Texas Tech University, Department of Range, Wildlife, and Fisheries Management: 25.

Racher, Brent; Britton, Carlton. 2003. Fire as part of an integrated management approach for saltcedar. In: Abstracts—Invasive plants in natural and managed systems: Linking science and management: Proceedings, 7th international conference on the ecology and management of alien plant invasions; 2003 November 3-7; Ft. Lauderdale, FL. [Lawrence, KS]: [Weed Science Society of America]. Abstract.

Racine, Charles H. 1979. The 1977 tundra fires in the Seward Peninsula, Alaska: effects and initial revegetation. BLM-Alaska Technical Report 4. Anchorage, AK: U.S. Department of the Interior, Bureau of Land Management, Alaska State Office. 51 p.

Racine, Charles H. 1981. Tundra fire effects on soils and three plant communities along a hill-slope gradient in the Seward Peninsula, Alaska. Arctic. 34(1): 71-84

Racine, Charles H.; Johnson, Lawrence A.; Viereck, Leslie A. 1987. Patterns of vegetation recovery after tundra fires in Northwestern Alaska, U.S.A. Arctic and Alpine Research. 19(4): 461-469

Rader, Laura Teresa. 2000. Biomass, leaf area and resource availability of kudzu dominated plant communities following herbicide treatments and induced competition from high density loblolly pine stands. Athens, GA: University of Georgia. 117 p. Thesis.

Randa, Lynda A.; Yunger, John A. 2001. A comparison of management techniques on plant species composition and biomass in a tallgrass prairie restoration. In: Bernstein, Neil P.; Ostrander, Laura J., eds. Seeds for the future; roots of the past: Proceedings of the 17th North American prairie conference; 2000 July 16-20; Mason City, IA. Mason City, IA: North Iowa Community College: 92-97.

Randall, J. 1996. Weed control for the preservation of biological diversity. Weed technology. 10: 370-383.

Randall, J. M.; Rejmánek, M. 1993. Interference of bull thistle (*Cirsium vulgare*) with growth of ponderosa pine (*Pinus ponderosa*) seedlings in a forest plantation. Canadian Journal of Forest Research. 23: 1507-1513.

Randall, J. M.; Rejmánek, M.; Hunter, J. C. 1998. Characteristics of the exotic flora of California. Fremontia. 26(4): 3-12.

Randall, John M. 1997. Defining weeds of natural areas. In: Luken, James O.; Thieret, John W., eds. Assessment and management of plant invasions. Springer Series on Environmental Management. New York: Springer-Verlag: 18-25.

Ranney, J. W.; Bruner, M. C.; Levenson, J. B. 1981. The importance of edge and the structure and dynamics of forest islands. In: Burgess, R. L.; Sharpe, D. M., eds. Forest island dynamics in man-dominated landscapes. New York: Springer-Verlag: 67-95.

Rasha, Renee. 2005. Fact sheet: Burma reed—*Neyraudia reynaudiana* (Kunth) Keng ex A.S. Hitchc, [Online]. In: Weeds gone wild: Alien plant invaders of natural areas. Plant Conservation Alliance, Alien Plant Working Group (Producer). Available at: http://www.nps.gov/plants/alien/fact/nere1.htm [2007, August 28].

Rasmussen, G. A. 1994. Prescribed burning considerations in sagebrush annual grassland communities. In: Monsen, Stephen B.; Kitchen, Stanley G., comps. Proceedings: Ecology and management of annual rangelands; 1992 May 18-22; Boise, ID. Gen. Tech. Rep. INT-GTR-313. Ogden, UT: U.S. Department of Agriculture, Forest Service, Intermountain Research Station: 69-70.

Ratzlaff, T. D.; Anderson, J. E. 1995. Vegetal recovery following wildfire in seeded and unseeded sagebrush steppe. Journal of Range Management. 48(5): 386-391.

Raunkiaer, C. 1934. The life forms of plants and statistical plant geography. Oxford: Clarendon Press. 632 p.

Raven, P. H. 1977. The California flora. In: Barbour, M. G.; Major, J., eds. Terrestrial vegetation of California. New York: Wiley Interscience: 109-137.

Raven, P. H.; Axelrod, D. I. 1978. Origin and relationships of the California flora. University of California Publications in Botany. 72: 1-134.

Rawinski, Thomas James. 1982. The ecology and management of purple loosestrife (*Lythrum salicaria* L.) in central New York. Ithaca, NY: Cornell University. 88 p. Thesis.

Redmann, R. E.; Romo, J. T.; Pylypec, B.; Driver, E. A. 1993. Impacts of burning on primary productivity of Festuca and Stipa-Agropyron grasslands in central Saskatchewan. The American Midland Naturalist. 130: 262-273.

Reed, H. E.; Seastedt, T. R.; Blair, J. M. 2005. Ecological consequences of C4 grass invasion of a C4 grassland: a dilemma for management. Ecological Applications. 15(5): 1560-1569.

Reed, R. A.; Johnson-Barnard, J.; Baker, W. L. 1996. Contribution of roads to forest fragmentation in the Rocky Mountains. Conservation Biology. 10(4): 1098-1106.

Rees, Daniel C.; Juday, Glenn Patrick. 2002. Plant species diversity on logged versus burned sites in central Alaska. Forest Ecology and Management. 155: 291-302.

Regan, Alan Chris. 2001. The effects of fire on woodland structure and regeneration of *Quercus garryana* at Fort Lewis, Washington. Seattle, WA: University of Washington. 78 p. Thesis.

Reichard, Sarah H.; Hamilton, Clement W. 1997. Predicting invasions of woody plants introduced into North America. Conservation Biology. 11: 193-203.

Reilly, Matthew J.; Wimberly, Michael C.; Newell, Claire L. 2006. Wildfire effects on plant species richness at multiple spatial scales in forest communities of the southern Appalachians. Journal of Ecology. 94(1): 118-130.

Reimer, D. N. 1973. Effects of rate, spray volume, and surfactant on the control of phragmites with glyphosate. Proceedings of the New England Weed Science Society. 27: 101-104.

Reinhardt, Elizabeth D.; Keane, Robert E.; Brown, James K. 1997. First Order Fire Effects Model: FOFEM 4.0, user's guide. Gen. Tech. Rep. INT-GTR-344. Ogden, UT: U.S. Department of Agriculture, Forest Service, Intermountain Research Station. 65 p.

Rejmánek, M. 1989. Invasibility of plant communities. In: Drake, J. A.; Mooney, H. A.; DiCastri, F.; Groves, R. H.; Kruger, F. J.; Rejmánek, M.; Williamson, M., eds. Biological Invasions: A Global Perspective. New York: John Wiley & Sons: 369-388.

Rejmánek, M.; Randall, J. M. 1994. Invasive alien plants in California: 1993 summary and comparison with other areas in North America. Madroño. 41: 161-177.

Rejmánek, M.; Richardson, D. M.; Higgins, S. I.; Pitcairn, M. J.; Grotkopp, E. 2005a. Ecology of invasive plants: state of the art. In: H. A. Mooney, R. N. Mack, J. A. McNeely, L. E. Neville, P. J. Schei, and J. K. Waage, eds.. Invasive Alien Species: A New Synthesis. SCOPE 63. Washington, DC: Island Press: 104-161.

Rejmánek, M.; Richardson, D. M.; Pyšek, P. 2005b. Plant invasions and invasibility of plant communities. In: van der Maarel, E., ed. Vegetation Ecology. Oxford, UK: Blackwell Publishing: 332-355.

Rejmánek, Marcel; Richardson, David M. 1996. What attributes make some plant species more invasive? Ecology. 77(6): 1655-1661.

Renne, I. J.; Spira, T. P. 2001. Effects of habitat, burial, age and passage through birds on germination and establishment of Chinese tallow tree in coastal South Carolina. Journal of the Torrey Botanical Society. 128: 109-119.

Rhoades, Chuck; Barnes, Thomas; Washburn, Brian. 2002. Prescribed fire and herbicide effects on soil processes during barrens restoration. Restoration Ecology. 10(4): 656-664.

Rhoads, Ann F.; Block, Timothy A. 2002. Invasive plant fact sheet II: Japanese barberry (*Berberis thunbergii* DC), [Online]. In: The global invasive species initiative. Arlington, VA: The Nature Conservancy (Producer). Available: http://tncweeds.ucdavis.edu/moredocs/berthu02.pdf [2006, August 25].

Rice, Barry Meyers; Randall, John, comps. 2004. Weed report: *Elaeagnus angustifolia*—Russian olive. In: Wildland weeds management and research: 1998-99 weed survey. Davis, CA: The Nature Conservancy, Wildland Invasive Species Program. 6 p. On file with: U.S. Department of Agriculture, Forest Service, Rocky Mountain Research Station, Missoula, MT.

Rice, Carol L. 1985. Fire history and ecology of the North Coast Range Preserve. In: Lotan, James E.; Kilgore, Bruce M.; Fisher, William C.; Mutch, Robert W., tech. coords. Proceedings, symposium and workshop on wilderness fire; 1983 November 15-18; Missoula, MT. Gen. Tech. Rep. INT-182. Ogden, UT: U.S. Department of Agriculture, Forest Service, Intermountain Forest and Range Experiment Station: 367-372.

Rice, Kevin J. 1989. Impacts of seed banks on grassland community structure and population dynamics. In: Leck, Mary Allessio; Parker, V. Thomas; Simpson, Robert L., eds. Ecology of soil seed banks. San Diego, CA: Academic Press, Inc: 211-230.

Rice, Peter M. 1993. Distribution and ecology of sulfur cinquefoil in Montana, Idaho and Wyoming. Final report: Montana Noxious Weed Trust Fund Project. Helena, MT: Montana Department of Agriculture. 11 p. On file with: U.S. Department of Agriculture, Forest Service, Rocky Mountain Research Station, Fire Sciences Laboratory, Missoula, MT.

Rice, Peter M. 2003. INVADERS Database System, [Online]. Missoula, MT: The University of Montana (Producer). Available: http://invader.dbs.umt.edu [2007, July 20].

Rice, Peter M. 2005. Fire as a tool for controlling nonnative invasive plants. Bozeman, MT: Center for Invasive Plant Management (Producer). http://www.weedcenter.org/management/burning_weeds.pdf. 52p.

Rice, Peter M.; Harrington, Michael G. 2003. The impact of prescribed fire following either herbicide or thinning treatments on bunchgrass and forested ecosystems in the Sawmill Creek Research Natural Area. Report Agreement Number 02-JV-11222048-141. Missoula, MT: U.S. Department of Agriculture, Forest Service, Rocky Mountain Research Station, Fire Sciences Laboratory. 43 p.

Rice, Peter M.; Harrington, Michael G. 2005a. Integration of herbicides and prescribed burning for plant community restoration. Rocky Mountain Research Station P.A. Nu. 01-PA-11222048-172.

Final Report PIAP Project Number RM-6. Missoula, MT: U.S. Department of Agriculture, Forest Service, Rocky Mountain Research Station, Fire Sciences Laboratory. 71 p.

Rice, Peter M.; Harrington, Michael G. 2005b. The impact of prescribed fire following either herbicide or thinning treatments on bunchgrass and forested ecosystems in the Sawmill Creek Research Natural Area. December 31, 2005 Progress Report. Agreement Number 02-JV-11222048-141. Missoula, MT: U.S. Department of Agriculture, Forest Service, Rocky Mountain Research Station, Fire Sciences Laboratory. 3 p.

Rice, Peter M.; Toney, J. C.; Bedunah, Donald J.; Carlson, Clinton E. 1997. Elk winter forage enhancement by herbicide control of spotted knapweed. Wildlife Society Bulletin. 25(3): 627-633.

Richards, R. T.; Chambers, J. C.; Ross, C. 1998. Use of native plants on federal lands: policy and practice. Journal of Range Management. 51: 625-632.

Richburg, Julie A. 2005. Timing treatments to the phenology of root carbohydrate reserves to control woody invasive plants. Amherst, MA: University of Massachusetts, Department of Natural Resources Conservation, 176 p. Dissertation.

Richburg, Julie A.; Dibble, Alison; Patterson, William A., III. 2001. Woody invasive species and their role in altering fire regimes of the northeast and mid-Atlantic states. In: Galley, K. E. M., and Wilson, T. P., eds. Proceedings of the Invasive Species Workshop: the Role of Fire in the Control and Spread of Invasive Species—Fire Conference 2000: the First National Congress on Fire Ecology, Prevention and Management San Diego, CA. Tallahassee, FL: Tall Timbers Research Station: 104-111.

Richburg, Julie A.; Patterson, William A., III. 2003a. Timing cutting and prescribed fire treatments for the control of northeastern woody invasive plants. Proceedings: Invasive plants in natural and managed systems: linking science and management [in conjunction with the] 7th international conference on the ecology and management of alien plant invasions Ft. Lauderdale, FL.

Richburg, Julie A.; Patterson, William A., III. 2003b. Can northeastern woody invasive plants be controlled with cutting and burning treatments? In: Bennett, Karen P., Dibble, Alison C., Patterson, William A. III, comps.. Proceedings: Using fire to control invasive plants: what's new, what works in the Northeast? Portsmouth, NH. Durham, NH: University of New Hampshire Cooperative Extension: 1-3.

Richburg, Julie A.; Patterson, William A., III; Ohman, Michael. 2004. Fire management options for controlling woody invasive plants in the northeastern and mid-Atlantic U.S, [Online]. Final report submitted to Joint Fire Science Program for Project Number 00-1-2-06. 59 p. Available: http://www.umass.edu/nrc/nebarrensfuels/publications/ [2007, July 12].

Richerson, Peter J.; Lum, Kwei-Lin. 1980. Patterns of plant species diversity in California: relation to weather and topography. American Naturalist. 116: 504-536.

Ricketts, T. H.; Dinerstein, E.; Olson, D. M.; Loucks, C. J.; Eichbaum, W.; DellaSala, D.; Kavanagh, K.; Hedao, P.; Hurley, P. T.; Carney, K. M.; Abell, R.; Walters, S. 1999. Terrestrial ecoregions of North America: A conservation assessment. Washington, DC: Island Press. [Pages unknown].

Risser, P. G.; Birney, E. C.; Blocker, H. D.; May, S. W.; Parton, W. J.; Wiens, J. A. 1981. The true prairie ecosystem. Stroudsburg, PA: Hutchinson Ross Publishing Company. [Pages unknown].

Rivas, Mercedes; Reyes, Otilia; Casal, Mercedes. 2006. Do high temperatures and smoke modify the germination response of Gramineae species? Forest Ecology and Management. 234S: S180.

Roberts, D. 1996. Climbing fern wreaks wetland havoc. Florida Department of Environmental Protection; Resource Management Notes. 8: 13.

Roberts, Fred H.; Britton, Carlton M.; Wester, David B.; Clark, Robert G. 1988. Fire effects on tobosagrass and weeping lovegrass. Journal of Range Management. 41(5): 407-409.

Roberts, M. R.; Gilliam, F. S. 1995. Patterns and mechanisms of plant diversity in forested ecosystems: implications for forest management. Ecological Applications. 5: 969-977.

Roberts, T. C., Jr. 1996. Is Utah Sahara bound? In: West, N. E., ed. Rangelands in a sustainable biosphere: Proceedings, 5th international rangelands congress; 1995 July 23-28; Salt Lake City, UT. Volume II: Invited presentations. Denver, CO: Society for Range Management: 475-476.

Roberts, Tom. 2005. [Personal communication]. July 18, Henderson, NV: U.S. Department of the Interior, Bureau of Land Management, Denver Federal Center.

Robertson, David J.; Antenen, Susan. 1990. Alien vine control in Piedmont forest restorations. Restoration & Management Notes. 8(1): 52.

Robertson, David J.; Robertson, Mary C.; Tague, Thomas. 1994. Colonization dynamics of four exotic plants in a northern Piedmont natural area. Bulletin of the Torrey Botanical Club. 121(2): 107-118.

Robertson, J. H.; Pearse, C. K. 1945. Artificial reseeding and the closed community. Northwest Science. 19: 59-66.

Robichaud, Peter R.; Beyers, Jan L.; Neary, Daniel G. 2000. Evaluating the effectiveness of postfire rehabilitation treatments. Gen. Tech. Rep. RMRS-GTR-63. Fort Collins, CO: U.S. Department of Agriculture, Forest Service, Rocky Mountain Research Station. 85 p.

Robinett, Dan. 1992. Lehmann lovegrass and drought in southern Arizona. Rangelands. 14: 100-103.

Robocker, W. C.; Miller, Bonita J. 1955. Effects of clipping, burning and competition on establishment and survival of some native grasses in Wisconsin. Journal of Range Management. 8: 117-120.

Roche, Ben F., Jr.; Roche, Cindy Talbott. 1999. Diffuse knapweed. In: Sheley, Roger L.; Petroff, Janet K., eds. Biology and management of noxious rangeland weeds. Corvallis, OR: Oregon State University Press: 217-230.

Rogers, G. F.; Vint, M. K. 1987. Winter precipitation and fire in the Sonoran Desert. Journal of Arid Environments. 13: 47-52.

Rogers, Garry F.; Steele, Jeff. 1980. Sonoran Desert fire ecology. In: Stokes, Marvin A.; Dieterich, John H., tech.l coords. Proceedings of the fire history workshop; 1980 October 20-24; Tucson, AZ. Gen. Tech. Rep. RM-81. Fort Collins, CO: U.S. Department of Agriculture, Forest Service, Rocky Mountain Forest and Range Experiment Station: 15-19.

Rogers, W. E.; Siemann, E. 2002. Effects of simulated herbivory and resource availability on native and invasive nonnative tree seedlings. Basic and Applied Ecology. 4: 297-307.

Rogers, W. E.; Siemann, E. 2003. Effects of simulated herbivory and resources on Chinese tallow tree (*Sapium sebiferum*, Euphorbiaceae) invasion of native coastal prairie. American Journal of Botany. 90: 241-247.

Rogers, W. E.; Siemann, E. 2004. Invasive ecotypes tolerate herbivory more effectively than native ecotypes of the Chinese tallow tree *Sapium sebiferum*. Journal of Applied Ecology. 41(3): 561-570.

Rogers, William E.; Siemann, Evan. 2005. Hervbivory tolerance and compensatory differences in native and invasive ecotypes of Chinese tallow tree (*Sapium sebiferum*). Plant Ecology. 181(1): 57-68.

Romme, W. H. 1982. Fire and landscape diversity in subalpine forests of Yellowstone National Park. Ecological Monographs. 52(5): 199-221.

Romme, W. H.; Bohland, Laura; Persichetty, Cynthia; Caruso, Tanya. 1995. Germination ecology of some common forest herbs in Yellowstone National Park, Wyoming, U.S.A. Arctic and Alpine Research. 27(4): 407-412.

Romme, W. H.; Floyd-Hanna, Lisa; Hanna, David D. 2003. Ancient piñon-juniper forests of Mesa Verde and the West: a cautionary note for forest restoration programs. In: Omi, Philip N.; Joyce, Linda A., tech. eds. Fire, fuel treatments, and ecological restoration: conference proceedings; 2002 April 16-18; Fort Collins, CO. Proceedings RMRS-P-29. Fort Collins, CO: U.S. Department of Agriculture, Forest Service, Rocky Mountain Research Station: 335-350.

Romme, W. H.; Knight, D. H. 1981. Fire frequency and subalpine forest succession along a topographic gradient in Wyoming. Ecology. 62(2): 319-326.

Romme, William. 1980. Fire history terminology: report of the ad hoc committee. In: Stokes, Marvin A.; Dieterich, John H., tech. coords. Proceedings of the fire history workshop. 1980 October 20-24; Tucson, AZ. Gen. Tech. Rep. RM-81. Fort Collins, CO: U.S. Department of Agriculture, Forest Service, Rocky Mountain Forest and Range Experiment Station: 135-137.

Rosburg, Thomas R. 2001. Iowa's non-native graminoids. Journal of the Iowa Academy of Science. 108(4): 142-153.

Rosburg, Thomas R.; Glenn-Lewin, David C. 1992. Effects of fire and atrazine on pasture and remnant prairie plant species in southern Iowa. In: Smith, Daryl D.; Jacobs, Carol A., eds. Recapturing a vanishing heritage: Proceedings, 12th North American prairie conference; 1990 August 5-9; Cedar Falls, IA. Cedar Falls, IA: University of Northern Iowa: 107-112.

Rossiter, N. A.; Setterfield, S. A.; Douglas, M. M.; Hutley, L. B. 2003. Testing the grass-fire cycle: alien grass invasion in the tropical savannas of northern Australia. Diversity and Distributions. 9: 169-176.

Rouget, M.; Richardson, D. M. 2003. Inferring process from pattern in plant invasions: a semimechanistic model incorporating propagule pressure and environmental factors. The American Naturalist. 162: 713-724.

Rowe, J. S. 1983. Concepts of fire effects on plant individuals and species. In: Wein, Ross W.; MacLean, David A., eds. The role of fire in northern circumpolar ecosystems. SCOPE 18. New York: John Wiley & Sons: 135-154.

Ruben, J. A.; Bolger, D. T.; Peart, D. R.; Ayres, M. P. 1999. Understory herb assemblages 25 and 60 years after clearcutting of a northern hardwood forest, USA. Biological Conservation. 90: 203-215.

Rudnicky, James L.; Patterson, William A., III; Cook, Robert P. 1997. Experimental use of prescribed fire for managing bird habitat at Floyd Bennett Field, Brooklyn, New York. In: Vickery, Peter, D.; Dunwiddie, Peter W., eds. Grasslands of northeastern North America: ecology and conservation of native and agricultural landscapes. Lincoln, MA: Massachusetts Audubon Society: 99-118.

Ruffner, C. M.; Abrams, M. D. 1998. Lightning strikes and resultant fires from archival (1912-1917) and current (1960-1997) information in Pennsylvania. Journal of the Torrey Botanical Society. 125: 249-252.

Ruggiero, Leonard F.; Jones, Lawrence L. C.; Aubry, Keith B. 1991. Plant and animal habitat associations in Douglas-fir forests of the Pacific Northwest: an overview. In: Ruggiero, Leonard F.; Aubry, Keith B.; Carey, Andrew B.; Huff, Mark H., tech. coords. Wildlife and vegetation of unmanaged Douglas-fir forests. Gen. Tech. Rep. PNW-GTR-285. Portland, OR: U.S. Department of Agriculture, Forest Service, Pacific Northwest Research Station: 447-462.

Rundel, P. W.; Vankat, J. L. 1989. Chaparral communities and ecosystems. In: Keeley, S. C., ed. The California chaparral: paradigms reexamined. Los Angeles, CA: Natural History Museum of Los Angeles County: 127-139.

Rupp, T. Scott; Chapin, F. Stuart, III; Starfield, Anthony M. 2000. Response of subarctic vegetation to transient climatic change on the Seward Peninsula in north-west Alaska. Global Change Biology. 6(5): 541-555.

Ruyle, G. B.; Roundy, B. A.; Cox, J. R. 1988. Effects of burning on germinability of Lehmann lovegrass. Journal of Range Management. 41(5): 404-406.

Ryan, K. C.; Frandsen, W. H. 1991. Basal injury from smoldering fires in mature *Pinus ponderosa* Laws. International Journal of Wildland Fire. 1(2): 107-118.

Ryan, Kevin C.; Noste, Nonan V. 1985. Evaluating prescribed fires. In: Lotan, James E.; Kilgore, Bruce M.; Fischer, William C.; Mutch, Robert W., tech. coords. Proceedings—symposium and workshop on wilderness fire; 1983 November 15-18; Missoula, MT. Gen. Tech. Rep. INT-182. Ogden, UT: U.S. Department of Agriculture, Forest Service, Intermountain Forest and Range Experiment Station: 230-238.

Sackett, Stephen S.; Haase, Sally M. 1998. Two case histories for using prescribed fire to restore ponderosa pine ecosystems in northern Arizona. In: Pruden, Teresa L.; Brennan, Leonard A., eds. Fire in ecosystem management: shifting the paradigm from suppression to prescription: Proceedings, Tall Timbers fire ecology conference; 1996 May 7-10; Boise, ID. No. 20. Tallahassee, FL: Tall Timbers Research Station: 380-389.

Safford, H.; Harrison, S. 2004. Fire effects on plant diversity in serpentine and sandstone chaparral. Ecology. 85: 539-548.

Sakai, Ann K.; Allendorf, Fred W.; Holt, Jodie S.; Lodge, David M.; Molofsky, Jane; With, Kimberly A.; Baughman, Syndallas; Cabin, Robert J.; Cohen, Joel E.; Ellstrand, Norman C.; McCauley, David E.; O'Neil, Pamela; Parker, Ingrid M.; Thompson, John N.; Weller, Stephen G. 2001. The population biology of invasive species. Annual Review of Ecology and Systematics. 32: 305-332.

Sala A.; Smith, S. D.; Devitt, D. A. 1996. Water use by *Tamarix ramosissima* and associated phreatophytes in a Mojave Desert floodplain. Ecological Applications. 6: 888-898.

Salo, Lucinda F. 2004. Population dynamics of red brome (*Bromus madritensis* subsp. *rubens*): times for concern, opportunities for management. Journal of Arid Environments. 57(3): 291-296.

Salo, Lucinda F. 2005. Red brome (*Bromus rubens* subsp. *madritensis*) in North America: possible modes for early introductions, subsequent spread. Biological Invasions. 7: 165-180.

Saltonstall, K. 2003. Microsatellite variation within and among North American lineages of *Phragmites australis*. Molecular Ecology. 12: 1689-1702.

Sampson, Arthur W. 1944. Plant succession on burned chaparral lands in northern California. Bull. 65. Berkeley, CA: University of California, College of Agriculture, Agricultural Experiment Station. 144 p.

Sampson, Arthur W.; Parker, Kenneth W. 1930. St. Johnswort on range lands of California. Bulletin 503. Berkeley, CA: University of California, College of Agriculture, Agriculture Experiment Station. 47 p.

Samuels, M. and Betancourt, J.L. 1982. Modeling the long-term effects of fuelwood harvests on pinyon-juniper woodlands. Environmental Management. 6(6): 505-515.

Sandberg, D. V.; Pierovich, J. M.; Fox, D. G.; Ross, E. W. 1979. Effects of fire on air: A state-of-knowledge review: Proceedings of the national fire effects workshop; 1978 April 10-14; Denver, CO. Gen. Tech. Rep. WO-9. Washington, DC: U.S. Department of Agriculture, Forest Service. 40 p.

Sands, Alan R.; Sather-Blair, Signe; Saab, Victoria. 2000. Sagebrush steppe wildlife: historical and current perspectives. In: Entwistle, P. G.; DeBolt, A. M.; Kaltenecker, J. H.; Steenhof, K., comps.. Sagebrush steppe ecosystems symposium: Proceedings; 1999 June 21-23; Boise, ID. Publ. No. BLM/ID/PT-001001+1150. Boise, ID: U.S. Department of the Interior, Bureau of Land Management, Boise State Office: 27-35.

Sapsis, David B. 1990. Ecological effects of spring and fall prescribed burning on basin big sagebrush/Idaho fescue--bluebunch wheatgrass communities. Corvallis, OR: Oregon State University. 105 p. Thesis.

Sarr, Daniel; Shufelberger, Amanda; Commons, Michael; Bunn, Wendy. 2003. Unpublished annual report for the Klamath Network Inventory Program: FY 2003 Non-Native Plant Inventory. Ashland, OR: U.S. Department of the Interior, Klamath Network-National Park Service, Inventory and Monitoring Program. [Number of pages unknown].

Sather, N. 1987a. Element stewardship abstract: *Bromus inermis* (awnless brome, smooth brome, [Online]. In: The global invasive species initiative. Arlington, VA: The Nature Conservancy (Producer). Available: http://tncweeds.ucdavis.edu/esadocs/documnts/bromine.pdf [2007, July 10].

Sather, N. 1987b. Element stewardship abstract: *Poa pratensis*, *Poa compressa* (Kentucky bluegrass, Canada bluegrass), [Online]. In: The global invasive species initiative. Arlington, VA: The Nature Conservancy (Producer). Available: http://tncweeds.ucdavis.edu/esadocs/documnts/poa_pra.pdf [2007, July 10].

Saunders, D.A.; Hobbs, R.J.; Margules, C.R. 1991. Biological consequences of ecosystem fragmentation: a review. Conservation Biology. 5(1): 18-32.

Sawyer, John O.; Keeler-Wolf, Todd. 1995. A manual of California vegetation. Misc. Report. Sacramento, CA: California Native Plant Society Press. 412 p.

Sawyer, John O.; Thornburgh, Dale A.; Griffin, James R. 1977. Mixed evergreen forest. In: Barbour, Michael G.; Major, Jack, eds. Terrestrial vegetation of California. New York: John Wiley and Sons: 359-381.

Sax, D. F.; Gaines, S. D.; Brown, J. H. 2002. Species invasions exceed extinctions on islands worldwide: a comparative study of plants and birds. American Naturalist. 160: 766-783.

Schacht, W.; Stubbendieck, J. 1985. Prescribed burning in the Loess Hills mixed prairie of southern Nebraska. Journal of Range Management. 38: 47-51.

Scherer G.; Zabowski, D.; Java, B.; Everett, R. 2000. Timber harvesting residue treatment . Part II understory vegetation response. Forest Ecology and Management. 126: 35-50.

Schirman, R. 1981. Seed production and spring seedling establishment of diffuse and spotted knapweed. Journal of Range Management. 34: 45-47.

Schlesginger, W. H.; Raikes, J. A.; Hartley, A. E.; Cross, A. E. 1996. On the spatial pattern of soil nutrients in desert ecosystems. Ecology. 77: 364-374.

Schlesinger, W. H.; Reynolds, Cunningham, G. L.; Huenneke, L. F.; Jarrell, W. M.; Virginia, R. A.; Whitford, W. G. 1990. Biological feedbacks in global desertification. Science. 247:1043-1048.

Schmalzer, P. A. 1995. Biodiversity of saline and brackish marshes of the Indian River lagoon: historic and current patterns. Bulletin of Marine Science. 57: 37-48.

Schmalzer, P. A.; Hinkle, C. R.; Mailander, J. L. 1991. Changes in community composition and biomass in *Juncus roemerianus* Scheele and *Spartina bakeri* Merr. marshes one year after a fire. Wetlands. 11: 67-86.

Schmid, M. K.; Rogers, G. F. 1988. Trends in fire occurrence in the Arizona Upland subdivision of the Sonoran Desert 1955-1983. Southwestern Naturalist. 33: 437-444.

Schmitz, D. C.; Hofstetter, R. H. 1999. Environmental, economic and human impacts. In: Laroche, F. B., ed. Melaleuca Management Plan. Florida Exotic Pest Plant Council: 17-21.

Schmitz, D. C.; Simberloff, D.; Hofstetter, R. H.; Haller, W.; Sutton, D. 1997. The ecological impacts of nonindigenous plants. In: Simberloff, Daniel; Schmitz, Don C.; Brown, Tom C., eds. Strangers in paradise, impact and management of nonindigenous species in Florida. Washington, DC: Island Press: 39-61.

Schoennagel, Tania L.; Waller, Donald M. 1999. Understory responses to fire and artificial seeding in an eastern Cascades Abies grandis forest, U.S.A. Canadian Journal of Forest Research. 29(9): 1393-1401.

Schoonmaker, Peter; McKee, Arthur. 1988. Species composition and diversity during secondary succession of coniferous forests in the western Cascade Mountains of Oregon. Forest Science. 34(4): 960-979.

Schramm, Peter. 1978. The „do's and don'ts" of prairie restoration. In: Glenn-Lewin, David C., and Landers, Roger Q. Jr., eds. Proceedings: Fifth Midwest prairie conference proceedings. Midwest prairie conference Ames, IA. Ames, IA: Iowa State University: 139-150.

Schreiner, Edward George. 1982. The role of exotic species in plant succession following human disturbance in an alpine area of Olympic National Park, Washington. Seattle, WA: University of Washington. 132 p. Dissertation.

Schuller, Reid. 1997. Changes in the south Puget prairie landscape. In: Dunn, Patrick; Ewing, Kern, eds. Ecology and conservation of south Puget Sound prairie landscape. Seattle, WA: The Nature Conservancy. 207-216.

Schultz, A. M.; Launchbaugh, J. L.; Biswell, H. H. 1955. Relationship between grass density and brush seedling survival. Ecology. 36(2): 226-238.

Schultz, Cheryl B.; Crone, Elizabeth E. 1998. Burning prairie to restore butterfly habitat: A modeling approach to management tradeoffs for the Fender's blue. Restoration Ecology. 6(3): 244-252.

Schultz, Gary E. 1993. Element stewardship abstract: Dioscorea bulbifera (air potato), [Online]. In: The global invasive species initiative. Arlington, VA: The Nature Conservancy (Producer). Available: http://tncweeds.ucdavis.edu/esadocs/documnts/diosbul.pdf [2004, December 29].

Schussman, Heather. 2006. [Review comments]. June 14. On file at: U.S. Department of Agriculture, Forest Service, Rocky Mountain Research Station, Fire Sciences Lab, Missoula, MT.

Schwartz, M. W.; Porter, D. J.; Randall, J. M.; Lyons, K. E. 1996. Impact of non-indigenous plants. In: Status of the Sierra Nevada. Sierra Nevada Ecosystem Project: Final report to Congress. Volume II: Assessments and scientific basis for management options. Wildland Resources Center Report No. 37. Davis, CA: University of California, Centers for Water and Wildland Resources: 1203-1226.

Schwartz, Mark W. 1997. Defining indigenous species: an introduction. In: Luken, James O.; Thieret, John W.; eds. Assessment and management of plant invasions. Springer Series on Environmental Management. New York: Springer-Verlag: 7-17.

Schwegman, John E.; Anderson, Roger C. 1986. Effect of eleven years of fire exclusion on the vegetation of a southern Illinois barren remnant. In: Clambey, Gary K.; Pemble, Richard H., eds. The prairie: past, present and future: Proceedings of the 9th North American prairie conference; 1984 July 29-August 1; Moorhead, MN. Fargo, ND: Tri-College University Center for Environmental Studies: 146-148.

Schwilk, D. W. 2003. Flammability is a niche construction trait; canopy architecture affects fire intensity. American Naturalist. 162: 725-733.

Scifres, C. J.; Hamilton, W. T. 1993. Prescribed burning for brushland management: The south Texas example. College Station, TX: Texas A&M University Press. 246 p.

Scott, J. H.; Reinhardt, E. D., comps.. 2007. FireWords Version 1.0: Fire Science Glossary [electronic]. U.S. Department of Agriculture, Forest Service, Rocky Mountain Research Station, Fire Sciences Laboratory (Producer). Available: http://www.firewords.net

Segelquist, C. A. 1971. Moistening and heating improve germination of two legume species. Journal of Range Management. 24: 393-394.

Seiger, Leslie. 1991. Element stewardship abstract for *Polygonum cuspidatum*, Japanese knotweed, Mexican bamboo, [Online]. Arlington, VA: The Nature Conservancy (Producer). 9 p. Available: www.tncweeds.ucdavis.edu [2007, July 12].

Selleck, G. W.; Coupland, R. T.; Frankton, C. 1962. Leafy spurge in Saskatchewan. Ecological Monographs. 32: 1-29.

Selmants, Paul C.; Knight, Dennis H. 2003. Understory plant species composition 30-50 years after clearcutting in southeastern Wyoming coniferous forests. Forest Ecology and Management. 185: 275-289.

Shafroth, P. B.; Stromberg, J. C.; Patten, D. T. 2002. Riparian vegetation response to altered disturbance and stress regimes. Ecological Applications. 12(1): 107-123.

Shafroth, Patrick B.; Auble, Gregor T.; Scott, Michael L. 1995. Germination and establishment of the native plains cottonwood (*Populus deltoides* ssp. *monilifera*) and the exotic Russian-olive (*Elaeagnus angustifolia* L.). Conservation Biology. 9(5): 1169-1175.

Sharp, Lee A.; Hironaka, M.; Tisdale, E. W. 1957. Viability of medusa-head (*Elymus caput-medusae* L.) seed collected in Idaho. Journal of Range Management. 10: 123-126.

Shaw, John D.; Steed, Brytten E.; DeBlander, Larry T. 2005. Forest inventory and analysis (FIA) annual inventory answers the question: what is happening to pinyon-juniper woodlands? Journal of Forestry. 2005: 280-285.

Shaw, R. B.; Castillo, J. M.; Laven, R. D. 1997. Impact of wildfire on vegetation and rare plants within the Kipuka Kalawamauna endangered plants habitat area—Pohakuloa Training Area, Hawaii. In: Greenlee, Jason M., ed. Proceedings, 1st conference on fire effects on rare and endangered species and habitats; 1995 November 13-16; Coeur d'Alene, ID. Fairfield, WA: International Association of Wildland Fire: 253-266.

Shearer, Raymond C.; Stickney, Peter F. 1991. Natural revegetation of burned and unburned clearcuts in western larch forests of northwest Montana. In: Nodvin, Stephen C.; Waldrop, Thomas A., eds. Fire and the environment: ecological and cultural perspectives: Proceedings of an international symposium; 1990 March 20-24; Knoxville, TN. Gen. Tech. Rep. SE-69. Asheville, NC: U.S. Department of Agriculture, Forest Service, Southeastern Forest Experiment Station: 66-74.

Sheeley, Scott E. 1992. Distribution and life history of *Vincetoxicum rossicum* (Asclepiadaceae): an exotic plant in North America. Syracuse, NY: State University of New York, College of Environmental Science and Forestry. 126 p. Thèsis.

Sheeley, Scott E.; Raynal, Dudley J. 1996. The distribution and status of species of *Vincetoxicum* in eastern North America. Bulletin of the Torrey Botanical Club. 123(2): 148-156.

Sheley, R. L.; Roche, B. F. Jr. 1982. Rehabilitation of spotted knapweed infested rangeland in northeastern Washington. Western Society of Weed Science. Abstract: 31.

Sheley, Roger L.; Jacobs, James J.; Carpinelli, Michael F. 1998. Distribution, biology, and management of diffuse knapweed (*Centaurea diffusa*) and spotted knapweed (*Centaurea maculosa*). Weed Technology. 12: 353-362.

Shelford, Victor E. 1963. The ecology of North America. Urbana, IL: University of Illinois Press. 610 p.

Shumway, Durland L.; Abrams, Marc D.; Ruffner, Charles M. 2001. A 400-year history of fire and oak recruitment in an old-growth oak forest in western Maryland, U.S.A. Canadian Journal of Forest Research. 31: 1437-1443.

Sieg, Carolyn Hull; Phillips, Barbara G.; Moser, Laura P. 2003. Exotic invasive plants. In Friederici, Peter, editor. Ecological Restoration of Southwestern Ponderosa Pine Forests. Washington, DC: Island Press: 251-267.

Siemann, E.; Rogers, W. E. 2003a. Herbivory, disease, recruitment limitation and the success of alien and native ecotypes of an non-native species. Ecology. 84: 1489-1505.

Siemann, E.; Rogers, W. E. 2003b. Reduced resistance of invasive varieties of the alien tree Sapium sebiferum to a generalist herbivore. Oecologia. 135: 451-457.

Silander, J. A.; Klepeis, D. M. 1999. The invasion ecology of Japanese barberry (*Berberis thunbergii*) in the New England landscape. Biological Invasions. 1: 189-201.

Simberloff, Daniel. 2003. Confronting introduced species: a form of xenophobia? Biological invasions. 5: 179-192.

Simms, P. L.; Risser, P. G. 2000. Grasslands. In: Barbour, M. G.; Billings, W. D., eds. North American terrestrial vegetation. New York: Cambridge University Press: 323-356.

Simonin, Kevin A. 2000. Euphorbia esula. In: Fire Effects Information System, [Online]. U.S. Department of Agriculture, Forest Service, Rocky Mountain Research Station, Fire Sciences Laboratory (Producer). Available: http://www.fs.fed.us/database/feis/ [2006, July 20].

Simpfendorfer, K. J. 1989. Trees, farms and fires. Land and Forests Bulletin No. 30. Victoria, Australia: Department of Conservation, Forests and Lands. 55 p.

Skinner, C. N.; Taylor, A. H. 2006. Fire in the Southern Cascade Bioregion. In: Sugihara, N.; van Wagtendonk, J. W.; Fites, J.; Thode, A., eds. Fire in California ecosystems. Berkeley, CA: University of California Press: 195-224.

Skinner, C. N.; Taylor, A. H.; Agee, J. K. 2006. Fire in the Klamath Mountains Bioregion. In: Sugihara, N.; van Wagtendonk, J. W.; Fites, J.; Thode, A., eds. Fire in California ecosystems. Berkeley, CA: University of California Press: 170-194

Skinner, Carl N.; Chang, Chi-ru. 1996. Fire regimes, past and present. In: Status of the Sierra Nevada. Sierra Nevada Ecosystem Project: Final report to Congress. Volume II: Assessments and scientific basis for management options. Wildland Resources Center Report No. 37. Davis, CA: University of California, Centers for Water and Wildland Resources: 1041-1069.

Skinner, Nancy G.; Wakimoto, Ronald H. 1989. Site preparation for rangeland grass planting—a literature review. In: Baumgartner, David M.; Breuer, David W.; Zamora, Benjamin A.; [and others], comps.. Prescribed fire in the Intermountain region: Symposium proceedings; 1986 March 3-5; Spokane, WA. Pullman, WA: Washington State University, Cooperative Extension: 125-131.

Slocum, M. G.; Platt, W. J.; Cooley, H. C. 2003. Effects of differences in prescribed fire regimes on patchiness and intensity of fires in subtropical savannas of Everglades National Park, Florida. Restoration Ecology. 11: 91-102.

Smeins, F. E.; Diamond, D. D.; Hanselka, C. W. 1992. Coastal prairie. In: Coupland, R. T., ed. Natural grasslands: Introduction and western hemisphere. Ecosystems of the World 8A. Amsterdam, Netherlands: Elsevier Science Publishers B. V.: 269-290.

Smith, Clifford W. 1985. Impact of alien plants on Hawaii's native biota. In: Stone, Charles P.; Scott, J. Michael, eds. Hawaii's terrestrial ecosystems: preservation and management. Honolulu, HI, University of Hawai`i, Cooperative National Park Resources Studies Unit: 180-243.

Smith, Clifford W; Tunison, J. Timothy. 1992. Fire and alien plants in Hawaii: research and management implications for native ecosystems. In: Stone, Charles P.; Smith, Clifford W.; Tunison, Timothy, eds. Symposium on alien plant invasions in native ecosystems of Hawaii: management and research. Honolulu, HI: University of Hawai`i, Cooperative Parks Resources Studies Unit: 394-408.

Smith, Jane Kapler, ed. 2000. Wildland fire in ecosystems: Effects of fire on fauna. Gen. Tech. Rep. RMRS-GTR-42-vol. 1. Ogden, UT: U.S. Department of Agriculture, Forest Service, Rocky Mountain Research Station. 83 p.

Smith, Karen A. 1985. Canada thistle response to prescribed burning (North Dakota). Restoration and Management Notes. 3(2): Note 94.

Smith, Larissa L.; DiTommaso, Antonio; Lehmann, Johannes; Greipsson, Sigurdur. 2006. Growth and reproductive potential of the invasive exotic vine *Vincetoxicum rossicum* in northern New York State. Canadian Journal of Botany. 84: 1771-1780.

Smith, Melinda D.; Knapp, Alan K. 1999. Nonnative plant species in a C4-dominated grassland: invasibility, disturbance, and community structure. Oecologia. 120(4): 605-612.

Smith, Melinda D.; Knapp, Alan K. 2001. Size of the local species pool determines invasibility of a C4-dominated grassland. Oikos. 92(1): 55-61.

Smith, Melinda D.; Wilcox, Julia C.; Kelly, Theresa; Knapp, Alan K. 2004. Dominance not richness determines invasibility of tallgrass prairie. Oikos. 106: 253-262

Smith, S. J.; Anderson, R. S. 1992. Late Wisconsin paleoecologic record from Swamp Lake, Yosemite National Park, California. Quartenary Research. 38: 91-102.

Smith, Stanley D.; Devitt, Dale A.; Sala, Anna; Cleverly, James R.; Busch, David E. 1998. Water relations of riparian plants from warm desert regions. Wetlands. 18(4): 687-696.

Smith, Stanley D.; Huxman, Travis E.; Zitzer, Stephen F.; Charlet, Therese N.; Housman, David C.; Coleman, James S.; Fenstermaker, Lynn K; Seemann, Jeffrey R.; Nowak, Robert S. 2000. Elevated CO_2 increases productivity and invasive species success in an arid ecosystem. Nature. 408: 79-81.

Smith, Stanley D.; Monson, Russell K.; Anderson, Jay E. 1997. Exotic plants. In: Smith, Stanley D.; Monson, Russell K.; Anderson, Jay E., eds. Physiological ecology of North American desert plants. New York: Springer-Verlag: 199-227.

Smith, Theodore A. 1970. Effects of disturbance on seed germination in some annual plants. Ecology. 51(6): 1106-1108.

Smith, W. Brad; Vissage, John S.; Darr, David R.; Sheffield, Raymond M. 2001. Forest Resources of the United States, 1997. Gen. Tech. Rep. NC-219. St. Paul, MN: U.S. Department of Agriculture, Forest Service, North Central Research Station. 109 p

Smith, W. K.; Gorz, H. J. 1965. Sweetclover improvement. Advances in Agronomy. 17: 163-231.

Smoliak, S.; Penney, D.; Harper, A. M.; Horricks, J. S. 1981. Alberta forage manual. Edmonton, AB: Alberta Agriculture, Print Media Branch. 87 p.

Snyder, J. R. 1991. Fire regimes in subtropical Florida. In: Hermann, Sharon, M., coord. High intensity fire in wildlands: management challenges and options: 17th Tall Timbers Fire Ecology Conference; 1989 May 18-21; Tallahassee, FL. Tallahassee, FL: Tall Timbers Research Station, The Nature Conservancy. 303-319.

Snyder, S. A. 1992a. *Elytrigia repens*. In: Fire Effects Information System, [Online]. U.S. Department of Agriculture, Forest Service, Rocky Mountain Research Station, Fire Sciences Laboratory (Producer). Available: http://www.fs.fed.us/database/feis/[2005, March 20].

Snyder, S. A. 1992b. *Phalaris arundinacea*. In: Fire Effects Information System, [Online]. U.S. Department of Agriculture, Forest Service, Rocky Mountain Research Station, Fire Sciences Laboratory (Producer). Available: http://www.fs.fed.us/database/feis/ [2007, May 17].

Sokal, Robert R.; Rohlf, F. James. 2000. Biometry. New York: W. H. Freeman and Co. 887 p.

Solecki, Mary Kay. 1997. Controlling invasive plants. In: Packard, Stephen; Mutel, Cornelia F., eds. The tallgrass restoration handbook: For prairies, savannas, and woodlands. Washington, DC: Island Press: 251-278.

Soll, Jonathan. 2004a. Controlling knotweed (*Polygonum cuspidatum, P. sachalinense, P. polystachyum* and hybrids) in the Pacific Northwest, [Online]. In: The global invasive species initiative. Arlington, VA: The Nature Conservancy (Producer). Available: http://tncweeds.ucdavis.edu/moredocs/polspp01.pdf[2005, January 15].

Soll, Jonathan. 2004b. Controlling Himalayan blackberry (*Rubus armeniacus* [*R. discolor, R. procerus*]) in the Pacific Northwest, [Online]. In: The global invasive species initiative. Arlington, VA: The Nature Conservancy (Producer). Available: http://tncweeds.ucdavis.edu/moredocs/rubarm01.pdf [2005, March 15].

Soll, Jonathan. 2004c. Controlling English ivy (*Hedera helix*) in the Pacific Northwest, [Online]. In: The global invasive species initiative. Arlington, VA: The Nature Conservancy (Producer). Available: http://tncweeds.ucdavis.edu/moredocs/hedhel02.pdf [2005, February 20].

Son, Yowhan; Lee, Yoon Young; Jun, Young Chul; Kim, Zin-Suh. 2004. Light availability and understory vegetation four years after thinning in a *Larix leptolepis* plantation of central Korea. Journal of Forest Research. 9: 133-139.

Soto, B.; Basanta, R.; Diaz-Fierros, F. 1997. Effects of burning on nutrient balance in an area of gorse (*Ulex europaeus* L.) scrub. Science of the Total Environment. 204(3): 271-281.

Spaeth, K. E.; Pierson, F. B.; Herrick, J. E.; Shaver, P. L.; Pyke, D. A.; Pellant, M.; Thompson, D.; Dayton, R. 2003. New proposed national resources inventory protocols on nonfederal rangelands. Journal of Soil and Water Conservation. 58(1): 18-21.

Sparks, Steven R.; West, Neil E.; Allen, Edith B. 1990. Changes in vegetation and land use at two townships in Skull Valley, western Utah. In: McArthur, E. Durant; Romney, Evan M.; Smith, Stanley D.; Tueller, Paul T., comps.. Proceedings--symposium on cheatgrass invasion, shrub die-off, and other aspects of shrub biology and management; 1989 April 5-7; Las Vegas, NV. Gen. Tech. Rep. INT-276. Ogden, UT: U.S. Department of Agriculture, Forest Service, Intermountain Research Station: 26-36.

Spears, B. M.; Rose, S. T.; Belles, W. S. 1980. Effect of canopy cover, seeding depth, and soil moisture on emergence of *Centaurea maculosa* and *C. diffusa*. Weed Research. 20: 87-90.

Speer, R. K.; Bailey, J. D. [In review]. Understory vegetation responses to group selection thinning and prescribed fire at the Southwest Plateau Fire and Fire Surrogate Sites. In: Van Ripper, C.; Sogge, M., eds. Proceedings of the 8th Biennial Conference of Research on the Colorado Plateau. U.S. Geological Survey/FRESC Report Series; [Meeting date unknown]; [Location unknown].

Sprenger, M. D.; Smith, L. M.; Taylor, John P. 2002. Restoration of riparian habitats using experimental flooding. Wetlands. 22: 49-57.

Springer, Judith D.; Waltz, Amy E. M.; Fulé, Peter Z.; Moore, Margaret M.; Covington, W. Wallace. 2001. Seeding versus natural regeneration: A comparison of vegetation change following thinning and burning in ponderosa pine. In: Vance, Regina K.; Edminster, Carleton B.; Covington, W. Wallace; Blake, Julie A., comps.. Ponderosa pine ecosystems restoration and conservation: steps toward stewardship: Conference proceedings; 2000 April 25-27; Flagstaff, AZ. Proceedings RMRS-P-22. Ogden, UT: U.S. Department of Agriculture, Forest Service, Rocky Mountain Research Station: 67-73.

Stanturf, J. A.; Wade, D. D.; Waldrop, T. A.; Kennard, D. K.; Achtemeier, G. L. 2002. Background paper: fire in southern forest landscapes. In: Wear, David N.; Greis, John G., eds. Southern forest resource assessment. Gen. Tech Rep. SRS-53. Asheville, NC: U.S. Department of Agriculture, Forest Service, Southern Research Station. 607-630.

Stebbins, G. L.; Major, J. 1965. Endemism and speciation in the California flora. Ecological Monographs. 35: 1-35.

Steen, Harold K. 1966. Vegetation following slash fires in one western Oregon locality. Northwest Science. 40(3): 113-120.

Stein, William I. 1995. Ten-year development of Douglas-fir and associated vegetation after different site preparation on Coast Range clearcuts. Res. Pap. PNW-RP-473. Portland, OR: U.S. Department of Agriculture, Forest Service, Pacific Northwest Forest and Range Experiment Station. 115 p.

Stephens, Scott L.; Fulé, Peter Z. 2005. Western pine forests with continuing frequent fire regimes: possible reference sites for management. Journal of Forestry. 103(7): 357-362.

Stephenson, John R.; Calcarone, Gena M. 1999. Mountain and foothills ecosystems. Albany, CA: U.S. Department of Agriculture, Forest Service, Pacific Southwest Research Station. 15-60.

Stephenson, N. L. 1998. Actual evapotranspiration and deficit: biologically meaningful correlates of vegetation distribution across spatial scales. Journal of Biogeography. 25: 855-870.

Steuter, Allen A. 1988. Wormwood sage controlled by spring fires (South Dakota). Restoration and Management Notes. 6(1): 35.

Stevens, Lawrence E. 1989. The status of ecological research on tamarisk (Tamaricaceae: *Tamarix ramosissima*) in Arizona. In: Kunzmann, Michael R.; Johnson, R. Roy; Bennett, Peter, technical coordinators. Tamarisk control in southwestern United States: Proceedings; 1987 September 2-3; Tucson, AZ. Special Report No. 9. Tucson, AZ: National Park Service, Cooperative National Park Resources Studies Unit, School of Renewable Natural Resources: 99-105.

Stevens, Sandy. 2002. Element stewardship abstract: *Lespedeza cuneata* (Dumont-Cours.) G. Don (sericea lespedeza, Chinese bush clover), [Online]. In: The global invasive species initiative. Arlington, VA: The Nature Conservancy (Producer). Available: http://tncweeds.ucdavis.edu/esadocs/documnts/lespcun.pdf[2003, November 17].

Stewart, G; Hull, A. C. 1949. Cheatgrass (*Bromus tectorum* L.)-an ecological intruder in southern Idaho. Ecology. 30: 58-74.

Stewart, Omer C. [Lewis, Henry T.; Anderson, M. Kat, eds]. 2002. Forgotten fires: Native Americans and the transient wilderness. Norman, OK: University of Oklahoma Press. 364 p.

Stewart, R. E. 1978. Origin and development of vegetation after spraying and burning in a coastal Oregon clearcut. Research Note PNW-317. Portland, OR: U. S. Department of Agriculture, Forest Service, Pacific Northwest Forest and Range Experiment Station. 10 p.

Stickney, P.F. 1990. Early development of vegetation following holocaustic fire in Northern Rocky Mountain forests. Northwest Science. 64: 243-246.

Stickney, Peter F. 1980. Data base for post-fire succession, first 6 to 9 years, in Montana larch-fir forests. Gen. Tech. Rep. INT-62. Ogden, UT: U.S. Department of Agriculture, Forest Service, Intermountain Forest and Range Experiment Station. 133 p.

Stickney, Peter F. 1986. First decade plant succession following the Sundance Forest Fire, northern Idaho. Gen. Tech. Rep. INT-197. Ogden, UT: U.S. Department of Agriculture, Forest Service, Intermountain Research Station. 26 p.

Stickney, Peter F.; Campbell, Robert B., Jr. 2000. Data base for early postfire succession in northern Rocky Mountain forests. Gen. Tech. Rep. RMRS-GTR-61-CD, [CD-ROM]. Fort Collins, CO: U.S. Department of Agriculture, Forest Service, Rocky Mountain Research Station. On file with: U.S. Department of Agriculture, Forest Service, Rocky Mountain Research Station, Fire Sciences Laboratory, Missoula, MT. Available: http://www.fs.fed.us/rm/pubs/rmrs_gtr61CD.htm [2008 April 4].

Stinson, Kenneth J.; Wright, Henry A. 1969. Temperatures of headfires in the southern mixed prairie of Texas. Journal of Range Management. 22(3): 169-174.

Stocker, G. C.; Mott, J. J. 1981. Fire in the tropical forests and woodlands of northern Australia. In: Gill, A. M.; Groves, R. H.; Noble, I. R., eds. Fire and the Australian biota. Canberra, Australia: Australian Academy of Science. 425-439.

Stocker, R. K.; Ferriter, A.; Thayer, D.; Rock, M.; Smith, S. 1997. L. microphyllum hitting south Florida below the belt. Wildland Weeds. Winter: 6-10.

Stocker, R. K.; Miller, R. E. Jr.; Black, D. W.; Ferriter, A. P.; Thayer, D. D. [In Press]. Using fire and herbicide to control *Lygodium microphyllum* and effects on a pine flatwoods plant community in south Florida. Natural Areas Journal [Volume unknown]. [Pages Unknown].

Stoddard, H. L. 1962. The use of fire in pine forests and game lands of the deep southeast. In: Proceedings 1st Annual Tall Timbers Fire Ecology Conference. Tallahassee, FL. Tallahassee, FL. 31-42.

Stoddard, H. L., Sr. 1931. The bobwhite quail: its habits, preservation, and increase. New York: Charles Scribner's Sons. 559 p.

Stohlgren, T. J.; Binkley, D.; Chong, G. W.; Kalkhan, M. A.; Schell, L. D.; Bull, K. A.; Otsuki, Y.; Newman, G.; Bashkin, M.; Son, Y. 1999a. Nonnative plant species invade hot spots of native plant diversity. Ecological Monographs. 69: 25-46.

Stohlgren, T. J.; Bull, K. A.; Otsuki, Y.; Villa, C.; Lee, M. 1998. Riparian zones as havens for nonnative plant species. Plant Ecology. 138: 113-125.

Stohlgren, T. J.; Schell, L. D.; Vanden Heuvel, B. 1999b. Effects of grazing and soil quality on native and nonnative plant diversity in Rocky Mountain grasslands. Ecological Applications. 9(1): 45-64.

Stohlgren, Thomas J.; Crosier, Catherine; Chong, Geneva W.; Guenther, Debra; Evangelista, Paul. 2005. Life-history habitat matching in invading non-native plant species. Plant and Soil. 277(1-2): 7-18.

Stoneberg, S. 1989. Goats make "cents" out of the scourge of leafy spurge. Rangelands. 11: 264-265.

Stransky, J. J. 1984. Forage yield of Japanese honeysuckle after repeated burning or mowing [*Lonicera japonica*]. Journal of Range Management. 37: 237-238.

Stransky, John J.; Huntley, Jimmy C.; Risner, Wanda J. 1986. Net community production dynamics in the herb-shrub stratum of a loblolly pine-hardwood forest: effects of clearcutting and site preparation. Gen. Tech. Rep. SO-61. New Orleans, LA: U.S. Department of Agriculture, Forest Service, Southern Forest Experiment Station. 11 p.

Streatfeild, Rosemary; Frenkel, Robert E. 1997. Ecological survey and interpretation of the Willamette Floodplain Research Area, W.L. Finley National Wildlife Refuge, Oregon, USA. Natural Areas Journal. 17: 346-354.

Streng, Donna R.; Glitzenstein, Jeff S.; Platt, William J. 1993. Evaluating effects of season of burn in longleaf pine forests: a critical literature review and some results from an ongoing long-term study. In: Hermann, Sharon M., ed. The longleaf pine ecosystem: ecology, restoration and management: Proceedings, 18th Tall Timbers fire ecology conference; 1991 May 30-June 2; Tallahassee, FL. No. 18. Tallahassee, FL: Tall Timbers Research, Inc: 227-263.

Stromberg, J. C.; Chew, M. 2002. Foreign visitors in riparian corridors of the American Southwest: is xenophobia justified? In: Tellman, B., ed. Invasive exotic species in the Sonoran Region. Tucson, AZ: University of Arizona Press: 195-219.

Stuart, J. D.; Grifantini, M. C.; Fox, L. 1993. Early successional pathways following wildfire and subsequent silvicultural treatment in Douglas-fir/hardwood forests, NW California. Forest Science. 39: 561-572.

Stuart, J.; Stephens, S. L. 2006. Fire in the North Coast bioregion. In: Sugihara, N.; van Wagtendonk, J. W.; Fites, J.; Thode, A., eds. Fire in California ecosystems. Berkeley, CA: University of California Press: 147-169.

Stuever, Mary C. 1997. Fire-induced mortality of Rio Grande cottonwood. Albuquerque, NM: University of New Mexico. 85 p. Thesis.

Stuever, Mary C.; Crawford, Clifford S.; Molles, Manuel C.; [and others]. 1997. Initial assessment of the role of fire in the Middle Rio Grande bosque. In: Greenlee, Jason M., ed. Proceedings, 1st conference on fire effects on rare and endangered species and habitats; 1995 November 13-16; Coeur d'Alene, ID. Fairfield, WA: International Association of Wildland Fire: 275-283.

Sturdevant, Nancy J.; Dewey, Jed. 2002. Evaluating releases of *Cyphocleonus achates* and *Agapeta zoegana* as potential field insectaries and effects of wildfire on previous releases. Forest Health Protection Report 02-6. Missoula, MT: U.S. Department of Agriculture, Forest Service, Northern Region. 5 p.

Stylinski, C. D.; Allen, E. B. 1999. Lack of native species recovery following severe exotic disturbance in southern Californian shrublands. Journal of Applied Ecology. 36: 544-554.

Sugihara, N. G.; van Wagtendonk, J. W.; Fites-Kaufman, J. 2006a. Fire as an ecological process. In: Sugihara, N. G.; van Wagtendonk, J. W.; Fites-Kaufman, J.; Shaffer, K. E.; Thode, A. E., eds. Fire in California ecosystems. Berkeley, CA: University of California Press. 58-74.

Sugihara, N. G.; van Wagtendonk, J. W.; Fites-Kaufman, J.; Shaffer, K. E.; Thode, A. E., eds. 2006b. Fire in California ecosystems. Berkeley, CA: University of California Press. 596 p.

Sullivan, Janet. 1992a. *Dactylis glomerata*. In: Fire Effects Information System, [Online]. U.S. Department of Agriculture, Forest Service, Rocky Mountain Research Station, Fire Sciences Laboratory (Producer). Available: http://www.fs.fed.us/database/feis/ [2005, May 22].

Sullivan, Janet. 1992b. *Lolium perenne*. In: Fire Effects Information System, [Online]. U.S. Department of Agriculture, Forest Service, Rocky Mountain Research Station, Fire Sciences Laboratory (Producer). Available: http://www.fs.fed.us/database/feis/ [2005, March 29].

Sullivan, John. 2005. Resource management planning efforts on the Bureau of Land Management's Snake River Birds of Prey National Conservation Area. In: Ralph, C. John; Rich, Terrell D., eds. Bird conservation implementation and integration in the Americas: proceedings of the 3rd international Partners in Flight conference: Vol. 2; 2002 March 20-24; Asilomar, CA. Gen. Tech. Rep. PSW-GTR-191. Albany, CA: U.S. Department of Agriculture, Forest Service, Pacific Southwest Research Station: 1184-1185.

Sumrall, L. Bradley; Roundy, Bruce A.; Cox, Jerry R.; Winkel, Von K. 1991. Influence of canopy removal by burning or clipping on emergence of *Eragrostis lehmanniana* seedlings. International Journal of Wildland Fire 1(1): 35-40.

Susko, David J.; Mueller, J. Paul; Spears, Janet F. 2001. An evaluation of methods for breaking seed dormancy in kudzu (*Pueraria lobata*). Canadian Journal of Botany. 79: 197-203.

Sutherland, Elaine Kennedy. 1997. History of fire in a southern Ohio second-growth mixed-oak forest. In: Pallardy, Stephen G.; Cecich, Robert A.; Garrett, H. Gene; Johnson, Paul S., eds. Proceedings, 11th central hardwood forest conference; 1997 March 23-26; Columbia, MO. Gen. Tech. Rep. NC-188. St. Paul, MN: U.S. Department of Agriculture, Forest Service, North Central Forest Experiment Station: 172-183.

Sutherland, S. 2004. What makes a weed a weed: life history traits of native and exotic plants in the USA. Oecologia. 141: 24-39.

Sutherland, Steve. 2006. [Personal communication]. April 1, Missoula, MT: U.S. Department of Agriculture, Forest Service, Rocky Mountain Research Station, Fire Sciences Laboratory.

Sutherland, Steve. 2007. Fire and weeds: general patterns. Unpublished paper on file at: U.S. Department of Agriculture, Forest Service, Rocky Mountain Research Station, Fire Sciences laboratory, Missoula, MT. 36 p.

Sutherland, Steve. 2008. Monitoring the impact of wildfire, fire suppression, and post-burn restoration on exotic weed invasion. Unpublished data on file at: U.S. Department of Agriculture, Forest Service, Rocky Mountain Research Station, Fire Sciences laboratory, Missoula, MT.

Sutton, R. F.; Tinus, R. W. 1983. Root and root system terminology. Forest Science Monograph 24. Washington, DC: Society of American Foresters. 137 p.

Svedarsky, W. D.; Buckley, P. E.; Feiro, T. A. 1986. The effect of 13 years of annual burning on an aspen-prairie ecotone in northwestern Minnesota. In: Clambey, Gary K., and Pemble, Richard H., eds. Proceedings: Proceedings of the ninth North American prairie conference—the prairie: past, present and future; Moorhead, MN. Fargo, ND: Tri-College University: 118-122.

Swan, Frederick R., Jr. 1970. Post-fire response of four plant communities in south-central New York State. Ecology. 51(6): 1074-1082.

Swanson, David K. 1996. Susceptibility of permafrost to deep thaw after forest fires in interior Alaska, U.S.A., and some ecological implications. Arctic and Alpine Research. 28(2): 217-227.

Swearingen, J. 2005. Alien plant invaders of natural areas, [Online]. Plant Conservation Alliance, Alien Plant World Group (Producer). Available: http://www.nps.gov/plants/alien/map/ardo1.htm [2006, March 7].

Swearingen, J.; Reshetiloff, K.; Slattery, B.; Zwicker, S. 2002. Plant invaders of Mid-Atlantic natural areas. Washington, DC: National Park Service and U.S. Fish & Wildlife Service. 82 p.

Sweeney, J. R. 1956. Responses of vegetation to fire: a study of herbaceous vegetation following chaparral fires. University of California Publications in Botany. 28: 143-250.

Swenson, Charles F.; LeTourneau, Duane; Erickson, Lambert C. 1964. Silica in medusahead. Weeds. 12: 16-18.

Swetnam, T. W. 1993. Fire history and climate change in giant sequoia groves. Science. 262: 885-889.

Swetnam, Thomas W.; Betancourt, Julio L. 1998. Mesoscale disturbance and ecological response to decadal climatic variability in the American Southwest. Journal of Climate. 11: 3128-3147.

Swezy, Michael; Odion, Dennis C. 1997. Fire on the mountain: a land manager's manifesto for broom control. In: Kelly, Mike; Wagner, Ellie; Warner, Peter, eds. Proceedings, California Exotic Pest Plant Council symposium; 1997 October 2-4; Concord, CA. Volume 3. Berkeley, CA: California Exotic Pest Plant Council: 76-81.

Szafoni, Robert E. 1991. Vegetation Management Guideline: Multiflora rose (*Rosa multiflora Thunb.*). Natural Areas Journal. 11(4): 215-216.

Taft, John B.; Solecki, Mary Kay. 1990. Vascular flora of the wetland and prairie communities of Gavin Bog and Prairie Nature Preserve, Lake County, Illinois. Rhodora. 92(871): 142-165.

Takeuchi, Wayne. 1991. Botanical survey of Puuwaawaa. Unpublished report on file at: Hawai`i Department of Land and Natural Resources, Division of Forestry and Wildlife, Hilo Branch, Hilo, HI. 62 p.

Tausch, Robin J. 1999. Transitions and thresholds: influence and implications for management in pinyon and Utah juniper woodlands. In: Monsen, Stephen B.; Stevens, R.; Tausch, Robin J.; Miller R.; Goodrich, S., eds. Proceedings: Ecology and management of pinyon-juniper communities within the Interior West. Ogden, UT: U.S. Department of Agriculture, Forest Service, Rocky Mountain Research Station: 61-65.

Tausch, Robin J.; West, Neil E. 1995. Plant species composition patterns with differences in tree dominance on a southwestern Utah piñon-juniper site. In: Shaw, Douglas W.; Aldon, Earl F.; LoSapio, Carol, tech. coords. Desired future conditions for piñon-juniper ecosystems: Proceedings of the symposium; 1994 August 8-12; Flagstaff, AZ. Gen. Tech. Rep. RM-258. Fort Collins, CO: U.S. Department of Agriculture, Forest Service, Rocky Mountain Forest and Range Experiment Station: 16-23.

Taverna, Kristin; Peet, Robert K.; Phillips, Laura C. 2005. Long-term change in ground-layer vegetation of deciduous forests of the North Carolina Piedmont, USA. Journal of Ecology. 93: 202-213.

Taylor, John P.; McDaniel, Kirk C. 1998a. Restoration of saltcedar (*Tamarix* spp.)-infested floodplains on the Bosque del Apache National Wildlife Refuge. Weed Technology. 12: 345-352.

Taylor, John P.; McDaniel, Kirk C. 1998b. Riparian management on the Bosque del Apache National Wildlife Refuge. New Mexico Journal of Science. 38: 219-232.

Tellman, Barbara, ed. 2002. Invasive exotic species in the Sonoran region. Arizona-Sonora Desert Museum Studies in Natural History. Tucson: University of Arizona Press. 424 p

Tesky, J. 1992. *Lespedeza bicolor*. In: Fire Effects Information System, [Online]. U.S. Department of Agriculture, Forest Service, Rocky Mountain Research Station, Fire Sciences Laboratory (Producer). Available: http://www.fs.fed.us/database/feis/ [2005, November 4].

Thilenius, John F. 1968. The *Quercus garryana* forests of the Willamette Valley, Oregon. Ecology. 49(6): 1124-1133.

Thomas, S. C.; Halpern, C. B.; Falk, D. A.; Liguori, D. A.; Austin, K. A. 1999. Plant diversity in managed forests: understory responses to thinning and fertilization. Ecological Applications. 9: 864-879.

Thompson, D. J.; Shay, J. M. 1985. The effects of fire on *Phragmites australis* in the Delta Marsh, Manitoba. Canadian Journal of Botany. 63: 1864-1869.

Thompson, D. J.; Shay, Jennifer M. 1989. First-year response of a *Phragmites* marsh community to seasonal burning. Canadian Journal of Botany. 67: 1448-1455.

Thompson, Daniel Q.; Stuckey, Ronald L.; Thompson, Edith B. 1987. Spread, impact, and control of purple loosestrife (*Lythrum salicaria*) in North American wetlands. Fish and Wildlife Research 2. Washington, DC: U.S. Department of the Interior, Fish and Wildlife Service. 55 p.

Thompson, Ralph L. 2001. Botanical survey of Myrtle Island Research Area, Oregon. Gen. Tech. Rep. PNW-GTR-507. Portland, OR: U.S. Department of Agriculture, Forest Service, Pacific Northwest Research Station. 27 p

Thompson, W. W.; Gartner, F. R. 1971. Native forage response to clearing low quality ponderosa pine. Journal of Range Management. 24: 272-277.

Thomsen, C. D.; Williams, W. A.; Vayssieres, M. P.; Turner. C. E.; Lanini, W. T. 1996. Yellow starthistle: biology and control. Resources Publication 21541. [Davis, CA]: University of California Division of Agriculture and Natural Resources: 333.

Thurber, G. 1880. Gramineae. In: Watson, S., ed. Geological survey of California: botany of California, Vol. II. Cambridge, MA: University Press. [Pages unknown].

Thysell, David R.; Carey, Andrew B. 2000. Effects of forest management on understory and overstory vegetation: a retrospective study. Gen. Tech. Rep. PNW-GTR-488. Portland, OR: U.S. Department of Agriculture, Forest Service, Pacific Northwest Research Station. 41 p.

Thysell, David R.; Carey, Andrew B. 2001a. Manipulation of density of *Pseudotsuga menziesii* canopies: preliminary effects on understory vegetation. Canadian Journal of Botany. 31: 1513-1525.

Thysell, David R.; Carey, Andrew B. 2001b. *Quercus garryana* communities in the Puget Trough, Washington. Northwest Science. 75(3): 219-235.

Tiedemann, Arthur R.; Conrad, Carol E.; Dieterich, John H.; [and others]. 1979. Effects of fire on water: A state-of-knowledge review: Proceedings of the national fire effects workshop; 1978 April 10-14; Denver, CO. Gen. Tech. Rep. WO-10. Washington, DC: U.S. Department of Agriculture, Forest Service. 28 p.

Timbrook, J.; Johnson, J. R.; Earle, D. D. 1982. Vegetation burning by the Chumash. Journal of California and Great Basin Anthropology. 4: 163-186.

Timmer, C. Elroy; Teague, Stanley S. 1991. Melaleuca eradication program: Assessment of methodology and efficacy. In: Center, Ted D.; Doren, Robert F.; Hofstetter, Ronald L.; Myers, Ronald; Whiteaker, Louis D., eds. Proceedings of the symposium on exotic pest plants; 1988 November 2-4; Miami, FL. Tech. Rep. NPS/NREVER/NRTR-91/06. Washington, DC: U.S. Department of the Interior, National Park Service: 339-351.

Timmins, S. M.; Williams, P. A. 1987. Characteristics of problem weeds in New Zealand's protected natural areas. In: Saunders, D. A.; Arnold, G. W.; Burridge, A. A.; Hopkins, A. J. M., eds. Nature conservation and the role of native vegetation. Chipping Norton, Australia: Surrey Beatty and Sons: [Pages unknown].

Tirmenstein, D. 1989a. Rubus discolor. In: Fire Effects Information System, [Online]. U.S. Department of Agriculture, Forest Service, Rocky Mountain Research Station, Fire Sciences Laboratory (Producer). Available: http://www.fs.fed.us/database/feis/ [2005, April 2].

Tirmenstein, D. 1989b. *Rubus laciniatus*. In: Fire Effects Information System, [Online]. U.S. Department of Agriculture, Forest Service, Rocky Mountain Research Station, Fire Sciences Laboratory (Producer). Available: http://www.fs.fed.us/database/feis/ [2005, April 2].

Tirmenstein, D. 1999. *Pascopyrum smithii*. In: Fire Effects Information System, [Online]. U.S. Department of Agriculture, Forest Service, Rocky Mountain Research Station, Fire Sciences Laboratory (Producer). Available: http://www.fs.fed.us/database/feis/ [2005, July 1].

Tisdale, E. W.; Hironaka, M.; Pringle, W. L. 1959. Observations on the autecology of *Hypericum perforatum*. Ecology. 40(1): 54-62.

Toorn, J. van der; Mook, J. H. 1982. The influence of environmental factors and management on stands of *Phragmites australis*. 1. Effect of burning, frost and insect damage on shoot density and shoot size. Journal of Applied Ecology. 19: 477-499.

Torell, Paul J.; Erickson, Lambert C.; Haas, Robert H. 1961. The medusahead problem in Idaho. Weeds. 9: 124-131.

Tosi, J.A., Jr.; Watson, V.; Bolaños, R.V.; 2002. Life zone maps of Hawaii. Joint venture of the Tropical Science Center, San Jose, Costa Rica and the Institute of Pacific Islands Forestry. CD containing electronic copies of maps as jpg and pdf files, and GIS shapefiles on file at: U.S. Department of Agriculture, Forest Service, Hilo, HI.

Tourn, G. M.; Menvielle, M. F.; Scopel, A. L.; Pidal, B. 1999. Clonal strategies of a woody weed: *Melia azedarach*. Plant and Soil. 217: 111-117.

Towle, Jerry C. 1982. Changing geography of Willamette Valley woodlands. Oregon Historical Quarterly. 83: 66-87.

Towne, E. Gene; Kemp, Ken E. 2003. Vegetation dynamics from annually burning tallgrass prairie. Journal of Range Management. 56(2): 185-192.

Towne, G.; Owensby, C. E. 1984. Long-term effects of annual burning at different dates in ungrazed Kansas tallgrass prairie. Journal of Range Management. 37: 392-397.

Trager, Matthew D.; Wilson, Gail W.; Hartnett, David C. 2004. Concurrent effects of fire regime, grazing and bison wallowing on tallgrass prairie vegetation. The American Midland Naturalist. 152(2): 237-247.

Trammell, T. L. E.; Rhoades, C. C.; Bukaveckas, P. A. 2004. Effects of prescribed fire on nutrient pools and losses from glades occurring within oak-hickory forests of central Kentucky. Restoration Ecology. 12(4): 597-604.

Transeau, Edgar Nelson. 1935. The Prairie Peninsula. Ecology. 16(3): 423-437.

Travnicek, Andrea J.; Lym, Rodney G.; Prosser, Chad. 2005. Fall-prescribed burn and spring-applied herbicide effects on Canada

thistle control and soil seedbank, in a northern mixed-grass prairie. Rangeland Ecology and Management. 58(4): 413-422.
Tu, Mandy. 2000. Element stewardship abstract: *Microstegium vimineum* (Japanese stilt grass, Nepalese browntop, Chinese packing grass), [Online]. In: The global invasive species initiative. Arlington, VA: The Nature Conservancy (Producer). Available: http://tncweeds.ucdavis.edu/esadocs/documnts/micrvim.pdf [2004, December 21].
Tu, Mandy. 2002a. Element stewardship abstract: *Cenchrus ciliaris* L. (African foxtail, buffelgrass, anjangrass), [Online]. In: The global invasive species initiative. Arlington, VA: The Nature Conservancy (Producer). Available: http://tncweeds.ucdavis.edu/esadocs/documnts/cenccil.pdf [2006, January 13].
Tu, Mandy. 2002b. Element stewardship abstract: *Dioscorea oppositifolia* L. syn. *Dioscorea batatas* (Decne) (Chinese yam, cinnamon vine), [Online]. In: The global invasive species initiative. Arlington, VA: The Nature Conservancy (Producer). Available: http://tncweeds.ucdavis.edu/esadocs/documnts/diosopp.pdf [2006, March 8].
Tu, Mandy. 2003. Element stewardship abstract: *Coronilla varia* L (crown vetch, trailing crown vetch), [Online]. In: The global invasive species initiative. Arlington, VA: The Nature Conservancy (Producer). Available: http://tncweeds.ucdavis.edu/esadocs/documnts/corovar.pdf [2006, January 13].
Tu, Mandy. 2004. Reed canary grass (*Phalaris arundinacea* L.) control & management in the Pacific Northwest, [Online]. In: The global invasive species initiative. Arlington, VA: The Nature Conservancy (Producer). Available: http://tncweeds.ucdavis.edu/moredocs/phaaru01.pdf [2005, January 15].
Tu, Mandy; Hurd, Callie; Randall, John M., eds. 2001. Weed control methods handbook: tools and techniques for use in natural areas. Davis, CA: The Nature Conservancy. 194 p. Available: http://tncweeds.ucdavis.edu/handbook.html [2007, September 5].
Tunison, J. Timothy; D'Antonio, Carla M.; Loh, Rhonda K. 2001. Fire and invasive plants in Hawai`i Volcanoes National Park. In: Galley, Krista E. M.; Wilson, Tyrone P., eds. Proceedings of the invasive species workshop: The role of fire in the control and spread of invasive species; Fire conference 2000: the first national congress on fire ecology, prevention, and management; 2000 November 27-December 1; San Diego, CA. Misc. Publ. No. 11. Tallahassee, FL: Tall Timbers Research Station: 122-131.
Tunison, J. Timothy; Leialoha, Julie A. K. 1988. The spread of fire in alien grasses after lightning strikes in Hawaii Volcanoes National Park. Hawaiian Botanical Society Newsletter. 27: 102-198.
Tunison, J. Timothy; Leialoha, Julie A. K.; Loh, Rhonda K.; Pratt, Linda W.; Higashino, Paul K. 1994. Fire effects in the coastal lowlands, Hawaii Volcanoes National Park. Technical Report 88. Honolulu, HI: University of Hawai`i, Cooperative Parks Resources Studies Unit. 42 p.
Tunison, J. Timothy; Loh, Rhonda K.; Leialoha, Julie A. K. 1995. Fire effects in the submontane seasonal zone, Hawaii Volcanoes National Park. Technical Report 97. Honolulu, HI: University of Hawai`i, Cooperative Parks Resources Studies Unit. 50 p.
Turkington, Roy A.; Cavers, Paul B.; Rempel, Erika. 1978. The biology of Canadian weeds. 29. *Melilotus alba* Ders. and *M. officinalis* (L.) Lam. Canadian Journal of Plant Science. 58: 523-537.
Turner, C. E.; Center, T. D.; Burrows, D. W.; Buckingham, G. R. 1998. Ecology and management of *Melaleuca quinquenervia*, an invader of wetlands in Florida, USA. Wetlands Ecology and Management. 5: 165-178.
Turner, Monica G.; Romme, William H. 1994. Landscape dynamics in crown fire ecosystems. Landscape Ecology. 9: 59-77.
Turner, Monica G.; Romme, William H.; Gardner, Robert H.; Hargrove, William W. 1997. Effects of fire size and pattern on early succession in Yellowstone National Park. Ecological Monographs. 67(4): 411-433.
Turner, Raymond M. 1974. Quantitative and historical evidence of vegetation changes along the Upper Gila River, Arizona. Geological Survey Professional Paper 655-H. Washington, DC: Department of the Interior, Geological Survey: H1-H20.
Tveten, R. K.; Fonda, R. W. 1999. Fire effects on prairies and oak woodlands on Fort Lewis, Washington. Northwest Science. 73(3): 145-158.
Tveten, Richard. 1997. Fire effects on prairie vegetation Fort Lewis, Washington. In: Dunn, Patrick; Ewing, Kern, eds. Ecology and conservation of south Puget Sound prairie landscape. Seattle, WA: The Nature Conservancy. 123-130.
Twedt, Daniel J.; Best, Chris. 2004. Restoration of floodplain forests for the conservation of migratory landbirds. Ecological Restoration. 22(3): 194-203.
Tyser, Robin W.; Worley, Christopher A. 1992. Alien flora in grasslands adjacent to road and trail corridors in Glacier National Park, Montana (U.S.A.). Conservation Biology. 6(2): 253-262.
U.S. Army, Sustainable Range Program. 2006. RTLA Technical Reference Manual: ecological monitoring on military lands draft, [Online]. Available: http://www.cemml.colostate.edu/itamtrm.htm. [2006, October 21].
U.S. Department of Agriculture (USDA) and U.S. Department of the Interior (USDI). 2006a. Interagency Burned Area Emergency Response Guidebook, Interpretation of Department of the Interior 620 DM3 and USDA Forest Service Manual 2523, for the emergency stabilization of federal and tribal trust lands, Version 4.0, [Online]. Available online at: http://fire.r9.fws.gov/ifcc/Esr/Planning/Assessment.htm [2007, August 1].
U.S. Department of Agriculture (USDA) and U.S. Department of the Interior (USDI). 2006b. Interagency Burned Area Rehabilitation Guidebook, Interpretation of Department of the Interior 620 DM3 for the burned area rehabilitation of federal and tribal trust lands, Version 1.3, [Online]. Available: http://fire.r9.fws.gov/ifcc/Esr/Planning/Assessment.htm [2007, August 1].
U.S. Department of Agriculture, Forest Service. 2001. Guide to noxious weed prevention practices, [Online]. Washington, DC: U.S. Department of Agriculture, Forest Service (Producer). 25 p. On file with: U.S. Department of Agriculture, Forest Service, Rocky Mountain Research Station, Fire Sciences Laboratory, Missoula, MT. Available: http://www.fs.fed.us/rangelands/ftp/invasives/documents/GuidetoNoxWeedPrevPractices_07052001.pdf [2007, August 1].
U.S. Department of Agriculture, Forest Service. 2003. Forest inventory and analysis national core field guide. Volume 1: field data collection procedure for phase 2 plots, Version 1.7.
U.S. Department of Agriculture, Forest Service. 2004. Avian, arthropod, and plant communities on unburned and wildfire sites along the Middle Rio Grande. 2004 annual report to Middle Rio Grande Conservancy District, Bosque del Apache National Wildlife Refuge. Albuquerque, NM: U.S. Department of Agriculture, Forest Service, Rocky Mountain Research Station. 32 p.
U.S. Department of Agriculture, Natural Resources Conservation Service (USDA, NRCS), Northeast Plant Materials Center. 2006. Plant fact sheet: Caucasian bluestem—*Bothriochloa bladhii* (Retz.) S.T. Blake, [Online]. In: Fact sheets and plant guides. In: The PLANTS Database. Baton Rouge, LA: National Plant Data Center (Producer). Available: http://plants.usda.gov/factsheet/pdf/fs_bobl.pdf [2007, August 22].
U.S. Department of the Interior, Fish and Wildlife Service. 1999. Fuel and fire effects monitoring guide, [Online]. Available: http://www.fws.gov/fire/downloads/monitor.pdf [2006, September 22].
U.S. Department of the Interior, National Park Service. 2003. Fire monitoring handbook. Boise, ID: National Interagency Fire Center, Fire Management Program Center. 274 p.
U.S. Department of the Interior (USDI); U. S. Forest Service (USFS). 2001. Managing the impacts of wildland fires on communities and the environment—the national fire plan, [Online]. Available: http//:www.fireplan.gov [2007, August 1].
U.S. Fish and Wildlife Service (USFWS), BayScapes Conservation Landscaping Program. 2004. [Online]. Available: http://www.nps.gov/plants/alien/pubs/midatlantic/eual.htm. [2006, October 24].
U.S. Geological Survey (USGS), Eastern Region Geography. 2001. Elevations and distances in the United States, [Online]. Available: http://erg.usgs.gov/isb/pubs/booklets/elvadist/elvadist.html
U.S. Geological Survey (USGS), National Wetlands Research Center. 2000. Coastal prairie—USGS FS-019-00, [Online]. In: NWRC Fact sheets. Lafayette, LA: U.S. Geological Survey, National Wetlands Research Center (Producer). Available: http://www.nwrc.usgs.gov/factshts/019-00.pdf [2007, August 21].
Uchytil, Ronald J. 1992a. *Melilotus alba*. In: Fire Effects Information System, [Online]. U.S. Department of Agriculture, Forest Service, Rocky Mountain Research Station, Fire Sciences Laboratory (Producer). Available: http://www.fs.fed.us/database/feis/ [2005, March 31].

Uchytil, Ronald J. 1992b. *Phragmites australis*. In: Fire Effects Information System, [Online]. U.S. Department of Agriculture, Forest Service, Rocky Mountain Research Station, Fire Sciences Laboratory (Producer). Available: http://www.fs.fed.us/database/feis/ [2006, August 28].

Uchytil, Ronald J. 1993. *Poa pratensis*. In: Fire Effects Information System, [Online]. U.S. Department of Agriculture, Forest Service, Rocky Mountain Research Station, Fire Sciences Laboratory (Producer). Available: http://www.fs.fed.us/database/feis/ 2005, March 31].

Uhl, Christopher; Kauffman J. Boone. 1990. Deforestation, fire susceptibility, and potential tree responses to fire in the Eastern Amazon. Ecology. 71: 437-449.

Underwood, E. C.; Klinger, R.; Moore, P. E. 2004. Predicting patterns of nonnative plant invasion in Yosemite National Park, California, USA. Diversity and Distributions. 10: 447-459.

Uresk, D. W.; Severson, K. E. 1998. Responses of understory species to changes in ponderosa pine stocking levels in the Black Hills. Great Basin Naturalist. 58: 312-327.

Ussery, Joel G.; Krannitz, Pam G. 1998. Control of Scot's broom (*Cytisus scoparius* (L.) Link.): The relative conservation merits of pulling versus cutting. Northwest Science. 72(4): 268- 273

Vail, Delmar. 1994. Symposium introduction: management of semi-arid rangelands—impacts of annual weeds on resource values. In: Monsen, Stephen B.; Kitchen, Stanley G., comps. Proceedings: Ecology and management of annual rangelands; 1992 May 18-22; Boise, ID. Gen. Tech. Rep. INT-GTR-313. Ogden, UT: U.S. Department of Agriculture, Forest Service, Intermountain Research Station: 3-4.

Vale, T. R. 1998. The myth of the humanized landscape: An example from Yosemite National Park. Natural Areas Journal. 18: 231-236.

Vallentine, J. F. 2001. Grazing management. San Diego, CA: Academic Press. 533 p.

Vallentine, John F. 1971. Range development and improvements. Provo, UT: Brigham Young University Press. 516 p.

van Andel, J.; Vera, F. 1977. Reproductive allocation in Senecio sylvaticus and *Chamaenerion angustifolium* in relation to mineral nutrition. Journal of Ecology. 65: 747-758.

Van Cleve, K.; Viereck, L. A.; Dyrness, C. T. 1987. Vegetation productivity and soil fertility in post-fire secondary succession in Interior Alaska. In: Slaughter, Charles W.; Gasbarro, Tony. Proceedings of the Alaska forest soil productivity workshop; 1987 April 28-30; Anchorage, AK. Gen. Tech. Rep. PNW-GTR-219. Portland, OR: U.S. Department of Agriculture, Forest Service, Pacific Northwest Station; Fairbanks, AK: University of Alaska, School of Agriculture and Land Resources Management: 101-102.

Van Dyke, E.; Holl, K. D. 2001. Maritime chaparral community transition in the absence of fire. Madroño. 48: 221-229.

Van Horn, Kent; Van Horn, Michele A.; Glancy, Robert C.; Goodwin, James; Driggers, Robert D. 1995. Fire management in wetland restoration [poster abstract]. In: Cerulean, Susan I.; Engstrom, R. Todd, eds. Proceedings of the 19th Tall Timbers fire ecology conference—fire in wetlands: a management perspective Tallahassee, FL. Tallahassee, FL: Tall Timbers Research Station: 169-170.

Van Lear, D. H.; Harlow, R. E. 2001. Fire in the eastern United States—influence on wildlife habitat. In: Ford, W. M.; Russell, K. R.; Moorman, C. E., eds. The role of fire in nongame wildlife management and community restoration: traditional uses and new directions: proceedings of a special workshop. Gen. Tech. Rep. NE-288. Newtown Square, PA: U.S. Department of Agriculture, Forest Service, Northeastern Research Station. 2-10.

Van Lear, David H.; Watt, Janet M. 1993. The role of fire in oak regeneration. In: Loftis, David L.; McGee, Charles E., eds. Oak regeneration: Serious problems, practical recommendations: Symposium proceedings; 1992 September 8-10; Knoxville, TN. Gen. Tech. Rep. SE-84. Asheville, NC: U.S. Department of Agriculture, Forest Service, Southeastern Forest Experiment Station: 66-78.

van Wagtendonk, Jan W. 1995. Large fires in wilderness areas. In: Brown, James K.; Mutch, Robert W.; Spoon, Charles W.; Wakimoto, Ronald H., tech. coords. Proceedings: symposium on fire in wilderness and park management; 1993 March 30-April 1; Missoula, MT. Gen. Tech. Rep. INT-GTR-320. Ogden, UT: U.S. Department of Agriculture, Forest Service, Intermountain Research Station: 113-116.

Van Wilgen, B.; Richardson, D. 1985. The effects of alien shrub invasions on vegetation structure and fire behaviour in South African fynbos shrublands. Journal of Applied Ecology. 22: 955-966.

Van, T. K.; Rayamajhi, M. B.; Center, T. D. 2005. Seed longevity of *Melaleuca quinquenervia*: a burial experiment in south Florida. Journal of Aquatic Plant Management. 43: 39-42.

Vankat, John L.; Snyder, Gary W. 1991. Floristics of a chronosequence corresponding to old field-deciduous forest succession in southwestern Ohio. I. Undisturbed vegetation. Bulletin of the Torrey Botanical Club. 118(4): 365-376.

Vavrek, Milan C.; Fetcher, Ned; McGraw, James B.; Shaver, G. R.; Chapin, F. Stuart, III; Bovard, Brian. 1999. Recovery of productivity and species diversity in tussock tundra following disturbance. Arctic, Antarctic, and Alpine Research. 31(3): 254-258.

Versfeld D. B.; Van Wilgen, B. W. 1986. Impacts of woody aliens on ecosystem properties. In: Macdonald, I. A. W.; Kruger, F. J.; Ferrar, A. A., eds. The ecology and control of biological invasions in South Africa. Cape Town, South Africa: Oxford University Press. 239-246.

Vickers, Gerald. 2003. [Remarks from panel discussion]. Highlights from the panel discussion, fire and invasive plants in the Northeast: What works? In: Using fire to control invasive plants: What's new, what works in the Northeast: 2003 Workshop Proceedings. 2003 January 24; Portsmouth, NH. Durham, NH: University of New Hampshire Cooperative Extension: 15-16.

Vickery, Peter D.; Zuckerberg, Benjamin; Jones, Andrea L.; Shriver, W. Gregory; Weik, Andrew P. 2005. Influence of fire and other anthropogenic practices on grassland and shrubland birds in New England. In: Saab, Victoria A.; Powell, Hugh D. W., eds. Fire and avian ecology in North America; Studies in Avian Biology No. 30. Ephrata, PA: Cooper Ornithological Society: 139-146.

Vickery, Peter, D.; Dunwiddie, Peter W., eds. 1997. Grasslands of northeastern North America: ecology and conservation of native and agricultural landscapes. Lincoln, MA: Massachusetts Audubon Society. 297 p.

Vickery, Peter. 2002. Effects of prescribed fire on the reproductive ecology of northern blazing star, *Liatris scariosa* var. *novae-angliae*. The American Midland Naturalist. 148: 20-27.

Viereck, L. A. 1983. The effects of fire in black spruce ecosystems of Alaska and northern Canada. In: Wein, Ross W.; MacLean, David A., eds. The role of fire in northern circumpolar ecosystems. New York: John Wiley and Sons Ltd.: 201-220.

Viereck, L. A.; Dyrness, C. T. 1979. Ecological effects of the Wickersham Dome Fire near Fairbanks, Alaska. Gen. Tech. Rep. PNW-90. Portland, OR: U.S. Department of Agriculture, Forest Service, Pacific Northwest Forest and Range Experiment Station. 71 p.

Viereck, Leslie A. 1973. Wildfire in the taiga of Alaska. Quaternary Research. 3: 465-495

Viereck, Leslie A.; Foote, Joan; Dyrness, C. T.; Van Cleve, Keith; Kane, Douglas; Seifert, Richard. 1979. Preliminary results of experimental fires in the black spruce type of interior Alaska. Res. Note PNW-332. Portland, OR: U.S. Department of Agriculture, Forest Service, Pacific Northwest Forest and Range Experiment Station. 27 p.

Vilà, M.; Burriel, J. A.; Pino, J.; Chamizo, J.; Llach, E.; Porterias, M.; Vives, M.; Salvador, R. 2005. Opuntia invasión and changes in land-use and fire frequencies in Cap de Creus, [Online]. Available: http://www.unesco.org/mab/EE/resources.htm [2005, December 15].

Vilà, Montserrat; Weiner, Jacob. 2004. Are invasive plant species better competitors than native plant species?—Evidence from pair-wise experiments. Oikos. 105(2): 229-238.

Vincent, Dwain W. 1992. The sagebrush/grasslands of the upper Rio Puerco area, New Mexico. Rangelands. 14(5): 268-271.

Vines, R. A. 1960. Trees, shrubs, and woody vines of the Southwest. Austin, TX: University of Texas Press. 1104 p.

Vinton, Mary Ann; Harnett, David C.; Finck, Elmer J.; Briggs, John M. 1993. Interactive effects of fire, bison (*Bison bison*) grazing and plant community composition in tallgrass prairie. The American Midland Naturalist. 129: 10-18.

Violi, H. 2000. Element stewardship abstract: *Paspalum notatum* Flüggé (Bahia grass, Bahiagrass), [Online]. In: The global invasive species initiative. Arlington, VA: The Nature Conservancy (Producer).

Available: http://tncweeds.ucdavis.edu/esadocs/documnts/paspnot.pdf [2005, January 4].
Virginia Department of Conservation and Recreation, Natural Heritage Program. 2002a. Species factsheet: autumn olive (*Elaeagnus umbellata* Thunberg) and Russian olive (*Elaeagnus angustifolia* L.), [Online]. In: Invasive alien plant species of Virginia. Virginia Native Plant Society (Producer). Available: http://www.dcr.virginia.gov/dnh/fselum.pdf [2004, December 21].
Virginia Department of Conservation and Recreation, Natural Heritage Program. 2002b. Species factsheet: Japanese stilt grass (*Microstegium vimineum*), [Online]. In: Invasive alien plant species of Virginia. Virginia Native Plant Society (Producer). Available: http://www.dcr.virginia.gov/dnh/fsmivi.pdf [2007, August 20]
Virginia Department of Conservation and Recreation, Natural Heritage Program. 2002c. Species factsheet: mile-a-minute (*Polygonum perfoliatum* L.), [Online]. In: Invasive alien plant species of Virginia. Virginia Native Plant Society (Producer). Available: http://www.dcr.virginia.gov/dnh/fspope.pdf [2004, December 21]
Virginia Department of Conservation and Recreation, Natural Heritage Program. 2002d. Species factsheet: Multiflora rose (*Rosa multiflora* Thunberg), [Online]. In: Invasive alien plant species of Virginia. Virginia Native Plant Society (Producer). Available: http://www.dcr.virginia.gov/dnh/fsromu.pdf [2004, December 21].
Virginia Department of Conservation and Recreation. 2005.The Natural Communities of Virginia, Classification of Ecological Community Groups (Version 2.2), Oak—Hickory Woodlands and Savannas, [Online]. Available: http://www.state.va.us/dcr/dnh/ncTIIIo.htm [2006, September].
Vitousek, P. M. 1986. Biological invasions and ecosystem properties: can species make a difference? In: Mooney, H. A.; Drake, J. A., eds. Ecology of biological invasions of North America and Hawaii. New York: Springer-Verlag. 163-178.
Vitousek, P. M. 1990. Biological invasions and ecosystem processes: toward and integration of population biology and ecosystem studies. Oikos. 57:7-13.
Vitousek, P. M.; Loope, L. L.; Stone, C. P. 1987. Introduced species in Hawaii, USA: biological effects and opportunities for ecological research. Trends in Ecology and Evolution. 2: 224-227.
Vitousek, P. M.; Walker, L. R. 1989. Biological invasion by Myrica faya in Hawaii [USA]: Plant demography, nitrogen fixation, ecosystem effects. Ecological Monographs. 59: 247-266.
Vitousek, Peter M.; D'Antonio, Carla M.; Loope, Lloyd L.; Westbrooks, Randy. 1996. Biological invasions as global environmental change. American Scientist. 84: 468-478.
Vitousek, Peter M.; Mooney, Harold A.; Lubchenco, Jane; Melillo, Jerry M. 1997. Human domination of earth's ecosystems. Science. 277: 494-499.
Vogl, R. J.; Schorr, P. K. 1972. Fire and manzanita chaparral in the San Jacinto mountains, California. Ecology. 53: 1179-1188.
Vogl, Richard J. 1969. The role of fire in the evolution of the Hawaiian flora and vegetation. In: Proceedings, annual Tall Timbers fire ecology conference; 1969 April 10-11; Tallahassee, FL. No. 9. Tallahassee, FL: Tall Timbers Research Station: 5-60.
Vogl, Richard J. 1974. Effects of fire on grasslands. Fire and Ecosystems. New York: Academic Press: 139-194.
Volland, Leonard A.; Dell, John D. 1981. Fire effects on Pacific Northwest forest and range vegetation. Portland, OR: U.S. Department of Agriculture, Forest Service, Pacific Northwest Region, Range Management and Aviation and Fire Management. 23 p.
Vollmer, Joseph and Vollmer, Jennifer. 2005. [Personal communication]. BASF Corporation, 2166 N 15th Street, Laramie, WY.
Wade, D. D. 1981. Some melaleuca-fire relationships, including recommendations for homesite protection. In: Geiger, K., ed. Proceedings of the Melaleuca Symposium; 1980 September 23-24; [Location unknown]. Tallahassee, FL: Florida Department of Agriculture and Consumer Services, Division of Forestry. 29-35.
Wade, Dale. 1988. Burning controls Brazilian pepper, brush on Sanibel Island (Florida). Restoration and Management Notes. 6(1): 49.
Wade, Dale. 2005. [Email to Alison Dibble]. January 13. Athens, GA: USDA Forest Service, Southern Research Station (retired). On file at U.S. Department of Agriculture, Forest Service, Rocky Mountain Research Station, Fire Sciences Laboratory, Missoula, MT.
Wade, Dale; Ewel, John; Hofstetter, Ronald. 1980. Fire in south Florida ecosystems. Gen. Tech. Rep. SE-17. Asheville, NC: U.S. Department of Agriculture, Forest Service, Southeastern Forest Experiment Station. 125 p.
Wade, Dale D.; Brock, Brent L.; Brose, Patrick H.; Grace, James B.; Hoch, Greg A.; Patterson, William A., III. 2000. Fire in eastern ecosystems. In: Brown, James K.; Smith, Jane Kapler, eds. Wildland fire in ecosystems: Effects of fire on flora. Gen. Tech. Rep. RMRS-GTR-42-vol. 2. Ogden, UT: U.S. Department of Agriculture, Forest Service, Rocky Mountain Research Station: 53-96.
Wade, M. 1997. Predicting plant invasions: making a start. In: Brock, J. H.; Wade, M.; Pyšek, P.; Green, D., eds. Plant invasions: studies from North America and Europe. Leiden, Netherlands: Backhuys Publishers: 1-18.
Wagener, W. W. 1961. Past fire incidence in Sierra Nevada forests. Journal of Forestry. 59: 739-748.
Wagner, W. L.; Herbst, D. H.; Sohmer, S. H. 1999. Manual of the flowering plants of Hawaii. Revised Ed. Bishop Museum Special Publication No. 97. Honolulu, HI: University of Hawai`i Press. 1919 p.
Wainwright, T. C.; Barbour, M. G. 1984. Characteristics of mixed evergreen forest in the Sonoma Mountains of California. Madroño. 31: 219-230.
Wakida, Charles K. 1997. Mauna Kea Ecosystem Wildland Fire Management Plan. Unpublished Report on file at: Hawai`i Department of Land and Natural Resources, Division of Forestry and Wildlife, Hilo Branch, Hilo, HI. 50 p.
Waldrop, Thomas A. (Supervisory Research Forester, USDA Forest Service Southern Research Station). 2006. [Letter to Molly E. Hunter]. March 27. 1 leaf. On file at: Colorado State University, Warner College of Natural Resources, Western Forest Fire Research Center, Fort Collins, CO.
Walker, J.; Peet, R. K. 1983. Composition and species diversity of pine-wiregrass savannas of the Green Swamp, North Carolina. Vegetatio. 55: 163-179.
Walker, Loren W. 2000. St. John's wort (*Hypericum perforatum* L. Clusiaceae): biochemical, morphological, and genetic variation within and among wild populations of the northwestern United States. Portland, OR: Portland State University. 162 p. Thesis.
Waloff, N.; Richards, O. W. 1977. The effect of insect fauna on growth mortality and natality of broom, *Sarothamnus scoparius*. Journal of Applied Ecology. 14(3): 787-798.
Wan, Shiqiang; Hui, Dafeng; Luo, Yiqi. 2001. Fire effects on nitrogen pools and dynamics in terrestrial ecosystems: a meta-analysis. Ecological Applications. 11(5): 1349-1365.
Ward, D. B. 1990. How many species are native to Florida? Palmetto. 9: 3-5.
Ward, P. 1968. Fire in relation to waterfowl habitat of the delta marshes. In: Proceedings, annual Tall Timbers fire ecology conference; 1968 March 14-15; Tallahassee, FL. No. 8. Tallahassee, FL: Tall Timbers Research Station: 255-267.
Warner, Peter J.; Bossard, Carla C.; Brooks, Matthew L.; DiTomaso, Joseph M.; Hall, John A.; Howald, Ann M.; Johnson, Douglas W.; Randall, John M.; Roye, Cynthia L.; Ryan, Maria M.; Stanton, Alison E. 2003. Criteria for categorizing invasive non-native plants that threaten wildlands, [Online]. California Exotic Pest Plant Council and Southwest Vegetation Management Association (Producer). 24 p. Available: http://www.werc.usgs.gov/lasvegas/pdfs/Warner_et_al_2003_Criteria%20for%20categorizing%20invasive.pdf [2007, August 28].
Washburn, Brian E.; Barnes, Thomas G.; Rhoades, Charles C.; Remington, Rick. 2002. Using imazapic and prescribed fire to enhance native warm-season grasslands in Kentucky, USA. Natural Areas Journal. 22(1): 20-27.
Washington State Noxious Weed Control Board. 2005. Written Findings, Gorse Ulex europaeus L., [Online] Available: http:www.nwcb.wa.gov/weed_info/Written_findings/Ulex_europaeus.html [2005, March 20].
Watkins, R.Z.; Chen, J.; Pickens, J.; Brosofske, K.D. 2003. Effects of forest roads on understory plants in a managed hardwood landscape. Conservation Biology. 17(2): 411-419.
Watson, A. K. 1980. The biology of Canadian weeds. 43. *Acroptilon* (Centaurea) *repens* (L.). D.C. Canadian Journal of Plant Science. 60: 993-1004.

Watson, A. K.; Renney, A. J. 1974. The biology of Canadian weeds. 6. *Centaurea diffusa* and *C. maculosa*. Canadian Journal of Plant Science. 54: 687-701.

Watterson, Nicholas. 2004. Exotic plant invasions from roads to stream networks in steep forested landscapes of western Oregon. Corvallis, OR: Oregon State University. 90 p. Thesis.

Weatherspoon, C. P. 2000. A long-term national study of the consequences of fire and fire surrogate treatments. In: Crossing the millennium: integrating spatial technologies and ecological principles for a new age in fire management: Proceedings of the Joint Fire Science conference and workshop; 1999 June 15-17; Boise, ID. Boise, ID: University of Idaho: 117-126

Weatherspoon, C. P.; Skinner, C. N. 1995. An assessment of factors associated with damage to tree crowns from the 1987 wildfires in northern California. Forest Science. 41: 430-451.

Weaver, H. 1951. Fire is an ecological factor in the Southwestern ponderosa pine forests. Journal of Forestry. 49: 93-98.

Weaver, Harold. 1959. Ecological changes in the ponderosa pine forest of the Warm Springs Indian Reservation in Oregon. Journal of Forestry. 57: 15-20.

Weaver, T.; Lichthart, J.; Gustafson, D. 1990. Exotic invasion of timberline vegetation, Northern Rocky Mountains, USA. In: Schmidt, Wyman C.; McDonald, Kathy J., comps.. Proceedings—symposium on whitebark pine ecosystems: ecology and management of a high-mountain resource; 1989 March 29-31; Bozeman, MT. Gen. Tech. Rep. INT-270. Ogden, UT: U.S. Department of Agriculture, Forest Service, Intermountain Research Station: 208-213.

Webb, Sara L.; Dwyer, Marc; Kaunzinger, Christina K.; Wyckoff, Peter H. 2000. The myth of the resilient forest: case study of the invasive Norway maple (*Acer platanoides*). Rhodora. 102: 332-354.

Webb, Sara L.; Kaunzinger, Christina Kalafus. 1993. Biological invasion of the Drew University (New Jersey) Forest Preserve by Norway maple. Bulletin of the Torrey Botanical Club. 120(3): 343-349.

Webb, Sara L.; Pendergast IV, Thomas H.; Dwyer, Marc E. 2001. Response of native and exotic maple seedling banks to removal of the exotic, invasive Norway maple (*Acer platanoides*). Journal of the Torrey Botanical Society. 128(2): 141-149.

Weddell, Bertie J. 2001. Fire in steppe vegetation of the northern intermountain region, [Online]. Boise, ID: U.S. Department of the Interior, Bureau of Land Management, Idaho State Office (Producer). Technical Bulletin No. 01-14. Available: http://www.id.blm.gov/techbuls/01_14/index.htm [2007, July 26].

Wein, Ross W. 1971. Fire and resources in the subarctic—panel discussion. In: Slaughter, C. W.; Barney, R. J.; Hansen, G. M., eds. Fire in the northern environment--a symposium; 1971 April 13-14; Fairbanks, AK. Portland, OR: U.S. Department of Agriculture, Forest Service, Pacific Northwest Forest and Range Experiment Station: 251-253.

Wein, Ross W.; Bliss, L. C. 1973. Changes in arctic Eriophorum tussock communities after fire. Ecology. 54(4): 845-852.

Wein, Ross W.; Wein, Gerold; Bahret, Sieglinde; Cody, William J. 1992. Northward invading non-native vascular plant species in and adjacent to Wood Buffalo National Park, Canada. Canadian Field-Naturalist. 106(2): 216-224.

Weis, Judith S.; Weis, Peddrick. 2003. Is the invasion of the common reed, *Phragmites australis*, into tidal marshes of the eastern US an ecological disaster? Marine Pollution Bulletin. 46: 816-820.

Weiss, J.; McLaren, D. 1999. Invasive assessment of Victoria's state prohibited, priority and regional priority weeds. Frankston, Victoria, Australia: Keith Turnbull Research Institute. 16 p.

Wells, C. G.; Campbell, Ralph E.; DeBano, Leonard F.; Lewis, Clifford E.; Fredriksen, Richard L.; Franklin, E. Carlyle; Froelich, Ronald C.; Dunn, Paul H. 1979. Effects of fire on soil: state-of-knowledge review: Proceedings of the national fire effects workshop; 1978 April 10-14; Denver, CO. Gen. Tech. Rep. WO-7. Washington, DC: U.S. Department of Agriculture, Forest Service. 34 p.

Wells, P. V. 1962. Vegetation in relation to geological substratum and fire in the San Luis Obispo Quadrangle, California. Ecological Monographs. 32: 79-103.

Weltzin, Jake F.; Archer, Steve; Heitschmidt, Rodney I. 1998. Defoliation and woody plant (*Prosopis glandulosa*) seedling regeneration: potential vs. realized herbivory tolerance. Plant Ecology. 138: 127-135.

Wendtland, Kyle J. 1993. Fire history and effects of seasonal prescribed burning on northern mixed prairie, Scotts Bluff National Monument, Nebraska Laramie, WY: University of Wyoming. 188 p.

West, Neil E. 1983. Intermountain salt desert shrubland. In: West, Neil E., ed. Temperate deserts and semi-deserts. Amsterdam: Elsevier: 375-397.

West, Neil E. 1988. Intermountain Desert, Shrub Steppes, and Woodlands. In: M.G. Barbour and W.D. Billings (eds.). North American Terrestrial Vegetation. NY, Cambridge University Press: 210-230.

West, Neil E. 1994. Effects of fire on salt-desert shrub rangelands. In: Monsen, Stephen B., and Kitchen, Stanley G., comps. Proceedings: ecology, management, and restoration of Intermountain annual rangelands. Gen. Tech. Rep. GTR-INT-313. Ogden, UT: U.S. Department of Agriculture, Forest Service, Intermountain Forest and Range Experiment Station: 71-74.

West, Neil E. 2000. Synecology and disturbance regimes of sagebrush steppe ecosystems. In: Entwistle, P. G.; DeBolt, A. M.; Kaltenecker, J. H.; Steenhof, K., comps.. In: Sagebrush steppe ecosystems symposium: Proceedings; 1999 June 21-23; Boise, ID. Publ. No. BLM/ID/PT-001001+1150. Boise, ID: U.S. Department of the Interior, Bureau of Land Management, Boise State Office: 15-26.

West, Neil E.; Chilcote, William W. 1968. *Senecio sylvaticus* in relation to Douglas-fir clear-cut succession in the Oregon Coast range. Ecology. 49(6): 1101-1107.

West, Neil E.; Hassan, M. A 1985. Recovery of sagebrush-grass vegetation following wildfire. Journal of Range Management. 38(2): 131-134.

West, Neil E.; Yorks, Terence P. 2002. Vegetation responses following wildfire on grazed and ungrazed sagebrush semi-desert. Journal of Range Management. 55(2): 171-181.

Westbrooks, Randy G. 1998. Invasive plants: changing the landscape of America. Fact Book. Washington, DC: Federal Interagency Committee for the Management of Noxious and Exotic Weeds. 109 p.

Westerling, A. L.; Hidalgo, H. G.; Cayan, D. R.; Swetnam, T. W. 2006. Warming and earlier spring increase Western U.S. forest wildfire activity. Science. 313(5789): 940-943.

Westman, W. E. 1979. A potential role of coastal sage scrub understories in the recovery of chaparral after fire. Madroño. 26: 64-68.

Westman, W. E. 1981a. Diversity relations and succession in California coastal sage scrub. Ecology. 62: 170-184.

Westman, W. E. 1981b. Factors influencing the distribution of species of Californian coastal sage scrub. Ecology. 65: 439-455.

Westman, W. E. 1983. Xeric Mediterranean-type shrubland associations of Alta and Baja California and the community continuum debate. Vegetatio. 52: 3-19.

Westman, W. E.; O'Leary, J. F. 1986. Measures of resilience: the response of coastal sage scrub to fire. Vegetatio. 65: 179-189.

Westman, W. E.; O'Leary, J. F.; Malanson, G. P. 1981. The effects of fire intensity, aspect and substrate on post-fire growth of Californian coastal sage scrub. In: Margaris, N. S.; Mooney, H. A., eds. Components of productivity of Mediterranean climate regions--basic and applied aspects. The Hague, The Netherlands: Dr. W. Junk Publishers: 151-179.

Westman, Walter E. 1982. Coastal sage scrub succession. In: Conrad, C. Eugene; Oechel, Walter C., tech. coords. Proceedings of the symposium on dynamics and management of Mediterranean-type ecosystems; 1981 June 22-26; San Diego, CA. Gen. Tech. Rep. PSW-58. Berkeley, CA: U.S. Department of Agriculture, Forest Service, Pacific Southwest Forest and Range Experiment Station: 91-99.

Wetzel, S. A. Fonda, R. W. 2000. Fire history of Douglas-fir forests in the Morse Creek drainage of Olympic National Park, Washington. Northwest Science. 74(4): 263-279.

Whelan, Christopher J.; Dilger, Michael L. 1992. Invasive, exotic shrubs: a paradox for natural area managers? Natural Areas Journal. 12: 109-110.

Whelan, Robert J. 1995. The ecology of fire. Cambridge, United Kingdom: Cambridge University Press. p 346.

Whisenant, S. G. 1989. Modeling Japanese brome population dynamics. Proceedings of the Western Society of Weed Science. 42: 176-185.

Whisenant, S. G. 1990a. Changing fire frequencies on Idaho's Snake River Plains: ecological and management implications. In: McArthur, E. Durant; Romney, Evan M.; Smith, Stanley D.; Tueller, Paul T., comps. Proceedings—symposium on cheatgrass invasion, shrub die-off, and other aspects of shrub biology and management; 1989 April 5-7; Las Vegas, NV. Gen. Tech. Rep. INT-276. Ogden, UT: U.S. Department of Agriculture, Forest Service, Intermountain Research Station: 4-10.

Whisenant, S. G. 1990b. Postfire population dynamics of *Bromus japonicus*. American Midland Naturalist. 123(2): 301-308.

Whisenant, S. G.; Bulsiewicz, W. R. 1986. Effects of prescribed burning on Japanese brome population dynamics. In: Proceedings of the XV international grassland conference Kyoto, Japan. Nishi-nasuno, Japan: Science Council of Japan and the Japanese Society for Grassland Science: p 803-804.

Whisenant, S. G.; Uresk, Daniel W. 1990. Spring burning Japanese brome in a western wheatgrass community. Journal of Range Management. 43(3): 205-208.

Whisenant, Steven G. 1985. Effects of fire and/or atrazine on Japanese brome and western wheatgrass. Proceedings of the Western Society of Weed Science. 38: 169-176.

White, Douglas W.; Stiles, Edmund W. 1992. Bird dispersal of fruits of species introduced into eastern North America. Canadian Journal of Botany. 70: 1689-1696.

White, L. L.; Zak, D. R. 2004. Biomass accumulation and soil nitrogen availability in an 87-year-old Populus grandidentatum chronosequence. Forest Ecology and Management. 191: 121-127.

White, Richard S.; Currie, Pat O. 1983. Prescribed burning in the northern Great Plains: yield and cover responses of 3 forage species in the Mixed Grass Prairie. Journal of Range Management. 36(2): 179-183.

White, S. D. 1995. Disturbance and dynamics in coastal sage scrub. Fremontia. 23: 9-16.

White, Seth M. 2004. Bridging the worlds of fire managers and researchers: lessons and opportunities from the Wildland Fire Workshops. Gen. Tech. Rep. PNW-GTR-599. Portland, OR: U.S. Department of Agriculture, Forest Service, Pacific Northwest Research Station. 41 p.

Whitson, T. D.; Koch, D. W. 1998. Control of downy brome (*Bromus tectorum*) with herbicides and perennial grass competition. Weed Technology. 12: 391-396.

Whittaker, R. H. 1954. The ecology of serpentine soils. IV. The vegetational response to serpentine soils. Ecology. 35: 275-288.

Wienk, Cody L.; Sieg, Carolyn Hull; McPherson, Guy R. 2004. Evaluating the role of cutting treatments, fire, and soil seed banks in an experimental framework in ponderosa pine forests of the Black Hills, South Dakota. Forest Ecology and Management. 192(2-3): 375-393.

Willard, E. Earl; Wakimoto, Ronald H.; Ryan, Kevin C. 1995. Vegetation recovery in sedge meadow communities within the Red Bench Fire, Glacier National Park. In: Cerulean, Susan I.; Engstrom, R. Todd, eds. Fire in wetlands: a management perspective: Proceedings, 19th Tall Timbers fire ecology conference; 1995 November 3-6; Tallahassee, FL. No. 19. Tallahassee, FL: Tall Timbers Research Station: 102-110.

Williams, C. E. 1994. Invasive alien plant species of Virginia. Richmond, VA: Department of Conservation and Recreation. [Pages unknown].

Williams, Charles E. 1993. The exotic empress tree, *Paulownia tomentosa*: an invasive pest of forests? Natural Areas Journal. 13(3): 221-222.

Williams, Charles E. 1998. History and status of Table Mountain Pine-pitch pine forests of the southern Appalachian Mountains (USA). Natural Areas Journal. 18(1): 81-90.

Williams, D. G.; Baruch, N. 2000. African grass invasions in the America: ecosystem consequences and the role of ecophysiology. Biological Invasions. 2(2): 123-140.

Williams, J. 1990. The coastal woodland of Hawaii Volcanoes National Park: vegetation recovery in a stressed ecosystem. Technical Report 72. Honolulu, HI: University of Hawai`i, Cooperative National Park Studies Unit. 78 p.

Williams, M. 1989. Americans and their forests: an historical geography. New York: Cambridge University Press. 411 p.

Williams, M. 1992. Americans and their forests. New York: Cambridge University Press. 599 p.

Williamson, M. 1993. Invaders, weeds and the risk from genetically manipulated organisms. Experimentia. 49: 219-224.

Williamson, M. 1996. Biological Invasions. London, UK: Chapman & Hall. 244 p.

Williamson, M.; Brown, K. 1986. The analysis and modelling of British invasions. Philisophical Transactions of the Royal Society of London. B314: 505-522.

Williamson, M.; Fitter, A. 1996. The characteristics of successful invaders. Biological Conservation. 78: 163-170.

Williamson, Mark. 1999. Invasions. Ecography. 22: 5-12.

Willis, B. D.; Evans, J. O.; Dewey, S. A. 1988. Effects of temperature and flaming on germinability of jointed goatgrass (*Aegilops cylindrica*) seed. Western Society of Weed Science. 41: 49-55.

Wills, R. 2001. Effects of varying fire regimes in a California native grasslands. Ecological Restoration. 19: 109.

Wills, R. 2006. Fire in the Central Valley Bioregion. In: Sugihara, N.; van Wagtendonk, J. W.; Fites, J.; Thode, A., eds. Fire in California ecosystems. Berkeley, CA: University of California Press: 447-480.

Wills, R. D. 1999. Effective fire planning for California native grasslands. In: Keeley, J. E.; Baer-Keeley, M.; Fotheringham, C. J., eds. 2nd interface between ecology and land development in California. Sacramento, CA: U.S. Geological Survey: 75-78.

Wills, Robin D.; Stuart, John D. 1994. Fire history and stand development of a Douglas-fir/hardwood forest in northern California. Northwest Science. 68(3): 205-211.

Wills, Robin. 2000. Effective fire planning for California native grasslands. Sacramento, CA: U.S. Geological Survey. Open-File Report 00-62.

Willson, G. D. 1991. Morphological characteristics of smooth brome used to determine a prescribed burn date. In: Smith, Daryl D., and Jacobs, Carol A., eds. Proceedings of the twelfth North American prairie conference: recapturing a vanishing heritage; Cedar Falls, IA. Cedar Falls, IA: University of Northern Iowa: 113-116

Willson, G. D.; Stubbendieck, J. 1996. Suppression of smooth brome by atrazine, mowing, and fire. The Prairie Naturalist. 28: 13-20.

Willson, G. D.; Stubbendieck, J. 1997. Fire effects on four growth stages of smooth brome (*Bromus inermis* Leyss.). Natural Areas Journal. 17: 306-312.

Willson, G. D.; Stubbendieck, J. 2000. A provisional model for smooth brome management in degraded tallgrass prairie. Ecological Restoration. 18: 34-38.

Willson, Gary D. 1992. Morphological characteristics of smooth brome used to determine a prescribed burn date. In: Smith, Daryl D., and Jacobs, Carol A., eds. Proceedings: Proceedings of the twelfth North American prairie conference: recapturing a vanishing heritage; Cedar Falls, IA. Cedar Falls, IA: University of Northern Iowa: 113-116.

Wilson, Mark V.; Ingersoll, Cheryl A.; Wilson, Mark G.; Clark, Deborah L. 2004. Why pest plant control and native plant establishment failed: A restoration autopsy. Natural Areas Journal. 24(1): 23-31.

Wilson, S. D. 1989. The suppression of native prairie by alien species introduced for revegetation. Landscape and Urban Planning. 17: 113-119.

Wimmer, S. Mark; Racher, Brent J.; Blair, Keith; Britton, Carlton M. 2001. Vegetation response to late-summer burning in the Hill Country of Texas. In: Zwank, Phillip J.; Smith, Loren M.; eds. Research highlights—2001: Range, wildlife, and fisheries management. Volume 32. Lubbock, TX: Texas Tech University, Department of Range, Wildlife, and Fisheries Management: 24.

Winne, J. C. 1997. History of vegetation and fire on the Pine Ridge pine grassland barrens of Washington County, Maine. In: Vickery, P. D.; Dunwiddie, P. W., eds. Grasslands of northeastern North America: ecology and conservation of native and agricultural landscapes. Lincoln, MA: Massachusetts Audubon Society: 25-52.

Winter, B. 1992. Leafy spurge control in a tallgrass prairie natural area. In: Masters, Robert A.; Nissen, Scott J.; Friisoe, Geir, eds. Leafy spurge symposium: Proceedings of a symposium; 1992 July 22-24; Lincoln, NE. Great Plains Agricultural Council Publication No. 144. [Manhattan, KS]: Great Plains Agricultural Council: 2-23 to 2-31.

Winter, Brian. 2005. [Email to Kris Zouhar]. March 7. Glyndon, MN: The Nature Conservancy, Bluestem Prairie Preserve. On file with: U.S. Department of Agriculture, Forest Service, Rocky

Mountain Research Station, Fire Sciences Laboratory, Missoula, MT: RWU 4403.

Winter, K.; Schmitt, M. R.; Edwards, G. E. 1982. *Microstegium vimineum*, a shade adapted C4 grass. Plant Science Letters. 24(3): 311-318.

Wirth, Troy A., Pyke, David A., 2007. Monitoring post-fire vegetation rehabilitation projects: current approaches, techniques, and recommendations for a common monitoring strategy in non-forested ecosystems: U.S. Geological Survey Scientific Investigations Report 2006-5048. [Washington, DC]: [U.S. Geological Survey]. 33 p.

Wisconsin Department of Natural Resources. 2004. Fact sheet: Reed canary grass (*Phalaris arundinacea*), [Online]. In: Invasive plant species. Madison, WI: Wisconsin Department of Natural Resources (Producer). Available: http://www.dnr.state.wi.us/invasives/fact/reed_canary.htm [2006, June 7].

Wiser, S. K.; Allen, R. B.; Clinton, P. W.; Platt, K. H. 1998. Community structure and forest invasion by an exotic herb over 23 years. Ecology. 79: 2071-2081.

Wolfson, B. A. S.; Kolb, T. E.; Sieg, C. H.; Clancy, K. M. 2005. Effects of post-fire conditions on germination and seedling success of diffuse knapweed in northern Arizona. Forest Ecology and Management. 216: 342-358.

Wolters, Gale L.; Sieg, Carolyn Hull; Bjugstad, Ardell J.; Gartner, F. Robert. 1994. Herbicide and fire effects on leafy spurge density and seed germination. Research Note RM-526. Fort Collins, CO: U.S. Department of Agriculture, Forest Service, Rocky Mountain Forest and Range Experiment Station. 5 p.

Wood, David M.; del Moral, Roger. 2000. Seed rain during early primary succession on Mount St. Helens, Washington. Madroño. 47(1): 1-9.

Woodall, S. L. 1979. Physiology of Schinus. In: Workman, R., ed.. Schinus-technical proceedings of techniques for control of Schinus in south Florida: a workshop for natural area managers. Sanibel, FL: The Sanibel-Captiva Conservation Foundation, Inc. 3-6.

Woodall, S. L. 1983. Establishment of Melaleuca quinquenervia seedlings in the pine/cypress ecotone of southwest Florida. Florida Scientist. 46: 65-71.

Woodall, Steven L. 1981. Site requirements for melaleuca seedling establishment. In: Geiger, R. K., ed. Proceedings of the Melaleuca symposium; 1980 September 23-24; [Ft. Meyers, FL]. Tallahassee, FL: Florida Department of Agriculture and Consumer Services, Division of Forestry: 9-15.

Woods, Kerry D. 1997. Community response to plant invasion. In: Luken, James O.; Thieret, John W.; eds. Assessment and management of plant invasions. Springer Series on Environmental Management. New York: Springer-Verlag: 56-68.

Wray, Jacilee; Anderson, M. Kat. 2003. Restoring Indian-set fires to prairie ecosystems on the Olympic Peninsula. Ecological Restoration. 21(4): 296-301.

Wright, Henry A.; Bailey, Arthur W. 1982. Fire ecology: United States and southern Canada. New York: John Wiley & Sons. 501 p.

Wright, Henry A.; Klemmedson, James O. 1965. Effect of fire on bunchgrasses of the sagebrush-grass region in southern Idaho. Ecology. 46(5): 680-688.

Wunderlin, R. P.; Hansen, B. 2003. Guide to the vascular plants of Florida, 2nd edition. Gainesville, FL: University Press of Florida. 787 p.

Xanthopoulos, Gavriil. 1986. A fuel model for fire behavior prediction in spotted knapweed (*Centaurea maculosa* L.) grasslands in western Montana. Missoula, MT: University of Montana. 100 p. Thesis.

Xanthopoulos, Gavriil. 1988. Guidelines for burning spotted knapweed infestations for fire hazard reduction in western Montana. In: Fischer, William C.; Arno, Stephen F., comps.. Protecting people and homes from wildfire in the Interior West: proceedings of the symposium and workshop; 1987 October 6-8; Missoula, MT. Gen. Tech. Rep. INT-251. Ogden, UT: U.S. Department of Agriculture, Forest Service, Intermountain Research Station: 195-198.

Xiong, Shaojun; Nilsson, Christer. 1999. The effects of plant litter on vegetation: A meta-analysis. Journal of Ecology. 87(6): 984-994.

Yost, Susan E.; Antenen, Susan; Harvigsen, Gregg. 1991. The vegetation of the Wave Hill Natural Area, Bronx, New York. Torreya. 118(3): 312-325.

Young, Frank L.; Ogg, Alex G. Jr.; Dotray, Peter A. 1990. Effect of postharvest field burning on jointed goatgrass (*Aegilops cylindrica*) germination. Weed Technology. 4: 123-127.

Young, J. A. 1991a. Cheatgrass. In: James, Lynn F.; Evans, John O., eds. Noxious range weeds. Westview Special Studies in Agriculture Science and Policy. Boulder, CO: Westview Press, Inc: 408-418.

Young, J. A. 1991b. Tumbleweed. Scientific American. 264(3): 82-87.

Young, J. A.; Budy, Jerry D. 1987. Energy crisis in 19th century Great Basin woodlands. In: Everett, Richard L., comp. Proceedings: pinyon-juniper conference; 1986 January 13-16; Reno, NV. Gen. Tech. Rep. INT-215. Ogden, UT: U.S. Department of Agriculture, Forest Service, Intermountain Research Station: 23-28.

Young, J. A.; Evans, R. A. 1973. Downy brome—intruder in the plant succession of big sagebrush communities in the Great Basin. Journal of Range Management. 26(6): 410-415.

Young, J. A.; Evans, R. A. 1974. Population dynamics of green rabbitbrush in disturbed big sagebrush communities. Journal of Range Management. 27: 127-132.

Young, J. A.; Evans, R. A. 1975. Germinability of seed reserves in a big sagebrush community. Weed Science. 23(5): 358-364.

Young, J. A.; Evans, R. A. 1978. Population dynamics after wildfires in sagebrush grasslands. Journal of Range Management. 31(4): 283-289.

Young, J. A.; Evans, R. A. 1981. Demography and fire history of a western juniper stand. Journal of Range Management. 34(6): 501-505.

Young, J. A.; Evans, R. A.; Major, J. 1972a. Alien plants in the Great Basin. Journal of Range Management. 25: 194-201.

Young, J. A.; Evans, R. A.; Robison, John. 1972b. Influence of repeated annual burning on a medusahead community. Journal of Range Management. 25(5): 372-375.

Young, J. A.; Evans, R. A.; Weaver, Ronald A. 1976. Estimating potential downy brome competition after wildfires. Journal of Range Management. 29(4): 322-325.

Youngblood, Andrew; Metlen, Kerry L.; Coe, Kent. 2006. Changes in stand structure and composition after restoration treatments in low elevation dry forests of northeastern Oregon. Forest Ecology and Management. 234(1-3): 143-163.

Zammit, C.; Zedler, P. H. 1994. Organization of the soil seed bank in mixed chaparral. Vegetatio. 111: 1-16.

Zar, Jerrold H. 1999. Biostatistical analysis. Upper Saddler River, NJ: Prentice Hall. 663 p. [plus appendices].

Zedler, J. G.; Kercher, S. 2004. Causes and consequences of invasive plants in wetlands: opportunities, opportunists, and outcomes. Critical reviews in plant sciences. 23: 431-452.

Zedler, J.; Loucks, O. L. 1969. Differential burning response of *Poa pratensis* fields and *Andropogon scopari* prairies in central Wisconsin. The American Midland Naturalist. 81: 341-352.

Zedler, P. H.; Zammit, C. A. 1989. A population-based critique of concepts of change in the chaparral. In: Keeley, S. C., ed. The California chaparral: paradigms reexamined. Los Angeles, CA: Natural History Museum of Los Angeles: 73-83.

Zedler, Paul H. 1995. Fire frequency in southern California shrublands: biological effects and management options. In: Keeley, Jon F.; Scott, Tom, eds. Brushfires in California: ecology and resource management: Proceedings; 1994 May 6-7; Irvine, CA. Fairfield, WA: International Association of Wildland Fire: 101-112.

Zedler, Paul H.; Gautier, Clayton R.; McMaster, Gregory S. 1983. Vegetation change in response to extreme events: the effect of a short interval between fires in California chaparral and coastal scrub. Ecology. 64(4): 809-818.

Zielke, Ken; Boateng, Jacob O.; Caldicott, Norm; Williams, Heather. 1992. Broom and gorse in British Columbia: A forestry perspective problem analysis. Province of British Columbia. Ministry of Forests. [Pages unknown].

Ziska, L. H.; Reeves, J. B., III; Blank, B. 2005. The impact of recent increases in atmospheric CO_2 on biomass production and vegetative retention of cheatgrass (*Bromus tectorum*): implications for fire disturbance. Global Change Biology. 11(8): 1325-1332.

Zouhar, Kris. 2001a. *Acroptilon repens*. In: Fire Effects Information System, [Online]. U.S. Department of Agriculture, Forest Service, Rocky Mountain Research Station, Fire Sciences Laboratory (Producer). Available: http://www.fs.fed.us/database/feis/ [2006, January 17].

Zouhar, Kris. 2001b. *Centaurea diffusa*. In: Fire Effects Information System, [Online]. U.S. Department of Agriculture, Forest Service, Rocky Mountain Research Station, Fire Sciences Laboratory (Producer). Available: http://www.fs.fed.us/database/feis/ [2006, January 17].

Zouhar, Kris. 2001c. *Centaurea maculosa*. In: Fire Effects Information System, [Online]. U.S. Department of Agriculture, Forest Service, Rocky Mountain Research Station, Fire Sciences Laboratory (Producer). Available: http://www.fs.fed.us/database/feis/ [2006, January 17].

Zouhar, Kris. 2001d. *Cirsium arvense*. In: Fire Effects Information System, [Online]. U.S. Department of Agriculture, Forest Service, Rocky Mountain Research Station, Fire Sciences Laboratory (Producer). Available: http://www.fs.fed.us/database/feis/ [2006, January 15].

Zouhar, Kris. 2002a. *Centaurea solstitialis*. In: Fire Effects Information System, [Online]. U.S. Department of Agriculture, Forest Service, Rocky Mountain Research Station, Fire Sciences Laboratory (Producer). Available: http://www.fs.fed.us/database/feis/ [2006, January 17].

Zouhar, Kris. 2002b. *Cirsium vulgare*. In: Fire Effects Information System, [Online]. U.S. Department of Agriculture, Forest Service, Rocky Mountain Research Station, Fire Sciences Laboratory (Producer). Available: http://www.fs.fed.us/database/feis/ [2005, April 1].

Zouhar, Kris. 2003a. *Bromus tectorum*. In: Fire Effects Information System, [Online]. U.S. Department of Agriculture, Forest Service, Rocky Mountain Research Station, Fire Sciences Laboratory (Producer). Available: http://www.fs.fed.us/database/feis/ [2006, January 18].

Zouhar, Kris. 2003b. *Linaria* spp. In: Fire Effects Information System, [Online]. U.S. Department of Agriculture, Forest Service, Rocky Mountain Research Station, Fire Sciences Laboratory (Producer). Available: http://www.fs.fed.us/database/feis/ [2005, March 29].

Zouhar, Kris. 2003c. *Tamarix* spp. In: Fire Effects Information System, [Online]. U.S. Department of Agriculture, Forest Service, Rocky Mountain Research Station, Fire Sciences Laboratory (Producer). Available: http://www.fs.fed.us/database/feis/ [2006, January 15].

Zouhar, Kris. 2004. *Hypericum perforatum*. In: Fire Effects Information System, [Online]. U.S. Department of Agriculture, Forest Service, Rocky Mountain Research Station, Fire Sciences Laboratory (Producer). Available: http://www.fs.fed.us/database/feis/ [2005, April 1].

Zouhar, Kris. 2005a. *Cytisus scoparius*, *C. striatus*. In: Fire Effects Information System, [Online]. U.S. Department of Agriculture, Forest Service, Rocky Mountain Research Station, Fire Sciences Laboratory (Producer). Available: http://www.fs.fed.us/database/feis/ [2006, August 28].

Zouhar, Kris. 2005b. *Elaeagnus angustifolia*. In: Fire Effects Information System, [Online]. U.S. Department of Agriculture, Forest Service, Rocky Mountain Research Station, Fire Sciences Laboratory (Producer). Available: http://www.fs.fed.us/database/feis/ [2006, January 17].

Zouhar, Kris. 2005c. *Genista monspessulana*. In: Fire Effects Information System, [Online]. U.S. Department of Agriculture, Forest Service, Rocky Mountain Research Station, Fire Sciences Laboratory (Producer). Available: http://www.fs.fed.us/database/feis/ [2006, May 1].

Zouhar, Kris. 2005d. *Ulex europaeus*. In: Fire Effects Information System, [Online]. U.S. Department of Agriculture, Forest Service, Rocky Mountain Research Station, Fire Sciences Laboratory (Producer). Available: http://www.fs.fed.us/database/feis/ [2006, September 21].

Zwolinski, Malcolm J. 1990. Fire effects on vegetation and succession. In: Krammes, J. S., technical coordinator. Effects of fire management of southwestern natural resources: Proceedings of the symposium; 1988 November 15-17; Tucson, AZ. Gen. Tech. Rep. RM-191. Fort Collins, CO: U.S. Department of Agriculture, Forest Service, Rocky Mountain Forest and Range Experiment Station: 18-24.

Appendix A: Glossary

These definitions were derived from chapters 1 through 4 of this volume, the Fire Effects Information System (FEIS), and the following sources: Agee (1993), Allaby (1992), Brown and others (1982), Brown and Smith (2000), Burke and Grime (1996), Helms (1998), Johnson (1992), Lincoln and others (1998), McPherson and others (1990), Neary and others (2005a), Romme (1980), Ryan and Noste (1985), Sakai and others (2001), Scott and Reinhardt (2007), Smith (2000), Sutton and Tinus (1983).

abundance: The number of individuals of a species in a given area; often used synonymously with "density"

adventitious: Structures or organs developing in an unusual position, such as roots originating on the stem

back fire: Fire set against an advancing fire to consume fuels and, as a consequence, prevent further fire

backing fire: A fire that is burning against the slope or wind, that is, moving down the slope or into the wind. This type of fire typically has the lowest fireline intensity on the fire's perimeter but may have long flame duration.

baseline fire regime: The fire regime needed to meet a specific goal, without necessarily requiring replication of past conditions

bud: A dormancy structure in shoots that consists of external protective scales and an internal embryonic shoot possessing meristem tissue

bulb: A short, solid, vertical underground stem with thin papery leaves

cambium: A layer of living, meristematic cells between the wood and the bark of a tree

caudex: The persistent and often woody base of an herbaceous perennial

condensation: The process by which water changes from gaseous to liquid phase and releases heat

controlled burn, controlled burning: See "prescribed fire"

corm: A short, solid, vertical underground stem with thin papery leaves

cover: The area of ground covered by a particular plant species, often expressed as a percent

crown fire: Fire that burns in the crowns of trees and shrubs, usually ignited by surface fire. Crown fires are common in coniferous forests and chaparral shrublands.

density: The number of individuals within a given area

depth of burn: Depth of ground fuels consumed by fire

dominance: The extent to which a given species predominates in a community because of its size, density, or cover

duff: Partially decomposed organic matter lying beneath the litter layer and above the mineral soil. It includes the humus and fermentation layers of the forest floor (Oa and Oe horizons, respectively).

duration of fire: The length of time that combustion occurs at a given point. Relates to downward heating and fire effects below the surface.

establishment (of a nonnative species): The process in which a species is able to grow and reproduce successfully in a new area

evapotranspiration: Evaporation from soils, plant surfaces, and water bodies, together with water losses from transpiring plants

fine fuels: Fast-drying, dead fuels, characterized by a high surface area to volume ratio, less than 1 cm in diameter. These fuels (for example, grasses, leaves, and needles) respond rapidly to changes in weather conditions. When dry, they ignite readily and are consumed rapidly.

fire-dependent ecosystems: Ecosystems where fire plays a vital role in determining the composition, structure, and landscape patterns

fire ecology: The study of relationships among fire, the environment, and living organisms

fire effects: The physical, chemical, and biological impacts of fire on the environment and ecosystem resources

fire exclusion: The policy and practice of excluding fire from wildlands. See also "fire suppression."

fire frequency: The recurrence of fire in a given area over time, often stated as number of fires per unit time

fire intensity: See "fireline intensity"

fire regime: Characteristic pattern of burning over large expanses of space and long periods of time. Fire regimes are described for a specific geographic area or vegetation type by the characteristic fire type (ground, surface, or crown fire), frequency, intensity, severity, size, spatial complexity, and seasonality.

fire-return interval: Number of years between fires at a given location; average number of years before fire reburns a given area

fire severity: The degree to which a site has been altered by fire; the effect of a fire on ecosystem properties, sometimes described by the degree of soil heating or mortality of vegetation

fire suppression: All work and activities connected with fire-extinguishing operations, beginning with discovery and continuing until the fire is completely extinguished. See also "fire exclusion."

fire type: Ground, surface, or crown

fireline intensity: Rate of heat release in the flaming front

flame length: The length of flames in the propagating fire front measured along the slant of the flame from the midpoint of its base to its tip

flame depth, flaming zone depth: Depth of the flaming front

flaming front: The zone in a spreading fire where combustion is primarily flaming. Behind this zone, combustion is primary glowing or involves the burning out of large fuels.

fuel: Living and dead vegetation that can be ignited. For descriptions of kinds of fuels and fuel classification, see Brown and Smith (2000).

fuel continuity: A qualitative description of the distribution of fuel both horizontally and vertically. Continuous fuels support fire spread better than discontinuous fuels.

fuel load, fuel loading: Weight of fuel per unit area. For descriptions of kinds of fuels and fuel classification, see Brown and Smith (2000).

grass/fire cycle: See "invasive plant/fire regime"

ground fire: Fire that burns in the organic material below the litter layer, mostly by smoldering combustion. Fires in duff, peat, dead moss and lichens, and partly decomposed wood are typically ground fires.

harm, ecological: In natural areas, occurs when an undesired species is abundant enough to cause significant, undesired changes in ecosystem composition, structure, or function

high-severity fire: Fire that alters soil properties and/or kills substantial amounts of underground plant tissue

invasibility: Susceptibility of a plant community to invasion

invasive: Species that can establish, persist, and spread in an area, and also cause—or have potential to cause—negative impacts or harm to native ecosystems, habitats, or species

invasive plant/fire regime: Occurs when a plant invasion alters fuelbed characteristics that alter the spatial and/or temporal distribution of fire on the landscape. These changes in fire regime, in turn, promote dominance of the invasive species.

ladder fuels: Shrubs, vines, forbs, young trees, and low branches on larger trees that provide continuous fine fuels from the surface into the crowns of dominant trees or shrubs

litter: Recently fallen plant material that is not decomposed or only partly decomposed; particles are still discernible

low-severity fire: Fire that causes little alteration to the soil and little mortality to underground plant parts or seed banks

meristem: Undifferentiated plant tissue with cells that can differentiate to form new tissues or organs

mesic: Pertaining to conditions of moderate moisture or water supply

mixed-severity fire regime: Pattern in which most fires either cause selective mortality of the overstory vegetation, depending on different species' susceptibility to fire, or sequential fires vary in severity. Applies only to forests, woodlands, and shrublands.

moderate-severity fire: Fire that causes moderate soil heating. Occurs where litter is consumed and duff is charred or consumed, but the underlying mineral soil is not visibly altered.

nonnative: A species that has evolved outside a particular area (for example, the United States) and has been transported to and disseminated in that area by human activities

organic soil: Soils with deep layers of organic matter that develop in poorly drained areas such as bogs, swamps, and marshes; and soils having more than 20 percent organic matter by weight

perennating tissues: See "meristem"

persistence (of a nonnative species): Establishment of a viable, self-sustaining population that maintains itself over time

prescribed fire: Fire burning with prescription, resulting from planned ignition, that meets management objectives; controlled application of fire to fuels in specified environmental conditions that allow the fire to be confined to a predetermined area and, at the same time, to produce fire behavior that will attain the planned management objectives

presettlement fire regime: Describes fire regimes before extensive settlement by European Americans, extensive conversion of wildlands for agriculture, and effective fire suppression. See also "baseline fire regime" and "reference fire regime."

propagule pressure: The availability, abundance, and mobility of propagules entering the plant community

reaction intensity: Rate of heat release per unit area of the flaming front

reference fire regime: See "baseline fire regime"

residence time: Time required for the flaming front of a fire to pass a stationary point at the surface of the fuel. The length of time that the flaming front occupies one point

rhizome: A horizontal underground stem with of a series of nodes that commonly produce roots

root crown: The point at which the root and stem of a plant meet and the vascular anatomy changes from that of a stem to that of a root; transition point between stem and root

seed bank: The community of viable seeds present in the soil and held in vegetation aboveground

species richness: The number of species present in a given area

spread (of a nonnative species): Increase in the size of existing populations and establishment of new, self-sustaining populations

stand-replacement fire regime: Pattern in which fire kills or top-kills the aboveground parts of the dominant vegetation. Using this definition, forests that routinely experience crown fire or severe surface fire have a stand-replacement fire regime; grasslands and many shrublands also have stand-replacement fire regimes because fire usually kills or top-kills the dominant vegetation layer.

stolon: A horizontal stem that creeps above ground and roots at the nodes or tips, giving rise to a new plant

succession: The gradual, somewhat predictable process of community change and replacement; the process of continuous establishment and extinction of populations at a particular site

surface fire: Fire that spreads in litter, woody material on or near the soil surface, herbs, shrubs, and small trees

tolerance: The capacity of an organism to subsist under a given set of environmental conditions

top-kill: Mortality of aboveground tissues of a plant without mortality of underground parts from which the plant can produce new stems and leaves

total fuel: The amount of biomass that potentially could burn

total heat release: The heat produced in the flaming front plus that released behind the flaming front through glowing and smoldering combustion

underburn: See "Understory fire"

understory fire: Fire that is not generally lethal to the dominant vegetation and does not substantially change the structure of the dominant vegetation. Applies only to forests, woodlands, and shrublands.

understory fire regime: Pattern in which most fires do not kill or top-kill the overstory vegetation and thus do not substantially change the plant community structure. Applies only to forests, woodlands, and shrublands.

vaporization: The process of adding heat to water until it changes from liquid to gaseous phase

wetlands: Areas that are saturated by surface water, groundwater, or a combination at a frequency and duration sufficient to support a prevalence of vegetation adapted to saturated soil conditions

wildfire: Fire that is not meeting management objectives and, therefore, requires a suppression response

wildland fire: Any nonstructural fire, other than prescribed fire, that occurs in a wildland setting

wildland fire use: Application of the appropriate management response to naturally ignited wildland fires to accomplish specific resource management objectives

xeric: Having very little moisture; tolerating or adapted to dry conditions

Index

A

absinth wormwood, 51
Acacia farnesiana, see klu
Acer platanoides, see Norway maple
Acroptilon repens, see Russian knapweed
adaptive management, 48, 60, 257, 259, 281, 294, 295
Aegilops cylindrica, see jointed goatgrass
Aegilops spp., see goatgrass
Aegilops triuncialis, see barbed goatgrass
Agropyron cristatum, see crested wheatgrass
Agropyron desertorum, see desert wheatgrass
Agrostis gigantea, see redtop
Ailanthus altissima, see tree-of-heaven
air potato, 95, 106, 107
Albizia julibrissin, see mimosa
alfalfa, 276
Alliaria petiolata, see garlic mustard
alsike clover, 219
Ampelopsis brevipedunculata, see porcelainberry
Amur honeysuckle, 63, 75
Angleton bluestem, 116
annual canarygrass, 24
annual ryegrass, 201
annual vernal grass, 22, 198, 209
Anthoxanthum aristatum, see annual vernal grass
Anthoxanthum odoratum, see sweet vernal grass
Arrhenatherum elatius, see tall oatgrass
Artemisia abinthium, see absinth wormwood
Arundo donax, see giant reed
Asian sword fern, 21, 227, 232, 233, 235, 240
autumn-olive, 20, 63, 68, 69, 81, 95, 100, 245–259
Avena barbata, see slender oat
Avena fatua, see wild oat
Avena spp., see oat

B

bahia grass, 55, 60, 95, 103, 105, 116
barbed goatgrass, 49, 58, 59, 180, 181, 182
barley, 179, 180, 184
Bell's honeysuckle, 80, 81
Bermudagrass, 116, 117, 137
Berberis thunbergii, see Japanese barberry
bigleaf periwinkle, 110
biological control, 55–56, 135, 138, 140
birdsfoot trefoil, 201
bird vetch, 198, 218–220 *passim*
blackberry, 21, 198, 214
black locust, 2, 63, 76–78 *passim*
black swallow-wort, 79, 80, 82, 83
Boer lovegrass, 21, 146
Bohemian knotweed, 217
Bothriochloa bladhii, see Caucasian bluestem
Bothriochloa ischaemum, see yellow bluestem
Brachypodium sylvaticum, see false brome
Brassica spp., see mustard
Brassica tournefortii, see Sahara mustard
Brazilian pepper, 12, 18, 27, 57, 93, 95, 97, 99, 101, 103, 104, 245–259
Brazilian satintail, 101, 245–259
Brazilian vervain, 116
bristly sheepburr, 219
brittlestem hempnettle, 219
brome fescue, 180
Bromus diandrus, see ripgut brome
Bromus hordeaceus, see soft chess
Bromus inermis, see smooth brome
Bromus japonicus, see Japanese brome
Bromus rubens, see red brome
Bromus tectorum, see cheatgrass
broomsedge, 227, 229, 231, 232, 234–237 *passim*
buckthorns, 70
buffelgrass, 41, 113, 116–118 *passim*, 126, 137, 144, 157, 158, 160, 227, 229, 234, 239
bull thistle, 16, 17, 21, 25, 27, 59, 122, 144, 148, 164, 167, 170, 199–206 *passim*, 245–259
bur clover, 184
Burned Area Emergency Response (BAER), 4, 274
bush beardgrass, 227, 229, 231–236 *passim*
bush honeysuckles, 20, 27, 60, 63, 66, 68, 70, 73–78 *passim*, 80, 81, 83, 88, 95, 100, 105, 107, 125, 245–259

C

Canada bluegrass, 52, 68, 73, 79
Canada thistle, 21, 25–27 *passim*, 52, 115, 116, 118, 122, 126, 144, 148, 149, 164–168 *passim*, 170, 199, 200, 202, 203, 205, 206, 216, 219, 220, 222, 245–259
Canadian horseweed, 165
Cardaria spp., see hoary cress
Carduus nutans, see musk thistle
Carpobrotus edulis, see hottentot fig
Carpobrotus spp., see iceplant
catclaw mimosa, 109
Caucasian bluestem, 115–120 *passim*, 129, 136
Celastrus orbiculatus, see Oriental bittersweet
Centaurea biebersteinii, see spotted knapweed
Centaurea diffusa, see diffuse knapweed
Centaurea maculosa, see spotted knapweed
Centaurea solstitialis, see yellow starthistle
Centaurea spp., see knapweed
Centaurea triumfettii, see squarrose knapweed
chaparral, 18, 25
  Interior West, 156
  Southwest coastal, 24, 26, 177, 182–187 *passim*, 265
cheatgrass, 11, 16, 19, 22, 24–26 *passim*, 28, 30, 35, 38–41 *passim*, 45, 50, 53–55 *passim*, 78, 115, 116, 118, 131, 142, 144, 147, 148, 152, 154, 156–160 *passim*, 162–168 *passim*, 170, 172, 179, 184, 191, 192, 220, 245–259, 265, 277, 279
Chenopodium album, see lambsquarters
chinaberry, 17, 20, 27, 95–98 *passim*, 100
Chinese privet, 51, 68, 75, 76, 97, 107, 116
Chinese silvergrass, 110
Chinese tallow, 12, 15, 18, 19, 27, 43, 51, 93, 95, 97–99 *passim*, 116–118 *passim*, 126, 127, 245–259
Chinese wisteria, 74, 110
Chinese yam, 95, 107
Chondrilla juncea, see rush skeletonweed
Cirsium arvense, see Canada thistle
Cirsium vulgare, see bull thistle
climate change, 28, 29–31, 34, 66, 88, 92, 111, 156, 194, 215, 221, 222, 295, 296

climbing fern species, 245–259
coastal scrub, 26, 41, 183–187, 265
cogongrass, 22, 93, 95, 101–103 *passim*, 105, 116, 118, 129, 245–259
colonial bentgrass, 209, 213
common buckthorn, 12, 27, 51, 60, 63, 68, 73, 75, 76, 78, 80, 81, 83, 87, 116, 118, 125
common dandelion, 21, 163, 165, 167, 168, 203, 205, 206, 215–217 *passim*, 219, 266
common groundsel, 17, 199, 205
common mullein, 116, 163, 165, 263
common pear, 21, 210–212 *passim*, 214
common pepperweed, 119, 219
common periwinkle, 110
common plantain, 219
common reed, 2, 83–86 *passim*
common sheep sorrel, 63, 78, 79, 166, 198, 205, 206, 210, 211, 217
common tansy, 219
common teasel, 57
common velvetgrass, 17, 21, 22, 191, 198, 209, 210, 212, 213
competition, 10, 121, 132, 133, 135, 140, 199, 200, 207–209 *passim*, 212, 214, 221, 256, 261
Convolvulus arvensis, see field bindweed
Coronilla varia, see crown vetch
Cortaderia spp., see pampas grass
creeping bentgrass, 205
creeping yellowcress, 219
crested wheatgrass, 16, 45, 116, 154, 158, 163, 277
crown fire, 3, 103, 104, 106, 110, 138, 212
crown vetch, 116, 118, 123
cutleaf blackberry, 51, 203, 206, 211, 212, 214
cutleaf filaree, 16, 17, 50, 53, 144, 157, 158, 161, 163, 184
cutleaf teasel, 57
Cynanchum louiseae, see black swallow-wort
Cynanchum rossicum, see pale swallow-wort
Cynodon dactylon, see Bermudagrass
Cytisus scoparius, see Scotch broom

D

Dactylis glomerata, see orchard grass
dallis grass, 229, 230, 237
Dalmatian toadflax, 11, 21, 22, 54, 115, 116, 144, 148, 150, 151, 153, 161, 163, 165, 166, 245–259
depth of burn, 255
Descurainia sophia, see flixweed tansymustard
desert wheatgrass, 16, 154
Dewey's sedge, 215
diffuse knapweed, 12, 42, 116, 133, 144, 148, 149, 157, 164, 245–259
Dioscorea alata, see water yam
Dioscorea bulbifera, see air potato
Dioscorea oppositifolia, see Chinese yam
Dipsacus fullonum, see common teasel
Dipsacus laciniatus, see cutleaf teasel
disturbance, 9, 10, 16, 22–25, 28, 58, 66, 77, 79, 83, 87, 88, 117, 120, 132, 134, 142, 154, 169, 170, 172, 173, 176, 180, 185, 191, 194, 198, 199, 200, 202, 203, 204, 206–208 *passim*, 215, 216, 219–223 *passim*, 261, 263, 265, 270, 273, 278, 294
diversity, 120, 130, 136, 202, 206, 208, 212, 221
duration of fire, 13, 22

E

Ehrharta stipoides, see meadow ricegrass
Elaeagnus angustifolia, see Russian-olive
Elaeagnus pungens, see thorny-olive
Elaeagnus umbellata, see autumn-olive
Elymus repens, see quackgrass
English ivy, 74, 95, 198, 207
Eragrostis chloromelas, see Boer lovegrass
Eragrostis curvula, see weeping lovegrass
Eragrostis lehmanniana, see Lehmann lovegrass
Erodium cicutarium, see cutleaf filaree
Erodium spp., see filaree
establishment (of a nonnative species), 93, 96, 105, 109, 117, 122, 148, 150, 167, 176, 191, 201, 206, 211, 213, 217, 234, 244, 250, 254, 255, 271, 277, 293, 295
eucalyptus, 188
Eucalyptus globulus, see Tasmaniam bluegum
Euonymus alatus, see winged euonymus
Euonymus fortunei, see winter creeper
European privet, 51, 63, 107
Euphorbia esula, see leafy spurge
Euryops multifidus, see sweet resinbush
evergreen blackberry, 210

F

false brome, 198, 203, 204, 212, 222
fennel, 54, 55, 176, 178, 179, 187, 195
Festuca filiformis, see fineleaved sheep fescue
Festuca rubra, see red fescue
fetid goosefoot, 165
field bindweed, 110, 219, 245–259
field foxtail, 215
filaree, 16, 176, 180, 184
fine fuels, 185
fine fuels, 39, 42, 74, 87, 93, 97, 99, 103, 104, 110, 128, 133, 145, 147, 148, 151, 155, 158, 159, 162, 163, 172, 208, 231, 240, 267
fineleaf sheep fescue, 73, 74, 78
fire behavior, 3, 33, 97, 102, 150, 203, 207, 212, 217, 220, 223, 250, 290
fire continuity, 115, 130, 185
fire exclusion, 4, 27, 42, 62, 67, 68, 73, 77, 80, 84, 87, 92, 93, 96, 101, 105, 107, 109, 115, 118–121 *passim*, 125, 126, 143, 146, 153, 162–164 *passim*, 166, 172, 176, 188, 190, 198, 204, 206, 208, 212, 214, 215, 222, 234, 244, 262
fire frequency, 4, 17–18, 33, 39, 51, 55, 58–59, 99, 101, 107, 108, 113, 115, 118, 119, 121–125 *passim*, 128, 130–133 *passim*, 136–138 *passim*, 142, 145–148 *passim*, 151, 154–156 *passim*, 159, 160, 163, 164, 166, 169, 170, 172, 176, 177, 181, 187, 198, 203, 206, 215, 218, 221, 222, 231, 233, 252, 257
fire intensity, see *fireline intensity*
fireline intensity, 4, 39, 138, 203, 207, 213
fire mortality, 127, 138, 147, 171, 207, 208, 214, 250, 291, 292
fire regime, 4, 9, 33, 36, 48, 56–59, 60, 61–62, 77, 78, 80, 94, 102, 106, 107, 110, 113, 117, 121, 131, 133, 137, 148–152, 158–160, 163, 166, 169, 170, 180, 185, 204, 205, 207, 212, 214, 217–218, 220, 221, 247, 250, 252, 255, 257, 262, 295, 296
  baseline, 5, 34, 80, 159, 177, 187, 247, 257, 294

mixed severity, 4, 67, 101, 154, 162, 164, 166, 198, 212, 218, 263–264, 266
presettlement, 4, 27–28, 67, 77, 82, 92, 105, 106, 108, 111, 115, 118, 130, 136, 139, 145, 146, 154, 155, 159, 162, 164, 166, 172, 176, 183, 184, 187, 189, 190, 198, 204, 206, 208, 215, 218, 230, 237, 247, 257
stand replacement, 4, 94, 108, 154, 155, 166, 167, 183, 184, 198, 204, 215, 218, 220, 264–266, 267
understory, 4, 101, 105, 164, 262–263, 266
fire-return interval, 38, 104, 105, 154, 155, 157–159 *passim*, 162, 166, 177, 183, 185, 187, 190, 230, 238
fire season, 33, 51–53 *passim*, 55, 56, 86, 87, 106, 115, 118–125 *passim*, 127, 128, 133, 135–138 *passim*, 145, 152, 154, 158, 181, 191, 199, 200, 209, 213–215 *passim*, 247, 252, 257
fire severity, 4, 12–17, 22, 33, 36, 42, 56, 79, 115, 122, 127, 130, 136, 152, 165–169 *passim*, 172, 191, 201, 202, 206, 208, 213, 214, 222, 247, 250, 252, 257, 283, 284, 290–292 *passim*
fire size, 18, 33, 40, 55, 58, 158, 163, 169, 170, 177, 185, 187, 231, 233
fire suppression, 23, 25, 168, 172, 177, 186, 191, 202, 218, 219, 221, 222, 269–280, 295
fire survival, 8, 12, 13–16, 18, 20, 43, 48, 50, 51, 67, 68–73, 75, 76, 78–82 *passim*, 84, 85, 96–98 *passim*, 100, 101, 103, 104, 106, 107, 127, 128, 146, 147, 153, 160, 171, 191, 234, 235, 238, 250, 291
firetree, 227, 237, 238
fire type, 9, 33
flame depth, 290
flame length, 138, 290
flammability, 207
flatspine stickseed, 165
flixweed tansymustard, 157, 163, 219
flooding, 55, 60, 82, 83, 96, 98, 99, 110, 137, 170–172 *passim*, 192
Foeniculum vulgare, see fennel
forest
Alaska hemlock-spruce, 198, 215–218, 266
boreal, 198, 218–220, 266
closed-canopy, 142, 166—168, 172
cypress swamp, 92, 95, 108–109
Hawai`i lowland, 228
Hawai`i montane, 230, 236, 240, 265
Hawai`i subalpine, 230
Hawai`i wet, 237–239
Interior West closed canopy, 25, 26, 142, 166–168
Interior West open canopy, 142, 164–166
lodgepole pine, 24, 266
Northeast coniferous, 61, 65, 69, 76–78
Northeast deciduous, 61, 65, 66, 68–76
Northeast mixed, 61, 65, 67, 68–76
Northwest coastal, 25
Northwest coastal Douglas-fir, 198–204
Northwest coastal montane, 198, 204–206, 263
ponderosa pine, 11, 24–27 *passim*, 53, 59, 262–263, 289
Southeast oak-hickory, 91, 95
Southeast pine, 91, 95, 100–105
Southeast tropical hardwood, 94, 95, 107–108
Southwest coastal, 25
Southwest coastal coniferous, 177, 188–192, 263
Southwest coastal mixed evergreen, 187–188, 263
fountain grass, 21, 22, 41, 116, 144, 157, 158, 160, 184, 227, 229, 232, 234, 235, 237, 239, 241, 265
foxtail fescue, 179, 180, 181, 184
Frangula alnus, see glossy buckthorn
French broom, 15, 17, 21, 27, 50, 54, 57, 179, 186, 188, 245–259
fuel, 58, 125, 130, 131, 137, 140, 148–152, 158–160, 163, 166, 170, 180, 185, 247, 250, 261, 296
fuel continuity, 35, 36, 40, 42, 57–58, 63, 74–75, 78, 80–81, 85–86, 87, 97, 102–103, 106, 108, 110, 133, 138, 139, 145, 146, 148, 151, 155, 158, 159, 160, 170, 180, 188, 192, 203, 207, 231, 233, 238, 240, 261, 263, 265, 277
fuel load, 35, 36, 38, 42, 50, 51, 57–58, 63, 74–75, 76, 78, 80–81, 85–86, 87, 97, 99, 102–103, 106, 110, 115, 118, 122, 123, 125, 127, 130, 137, 138, 145, 146, 152, 154, 155, 158, 159, 161, 163, 170, 187, 188, 191, 192, 204, 207, 208, 212–215 *passim*, 217, 222, 231, 233, 234, 238, 240, 241
fuel treatment, 23, 24, 25, 261–268, 276

G

garden cornflower, 209, 211, 219
garlic mustard, 16, 48, 71, 73, 75, 76, 80, 83, 95, 116, 245–259
Genista monspessulana, see French broom
giant knotweed, 217
giant reed, 36, 105, 110, 116–118 *passim*, 140, 179, 192, 193, 245–259
giant sugarcane plumegrass, 116
Glechoma hederacea, see ground-ivy
glossy buckthorn, 20, 27, 51, 60, 63, 68, 73, 75, 78, 83–85 *passim*, 87, 125
goatgrass, 179
golden bamboo, 110
gorse, 15, 17, 27, 54, 198, 203, 216, 245–259
grain barley, 58
grass/fire cycle, see *invasive plant/fire regime*
grassland
California, 27, 49, 52, 53, 58, 59, 177–182
desert, 49, 145, 146–147, 151, 152
Hawai`i lowland, 228, 234, 240
mountain, 19, 54, 57, 142, 145–146, 147, 151–152
Northeast, 65, 78–82
Northwest montane meadow, 204–206
Pacific northwest montane meadow, 198
shortgrass, 114, 118, 122, 131, 132, 136–137, 138
Southeast wet, 27, 94–100, 95, 109
Southwest coastal, see *grassland, California*
grazing, 23, 25, 42, 50, 53, 58, 81, 114, 115, 119, 120, 122–124 *passim*, 126, 128, 130, 132, 141, 142, 145, 146, 151, 153–156 *passim*, 160, 162, 163, 169, 171, 172, 176, 180, 208, 277
ground fire, 3, 14, 15, 16, 50, 96, 97, 98, 107, 110
ground-ivy, 68, 72, 73, 83
guineagrass, 115–118 *passim*, 129, 136, 137, 234

H

hairy catsear, 21, 198, 199, 205, 209–211 *passim*, 213, 216
halogeton, 157
harm
ecological, 2, 20, 37, 45, 117, 255, 277

hawkweed, 142, 167
heath woodrush, 210, 211
Hedera helix, see English ivy
herbicide, 51, 53, 54, 60, 81, 86, 87, 99, 100, 105, 106, 125, 127, 128, 135, 137, 138, 148, 152, 153, 160, 161, 171, 177, 178, 195, 200, 208, 214, 239, 265, 275, 284, 289
Hieracium spp., see hawkweed
high severity fire regime, see *fire regime, stand replacement*
Hilo grass, 227, 238
Himalayan blackberry, 22, 26, 51, 202, 203, 206–208 *passim*, 210–212 *passim*, 214
hoary cress, 148, 149, 245–259
Holcus lanatus, see common velvet grass
Hordeum murinum, see mouse barley
Hordeum spp., see barley
Hordeum vulgare, see grain barley
hottentot fig, 179, 187
houndstongue, 148, 245–259
Hyparrhenia rufa, see thatching grass
Hypericum perforatum, see St. Johnswort
Hypochaeris radicata, see hairy catsear

I

iceplant, 41
impact, 128, 134, 142, 145, 165, 170, 173, 190, 201, 207, 211, 255, 256, 296
Imperata brasiliensis, see Brazilian statintail
Imperata cylindrica, see cogongrass
invasibility, 2, 8, 9, 18, 27, 92, 96, 109, 119, 172, 206, 209, 214, 221, 222, 244, 255, 256, 270, 279, 294, 296
invasive plant, 2
invasive plant/fire regime, 19, 26, 28, 37–40 *passim*, 43, 44, 93, 97, 103, 110, 142, 158, 159, 172, 185–186, 203–204, 231, 232, 234, 240, 255, 265, 267
Ipomoea coccinea, see redstar
Italian ryegrass, 11, 58, 179, 180, 184
itchgrass, 116
Iva axillaris, see poverty weed

J

Japanese barberry, 53, 63, 66, 68, 69, 74–78 *passim*, 80, 83, 87, 88
Japanese brome, 11, 16, 19, 25, 53, 116, 118, 131, 144, 147, 148, 162–164 *passim*
Japanese climbing fern, 41, 95, 96, 101, 116
Japanese honeysuckle, 54, 63, 66, 68, 71, 74–77 *passim*, 80, 81, 95, 100, 101, 104–106 *passim*, 116
Japanese knotweed, 41, 83, 85, 87, 207, 217
Japanese privet, 68, 107
Japanese stiltgrass, 41, 63, 66, 68, 72, 73, 74, 76, 77, 80, 83, 88, 95, 100, 245–259
Japanese wisteria, 110
Johnson grass, 17, 41, 116–118 *passim*, 129, 137, 144, 146, 160, 245–259
jointed goatgrass, 49, 157

K

Kentucky bluegrass, 11, 18, 52, 55, 116–121 *passim*, 123, 144, 152, 153, 157, 161, 165, 191, 205, 210, 211, 217, 221, 271
kiawe, 227, 228, 233
kikuyu grass, 227
Kleberg bluestem, 116
klu, 227, 233
knapweeds, 27, 118
knotweeds, 198, 207, 217
koa haole, 227, 228, 233
kochia, 116
kudzu, 17, 20, 54, 55, 66, 68, 71, 74–76 *passim*, 95, 103, 104–107 *passim*, 245–259

L

Lactuca serriola, see prickly lettuce
Lactuca spp., see wild lettuce
lambsquarters, 165, 263
leafy spurge, 54–56 *passim*, 113, 115, 116, 118, 120, 134–136, 140, 144, 148–150 *passim*, 153, 157, 164, 284, 286, 287
Lehmann lovegrass, 16, 17, 19, 21, 54, 116, 136, 144–146 *passim*, 151, 156, 161
Lepidium latifolium, see perennial pepperweed
lespedeza, 95
Lespedeza bicolor, see lespedeza
Lespedeza cuneata, see sericea lespedeza
Leucaena leucocephala, see koa haole
Ligustrum japonicum, see Japanese privet
Ligustrum sinense, see Chinese privet
Ligustrum spp., see privet
Ligustrum vulgare, see European privet
Linaria dalmatica, see Dalmatian toadflax
Linaria vulgaris, see yellow toadflax
litter, 40, 42, 44, 52–53, 55, 74, 78, 85, 86, 88, 104, 106, 120, 131, 132, 135, 137, 148, 151, 159, 170, 192, 203, 211, 214, 290, 291
little quakinggrass, 209
Lolium arundinaceum, see tall fescue
Lolium multiflorum, see Italian ryegrass
London rocket, 11, 144, 147, 157, 163
Lonicera japonica, see Japanese honeysuckle
Lonicera spp., see bush honeysuckles
lupine, 78, 208
Lupinus spp., see lupine
Lygodium japonicum, see Japanese climbing fern
Lygodium microphyllum, see Old World climbing fern
Lythrum salicaria, see purple loosestrife

M

Macartney rose, 54, 116–118 *passim*, 126, 128
meadow foxtail, 221
meadow ricegrass, 227, 230, 237
Medicago polymorpha, see bur clover
Medicago spp., see alfalfa
Mediterranean grass, 17, 39, 50, 157–159 *passim*, 161
Mediterranean sage, 157
medusahead, 12, 16, 17, 38, 49, 53, 54, 58, 144, 147, 148, 152, 156–158 *passim*, 162, 163, 179–182 *passim*, 265
melaleuca, 12, 15, 18–20 *passim*, 27, 53, 93, 95–99 *passim*, 101, 103, 107–110 *passim*, 245–259
Melaleuca quinquenervia, see melaleuca
mile-a-minute, 71, 83, 85, 87
Melia azedarach, see chinaberry
Melilotus alba, see white sweetclover

Melilotus officinalis, see yellow sweetclover
Melilotus spp., see sweetclover
Melinis minutiflora, see molasses grass
Melinis repens, see Natal redtop
mimosa, 17, 95, 105, 106
Mesembryanthemum spp., see iceplant
Microstegium vimineum, see Japanese stiltgrass
Mimosa pigra, see mimosa
Miscanthus sinensis, see Chinese silvergrass
Missouri bladderpod, 116
molasses grass, 227, 229, 231–237 *passim*, 240
monitoring, 3, 29, 31, 47, 48, 60, 81, 88, 89, 105, 172, 173, 256, 257, 259, 263, 266, 275, 277, 281–292, 294
Morella faya, see firetree
Morrow's honeysuckle, 80, 81, 87
mouse barley, 52, 58
multiflora rose, 63, 68, 70, 76, 79–81 *passim*, 83, 110, 116, 245–259
musk thistle, 21, 26, 115, 116, 122, 144, 148, 245–259
mustard, 179, 184
Mycelis muralis, see wall-lettuce
Mycelis spp., see wild lettuce

N

narrowleaf hawksbeard, 219, 220
Natal redtop, 227, 233, 239
Nephrolepis multiflora, see Asian sword fern
Neyraudia reynaudiana, see silkreed
nonnative, 2
nonnative pathogens, 66, 77, 88
North Africa grass, 25
Norway maple, 63, 66, 68, 69, 73, 75, 77, 80, 83, 245–259

O

oat, 58, 179, 219
oatgrass, 211, 212
old field, 78–82, 105
Old World climbing fern, 41, 95–98 *passim*, 100–102 *passim*, 104, 107, 108
oneseed hawthorn, 212
orchard grass, 116, 164, 165, 201, 207, 216, 217
organic soil, 97, 98, 107, 110
Oriental bittersweet, 63, 66, 68, 71, 74, 75, 77, 78, 80, 83, 88, 245–259
oxeye daisy, 116, 219

P

pale madwort, 163
pale swallow-wort, 79, 82, 83
pampas grass, 179, 184, 185
paradise apple, 212
Paspalum conjugatum, see Hilo grass
Paspalum notatum, see bahia grass
Paulownia tomentosa, see princesstree
Penn sedge, 209, 213
Pennisetum ciliare, see buffelgrass
Pennisetum clandestinum, see Kikuyu grass
Pennisetum setaceum, see fountain grass
perennial pepperweed, 54, 179, 245–259
perennial ryegrass, 201, 203, 206, 217, 221
perennial sowthistle, 219, 245–259
persistence (of a nonnative species), 117, 148, 152, 153, 156, 163, 172, 186, 191, 234, 235, 244, 254, 256, 264, 266, 294
Phalaris arundinacea, see reed canarygrass
Phalaris canariensis, see annual canarygrass
Phleum pratense, see timothy
Phragmites australis, see common reed
Phyllostachys aurea, see golden bamboo
pineapple weed, 219
Poa compressa, see Canada bluegrass
Poa nemoralis, see wood bluegrass
Poa pratensis, see Kentucky bluegrass
Polygonum cuspidatum, see Japanese knotweed
Polygonum perfoliatum, see mile-a-minute
Polygonum spp., see knotweeds
porcelainberry, 66, 70, 74, 79, 80, 83
Portuguese broom, 245–259
Potentilla recta, see sulfur cinquefoil
poverty weed, 58
prairie
  northern and central tallgrass, 49, 52, 58, 114, 117–125 *passim*
  northern mixedgrass, 51, 55, 58, 114, 118, 122, 123, 131–136, 139
  palmetto, 94, 95
  southern mixedgrass, 114, 118, 129, 136, 138
  southern tallgrass, 27, 43, 53, 114, 117–119 *passim*, 125–130, 132
prescribed fire, 62, 67, 75–79 *passim*, 81–84 *passim*, 86–87, 92, 93, 98–100, 102–108 *passim*, 110, 111, 113, 115, 118–125 *passim*, 127, 128, 130–139 *passim*, 150, 152–153, 160–161, 163–164, 166, 171, 173, 177, 180, 187, 199, 204, 205, 208, 209–211, 212–215, 218, 220, 221, 223, 239–240, 241, 244, 247, 252, 257, 261, 265, 289, 291, 295
prickly lettuce, 11, 17, 22, 25, 26, 144, 163, 165, 167, 168, 200, 201, 263
prickly Russian-thistle, 26
princesstree, 22, 66, 68, 69, 77, 110
privet, 66, 68, 70, 75–77 *passim*, 83, 95, 105, 245–259
propagule pressure, 9, 22, 24, 28, 79, 83, 109, 110, 119, 137, 180, 185, 186, 188, 191, 250, 261, 263, 270–275 *passim*, 277–279, 294, 295
Prosopis pallida, see kiawe
prostrate knotweed, 119
Psathrostachys juncea, see Russian wildrye
Pueraria montana var. lobata, see kudzu
puncture vine, 144, 147
purple loosestrife, 83, 84, 86, 87, 116, 117, 245–259
Pyrus communis, see common pear

Q

quackgrass, 52, 116, 199, 205

R

rattail sixweeks grass, 16, 58
reaction intensity, 4
red brome, 11, 17, 30, 38, 39, 41, 49, 50, 52, 142, 144, 155, 157–159 *passim*, 161–163 *passim*, 172, 179, 180, 184, 265
red fescue, 219, 221, 227, 235
red-horned poppy, 116

redstar, 146
redtop, 16, 68, 79, 167
reed canarygrass, 2, 72, 76, 83, 84, 86, 198, 208, 218, 219, 221
reference fire regime. *See* baseline fire regime
residence time, 35
resource availability, 10–11, 24, 270–273 *passim*, 275–278 *passim*, 294
Rhamnus cathartica, see common buckthorn
riparian, 35, 36, 44, 55, 65, 82–87, 114, 117, 118, 126, 128, 130, 132, 137–140, 168–171, 191–192, 198, 202, 206–208
ripgut brome, 17, 35, 54, 58, 179, 180, 184, 186, 271
Robinia pseudoacacia, see black locust
Rosa bracteata, see Macartney rose
Rosa eglanteria, see sweetbriar rose
Rosa multiflora, see multiflora rose
rough bluegrass, 199
rough hawkbit, 210
Rubus discolor, see Himalayan blackberry
Rubus lacinatus, see cutleaf blackberry
Rubus phoenicolasius, see wineberry
Rumex acetosella, see common sheep sorrel
rush skeletonweed, 144, 148, 149, 157, 245–259
Russian knapweed, 116, 134, 144, 148, 149, 165, 245–259
Russian-olive, 20, 68, 116–118 *passim*, 139, 141, 144, 169, 170, 172, 191, 245–259
Russian-thistle, 144, 147, 157, 159, 163, 165
Russian wildrye, 277
ryegrass, 116

S

safety, 45, 59, 88
Sahara mustard, 54, 144, 157, 158
salsify, 166
Salsola kali, see Russian-thistle
Salsola tragus, see prickly Russian-thistle
saltcedar, 44, 113, 118, 179, 245–259
savanna
  oak, 27, 62, 67, 75, 76, 119, 125
  Southeast pine, 91, 95, 100–105
Schinus terebinthifolius, see Brazilian pepper
Schismus spp., see Mediterranean grass
Schizachyrium condensatum, see bush beardgrass
Scotch broom, 11, 15–17 *passim*, 19, 21, 22, 26, 27, 50, 54, 56, 57, 59, 79–81 *passim*, 179, 186, 188, 191, 198, 202, 203, 206–209 *passim*, 211–214 *passim*, 216, 245–259
Scotch thistle, 116
seed
  reproduction by, 14, 16, 26, 47, 48, 58, 69, 96, 99, 100, 101, 106, 132, 134, 146, 149, 150, 152, 162, 202, 204, 213, 216, 232, 246, 248, 250, 254, 255, 257, 261
seed bank, 8, 10, 18, 21, 22, 24, 49, 50, 55, 100, 107, 139, 148, 159, 162, 181, 235, 246, 248, 255, 263
  aerial, 12
  soil, 16–17, 27, 68–73, 76, 77, 80, 81, 84, 85, 98, 122–124 *passim*, 126, 130, 132–134 *passim*, 153, 167, 199, 200, 202, 203, 213, 214, 219
seed dispersal, 18, 21, 63, 83, 96, 99, 104, 125, 126, 134, 137–139 *passim*, 149, 167, 185, 188, 199, 200, 205, 207, 218, 220, 222, 246, 248, 255, 295
seed production, 22, 48, 49, 120, 123, 126, 131, 134, 138, 139, 150, 169, 200, 246–248 *passim*, 255
seed scarification, 17, 21, 80, 105, 106, 150, 203
Senecio jacobaea, see tansy ragwort
Senecio sylvaticus, see woodland groundsel
Senecio vulgaris, see common groundsel
sericea lespedeza, 54, 95, 100, 116–118 *passim*, 123, 126, 245–259
Shepherd's purse, 219
shrubby lespedeza, 100
shrubland
  desert, 17, 19, 35, 38, 39, 50, 137, 155–157 *passim*, 159–161 *passim*
  Hawai`i lowland, 228, 236
  Hawai`i montane, 229, 234, 236, 237
  Hawai`i subalpine, 229, 233, 234
  mountain, 26
  sagebrush, 11, 19, 21, 38, 40, 50, 53, 55, 132, 154, 156, 158, 160, 265
Sida abutifolia, see spreading fanpetals
silkreed, 100, 103
Sisymbrium altissimum, see tumble mustard
Sisymbrium irio, see London rocket
slender oat, 52, 180, 184
smooth brome, 11, 52, 55, 59, 113, 115, 116, 118, 120, 122, 152, 163–165 *passim*, 219
soft brome, 180, 186
soft chess, 11, 53, 58
soil heating, 12–14 *passim*, 106, 152, 161, 201, 202, 208, 211, 212, 218, 263, 290, 292
soil moisture, 121, 271
Solanum viarum, see tropical soda apple
Sonchus arvensis, see perrenial sowthistle
Sorghum halepense, see Johnson grass
Spanish broom, 21, 27, 245–259
spotted knapweed, 16, 17, 21, 54, 57, 58, 79, 115, 116, 132, 142–144 *passim*, 148, 149, 151, 153, 157, 164–168 *passim*, 170, 216, 245–259, 273–275 *passim*, 283, 284, 286– 290 *passim*
spreading fanpetals, 146
spread (of a nonnative species), 93, 96, 99, 100, 106, 108, 109, 117, 146, 148, 157, 158, 169, 181, 206, 234, 235, 238, 240, 244, 254, 256, 261, 271, 277, 293
squarrose knapweed, 54, 116, 160
St. Johnswort, 16, 17, 21, 22, 25, 54, 142, 144, 148, 150, 153, 167, 198, 200–202 *passim*, 206, 210–212 *passim*, 216, 245–259
sulfur cinquefoil, 19, 21, 144, 148, 150, 153, 245–259
surface fire, 3, 67, 162
survival, 216
swallow-worts, 71
sweetbriar rose, 21, 51, 210–212 *passim*, 214
sweet cherry, 212
sweetclovers, 18, 115, 124, 198, 276
sweet resinbush, 144, 147
sweet vernal grass, 73, 74, 209, 227, 229, 235, 237

T

Taeniatherum caput-medusae, see medusahead
tall fescue, 54, 95, 97, 100, 207
tallgrass prairie, 17, 18, 24, 27

tall oatgrass, 198, 211
tamarisk, 15, 20, 22, 54, 55, 57, 60, 110, 113, 115–118 *passim*, 126, 128, 136, 138, 141, 144, 169–172 *passim*, 179, 191, 192, 245–259
Tamarix spp., see tamarisk
tansy ragwort, 16, 21, 22, 25, 200, 202, 203, 205, 210, 211, 213, 216
Taraxacum officinale, see common dandelion
Tasmanian bluegum, 179
Tatarian honeysuckle, 63
temperature
  soil, 50, 51
thatching grass, 227, 233, 239
thorny-olive, 95, 105
timothy, 167, 198, 205, 207, 215, 217, 219, 221
total heat release, 4
tree-of-heaven, 20, 66, 68, 69, 75, 77, 83, 88, 95, 105, 106, 179, 188, 191, 245–259
Triadica sebifera, see Chinese tallow
Tribulus terrestris, see puncture vine
tropical soda apple, 95, 100
tumble mustard, 16, 144, 148, 157, 163, 219
tundra, 198, 220–222

U

Ulex europeus, see gorse
underburn, see *fire regime, understory*
Urochloa maxima, see guineagrass

V

vaseygrass, 116
vegetative regeneration, 8, 14–16, 20, 43, 44, 67–73 *passim*, 76, 79, 80, 84, 85, 96, 100, 101, 104–107 *passim*, 120, 122, 123, 125–128 *passim*, 130, 132, 134–140 *passim*, 146, 149, 150, 169, 171, 195, 202, 204, 211, 213–215 *passim*, 217, 232, 234, 235, 238, 246, 248, 250, 254, 255, 257
velvetgrass, 211, 227, 229, 235, 237
Ventenata dubia, see North Africa grass
Verbascum thapsus, see common mullein
Vicia cracca, see bird vetch
Vinca major, see bigleaf periwinkle
Vinca minor, see common periwinkle
Vulpia myuros, see rattail sixweeks grass

W

wall-lettuce, 25, 200, 201, 205
water yam, 95, 107
weeping lovegrass, 21, 144, 145, 146, 152
western wheatgrass, 216
wetland, 108
  Northeast fresh, 65, 82–87
  Northeast tidal, 65, 82–87
  Southeast, 53, 92, 94–100, 117
  Southeast cypress swamp, see *forest, cypress swamp*
  Southwest coastal, 191–192
white sweetclover, 11, 17, 51, 116, 118, 124, 206, 207, 216, 219
whitetop, 116
wild lettuce, 21, 25
wildlife, 137
  effects on, 38, 40, 59, 60, 78, 79, 88, 119, 201
wild oat, 52, 180, 184, 186, 219, 265
wineberry, 76
winged euonymus, 63, 70, 76, 77, 95, 105
winter creeper, 95, 105
Wisteria floribunda, see Japanese wisteria
Wisteria sinensis, see Chinese wisteria
wood bluegrass, 78
woodland
  Hawai`i lowland, 11, 21, 228, 231, 233–235 *passim*, 240
  Northwest oak, 22, 27, 59, 198, 208–215
  oak-hickory, 105–107
  piñon-juniper, 21, 26, 27, 132, 161–164, 265
woodland groundsel, 17, 25, 200, 203, 205

Y

yellow bluestem, 116, 118, 129, 136
yellow glandweed, 210, 211
yellow hawkweed, 219
yellow salsify, 205
yellow starthistle, 16, 19, 22, 49, 54, 116, 134, 144, 148, 176, 179–182 *passim*, 184, 245–259
yellow sweetclover, 17, 51, 116, 118, 124, 163, 165, 216, 217, 219
yellow toadflax, 116, 144, 148, 150, 198, 216, 219, 245–259

U.S. GOVERNMENT PRINTING OFFICE: 2008 — 747-256